Wave Theory Foundations, Analysis and Algorithms

ristopher F. Barnes, PhD.

gress Cataloging-in-Publication Data

her F.
rture radar, wave theory foundations, analysis and algorithms
ographical references and index.
92-31373-2

erture radar. 2. Wave theory. 3. Analysis and algorithms. I.

United States of America
3 2 1

by Tyler D. Thompson.

Synthetic

Wave Theo

Analysis a

Christopher F

Georgia Institute

Synthetic Aperture Radar

Library of Cong

Barnes, Christop
Synthetic ape
Includes bibl
ISBN 978-0-
First Edition
1. Synthetic a
Title.

Printed in the
10 9 8 7 6 5 4

Cover design

Preface

Synthetic aperture radar (SAR) analysis, algorithms and applications have undergone continual development over a span of more than six decades. The SAR image formation processing options have fanned into an evolutionary tree of algorithmic choices that still continues to grow.

An awareness of the broad historical background of SAR and familiarity with a diverse set of rather sophisticated mathematics is required to obtain a workable understanding of SAR. The depth and breadth of the subject matter, along with SAR's varied history, makes SAR theory and practice challenging to master.

Learning SAR by an in-depth study of the scientific journal and technical literature has its own challenges. SAR literature is not readily assessable for various reasons, including inconsistent assumptions of specialized radar hardware subsystems (e.g., stretch receivers versus chirp receivers), inconsistent terminology, inconsistent mathematical notation, multiple coordinate system definitions and possibly, lack of full technical disclosures required to protect intellectual property. The highest hurdle is probably the existence of multiple simplifying signal analysis assumptions at the foundations of several variant SAR "theories."

My purpose in writing this book is to help the serious student obtain a solid understanding of a unified SAR theory, learn methods of SAR analysis and become more familiar with the many types of SAR algorithms. I seek to explain theoretical concepts in an intuitive way. Background radar concepts and foundational mathematical tools are covered. Special attention is afforded to Huygens-Fresnel wave theory.

This work contains measures of technical innovation in the areas of volumetric SAR analysis and image formation processing, and in methods used for the analysis of temporal chirp waveforms and spatial chirp phase curtains.

The pedagogical approach of this work was developed over about a decade and a half of teaching SAR at the professional education and graduate level at the Georgia Institute of Technology. Being an engineer, I make no pretense at mathematical rigor. I seek to physically motivate insights into key concepts by conveying conceptual and geometric interpretations sufficient for sound engineering. I strive to be sufficiently didactic in relevant details. I strive to not be overly pedantic in what I consider to be mathematical minutiae (with the caveat that one's minutiae may be another's opus). I introduce or import terminology from related wave theory and computed imaging disciplines that I believe to be helpful for teaching SAR. I also permit myself the flexibility of enhancing and developing mathematical notations that I believe to be *relatively* concise and clarifying.

I hope you find this work helpful in your mastery of the scientific foundations of SAR and useful in your adaptations and applications of SAR image formation processing algorithms.

– Christopher F. Barnes
June 2014

About the Author

Christopher F. Barnes was awarded the Ph.D. degree by Brigham Young University in 1989. He was employed by the Georgia Tech Research Institute (GTRI) Sensors and Electromagnetic Applications Laboratory and worked there in basic and applied research for thirteen years. Dr. Barnes transferred within the Georgia Institute of Technology from GTRI to the School of Electrical and Computer Engineering in 2002, with initial assignment to the Savannah campus. In 2013, he transferred to the Atlanta campus.

Professor Barnes is an avid saltwater fisherman. On most outings, he almost always catches fresh air and sunshine. He occasionally catches other intended quarry.

The charm of fishing is that it is the pursuit of what is elusive but obtainable, a perpetual series of occasions for hope[1].

The joys of engineering are much the same.

[1] Quote attributed to Scottish politician John Buchan (1875–1940).

Contents

List of Figures

Synthetic Aperture Radar

Wave Theory Foundations

Analysis and Algorithms

Christopher F. Barnes, Ph.D.

Georgia Institute of Technology

Chapter 1

Synthetic Aperture Radar

Synthetic aperture radar (SAR) signal analysis and system engineering have been approached over the years with various theoretical tactics and technological tools. The first investigators of SAR formulated their understanding from a Doppler[1] perspective [990]. As SAR system development and theoretical understanding progressed, the science of SAR was recognized as related to other computed imaging disciplines, including holography [309,433,638], ray tomography [415,416] and diffraction tomography [187, 565]. More recently, SAR imaging has been identified as related to seismic imaging [137, 728, 790]. The common thread that runs through all of these computed imaging disciplines is a wave theory foundation.

We build a framework for SAR signal analysis and algorithm development based on wave propagation theories. We assert that all SAR image formation algorithms are related in some way to the wave theories of Huygens[2] and Fresnel[3].

Additionally, we describe the early SAR signal models and processing methods initially developed by the inventors of SAR as related to either the Fresnel (truncated series) approximation [960] or Fraunhofer[4] (plane wave) approximation of Huygens-Fresnel spherical wave field analysis.

The more recently developed "range migration" or "omega-k" image formation algorithm is free of the approximations of Fraunhofer and Fresnel [885]. We show that omega-k processing is relatable to special yet simple frequency-wavenumber domain functions that we call *Huygens-Fresnel transfer functions* [150, 395, 723, 993]. Huygens-Fresnel transfer functions are dimensionally separable frequency-wavenumber domain descriptions of spherical wave propagation that occurs in time and space.

We seek to coalesce various perspectives for understanding SAR into a wave theory framework based on concepts postulated or developed by Huygens, Fres-

[1]Christian Doppler (1803–1857) was an Austrian physicist who first described the frequency shift effect associated with relative motion between signal source and signal sensor.

[2]Christiaan Huygens (1629–1696) was a Dutch physicist who proposed a model for light wave propagation based on point sources that radiate impulsive spherical wavelets.

[3]Augustin-Jean Fresnel (1788–1827) was a French physicist who extended Huygens' model to include point sources that produce monochromatic sinusoidal spherical wave fields.

[4]Joseph von Fraunhofer (1787–1826) was a German physicist who was first to extensively explore the effects of a diffraction grating on light.

nel, Helmholtz[5] and Kirchhoff[6]. We use differential and integral forms of wave motion equations to set up scattered wave field models described in the time and space domains. These models are then subjected to partial or full Fourier[7] transforms, leading to analysis methods and associated SAR image formation algorithms performed in joint time-space, frequency-space, frequency-wavenumber or time-wavenumber domains.

This introductory chapter is outlined as follows: Section 1.1 addresses differences of radar imaging with real and synthetic apertures. Section 1.2 gives a short history of the invention and development of SAR. Section 1.3 introduces the operational concepts of monostatic SAR systems. Section 1.4 describes multistatic SAR configurations, including interferometric modalities and multiple pass flight operations.

1.1 Imaging Radars

We give a short primer on real aperture radar and the limitations of real aperture imaging that motivated the development of synthetic aperture radar imaging methods [784, 811, 842].

Real Aperture Imaging

The limiting factor in the cross-range or along-track (y dimension) resolution Δ_y^{ant} of a real aperture imaging radar is the range dependent width of the along-track antenna beam pattern,

$$\Delta_y^{\text{ant}}(r) \propto r \cdot \theta_y^{\text{ant}}, \tag{1.1}$$

where r is down-range distance and θ_y^{ant} is the along-track angular beamwidth of the antenna [23, 114]. Since the resolution $\Delta_y^{\text{ant}}(r)$ grows coarser with increasing r, a real aperture imaging radar has an along-track resolution that degrades with increasing range.

Typically, the angular beamwidth of an antenna with along-track physical diameter D_y is dependent on the wavelength λ of the transmitted signal,

$$\theta_y^{\text{ant}}(\lambda) \propto \frac{\lambda}{D_y}.$$

Hence, a real aperture imaging radar has cross-range resolution that also degrades with decreasing signal frequency (i.e., increasing wavelength) and with decreasing antenna size,

$$\Delta_y^{\text{ant}}(r, \lambda) \propto r \cdot \frac{\lambda}{D_y}.$$

[5] Hermann Ludwig Ferdinand von Helmholtz (1821–1894) was a German physician and physicist who made significant contributions to multiple areas of modern science, including wave theory.

[6] Guastav Robert Kirchhoff (1824–1887) was a German physicist known for his work in blackbody radiation, electricity and thermochemistry.

[7] Jean-Baptiste-Joseph Fourier (1768–1830) was a French mathematician who is widely known for his many contributions in analytic mathematics.

Synthetic Aperture Imaging

Real aperture imaging, with its characteristic of along-track resolution that degrades with increasing range, does not exploit the spatially induced phase variations that occur within sets of transmitted signal echoes collected by a moving radar [910]. Synthetic aperture imaging, as a form of spatially coherent imaging, utilizes phase information gathered by a movable radar system to improve along-track resolution through batch processing of sets of pulse echoes [112, 175, 180, 226, 332, 406, 440, 471, 472, 541, 542, 719, 885].

Resolution Limits of Point Reference and Line Reference Systems

The achievable resolution of a SAR system depends on the operational mode (see Section 1.3) and on attributes of a radar's motion compensation system. The purpose of motion compensation is spatial phase referencing and stabilization. Motion compensation measurements are based on a suite of onboard motion sensors that collectively provide data to support accurate estimates of true platform position, velocity and acceleration [123, 340, 439, 524, 531, 533, 1005].

In a stripmap mode SAR that employs phase stabilization to a reference range line, the achievable along-track resolution is

$$\Delta_y^{\text{strip:line}} = \frac{D_y}{4}. \tag{1.2}$$

In a spotlight mode SAR that employs phase stabilization to a reference ground point, the achievable along-track resolution is the same,

$$\Delta_y^{\text{spot:point}} = \frac{D_y}{4}.$$

In both of these cases, the along-track resolution Δ_y is independent of range and depends only on antenna size D_y. In a spotlight mode SAR that employs phase stabilization to a reference range line, the ultimate along-track resolution is

$$\Delta_y^{\text{spot:line}} = \frac{\lambda}{4},$$

which is only achieved in the limit of an infinitely long synthetic aperture. In this case, the lower bound to the along-track resolution depends only on the transmitted (continuous wave) signal wavelength λ. This lower bound to resolution is the (round trip) *diffraction limit.*

Stripmap Mode Resolution Limit

Even if $D_y < \lambda$, the line phase referenced stripmap mode resolution $\Delta_y^{\text{strip:line}}$ and point phase referenced spotlight mode resolution $\Delta_y^{\text{spot:point}}$ cannot be improved beyond the diffraction limit $\lambda/4$.

Synthetic Array and Passive Array Resolution Limit

The achievable resolution of an infinitely long synthetic array operating in an active sensing mode is double the achievable resolution of an infinitely long real array operating in a passive sensing mode. The resolution limit of the line referenced spotlight mode system,

$$\Delta_y^{\text{spot:line}} = \frac{\lambda}{4}$$

is lower than the (one way) diffraction limit

$$\Delta_y^{\text{array:passive}} = \frac{\lambda}{2}$$

of a passive real array by a factor of one half. This reduction in the resolution limit is due to round trip signal propagation of the active system. The signal wavelength λ is in effect reduced to $\lambda/2$ by the doubling of the total number of phase cycles experienced in two way signal travel, which is equivalent to an effective doubling of the frequency of the transmitted signal [493].

1.2 Historical Overview

We give an overview of the history of the invention and development of SAR. We point out major technological and theoretical milestones (and stumbling stones) that mark the pathway of the evolution of SAR.

Doppler Beam Sharpening

C. A. Wiley is viewed as the inventor of an early form of stripmap mode SAR known as *Doppler beam sharpening* (DBS). Wiley's ideas were based on observations of the spectral properties of periodic pulse echoes of a squinted airborne radar [120, 221, 832, 990]. Wiley noticed the spacing of spectral lines by the pulse repetition frequency (PRF). He hypothesized that spectral shifts of different along-track scatterers should be resolvable with a resolution given by

$$\Delta_y^{\text{DBS}}(r) \propto r \cdot \theta_y^{\text{eff}}, \tag{1.3}$$

where $\theta_y^{\text{eff}} \ll \theta_y^{\text{ant}}$ is the effective beamwidth suggested by the *angular* resolution of the spectral analysis [42].

The first Doppler beam sharpening system was demonstrated early in 1952. This DBS system reduced the effective beamwidth of a real aperture Yagi antenna from 100 degrees to 1 degree (at a transmitted signal frequency of 930 MHz) [990]. The first Doppler beam sharpened images were obtained in 1953.

Wiley filed for patent protection of the DBS system (disclosed in 1951, filed in 1954). Wiley received a secrecy order from the U.S. Patent Office in 1955. The secrecy order was rescinded in 1964.

Doppler Beam Sharpening as Unfocused SAR

Doppler beam sharpening has sometimes been called "unfocused SAR" [26, 440]. In an "unfocused SAR" system, the along-track resolution, although enhanced, retains an azimuth focusing dependency on range. Although (1.3) is an improvement over (1.1), the along-track resolution achievable by Doppler beam sharpening remains range dependent. Furthermore, DBS systems are operative only in the far field where "unfocused" beamwidths are wide [990]. In this sense, Wiley's Doppler based concepts did not provide complete (range independent) aperture synthesis.

Stripmap Mode Synthetic Aperture Radar

Independently of Wiley, C. W. Sherwin of the University of Illinois developed in 1952 the foundations of "fully focused" stripmap mode aperture synthesis. Sherwin postulated in 1953 that a fully focused synthetic aperture should improve along-track resolution beyond that achievable with Doppler beam sharpening [223, 638]. Sherwin's ideas were applicable in both the near field and far field of the antenna beam.

A joint research project that included Sherwin and L. J. Cutrona of the University of Michigan was initiated in the summer of 1953. This U.S. Army research project, conducted under the code name of "Project Wolverine," demonstrated in 1953 early forms of SAR focusing [120].

Optical Processing of Film Recorded Data

Because the data production rates of the first SAR demonstration systems exceeded the ability to achieve real-time data recording in analog electromagnetic formats, these first prototype systems used transduced optical film recordings of coherent radar data. Optical post processing was used to focus the film recorded data [942]. Work on optical processors for SAR was significantly advanced in 1954 [990]. A prototype SAR system with an X-band radar and an optical computer produced a form of focused stripmap mode imagery in the summer of 1957 [26].

Development work continued and research results led Cutrona to postulate in 1961 that a fully focused synthetic aperture should provide along-track resolution *independent* of range [229, 832]. Eventually, a Cutrona-Sherwin led project was able to demonstrate that azimuth resolution independent of range can indeed be achieved in stripmap mode SAR.

Early SAR Radars

The Cutrona-Sherwin research group, subsequently working with Texas Instruments, advanced the SAR concept to a technical maturity level that resulted in the AN/UPD-1 radar system [26]. A total of five radars were built and tested [229]. Delivery of these SAR systems continued through the AN/UPD-4, which was built by Goodyear Aerospace in 1972 [505]. A multi-frequency SAR was included in the Apollo lunar mission in 1972 [26]. The SEASAT SAR system was launched in 1975 [23, 58, 214, 507, 530, 590, 1000]. A description of various SAR system developments that occurred after SEASAT through the early 1990s can be found in [981].

Spotlight Mode Synthetic Aperture Radar

Operational concepts for spotlight mode SAR emerged in the mid to late 1970s, about twenty years after the inception of stripmap mode concepts [115, 190, 488, 976, 977, 987].

J. L. Walker and W. M. Brown built the first prototype of a form of spotlight mode SAR called *inverse SAR* (ISAR). Their prototype optically processed coherent radar data scattered by objects placed on a rotating platform [115, 411]. This ISAR prototype was configured to simulate an airborne radar flying in a circle around a reference ground point [679].

This was the genesis of motion compensation to a point concepts eventually employed for phase stabilization in polar format spotlight mode systems. The circular flight profile suggested by a spotlight mode turn table model represented a major difference compared to the flight profile employed in stripmap mode, where the flight path is required to be a straight line and where phase is stabilized in reference to a fixed range value.

Evolution of SAR Concepts and Methods

In addition to the invention, development and demonstration of SAR concepts and systems, the evolution of the theoretical basis and mathematical foundations of SAR has an interesting and meandering history.

SAR and Holography

Some forms of SAR analysis and image formation processing can be equated with the methods of optical holography invented by Gabor[8] [369–371]. Gabor's holographic imaging methods are based on inversion of spherically scattered coherent optical wave fields [89, 665]. His methods have found applications in various computed imaging disciplines, including diffraction tomography [251] and seismology [77].

The applicability of holographic wave field inversion methods to synthetic aperture radar algorithm development was recognized in the 1960s [213, 502, 562, 572]. However, implementation challenges that existed before the age of high speed digital computing hindered this approach in the advancement of SAR. As a result, the evolution of SAR systems detoured into various disciplines empowered (and constrained) by the technical practicalities of the time. More recently, the age of high performance computing has allowed SAR algorithm development to return to the theoretical foundations of holography.

Fresnel Approximation

Many pioneering developers of stripmap mode SAR formulated their concepts within the context of Doppler processing of radar signals [117, 976]. Their SAR signal models usually utilized a second or third order truncated Taylor[9] series

[8] Dennis Gabor (Gabor Denes) (1900–1979) was a Hungarian born British electrical engineer who invented holography, for which he received the 1971 Nobel Prize in physics.

[9] Sir Brook Taylor (1685–1731) was an English mathematician who is best known for the Taylor's theorem and the Taylor series.

expansion to simplify the analysis of the time evolution of sensor-scatterer slant-range. An approximation that truncates a slant-range expression in wave field models to second order is called a *Fresnel approximation* [607]. A Fresnel approximation provides a parabolic approximation of the true hyperbolic spatial geometry [648] that determines pulse signal echo delay. A change from hyperbolic to parabolic descriptions of sensor-scatterer imaging geometries makes spatial Fourier transforms easier to evaluate. The simplified Fourier analysis of second order systems yielded tractable SAR spectral models and associated inversion methods for early SAR systems [751]. This evolutionary tack of SAR algorithm development led to the family of *range-Doppler algorithms* [117, 218, 221].

Early Range-Doppler Algorithms and Systems

A major challenge in early range-Doppler algorithm implementation was dealing with a multiplicity of range dependent "focusing filters." A different cross-range focusing filter was allocated to each subswath of the full range swath. Sometimes, especially in early versions of range-Doppler algorithms, a truncated series representation of a slant-range hyperbola specific to each subswath was used. Modern range-Doppler algorithms utilize higher order signal models in developing cross-range focusing filters for along-track image formation.

Typically, in range-Doppler imaging systems, *down-range focusing* was performed by the receiver hardware subsystem. Pulse compressed electromagnetic echo signals were transduced and photographically recorded on film. The *along-track focusing* of film recorded data was subsequently performed by an optical computing system comprised of lenses and optical filters [120, 228, 301, 433, 942]. A bank of range variant along-track focusing filters could be implemented with a single conical lens.

Modern Range-Doppler Algorithms and Systems

Range-Doppler algorithms have persisted in SAR practice, especially in space based SAR systems [763]. Range-Doppler designs and implementations have become quite sophisticated [218], including enhancements based on chirp scaling. The invention of the modern version of the range-Doppler algorithm is attributed to C. Wu [217, 500, 999, 1001].

Digital Processing

Analog optical processors persisted until the late 1960s [120, 227, 228], after which hybrids of optical and digital processing systems began to emerge [221]. Fully digital computing systems were first used in SAR in 1975 for processing SEASAT data [24, 25, 59, 60, 136, 966].

Fraunhofer Approximation

Fundamental differences in the signal analysis between stripmap and spotlight mode SARs caused the Taylor series signal modeling methods used in stripmap SAR to be abandoned in the early analysis of spotlight mode signals. The reduced order simplification achieved by a truncated Taylor series was replaced with another simplifying approach based on far field plane wave approximations of spherical wavefronts. The tangential plane wave approximation of a spherical wavefront is called the *Fraunhofer approximation* [607]. The primary driver for the continued use of a simplifying assumption (the Fraunhofer approximation) for spotlight mode algorithm development was to make spatial analysis of the SAR signal spectrum more tractable with more easily evaluated spatial Fourier transforms.

Polar Formatting Algorithms and the Fourier Projection Theorem

The plane wave Fraunhofer model for spotlight mode SAR led to *polar formatting algorithms* [490, 679]. Polar formatting methods are explainable in terms of the *Fourier projection theorem* of ray tomographic signal processing [187, 565]. The Fourier projection theorem requires a planar illuminating wavefront. This requirement is met in SAR if the range is mandated to be large and the illuminated spot size is constrained to be small.

Stolt Formatting Algorithms and the Fourier Diffraction Theorem

The Fourier projection theorem is sometimes viewed as a constrained special case of a more general theorem called the *Fourier diffraction theorem* [40]. The Fourier diffraction theorem is not subject to long range and small spot constraints. Hence, if range is short or the scene is large, then the Fourier diffraction theorem can be used as a basis for image formation algorithm development. The Fourier diffraction theorem is not compatible with a plane wave assumption, and requires analysis that accommodates spherical waves. The Fourier diffraction theorem is a theoretical foundation in holography, diffraction tomography, seismic imaging and radio astronomy. *Stolt formatting algorithms* are based on a passband form of the Fourier diffraction theorem.

Motion Through Resolution Cells

When the accuracy of a Fraunhofer plane wave approximation begins to fail as ground scene size becomes large or range becomes short, a scatterer's echo can occur in a range bin other than that suggested by Fraunhofer's plane wave model. This effect is called *range cell migration* or *motion through resolution cells* [26, 117, 526].

Temporal Doppler

As polar format algorithm development for SAR image formation processing advanced, various insights and other perspectives for understanding SAR were identified. For example, the "stop-and-hop" assumption of D. C. Munson [679] showed, as a thought experiment, that the SAR platform need not be moving during the transmission and reception of each pulse echo. This insight can lead one to ask that if there is no requirement for relative movement between sensor and scatterer, can the Doppler effect [664] be the underlying phenomenon exploited by SAR? Although sometimes debated, the answer is no [493, 683]. A physical array does not rely on Doppler phenomenology for beamforming and neither does a synthetic array.

Spatially Induced Phase Shifts

SAR exploits phase shifts associated with *inter-pulse* signal attributes caused by changing sensor *position* between pulses. Any *intra-pulse* Doppler shift sensed within the pulse duration and caused by a range closure rate between sensor and scene is usually ignored in the formulation of SAR algorithms [407, 683]. However, SAR signals can be *formally* analyzed from a time domain Doppler perspective. This is a common approach used to set up range-Doppler algorithm development [26, 218, 445].

SAR and Angular Diversity

The true physical basis of SAR imaging is angular diversity created by changes of sensor position, rather than Doppler induced frequency diversity [114, 638, 679, 976]. View angle diversity in SAR can be characterized by an angular sweep of the line of sight plane wave element of the incident and reflected waves between the sensor and scatterers. An angular sweep of these constituent directed plane waves induces an *angular spectrum manifold* of SAR phase history data when described in the frequency-wavenumber domain.

Spatial Doppler

It has been suggested that the spatial or angular modulation of SAR should be called the *spatial Doppler effect* [407, 885, 1007]. The terminology of "spatial Doppler" indicates both similarities and differences with respect to temporal Doppler. However, the wording of "spatial Doppler" may not be sufficiently clarifying in regards to the true foundational phenomenology of angular diversity of pulse echo measurements.

Stolt Formatting Algorithms

Beginning in the late 1980s, the advances of high performance computing enabled digital processing methods to liberate computationally constrained formulations of SAR into "approximation free" zones of spherical wave theory [136, 137, 885]. SAR imaging via spherical wave field inversion [865] based on Huygens-Fresnel principles opens the door to exploit wave field concepts developed by Helmholtz, Kirchhoff, Born[10] and others. A new evolutionary path free of the approximations of Fresnel and Fraunhofer for SAR development emerged. These new algorithms were typically called "omega-k" algorithms because processing is performed in a combined temporal-frequency and spatial-wavenumber domain.

Because the SAR omega-k algorithm of the 1980s first appeared to be a leverage of literature related to seismic inversion [134, 137, 218, 790, 791], its relationship to previously established holographic methods was not widely recognized.

[10]Max Born (1882–1970) was a German born British mathematician who worked in quantum mechanics and optics.

Omega-k processing is a reemergence of the early SAR holography ideas established by E. N. Leith and J. Upatnieks in the 1960s [572, 573, 575–578]. Rather than having a basis in seismology, this late evolutionary branch of omega-k processing in SAR practice can be viewed as a return to the roots of Gabor's work in holography [369, 370, 374, 748, 887].

Some versions of omega-k processing also emerged about this time without reference to seismic processing [40, 448]. Still others advanced the development of omega-k algorithms based on related prior work and experience in medical ultrasonic tomographic imaging [864, 865, 868, 898].

There was a degree of initial scepticism by some towards omega-k's suitability [241] and computationally feasibility [495] in SAR.

The omega-k algorithm has many names and has been called the *wavenumber algorithm* [584], the *range migration algorithm* [146], the *wavenumber interpolation algorithm* [648], the *F-K algorithm* [376] and the *wavefront reconstruction algorithm* [885]. Another name for this algorithm, which suggests similarities and differences with the polar format algorithm, is the *Stolt format algorithm* [913, 914].

Stolt Formatted Angular Spectrums

The omega-k algorithm is an image formation process most often implemented in a frequency-wavenumber domain called the *angular spectrum* [84, 196]. The angular spectrum is obtained by *Stolt formatting* of *rectilinear* frequency-wavenumber data. An angular spectrum is an expression of a spherical wave field decomposed into a collection of directed plane waves that propagate in all directions [84, 195, 289, 322, 354, 466, 666, 810, 835, 988]. The descriptive mix of the amplitude and phase coefficients of these constituent directed plane waves comprises the angular spectrum of Stolt formatted data.

Range-Doppler Perspective of Omega-K Processing

Assertions exist that the omega-k algorithm is subject to approximations that hinder application to large range swaths. Beyond constraints associated with the possible use of *stretch* or *deramp* processing for the pulse compression of chirp waveforms (see Section 12.5), these assertions are yet to be validated. Such suspicions are likely a result of viewing the Stolt mapping as a form of *differential range-Doppler processing,* where the Huygens-Fresnel transfer (the omega-k "matched filter") is viewed as a process of "bulk focusing" the data relative to a reference hyperbola through scene or swath center [40]. In reality, Stolt formatting provides exact "correction" of all differential elements of hyperbolic range walk across the entire range swath, no matter how large the range swath [216, 218, 397]. Moreover, we shall see that Stolt formatting can be performed *without* Huygens-Fresnel transfers.

Slow-Time Matched Filtering Not Required in Stolt Formatting

There was a period of time in the development of SAR where a widely held view stipulated that along-track processing was inherently dependent upon the matched filtering concept. This view asserted that a correlation or matched filtering process with a slow-time reference signal was required to form the image in the along-track dimension.

Different along-track matched filters were typically used for different range subswaths. This bank of slow-time matched filters had range dependencies that seemed inescapable [752]. This view point, like that of the Doppler formulation, has utility in certain contexts, but can be somewhat misleading in other contexts.

We develop a generalized wave theory framework where a spatial matched filter need not always be employed in SAR image formation processing. The core "matched filter" of along-track "focusing" in omega-k processing is a wave field inversion mechanism definable as a Huygens–Fresnel transfer function. A Huygens-Fresnel transfer achieves spherical wave field inversion in the time-space domain with a phase-only multiply operation performed in the frequency-wavenumber domain.

Huygens-Fresnel Transfers in Stolt Formatting are Optional

Moreover, wavenumber domain Huygens-Fresnel transfers are an *optional* processing step of omega-k processing based on Stolt formatting. With Huygens-Fresnel transfers, Fourier inversion of Stolt formatted data produces a SAR image in a coordinate system referenced to scene center. Without Huygens-Fresnel transfers, Fourier inversion of Stolt formatted data produces a SAR image in a coordinate system referenced to the synthetic aperture flight path. In either case, with or without Huygens-Fresnel transfers, Stolt formatting induces no range dependent depth of focus and leaves no residual depth of focus blurring.

If Huygens-Fresnel transfers are not performed, then it may appear as though SAR image formation processing is being achieved without slow-time matched filtering and indeed, it is. Along-track image formation processing based in wave field inversion is not a chore inherently requiring matched filtering or correlation processing.

Omega-K Processing Without Stolt Formatting

Moreover, variants of omega-k processing exist that can be performed with or without Stolt formatting. One approach performs Fourier migration via Stolt transformation of the SAR signal spectrum. Another approach executes multiple Huygens-Fresnel transfers, performing one Huygens-Fresnel transfer for each cross-range line contained in the SAR image. Stolt formatting is not performed in this later form of omega-k processing (see Section 12.8).

Competing SAR "Theories"

By the time the essentially approximation free Huygens-Fresnel formulation of SAR image formation processing (re)emerged in the 1990s, the Fresnel and Fraunhofer approximation based methods, dominant in the practice of stripmap and spotlight mode SAR, had become entrenched in various expertly practicing SAR communities. The strategic space based SAR practitioners had their finely tuned range-Doppler (Fresnel approximation) methods and specialized radar system hardware. The tactical airborne SAR practitioners had their efficient polar format

(Fraunhofer approximation) methods and their own specialized radar hardware subsystems. The differences in these algorithms, tailored to specific operational concepts and system hardware architectures, did little to foster interaction between these different SAR communities and related theoretical subdisciplines.

The three formulations of SAR signal models (range-Doppler Fresnel, polar format Fraunhofer and Stolt format Huygens-Fresnel) and supporting radar hardware system architectures do not always agree with each other in the derivation of key system engineering requirements, let alone details of the different image formation processing algorithms.

Balkanization of the practicing SAR communities has been further exacerbated by the high costs of SAR hardware systems. High costs tend to extend legacy system life cycles and delay technology refresh. Such delays hinder cross subdiscipline interactions. It may even be fair to say that the fragmented world of SAR practice (as finely tuned as each subdiscipline may be) has led to niche SAR practicum. In some cases, these niches are somewhat mired in technological and theoretical ruts.

A situation now exists for those who are new to SAR to enter into only a subspace of the conversation and practice. Unfortunately, this makes it possible to misapply newly learned SAR concepts of one niche into an un-accommodating domain of another niche [899].

Naive selection and blending of SAR conceptual frameworks can cause erroneous top level system requirements to be defined, especially in regards to data acquisition parameters such as pulse repetition frequency and motion compensation designs. Top level, front end hardware engineering design decisions must flow into compatible back end, computational subsystem and image formation algorithm choices.

SAR System Requirement Specifications

SAR system designs are specific to underlying SAR signal model formulations. SAR system design choices imply concomitant selections of image formation methods and vice versa.

For example, the modeled bandwidth of the along-track signal determines the required pulse repetition frequency. The Fraunhofer (plane wave) approximation based polar format algorithm (and associated motion compensation to a point data collection subsystems) employ signal analysis methods that model along-track signal bandwidths as proportional to along-track ground scene size. Ground scene size must be constrained to maintain the validity of the Fraunhofer approximation. On the other hand, the Huygens-Fresnel formulation and associated Stolt formatting data collection subsystem designs are associated with signal models where the along-track signal bandwidth is characterized as proportional to the *sum* of along-track ground scene size (unconstrained) and the length of the synthetic aperture (also unconstrained).

These different signal bandwidth models of polar formatting and Stolt formatting systems lead to different pulse repetition frequency requirements. Naive blending of various design principles may lead to erroneous specification of radar system requirements. Need we say that an incorrectly specified pulse repetition frequency for a radar system is a serious design flaw?

SAR Data Incompatibilities

Image formation processing of SAR data with an algorithm different than that intended during SAR system design can be problematic. The data acquired with one type of SAR hardware system are not amenable to processing by image formation algorithms intended as a backend for a different type of SAR system hardware. A situation now exists where sharing of raw SAR data, without broad understanding of the SAR cross subdiscipline technological and theoretical issues, is fraught with challenges.

Anecdotes [885] can be shared where poorly postulated SAR cross subdiscipline experiments led to initial skepticism of omega-k inversion methods during the early 1990s [241]. The newer omega-k SAR algorithms did not at first appear to perform well when applied to legacy system data sets [378, 899]. The poor results were eventually correctly understood by some as not inherent in the omega-k algorithm, but attributable to the underlying differences in data acquisition requirements [885].

SAR Image "Enhancements?"

Unfortunately, these early missteps have led to various reparative, but misguided image enhancement techniques to be proposed for SAR. For example, the SAR literature contains proposals for the use of low pass "Doppler filtering" of collected SAR data to improve image quality [885]. Why expend the effort and expense to collect high bandwidth SAR data for high resolution imaging if the high frequency portion of the Doppler data spectrum is going to be intentionally low pass filtered away?

A closer examination shows that some of these SAR "enhancements" are not helpful from the perspective of foundational SAR theory, but seek to compensate latent (and possibly forgotten) signal modeling approximations and associated limitations.

Reconditioning of SAR Data

SAR data acquired from different types of hardware systems can sometimes be made compatible with post processing. On such example is M. Soumekh's "time domain decompression" method [885], which we call *unstretch processing*. Unstretch processing is a *reramping* method applicable when linear frequency modulated chirp waveforms and stretch receiver hardware subsystems are used.

An analogous slow-time process that we call *rechirp processing* is applicable in the along-track spatial dimension for "undoing" the motion compensation to a ground reference point [397]. Motion compensation to a point is required in polar format processing. Motion compensation to a point is not compatible with (slant-plane) Stolt formatting[11].

Once polar format data has been reconditioned via unstretch and rechirp preprocessing, legacy polar format data can be safely exchanged for subsequent processing with algorithms that perform Stolt formatting.

[11]Phase referencing to a ground point is required in three dimensional Stolt formatting (see Chapter 13).

1.3 Monostatic SAR Modes and Submodes

Synthetic aperture radars that transmit and receive signals with a single antenna are *monostatic.* Flight concepts, together with various antenna steering strategies, comprise operational concepts called *SAR modes.* The most common modes are *stripmap* and *spotlight* modes.

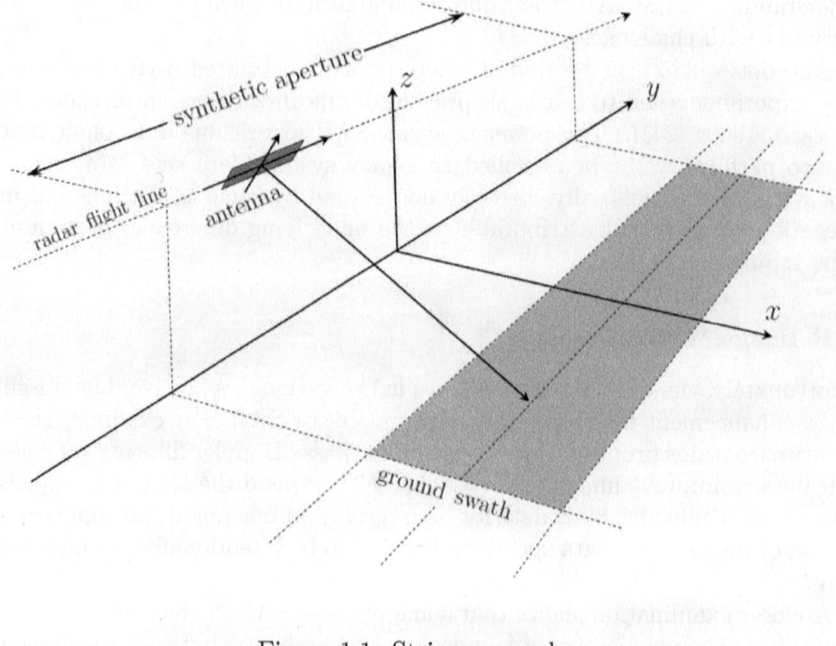

Figure 1.1: Stripmap mode.

SAR modes can also be qualified by *submodes.* Example submodes include linear flight SAR with broadside [146, 720], squint [344, 490], forward looking [132, 133, 230, 613, 615, 872], low altitude [611], fly over or down looking [125, 383, 385, 985] configurations.

Fundamental characteristics of SAR imagery are determined by the operational mode or submode used to collect the SAR data.

Stripmap Mode

A stripmap mode SAR [114, 120, 160, 496, 544, 572, 675] holds the antenna in a fixed position relative to the platform's air frame along a linear flight path as shown in Fig. 1.1. If the antenna's fixed boresight is perpendicular to the flight vector, then the stripmap mode is in a *broadside* geometry. Any other fixed antenna boresight angle results in a *squint* geometry.

Stripmap Mode Ground Swath

A stripmap system produces imagery of a continuous ground swath parallel to the radar flight line. The ground swath is offset by a fixed down-range distance.

Stripmap Mode Along-Track Resolution

The along-track resolution of a pulse echo stripmap mode SAR is

$$\Delta_y^{\text{strip}} \propto \frac{D_y}{4}, \tag{1.4}$$

where D_y is the width of the antenna in the along-track flight dimension y [332, 683, 751].

Compared to other SAR modes, stripmap mode SARs typically provide higher area coverage rates with a compromised trade off of lower along-track resolution. The along-track resolution is lower due to the dwell time restriction imposed by the fixed mounting of the antenna (no antenna steering).

Stripmap Mode Along-Track Resolution Antenna Size Dependencies

The along-track resolution (1.4) of stripmap mode SAR is *directly* proportional to the antenna diameter. A *smaller* antenna is required to provide higher resolution in stripmap mode synthetic aperture imaging. Unlike real aperture imaging, where a smaller antenna degrades resolution, a smaller along-track antenna diameter D_y actually improves the along-track resolution in stripmap mode. The broadened beam pattern produced by a smaller antenna allows more pulse echoes to be gathered from each scatterer by the passing SAR sensor. The broadened beam extends the SAR sensor's angular dwell on each scatterer, thereby increasing along-track resolution through stripmap mode aperture synthesis.

Stripmap Mode Pulse Counts

In stripmap mode, the number of pulse echoes collected from each scatterer depends on each scatterer's range. The beamwidth is narrower at near ranges and broader at far ranges. Scatterers at far ranges are illuminated longer than scatterers at near ranges. Hence, the "length" of the synthetic aperture in stripmap mode has a range dependent definition. Different scatterers at different ranges have different slow-time synthetic aperture dwell periods and different observational (impinging) pulse counts.

Measures of "Dwell"

The observational "dwell" of a SAR system can be measured in terms of time, distance or angle. Each of these dwell measures may or may not be scatterer specific, depending upon the operational mode of the SAR system.

In stripmap mode, a temporal dwell is determined by the time interval a scatterer is illuminated by the passing antenna beam. A spatial dwell is determined by the length of the flight path experienced during the temporal dwell. An angular dwell is measured by the range of view angles subtended between sensor and scatterer as the radar flies by.

Angular Dwell Independence of Range in Stripmap Mode

The observational dwell angle in stripmap mode is determined by the SAR antenna's angular field of view, i.e., by the antenna's angular beamwidth. The antenna's angular field of view is independent of range. Hence, all scatterers in stripmap mode experience the same angular dwell, regardless of each scatterer's down-range position. Stripmap mode's along-track point spread function has no dependence on range and is spatially invariant within the SAR image. We shall see that this is not the case in spotlight mode SAR, where the image point spread function is inescapably spatially variant.

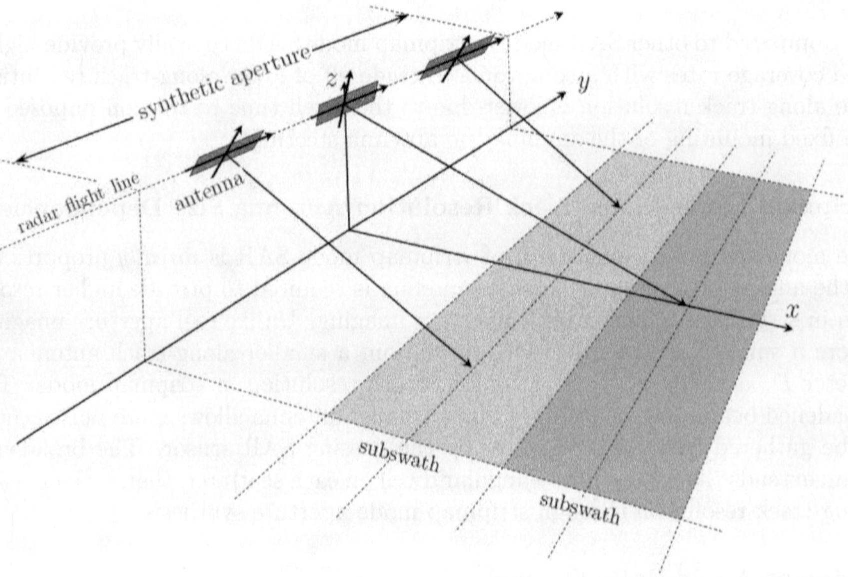

Figure 1.2: Stripmap scan submode.

Stripmap Scan Submode

A stripmap mode radar may operate in a *stripmap scan submode* as shown in Fig. 1.2. This submode seeks to expand the extent of the radar ground coverage in the down-range dimension. The scan submode is created by using multiple beams [225, 485] or by varying the antenna's depression angle on a pulse to pulse, or group of pulses basis [222, 943]. The antenna's depression angle is staggered to create multiple subswaths centered at different down-range distances. The subswaths of the composite strip are illuminated in round robbin fashion.

Increased down-range coverage comes at the expense of reduced along-track resolution or degraded along-track point spread function quality [1024]. Distributing pulses among multiple down-range subswaths reduces the supportable along-track signal bandwidth because of the lower along-track spatial sampling rate. Additionally, allocating slow-time dwell in terms of groups of pulses creates an on-off modulation or scalloping of the collected along-track phase history. This scalloping degrades the quality of the along-track point spread function.

A spoiled beam on transmit with digital beamforming on receive, where a phased array receiving antenna is divided into multiple subapertures can also be used to increase range swath size [546].

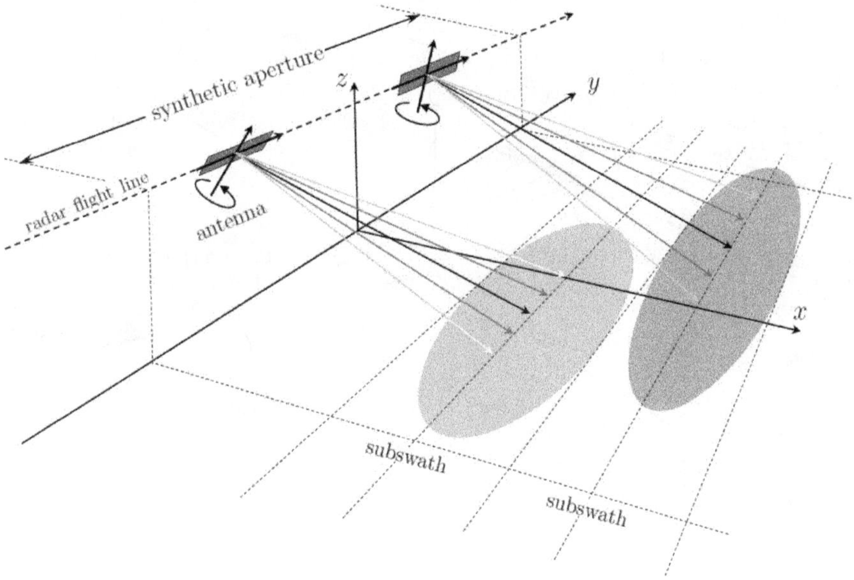

Figure 1.3: Stripmap sweep submode.

Stripmap Sweep Submode

In one variant of scan mode SAR, the antenna is swept through each subswath as shown in Fig. 1.3. The sweep direction is in a direction opposite of the sweep direction performed in spotlight mode [1024]. A cycle of backward to forward antenna sweeps gives opportunity to switch the antenna depression angle to point at different subswaths. A key advantage in this antenna steering strategy is that each scatterer is illuminated by the full along-track beamwidth pattern of the antenna that has both translational movement (platform flight) and angular sweep (antenna steering without staring). This eliminates the amplitude modulation scalloping caused by on-off revisits during the along-track dwell with different parts of the antenna beam pattern. A drawback is that the effective antenna beamwidth with respect to time-space parameters is shrunk by a factor determined by the frequency of the sweep rate. As expected for a scan submode, the along-track resolution is reduced. The loss is incurred by reducing the effective along-track beamwidth.

Spotlight Mode

As a means to escape the along-track resolution limit imposed by the beamwidth restriction on angular dwell of a stripmap mode SAR, a spotlight mode SAR steers and stares the antenna towards a scene of interest as shown in Fig. 1.4 [105, 529–531, 543, 650, 679, 805]. This control of the antenna aim point extends the observational dwell angle of the data collection. A spotlight mode SAR seeks to provide finer along-track resolution with a compromised trade off of smaller area coverage. Typically, stripmap mode is used strategically for wide area surveil-

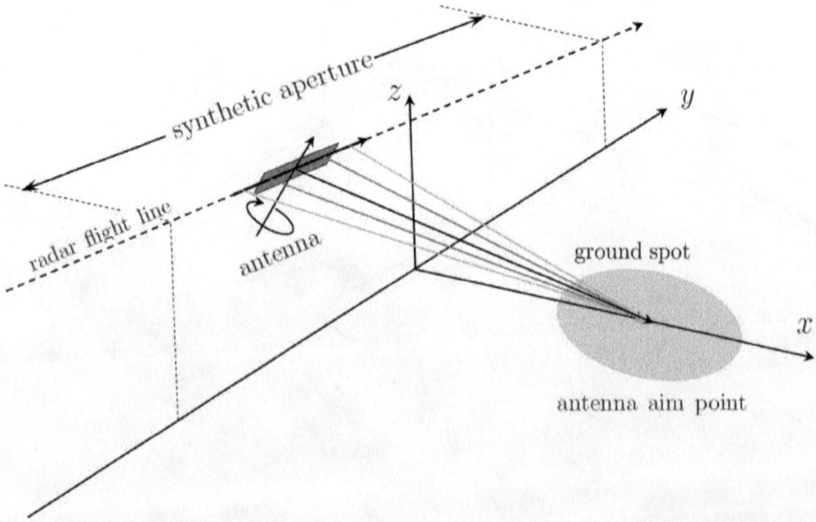

Figure 1.4: Spotlight mode.

lance. Spotlight mode operational concepts are more tactical in nature, providing finer detail on a spot of known or suspected interest.

Spotlight Mode Angular Dwell

Each scatterer in the ground scene of a spotlight mode SAR experiences an unique angular dwell. Hence, the point spread function of spotlight mode imaging is spatially variant.

Inverse Spotlight Submode

Inverse SAR (ISAR) is analogous to spotlight mode SAR, but the steered antenna remains stationary and tracks a moving object such as an aircraft or satellite as shown in Fig. 1.5 [639, 640, 755, 854]. ISAR can also be used to form an image of a rotating object on a turntable [26, 168] or can be used to image celestial bodies like the Moon and planets [744].

ISAR imaging algorithms were initially developed using the range-Doppler formulation of SAR [115, 976]. ISAR image formation processing can support three dimensional imaging [26]. Interferometric ISAR concepts can also be formulated [872].

Spotlight Spotmap Submode

A spotlight mode radar may be employed in an operational concept sometimes called *spotlight mosaic mode.* This submode seeks to expand the coverage rate of the primary spotlight mode by imaging multiple spots as shown in Fig. 1.6. Multiple spot images are combined in post processing to form an imaged mosaic

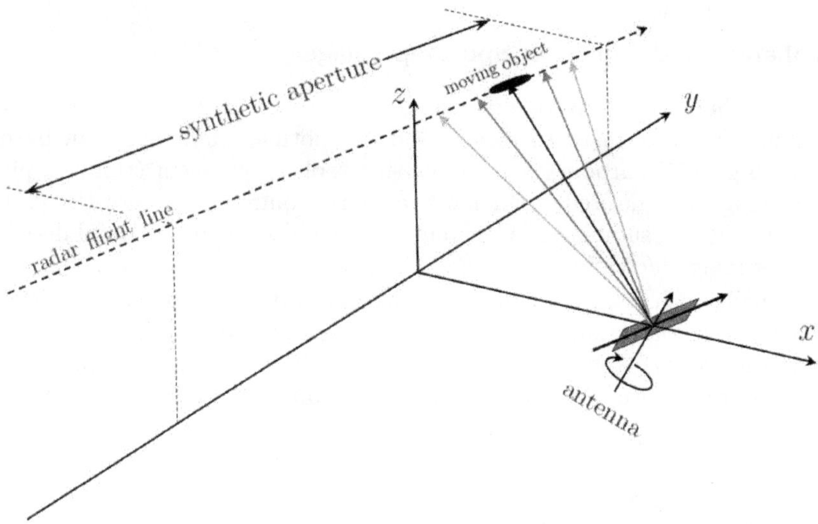

Figure 1.5: Inverse spotlight mode.

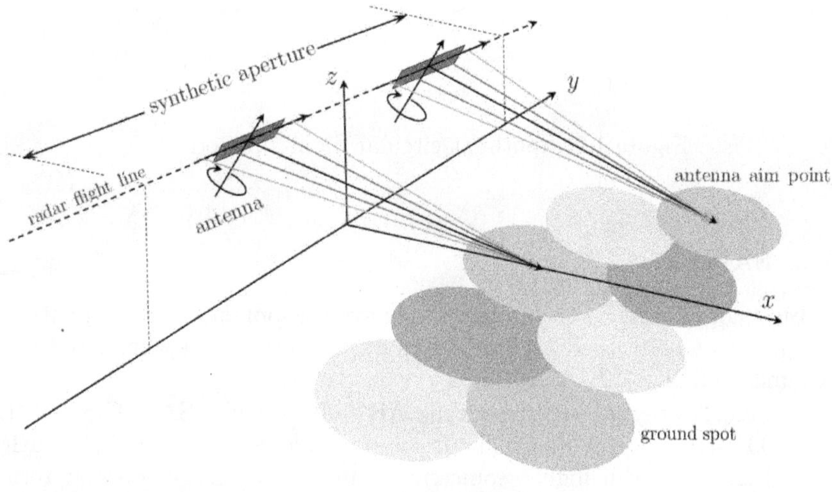

Figure 1.6: Spotlight spotmap submode.

of a larger scene of interest. Spotlight mosaic mode may also be reasonably called *spotmap mode* (a blending of the spot-light/strip-map terms).

Non-coherent and Coherent Spotmap Imagery

If only non-coherent *detected* (real valued pixel) spotmap imagery is desired, then the seaming operation used to blend together spotmap subimages can be performed with general purpose image processing tools. If coherent (complex pixel) spotmap imagery is desired, then much care is required to ensure the correct seaming of coherent subimages. Coherent seaming also requires careful design of the subspot image formation processing algorithm.

High quality coherent spotmap SAR imagery is difficult to achieve with polar format image formation of spot images. Coherent spotmap imagery is more readily obtained with spot images formed by Stolt formatting, where coherent integrations are performed in either the complex pixel or wavenumber domain.

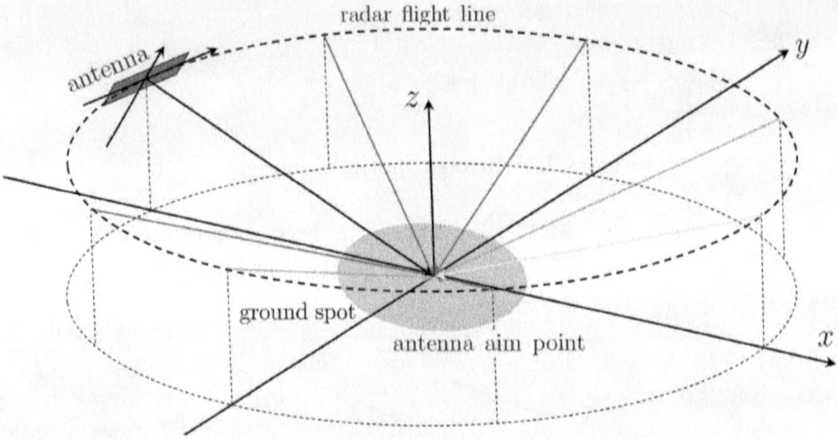

Figure 1.7: Spotlight circular flight submode.

Circular SAR

Some objects of interest on the ground may have inopportunely poor orientations with respect to a selected flight line. Loss of non-returning specular reflections reduces image quality [10, 165, 683].

A more recent operational concept in SAR called circular SAR (C-SAR) [127, 129, 140, 499, 804, 885] forms a synthetic aperture along a circular flight path as shown in Fig. 1.7. Such flight geometries counter low energy specular returns by performing data collections over a partial or full 360 degree sweep of aspect angles.

C-SAR images typically do not contain shadows of objects within the scene. Depending on the type of algorithm used for C-SAR image formation, artifacts such as layover may also be eliminated.

C-SAR enables higher "along-track" resolutions to be achieved because of increased angular dwells [27]. Note that it is not possible to define an "along-track" sensor referenced Cartesian[12] coordinate system in C-SAR because of the ever changing flight path direction of C-SAR operations. Hence, there is no notion of a Cartesian aperture coordinate system nor is there a fixed "slant-plane" in a C-SAR imaging geometry.

Volumetric SAR

When the synthetic aperture has sufficient degrees of freedom from flight operations or from multiple sensor configurations, then three dimensional volumetric synthetic aperture radar (VolSAR) imagery can be produced. "Three dimensional" can mean different things in SAR imagery. For one, it can mean the extraction of terrain elevation data [284], for another, it can mean volumetric imaging. Depending on sensor wavelength, volumetric imaging may provide either three dimensional penetrative imaging or three dimensional surface imaging.

VolSAR image formation processing is best suited to meet the challenging demands (and opportunities) of wide beam, large ground spot and short range operational scenarios. Layover and foreshortening effects should not exist in properly formed VolSAR image products.

Three Dimensional SAR Image Formation Processing

Various three dimensional SAR image formation processing methods exist for VolSAR. Monostatic synthetic aperture radars with three dimensional imaging capability have been proposed with different SAR data acquisition concepts and a variety of algorithms for VolSAR image formation. These include C-SAR, elevation C-SAR [885] and multi-dimensional aperture synthesis systems.

The possibility of three dimensional imaging for rotating objects has been long recognized and approached from a subaperture/subpatch form of range-Doppler processing [26].

Another approach is related to seismically motivated inversion methods [77, 182]. Yet another is a three dimensional generalization of two dimensional Fourier-Hankel[13] methods [324]. Other approaches for multi-dimensional SAR include concepts of multiple pass stereo SAR with different squint angles [247, 274, 284, 491], three dimensional polar formatting [490] and three dimensional tomographic approaches [933].

Meandering Curvilinear Flight SAR

An advanced and emerging concept in SAR is meandering curvilinear flight SAR [687], as illustrated by the example flight path shown in Fig. 1.8. The opportunity afforded by this operational concept is expansion of aperture synthesis processing into down-range and above-ground dimensions. The challenge encountered by

[12] Rene Descartes (1569–1650) was a French philosopher, mathematician, physicist and writer who is known for his work in analytical geometry.

[13] It is more appropriate to view cylindrical Fourier-Hankel formulations of SAR as a specialization of spherical Huygens-Fresnel formulations, not the other way around.

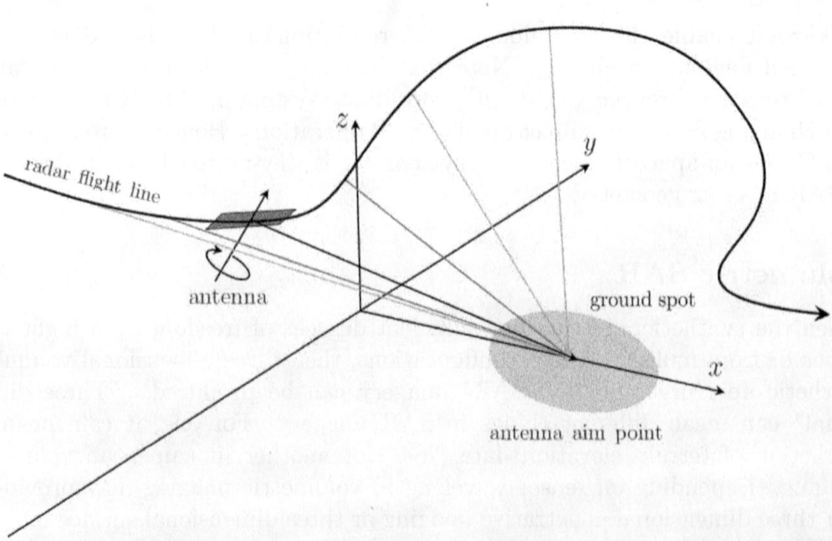

Figure 1.8: Spotlight curvilinear flight submode.

allowing a meandering curvilinear SAR flight path is algorithm development for three dimensional SAR image formation.

1.4 Multistatic and Multipass SAR Modes

Multiple antennas on one or more SAR platforms provide *bistatic* or *multistatic* SAR capabilities [14, 548, 604, 799, 882, 978]. Multistatic SAR enables interferometric measurements that provide forms of three dimensional SAR imagery or generation of motion profiles of moving scene content.

Multiple antenna configurations on a single platform can provide *single pass interferometry.* Multistatic SAR modality can also be achieved by a single platform with a single antenna that makes multiple passes over an area of interest in tailored flight operations [364, 465, 825]. Multipass monostatic interferometry enables applications such as detection of scene changes that occur over time.

Interferometric SAR

Interferometric SAR processing typically registers two complex SAR images and forms an interferogram by multiplying one registered image with the conjugate of the other [1030]. This produces a spatial map of phase differences. The principal phase value of the interferogram is unwrapped and a physical attribute such as scatterer elevation is estimated based on phase differences [823, 824]. Interferometric SAR can be used to achieve either *along-track interferometry* [145] or *down-range interferometry* [749].

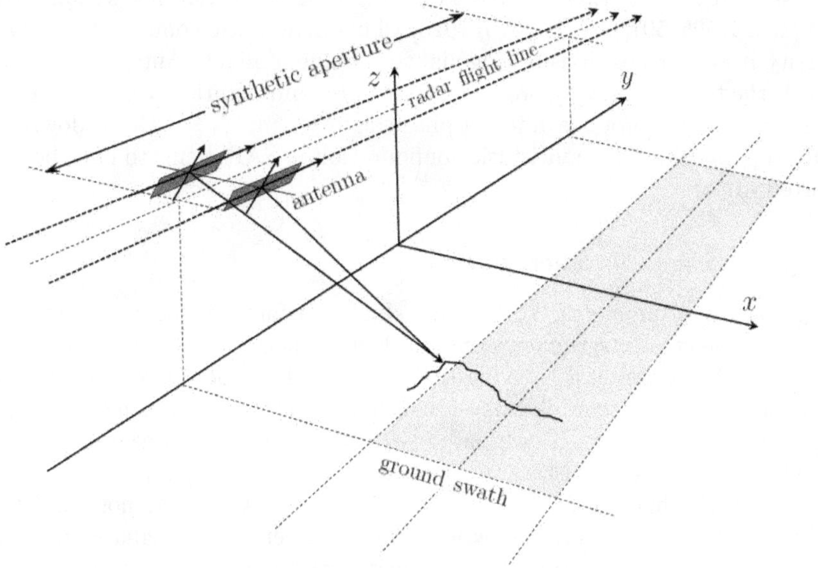

a. Terrain elevation estimation with down-range interferometric SAR.

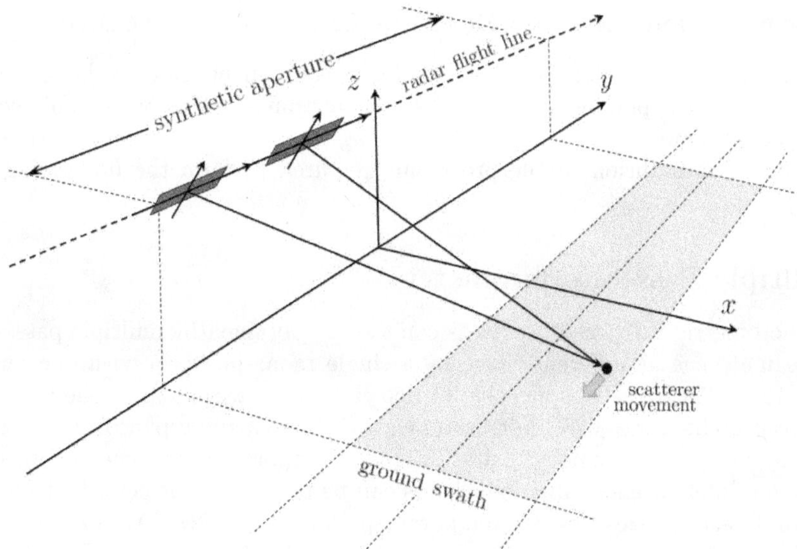

b. Scatterer movement estimation with along-track interferometric SAR.

Figure 1.9: Interferometric configurations.

Down-Range Interferometry

Down-range interferometric SAR from a single platform in single pass operations, as illustrated in Fig. 1.9a, is able to estimate height variations within a scene [87,191,384,596,597,609,620,737,1028]. This multistatic configuration requires diversity in sensor positions orthogonal to the line of flight. Antennas positioned towards the tip of each wing produce the required multistatic diversity. Scatterers at different heights produce different phase history signatures with the down-range mount separation. This multistatic configuration in SAR seems to have been first reported in [408].

Along-Track Interferometry

Along-track interferometric SAR from a single platform in single pass operations is able to detect and estimate along-track movement of scatterers as shown in Fig. 1.9b. The notion of a "moving scatterer" need not be solely associated with a moving object, but can describe a continuous distribution of moving scatterers, such as surface motion associated with ocean currents, glaziers, lava fields, oil spills, etc. This multistatic configuration requires sensor position diversity in line with the flight path (e.g, antennas located towards the nose and tail of the fuselage). Moving scatterers will produce different phase signatures with the along-track mount separation. Along-track interferometry is able to sense components of scatterer motion aligned with the flight line.

Movement Detection with Down-Range Interferometric SAR

Moving scatterers with range closure velocity components can also be sensed and detected with a pair of wing tip mounted antennas. The processing required for detecting moving objects with both along-track and down-range motion components is in addition to the processing required to form the underlying SAR images.

Multiple Pass Interferometry

Interferometric SAR measurements can also be obtained by multiple passes over the same scene at different times by a single radar platform with one antenna [121,197]. These systems are able to detect changes within the scene that occur between flight times [626]. For example, a C-SAR carrying platform can repeatedly fly in a circle around a spot of interest. Temporal interferometric processing between different encircling flight loops can be performed for persistent monitoring to detect changes that are temporal, spatial or both [27]. An important form of change detection processing with complex SAR imagery is *coherent change detection* [490].

Fusion of Multiple Pass Imagery

Data fusion can utilize a variety of multiple pass flight configurations as illustrated in Fig. 1.10 [61,311,410,463,464]. Non-coherent integration of SAR data is one approach to data fusion [667,669]. A potentially more interesting and more

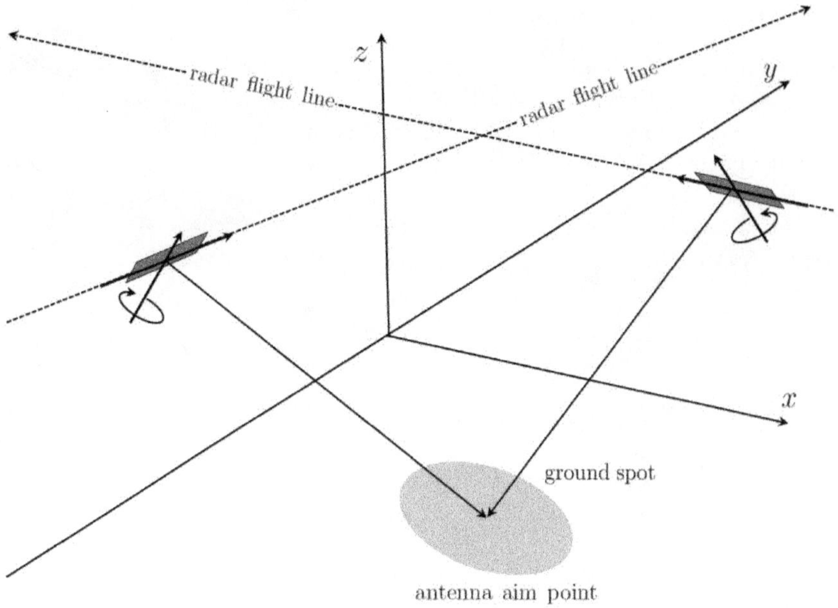

Figure 1.10: Multiple pass SAR flight geometries.

powerful approach is to perform SAR data fusion coherently. Coherently fused data supports the formation of various types of volumetric imagery [341].

Chapter 2

Signal Analysis Preliminaries

Background material helpful for analyzing SAR systems and signals is presented in this chapter. Section 2.1 sets out the importance of waves defined in time and space. Section 2.2 defines multiple coordinate systems useful in SAR signal analysis. Section 2.3 takes a close look at the role of sensor-scatterer distance in SAR imaging geometries. Section 2.4 looks at direction as another facet of spatial geometry that is of primary importance. Section 2.5 describes the various meanings of the word "phase" as used in SAR signal analysis. Section 2.6 defines window and weight functions used to characterize amplitude modulations. Section 2.7 introduces the important classes of baseband and passband signals. Section 2.8 gives an overview of the structure and function of coherent receivers. Section 2.9 previews the various wavenumber domains that will be encountered. Section 2.10 provides a quick review of useful elementary signals. Section 2.11 describes the applicability of the Heisenberg uncertainty principle in SAR signal analysis. Section 2.12 explains Nyquist sampling requirements. Section 2.13 takes a careful look at various definitions of sinc functions. Finally, Section 2.14 covers the theory of signal reconstructions via Shannon's sampling theorem.

2.1 Waves in Time and Space

Waves that propagate through space as a function of time are a key element in SAR signal modeling. SAR signal analysis involves working with both temporal and multi-dimensional spatial signals. Temporal wave analysis is typically done at a fixed point or region in space where time is a free parameter. Spatial wave analysis is typically done at a fixed point in time where spatial parameters are free variables.

Temporal Sinusoidal Waves

Temporal waves of interest include the *monochromatic sinusoidal wave* of a fixed frequency. The relationship between a temporal sinusoidal wave's period T and the corresponding signal frequency f is conventionally described with these expressions,

$$\underset{\substack{\text{wave period}\\ \text{in time}}}{T} \quad \Leftrightarrow \quad \underset{\substack{\text{cyclic}\\ \text{frequency}}}{\tfrac{1}{T} = f} \quad \Leftrightarrow \quad \underset{\substack{\text{angular}\\ \text{frequency}}}{2\pi f = \omega,} \tag{2.1}$$

where $\Leftrightarrow$ indicates relationships between different, but dependent, signal parameters. The relationship $f = \frac{1}{T}$ defines the signal's *cyclic frequency* in hertz[1]. The parameter $\omega = 2\pi f$ is the signal's *angular frequency* measured in radians per second.

Spatial Sinusoidal Waves

Temporal sinusoidal signals are quite familiar. Spatial sinusoidal wave fields may be less so. There are a few inconsistencies in the conventional notation of spatial sinusoidal waves that arise when compared to sinusoidal waves defined in time. Comparing the notations for parameters of spatial sinusoidal waves,

$$\underset{\substack{\text{wavelength}\\ \text{in space}}}{\lambda} \quad \Leftrightarrow \quad \underset{\substack{\text{spatial}\\ \text{frequency}}}{\tfrac{1}{\lambda} = ?} \quad \Leftrightarrow \quad \underset{\text{wavenumber}}{\tfrac{2\pi}{\lambda} = k} \tag{2.2}$$

to those used for temporal waves (2.1) reveals a discrepancy. There is no standard symbol for spatial frequencies in (2.2) at the place holder of "?". This missing spatial wave parameter should correspond to the standard symbol "f" used for the cyclic frequencies of temporal waves. Notational convention is that the parameter,

$$k = \frac{2\pi}{\lambda},$$

which is the spatial sinusoid signal's *wavenumber*, is typically used to specify spatial frequency. Since SAR signal processing is dominated by analysis of spatial waves, this discrepancy in conventional notation and nomenclature for waves in space creates a slight hurdle that makes expressions of SAR mathematics a little more challenging to consume on first exposure. The physical attribute associated with the temporal wave parameter "f" may be intuitive, but it is often the case that the wavenumber "k" is less so. Moreover, temporal frequency is a one dimensional quantity. Spatial frequency or wavenumber may be multi-dimensional [502].

Wave Fields

A region of space over which a standing wave is defined or a region of space over which a wavelet propagates with time is a *wave field.* Wave fields are characterized

[1] Heinrich Rudolf Hertz (1857–1894) was a German physicist who first proved the existence of electromagnetic waves as theorized by Maxwell.

by a spatial signal's amplitude and phase variations. Wave fields can also be characterized by the real and imaginary parts of complex expressions.

Phase Fields

If only the phase variation over space is of interest, then the field is referred to as a *phase field.* Phase fields are often analyzed with the assumption of constant peak wave amplitude.

Dynamic and Static Wave Fields

A wave field over space with wave motion that varies with time is *dynamic.* Otherwise, the wave field is *static.*

Monochromatic and Polychromatic Harmonic Wave Fields

A wave or phase field in time or space that is sinusoidal is *harmonic.* A harmonic field comprised of a single frequency sinusoid is *monochromatic.* If multiple sinusoids of different frequencies exist, the field is *polychromatic.*

Electromagnetic Harmonic Plane Waves

Dynamic electromagnetic sinusoidal plane wave fields have a relationship between the spatial wavenumber k and the temporal angular frequency ω given by

$$\begin{aligned} k &= \frac{\omega}{c}, \\ &= k_\omega, \end{aligned}$$

where c is the speed of light. The wavenumber k_ω is a measure, at a fixed time instant, of the change of phase of a propagating sinusoidal electromagnetic wave as a function of distance. A spatial signal's wavenumber is quantified in terms of radians per meter.

For example, a dynamic electromagnetic wave field with temporal cyclic frequency of 300 MHz (and temporal angular frequency of $2\pi \times 300 \times 10^6$ radians per second) has a wavelength of about 1 meter and a spatial wavenumber of about $2\pi \approx 6.3$ radians per meter.

Some of the cyclic frequencies, wavelengths and wavenumbers commonly encountered in radar are presented in Table 2.1.

Wave Field Manifolds

Wave and phase fields of interest in SAR signal analysis can be one, two or three dimensional. Wave and phase fields can be defined over linear, curvilinear, surface or volume subregions of three dimensional space. We refer to localized regions of wave field support as *manifolds.*

Phase Map Visualizations

The *phase map* is a useful way of visualizing wave or phase fields on a manifold.

Table 2.1: Electromagnetic Spectrum Wavenumber Parameters

Frequency f	Wavelength λ	Wavenumber k_ω			
0.1 GHz	300.0 cm	2.1	=	0.67 π	radians/meter
0.3 GHz	100.0 cm	6.3	=	2.00 π	radians/meter
0.5 GHz	60.0 cm	10.4	=	3.33 π	radians/meter
1.0 GHz	30.0 cm	20.9	=	6.67 π	radians/meter
3.0 GHz	10.0 cm	62.9	=	20.00 π	radians/meter
5.0 GHz	6.0 cm	104.7	=	33.33 π	radians/meter
10.0 GHz	3.0 cm	209.4	=	66.67 π	radians/meter
15.0 GHz	2.0 cm	314.2	=	100.00 π	radians/meter
35.0 GHz	0.9 cm	733.0	=	233.33 π	radians/meter
95.0 GHz	0.3 cm	1,989.7	=	633.33 π	radians/meter

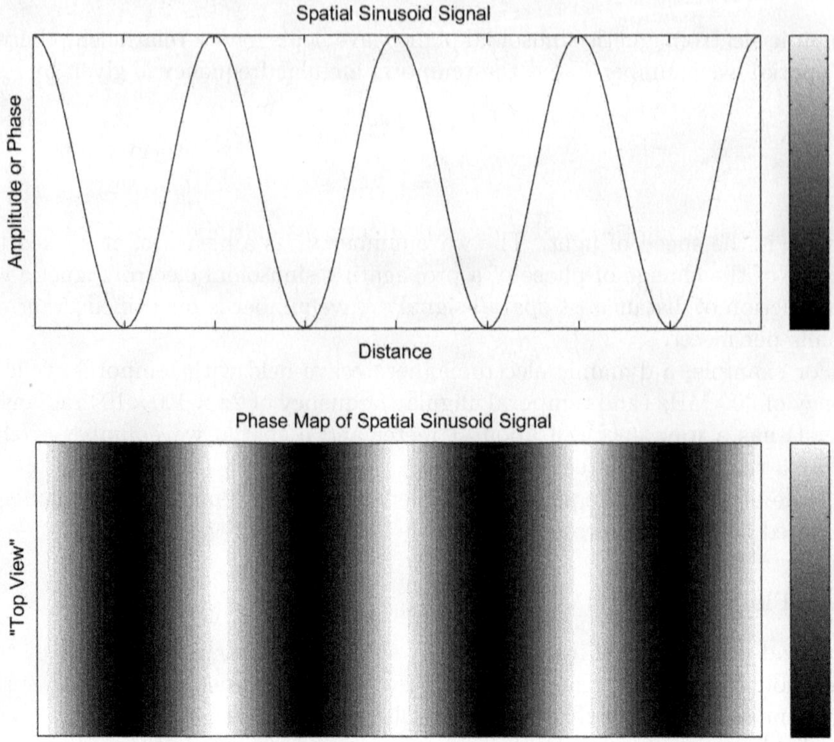

Figure 2.1: Sinusoidal wave in one dimension.

Phase Map of One Dimensional Wave Field

An example phase map of a one dimensional spatial wave field is shown in Fig. 2.1. Gray scale value is used to visually convey spatial signal variations. The phase map of Fig. 2.1 provides a "top view" of a one dimensional wave or phase field.

a. Sinusoidal wave fields in two dimensions.

b. Sinusoidal wave fields in three dimensions on square manifolds.

Figure 2.2: Multi-dimensional phase maps.

Phase Map of Two Dimensional Wave Fields

A two dimensional phase map can convey information about wave amplitude or phase as a function of two spatial variables. Example two dimensional phase maps are shown in Fig. 2.2a. A phase map can also be used to show the values of the real or imaginary part of a complex signal defined over space.

Phase Map of Three Dimensional Wave Fields

A phase map over the surface of some three dimensional manifold can be used to convey information about a three dimensional wave or phase field. In the phase maps of Fig. 2.2b, for example, the wave field extends throughout three dimensional space. The phase field is painted only on the surface of the manifold, which is a cube in the examples of Fig. 2.2b.

2.2 Coordinate Systems

The use of multiple coordinate systems defined over various parameter spaces facilitates analysis of SAR signals. We define five coordinate systems used in various contexts in our study of SAR. These are the *time coordinate system*, the *aperture coordinate system*, the *scene coordinate system*, the *slant-plane coordinate system* and the *image coordinate system.*

Time Coordinate System

There are two "time" variables useful in SAR signal analysis,

$$\begin{aligned} \text{fast-time} \quad &\rightarrow \quad t, \\ \text{slow-time} \quad &\rightarrow \quad \tau. \end{aligned}$$

Fast-time t measures signal propagation time and is called "fast" because of the speed of light propagation of electromagnetic signals. *Slow-time* τ measures signal characteristics related to the position and relatively slow movement of the radar carrying platform.

For linear flight SAR, the along-track antenna position u_y along the flight line y is a function of slow-time τ,

$$u_y(\tau) = \tau \cdot v_{\text{radar}},$$

where v_{radar} is the speed of the moving radar. More often than not, the radar's along-track position u_y, instead of τ, is referred to as the "slow-time" parameter. This terminology is imprecise because u_y is a spatial, not temporal variable.

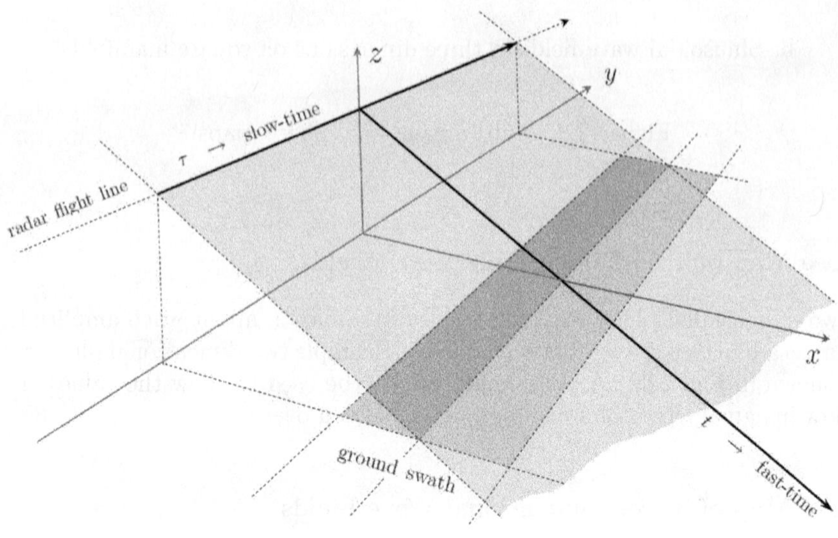

Figure 2.3: Time coordinate system.

The purpose of the two dimensional coordinate system (t, u_y) is to describe a SAR signal in the time-space domain, i.e., in a plane formed by the fast-time and slow-time dimensions. The time coordinate system has at its origin the zero value $t = 0$ of fast-time and the zero value $\tau = 0$ of slow-time. The slow-time origin $\tau = 0$ coincides with the position $u_y = 0$ at the center of a linear synthetic aperture. As shown in Fig. 2.3, these two time-space parameters can be viewed as forming a *slant-plane* where the fast-time dimension is aligned with the *down-range dimension.* The slow-time dimension is called the *along-track dimension.* The along-track direction is orthogonal to the fast-time dimension and is aligned with the flight vector of linear flight SAR.

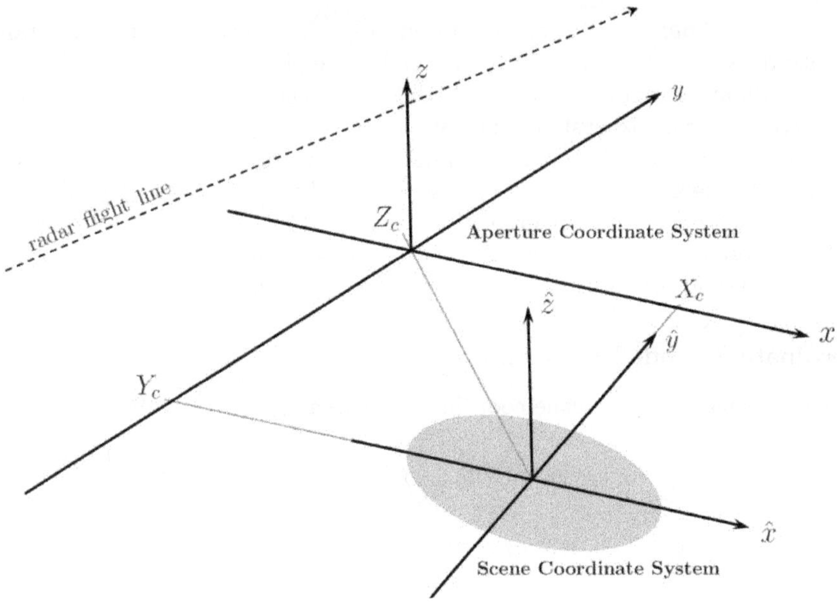

Figure 2.4: Aperture and scene coordinate systems.

Aperture Coordinate System

The role of the *aperture coordinate system* is to describe spatial parameters with respect to the radar carrying platform's flight path [407]. The aperture coordinate system has at its origin the center point of the synthetic aperture flight line. The *along-track axis* of the aperture coordinate system is parallel to the platform's flight vector and is labeled as the y axis. The horizontal x axis is perpendicular to the y axis and is called the *down-range axis.* The z axis finishes out a three dimensional right handed Cartesian coordinate system as illustrated in Fig. 2.4 and measures height above the ground plane. We sometimes call the height dimension the *elevation* or *above-ground dimension.*

Locations expressed in the aperture coordinate system are represented by a triplet

$$(x, y, z)$$

or by the vector

$$\boldsymbol{x} = [x\ y\ z].$$

The position of the SAR antenna in the aperture coordinate system is expressed as a triplet (u_x, u_y, u_z) or by the vector

$$\boldsymbol{u} = [u_x \; u_y \; u_z].$$

The coordinates of the ground scene center point in the aperture coordinate system are (X_c, Y_c, Z_c).

Scene Coordinate System

The *scene coordinate system,* as illustrated in Fig. 2.4, describes spatial parameters with respect to the center of the ground scene [40, 407, 868]. The axes of the scene coordinate system are aligned with, but typically offset from the axes of the aperture coordinate system. The scene coordinate system and the aperture coordinate system are offset from each other by (X_c, Y_c, Z_c). The coordinates of the center of the ground scene in the scene coordinate system are $(0, 0, 0)$. To distinguish between locations (x, y, z) expressed in the aperture coordinate system from those expressed in the scene coordinate system, the triplets of the scene coordinate system are annotated with carets, e.g., $(\hat{x}, \hat{y}, \hat{z})$.

Coordinate System Transformation

The transformation from scene coordinates to aperture coordinates is given by

$$\begin{aligned} x &= X_c + \hat{x}, \\ y &= Y_c + \hat{y}, \\ z &= Z_c + \hat{z} \end{aligned}$$

and, vice versa, by

$$\begin{aligned} \hat{x} &= x - X_c, \\ \hat{y} &= y - Y_c, \\ \hat{z} &= z - Z_c. \end{aligned}$$

Slant-Plane Coordinate System

SAR signal analysis is often reduced from a three dimensional problem to a two dimensional problem. The *slant-plane coordinate system* is one possible choice for this two dimensional coordinate system. The *ground plane* is another option. The radar slant-plane is defined by the nominal flight vector of the radar carrying platform and the line of sight *slant-range vector* to scene center as shown in Fig. 2.5.

In much of our signal analysis, we do not compress the three dimensional problem to a two dimensional problem by a projection down into the ground plane or up into the slant-plane. Rather, we assume scatterers are located at two dimensional points (x, y) resident in the slant-plane. With this simple assumption, we proceed with our analysis in this flat subspace of the truer three dimensional problem [683].

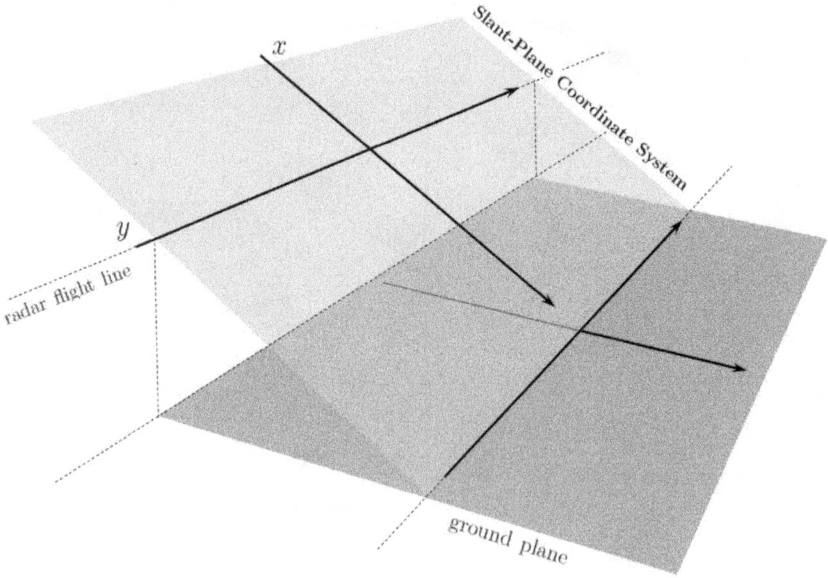

Figure 2.5: Linear flight slant-plane geometry.

Image Coordinate System

Another useful coordinate system is the discrete *image array of pixels* or the *image coordinate system.* The pixel coordinate system can be used to define the relationship between a pixel index pair $[n_x, n_y]$ in the image domain and the corresponding ground position $(\hat{x}[n_x], \hat{y}[n_y])$ within the scene coordinate system. If the image resides in the aperture coordinate system, then the ground position corresponding to indexed location is $(x[n_x], y[n_y])$.

2.3 Sensor-Scatterer Distance

Arguably, the two most important parameters of SAR signal analysis are *distance* and *direction.* Gathered data about the distance and direction between the moving radar sensor and each and every scatterer in the illuminated ground scene empowers SAR imaging.

Various ways that distance is described in SAR signal analysis is presented in this section. Various ways that direction is described in SAR signal analysis in both spatial and wavenumber domains is presented in the next section.

Linear Flight Slant-Plane Sensor-Scatterer Distance

With the linear flight path $\boldsymbol{u}$ aligned with the y axis, the distance between a point antenna a_p at aperture position

$$\boldsymbol{u} = (0, u_y)$$

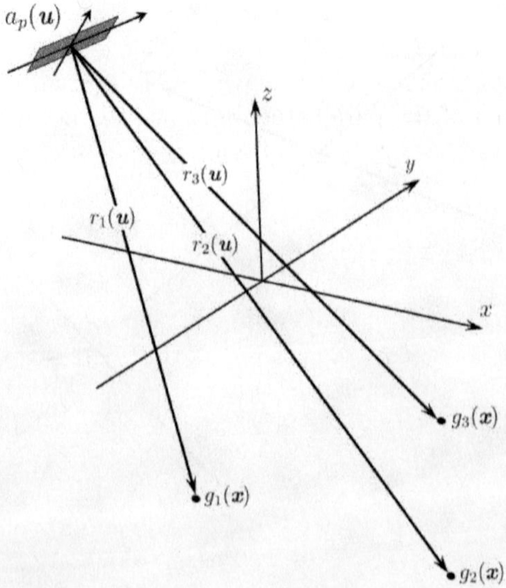

Figure 2.6: Sensor-scatterer distance.

in the slant-plane and a scatterer g_n located at

$$\boldsymbol{x} = (x_n, y_n)$$

in the slant-plane is given by

$$r_n(u_y) = \sqrt{x_n^2 + (u_y - y_n)^2}. \tag{2.3}$$

This is the *sensor-scatterer slant-range distance.*

Non-Linear Flight Sensor-Scatterer Slant-Range Distance in Three Dimensions

When we consider non-linear flight paths, the instantaneous radar position in three dimensions is described by the point $\boldsymbol{u}$ that varies with changing radar platform position within a possibly curvilinear, surface or volumetric synthetic aperture. The corresponding slant-range distance between a point antenna a_p at $\boldsymbol{u} = (u_x, u_y, u_z)$ and a single scatterer g_n at $\boldsymbol{x} = (x_n, y_n, z_n)$ is

$$\begin{aligned} r_n(\boldsymbol{u}) &= |a_p(\boldsymbol{u}) - g_n(\boldsymbol{x})|, \\ &= \sqrt{(u_x - x_n)^2 + (u_y - y_n)^2 + (u_z - z_n)^2}. \end{aligned}$$

Slant-range is illustrated for three scatterers in Fig. 2.6. The slant-range distance is different than the down-range distance x_n, which is invariant in slow-time.

Notational Rule for Ordered Distance and Difference Parameters

A notational rule we use is that the variables within distance and squared difference expressions such as $(u_y - y_n)^2$ are ordered such that the free parameter

(u_y in this instance) comes first and the fixed parameter (y_n in this instance) is placed last. The distance $|a_p(\boldsymbol{u}) - g_n(\boldsymbol{x})|$ is another example of this notational rule and represents measurement of the wave field scattered by $g_n(\boldsymbol{x})$ and sensed at $a_p(\boldsymbol{u})$. With a switch of the parameter positions, the notation $|g_n(\boldsymbol{x}) - a_p(\boldsymbol{u})|$ implies illumination of $g_n(\boldsymbol{x})$ by the point antenna $a_p(\boldsymbol{u})$.

2.4 Sensor-Scatterer Direction

A key element in SAR signal analysis is the variation of sensor-scatterer view angle as a function of sensor position in slow-time. In this section, we lay a foundation for directional analysis by reviewing *direction angles* and *direction cosines* from Cartesian geometry and elaborate on their application in SAR signal analysis.

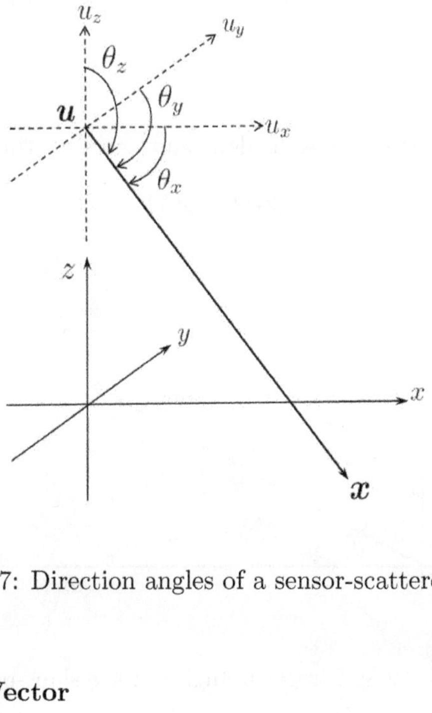

Figure 2.7: Direction angles of a sensor-scatterer vector.

Sensor-Scatterer Vector

A vector

$$\overrightarrow{\boldsymbol{ux}} \tag{2.4}$$

that originates at the sensor position

$$\boldsymbol{u} = [u_x \ u_y \ u_z]$$

and which extends to a scatterer positioned at

$$\boldsymbol{x} = [x \ y \ z],$$

as shown in Fig. 2.7, is called the *sensor-scatterer vector* or *line of sight vector.*

Sensor-Scatterer Direction Angles

The line of sight vector between sensor and scatterer define *sensor-scatterer direction.* In the definition of *direction angles*

$$\theta_i : i \in \{x, y, z\}$$

between sensor and scatterer, the origin of the coordinate system used to describe direction is assumed to continually move with the sensor's changing position $\boldsymbol{u}$ within the synthetic aperture. This geometry is illustrated in Fig. 2.7. The vector of *dimensional direction angles,*

$$\boldsymbol{\theta} = [\theta_x \ \theta_y \ \theta_z] \tag{2.5}$$

that a scatterer's line of sight vector makes with the positive directions of the u_x, u_y and u_z axes of the translated coordinate system are called the *sensor-scatterer direction angles.*

Slow-Time Direction Angles

We can write the direction angles as dependent on slow-time position $\boldsymbol{u}$,

$$\boldsymbol{\theta}(\boldsymbol{u}) = [\theta_x(\boldsymbol{u}) \ \theta_y(\boldsymbol{u}) \ \theta_z(\boldsymbol{u})].$$

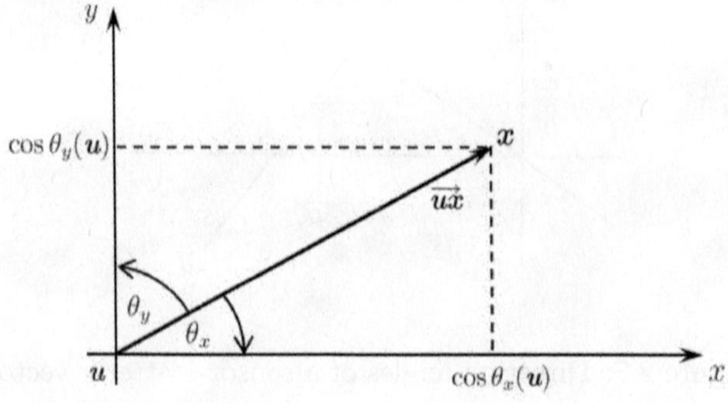

Figure 2.8: Direction angles in the slant-plane.

Slant-Plane Sensor-Scatterer Direction Angles

In the two dimensional slant-plane of linear flight SAR, sensor-scatterer direction can be characterized by the direction angle $\theta_x(u_y)$ measured with respect to the x axis at the radar position $(0, u_y)$. The slant-plane line of sight vector $\overrightarrow{ux}$ to the scatterer at (x, y) is shown in Fig. 2.8. The sensor-scatter direction angle θ_x with respect to the down-range x axis is a function of the radar's along-track position u_y,

$$\theta_x(u_y) \quad = \quad \arccos\left[\frac{x}{\sqrt{x^2 + (u_y - y)^2}}\right].$$

Equivalently, direction in the slant-plane can be characterized with the angle $\theta_y(u_y)$ made with respect to the along-track y axis. Only one of the direction angles $\theta_x(u_y)$ or $\theta_y(u_y)$ need be specified. Knowledge of one implies the value of the other.

Direction could also be quantified by *direction sines,* rather than with *direction cosines,* where in this case,

$$\theta_x(u_y) \quad = \quad \arcsin\left[\frac{u_y - y}{\sqrt{x^2 + (u_y - y)^2}}\right].$$

We shall usually employ direction cosines.

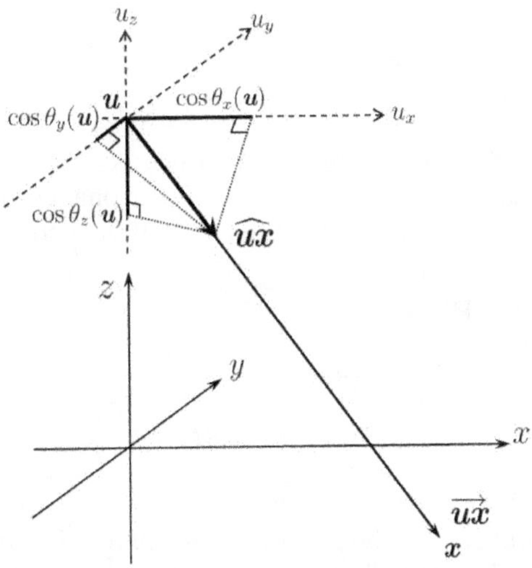

Figure 2.9: Direction cosines of a sensor-scatterer vector.

Sensor-Scatterer Direction Cosines

The *direction cosines* of the slow-time sensor-scatterer direction angles defined in three dimensions are

$$\begin{aligned}
\cos\theta_x(\boldsymbol{u}) &= \frac{u_x - x}{|\boldsymbol{u} - \boldsymbol{x}|}, \\
\cos\theta_y(\boldsymbol{u}) &= \frac{u_y - y}{|\boldsymbol{u} - \boldsymbol{x}|}, \\
\cos\theta_z(\boldsymbol{u}) &= \frac{u_z - z}{|\boldsymbol{u} - \boldsymbol{x}|}.
\end{aligned}$$

The lengths of the slow-time direction cosines are determined by the dimensional projections of the sensor-scatterer unit vector $\widehat{\boldsymbol{ux}}$ as shown in Fig. 2.9.

Pythagorean Triplet of Direction Cosines

The slow-time direction cosines are not independent. Collectively they satisfy a *Pythagorean triplet* condition,

$$\cos^2\theta_x(\boldsymbol{u}) + \cos^2\theta_y(\boldsymbol{u}) + \cos^2\theta_z(\boldsymbol{u}) = 1.$$

This Pythagorean triplet condition provides a constraint among direction parameters. In three dimensional contexts, knowledge of any two direction cosine angles infers the value of the remaining one. In two dimensional contexts, knowledge of any one infers the other. We shall see that the Pythagorean relationship is key to a process called *Fourier migration*, and more generally, in a process called *angular spectrum processing*.

2.5 Phase History

The notion of a *phase history* is central in SAR analysis. We define various meanings of the word "phase" as used in discussions about SAR. We shall define exactly what we mean by "instantaneous phase," "phase shift," "phase center," "phase function," "phase factor," "phase field," "phase history" and "phase map."

Sinusoid Signals and Phase

Consider a real sinusoidal waveform,

$$w(t) = \cos\left[\phi(t)\right]. \tag{2.6}$$

The *instantaneous phase* of the transmitted waveform $w(t)$ is the real number value of $\phi(t)$ at the instant t. The instantaneous phase $\phi(t)$ as a function of time is the *phase function*. If $\phi(t)$ is linear, e.g.,

$$\phi(t) = \omega_c t,$$

then the signal (2.6) is said to be a *continuous wave* signal. We sometimes indicate a monochromatic continuous wave tone signal with a "cw" superscript,

$$w(t)^{\text{cw}} = \cos\left(\omega_c t\right).$$

When a tone signal $\cos(w_c t)$ of fixed frequency ω_c is transmitted, the received echo signal $s(t)^{\text{cw}}$ can be modeled as both *amplitude modulated* by a real function $g(t)$ and *phase modulated* by a real function $\theta(t)$,

$$s(t)^{\text{cw}} = g(t)\cdot\cos\left[\omega_c t + \theta(t)\right]. \tag{2.7}$$

Echo Signal Amplitude Modulation

The amplitude modulation $g(t)$ of a received echo signal (2.7) conveys information about the reflective strength of a scatterer. The Fourier transform of the time dependent amplitude response function $g(t)$ reveals the frequency dependent scattering properties of the scatterer.

Echo Signal Phase Modulation

The *phase modulation* $\theta(t)$ of (2.7) informs, in part, about the dielectric properties of the scatterer. The angle $\theta(t)$ is the *instantaneous phase shift* of the monochromatic sinusoid (2.7) at time t. The time rate of change

$$\partial_t \{\phi(t)\} = \omega_c + \partial_t \{\theta(t)\}$$

of the phase function

$$\phi(t) = \omega_c t + \theta(t)$$

determines the *instantaneous frequency* of the continuous wave echo signal $s(t)^{\text{cw}}$.

Real and Complex Signals

The mathematical notation of SAR signal analysis is often less cumbersome when complex forms of functions are used.

Complex Phase Signals

We express a *complex phase signal* as an exponential function $\exp(\cdot)$ that has a purely imaginary argument comprised of a real phase function $\phi(t)$,

$$\exp\left[j\phi(t)\right].$$

Euler's Formula

A complex phase function is structured with real and imaginary parts as expressed by Euler's[2] formula,

$$\exp(j\theta) = \cos\theta + j\sin\theta. \tag{2.8}$$

Complex Tone Signals

A *complex tone signal*

$$\exp(j\omega_c t) = \cos(\omega_c t) + j\sin(\omega_c t) \tag{2.9}$$

consists of an "in-phase" real component $\cos(j\omega_c t)$ and an imaginary component $\sin(j\omega_c t)$ in relative quadrature phase, sometimes called the "quad-phase" element. We sometimes refer to the complex tone signal (2.9) as a *complex sinusoid.*

Phasors

The complex sinusoid (2.9) is often used to *up convert* baseband signals into passband signals. The *carrier* $\exp(j\omega_c t)$ can be viewed as a counter clockwise spinning unit *phasor* rotating with angular frequency ω_c. In this context, the "c" subscript of ω_c indicates "carrier." For continuous wave signals, this subscript can be interpreted as meaning "constant."

---

[2]Leonhard Euler (1707–1783) was a Swiss mathematician and physicist who worked in mechanics, fluid dynamics, optics and astronomy.

Notational Rule: Syntax for Exponential Signals

We seek to consistently write exponential functions that have constant or real arguments with superscript syntax, e.g., $e^{j\theta}$ and e^{-t}. Complex phase signals that have temporally or spatially dependent arguments are written without superscript notation, for example, we write $\exp(j\omega t)$ instead of $e^{j\omega t}$ and we write $\exp[-jk_\omega r(u_y)]$ instead of $e^{-jk_\omega r(u_y)}$. Consistent use of such notational rules makes the core of the mathematics easier to see at the equation's notational baseline, rather than having details of complicated SAR phase signals relatively hidden in the fine print of exponential function superscripts.

Notational Rule: Phase Factor Parameters

We use a notational convention that the free parameters of $\exp(j\cdot)$ functions are placed last in multiplicative argument chains. For example, we write $\exp(j\omega_c t)$ with time t as the free parameter and with ω_c a constant, and we write $\exp(jT_c\omega)$ with ω as the free parameter and with T_c a constant.

Notational Rule: Explicit Multiplication Operators

Bullets $(\cdot)$ are used for clarity at times where confusion between multiplications and other functional notations is possible, e.g., we write

$$\exp\left[j \cdot \overline{\alpha \cdot (t - t_o)} \cdot t\right],$$

instead of

$$\exp\left[j \cdot \overline{\alpha(t - t_o)} \cdot t\right]$$

to emphasize that α is a fixed parameter, rather than a functional.

Additive Phase Functions and Multiplicative Phase Factors

Among the wonderful properties of the complex exponential function (2.8) is the key property [803] that provides a bridge between additive phase functions and multiplicative phase factors,

$$\exp(-jx_n k_x - jy_n k_y) = \exp(-jx_n k_x) \cdot \exp(-jy_n k_y).$$

Phase History

One or more complex *phase signals* $\exp(j\cdot)$ multiplied together are said to form a *phase history*. For example, the phase history

$$\exp\left[j\frac{\pi}{4} + j\omega_c t - jk_\omega r(u_y)\right] = e^{j\frac{\pi}{4}} \cdot \exp(j\omega_c t) \cdot \exp[-jk_\omega r(u_y)] \tag{2.10}$$

has three *phase factors*. The phase factor $e^{j\frac{\pi}{4}}$ is a constant phase shift, $\exp(j\omega_c t)$ is a *temporal phase factor* and $\exp[-jk_\omega r(u_y)]$ is a *spatial phase factor*. Phase histories exist in temporal domains, spatial domains or combinations of both. Phase histories can also exist in wavenumber domains, either full or in part.

Phase History Decompositions

A phase history factorization into temporal, spatial and wavenumber phase factors is called a *phase history decomposition.*

Temporally and Spatially Separable Phase Histories

A study of signals that vary in time and space is the beginning of SAR signal analysis. SAR temporal phase elements are often separable (through phase factor factorization) from spatial phase elements as shown by the example of (2.10). Such phase histories are said to be *time-space separable.*

Tonal Wavenumber Phase Histories

A particularly important *wavenumber phase history* in SAR analysis is the *tonal wavenumber phase history*

$$\exp(-jx_n k_x - jy_n k_y - jz_n k_z) = \exp(-jx_n k_x)\cdot\exp(-jy_n k_y)\cdot\exp(-jz_n k_z), \quad (2.11)$$

where x_n, y_n and z_n are spatial parameters and where k_x, k_y and k_z are *dimensional wavenumbers.*

Dimensionally Separable Spatial Phase Histories

When a wavenumber phase history is decomposable as demonstrated by (2.11), the phase history is said to be *dimensionally separable.* Dimensionally separable wavenumber phase factors do not imply separability of the supporting wavenumber manifold (the support region where the phase history data are non-zero). The supporting wavenumber manifold must be rectilinear or box shaped to be dimensionally separable. Any other manifold shape is non-separable.

Phase Factor Notation

We write a prefix $\measuredangle$ to indicate a phase factor. We write a prefix $\angle$ to indicate a phase function. Hence, given a phase modulated signal

$$s(t) = g(t) \cdot \exp\left[j\theta(t)\right],$$

we can write

$$\begin{aligned} \angle s(t) &= \theta(t), \\ \measuredangle s(t) &= \exp\left[j\theta(t)\right], \end{aligned}$$

and combining $\angle s(t)$ and $\measuredangle s(t)$,

$$\measuredangle s(t) = \exp\left[j\angle s(t)\right].$$

2.6 Windows and Weights

We give an overview of tools and methods for SAR signal analysis that involve *window* and *weight* functions descriptive of amplitude modulations.

Window Functions

Window functions can be used to define pulse boundaries in time or to specify spectral boundaries in frequency or wavenumber domains. The conventional notation [997] for a rectangular window function is

$$\mathsf{rect}\left(\frac{t}{T_p}\right) = \begin{cases} 1 & \text{for} \quad |t| < \frac{T_p}{2}, \\ 0 & \text{otherwise.} \end{cases}$$

We prefer a notation that is in some ways more compact and illustrative [638],

$$\Pi_{(T_p)}(t) = \begin{cases} 1 & \text{for} \quad |t| < \frac{T_p}{2}, \\ 0 & \text{otherwise.} \end{cases} \tag{2.12}$$

The uppercase Greek symbol Π of (2.12) provides a notational pictograph that is similar to the shape ⊓ of a rectangular window function as shown in Fig. 2.10.

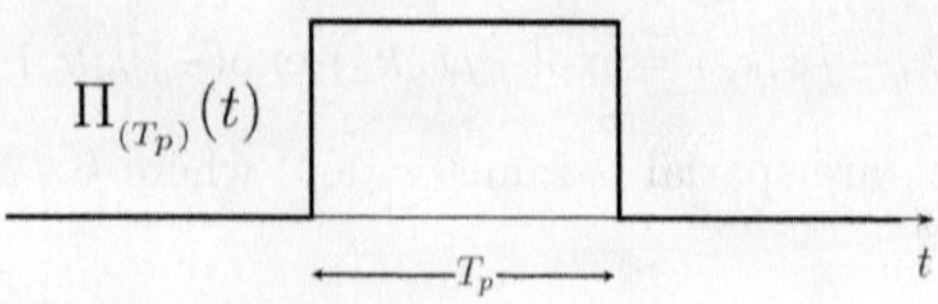

Figure 2.10: Window function.

Window Functions and Pulse Signals

Window functions (2.12) are used to represent *pulse signals* with a time duration of T_p. The subscripted notation

$$\Pi_{(T_p)}(t)$$

indicates a window function with "on" duration of T_p seconds. The pulse is assumed centered about $t = 0$ with time extent $[-T_p/2, +T_p/2]$. We also use a notational variant

$$\Pi_{[T_s,T_f]}(t),$$

where $[T_s, T_f]$ indicates an *on-off window* that turns on at $t = T_s$ and turns off at $t = T_f$. Hence,

$$\Pi_{(T_p)}(t) = \Pi_{[-T_p/2,+T_p/2]}(t). \tag{2.13}$$

We are sometimes constrained by the causality of only positive time measurements and must consider a temporal window function of the form

$$\Pi_{[0,T_p]}(t),$$

which starts at $t = 0$ and ends at $t = T_p$. We sometimes write this as $\Pi_{(T_p)}(t)$, where it can be determined from context that

$$\Pi_{(T_p)}(t) = \Pi_{[0,T_p]}(t),$$

instead of the definition of (2.13).

Weight Functions

When a window function provides an amplitude modulation in addition to an on-off time or frequency gating, we call the function a *weight function*. We add a caret to "Π" to indicate the existence of amplitude modulation,

$$\widehat{\Pi}_{(T_p)}(t) = \text{weight}\left(\frac{t}{T_p}\right).$$

The duration of a weight function can sometimes only be nominally (not precisely) defined. For example, when a weight function is used to describe the spectrum of a time limited (pulse) signal. In such cases, the weight function width is often defined as the span between first nulls of the (baseband) spectrum.

2.7 Baseband and Passband Signals

We give an overview of tools and methods for SAR signal analysis that involve *baseband* and *passband* signal definitions.

Baseband Pulses

Consider the general *baseband pulse,*

$$p_{\Pi(T_p)}(t).$$

This pulse is time limited to an interval of T_p seconds centered at time $t = 0$. This pulse has a Fourier transform represented by

$$P_{\Pi(2\omega_o)}(\omega_b). \tag{2.14}$$

Baseband angular frequencies are indicated by ω_b and baseband angular frequency bandwidth by β_b. The general *baseband pulse spectrum* (2.14) is centered about $\omega_b = 0$ and is assumed to be essentially bandlimited to a bandwidth of

$$\beta_b = 2\omega_o.$$

We call ω_o the *half-width* of the pulse spectrum.

Baseband Pulse Fourier Transform Pairs

The baseband pulse signal Fourier transform pair is

$$p_{\Pi(T_p)}(t) \;{}_t\!\Leftrightarrow_{\omega_b}\; P_{\Pi(2\omega_o)}(\omega_b).$$

We seek to consistently describe baseband pulse signals with the notation "$p(\cdot)$" and their Fourier transforms with "$P(\cdot)$."

Passband Pulse Signals

Carrier modulation or *up conversion* of a baseband pulse $p_{\Pi(T_p)}(t)$ by a time domain *carrier* (complex) sinusoid $\exp(j\omega_c t)$ with angular frequency ω_c produces a *passband pulse signal,*

$$s_{\Pi(T_p)}(t) = \exp(j\omega_c t) \times p_{\Pi(T_p)}(t).$$

We seek to consistently describe a *passband signal* with "$s(\cdot)$" and the *passband pulse spectrum* with "$S(\cdot)$." Implicit in the "$s(\cdot)$" notation for passband pulse signals is the existence of the carrier $\exp(j\omega_c t)$.

We often use scripted notation for indicating the presence of shifts and carriers in time and frequency domains. We express a passband signal in terms of the baseband pulse with scripted notation as

$$s_{\Pi(T_p)}(t) = p_{\Pi(T_p)}^{\text{carrier}(\omega_c)}(t).$$

Careful and consistent use of scripted notations provides informative and compact signal state annotations.

Angular Frequency Parameters

We often subscript various angular frequency variables ω. We indicate the carrier frequency as ω_c, the baseband frequency as ω_b and the passband frequency as ω_p. The carrier, baseband and passband angular frequencies are related as

$$\omega_p = \omega_c + \omega_b.$$

Passband Pulse Signal Spectrum

The passband pulse signal spectrum is the same as the baseband signal spectrum, but right shifted by a constant frequency offset of ω_c,

$$S_{\Pi(2\omega_o)}(\omega_p) = P_{\Pi(2\omega_o)}(\omega_p - \omega_c). \tag{2.15}$$

The passband pulse signal spectrum is essentially bandlimited to the passband window function,

$$\Pi_{[\omega_c-\omega_o,\omega_c+\omega_o]}(\omega_p).$$

The passband pulse signal has bandwidth $2\omega_o$, which is indicated in (2.15) by the subscript

$$\Pi(2\omega_o).$$

When we want to be more explicit about *passband center frequency* location in the ω_p domain, we sometimes write the subscript as

$$S_{\Pi(2\omega_o @ \omega_c)}(\omega_p).$$

Fourier transforms of passband pulse signals in terms of the up (right) shifted baseband pulse spectrum are expressible in scripted notation,

$$S_{\Pi(2\omega_o)}(\omega_p) = P_{\Pi(2\omega_o)}^{\text{R.shift}(\omega_c)}(\omega_b),$$

where the R.shift(ω_c) superscript indicates up conversion of $P_{\Pi(2\omega_o)}(\omega_b)$.

Passband Pulse Fourier Transform Pair

In summary, the Fourier transform pair of a passband pulse signal has multiple notational forms,

$$\begin{array}{rcl} s_{\Pi(T_p)}(t) & t \Leftrightarrow \omega_p & S_{\Pi(2\omega_o)}(\omega_p), \\ p_{\Pi(T_p)}^{\text{carrier}(\omega_c)}(t) & t \Leftrightarrow \omega_b & P_{\Pi(2\omega_o)}^{\text{R.shift}(\omega_c)}(\omega_b), \\ \exp(j\omega_c t) \times p_{\Pi(T_p)}(t) & t \Leftrightarrow \omega_b & P_{\Pi(2\omega_o)}(\omega_p - \omega_c). \end{array}$$

Both algebraic and scripted syntax options are used, depending on whether clarity by conciseness or the need to expose algebraic details is desired.

2.8 Coherent Receivers

The pair of scatterer informative parameters $g(t)$ and $\theta(t)$ of a continuous wave echo signal,

$$s(t)^{\text{cw}} = g(t) \cdot \cos\left[\omega_c t + \theta(t)\right],$$

can be extracted by a *coherent receiver* [119]. To explain the function of a coherent receiver, consider a real sinusoid signal that is expressed and manipulated in complex analytic form [202],

$$\begin{aligned} s(t)^{\text{cw}} &= g(t) \cdot \cos\left[\omega_c t + \theta(t)\right], \\ &= \mathsf{Real}\left\{g(t) \cdot \cos\left[\omega_c t + \theta(t)\right] + jg(t) \cdot \sin\left[\omega_c t + \theta(t)\right]\right\}, \\ &= \mathsf{Real}\left\{g(t) \cdot \exp\left[j\omega_c t + j\theta(t)\right]\right\}, \\ &= \mathsf{Real}\left\{g(t) \cdot \exp(j\omega_c t) \cdot \exp\left[j\theta(t)\right]\right\}. \end{aligned} \tag{2.16}$$

To simplify notation, we suppress the $\mathsf{Real}\{\cdot\}$ operator and use a short hand for (2.16) by writing

$$s(t)^{\text{cw}} = g(t) \cdot \exp(j\omega_c t) \cdot \exp\left[j\theta(t)\right]. \tag{2.17}$$

The parameters $g(t)$ and $\theta(t)$ embedded in the echo signal express amplitude and phase properties of the illuminated scatterer. The embedded carrier $\exp(jw_c t)$ within the received echo signal (2.17) provides no information about scatterer properties.

Carrier Demodulation of a Continuous Wave Echo Signal

The purpose of *carrier demodulation* or *down conversion* of the received continuous wave echo signal is to remove the carrier element

$$\exp(jw_c t) = \cos(w_c t) + j\sin(w_c t) \tag{2.18}$$

from (2.17). The remaining scatterer informative parameters $g(t)$ and $\theta(t)$ can be extracted by the in-phase and quad-phase channels of the coherent receiver shown in Fig. 2.11 [984].

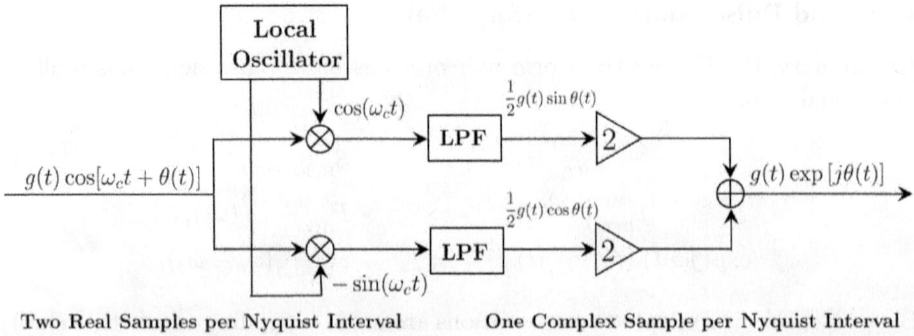

Figure 2.11: Coherent receiver for demodulation and real to complex signal conversion.

The operation of a coherent receiver is based on the temporal orthogonality of sine and cosine waveforms. Consider the trigonometric identity,

$$\begin{aligned} s(t)^{\text{cw}} &= \mathsf{Re}\left\{g(t)\cdot\exp(j\omega_c t)\cdot\exp[j\theta(t)]\right\}, \\ &= g(t)\cos[\omega_c t + \theta(t)], \\ &= g(t)\cos\theta(t)\cos(\omega_c t) - g(t)\sin\theta(t)\sin(\omega_c t). \end{aligned} \tag{2.19}$$

The local oscillator and low pass filters of the coherent receiver demodulate the carrier elements $\cos(\omega_c t)$ and $\sin(\omega_c t)$ from (2.19). The top path of the coherent receiver system's signal flow diagram outputs the cosine dependent in-phase signal element,

$$\cos(\omega_c t) \times \{g(t)\cos\theta(t)\cos(\omega_c t) - g(t)\sin\theta(t)\sin(\omega_c t)\} = \frac{1}{2}g(t)\cos\theta(t).$$

Likewise, the lower path of the coherent receiver outputs the sine dependent quad-phase signal element,

$$-\sin(\omega_c t) \times \{g(t)\cos\theta(t)\cos(\omega_c t) - g(t)\sin\theta(t)\sin(\omega_c t)\} = \frac{1}{2}g(t)\sin\theta(t).$$

After amplitude amplification by a factor of two, the coherent receiver has two channels with real outputs,

$$\begin{aligned} \text{In-Phase Channel:} &\Rightarrow g(t)\cos\theta(t), \\ \text{Quad-Phase Channel:} &\Rightarrow g(t)\sin\theta(t). \end{aligned}$$

The digital data of these channels are collected and combined in a complex data representation,

$$g(t)\cdot\exp[j\theta(t)] = g(t)\cos\theta(t) + j\cdot g(t)\sin\theta(t).$$

This process of analog signal to complex digital data conversion is called *coherent detection* [984].

With coherent detection, Nyquist rate requirements can be expressed in terms of one complex data sample per Nyquist interval or a rate based on two real samples per Nyquist interval.

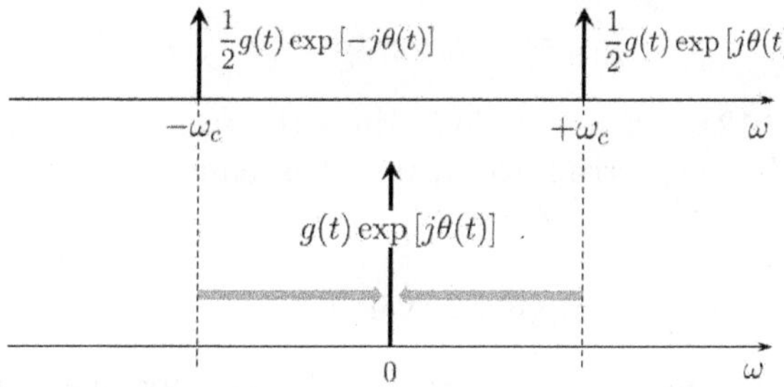

a. Double sideband real signal spectra before and after down conversion.

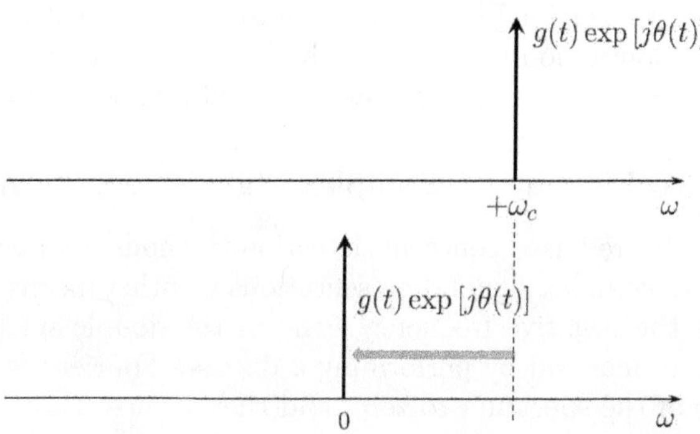

b. Single sideband complex signal spectra before and after down conversion.

Figure 2.12: Down converted double and single sideband signal spectra.

Spectral Representations of Signal Demodulation

As expressed by an inverse Euler equation,

$$g(t) \cdot \cos[w_c t + \theta(t)] \quad = \quad g(t) \cdot \frac{\exp\{j[w_c t + \theta(t)]\} + \exp\{-j[w_c t + \theta(t)]\}}{2},$$

amplitude and phase modulated sinusoid signals are *double sideband signals.* The spectral description of down conversion of a real sinusoidal signal

$$s(t)^{\text{cw}}_{\text{real}} = g(t)_{\text{real}} \cdot \cos[\omega_c t + \theta(t)]$$

is shown in Fig. 2.12a. Demodulation brings the two parts of the real signal's double sided passband spectrum together at baseband. Analytic complex representations of signals

$$s(t)^{\text{cw}}_{\text{complex}} = g(t)_{\text{real}} \cdot \exp(j\omega_c t) \cdot \exp\left[j\theta(t)\right]$$

are *single sideband signals* as shown Fig. 2.12b. The operation of demodulation is expressed in complex form by multiplication of $s(t)^{\text{cw}}$ with $\exp(-j\omega_c t)$,

$$\begin{aligned} g(t)_{\text{complex}} &= \exp(-j\omega_c t) \times \left\{ s(t)^{\text{cw}}_{\text{complex}} \right\}, \\ &= g(t)_{\text{real}} \cdot \exp\left[j\theta(t)\right]. \end{aligned}$$

An illustration of the spectral description of baseband conversion of a complex sinusoidal signal is shown in Fig. 2.12b. Comparing this with Fig. 2.12a, the result of baseband conversion is the same whether the signal was actually real or formally expressed with the complex analytic signal representation.

Conversion of Real Signals to Complex Signals in Software

In addition to hardware based coherent detection methods, real baseband signals can be converted to complex signal representations by other means [621,653]. One way is to remove the negative frequency band of the double sideband spectrum in software. This is achieved by performing a discrete Fourier transform, setting the negative side of the spectrum to zero, and then inverse Fourier transforming to a complex representation. Since removing half of the spectrum reduces the signal power by two, the result should be multiplied after conversion to complex format by a factor of two. Other ways include using the Hilbert transform (see Section 14.5) or by processing with quadrature filters to produce analytic signal descriptions [309, 782].

2.9 Wavenumber Domains

Bandwidth analysis in the frequency-wavenumber or "omega-k" (ω, k) domain is required to determine Nyquist sampling constraints of SAR signals that exist in the time-space (t, x) domain. Fourier analysis of SAR signals has essential utility for determining Nyquist sampling requirements, even if a particular SAR image formation method is not performed in a frequency or wavenumber domain.

Temporal Wavenumbers

The wavenumber associated with a propagating monochromatic sinusoidal wave moving at the speed of light c with temporal angular frequency ω is [91]

$$k_\omega = \frac{\omega}{c}.$$

The *temporal wavenumbers* related to baseband and passband angular frequency variables are

$$\begin{array}{lcll} \omega & \rightarrow & k_\omega & \text{temporal wavenumber,} \\ \omega_c & \rightarrow & k_{\omega_c} & \text{carrier wavenumber,} \\ \omega_b & \rightarrow & k_{\omega_b} & \text{baseband wavenumber,} \\ \omega_p & \rightarrow & k_{\omega_p} & \text{passband wavenumber.} \end{array}$$

We often refer to a "temporal wavenumber" k_ω as a "frequency" whether it is expressed with ω or k_ω notation.

Spatial Wavenumbers

There are also many different *spatial wavenumber* variables encountered in radar signal analysis [306, 937].

The wavenumber k_{u_y} is the Fourier transform variable corresponding to the aperture position variable u_y.

Fourier transforms of temporal-spatial (t, u_y) domain signals lead to spectra described in the *k-space* (k_ω, k_{u_y}) domain, which is the frequency-wavenumber domain of the *SAR signal spectrum.*

If a terrain scattering function is expressed in terms of spatial coordinates (x, y), then the corresponding spatial wavenumber pair is (k_x, k_y).

Here is a list of many of the k-space variables (and their corresponding Fourier transform variables) associated with SAR system analysis:

$$\begin{array}{lcll} u_y & \Leftrightarrow & k_{u_y} & \text{aperture wavenumber,} \\ \\ x & \Leftrightarrow & k_x & \text{down-range wavenumber in aperture coordinate system,} \\ y & \Leftrightarrow & k_y & \text{along-track wavenumber in aperture coordinate system,} \\ z & \Leftrightarrow & k_z & \text{above-ground wavenumber in aperture coordinate system,} \\ \hat{x} & \Leftrightarrow & k_{\hat{x}} & \text{down-range wavenumber in scene coordinate system,} \\ \hat{y} & \Leftrightarrow & k_{\hat{y}} & \text{along-track wavenumber in scene coordinate system,} \\ \hat{z} & \Leftrightarrow & k_{\hat{z}} & \text{above-ground wavenumber in scene coordinate system,} \\ \\ x & \Leftrightarrow & \breve{k}_x & \text{Stolt formatted down-range wavenumber,} \\ y & \Leftrightarrow & \breve{k}_y & \text{Stolt formatted along-track wavenumber,} \\ z & \Leftrightarrow & \breve{k}_z & \text{Stolt formatted above-ground wavenumber,} \\ x & \Leftrightarrow & \check{k}_x & \text{polar formatted down-range wavenumber,} \\ y & \Leftrightarrow & \check{k}_y & \text{polar formatted along-track wavenumber,} \\ z & \Leftrightarrow & \check{k}_z & \text{polar formatted above-ground wavenumber.} \end{array}$$

We seek to prevent confusion by consistently subscripting k-space wavenumber variables. We almost never simply write "k," except in reference to "k-space."

2.10 Elementary Functions

Certain elementary functions occur often in SAR signal analysis. Some of the more important of these include:

·impulse function at (x_n, y_n) $\rightarrow$ $\delta(x - x_n, y - y_n)$,
·sinc function at (x_n, y_n) $\rightarrow$ $\text{sinc}(x - x_n, y - y_n)$,
·hyperbola as a function of u_y at (x_n, y_n) $\rightarrow$ $[\sqrt{x_n^2 + (u_y - y_n)^2}, u_y]$,
·parabola as a function of u_y at (x_n, y_n) $\rightarrow$ $\left[x_n + (u_y - y_n)^2/(2x_n), u_y\right]$,
·sinusoid in time $\rightarrow$ $\exp(j\omega_c t)$,
·sinusoid in space (one dimensional) $\rightarrow$ $\exp(jk_x x)$,
·sinusoid in spatial (x, y) plane $\rightarrow$ $\exp(jk_x x + jk_y y)$,
·sinusoid in wavenumber (k_x, k_y) plane $\rightarrow$ $\exp(jx_n k_x + jy_n k_y)$,
·harmonic spherical wave field $\rightarrow$ $\exp(-jk_{\omega_c} r)$.

We review some of these elementary signals and their Fourier transforms.

Impulse Function in a Plane

The Fourier transform of a Dirac[3] impulse function

$$\delta(x - x_n, y - y_n)$$

located at (x_n, y_n) in the (x, y) plane is a pair of complex sinusoids,

$$\exp(-jx_n k_x) \cdot \exp(-jy_n k_y)$$

defined over the infinite k-space plane (k_x, k_y). The phase factor $\exp(-jx_n k_x)$ is a complex sinusoid with a frequency of $-x_n$ in the k_x wavenumber dimension. The phase factor $\exp(-jy_n k_y)$ is a complex sinusoid with a frequency of $-y_n$ in the k_y wavenumber dimension.

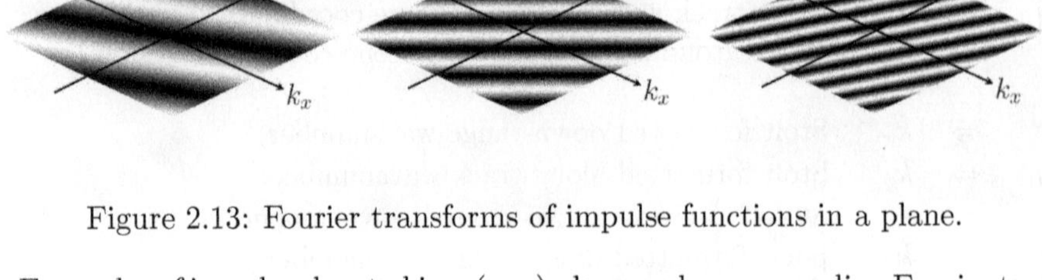

Figure 2.13: Fourier transforms of impulse functions in a plane.

Examples of impulses located in a (x, y) plane and corresponding Fourier transforms in the (k_x, k_y) domain are shown in Fig. 2.13. Note that the spatial domain

[3]Paul Adrien Maurice Dirac (1902–1984) was a British theoretical physicist who is known for his early contributions to quantum mechanics and quantum electrodynamics, including his development and use of generalized functions [377].

vector directed from the origin to the location of the singularity (i.e., impulse) corresponds to the direction of the directed tonal (sinusoidal) wave field in the k-space plane. Additionally, the distance of the impulse from the origin determines the frequency of the directed sinusoidal wave in the k-space plane. The larger the distance, the higher the frequency of the tonal wave field.

Expressions of Fourier Transform Pairs

We sometimes express a nominal (ignoring constants) Fourier transform pair with the notation

$$\delta(x - x_n, y - y_n) \rightleftharpoons \exp(-jx_n k_x) \cdot \exp(-jy_n k_y), \tag{2.20}$$

where $\rightleftharpoons$ indicates an informally expressed Fourier relationship where amplitude functions or constants are suppressed. A completely expressed Fourier transform pair is indicated with $\Leftrightarrow$ and since (2.20) is fully expressed, we write

$$\delta(x - x_n, y - y_n) \Leftrightarrow \exp(-jx_n k_x) \cdot \exp(-jy_n k_y).$$

Plane Waves

Plane wave models are important in some approaches to SAR analysis and image formation processing. For example, Stolt formatting can be viewed as essentially a plane wave decomposition obtained by a spatial Fourier transform of spherical wave field structures that exist in the time-space domain [812]. As another example, the structure of plane waves is a key assumption in the far field approximation used by polar formatting. We provide background material on directed plane waves.

Directed Plane Wave

Consider a plane wave

$$pp(t, \boldsymbol{x})_{\hat{\boldsymbol{r}}} = pp(ct - \boldsymbol{x} \cdot \hat{\boldsymbol{r}})$$

traveling in time and space that is moving in the direction of propagation defined by the unit vector,

$$\hat{\boldsymbol{r}} = \frac{\boldsymbol{r}}{|\boldsymbol{r}|},$$

where $\hat{\boldsymbol{r}} = [\hat{r}_x \ \hat{r}_y \ \hat{r}_z]$. The unit vector $\hat{\boldsymbol{r}}$ is aligned with the direction of plane wave motion and is normal to the propagating plane wave as illustrated in Fig. 2.14.

Directed Monochromatic Sinusoidal Plane Wave Field

Of frequent interest in SAR signal analysis is the *directed monochromatic sinusoidal plane wave field,* which in real form is given by

$$pp(ct - \boldsymbol{x} \cdot \hat{\boldsymbol{r}})_\omega = \cos[k_\omega(ct - \boldsymbol{x} \cdot \hat{\boldsymbol{r}})],$$

where the temporal wavenumber is

$$k_\omega = \frac{\omega}{c}.$$

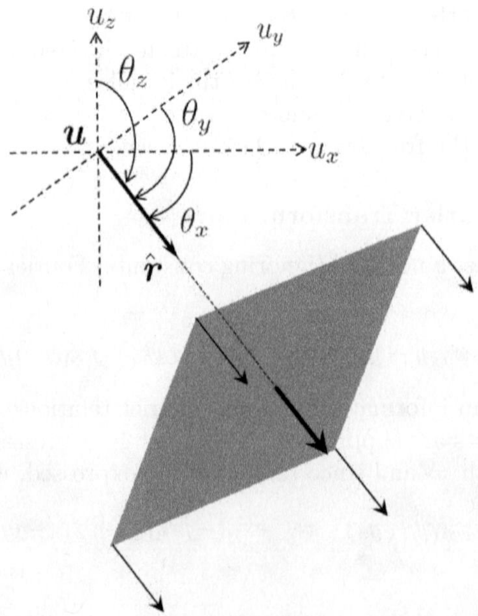

Figure 2.14: Directed plane wave.

Sinusoidal Plane Wave Moving Along a Coordinate Axis

Consider a sinusoidal plane wave field directed along the x axis. In this case,

$$\begin{aligned} \hat{\boldsymbol{r}} &= [1\ 0\ 0], \\ \boldsymbol{x} \cdot \hat{\boldsymbol{r}} &= x \end{aligned}$$

and

$$\begin{aligned} pp(ct - \boldsymbol{x} \cdot \hat{\boldsymbol{r}})_\omega &= \cos[k_\omega(ct - \boldsymbol{x} \cdot \hat{\boldsymbol{r}})], \\ &= \cos[k_\omega(ct - x)], \\ &= \cos(\omega t - k_\omega x). \end{aligned} \tag{2.21}$$

Wavenumber Mappings Between Time and Space

If we define a spatial wavenumber k_x as suggested by the form of (2.21), and if we define an equivalence relationship between temporal and spatial wavenumbers

$$k_x = k_\omega, \tag{2.22}$$

then the three dimensional sinusoidal plane wave propagating in the direction of x is expressed as

$$\cos(\omega t - k_\omega x) = \cos(\omega t - k_x x).$$

Wavelength Mappings Between Time and Space

In addition to the relationship between temporal and spatial wavenumbers, there exists a relationship between temporal and spatial wavelengths of propagating plane waves. Explicitly, for a sinusoidal plane wave with frequency ω propagating in the x direction, from (2.22),

$$\frac{2\pi}{\lambda_x} = \frac{2\pi}{\lambda_\omega}.$$

Hence, a sinusoidal plane wave moving along a coordinate axis induces a direct mapping from the temporal wavelength λ_ω to the spatial wavelength λ_x,

$$\lambda_x = \lambda_\omega.$$

Plane Waves Moving in Arbitrary Directions

In the general case, where $\hat{\boldsymbol{r}}$ is not aligned with a coordinate axis, we need to deal with the projections of the temporal wavenumber onto dimensional wavenumbers (and associated mappings of wavelengths) corresponding to more than just one spatial dimension. A temporal wavenumber k_ω projected onto dimensional wavenumbers describes a sinusoidal plane wave traveling in some general direction. For example, a sinusoidal plane wave of temporal frequency ω propagating in the direction of $\hat{\boldsymbol{r}}$ can be expressed as

$$\begin{aligned} pp(ct - \boldsymbol{x}\cdot\hat{\boldsymbol{r}})_\omega &= \cos[k_\omega(ct - \boldsymbol{x}\cdot\hat{\boldsymbol{r}})], \\ &= \cos\left[\omega t - k_\omega(\boldsymbol{x}\cdot\hat{\boldsymbol{r}})\right], \\ &= \cos(\omega t - \boldsymbol{k}_{\hat{\boldsymbol{r}}}\cdot\boldsymbol{x}), \end{aligned}$$

where, in a generalized version of (2.22),

$$\boldsymbol{k}_{\hat{\boldsymbol{r}}} = k_\omega\hat{\boldsymbol{r}}.$$

This plane wave has movement in time and space in the direction identified by the *directed wavenumber vector,*

$$\begin{aligned} \boldsymbol{k}_{\hat{\boldsymbol{r}}} &= [k_{\hat{r}_x}\ k_{\hat{r}_y}\ k_{\hat{r}_z}], \\ &= k_\omega\cdot\left[\hat{r}_x\ \hat{r}_y\ \hat{r}_z\right], \\ &= k_\omega\hat{\boldsymbol{r}}. \end{aligned}$$

The direction of the wavenumber vector $\boldsymbol{k}_{\hat{\boldsymbol{r}}}$ is determined by $\hat{\boldsymbol{r}}$. Hence, $\boldsymbol{k}_{\hat{\boldsymbol{r}}}$ is a vector normal to the planar “wavefront” of the directed sinusoidal plane wave. The magnitude of $\boldsymbol{k}_{\hat{\boldsymbol{r}}}$ is determined by the temporal wavenumber k_ω.

Spherical Waves

Models of spherical wavelets and spherical wave fields are key in SAR signal analysis.

Spherical Impulsive Wavelets

The most important wave in time-space domain SAR signal analysis is the spherical impulsive wavelet. The wavefront of an expanding impulsive spherical wavelet is infinitesimally thin.

Spherical Sinusoidal Wave Fields

An important wave field in frequency-space domain SAR signal analysis is the spherical sinusoidal wave field. Directed sinusoidal (planar) wave fields are also important.

Monochromatic and Polychromatic Spherical Sinusoidal Wave Fields

Often in the frequency-space domain analysis of SAR, a monochromatic spherical sinusoidal wave field is considered, especially when the transmitted waveform is a sinusoidal continuous wave signal. When the analysis is extended to polychromatic pulsed radar systems, all constituent frequencies (whether continuous or discrete) within the passband of the sensor are considered.

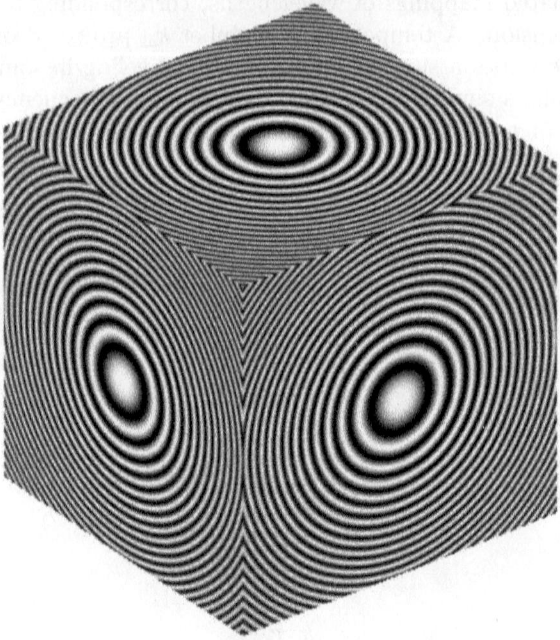

Figure 2.15: Fresnel zone plates on a cubic manifold centered on a spherical sinusoidal phase field.

Spherical Phase Field Sensed on Fresnel Zone Plate

The *Fresnel zone plate* of Fourier optics is the phase pattern that results from sampling a monochromatic spherical sinusoidal wave field on a planar surface.

Fig. 2.15 shows Fresnel zone plate phase patterns on the sides of a cube centered about a spherical sinusoidal wave field's source point. A linear flight SAR can be viewed as collecting a linear sample of the phase pattern of a Fresnel zone plate.

2.11 Heisenberg Uncertainty Principle

Heisenberg's uncertainty principle[4] establishes bounds in quantum mechanics to an observer's certainty or confidence in co-measurements of pairs of conjugate physical properties. A moving particle's position and momentum are example properties that comprise such a conjugate pair[5] of measurable attributes.

Heisenberg's uncertainty principle is also operative in the study of signals that are both time limited and band limited in some sense [845–848].

In the signal analysis of SAR, for example, Heisenberg's uncertainty principle establishes the impossibility of a short observation (e.g., a short SAR sensor dwell) from producing high resolution (i.e., precise localization of a scatterer's location in the SAR image domain). Spatial window parameters such as the synthetic aperture length Π_{u_y} are descriptive of observation dwell. Parameters such as resolution Δ_{u_y} are descriptive of limits of certainty.

The Heisenberg uncertainty principle, as applied in SAR signal analysis, mandates that localization or high precision (small Δ) in a temporal or spatial domain requires an inversely proportionate observation or measurement dwell (large Π) in the corresponding frequency or wavenumber domain.

Four Attributes of Heisenberg Pairs in Signal Analysis

The Heisenberg uncertainty principle constraint in temporal signal analysis[6] can be expressed as

$$\Delta_t \times \Pi_\omega \geq 2\pi,$$

which precludes a small time localization Δ_t without a concomitant angular frequency support of at least

$$\Pi_\omega \geq \frac{2\pi}{\Delta_t}.$$

Similarly, angular frequency resolution Δ_ω, as constrained by the Heisenberg uncertainty relationship,

$$\Pi_t \times \Delta_\omega \geq 2\pi,$$

requires time observation intervals of at least

$$\Pi_t \geq \frac{2\pi}{\Delta_\omega}.$$

[4] Werner Heisenberg (1901–1976) was a German theoretical physicist who made foundational contributions to quantum mechanics and established its uncertainty principle.

[5] Given Π_x as a quantity of error in observations of a particle's position, an uncertainty level below Δ_p of the particle's momentum cannot be escaped where $\Pi_x \Delta_p \geq h$, with h being Planck's constant. Hence, a reduction in the error of the position observation Π_x requires a concomitant increase in the observer's uncertainty (error range) of the momentum measurement Δ_p.

[6] Time and frequency as conjugate physical attributes comes from the energy-time uncertainty conjugate pair $\Delta_\epsilon \Pi_t \geq h$, with h being Planck's constant and energy defined as $\epsilon = hf$. Hence, $\Delta_{(hf)}\Pi_t \geq h$ leads to $\Delta_f \Pi_t \geq 1$, or in terms of angular frequency, $\Delta_\omega \times \Pi_t \geq 2\pi$.

Similar expressions hold for other spatial and wavenumber Fourier transform pair variables. For example, for the down-range spatial parameter x and associated wavenumber k_x, there exists a pair of Heisenberg relationships,

$$\begin{array}{ccccc} \Delta_x & \times & \Pi_{k_x} & \geq & 2\pi, \\ \Pi_x & \times & \Delta_{k_x} & \geq & 2\pi. \end{array} \tag{2.23}$$

For the along-track ground scene spatial parameter y and associated wavenumber k_y, there exist the Heisenberg requirements of

$$\begin{array}{ccccc} \Delta_y & \times & \Pi_{k_y} & \geq & 2\pi, \\ \Pi_y & \times & \Delta_{k_y} & \geq & 2\pi. \end{array}$$

For the synthetic aperture spatial parameter u_y and associated wavenumber k_{u_y},

$$\begin{array}{ccccc} \Delta_{u_y} & \times & \Pi_{k_{u_y}} & \geq & 2\pi, \\ \Pi_{u_y} & \times & \Delta_{k_{u_y}} & \geq & 2\pi, \end{array}$$

and so on. In each case, a Fourier transform pair of variables, e.g., u_y and k_{u_y}, is qualified by four Heisenberg uncertainty related parameters, e.g., Δ_{u_y}, $\Delta_{k_{u_y}}$, Π_{u_y} and $\Pi_{k_{u_y}}$.

Specification and Derivation of Heisenberg Parameters

Specification of certain pairs of parameters is sufficient to cover the degrees of freedom among the four attributes of conjugate properties. For example, consider the four attributes related to sample density and observation support of x and k_x,

$$\{\Delta_x, \Delta_{k_x}, \Pi_x, \Pi_{k_x}\}. \tag{2.24}$$

Specification of all members of the quadruple (2.24) can be done with specification of the pair Δ_x and Π_x, the pair Δ_x and Δ_{k_x}, the pair Π_x and Π_{k_x} or the pair Π_x and Δ_{k_x}. The remaining unspecified attributes are derived from the specified pair per the Heisenberg relationships (2.23).

Heisenberg Uncertainty and Chirp Signals: Preview of a Special Case

Heisenberg relationships hold without qualification in familiar classes of waveforms and signals, except as we shall see in Chapter 6, for the window functions of temporal and spatial *chirp signals.* Various aspects of chirp signal window functions remain rather amphibious (half-in/half-out) with respect to the Heisenberg uncertainty principle.

Furthermore, we shall see that the multiplicative window functions of chirp signals do not map (as normally expected with Fourier signal analysis) to convolutional window functions in the corresponding Fourier domain.

We shall explore these interesting exceptions to Heisenberg constraints and conventional signal analysis in Section 6.4.

2.12 Nyquist Sampling

Heisenberg uncertainty intervals determine the Nyquist[7] rate required for unaliased sampling to occur. A Nyquist sampling rate is sufficient to enable a Shannon[8] reconstruction. Shannon reconstructions empower discrete data to represent analog data without suffering loss of "information" [821, 844]. Nyquist-Shannon sampling and reconstruction are briefly explained.

Nyquist Sampling of Temporal Signals

Given a continuous time signal $f(t)$, a sampled version is

$$\begin{aligned} f[n] &= f(t) \cdot \sum_{n=-\infty}^{\infty} \delta\left(t - n \cdot \Delta_t\right), \\ &= \sum_{n=-\infty}^{\infty} f(n \cdot \Delta_t) \cdot \delta\left(t - n \cdot \Delta_t\right), \end{aligned}$$

where Δ_t is the sampling interval.

Shannon and Nyquist established conditions under which the continuous signal $f(t)$ is fully characterized by discrete samples $f[n]$ [356, 606, 809, 821]. The primary requirement is that if the bandwidth of $f(t)$ spans $(-f_o, f_o)$, then Nyquist sampling (real valued samples) requires

$$\Delta_t = \frac{1}{2f_o}.$$

An analog-to-digital converter cannot operate at a sampling (cyclic) frequency less than

$$f_s = \frac{1}{\Delta_t}$$

without causing frequency aliasing and loss of information.

Complex and Real Sample Counts

If complex samples are desired, then the required sampling rate is reduced by half. The sample count obtained by complex sampling remains the same as real sampling because both a real and imaginary part comprise each complex sample.

Nyquist Rates and Heisenberg Uncertainty

The upper bound for the duration of a Nyquist sampling interval of a temporal signal is

$$\Delta_t \leq \frac{2\pi}{\Pi_\omega},$$

where Π_ω characterizes the angular frequency bandwidth of the signal being sampled.

[7] Harry Nyquist (1889–1976) was a Swedish electrical engineer who made important contributions to digital and analog signal theory, communication theory and information theory.

[8] Claude Elwood Shannon (1916–2001) was an American electronics engineer and mathematician who is considered the father of modern communication theory and the founder of the discipline of information theory.

Nyquist Sampling of Spatial Signals

Similar expressions exist for Nyquist sampling densities in spatial wavenumber domains. For example,

$$\Delta_{k_x} \leq \frac{2\pi}{\Pi_x}$$

shows that a wavenumber sample density of Δ_{k_x} is required to support an x domain spatial extent of Π_x. The Nyquist sample density of a spatial wavenumber supports representation of scene extent. We typically characterize the down-range scene size as

$$\begin{aligned} \Pi_x &= 2X_o, \\ &= \Pi_{(2X_o)}(x). \end{aligned}$$

A similar requirement holds for Δ_{k_y} given the along-track scene size $\Pi_y = 2Y_o$,

$$\Delta_{k_y} \leq \frac{2\pi}{2Y_o}$$

Analog-to-Digital Conversion Hardware

Beyond idealized notions of sampling of continuous time signals to provide discrete time samples, efficient and effective sampling of signals with realizable hardware systems has other factors that should be considered in developing data acquisition systems [32].

Over Sampling Factor

Often a signal is intentionally sampled at a rate faster than the Nyquist rate. The *over sampling ratio* is the factor that the selected sampling rate exceeds the Nyquist rate. Over sampling factors of between 1.1 and 1.2 have been suggested for temporal data acquisition. Over sampling factors of between 1.1 and 1.4 have been suggested for pulse repetition frequencies [218].

2.13 Sinc Functions

Sinc functions play a major role in SAR signal analysis. There are subtleties involved with scalings and normalizations of various forms of sinc functions.

Sinc Function

The *sinc function* is a sine function with a linear rate of attenuation,

$$\text{sinc}(t) = \frac{\sin(t)}{t}. \tag{2.25}$$

The sinc function has a magnitude of unity at $t = 0$. The sinc function has a Fourier transform that is the square window function with unit half-width,

$$\text{sinc}(t) \ {}_t\!\Leftrightarrow_\omega \ \pi \cdot \Pi_{[-1,+1]}(\omega).$$

The distance between zero crossings away from the mainlobe of (2.25) is

$$\Delta_t = \pi.$$

The width of the mainlobe of the sinc function (2.25) is 2π.

Sinc Function with Time Scaling

The *sinc function with time scaling* [734] is defined as

$$\text{sinc}(\omega_o t) = \frac{\sin(\omega_o t)}{\omega_o t}.$$

The Fourier transform of $\text{sinc}(\omega_o t)$ is a scaled square window function with a half-width of ω_o,

$$\text{sinc}(\omega_o t) \;\; {}_t\!\Leftrightarrow_\omega \;\; \frac{2\pi}{2\omega_o} \cdot \Pi_{[-\omega_o,+\omega_o]}(\omega), \tag{2.26}$$

where

$$\Pi_{[-\omega_o,+\omega_o]}(\omega) = \Pi_{[-1,+1]}\left(\frac{1}{\omega_o} \cdot \omega\right).$$

The baseband angular frequency bandwidth of the sinc function with ω_o time scaling is

$$\begin{aligned} \beta &= 2\omega_o, \\ &= \Pi_{(2\omega_o)}(\omega). \end{aligned}$$

The distance between zero crossings of the sinc function with ω_o time scaling is

$$\Delta_t = \frac{2\pi}{2\omega_o}.$$

The parameters Δ_t and $\Pi_{(2\omega_o)}(\omega)$ of this form of the sinc function satisfy the Heisenberg uncertainty relationship,

$$\Delta_t = \frac{2\pi}{\Pi_{(2\omega_o)}(\omega)}.$$

Normalized Sinc Function

The *normalized sinc function* [997] has a time scaling parameter set to $\omega_o = \pi$,

$$\text{sinc}(\pi t) = \frac{\sin(\pi t)}{\pi t}.$$

We write this special form of the sinc function as

$$\mathsf{sinc}(t) = \text{sinc}(\pi t).$$

This sinc function is viewed as normalized because its integral over all t yields a finite area of one. The Fourier transform of the normalized sinc function is the square window function with a half-width of π,

$$\mathsf{sinc}(t) \;\; {}_t\!\Leftrightarrow_\omega \;\; \Pi_{[-\pi,+\pi]}(\omega).$$

The distance between zero crossings of the normalized sinc function is

$$\Delta_t = 1.$$

Normalized Sinc Function with Time Scaling

Of particular utility in SAR signal analysis expressed in terms of the cyclic (not angular) frequency variable is the normalized sinc function with time scaling given by

$$\mathsf{sinc}(Bt) = \operatorname{sinc}(\pi Bt).$$

With this choice of the time scaling parameter, we can equate the angular and cyclic frequency bandwidth parameters,

$$\omega_o = \pi B,$$

where ω_o is a half bandwidth parameter and B is a full bandwidth parameter. The corresponding Fourier transform pair is

$$\mathsf{sinc}(Bt) \;{}_t\!\Leftrightarrow_\omega\; \frac{1}{B} \cdot \Pi_{[-\pi B, +\pi B]}(\omega). \tag{2.27}$$

The Fourier transform pair (2.27) can be expressed in terms of cyclic frequency as

$$\mathsf{sinc}(Bt) \;{}_t\!\Leftrightarrow_f\; \frac{1}{B} \cdot \Pi_{[-B/2, +B/2]}(f)$$

and has a cyclic frequency bandwidth of B. The angular frequency bandwidth is

$$\beta = 2\pi B.$$

The distance between zero crossings of the normalized sinc function with time scaling is

$$\Delta_t = \frac{1}{B}. \tag{2.28}$$

Using (2.27) and (2.28),

$$\mathsf{sinc}\left(\frac{t}{\Delta_t}\right) \;{}_t\!\Leftrightarrow_f\; \frac{1}{B} \cdot \Pi_{[-B/2, +B/2]}(f)$$

or in angular frequency form,

$$\mathsf{sinc}\left(\frac{t}{\Delta_t}\right) \;{}_t\!\Leftrightarrow_\omega\; \frac{2\pi}{\beta} \cdot \Pi_{[-\beta/2, +\beta/2]}(\omega).$$

For spatial systems,

$$\mathsf{sinc}\left(\frac{x}{\Delta_x}\right) \;{}_x\!\Leftrightarrow_{k_x}\; \frac{2\pi}{\beta_x} \cdot \Pi_{[-\beta_x/2, +\beta_x/2]}(k_x)$$

with

$$\Delta_x = \frac{2\pi}{\beta_x}$$

being spatial resolution, where

$$\beta_x = 2k_{x_o}$$

is the extent of the spatial wavenumber manifold.

Sinc Functions with Nyquist Parameters

A set of useful Fourier transform pairs in terms of temporal or spatial resolution, and associated bandwidth parameters, i.e., in terms of Nyquist parameters, is

$$\begin{aligned} \mathsf{sinc}\left(\frac{t}{\Delta_t}\right) \quad & {}_{t \Leftrightarrow \omega} \quad \Delta_t \cdot \Pi_{(\beta)}(\omega), \\ \mathsf{sinc}\left(\frac{x}{\Delta_x}\right) \quad & {}_{x \Leftrightarrow k_x} \quad \Delta_x \cdot \Pi_{(\beta_x)}(k_x). \end{aligned} \tag{2.29}$$

The Heisenberg uncertainty principle is clearly evident in the expressions of (2.29), where

$$\begin{aligned} \Delta_t &= \frac{2\pi}{\Pi_{(\beta)}(\omega)}, \\ \Delta_x &= \frac{2\pi}{\Pi_{(\beta_x)}(k_x)}. \end{aligned}$$

Sinc Functions and Pulse Fourier Transforms

By the time-frequency duality property of the Fourier transform[9] and from (2.26),

$$\begin{aligned} \Pi_{(\Delta_t)}(t) \quad & {}_{t \Leftrightarrow \omega} \quad \Delta_t \cdot \operatorname{sinc}\left[(\Delta_t/2)\omega\right], \\ \Pi_{(\Delta_t)}(t) \quad & {}_{t \Leftrightarrow f} \quad \Delta_t \cdot \mathsf{sinc}\left(\Delta_t f\right), \end{aligned} \tag{2.30}$$

which are given in terms of both angular and cyclic frequency parameters. If the temporal resolution is written in terms of a half-width T_s, i.e., in terms of a Nyquist sample interval T_s density requirement of two real samples per resolution cell,

$$\Delta_t = 2T_s$$

then (2.30) are expressed in more familiar form,

$$\begin{aligned} \Pi_{(2T_s)}(t) \quad & {}_{t \Leftrightarrow \omega} \quad 2T_s \cdot \operatorname{sinc}\left(T_s \omega\right), \\ \Pi_{(2T_s)}(t) \quad & {}_{t \Leftrightarrow f} \quad 2T_s \cdot \mathsf{sinc}\left(2T_s f\right). \end{aligned}$$

2.14 Shannon Reconstructions

Given a set of uniformly spaced samples $f[n]$, Shannon interpolation reconstructs analog values between discrete sampling points. Shannon's sampling theorem establishes under what conditions a general class of continuous time signals can be reconstructed exactly by using a sequence of discrete time samples.

Interpolation with Uniformly Spaced Samples

A continuous time normalized sinc function is scaled to each discrete sample value's $f[n]$ and shifted to the position of each sample to reconstruct $f(t)$. If the spectrum $F(\omega)$ of the baseband signal $f(t)$ is such that

$$F(\omega) = 0$$

[9] If $f(t) \Leftrightarrow F(\omega)$, then $F(t) \Leftrightarrow 2\pi f(-\omega)$.

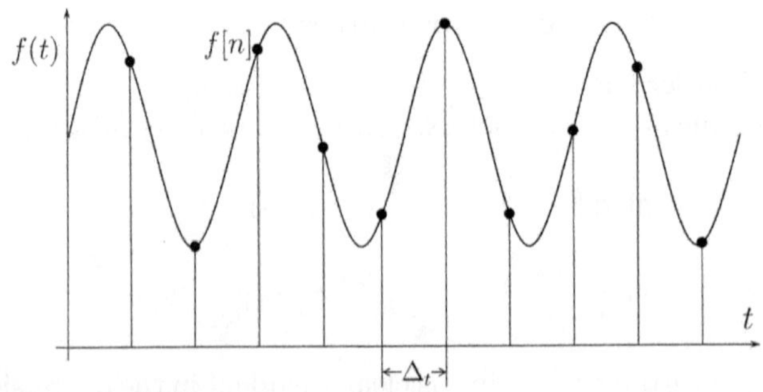

Figure 2.16: Interpolation from uniformly spaced samples.

for $\omega \geq \omega_o$ and

$$\Delta_t = \frac{2\pi}{2\omega_o},$$

then exact reconstruction is obtained by

$$f(t) = \sum_{n=-\infty}^{\infty} f[n] \cdot \mathsf{sinc}\left(\frac{t - n \cdot \Delta_t}{\Delta_t}\right).$$

The normalized sinc function with time scaling $1/\Delta_t$ has unit value at zero and zero values at integral multiples of the sampling interval Δ_t. Values of the normalized and scaled sinc functions, i.e., $f[n] \cdot \mathsf{sinc}\,[(t - n \cdot \Delta_t)/\Delta_t]$, are summed to determine the analog waveform value at any point t between discrete sample points as shown in Fig. 2.16.

Shannon Reconstructions with Truncated Sinc Functions

The interpolation sinc function can be truncated to finite length in applications, typically four to eight samples on each side of the sinc mainlobe, to reduce computational costs with an associated loss of fidelity. Finite sums give approximate reconstructions. Reconstructions with finite kernel interpolations can be performed to arbitrary accuracy by increasing the order of the finite sum and the extent of the sinc interpolator [960]. Exact reconstructions require infinite sums.

Interpolation with Non-Uniformly Spaced Samples

Interpolation between discrete samples can also be achieved, as illustrated in Fig. 2.17, with non-uniformly spaced sample sets [139,1011]. If a uniformly spaced set of samples is desired by interpolation from non-uniformly spaced samples, then a set of uniformly spaced objective output sample points is first defined. The interpolator sinc kernel functions are scaled by each input sample value and are positioned at the location of each of the original, non-uniformly spaced input samples. The values of the input samples extrapolated by the sinc functions to each of the desired uniformly sampled output points are summed. For finite length sinc interpolation kernels, only those non-uniformly spaced input samples

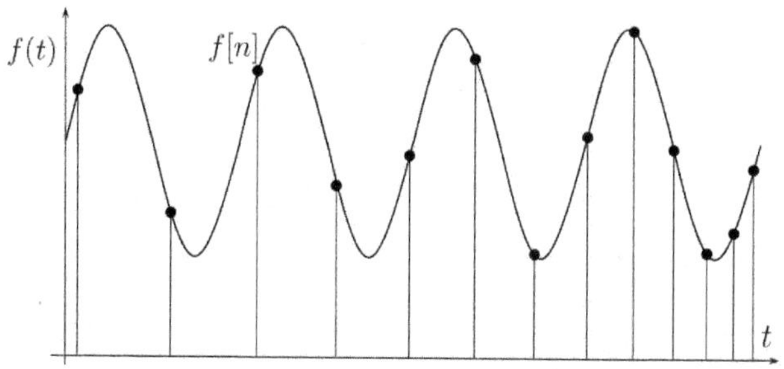

Figure 2.17: Interpolation from non-uniformly spaced samples.

sufficiently near each desired output sample are extrapolated and included in the sum.

To avoid loss of information, the largest (non-uniform) sample spacing cannot exceed the Nyquist interval.

A Jacobian[10] determinant must be used in scaling the terms of the summation to prevent changes in the energy level between the original non-uniform input sample set and the desired uniform output sample set [860].

[10] Car Gustav Jacob Jacobi (1804–1851) was a German mathematician.

Chapter 3

Wave Theory Framework

We present a wave theory framework for SAR [198, 424, 812]. The key concepts are the *expanding impulsive spherical wavelet* of Huygens and the *monochromatic sinusoidal spherical wave field* of Fresnel. Both of these originate at a point singularity. Huygens' wavelet is a solution to a time-space domain wave motion equation derived from Maxwell's equations. Fresnel's wave field is a solution to a frequency-space domain wave motion equation called the Helmholtz equation. These are related to each other by temporal Fourier transformation. A Fresnel wave field is a monochromatic sample of the infinite bandwidth temporal Fourier transform of Huygens' expanding spherical wavelet.

Huygens and Fresnel wave theories provide a framework for models of SAR signal transmission, scattering and reception. Huygens' wavelet is key in time-space analysis. Fresnel's wave field is key in frequency-space analysis. The spatial Fourier transform of Fresnel's wave field provides a pathway to frequency-wavenumber domain analysis, where the key entity is the Huygens-Fresnel transfer function.

This chapter is outlined as follows: Section 3.1 covers the historical background of the development of Huygens-Fresnel wave theories. Section 3.2 reviews Maxwell's equations and derives time-space domain expressions of electromagnetic wave motion. Section 3.3 describes plane wave and spherical wave motion in free space. Section 3.4 gives Helmholtz's equation of wave motion expressed in the frequency-space domain. Section 3.5 presents Fresnel's sinusoidal spherical wave field. Section 3.6 derives Huygens' expanding impulsive spherical wavelet as the inverse Fourier transform of an infinite bandwidth collection of Fresnel wave fields. Section 3.7 introduces Huygens-Fresnel transfer functions of the frequency-wavenumber domain. Section 3.8 shows how Huygens-Fresnel transfer functions are naturally embedded in the spectral description of a terrain function expressed in the aperture coordinate system. Section 3.9 introduces the concept of indirectly performing wave field inversion in the time-space domain through application of inverse Huygens-Fresnel transfers in the frequency-wavenumber domain. Finally, Section 3.10 describes the tonal wave fields that exist on spherical manifolds in the wavenumber domain. The dimensional frequencies of these wavenumber domain tonal wave fields are descriptive of point scatterer positions within the scene's terrain function.

3.1 Wave Theory Concepts

We begin our presentation of wave theory concepts by providing a historical overview of the development of various theories postulated to explain the propagation of light.

Early theories for light propagation, based on observations of assumed analogous mechanical wave motions in media such as gases, liquids and solids, were motivated by a dynamic analysis of longitudinal (back and forth) waves in matter. These early wave motion models were later enhanced by incorporation of aspects of waves that travel with both longitudinal and transverse (up and down) motion.

Light, as an electromagnetic wave, propagates without the need of a supporting medium. Light can traverse "free space." Light traverses only by transverse wave motion. Electromagnetic wave propagation is a vector phenomenon.

Mechanical waves need supporting media for propagation. Sound is an example of a mechanical wave. Acoustic wave propagation is a scalar phenomenon.

Although a classical mechanical analysis of the physics of light is impossible (the tools of modern physics are required), there is much value in a study of analogous mechanical scalar acoustic waves. The mechanical analysis of sound especially aids our understanding of descriptions of scalar elements of vector electromagnetic (light) waves.

Huygens Principles of Light Wave Motion

Principles postulated by Huygens provided an early formulation of a theory of wave propagation for light. Although intuitive and easily motivated by observations of wave phenomena in a physical medium such as water, there were certain curious aspects of the Huygens principles that did not seem to be justified by the limited mechanical analysis of his time. These suspect aspects have been elaborated and clarified with additional study by Fresnel, Poisson[1], Helmholtz, Kirchhoff and others. These additional insights were obtained, in part, by a mechanical analysis of a combined potential and kinematic model for acoustic wave propagation [35].

Huygens' Expanding Impulsive Spherical Light Wavelet

In his *Traite de la Lumiere*, published in 1690, Huygens offered an explanation for the propagation of light that included several key concepts. Similar to waves caused by tossing a pebble into still water, Huygens envisioned a point disturbance of some medium causing a spherical wavelet of light to propagate out in all directions. This imagined medium was called "aether" [296].

The concept of Huygens' expanding impulsive spherical wavelet is illustrated by a sequence of expanding circles in Fig. 3.1a. The wavefront of this wavelet is infinitesimally thin. Hence, the wavelet is described as "impulsive" (not only because of the singular nature of its source, but also because of the infinitesimal thinness of its shell). The intensity of the spherical impulsive wavelet decays at a rate proportional to the wavelet's distance from the source point (intensity decay is indicated in Figs. 3.1a-b by decreasing gray scale value).

[1]Siméon Denis Poisson (1781–1840) was a French mathematician and physicist.

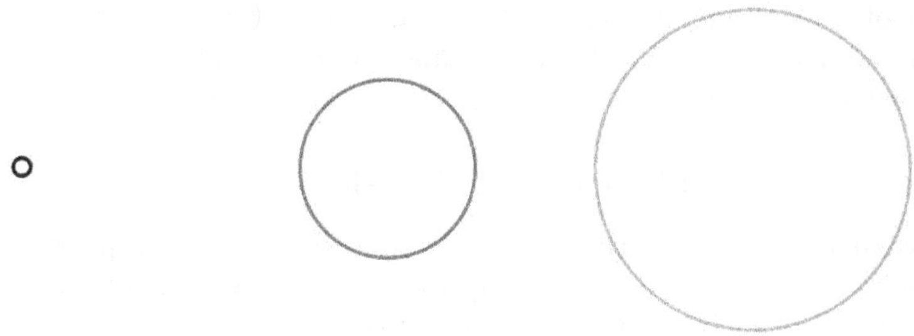

a. Huygens' expanding spherical wavelet.

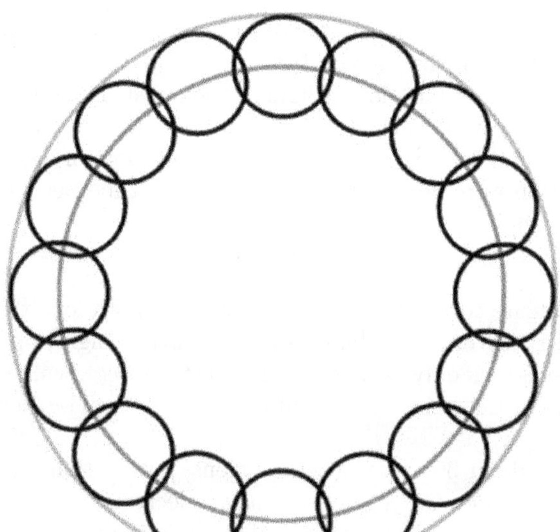

b. Huygens' concept of secondary spherical wavelets.

Figure 3.1: Huygens' wavelet propagation concepts.

Huygens' Primary and Secondary Wavelets

Huygens also postulated the concept of secondary wavelets caused by the disturbance of a passing primary wavelet of light. Each point on the primary wavefront is viewed as causing new impulsive disturbances that lead to secondary expanding spherical wavelets. These secondary wavelets empower the continuing propagation of the primary wavelet. Within Huygens' conceptual framework, the primary propagating wavelet was imagined as the amplitude envelop of secondary wavelets [35]. This concept is illustrated in Fig. 3.1b.

Challenges Faced by Huygens' Wavelet Model

When delved more deeply, Huygens' impulsive wavelet concepts suffer from several difficulties. For one, dealing with non-spherical (e.g., plane wave) propagation presents challenges. For another, the secondary wavelet concept would seem to imply that the secondary light disturbances should cause waves that propagate inwards as well as outwards.

Particle versus Wave Theories of Light Movement

These conceptual gaps, in part, motivated others, including Sir Isaac Newton[2], to abandon a wave theory framework for light. Newton turned his attention to a corpuscular (particle) explanation of light. Over most of the next two centuries, advocates of wave theories and advocates of particle theories debated the relative merits of their points of view. The arguments in support of a corpuscular theory of light propagation were dominant for most of that time period.

Study of Light is Beyond Classical Physics

Waves that propagate through matter can consist of both longitudinal (e.g., pressure) waves and transverse (e.g., shear) *scalar waves.* Light, on the other hand, propagates in free space only as a transverse *vector wave.* Scalar waves in matter and vector waves in free space have analogous properties, but are fundamentally different.

It wasn't until about 300 years after Huygens' publication that light was eventually identified by James Clerk Maxwell[3] as an electromagnetic phenomenon and as such, outside the bounds of a mechanical analysis grounded in classical physics. The wave theory and particle perspectives of light propagation were eventually blended in the tenets of quantum mechanics of modern physics.

The Empirical Study of Light

The road to a complete theory for the propagation of light is marked with major milestones, both experimental and theoretical (see Section 11.1 for additional material on a historical overview of the study of light). There is value in returning to the topic of early wave theories for light propagation based on observations of

[2] Sir Isaac Newton (1642–1727) was an English physicist, mathematician, astronomer, natural philosopher and theologian.

[3] James Clerk Maxwell (1831–1879) was a Scottish theoretical physicist and mathematician who made foundational contributions to the theory of classical electromagnetics.

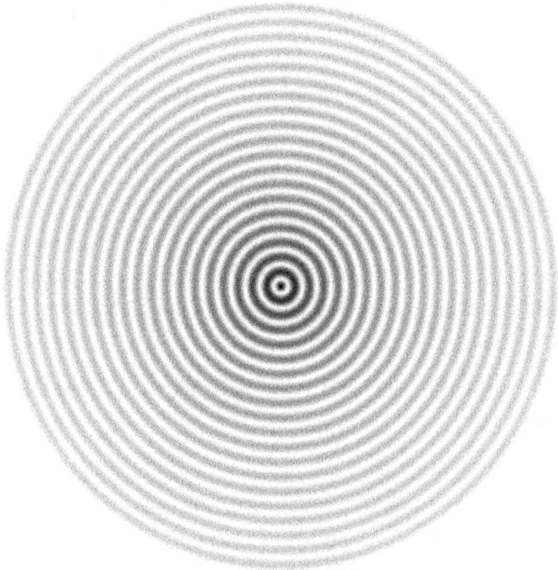

a. Fresnel's sinusoidal wave field.

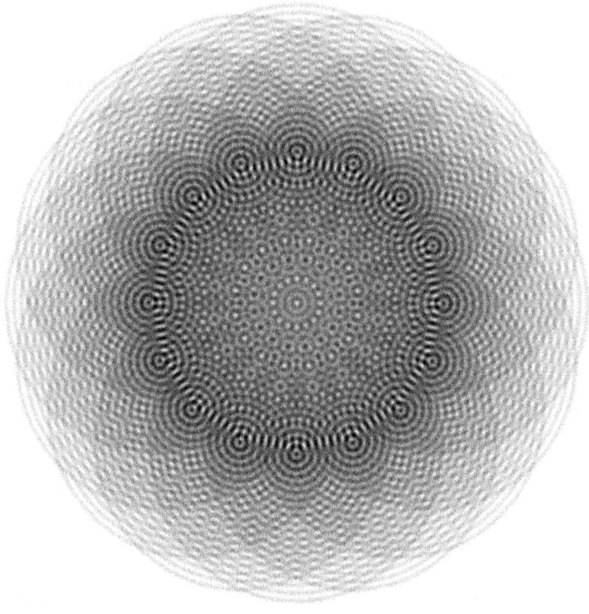

b. Secondary Fresnel wave fields.

Figure 3.2: Fresnel's sinusoidal spherical wave field concepts.

dynamic mechanical properties of waves in matter. We will point out the most important advances of this empirical study and the supporting theories that relate to the wave theory framework we develop for SAR signal analysis.

Thomas Young's Experiments in Light Wave Diffraction

Thomas Young[4] demonstrated in a set of double slit diffraction experiments in 1801 that "light added to light can produce darkness" [1020–1022]. Young subsequently postulated a wave theory basis for light diffraction. Young's wave theory hypothesis was in contrast to a century long held view, established by Sir Isaac Newton [695], that light is comprised of moving particles. Young's experiments empirically established that light also has wave like behaviors.

Fresnel's Monochromatic Sinusoidal Spherical Wave Field

Building on observations from the experimental work of Young, Fresnel [360] extended the Huygens principles by replacing the notion of an expanding impulsive spherical wavelet with the concept of a continuously radiating monochromatic sinusoidal spherical wave field as shown in Fig. 3.2a. Fresnel's wave field also "emerges" from a point singularity [35]. Fresnel accepted Huygens' concepts of primary and secondary waves as illustrated with sinusoidal wave fields in Fig. 3.2b. Fresnel also proposed that light moves along lines of constructive sinusoidal phase interactions, rather than along ridges of wavelet amplitude interactions. Fresnel's work proved to be a rather robust addition to Huygens' wave theory framework. Similar to Fourier synthesis in harmonic signal analysis, Fresnel's spatial harmonic models have the ability to express the synthesis of arbitrarily shaped wavefronts in terms of coherent elements of constituent monochromatic sinusoidal spherical wave fields.

Helmholtz's Formulation

Fresnel, in his analysis of acoustic wave fields, modeled the physics of acoustic wave motion as involving two parts. Fresnel showed that each secondary wave's point source can be viewed as giving rise to both an acoustic compression wave (causing local changes of propagation medium density, i.e., wave potential) and an acoustic velocity wave (the conventional notion of longitudinal wave motion, i.e., wave kinematics). However, Fresnel had to make assumptions that seemed to have little physical basis for his analysis to remain mathematically cohesive. One such assumption was that the amplitudes of secondary waves are produced relative to the amplitudes of primary waves at a ratio of 1 to λ, where λ is the harmonic field's wavelength. Another strange assumption was that the secondary waves are oscillating a quarter of a period ahead of the primary wave. This seemingly non-causal element of his analysis led many to regard Fresnel's modifications of Huygens' concepts as merely a method of calculation, rather than physical modeling of mechanical wave dynamics.

[4]Thomas Young (1773–1829) was an English scientist who first demonstrated the diffraction of light.

Helmholtz's Integral Analysis of Huygens Wavelets

In follow on work to Fresnel's differential analysis, Helmholtz showed, with a careful integral formulation of Fresnel's ideas, that the oddities of secondary wave scaling and quarter wavelength phase advance can be physically justified in multimodal longitudinal and transverse wave mechanics. Key to Helmholtz's solution was an imagined small volume and its closed surface. Helmholtz developed integral descriptions based on analysis of the wave field and its derivatives on the closed surface containing a point singularity.

The analytical extension of Helmholtz's methods gave a way to rigorously deal with Huygens' original notion of an infinitesimally small point source. Helmholtz's approach changed the differential analysis of a point singularity to an integrative analysis over a closed surface that contains the singularity.

The Notion of "State" in Wave Fields

A key element of the wave theory framework of Huygens and Fresnel is that full knowledge of the *wave field state,* as measured over an enclosing surface that contains source singularities, is sufficient to estimate the state of the wave field at any other point in space or time. A requirement for explicit knowledge about a field's source singularity is eliminated if sufficient information about the field on a surrounding closed surface is known.

The mathematics of wave state can be reformulated away from the problematic differential analysis of point singularities to an integrative analysis over a closed surface containing the singularity. The ability to express wave state at a time after its impulsive temporal origin gives one the option of not dealing directly with singularities in time and space.

Interestingly, this ability is lost if the problem dimensionality, even with the integrative formulation, is reduced from three dimensions to two dimensions (see Chapter 11).

Equivalent Surface Sources

Helmholtz's integral analysis can be over closed surfaces that may or may not contain point sources. When a closed surface contains point sources, a Helmholtz analysis replaces the point singularities with a distribution of sources over the closed surface that maintains the given field exterior to the closed surface and which produces a null field interior to the closed surface. When the field's sources are exterior to the closed surface, then the distribution of surface sources replaces the original exterior field with a null field and induces a field in the interior that is identical to the original given field.

Kirchhoff's Integral Formula

The analytical approach of Helmholtz opened the door for others to use surface integrals and volume integrals in wave field analysis. In particular, this led Kirchhoff to define various boundary condition requirements motivated by analysis of a general class of mechanical waves that propagate in matter. Kirchhoff's boundary conditions proved consistent with Huygens' spherical wavelet premises.

Kirchhoff's model is applicable to mechanical waves of any structure and origin [35, 89]. Fresnel's and Helmholtz's enhancements of the Huygens principles are special cases of the general formulation of Kirchhoff. Kirchhoff's integral theorem has found many applications in computing imaging with sensor systems that employ mechanical waves, such as in ultrasonic medical imaging and in seismic imaging [376]. Kirchhoff's methods, however, do not provide a satisfactory description of electromagnetic vector wave fields.

Diffracted Wave Fields

Wave and matter interactions can be modeled by classical approaches using ray theory, or more generally, by a framework of wave diffraction as provided by Huygens-Fresnel principles [337]. A diffracted wave field produced by a two slit diffraction grating is shown in Fig. 3.3a.

Wave diffraction models also cover the case where there is a semi-transparent film for light to pass through [150]. Such a diffracting film and surrounding gap aperture is a suitable model for a SAR ground scene with a continuum of scatterers of variable strength. Wave fields produced by scatterers in SAR and the wave fields of light refracted or diffracted through a semi-translucent film can be considered similar. The SAR ground scene is imagined as a semi-transparent film that covers the classical gap aperture of a diffraction experiment [45,345,576–578,580], with an imagined plane wave illumination source behind (underneath) the SAR ground scene as illustrated in Fig. 3.3b.

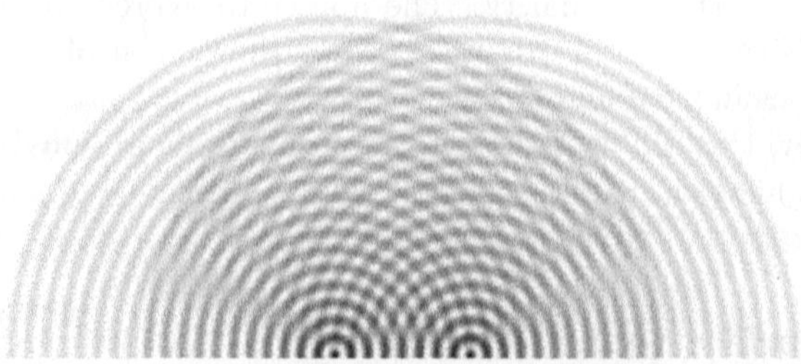

a. Diffracted wave field of two slits separated by eight wavelengths.

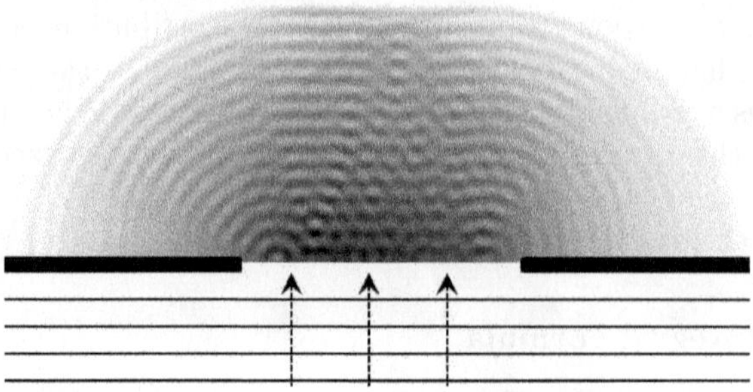

b. Diffracted wave field of translucent film and a gap aperture.

Figure 3.3: SAR scattering modeled as a diffracted wave field.

Wave Field Inversion

If a monochromatic continuous wave illuminates a SAR ground scene, then the scattered wave field can be viewed as produced by a set of variable strength Fresnel radiators. The wave field inversion problem is to estimate scatterer strength and position from remote, possibly incomplete, measurements of the wave field [433, 481, 580, 953].

Similarities between SAR image formation processing and the diffracted wave field inversion problem are based on the following comparison:

Diffracted Wave Field Inversion: Diffraction of monochromatic waves by a restrictive, semi-transmissive gap aperture produces a wave field remotely measurable in time and space. Wave field inversion seeks to estimate the perturbing aperture (shape and translucence) based on remote measurements of diffracted wave or phase data.

Remotely Sensed SAR Processing: Scattering of waves from an active sensor by a terrain function produces a wave field remotely measurable in time and space. SAR image formation processing seeks to estimate the terrain scattering function based on remotely sensed scattered wave field measurements.

Wave Field Inversion in Various Domains

Although data collection occurs in the time-space domain, processing to achieve wave field inversion may be performed not only in the time-space domain, but possibly in the frequency-space domain, in the time-wavenumber domain or in the frequency-wavenumber domain. Computed imaging algorithms based on wave field inversion exist for all these domain combinations.

Isotropic Radiation, Scattering and Reception in SAR

We begin our SAR signal analysis by describing omnidirectional electromagnetic radiation. We find Huygens-Fresnel wave field point sources useful for modeling the following elements of SAR analysis problems:

- isotropic electromagnetic radiation from a point antenna in transmit mode,
- isotropic scattering from ground points and
- isotropic reception of backscatter by a point antenna in receive mode.

Isotropic Radiation from a Point Antenna

In our initial SAR analysis, we assume the antenna is a point radiator that radiates a sinusoidal continuous wave in all directions. The transmitted electromagnetic signal is viewed as creating a primary spherical wave field in a way that is consistent with Fresnel's extension of Huygens principles [431].

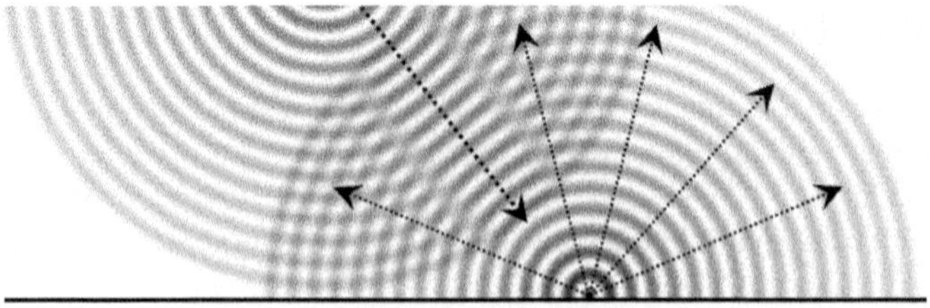

Figure 3.4: Isotropic scatter of a ground point.

Isotropic Scattering from a Ground Point

For the purposes of developing a baseline SAR theory, each scatterer in the illuminating wave field is viewed, as illustrated in Fig. 3.4, as an energized Fresnel radiator producing its own secondary spherical sinusoidal wave field. This perspective is useful for wave field signal modeling whether the illuminating wave field is planar, spherical or otherwise.

Weak Scattering Assumption

Each scatterer is imagined to react only to the incident field and not to the secondary fields produced by nearby scatterers. This assumption is called the weak scattering or first Born approximation [89].

Impulse Function Model of Point Scatterers

We assume each scatterer on the ground can be modeled by a point singularity described as an impulse function,

$$g_n(\boldsymbol{x}) = g_n \cdot \delta(x - x_n, y - y_n).$$

Each isotropic scatterer scatters electromagnetic wave energy with equal strength in all directions. Each scatterer may also induce a phase shift, since g_n is complex.

Isotropic Reception at a Point Antenna

We will employ a model of a directed antenna beam rather late in our analysis of SAR signals (Chapter 9.) Until then, we assume the antenna operates as a point receiver that receives signals equally well from all directions. At each sampling instant, the received echo signal is the superposition of echoes from scatterers with equidistant positions from the antenna. Our model of the received echo signal from a ground scene is an *integrative measure* of multiple scatterers illuminated by a passing spherical wavefront [80, 679].

3.2 Maxwell's Equations of Electromagnetics

Maxwell developed a set of foundational equations that describe the time-space behavior of electromagnetic fields. We use Maxwell's equations of electromagnetic vector fields to develop vector and scalar wave motion equations useful in formulating the wave theory framework for SAR [238].

Maxwell's Equations in Differential Form

Maxwell's equations in differential form are

$$\begin{array}{rrrcrl} \nabla & \cdot & \boldsymbol{D}(t,\boldsymbol{x}) & = & & \rho(t,\boldsymbol{x}), \\ \nabla & \cdot & \boldsymbol{B}(t,\boldsymbol{x}) & = & & 0, \\ \nabla & \times & \boldsymbol{E}(t,\boldsymbol{x}) & = & - & \partial_t\left\{\boldsymbol{B}(t,\boldsymbol{x})\right\}, \\ \nabla & \times & \boldsymbol{H}(t,\boldsymbol{x}) & = & & \partial_t\left\{\boldsymbol{D}(t,\boldsymbol{x})\right\} + \boldsymbol{J}(t,\boldsymbol{x}), \\ & & \boldsymbol{D}(t,\boldsymbol{x}) & = & & \epsilon_o \boldsymbol{E}(t,\boldsymbol{x}), \\ & & \boldsymbol{B}(t,\boldsymbol{x}) & = & & \mu_o \boldsymbol{H}(t,\boldsymbol{x}), \end{array} \tag{3.1}$$

where vector quantities are indicated with bold fonts and where

- $\boldsymbol{E}$ is the electric field intensity vector (units[5] of V/m),
- $\boldsymbol{D}$ is the electric flux density vector (units[6] of C/m^2),
- $\boldsymbol{H}$ is the magnetic field intensity vector (units[7] of A/m),
- $\boldsymbol{B}$ is the magnetic flux density vector (units[8] of Wb/m^2),
- ρ is the charge density (units of C/m^3),
- $\boldsymbol{J}$ is the electric current density vector (units of A/m^2),
- ϵ_o is the free space electric permittivity (units[9] of F/m),
- μ_o is the free space magnetic permeability (units[10] of H/m)

and where $\nabla \cdot \{\cdot\}$ is the divergence operator and $\nabla \times \{\cdot\}$ is the curl operator.

Speed of Electromagnetic Wave Propagation in Free Space

The speed of light in terms of the permittivity and permeability parameters of Maxwell's equations is given by

$$c = \frac{1}{\sqrt{\mu_o \epsilon_o}}.$$

---

[5] Alessandro Volta (1745–1827) was an Italian physicist known for the invention of the battery and for whom the *volt* unit is named.

[6] Charles-Augustin de Coulomb (1736–1806) was a French physicist known for his study of electrostatics and for whom the *coulomb* unit is named.

[7] Andre-Marie Ampere (1775–1836) was a French physicist and mathematician known for his work in electromagnetism and for whom the *ampere* unit is named.

[8] Wilhelm Eduard Weber (1804–1891) was a German physicist who made significant contributions in the study of electromagnetism and for whom the *weber* unit is named.

[9] Michael Faraday (1791–1867) was an English chemist and physicist who contributed much to the fields of electromagnetism and electrochemistry and for whom the *farad* unit is named.

[10] Joseph Henry (1797–1878) was an American scientist who made early discoveries in the field of electromagnetics and for whom the *henry* unit is named.

Electromagnetic Wave Motion Equations

We develop a general expression of electromagnetic vector wave motion in the time-space domain $(t, \boldsymbol{x})$. Points in three dimensional Cartesian space are indicated by the vector

$$\boldsymbol{x} = [x\ y\ z].$$

Scalar quantities are usually expressed in terms of Cartesian coordinates with the scalar variable x. The scalar r and the vector $\boldsymbol{r}$ are also used when radial distance

$$r = |\boldsymbol{r}|$$

in a three dimensional space is of primary interest, such as when spherical symmetry exists.

Electric Field Vector Wave Equation

A wave motion equation for the electric field intensity vector $\boldsymbol{E}$ can be derived by manipulating Maxwell's equations. Begin by eliminating $\boldsymbol{D}$ and $\boldsymbol{B}$ from the dynamic (time dependent) expressions through substitution of $\boldsymbol{H} = (1/\mu_o)\boldsymbol{B}$ and $\boldsymbol{D} = \epsilon_o \boldsymbol{E}$ into

$$\nabla \times \boldsymbol{H}(t, \boldsymbol{x}) = \partial_t \{\boldsymbol{D}(t, \boldsymbol{x})\} + \boldsymbol{J}(t, \boldsymbol{x}),$$

to obtain

$$\nabla \times \boldsymbol{B}(t, \boldsymbol{x}) = \mu_o \partial_t \{\epsilon_o \boldsymbol{E}(t, \boldsymbol{x})\} + \mu_o \boldsymbol{J}(t, \boldsymbol{x}). \tag{3.2}$$

Next, take the curl of both sides of

$$\nabla \times \boldsymbol{E}(t, \boldsymbol{x}) = -\partial_t \{\boldsymbol{B}(t, \boldsymbol{x})\}$$

to obtain

$$\begin{aligned} \nabla \times \nabla \times \boldsymbol{E}(t, \boldsymbol{x}) &= \nabla \times [-\partial_t \{\boldsymbol{B}(t, \boldsymbol{x})\}], \\ &= -\partial_t [\nabla \times \boldsymbol{B}(t, \boldsymbol{x})] \end{aligned} \tag{3.3}$$

and substitute (3.2) for $\nabla \times \boldsymbol{B}(t, \boldsymbol{x})$ in (3.3) to get

$$\nabla \times \nabla \times \boldsymbol{E}(t, \boldsymbol{x}) = -\partial_t^2 \{\mu_o \epsilon_o \boldsymbol{E}(t, \boldsymbol{x})\} - \partial_t \{\mu_o \boldsymbol{J}(t, \boldsymbol{x})\}. \tag{3.4}$$

Next, use an identity that describes the curl of a curl,

$$\nabla \times \nabla \times \boldsymbol{E} = \nabla \{\nabla \cdot \boldsymbol{E})\} - \nabla^2 \boldsymbol{E},$$

to rewrite (3.4) as

$$\nabla^2 \{\boldsymbol{E}(t, \boldsymbol{x})\} - \partial_t^2 \{\mu_o \epsilon_o \boldsymbol{E}(t, \boldsymbol{x})\} = \partial_t \{\mu_o \boldsymbol{J}(t, \boldsymbol{x})\} + \nabla \{\nabla \cdot \boldsymbol{E}(t, \boldsymbol{x})\},$$

where $\nabla\{\cdot\}$ is the gradient operator and where

$$\nabla^2 \{\cdot\} = \partial_x^2 \{\cdot\} + \partial_y^2 \{\cdot\} + \partial_z^2 \{\cdot\}$$

is the Laplacian[11] operator. With the assumption of charge free space, $\nabla \cdot \boldsymbol{E} = 0$ and we obtain the *vector wave equation* for the electric field intensity vector [665],

$$\nabla^2 \{\boldsymbol{E}(t, \boldsymbol{x})\} - \partial_t^2 \{\mu_o \epsilon_o \boldsymbol{E}(t, \boldsymbol{x})\} = \partial_t \{\mu_o \boldsymbol{J}(t, \boldsymbol{x})\}.$$

A similar development exist for the magnetic field intensity vector $\boldsymbol{H}(t, \boldsymbol{x})$.

[11] Pierre-Simon Marguis de Laplace (1749–1827) was a French mathematician and astronomer.

Electric Field Scalar Wave Motion Equation

Each scalar component $E \in \{E_x, E_y, E_z\}$ of a vector electric field $\boldsymbol{E} = [E_x\ E_y\ E_z]$ satisfies a *scalar wave motion equation.* The scalar wave motion equation for the electric field in the x dimension is

$$\partial_x^2 \{E(t,x)\} - \partial_t^2 \{E(t,x)/c^2\} = j'(t,x), \tag{3.5}$$

where $E = E_x$ and where $j'(t,x)$ indicates a scalar component of the time derivative of the vector current density,

$$j'(t,x) = \partial_t\{\mu_o \boldsymbol{J}\}.$$

Scalar Electric Field of an Impulsive Time-Space Current Density

Assuming that the changing current density of (3.5) is impulsively produced by a point source located at $x = x_o$ at the time instant $t = t_o$, we set

$$j'(t - t_o, x - x_o) \quad = \quad \delta(t - t_o, x - x_o)$$

and we have

$$\partial_x^2 \{E(t - t_o, x - x_o)\} - \partial_t^2 \{E(t - t_o, x - x_o)/c^2\} = \delta(t - t_o, x - x_o). \tag{3.6}$$

Equation (3.6) is an impulsively driven *time-space domain inhomogeneous scalar wave motion equation* in differential form. The forcing function on the right hand side represents a source singularity in time and space that is viewed, in the sense of Huygens' concepts, as creating the field $E(t - t_o, x - x_o)$.

Scalar Electric Field of Free Space

In the absence of charge distributions or currents, the scalar wave motion equation for a scalar component of an electric vector field in free space is a *time-space domain homogeneous scalar wave motion equation,*

$$\partial_x^2 \{E(t,x)\} - \partial_t^2 \{E(t,x)/c^2\} = 0. \tag{3.7}$$

Electric Field Planar Wave Motion Equations

A general solution of the scalar wave motion equation (3.7) is

$$E(t,x) = pp(ct - x) + pp(ct + x),$$

where $pp(ct - x)$ is a plane wave that propagates to the right along x and where $pp(ct + x)$ is a plane wave that propagates to the left [35].

Spherical Electric Field Wave Motion Equations

If the homogeneous scalar electric wave motion equation of free space (3.7) is expressed in terms of the spherical coordinate parameters descriptive of three dimensional space,

$$E(t,x) \mapsto E(t,r,\theta,\phi),$$

then the free space scalar wave motion equation corresponding to (3.7) becomes

$$\begin{aligned}\partial_r^2\{E\} + 2\partial_r\{E\}/r + \partial_\theta^2\{E\}/r^2 \\ + \cot\theta \cdot \partial_\theta\{E\} + \partial_\phi^2\{E\}/r^2 \sin^2\theta - \partial_t^2\{E/c^2\} = 0. \qquad (3.8)\end{aligned}$$

If the scalar wave $E(t, r, \theta, \phi)$ of (3.8) only depends on spherical distance r and time t, i.e., if $E(t, r, \theta, \phi)$ is invariant to θ and ϕ, then we can make the notational change,

$$E(t, r, \theta, \phi) \mapsto E(t, r)$$

and reduce (3.8) to

$$\partial_r^2\{E(t, r)\} + 2\partial_r\{E(t, r)\}/r - \partial_t^2\{E(t, r)/c^2\} = 0. \qquad (3.9)$$

Moreover, the wave motion equation (3.9) can be expressed[12] as

$$\partial_r^2\{r \cdot E(t, r)\} - \partial_t^2\{r \cdot E(t, r)\}/c^2 = 0. \qquad (3.10)$$

General Solution of Spherical Wave Motion Equation

Equations (3.9) and (3.10) are scalar expressions of spherical wave motion in three dimensional space [35]. Additionally, the combination $r \cdot E(t, r)$ of (3.10) satisfies the equation of plane wave motion as given by (3.7) and (3.2) [35]. Hence, a general solution of the spherical wave field equation (3.10) is

$$r \cdot E(t, r) = ss(ct - r) + ss(ct + r), \qquad (3.11)$$

where the spherical wave $ss(ct - r)$ travels outwardly with time and where the spherical wave $ss(ct + r)$ travels inwardly with time. For the particular solution

$$E(t, r) = \frac{1}{r} \cdot ss(ct - r),$$

the electric field $E(t, r)$ is an outwardly propagating spherical wave that attenuates at a rate of $1/r$.

Impulsively Driven Expanding Spherical Wave Field

The part of the general solution (3.11) of most interest is the outwardly expanding field,

$$\begin{aligned} r \cdot E(t, r) &= ss(ct - r), \\ &= ss(t, r). \qquad (3.12)\end{aligned}$$

We seek to consistently use the notation "$ss(t, r)$" for an outwardly expanding spherical wave that has been normalized to remove the $1/r$ attenuation factor.

[12] Since the double derivative of rE with respect to r gives $(rE)'' \mapsto (rE' + E)' \mapsto (rE'' + 2E')$.

Spherical Electrical Field Wave Motion Equation

Considering the electromagnetic wave field equation (3.6) of a point source at t_o and r_o, we obtain from (3.10) and (3.12) an *inhomogeneous spherical wave motion equation* driven by an impulse function in the time-space domain,

$$\partial_r^2 \{ss(t-t_o, r-r_o)\} - \partial_t^2 \{ss(t-t_o, r-r_o)/c^2\} \quad = \quad \delta(t-t_o, r-r_o). \tag{3.13}$$

We often assume $x_o = 0$ and $t_o = 0$ to shorten such expressions.

3.3 Time-Space Scalar Wave Motion Equations

In the next few sections we generalize the wave motion equations of the time-space domain and develop wave motion equations in the frequency-space and frequency-wavenumber domains. We do this by performing marginal and full Fourier transforms of the time-space domain expressions.

Time-Space Wave Motion Equations

A partial differential equation obtained from Maxwell's equations and descriptive of general time-space wave motion $ww(t, \boldsymbol{x})$ in rectilinear form in three dimensional free space is

$$\nabla^2 \{ww(t, \boldsymbol{x})\} - \partial_t^2 \{ww(t, \boldsymbol{x})/c^2\} = 0. \tag{3.14}$$

The absence of wave sources, point radiators or otherwise, is indicated by the zero on the right hand side of (3.14). In essence, this equation indicates that second order effects over space $\nabla^2 \{\cdot\}$ are countered by second order effects in time $\partial_t^2 \{\cdot\}$ — leading to a never ending interplay that propagates the electromagnetic wave $ww(t, \boldsymbol{x})$ throughout time and space.

Scalar Wave Motion in Free Space

A general time-space scalar function $ww(t, x)$ satisfies a one dimensional element of the three dimensional wave motion equation (3.14),

$$\partial_x^2 \{ww(t, x)\} - \partial_t^2 \{ww(t, x)/c^2\} = 0. \tag{3.15}$$

Planar and Spherical Wave Solutions

Any solution $ww(t, x)$ of (3.15) is a *wave function.* We have already identified two special solutions: the plane wave solution

$$ww(t, x) = pp(ct - x)$$

for a three dimensional plane wave moving in the direction of x and the non-attenuating spherical wave solution

$$ww(t, r) = ss(ct - r)$$

for an expanding spherical wave with radius r. The wave $ss(ct-r)$ is a solution of the impulsively driven inhomogeneous wave motion equation

$$\partial_r^2 \{ww(t,r)\} - \partial_t^2 \{ww(t,r)/c^2\} = \delta(t,r). \tag{3.16}$$

Attenuating and Non-Attenuating Expanding Spherical Waves

A solution of spherical wave motion that satisfies (3.16) is of the form [35],

$$\begin{aligned} ss(t,r) &= ss(ct-r), \\ &= r \cdot hh(ct-r). \end{aligned}$$

The outwardly expanding spherical wave

$$hh(t,r) = hh(ct-r)$$

attenuates in all directions from the origin at the rate of $1/r$. The expanding spherical wave field

$$ss(t,r) = ss(ct-r)$$

suffers no attenuation.

Notational Rules for Spherical Waves

We use the notation $hh(t,r)$ for spherical waves with attenuation. We use the notation $ss(t,r)$ for spherical waves with no attenuation. We also express a spherical wave in the time-space domain as $hh(t,\boldsymbol{r})$, where use of the radial vector $\boldsymbol{r}$ is a reminder that the wave is in three dimensional space and has spherical symmetry.

Seeking Solutions to Time-Space Wave Motion Equations

We have not specifically identified expressions for $ss(t,r)$ or $hh(t,r)$ that are solutions of (3.16). Rather than seek to directly determine these solutions of the time-space wave motion equation, we use a round about approach that first involves a frequency-space domain description of time-space wave motion. We then use the solution of the frequency-space domain wave motion equation to find a particular time-space domain solution.

3.4 Frequency-Space Scalar Wave Motion Equations

We next employ a temporal Fourier transform to convert the time-space domain wave motion equation

$$\nabla^2 \{ss(t,\boldsymbol{x})\} - \partial_t^2 \{ss(t,\boldsymbol{x})/c^2\} = \delta(t,\boldsymbol{x}). \tag{3.17}$$

into a frequency-space domain expression. The result is Helmholtz's equation. We seek the temporal Fourier transform of (3.17).

Helmholtz's Equation

For the moment, assume $\boldsymbol{x}$ is a vector in a rectilinear coordinate system. Let $Ss(\omega, \boldsymbol{x})$ be the marginal (temporal) Fourier transform of $ss(t, \boldsymbol{x})$,

$$ss(t, \boldsymbol{x}) \;{}_t\!\Leftrightarrow_\omega\; Ss(\omega, \boldsymbol{x}).$$

By linear system theory, we can take the temporal Fourier transform of all parts of (3.17). Consider first the temporal Fourier transform of the second term,

$$\frac{1}{c^2} \cdot \partial_t^2 \left\{ss(t, \boldsymbol{x})\right\}.$$

The Fourier transform pair of the first partial derivative $\partial_t \left\{ss(t, \boldsymbol{x})\right\}$ is

$$\partial_t \left\{ss(t, \boldsymbol{x})\right\} \;{}_t\!\Leftrightarrow_\omega\; j\omega \cdot Ss(\omega, \boldsymbol{x}).$$

The second partial derivative of $\partial_t^2 \left\{ss(t, \boldsymbol{x})\right\}$ yields the Fourier transform pair,

$$\partial_t^2 \left\{ss(t, \boldsymbol{x})\right\} \;{}_t\!\Leftrightarrow_\omega\; -\omega^2 \cdot Ss(\omega, \boldsymbol{x}).$$

The temporal Fourier transform of all of (3.17) yields *Helmholtz's equation,*

$$\nabla^2 \left\{Ss(\omega, \boldsymbol{x})\right\} + k_\omega^2 \cdot Ss(\omega, \boldsymbol{x}) = 1 \cdot \delta(\boldsymbol{x}), \tag{3.18}$$

where

$$k_\omega = \frac{\omega}{c}$$

is the temporal (angular frequency) wavenumber. The result (3.18) also engages the Fourier transform pair

$$\delta(t) \;{}_t\!\Leftrightarrow_\omega\; 1,$$

where we have used

$$\delta(t, \boldsymbol{x}) = \delta(t) \cdot \delta(\boldsymbol{x}).$$

A Green's Function Solution

A solution of a wave motion equation is called a *Green's*[13] *function*[14].

A Greens Function for Helmholtz's Wave Motion Equation

An important Green's function, which we denote as $Hh(\omega, r)$, is a solution of Helmholtz's equation.

We will show that this Green's function is equivalent to Fresnel's spherical wave field. Then, given Fresnel's spherical wave field solution of the frequency-space domain wave motion equation, the time-space domain solution is obtained by an inverse temporal Fourier transform of the Green's function $Hh(\omega, r)$. We shall see that the resulting time-space domain solution is Huygens' expanding spherical wavelet.

---

[13]George Green (1793–1841) was a British mathematician and physicist who is known for his foundational work in potential functions of physics, and his early theories of electricity and magnetism.

[14]As noted in [260] and as eloquently stated by F. Rohrlich [792], "The commonly used expressions 'the Green's function' and 'a Green's function' represent an atrocity to the English language. I doubt that those who use them ever refer to 'a Shakespeare's sonnet.' " To the reader of this text, with apology, the author perpetuates this atrocity.

Hinge Pin Between Time, Space, Frequency and Wavenumber

It can be shown that

$$Ss(\omega, \boldsymbol{x}) = r \cdot Hh(\omega, r)$$

is a solution to Helmholtz's equation (3.18), where $r = |\boldsymbol{x}|$ and where the Green's function solution $Hh(\omega, r)$ has spherical symmetry. Hence,

$$Hh(\omega, r) = Hh(\omega, \boldsymbol{r}). \tag{3.19}$$

where $\boldsymbol{r}$ is a radial vector. The Green's function solution (3.19) will prove useful as an analytic hinge pin in at least three ways.

1. We shall seek to determine $hh(t, \boldsymbol{r})$ by finding the inverse temporal Fourier transform of $Hh(\omega, \boldsymbol{r})$,

   $$hh(t, \boldsymbol{r}) = \mathsf{iFT}^{\omega}_{\text{1D}} \left\{ Hh(\omega, \boldsymbol{r}) \right\}.$$

   We shall see that $hh(t, \boldsymbol{r})$ is an expanding Huygens impulsive spherical wavelet in time and space. Knowledge of $hh(t, \boldsymbol{r})$ is key to wave field inversion approaches in the time-space domain. In particular, $hh(t, \boldsymbol{r})$ is useful in formulating various time-space domain SAR image formation algorithms, including algorithms based on hyperbolic range walk coherent summing (see Section 12.2) in the signal domain and spherical direct back-propagation coherent summing in the image domain (see Section 12.3).

2. We shall seek to determine $HH(\omega, \boldsymbol{k_r})$ by performing the forward spatial Fourier transforms of $Hh(\omega, \boldsymbol{r})$ with respect to the *rectilinear* dimensions of $\boldsymbol{r}$,

   $$HH(\omega, \boldsymbol{k_r}) = \mathsf{FT}^{\boldsymbol{r}}_{\text{3D}} \left\{ Hh(\omega, \boldsymbol{r}) \right\},$$

   where $\boldsymbol{k_r} = [k_{r_x} \ k_{r_y} \ k_{r_z}]$. Knowledge of $HH(\omega, \boldsymbol{k_r})$ is key to wave field inversion approaches in the frequency-wavenumber domain. We shall see that $\measuredangle HH(\omega, \boldsymbol{k_r})$ is composed of Huygens-Fresnel transfer functions,

   $$\measuredangle HH(\omega, \boldsymbol{k_r}) = H_{r_x}(\omega, k_{r_x}) \cdot H_{r_y}(\omega, k_{r_y}) \cdot H_{r_z}(\omega, k_{r_z}).$$

   We shall see that $HH(\omega, \boldsymbol{k_r})$ is particularly useful in the class of Stolt formatting algorithms (see Chapter 7).

3. One way to determine $hH(t, \boldsymbol{k_r})$ is by performing the temporal inverse Fourier transform of $HH(\omega, \boldsymbol{k_r})$,

   $$hH(t, \boldsymbol{k_r}) = \mathsf{iFT}^{\omega}_{\text{1D}} \left\{ HH(\omega, \boldsymbol{k_r}) \right\}.$$

   We shall see that $hH(t, \boldsymbol{k_r})$ is key to wave field inversion approaches in the time-wavenumber domain, which is the domain of the class of range-Doppler (see Section 12.9) and chirp scaling image formation algorithms (see Section 12.10).

4. We shall see that knowledge of $Hh(\omega, \boldsymbol{r})$ itself is useful in developing SAR image formation processing algorithms implemented within the frequency-space domain.

3.5 Fresnel Wave Fields

We now present the important Green's function [902] that we denote as $Hh(\omega, \boldsymbol{r})$. This Green's function is a solution of Helmholtz's wave motion equation (3.18) for unbounded free space [35],

$$Hh(\omega, \boldsymbol{r}) = \frac{1}{4\pi} \cdot \frac{1}{|\boldsymbol{r}|} \cdot \exp\left(-jk_\omega |\boldsymbol{r}|\right). \tag{3.20}$$

Since this solution only involves the magnitude of $|\boldsymbol{r}|$, the wave field $Hh(\omega, \boldsymbol{r})$ is a monochromatic complex sinusoidal spherical wave field in the frequency-space domain $(\omega, \boldsymbol{r})$ that originates at a point singularity positioned at the origin. We recognize (3.20) as a *Fresnel wave field.* The importance of (3.20) in wave motion analysis cannot be overstated. Phase maps of example Fresnel wave fields over $\boldsymbol{r}$ for various values of k_ω are shown in Fig. 3.5a.

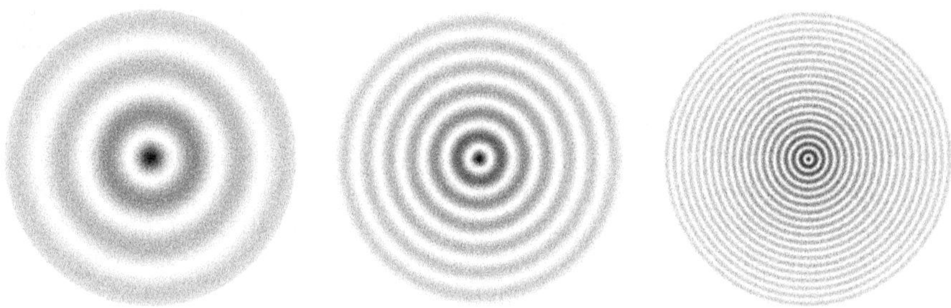

a. Fresnel wave fields.

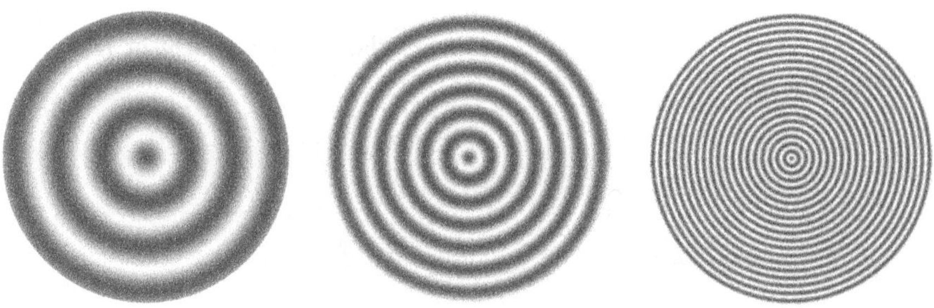

b. Fresnel phase fields.

Figure 3.5: Fresnel wave and phase fields.

Fresnel Phase Field

We indicate the phase factor element of Fresnel's wave field (3.20) by

$$\begin{aligned} Ss(\omega, \boldsymbol{r}) &= \measuredangle Hh(\omega, \boldsymbol{r}), \\ &= \exp\left(-jk_\omega |\boldsymbol{r}|\right), \\ &= 4\pi \cdot |\boldsymbol{r}| \cdot Hh(\omega, \boldsymbol{r}). \end{aligned}$$

The field $Ss(\omega, \boldsymbol{r})$ describes a monochromatic complex sinusoidal spherical wave field with no amplitude decay, i.e., a phase field. Phase maps of example Fresnel phase fields over $\boldsymbol{r}$ for various values of k_ω are shown in Fig. 3.5b.

Dynamic Fresnel Phase Fields

Dynamic Fresnel wave and phase fields are created by adding a temporal sinusoidal phase factor $\exp(j\omega t)$ to the spatial element $\exp(-jk_\omega r)$,

$$\begin{aligned} Ss(\omega, ct-r) &= \exp(j\omega t) \cdot Ss(\omega, r), && (3.21)\\ &= \exp(j\omega t) \cdot \exp(-jk_\omega r), \\ &= \exp[j(\omega t - k_\omega r)], \\ &= \exp[jk_\omega (ct - r)]. && (3.22) \end{aligned}$$

A dynamic sinusoidal Fresnel wave field emanates from the source singularity with monochromatic radial variability given by $k_\omega = \omega / c$ and propagates through space with expanding radius $ct - r$ and with sinusoidal temporal variability given by $\exp(j\omega t)$.

The form of (3.22) matches the general solution (3.11) of the wave motion equation descriptive of outwardly expanding spherical wave fields.

The spatial-temporal separability of (3.21) is key in analytic methods that seek harmonic wave fields solutions to inhomogeneous forms of Helmholtz's wave motion equation [35].

3.6 Huygens Wavelets

Consider the marginal (temporal only) inverse Fourier transform of $Hh(\omega, \boldsymbol{r})$ in the Fourier transform pair,

$$hh(t, \boldsymbol{r}) \;\; {}_t\Leftrightarrow_\omega \;\; Hh(\omega, \boldsymbol{r}),$$

where

$$Hh(\omega, \boldsymbol{r}) = \frac{1}{4\pi} \cdot \frac{1}{|\boldsymbol{r}|} \exp(-jk_\omega |\boldsymbol{r}|). \qquad (3.23)$$

The inverse Fourier transform is evaluated by

$$\begin{aligned} hh(t, \boldsymbol{r}) &= \mathsf{iFT}^{\omega}_{1D} \{Hh(\omega, \boldsymbol{r})\}, \\ &= \frac{1}{2\pi} \int_{-\infty}^{+\infty} Hh(\omega, \boldsymbol{r}) \cdot \exp(j\omega t) \cdot d\omega, \\ &= \frac{1}{2\pi} \int_{-\infty}^{+\infty} \frac{1}{4\pi} \cdot \frac{1}{|\boldsymbol{r}|} \cdot \exp(-jk_\omega |\boldsymbol{r}|) \cdot \exp(j\omega t) \cdot d\omega. && (3.24) \end{aligned}$$

We interpret (3.24) as the temporal inverse Fourier transform of the constant "1" unity function,

$$hh(t, \boldsymbol{r}) = \frac{1}{4\pi} \cdot \frac{1}{|\boldsymbol{r}|} \cdot \frac{1}{2\pi} \int_{-\infty}^{+\infty} 1 \cdot \exp\left(-j\frac{1}{c}|\boldsymbol{r}|\omega\right) \cdot \exp(j\omega t) \cdot d\omega, \quad (3.25)$$

where the phase factor $\exp(-jk_\omega|\boldsymbol{r}|)$ is a frequency domain description of the time delay

$$t_r = \frac{1}{c}|\boldsymbol{r}|.$$

Since

$$\frac{1}{2\pi}\int_{-\infty}^{+\infty} 1 \cdot \exp(j\omega t) \cdot d\omega = \delta(t),$$

the inverse Fourier transform (3.25) of the Fresnel wave field evaluates to [181]

$$hh(t, \boldsymbol{r}) = \frac{1}{4\pi} \cdot \frac{1}{|\boldsymbol{r}|} \cdot \delta(t - t_r). \tag{3.26}$$

Equation (3.26) is Huygens' expanding impulsive spherical wavelet.

Huygens Wavelet Time-Space Source Points

The expanding impulsive (the wavefront is impulsively thin) wavelet (3.26) is produced by a source temporally impulsive at time $t = 0$ and spatially impulsive at $\boldsymbol{r} = \boldsymbol{0}$ [557]. More generally, a Huygens wavelet with an impulsive source at $(t_o, \boldsymbol{r}_o)$ is [175],

$$hh(t - t_o, \boldsymbol{r} - \boldsymbol{r}_o) = \frac{1}{4\pi} \cdot \frac{1}{|\boldsymbol{r} - \boldsymbol{r}_o|} \cdot \delta(t - t_o - t_{r_o}), \tag{3.27}$$

where

$$t_{r_o} = \frac{1}{c}|\boldsymbol{r} - \boldsymbol{r}_o|.$$

3.7 Huygens-Fresnel Transfer Functions

At this point of our analysis, we have knowledge of $Hh(\omega, \boldsymbol{r})$ as given by (3.20) and we have knowledge of $hh(t, \boldsymbol{r})$ as given by (3.26). However, we do not yet know the desired frequency-wavenumber domain function $HH(\omega, \boldsymbol{k}_r)$.

Spatial Fourier Transform of Fresnel Spherical Wave Field

We determine $HH(\omega, \boldsymbol{k}_r)$ via forward Fourier transforms of a rectilinear form of $Hh(\omega, \boldsymbol{r})$ in the dimensions of

$$\boldsymbol{r} = [r_x \; r_y \; r_z].$$

This three dimensional spatial Fourier transform is

$$\begin{aligned} & HH(\omega, \boldsymbol{k}_r) \\ &= \mathsf{FT}^{r}_{3D}\left\{Hh(\omega, \boldsymbol{r})\right\}, \\ &= \int_{-\infty}^{+\infty}\int_{-\infty}^{+\infty}\int_{-\infty}^{+\infty} \frac{1}{4\pi} \cdot \frac{1}{|\boldsymbol{r}|} \exp(-jk_\omega|\boldsymbol{r}|) \\ &\qquad \cdot \exp(-jk_{r_x} r_x)\exp(-jk_{r_y} r_y)\exp(-jk_{r_z} r_z) \cdot dr_x dr_y dr_z, \\ &= \int_{-\infty}^{+\infty}\int_{-\infty}^{+\infty}\int_{-\infty}^{+\infty} \frac{1}{4\pi} \cdot \frac{1}{|\boldsymbol{r}|} \exp(-jk_\omega|\boldsymbol{r}|) \cdot \exp(-j\boldsymbol{k}_r \cdot \boldsymbol{r}) \cdot d\boldsymbol{r}, \end{aligned} \tag{3.28}$$

where the wavenumber variables introduced by the Fourier transform (3.28) comprise a wavenumber vector [864],

$$\boldsymbol{k_r} = [k_{r_x}\ k_{r_y}\ k_{r_z}]$$

and where the phase function of the kernel of the multi-dimensional spatial Fourier transform (3.28) is expressed as an inner product,

$$k_r \cdot \boldsymbol{r} = k_{r_x} \cdot r_x + k_{r_y} \cdot r_y + k_{r_z} \cdot r_z.$$

The sought after $HH(\omega, \boldsymbol{k_r})$, as expressed by (3.28), was non-trivial to derive. Without providing the details of a derivation here, $HH(\omega, \boldsymbol{k_r})$ of (3.28) is given by [665, 734, 816]

$$HH(\omega, \boldsymbol{k_r}) \quad = \quad a(\omega, \boldsymbol{k_r})^{3\text{D}}_{\text{FT}} \cdot H_{r_x}(\omega, k_{r_x}) \cdot H_{r_y}(\omega, k_{r_y}) \cdot H_{r_z}(\omega, k_{r_z}), \quad (3.29)$$

where

$$a(\omega, \boldsymbol{k_r})^{3\text{D}}_{\text{FT}}$$

is a frequency-wavenumber domain weight function of secondary importance in our analysis and where

$$\begin{aligned} H_{r_x}(\omega, k_{r_x}) &= \exp[-jr_x \cdot k_{r_x}(\omega)], \\ H_{r_y}(\omega, k_{r_y}) &= \exp[-jr_y \cdot k_{r_y}(\omega)], \\ H_{r_z}(\omega, k_{r_z}) &= \exp[-jr_z \cdot k_{r_z}(\omega)], \end{aligned} \quad (3.30)$$

are dimensionally separable *Huygens-Fresnel transfer functions.*

Pythagorean Coupling of Huygens-Fresnel Transfers

The dimensional spatial wavenumbers k_{r_x}, k_{r_y} and k_{r_z} of (3.30) are interdependent and are functionally dependent on the temporal angular wavenumber $k_\omega = \omega/c$ per a Pythagorean constraint,

$$\begin{aligned} k_{r_x}(\omega) &= \sqrt{k_\omega^2 - k_{r_y}^2 - k_{r_z}^2}, \\ k_{r_y}(\omega) &= \sqrt{k_\omega^2 - k_{r_x}^2 - k_{r_z}^2}, \\ k_{r_z}(\omega) &= \sqrt{k_\omega^2 - k_{r_x}^2 - k_{r_y}^2}. \end{aligned} \quad (3.31)$$

Fourier-Wavenumber Manifold

The Fourier transform (3.29) is non-zero only where (3.31) are satisfied for a given k_ω value. In this sense, (3.31) defines the supporting wavenumber domain manifold of the slow-time Fourier transform $HH(\omega, \boldsymbol{k_r})$ of $Hh(\omega, \boldsymbol{r})$.

Ewald Sphere

The Pythagorean relationship between the temporal wavenumber k_ω and spatial wavenumbers of $\boldsymbol{k_r}$ has a simple geometric interpretation. Equations (3.31) bound the solution of the slow-time (spatial) Fourier transform (3.29) to the surface of a special sphere in k-space called the *Ewald*[15] *sphere* [315,316]. Per the Pythagorean constraint (3.31), the Ewald sphere of k-space has a radius of k_ω and the data resident on its surface are given by (3.29). The Ewald sphere is the frequency-wavenumber domain manifold of the spatial Fourier transform of a Fresnel wave field of a given frequency.

Ewald Sphere as an Angular Spectrum Manifold

A harmonic spherical wave field of frequency ω can be decomposed into an *angular spectrum* of harmonic plane waves with directions that sweep the unit sphere. The associated spherical manifold of radius k_ω of coefficient weights $a(\omega, \boldsymbol{k_r})_{\text{FT}}^{\text{3D}}$ in the $(k_{r_x}, k_{r_y}, k_{r_z})$ wavenumber domain reside on the Ewald sphere.

Much more will be said about the attributes of an Ewald sphere as a descriptor of a SAR phase history in the frequency-wavenumber domain.

The Ewald sphere is emergent in the analysis of a variety of computed imaging problems, including crystallography [315, 316], diffraction tomography [252, 674, 837], seismology, microscopy [917, 918], microwave imaging, radio astronomy and holography [38, 993].

Spatial Fourier Transform Attributes

We summarize key attributes of the spatial Fourier transform pair,

$$Hh(\omega, \boldsymbol{r}) \quad \Leftrightarrow \quad HH(\omega, \boldsymbol{k_r}).$$

On one side of the Fourier transform pair is a monochromatic sinusoidal Fresnel spherical wave field $Hh(\omega, \boldsymbol{r})$. On the other side is a tonal wave field $HH(\omega, \boldsymbol{k_r})$ resident on the surface of the Ewald sphere. The radial spatial frequency of the spherical wave field $Hh(\omega, \boldsymbol{r})$ is k_ω. The wavenumber wave field $HH(\omega, \boldsymbol{k_r})$ corresponds to dimensionally separable tonal phase factors,

$$\exp(-jx_n k_x) \cdot \exp(-jy_n k_y) \cdot \exp(-jz_n k_z)$$

with frequencies in the k_{r_x}, k_{r_y} and k_{r_z} dimensions determined by the position parameters (x_n, y_n, z_n) of the point source of the spatial spherical wave field $Hh(\omega, \boldsymbol{r})$ (see Section 3.10).

3.8 Huygens-Fresnel Transfers in Terrain Spectra

Most of our SAR signal analysis employs discrete models of terrain scattering. The wavenumber domain phase histories of discrete terrain scattering functions, once properly formatted, are essentially comprised of Huygens-Fresnel transfer functions. These are the same Huygens-Fresnel transfer functions that arise from the spatial Fourier transform of a Fresnel spherical wave field.

[15] Paul Peter Ewald (1888–1985) was a German born American physicist who is known for his pioneering work in x-ray diffraction.

Terrain Point Scattering Functions

Consider the point scatterer located at $\boldsymbol{x}_n$ in a three dimensional terrain function,

$$\begin{aligned} \boldsymbol{g}_n(\boldsymbol{x}) &= g_n \cdot \delta(\boldsymbol{x} - \boldsymbol{x}_n), \\ &= ggg_n(x, y, z). \end{aligned}$$

Terrain Point Scattering Spectrum

The three dimensional terrain spectrum is obtained via three Fourier transforms in the component directions of $\boldsymbol{x}$,

$$\begin{aligned} \boldsymbol{G}_n(\boldsymbol{k_x}) &= \mathsf{FT}^{\boldsymbol{x}}_{3\mathrm{D}} \left\{\boldsymbol{g}_n(\boldsymbol{x})\right\}, \\ &= \mathsf{FT}^{\boldsymbol{x}}_{3\mathrm{D}} \left\{g_n \cdot \delta(\boldsymbol{x} - \boldsymbol{x}_n)\right\}, \\ &= \exp\left(-jx_n k_x\right) \cdot \exp\left(-jy_n k_y\right) \cdot \exp\left(-jz_n k_z\right), \end{aligned}$$

where in this case, we have assumed $|g_n(\boldsymbol{x})| = 1$. We sometimes write

$$\boldsymbol{G}_n(\boldsymbol{k_x}) = GGG_n(k_x, k_y, k_z)$$

for the spectrum of a three dimensional terrain function. Use of upper case letters indicates that a marginal Fourier transform has been performed along the corresponding spatial dimensions of

$$\boldsymbol{g}_n(\boldsymbol{x}) = ggg_n(x, y, z).$$

Terrain Spectrum Dimensional Phase Factors

The dimensionally separable phase factors of the spectrum $\boldsymbol{G}_n(\boldsymbol{k_x})$ of the nth scatterer, as described in the aperture coordinate system, are the following:

- Down-range scatterer position phase factor

  $$\exp\left(-jx_n k_x\right) = G_n(k_x)_{\mathrm{ACS}}$$

  with down-range positional frequency x_n in the down-range wavenumber dimension k_x.

- Along-track scatterer position phase factor

  $$\exp\left(-jy_n k_y\right) = G_n(k_y)_{\mathrm{ACS}}$$

  with along-track positional frequency y_n in the along-track wavenumber dimension k_y.

- Above-ground scatterer position phase factor

  $$\exp\left(-jz_n k_z\right) = G_n(k_z)_{\mathrm{ACS}}$$

  with above-ground positional frequency z_n in the elevation wavenumber dimension k_z.

Terrain Spectrum in Aperture Coordinate System

The terrain spectrum in the aperture coordinate system is

$$\begin{aligned} \boldsymbol{G}_n(\boldsymbol{k_x})_{\text{ACS}} &= \sum_n \exp\left[-j\left(x_n k_x + y_n k_y + z_n k_z\right)\right], \\ &= \sum_n G_n(k_x)_{\text{ACS}} \cdot G_n(k_y)_{\text{ACS}} \cdot G_n(k_z)_{\text{ACS}}, \\ &= GGG_n(k_x, k_y, k_z)_{\text{ACS}}. \end{aligned}$$

Terrain Spectrum in Scene Coordinate System

The separable terrain spectrum phase factors of a scatterer at $(\hat{x}_n, \hat{y}_n, \hat{z}_n)$ are expressible in the scene coordinate system as

$$\begin{aligned} G_n(k_{\hat{x}})_{\text{SCS}} &= \exp\left(-j\hat{x}_n k_{\hat{x}}\right), \\ G_n(k_{\hat{y}})_{\text{SCS}} &= \exp\left(-j\hat{y}_n k_{\hat{y}}\right), \\ G_n(k_{\hat{z}})_{\text{SCS}} &= \exp\left(-j\hat{z}_n k_{\hat{z}}\right), \end{aligned}$$

where the $\hat{k}_i$ with $i \in \{\hat{x}, \hat{y}, \hat{z}\}$ are the wavenumbers of the scene coordinate system. The terrain spectrum expressed in the scene coordinate system is

$$\begin{aligned} \boldsymbol{G}_n(\boldsymbol{k}_{\hat{\boldsymbol{x}}})_{\text{SCS}} &= \sum_n \exp\left[-j\left(\hat{x}_n k_{\hat{x}} + \hat{y}_n k_{\hat{y}} + \hat{z}_n k_{\hat{z}}\right)\right], \\ &= \sum_n G_n(k_{\hat{x}})_{\text{SCS}} \cdot G_n(k_{\hat{y}})_{\text{SCS}} \cdot G_n(k_{\hat{z}})_{\text{SCS}}, \\ &= GGG_n(k_{\hat{x}}, k_{\hat{y}}, k_{\hat{z}})_{\text{SCS}}. \end{aligned}$$

In Situ Fields of the Terrain Spectrum

A key insight to understanding SAR wave field inversion is the view that

$$\boldsymbol{G}_n(\boldsymbol{k}_{\hat{\boldsymbol{x}}})_{\text{SCS}}$$

is representative of the terrain scattered wave field as it exist at the ground scene and

$$\boldsymbol{G}_n(\boldsymbol{k_x})_{\text{ACS}}$$

is representative of the terrain scattered wave field as it is remotely sensed along the SAR flight line. To see this, consider the following arguments, which may seem like a mere tautology of notation, but a key element of wavenumber domain SAR image formation processing methods is resident in these wave field expressions.

Huygens-Fresnel Transfers Between Aperture and Scene Coordinate Systems

By analyzing terrain spectra expressed in both aperture and scene coordinate systems,

$$(x, y, z) = (X_c + \hat{x}, Y_c + \hat{y}, Z_c + \hat{z}),$$

we can identify the Huygens-Fresnel transfer functions that represent coordinate shifts, or equivalently, propagations of the corresponding wave fields in the spatial domain. Proceeding, the relationships between the dimensional phase factors of the ground scene spectrum $\boldsymbol{G}_n(\boldsymbol{k}_x)_{\text{ACS}}$ and those expressed in the scene coordinate system $\boldsymbol{G}_n(\boldsymbol{k}_{\hat{x}})_{\text{SCS}}$ can be equated,

$$\sum_n \exp\left[-j\left(x_n k_x + y_n k_y + z_n k_z\right)\right]$$
$$= \sum_n \exp\left\{-j\left[(X_c + \hat{x}_n)k_{\hat{x}} + (Y_c + \hat{y}_n)k_{\hat{y}} + (Z_c + \hat{z}_n)k_{\hat{z}}\right]\right\}.$$

Hence,

$$\begin{aligned} G_n(k_x)_{\text{ACS}} &= \exp(-jX_c k_{\hat{x}}) \cdot \exp(-j\hat{x}_n k_{\hat{x}}), \\ &= H_{X_c}(k_{\hat{x}}) \cdot G_n(k_{\hat{x}})_{\text{SCS}}, \end{aligned}$$

$$\begin{aligned} G_n(k_y)_{\text{ACS}} &= \exp(-jY_c k_{\hat{y}}) \cdot \exp(-j\hat{y}_n k_{\hat{y}}), \\ &= H_{Y_c}(k_{\hat{y}}) \cdot G_n(k_{\hat{y}})_{\text{SCS}}, \end{aligned}$$

$$\begin{aligned} G_n(k_z)_{\text{ACS}} &= \exp(-jZ_c k_{\hat{z}}) \cdot \exp(-j\hat{z}_n k_{\hat{z}}), \\ &= H_{Z_c}(k_{\hat{z}}) \cdot G_n(k_{\hat{z}})_{\text{SCS}}, \end{aligned}$$

where

$$\begin{aligned} H_{X_c}(k_{\hat{x}}) &= \exp\left(-jX_c k_{\hat{x}}\right), \\ H_{Y_c}(k_{\hat{y}}) &= \exp\left(-jY_c k_{\hat{y}}\right), \\ H_{Z_c}(k_{\hat{z}}) &= \exp\left(-jZ_c k_{\hat{z}}\right) \end{aligned}$$

are the Huygens-Fresnel transfer functions. Hence, the "forward model" of the wave field, as measured by a SAR sensor, is

$$H_{X_c}(k_{\hat{x}}) \cdot H_{Y_c}(k_{\hat{y}}) \cdot H_{Z_c}(k_{\hat{z}}) \cdot GGG(k_{\hat{x}}, k_{\hat{y}}, k_{\hat{z}})_{\text{SCS}} \quad = \quad GGG(k_x, k_y, k_z)_{\text{ACS}}.$$

The functions $H_{X_c}(k_{\hat{x}})$, $H_{Y_c}(k_{\hat{y}})$ and $H_{Z_c}(k_{\hat{z}})$ are the separable dimensional elements of the spatial Fourier transform

$$HH(\omega, \boldsymbol{k}_r) = \mathsf{FT}^{r}_{\text{3D}}\left\{Hh(\omega, \boldsymbol{r})\right\}$$

of a Fresnel wave field with a driving impulsive source located at (X_c, Y_c, Z_c) [74, 124, 137, 174, 309, 395, 407].

Spherical wave field inversion in the time-space domain is synonymous with Huygens-Fresnel transfers in the frequency-wavenumber domain [96, 374, 395, 438, 648, 649]. Huygens-Fresnel transfers perform coordinate system shifts with operations executed in the wavenumber domain. Furthermore, these wavenumber domain "shift operators" can be applied independently to each dimension within the wavenumber domain.

Terrain Scattered Waves as Secondary Huygens Wavelets or Fresnel Wave Fields

In the case of terrain scattered wave fields, the field's "source" is ground scattering resulting from the impinging field from the radar. If the transmitted signal is an impulse, then viewing the scattered field as secondary scattering from the primary illumination is consistent with the wavelet propagation concept as imagined by Huygens. If the illuminating field consists of a sinusoidal continuous wave signal, then the scattered "secondary" fields are consistent with the wave field propagation concepts as imagined by Fresnel. When chirp waveforms are transmitted, these concepts extend such that each point scatterer is viewed as radiating a secondary chirp waveform (see Chapter 6).

3.9 Huygens-Fresnel Inversion

Remote sensing and imaging based on wave field inversion is processing that seeks to reverse the forward wave propagation model to estimate characteristics of the cause of forward scattering. Information about scatterers (viewed as Huygens-Fresnel secondary point sources) can be estimated from remotely observed wave field measurements. Scatterer information such as position, scatterer strength, dielectrically induced phase shifts, etc. can be discovered through wave field inversion. Wave field inversion, in essence, performs a backwards propagation of wave field measurements to recover an estimate of the "in situ" field as it exists at the source area of interest, i.e., at the illuminated ground scene.

Spherical Wave Field Inversion

Back propagation of the scattered wave field, as measured along the sensor's flight path, to the in situ wave field at the ground scene is the idea of wave field inversion applied to SAR. These shifts between the wave field at the sensor and the wave field at the scene can be characterized in either spatial or wavenumber domains.

When described in wavenumber domains, forward wave field propagations are described by dimensionally separable complex sinusoids, i.e, by tonal fields with wavenumber domain frequencies determined by the distance of the scatterer from the origin of the sensor (aperture) coordinate system. Back propagation, as described in a wavenumber domain, is nothing more than multiplying these complex sinusoidal tonal phase factors with conjugated tonal phase factors, i.e., with inverse Huygens-Fresnel transfer functions. Hence, to convert from the aperture to scene coordinate system, we apply inverse Huygens-Fresnel transfer functions,

$$H^*_{X_c}(k_x) \cdot H^*_{Y_c}(k_y) \cdot H^*_{Z_c}(k_z) \cdot GGG(k_x, k_y, k_z)_{\text{ACS}} \quad = \quad GGG(k_{\hat{x}}, k_{\hat{y}}, k_{\hat{z}})_{\text{SCS}}.$$

Huygens-Fresnel transfer functions are pure phase operators that exactly describe spherical wave dispersion. Huygens-Fresnel transfer functions, forward and inverse, provide means to extrapolate spherical wave fields in the spatial domain with a wavenumber domain operator. This approach to backwards wave propagation can be viewed as a rigorous inverse wave propagation process subject only to the limitations of the scalar wave equation assumption [813].

Equation (3.29) establishes that the spatial Fourier transform of Fresnel's spherical wave field solution of Helmholtz's wave motion equation is decomposable into Huygens-Fresnel transfer functions. Huygens-Fresnel transfer functions are tonal in nature and are dimensionally separable in k-space. Huygens-Fresnel transfers can be viewed as frequency-wavenumber domain expressions of Huygens' spherical wavelet propagation and related monochromatic Fresnel wave fields.

Inverse Huygens-Fresnel transfer functions are wave field back-propagation operators [649]. These wavenumber domain operations of Huygens-Fresnel transfers (3.30) have been called "Fourier propagations" [849].

Inversion with Huygens-Fresnel Transfers: Summary

We have presented multiple expressions of Huygens-Fresnel concepts in various domains. These lead to useful SAR signal analysis and modeling tools, including the following:

- Huygens' expanding spherical wavelets $hh(t, \boldsymbol{r})$ in the time-space domain.
- Fresnel sinusoidal spherical wave fields $Hh(\omega, \boldsymbol{r})$ in the frequency-space domain.
- Huygens-Fresnel transfers $HH(\omega, \boldsymbol{k_r})$ in the frequency-wavenumber domain.

To aid in clarity of discussion, we shall seek to be consistent in the use of the following notation and terminology:

$hh(t, \boldsymbol{r})$	–	time-space domain Huygens wavelet,
$Hh(\omega, \boldsymbol{r})$	–	frequency-space domain Fresnel wave field,
$HH(\omega, \boldsymbol{k_r})$	–	frequency-wavenumber domain Huygens-Fresnel transfer functions.

All of these entities, either in exact or approximate forms, provide the basis of most SAR image formation algorithms.

3.10 Wavenumber Domain Tonal Wave Fields

We take this opportunity to preview and illustrate the wavenumber domain tonal wave fields of the Huygens-Fresnel wave theory framework. We illustrate how the position of a point singularity that produces a spherical wavelet or spherical harmonic wave field in the spatial domain determines the direction and frequency of this tonal wavenumber wave field, and how this wave field varies if expressed in the scene or aperture coordinate system.

Wavenumber Wave Fields

Wave fields that are impulsively spherical in the time-space domain become harmonic and infinitely (temporal) broadband (polychromatic) in the frequency-space domain after subjection to a temporal Fourier transform. Furthermore, the constituent harmonic spherical wave fields (that span all temporal frequencies) in the frequency-space domain become spatially tonal, once subjected to a spatial

Fourier transform. The dimensional frequencies of these tonal wavenumber wave fields in the $\boldsymbol{k}_{\hat{\boldsymbol{x}}}$ domain are determined by the scatterer position parameters

$$\hat{\boldsymbol{x}}_n = [\hat{x}_n \; \hat{y}_n \; \hat{z}_n]$$

of the scene coordinate system, or determined by the scatterer position parameters

$$\boldsymbol{x}_n = [x_n \; y_n \; z_n]$$

of the aperture coordinate system.

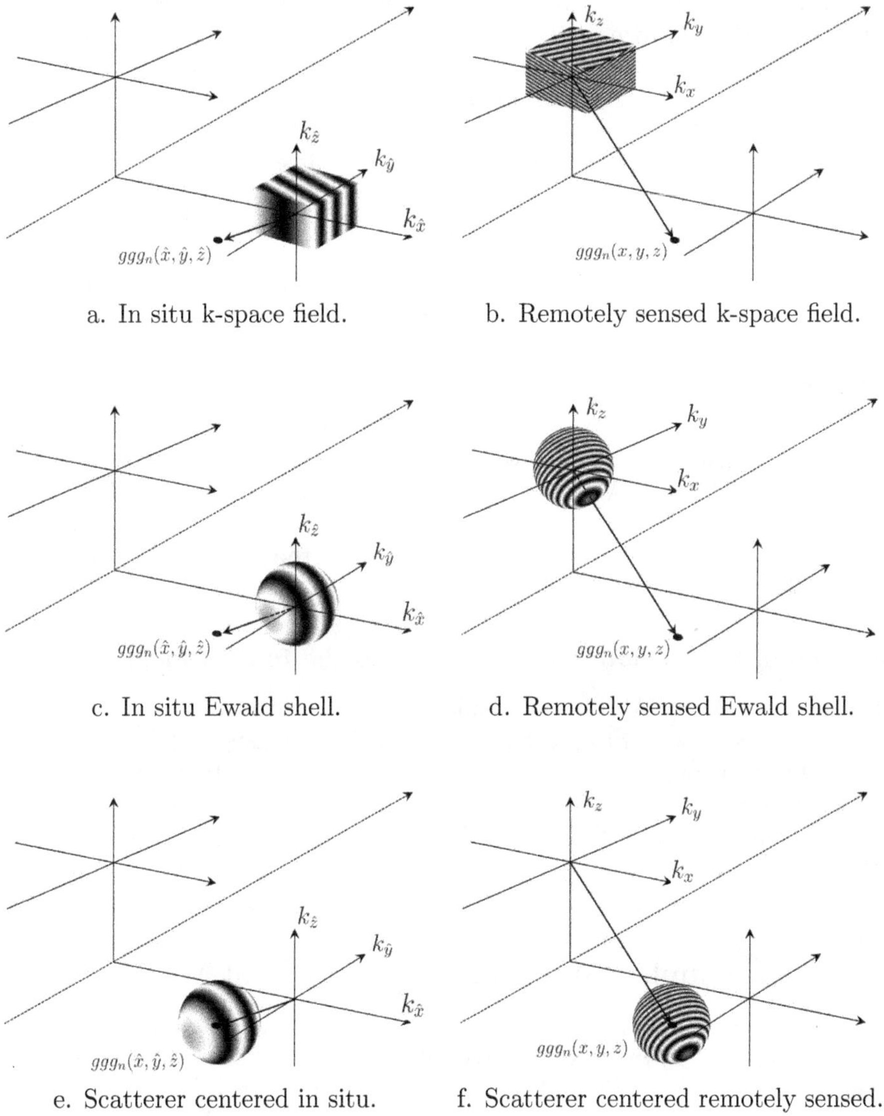

a. In situ k-space field.

b. Remotely sensed k-space field.

c. In situ Ewald shell.

d. Remotely sensed Ewald shell.

e. Scatterer centered in situ.

f. Scatterer centered remotely sensed.

Figure 3.6: Ewald sphere tonal wavenumber wave field.

In Situ Tonal Wavenumber Wave Field on Cubic Manifold

The tonal wave field of the spatial Fourier transform of an impulsively modeled scatterer extends throughout the wavenumber domain without bound. Fig. 3.6a shows the tonal wavenumber wave field of a single scatterer g_n as described in the scene coordinate system. The direction and frequency of this tonal wave field in the $\boldsymbol{k}_{\hat{\boldsymbol{x}}}$ domain are determined by the scatterer position parameters

$$\hat{\boldsymbol{x}}_n = [\hat{x}_n \ \hat{y}_n \ \hat{z}_n].$$

This is the terrain function's in situ k-space wave field. A sample of the in situ k-space wave field is drawn as a phase map on an arbitrarily placed cubic manifold in Fig. 3.6a. There is a blending of spatial and k-space entities in this figure. The cubic manifold is shown at the origin of the scene coordinate system in Fig. 3.6a. No matter where this cubic manifold is positioned in $\boldsymbol{k}_{\hat{\boldsymbol{x}}}$-space, the tonal form of the phase map on its surface is fundamentally the same.

Remotely Sensed Tonal Wavenumber Wave Field on Cubic Manifold

Fig. 3.6b shows the tonal wave field of the same scatterer, but as described in the aperture coordinate system of the SAR sensor. The direction and frequency of this tonal wave field in the $\boldsymbol{k}_{\boldsymbol{x}}$ domain are determined by the scatterer position parameters

$$\boldsymbol{x}_n = [x_n \ y_n \ z_n].$$

As with the in situ field, the tonal wave field extends throughout the wavenumber domain without bound. A sample of the tonal wavenumber wave field is drawn as a phase map on an arbitrarily placed cubic manifold. No matter where this cubic manifold is positioned in $\boldsymbol{k}_{\boldsymbol{x}}$-space, the form of the phase map on its surface is essentially the same.

Inversion of the Remotely Sensed Field to the In Situ Field

A primary task of the SAR imaging challenge is to invert the wave field as remotely sensed by the SAR sensor to the in situ wave field as it exists at the ground scene. Wave field inversion performed in k-space is preparatory to Fourier inversion that yields a SAR image centered in the scene coordinate system. However, we must deal with the bandwidth limitations of a practical SAR sensor system to better understand the supporting wavenumber manifolds available for wave field inversion operations.

Tonal Wavenumber Wave Field on Spherical Manifold

A SAR system is a passband sensor. As such, the supporting manifold of measured data is limited to the sensor's passband. When this manifold is portrayed in the wavenumber domain, the manifold is a k-space sphere with radius determined by the passband of the temporal angular frequency wavenumber k_ω as explained by the constraints of (3.31). Hence, to be more descriptive of data collections performed by SAR, the portions of the tonal wavenumber wave fields of Figs. 3.6a-b that are obtainable with a monochromatic continuous wave SAR sensor are only

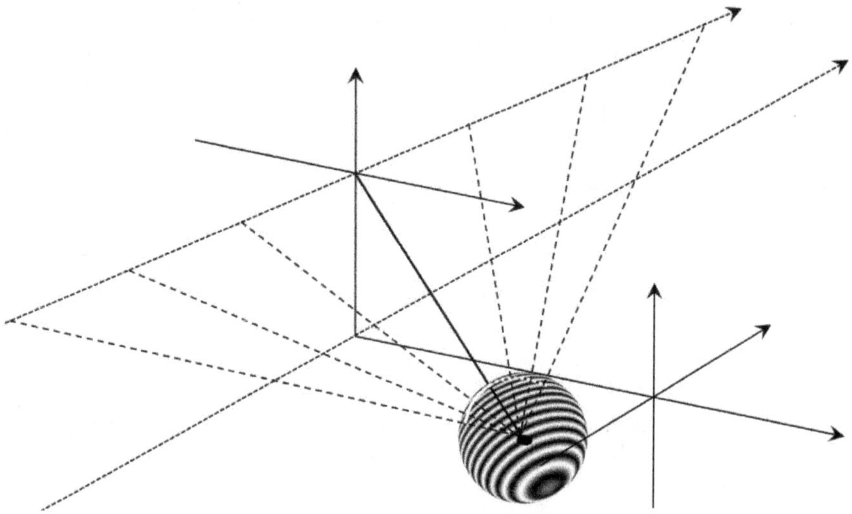

a. Linear flight k-space trace on Ewald sphere.

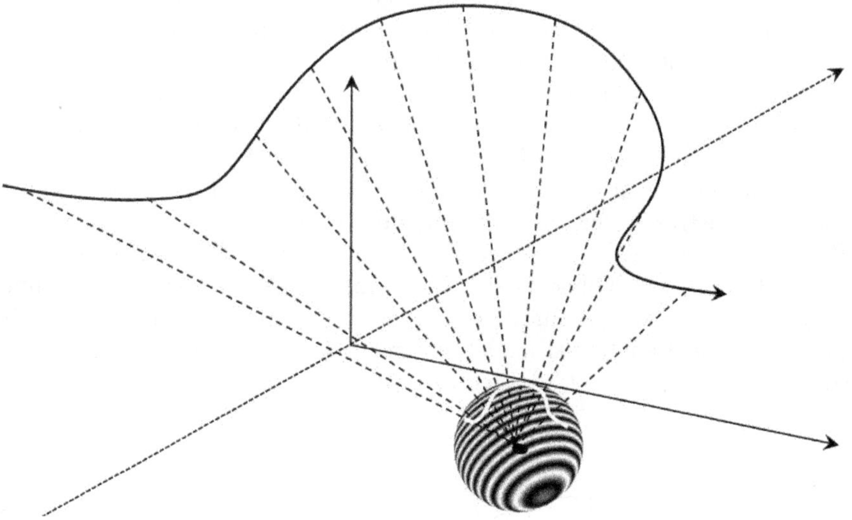

b. Non-linear flight k-space trace on Ewald sphere.

Figure 3.7: Flight induced traces on Ewald sphere.

on the surface of a k-space Ewald sphere with wavenumber radius given by k_ω as shown in Figs. 3.6c-d. The measurable wave field of the in situ wavenumber domain is not boundless, but is constrained to reside only on a manifold defined by (3.31).

The wavenumber representation of data collected by a SAR sensor that employs a continuous wave signal resides on a spherical shell manifold. This spherical shell manifold is the shell of the Ewald sphere. The *Ewald shell* is centered at the origin of the aperture or scene wavenumber coordinate system, whichever is used to describe the collected data, as illustrated in Fig. 3.6c-d.

Flight Geometry Induced Manifolds on Ewald Sphere

The tonal wavenumber wave field of each scatterer within the ground scene induces its own tonal phase function on a common Ewald shell shared by all scatterers. SAR sensors only collect submanifolds of data based on scatterer specific traces of angular dwell between sensor and scatterers. These submanifold traces are determined by the imaging geometry between sensor and scatterers, as empowered by the geometry of the flight profile. The full Ewald shell manifold of tonal wavenumber domain data would only be obtainable if the SAR sensor did a data collection from all possible sensor-scatterer view angles.

Example Ewald shell submanifold traces are illustrated in Figs. 3.7a-b. Slant-range vectors between a sensor and scatterers induce curvilinear traces on the Ewald shell as the sensor moves. Each scatterer within the terrain function induces its own unique trace on the shared Ewald sphere. If more than one scatterer exists, then the induced signatures of all scatterers create a complicated composite phase field on the shared Ewald shell. Collectively, all scatterers of the entire ground scene induce a fused manifold on the surface of the Ewald sphere. To simplify the visual presentation of these complicated composite phase maps, the induced k-space phase history of each scatterer can be shown on its own Ewald shell, with the individual Ewald shells centered on the locations of the individual scatterers as illustrated (for one scatterer) in Figs. 3.6f-e.

The traces of all scatterers induced by linear flight SAR lie on a slant-plane slice of the Ewald shell as illustrated in Fig. 3.7a. Traces obtained with non-linear flight SARs meander on the surface of the Ewald sphere as shown in Fig. 3.7b.

The Ewald shell analytical model and geometric perspective of scatterer induced wavenumber domain traces reduce SAR signal modeling down to a simple and easily visualized concept.

Chapter 4

Cross-Range Profiling with Aperture Synthesis

We now lay the ground work for understanding SAR image formation in the cross-range dimension by aperture synthesis. We examine the one dimensional along-track profiling problem where scatterers are constrained to share a fixed down-range distance from the radar. The scatterers are modeled as dispersed only in the cross-range dimension. We assume an isotropic point antenna. We only consider single bounce echoes that return to the sensor directly from discrete scatterers [12, 324, 448].

Understanding key concepts is easier if we make two assumptions. First, we assume the transmitted signal is a continuous wave sinusoid [186, 648]. Second, we assume the synthetic aperture is infinitely long [885]. The first assumption allows us to focus on a monochromatic waveform with a spectrum comprised of a single frequency. The transmitted, scattered and received wave fields produced by this waveform are closely aligned with harmonic spherical Fresnel wave fields. The second assumption allows us to start with the simpler Fourier analysis of the unwindowed spatial chirp signal that comprises the spatial phase curtain of each scatterer. The phase curtain is modeled as unwindowed because of the assumed infinite length of the synthetic aperture.

The cross-range profiling method examined here is part of the slant-plane Stolt formatting algorithm discussed in more detail in Chapter 7. The down-range offset between the terrain function and the radar flight line leads to a forward Huygens-Fresnel transfer embedded in the frequency-wavenumber phase history. The key element of this phase history is a weighted portion of the terrain spectrum. The remotely sensed portion of the terrain spectrum is prepared for Fourier inversion by Fourier migration and by interpolation from a non-uniform data raster into an uniform grid of angular spectrum data. Fourier inversion results in the cross-range profile.

This chapter is outlined as follows: Section 4.1 establishes the geometry of the along-track profiling problem and presents the along-track terrain function expressed in spatial and wavenumber domains. Section 4.2 describes Huygens-Fresnel decompositions of broadside and squint ground scenes. Section 4.3 gives a model for the received echo signal when a continuous wave tone signal is transmitted and describes the gathered time-space phase history. Section 4.4 analyzes the frequency-space phase history. Section 4.5 explores the rectilinear spectrum of the infinite synthetic aperture. Section 4.6 introduces the notion of miter cuts of wavelength and the associated dimensional spatial wavenumber projections of a directed temporal wavenumber vector. Section 4.7 develops the finite frequency-wavenumber window function associated with an infinite synthetic aperture. Section 4.8 describes Fourier migration from the rectilinear spectrum into the angular spectrum via Stolt formatting. Section 4.9 explores the geometry of the angular spectrum and demonstrates the synthesis of a spherical wave field with a full spherical sweep of directed plane waves. Section 4.10 develops the Huygens-Fresnel transfer decomposition of the angular spectrum. Section 4.11 formulates wave field inversion by inverse Huygens-Fresnel transfers and deals with uniform resampling of Fourier migrated data. Finally, Section 4.12 examines characteristics of the cross-range profile.

4.1 Cross-Range Profiling

We next establish spatial and spectral descriptions of the terrain function in cross-range profiling. We use idealized point scatterer functions. We shall develop the corresponding wavenumber descriptions in both the aperture and scene coordinate systems.

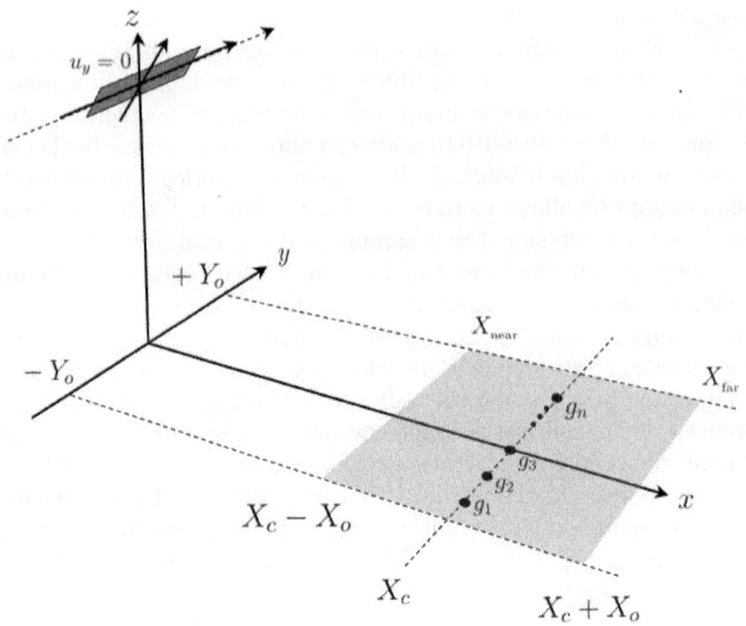

Figure 4.1: Imaging geometry for broadside along-track profiling.

Cross-Range Profiling Geometry

Assume the existence of a collection of discrete scatterers dispersed in a cross-range terrain function centered at (X_c, Y_c) in the aperture coordinate system as shown in Fig. 4.1. The index n identifies point scatterers g_n located at along-track positions y_n. The along-track extent of the single line of scatterers that comprises this simple terrain function is bounded to a segment of length $2Y_o$ centered at Y_c. The two dimensional ground patch that contains this along-track line of scatterers has a nominal down-range swath width of $2X_o$. All scatterers share the same down-range distance $x_n = X_c$ centered within the range swath.

The scatterer reflectivities g_n are assumed to be omnidirectional and independent of illuminating signal frequency. Scatterer specific radar cross section g_n is assumed to be complex valued [23]. A sinusoid signal reflected from the point scatterer g_n has an echo strength of $|g_n|$ and a phase shift of $\angle g_n$.

Broadside and Squint Cross-Range Terrain Functions

If the scene's along-track center position is $Y_c = 0$ in the aperture coordinate system, then the scene is a *broadside scene* as shown in Fig. 4.1. Scatterers in a broadside scene have along-track positions that satisfy

$$y_n \in [-Y_o, Y_o] .$$

Any cross-range terrain function with a non-zero along-track center point ($Y_c \neq 0$) is a *squint scene.* Scatterers in a squint scene have along-track positions that satisfy

$$y_n \in [Y_c - Y_o, Y_c + Y_o] ,$$

as illustrated in Fig. 4.2.

Boresight and Off Boresight Scatterers

We define terminology for the relative positions of individual scatterers with respect to the along-track center point Y_c. The single scatterer at $y_n = Y_c$ (if it exists) is the *boresight scatterer.* All other scatterers with along-track positions $y_n \neq Y_c$ are *off boresight.* Off boresight scatterers may exist in a broadside scene and a boresight scatterer may exist in a squint scene.

Terrain Function Notation for Along-Track and Down-Range Profiles

In general, we write $gg(x, y)$ for a two dimensional terrain function. If a terrain function only contains an along-track line of scatterers, then we write $ga(X_c, y)$. If a terrain function only contains a down-range line of scatterers, then we write $rg(x, Y_c)$.

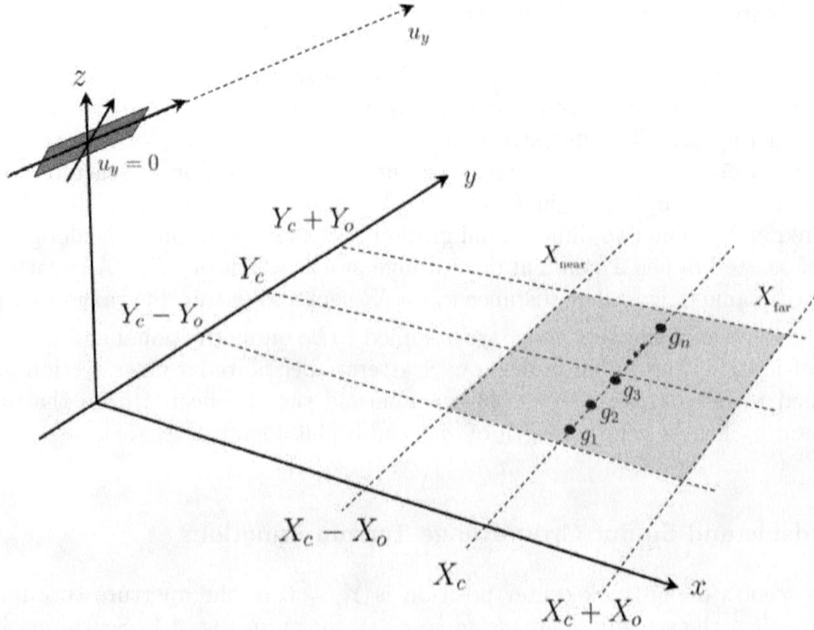

Figure 4.2: Imaging geometry for squint along-track profiling.

Along-Track Terrain Function in Aperture Coordinate System

The terrain function for an along-track, broadside, cross-range line of scatterers located at a down-range distance of X_c is

$$\begin{aligned} g_{\Pi(2X_o@X_c)}a_{\Pi(2Y_o@0)}(x,y)_{\mathrm{ACS}} &= \sum_n g_n \cdot \delta(x - X_c, y - y_n), \\ &= \delta(x - X_c) \cdot \underbrace{\sum_n g_n \cdot \delta(y - y_n)}_{a_{\Pi(2Y_o@0)}(y)_{\mathrm{ACS}}}, \end{aligned} \tag{4.1}$$

where the along-track line of scatterers that we seek to profile is

$$a_{\Pi(2Y_o@0)}(y)_{\mathrm{ACS}} = \sum_n g_n \cdot \delta(y - y_n).$$

The first subscript $\Pi(2X_o@X_c)$ of

$$g_{\Pi(2X_o@X_c)}a_{\Pi(2Y_o@0)}(x,y)_{\mathrm{ACS}} \tag{4.2}$$

indicates a down-range distance of X_c and a nominal range swath of width $2X_o$. The second subscript $\Pi(2Y_o@0)$ of (4.2) indicates that the along-track extent of the line of scatterers $a_{\Pi(2Y_o@0)}(y)_{\mathrm{ACS}}$ is restricted to a line segment of length $2Y_o$ centered at $Y_c = 0$.

Broadside Terrain Spectrum

The two dimensional Fourier transform of (4.1) yields the *terrain spectrum*,

$$\mathsf{FT}_{2D}\left\{g_{\Pi(2X_o@X_c)}a_{\Pi(2Y_o@0)}(x,y)_{\text{ACS}}\right\} = \exp(-jX_ck_x)\cdot\sum_n g_n\cdot\exp(-jy_nk_y). \tag{4.3}$$

Since all along-track scatterers share the same down-range distance X_c, the phase factor $\exp(-jX_ck_x)$ of (4.3) is independent of n and moves to the outside of the terrain function's summation. The remaining set of scaled and summed phase factors,

$$\sum_n g_n\cdot\exp(-jy_nk_y),$$

constitutes the along-track terrain spectrum.

Along-Track Terrain Function in Scene Coordinate System

The along-track terrain function in the scene coordinate system is

$$g_{\Pi(2X_o@0)}a_{\Pi(2Y_o@0)}(\hat{x},\hat{y})_{\text{SCS}} = \sum_n g_n\cdot\delta(\hat{x}-\hat{x}_n,\hat{y}-\hat{y}_n).$$

The scene is centered at $(0,0)$. Since the point (X_c,Y_c) in the aperture coordinate system coincides with the origin of the scene coordinate system, we have $\hat{x}_n=0$ for all scatterers on the cross-range line of interest. Hence,

$$g_{\Pi(2X_o@0)}a_{\Pi(2Y_o@0)}(\hat{x},\hat{y})_{\text{SCS}} = \delta(\hat{x})\cdot\sum_n g_n\cdot\delta(\hat{y}-\hat{y}_n). \tag{4.4}$$

Along-Track Terrain Spectrum in Scene Coordinate System

The two dimensional Fourier transform of (4.4) gives the along-track terrain spectrum described in the scene coordinate system,

$$\begin{aligned}\mathsf{FT}_{2D}\left\{g_{\Pi(2X_o@0)}a_{\Pi(2Y_o@0)}(\hat{x},\hat{y})_{\text{SCS}}\right\} &= \exp(-j\cdot 0\cdot k_{\hat{x}})\cdot\sum_n g_n\cdot\exp(-j\hat{y}_nk_{\hat{y}}),\\ &= 1\cdot\sum_n g_n\cdot\exp(-j\hat{y}_nk_{\hat{y}}).\end{aligned}$$

We write this as

$$G_{\Pi(\infty)}A_{\Pi(\infty)}(k_{\hat{x}},k_{\hat{y}})_{\text{SCS}} = 1\cdot\underbrace{\sum_n g_n\cdot\exp(-j\hat{y}_nk_{\hat{y}})}_{A_{\Pi(\infty)}(k_{\hat{y}})_{\text{SCS}}}, \tag{4.5}$$

where the spectrum of the along-track profile is

$$A_{\Pi(\infty)}(k_{\hat{y}})_{\text{SCS}} = \sum_n g_n\cdot\exp(-j\hat{y}_nk_{\hat{y}}).$$

Terrain Spectrum Notation

In general, where scatterers are dispersed in two dimensions, we write

$$G_{\Pi(\infty)}G_{\Pi(\infty)}(k_{\hat{x}}, k_{\hat{y}})_{\text{SCS}} = \sum_n g_n \cdot \exp(-j\hat{x}_n k_{\hat{x}}) \cdot \exp(-j\hat{y}_n k_{\hat{y}}).$$

The $\Pi(\infty)$ subscripts of $G_{\Pi(\infty)}G_{\Pi(\infty)}(\cdot,\cdot)$ indicate infinite support for the down-range wavenumber $k_{\hat{x}}$ and the along-track wavenumber $k_{\hat{y}}$. These infinite bandwidths in the wavenumber domain result from use of the impulse function $\delta(\cdot,\cdot)$ as a spatial domain scatterer model. Infinite bandwidth of the terrain spectrum is not a result of the assumed infinite synthetic aperture length.

Double Scripted Notation and Marginal Fourier Transforms

The use of double scripted notation helps in the expression of signals where we are interested in both full and marginal multi-dimensional Fourier transforms [407, 770]. For example, $GG(k_x, k_y)$ is the Fourier transform of $gg(x, y)$ in both x and y dimensions. Only the down-range marginal Fourier transform has been applied to x in $Gg(k_x, y)$. Only the along-track marginal Fourier transform has been applied to y in $gG(x, k_y)$.

4.2 Along-Track Profile Huygens-Fresnel Analysis

To develop a wavenumber domain description of a transfer from the scene coordinate system to the aperture coordinate system, consider

$$g_{\Pi(2X_o @ X_c)} g_{\Pi(2Y_o @ Y_c)}(x, y)_{\text{ACS}} = \sum_n g_n \cdot \exp[-j(X_c + \hat{x}_n)k_x] \cdot \exp[-j(Y_c + \hat{y}_n)k_y]$$

and since (with $\hat{x}_n = 0$ and $Y_c = 0$ in broadside along-track profiling),

$$g_{\Pi(2X_o @ X_c)} a_{\Pi(2Y_o @ 0)}(x, y)_{\text{ACS}} = \sum_n g_n \cdot \exp\left(-jX_c k_x\right) \cdot \exp\left(-j\hat{y}_n k_y\right).$$

Moreover, since the orientation of the aperture and scene coordinate systems coincide, we can equate the aperture and scene coordinate system wavenumbers,

$$\begin{aligned} k_x &= k_{\hat{x}}, \\ k_y &= k_{\hat{y}}. \end{aligned}$$

The scene in the aperture coordinate system, but expressed using scene coordinate system parameters, is

$$g_{\Pi(2X_o @ X_c)} a_{\Pi(2Y_o @ 0)}(x, y)_{\text{ACS}} = \exp(-jX_c k_{\hat{x}}) \cdot \sum_n g_n \cdot \exp\left(-j\hat{y}_n k_{\hat{y}}\right). \tag{4.6}$$

Along-Track Profile Forward Huygens-Fresnel Transfers

We identify the phase factor $\exp(-jX_c k_{\hat{x}})$ of (4.6) as the forward Huygens-Fresnel transfer function in the down-range dimension,

$$H_{X_c}(k_{\hat{x}}) = \exp(-jX_c k_{\hat{x}}).$$

The forward Huygens-Fresnel transfer $H_{X_c}(k_{\hat{x}})$ describes the difference between the wavenumber function of the phase field produced (scattered) at its source (in situ at the ground scene) and the wavenumber function of the phase field as measured (remotely sensed) along the radar's flight line. This Huygens-Fresnel transfer function indicates the flight path is offset from the line of along-track scatterers in the down-range dimension by a distance of X_c.

Broadside Huygens-Fresnel Analysis

From (4.5) and (4.6), the Huygens-Fresnel decomposition of the terrain spectrum is

$$G_{\Pi(\infty)} A_{\Pi(\infty)}(k_x, k_y)_{\mathrm{ACS}} = H_{X_c}(k_{\hat{x}}) \cdot G_{\Pi(\infty)} A_{\Pi(\infty)}(k_{\hat{x}}, k_{\hat{y}})_{\mathrm{SCS}}.$$

The wavenumber phase history of the along-track ground scene

$$G_{\Pi(\infty)} A_{\Pi(\infty)}(k_x, k_y)_{\mathrm{ACS}},$$

as described in the aperture coordinate system, is equal to

$$G_{\Pi(\infty)} A_{\Pi(\infty)}(k_{\hat{x}}, k_{\hat{y}})_{\mathrm{SCS}}$$

of the scene coordinate system subjected to the down-range forward Huygens-Fresnel transfer $H_{X_c}(k_{\hat{x}})$.

Squint Huygens-Fresnel Analysis

Since $y = Y_c + \hat{y}$ with $Y_c \neq 0$ for a squint scene, the Fourier transform pair that describes a squint scene in the aperture coordinate system is

$$g_{\Pi(2X_o @ X_c)} a_{\Pi(2Y_o @ Y_c)}(x, y)_{\mathrm{ACS}} \quad \Leftrightarrow \quad G_{\Pi(\infty)} A_{\Pi(\infty)}(k_x, k_y)_{\mathrm{ACS}},$$

where

$$G_{\Pi(\infty)} A_{\Pi(\infty)}(k_x, k_y)_{\mathrm{ACS}} \quad = \quad H_{X_c}(k_{\hat{x}}) \cdot H_{Y_c}(k_{\hat{y}}) \cdot G_{\Pi(\infty)} A_{\Pi(\infty)}(k_{\hat{x}}, k_{\hat{y}})_{\mathrm{SCS}}.$$

Both along-track $H_{Y_c}(k_{\hat{y}})$ and down-range $H_{X_c}(k_x)$ Huygens-Fresnel transfers are required to describe the squint scene in the aperture coordinate system, where

$$H_{Y_c}(k_{\hat{x}}) = \exp(-jY_c k_{\hat{y}})$$

is the forward Huygens-Fresnel transfer that describes the along-track position of the squint scene.

Terrain Function Transform Pair in Scene Coordinate System

The Fourier transform pair of a broadside or squint scene in the scene coordinate system is

$$g_{\Pi[-X_o,X_o]}\, a_{\Pi[-Y_o,Y_o]}(\hat{x},\hat{y})_{\text{SCS}} \quad \Leftrightarrow \quad G_{\Pi(\infty)} A_{\Pi(\infty)}(k_{\hat{x}},k_{\hat{y}})_{\text{SCS}}. \tag{4.7}$$

No forward Huygens-Fresnel transfers exist. The scene coordinate system Fourier transform pair (4.7) does not indicate if the scene is broadside or squint. This distinction can only be expressed in the aperture coordinate system.

4.3 Along-Track Profile Time-Space Phase History

We begin by describing phase data collected in the time-space domain along a linear flight path that passes through spherically scattered phase fields when the transmitted waveform is a sinusoidal continuous wave signal.

Isotropic Radiation, Scattering and Reception

We assume the radar is a point antenna that radiates in all directions. We assume isotropic scattering from discrete scatterers on the ground.

Transmitted Continuous Wave Tone Signal

Consider a continuous wave tone signal at a frequency of ω_c. The infinite duration transmitted waveform in complex form is

$$w(t)^{\text{cw}} = \exp(j\omega_c t).$$

Received Echo Signal

The echo signal received from scatterer g_n located at (x_n, y_n) (where $x_n = X_c$ for cross-range profiling) is a scaled and delayed echo of the transmitted waveform $w(t)^{\text{cw}}$,

$$ss(t,u_y)_n^{\text{cw}} = g_n \cdot w\left[t - t_n(u_y)\right]^{\text{cw}}. \tag{4.8}$$

The double scripted notation $ss(t,u_y)_n^{\text{cw}}$ of (4.8) indicates the received echo signal depends on both the fast-time and slow-time parameters t and u_y. The echo strength is scaled by the scatterer's radar cross section coefficient g_n. Other less germane and deterministic factors (such as round trip attenuation) are ignored for the sake of notational simplicity.

Slant-Range Distance

The along-track position of the radar's point antenna a_p is $(0, u_y)$ in a two dimensional (slant-plane) aperture coordinate system. Antenna position varies as a function of slow-time u_y,

$$a_p(0, u_y).$$

The slant-range distance between the sensor and the nth scatterer is

$$\begin{aligned} r_n(u_y) &= |a_p(0, u_y) - g_n(X_c, y_n)|, \\ &= \sqrt{(0 - X_c)^2 + (u_y - y_n)^2}. \end{aligned} \tag{4.9}$$

Echo Signal of Entire Along-Track Ground Scene

The superposition of echoes from all scatterers within the ground scene describes the terrain scattered signal received by the radar,

$$\begin{aligned} ss(t, u_y)^{\text{cw}}_{\text{ACS}} &= \sum_n g_n \cdot w\left[t - t_n(u_y)\right]^{\text{cw}}, \\ &= \sum_n g_n \cdot w\left[t - \frac{1}{c} \cdot 2r_n(u_y)\right]^{\text{cw}}. \end{aligned} \tag{4.10}$$

Along-Track Profile Time-Space Phase History

With substitution of the complex tone signal waveform

$$w(t)^{\text{cw}} = \exp(j\omega_c t)$$

into (4.10), we obtain a factorized *time-space phase history,*

$$\begin{aligned} ss(t, u_y)^{\text{cw}}_{\text{ACS}} &= \sum_n g_n \cdot \exp\left\{j\omega_c \left[t - \frac{2}{c} r_n(u_y)\right]\right\}, \\ &= \sum_n g_n \cdot \exp(j\omega_c t) \cdot \exp\left[-j\omega_c \frac{2}{c} r_n(u_y)\right]. \end{aligned} \tag{4.11}$$

Time-Space Phase Measurements in Fresnel Phase Fields

The time-space phase history (4.11) contains scatterer informative phase factors,

$$\exp\left[-j\omega_c \frac{2}{c} r_n(u_y)\right] = \exp\left[-jk_{\omega_c} \cdot 2r_n(u_y)\right], \tag{4.12}$$

where $k_{\omega_c} = \omega_c / c$.

Scatterer Independent Phase Factors

We see in (4.11) a phase factor $\exp(j\omega_c t)$ that is independent of the scatterer index n. This independence allows us to write,

$$ss(t, u_y)^{\text{cw}} = \underbrace{\exp(j\omega_c t)}_{\substack{\text{no scatterer} \\ \text{information}}} \cdot \sum_n g_n \cdot \underbrace{\exp\left[-jk_{\omega_c} \cdot 2r_n(u_y)\right]}_{\substack{\text{scatterer informative} \\ \text{phase curtains}}}. \tag{4.13}$$

The fast-time dependent phase factor $\exp(j\omega_c t)$ provides no information about echo strength or information about the positions of along-track scatterers within the cross-range terrain function.

Temporal-Spatial Separability

The phase factor $\exp(j\omega_c t)$ is identical to the transmitted continuous wave signal $w(t)^{^{\mathrm{cw}}}$. The temporally dependent waveform phase factor

$$w(t)^{^{\mathrm{cw}}} = \exp(j\omega_c t)$$

in (4.13) is separable from the spatially dependent phase factors,

$$\sum_n g_n \cdot \exp\left[-jk_{\omega_c} \cdot 2r_n(u_y)\right].$$

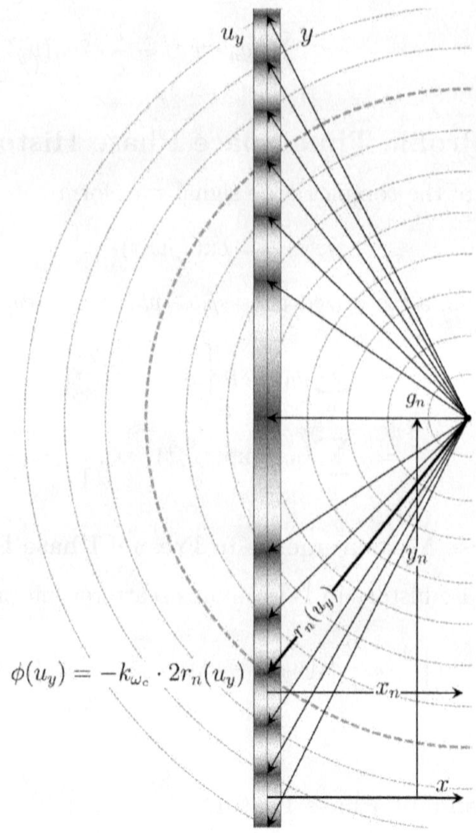

Figure 4.3: Slow-time phase curtain of a linear synthetic aperture.

Spherical Phase Field Sensed Along Flight Line

The phase factor (4.12) represents phase measurements collected by a linear flight SAR within the spherical phase field of the nth scatterer. Shown in Fig. 4.3 is a set of concentric circles (dotted lines) that represent spherical surfaces of constant phase that cut the slant-plane within the scattered field of a point scatterer. The scatterer g_n is located at a down-range distance of $x_n = X_c$. The scatterer is off boresight with an along-track offset of y_n. The slant-plane circles of constant phase that intersect the flight line are determined by sensor-scatterer slant-range values

$$r_n(u_y) = \sqrt{X_c^2 + (u_y - y_n)^2}$$

and by the temporal wavenumber

$$k_{\omega_c} = \frac{\omega_c}{c}.$$

The SAR sensor makes measurements of the phase variation of (4.12) along the flight line u_y as illustrated in Fig. 4.3. The intersections of the linear flight path with the spherical phase field (represented by circular lines of constant phase) determine the data of the *slow-time phase curtain* gathered by the SAR sensor. The slow-time phase curtain is shown as a phase map in Fig. 4.3. Slow-time phase curtains have been given various names, including "Doppler history" [137] and "spatial Doppler" [885].

Canonical Form of Time-Space Phase History

We define a canonical form of a time-space phase history as

$$ss(t, u_y)^{\text{cw}} = \underbrace{w(t)^{\text{cw}}}_{\substack{\text{fast-time} \\ \text{waveform}}} \cdot \sum_n g_n \cdot \underbrace{\exp\left[-jk_{\omega_c} \cdot 2r_n(u_y)\right]}_{\substack{\text{slow-time} \\ \text{phase curtains}}}. \tag{4.14}$$

The structure of (4.14) will occur often in our analysis of SAR time-space phase histories.

Range Walk Phase Curtains

The slant-range $r_n(u_y)$ is the radius of a sphere centered at the nth scatterer that extends to the point antenna at aperture position u_y. The phase function of a scatterer's spherical phase field is

$$\phi(u_y) = -k_{\omega_c} \cdot 2r_n(u_y). \tag{4.15}$$

The spatial spherical phase function (4.15) is a covert expression of fast-time round trip signal delay,

$$\phi(u_y) = -\omega_c \cdot t_n(u_y), \tag{4.16}$$

where the fast-time echo delay $t_n(u_y)$ is dependent on the round trip distance $2r_n(u_y)$ between the scatterer and aperture position u_y,

$$t_n(u_y) = \frac{1}{c} \cdot 2r_n(u_y).$$

Signal Delay Hyperbolas

When fast-time echo delay t_n is plotted as a function of u_y, the fast-time/slow-time function $t_n(u_y)$ maps out a hyperbola [648] as shown in Fig. 4.4a. The hyperbolic trace of the corresponding slant-range distance $r_n(u_y)$ shown in Fig. 4.4b is called *range walk.* Note that the fast-time axis has been flipped in Figs. 4.4a-b so that movement to the left on the horizontal time axis represents an increase in signal delay. At data collection range points $r_n(u_y)$ along the range walk hyperbola, the $t_n(u_y)$ echo delay phase measurements

$$\exp\left[-jk_{\omega_c} \cdot 2r_n(u_y)\right] = \exp\left[-j\omega_c \cdot t_n(u_y)\right] \tag{4.17}$$

are gathered by the SAR sensor moving along the flight axis $(0, u_y)$.

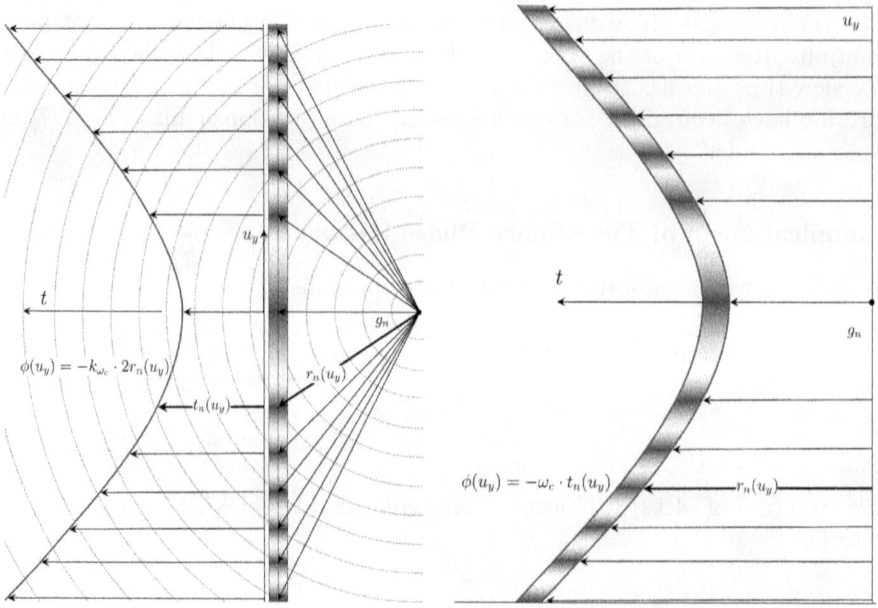

a. Signal delay hyperbola. b. Range walk phase curtain.

Figure 4.4: Hyperbolic range walk to signal delay mapping.

Slow-Time to Range Walk Phase Curtain Mapping

The slow-time phase curtain of Fig. 4.4a maps to a *range walk phase curtain* as shown in Fig. 4.4b.

Range Walk Signals of Different Along-Track Offsets

Since we assume $x_n = X_c$ in our discussion of along-track profiling, the hyperbolic range walk of different scatterers dispersed in the along-track dimension are identical in shape, but have different along-track positions as illustrated in Fig. 4.5a.

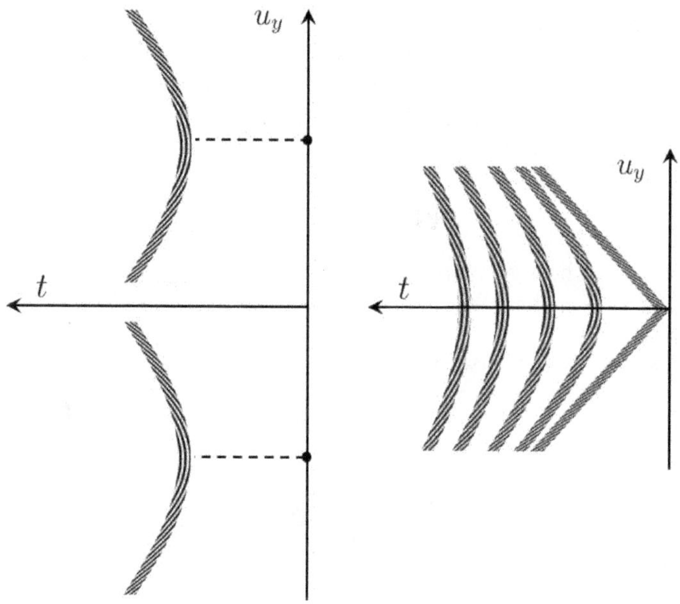

a. Same-range range walks. b. Same-azimuth range walks.

Figure 4.5: Range walk phase curtains as a function of scatterer position.

Range Walk Phase Curtains of Different Down-Range Offsets

Although somewhat off topic, consider the hyperbolic slant-range geometry of scatterers located at different down-range distances. Differences in down-range offset distances between the sensor and scatterers induce different hyperbolic range walk shapes as shown in Fig. 4.5b. This figure contains slant-range hyperbolas of five scatterers. One scatterer among this set has no down-range offset ($x_n = 0$).

The phase maps of Figs. 4.5a-b show the full time-space echo signal (4.14), not just the phase curtain (4.17) as shown in Figs. 4.4a-b.

Time-Space Phase History Temporal-Spatial Phase Factor Separability

Because of the separability of the time-space phase history into temporally and spatially dependent elements as expressed in (4.14), it is possible to produce separate phase maps of these different phase factors. Shown in Figs. 4.6a-c are elements of the time-space phase history decomposition of the range walk hyperbolas of the five scatterers shown in Fig. 4.5b.

The phase maps of Fig. 4.6a show the full time-space phase history

$$ss(t, u_y)^{\text{cw}} \tag{4.18}$$

of each scatterer. The phase maps of Fig. 4.6b display the real part of the temporal waveform phase factors

$$w(t)^{\text{cw}} = \exp(j\omega_c t) \tag{4.19}$$

as measured over a short fast-time window $[T_s, T_f]$, with $T_s = t_n(u_y)$ for aperture position u_y. Such a window does not really exist with non-pulsed, infinite duration, continuous wave signals. This window is used for concept illustration (to allow us to see signal delay in Figs. 4.6a-c. The phase maps of Fig. 4.6c show the real part of the range walk phase curtain

$$\exp\left[-jk_{\omega_c} \cdot 2r_n(u_y)\right] \tag{4.20}$$

of each scatterer. The two phase factors (4.19) and (4.20) combine to give the composite time-space phase history (4.18) as shown in Fig. 4.6a.

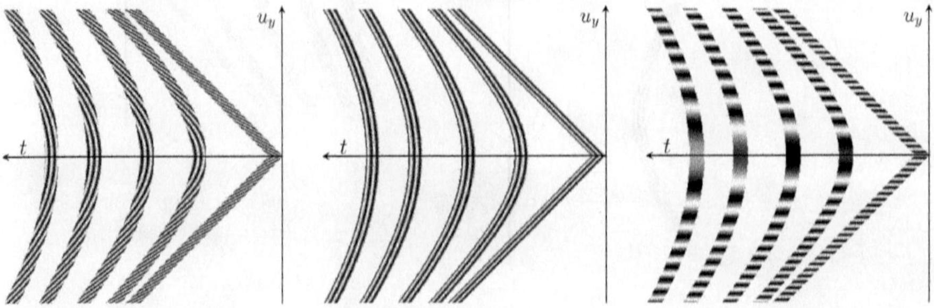

a. Time-space phase history. b. Temporal phase factor. c. Spatial phase factor.

Figure 4.6: Continuous wave time-space phase history decomposition.

4.4 Frequency-Space Phase History

Our next task is to develop the frequency-space domain description of the data collected in the time-space domain.

Continuous Wave Echo Signal Down Conversion

We formulate an along-track imaging approach for cross-range profiling by first removing from the measured time-space phase history the phase factor that is not scatterer informative. Down conversion of the received echo signal achieves this objective:

$$\begin{aligned} Ss(\omega_c, u_y) &= \underbrace{\exp(-j\omega_c t)}_{\substack{\text{local} \\ \text{oscillator}}} \underbrace{\times}_{\substack{\text{down} \\ \text{conversion}}} \underbrace{ss(t, u_y)^{\text{cw}}}_{\substack{\text{received} \\ \text{signal}}}, \qquad (4.21) \\ &= \underbrace{\exp(-j\omega_c t)}_{\substack{\text{cancels} \\ \text{phase factor}}} \times \underbrace{\exp(j\omega_c t)}_{\substack{\text{no scatterer} \\ \text{information}}} \sum_n g_n \cdot \underbrace{\exp\left[-jk_{\omega_c} \cdot 2r_n(u_y)\right]}_{\substack{\text{scatterer informative} \\ \text{phase factors}}}, \end{aligned}$$

which results in

$$Ss(\omega_c, u_y) = \sum_n g_n \cdot \underbrace{\exp\left[-jk_{\omega_c} \cdot 2r_n(u_y)\right]}_{\substack{\text{frequency-space domain} \\ \text{phase curtains}}} .$$

Down conversion of the received echo signal, achieved with multiplication of $ss(t, u_y)^{\text{cw}}$ by $\exp(-j\omega_c t)$, removes the temporal element $\exp(j\omega_c t)$ embedded in the time-space phase history. We call

$$Ss(\omega_c, u_y) = \sum_n g_n \cdot \exp\left[-jk_{\omega_c} \cdot 2r_n(u_y)\right] \tag{4.22}$$

the *continuous wave frequency-space phase history.*

The frequency-space phase history (4.22) is composed solely of phase curtains, one for each scatterer.

Since the phase curtains $\exp\left[-jk_{\omega_c} \cdot 2r_n(u_y)\right]$ are independent of time, down conversion does not cause a left shift of the spectrum of these passband phase factors. Hence, there is no down conversion related "L.shift" superscript annotation added to $Ss(\omega_c, u_y)$.

Double Scripted Notation for CW Frequency-Space Phase History

A demodulated signal's conversion from the time-space domain to the frequency-space domain is indicated with the uppercase-lowercase double scripted notation $Ss(\omega_c, u_y)$. This notation usually indicates when a temporal Fourier transform converts the time-space signal $ss(t, u_y)^{\text{pulse}}$ to the frequency-space phase history, $Ss(\omega, u_y)^{\text{pulse}}$ [407, 770]. In the case at hand, however, where a continuous waveform instead of a pulse waveform is used, a Fourier transform is not performed to obtain $Ss(\omega_c, u_y)$ from $ss(t, u_y)^{\text{cw}}$. Down conversion eliminates the temporal dependence of the time-space phase history when continuous wave tone signals are transmitted. The remaining phase history elements depend on the tone signal's frequency parameter and hence, reside in the frequency-space domain.

Phase Curtains in the Frequency-Space Phase History

Consider the phase curtain $\exp\left[-jk_{\omega_c} \cdot 2r_n(u_y)\right]$ of the nth scatterer. This phase factor represents a spatial sample along u_y of the static element of a monochromatic Fresnel phase field. Hence, it is appropriate to use the notation,

$$\measuredangle Hh^2[\omega_c, r_n(u_y)] = \exp\left[-jk_{\omega_c} \cdot 2r_n(u_y)\right]. \tag{4.23}$$

The double scripted notation $Hh(\cdot,\cdot)$ indicates residency in the frequency-space domain of Fresnel's harmonic spherical phase field.

Canonical Form of Frequency-Space Phase Curtains

We can now write the frequency-space phase history in a useful canonical form,

$$Ss(\omega_c, u_y) = \sum_n g_n \cdot \measuredangle Hh^2\left[\omega_c, r_n(u_y)\right]. \tag{4.24}$$

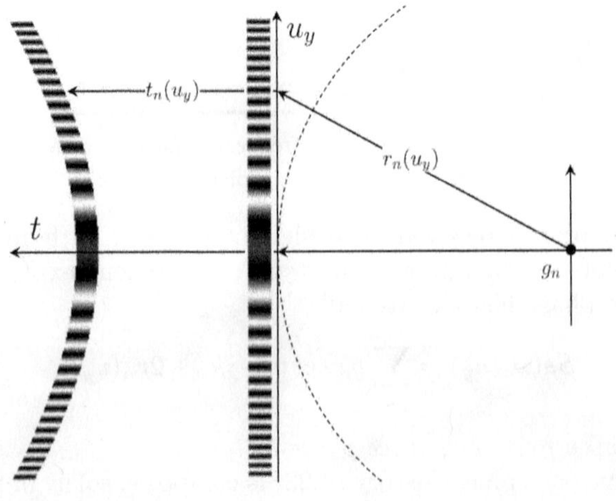

Figure 4.7: Slow-time to range walk phase curtain mapping.

The structure of (4.24) will occur often in our analysis of a monochromatic SAR frequency-space phase history.

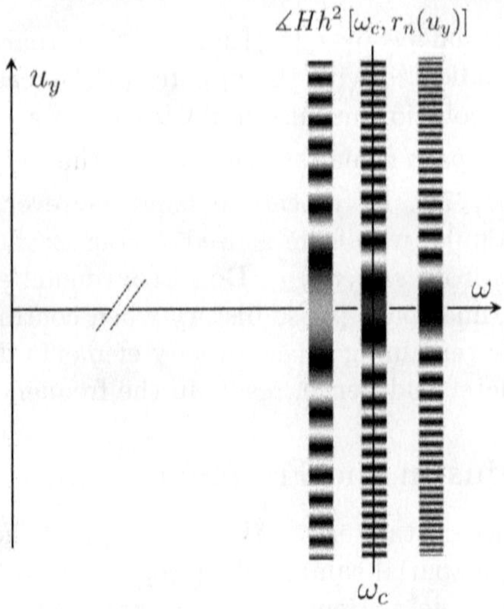

Figure 4.8: Frequency-space phase curtains.

Finite Aperture Frequency-Space Phase History Characteristics

To see the difference between time-space (4.14) and frequency-space (4.23) phase map presentations of scatterer specific phase curtains, the time-space domain phase curtains of a single scatterer in both slow-time and range walk forms are shown in Fig. 4.7. These phase maps are plotted as a function of time and space, i.e., in the (t, u_y) domain and in the $[t, r_n(u_y)]$ domain for one continuous wave frequency value ω_c.

Range walk is a time domain, not a frequency domain attribute of SAR phase histories and hence, is not evident in frequency-space domain phase histories. When producing phase maps in the frequency-space domain, the phase maps are drawn as a function of u_y versus ω_c or ω_p as shown in Fig. 4.8. The center phase map of Fig. 4.8 is the slow-time phase curtain of Fig. 4.7, but replotted as a function of (ω_c, u_y).

For systems that use continuous wave waveforms, the phase curtain of the frequency-space domain is identical to the slow-time phase curtain of the time-space domain.

The phase maps on the right and left of the center frequency-space phase curtain shown in Fig. 4.8 have lower and higher continuous wave ω_c frequency values. The three phase maps of Fig. 4.8 illustrate how frequency-space and time-space slow-time phase curtains change when the continuous wave frequency ω_c is decreased or increased.

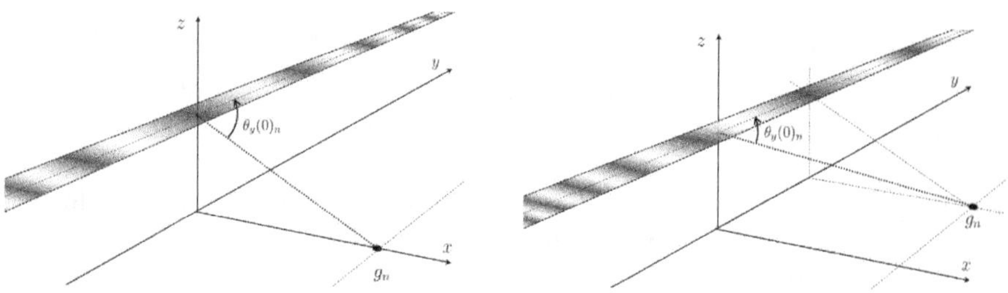

a. Boresight phase curtain. b. Off boresight phase curtain.

Figure 4.9: Boresight and off boresight slow-time phase curtains.

Boresight and Off Boresight Frequency-Space Phase Histories

Consider the slow-time phase curtains (equivalent to the frequency-space phase curtains) shown in Figs. 4.9a-b. These phase curtains are illustrated along SAR flight lines. The scatterer of Fig. 4.9a is at boresight and the scatterer of Fig. 4.9b is off boresight.

4.5 Frequency-Wavenumber Phase History

A spatial Fourier transform converts the phase history $Ss(\omega_c, u_y)_n$ from the frequency-space domain to the frequency-wavenumber domain,

$$Ss(\omega_c, u_y)_n \mapsto SS(\omega_c, k_{u_y})_n.$$

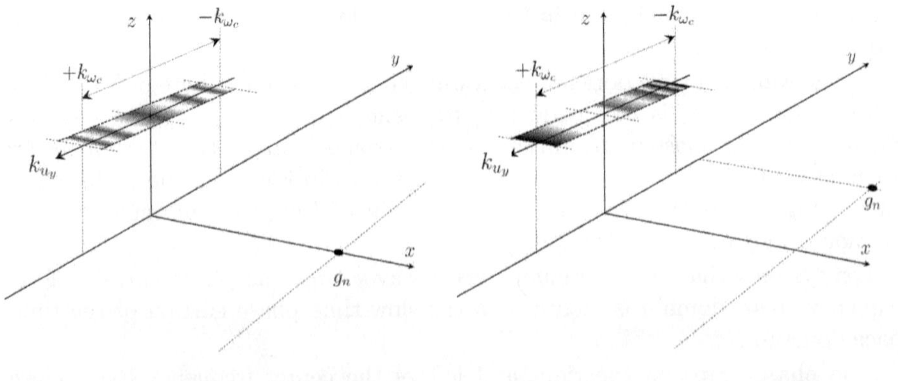

a. Boresight scatterer. b. Off boresight scatterer.

Figure 4.10: Boresight and off boresight frequency-wavenumber phase histories.

Boresight and Off Boresight Frequency-Wavenumber Phase Histories

Shown in Figs. 4.10a-b are the along-track Fourier transforms of the infinitely long frequency-space phase curtains of the boresight and off boresight scatterers of Figs. 4.9a-b. The frequency-space phase history of the infinite aperture is infinitely long in space, but the corresponding frequency-wavenumber phase history is band limited as a function of aperture wavenumber k_{u_y}.

The Fourier transform of the phase curtain of the boresight scatterer at broadside in Fig. 4.10a, returns a *slow-time baseband aperture wavenumber phase history.* If the scatterer is off boresight, as shown in Fig. 4.10b, then the frequency-wavenumber phase history is a *slow-time passband aperture wavenumber phase history* [885].

Slow-Time Fourier Transform Expression

We seek analytic expression of the spatial Fourier transform $SS(\omega_c, k_{u_y})$. We begin by writing the spatial Fourier transform of the frequency-space phase curtain of a single scatterer

$$
\begin{aligned}
& SS(\omega_c, k_{u_y})_n \\
&= \mathsf{FT}_{1\text{D}}^{u_y}\left\{Ss(\omega_c, u_y)_n\right\}, \\
&= \int_{-\infty}^{\infty} Ss(\omega_c, u_y)_n \cdot \exp(-jk_{u_y} u_y) \cdot du_y, \\
&= \int_{-\infty}^{\infty} g_n \cdot \exp\left[-jk_{\omega_c} \cdot r_n(u_y)\right] \cdot \exp(-jk_{u_y} u_y) \cdot du_y, \\
&= g_n \cdot \int_{-\infty}^{\infty} \exp\left[-jk_{\omega_c} \cdot \sqrt{x_n^2 + (u_y - y_n)^2}\right] \cdot \exp(-jk_{u_y} u_y) \cdot du_y.
\end{aligned}
\tag{4.25}
$$

We have generalized the frequency-space phase history $Ss(\omega_c, u_y)_n$ by writing x_n instead of X_c. We call the Fourier transform variable k_{u_y} introduced in (4.25) the *aperture wavenumber* or the *slow-time wavenumber*.

Slow-Time Fourier Transform Evaluation

Historically, evaluation of the spatial Fourier transform (4.25) of the hyperbolic chirp signal

$$\exp\left[-jk_{\omega_c} \cdot \sqrt{x_n^2 + (u_y - y_n)^2}\right]$$

proved difficult. Methods based on the principle of stationary phase (see Section 15.4) were required. Application of these methods (see Section 15.6) allows the Fourier transform of (4.25) to be approximately evaluated as [374]

$$\begin{aligned} SS(\omega_c, k_{u_y})_n \\ \approx\ & g_n \cdot a(\omega_c, k_{u_y})_{\text{psp}} \cdot \exp(-jx_n \cdot 2\sqrt{k_{\omega_c}^2 - k_{u_y}^2}\ ) \cdot \exp(-jy_n \cdot 2k_{u_y}), \quad (4.26) \end{aligned}$$

where

$$a(\omega_c, k_{u_y})_{\text{psp}} = e^{-j\frac{\pi}{4}} \cdot \lambda_x(\omega_c, k_{u_y}). \quad (4.27)$$

SAR Signal Spectrum

Equation (4.26) is the continuous wave echo signal's *frequency-wavenumber phase history*, or more concisely, the *SAR signal spectrum*. The frequency-wavenumber phase history is the oft noted "omega-k" phase history of SAR.

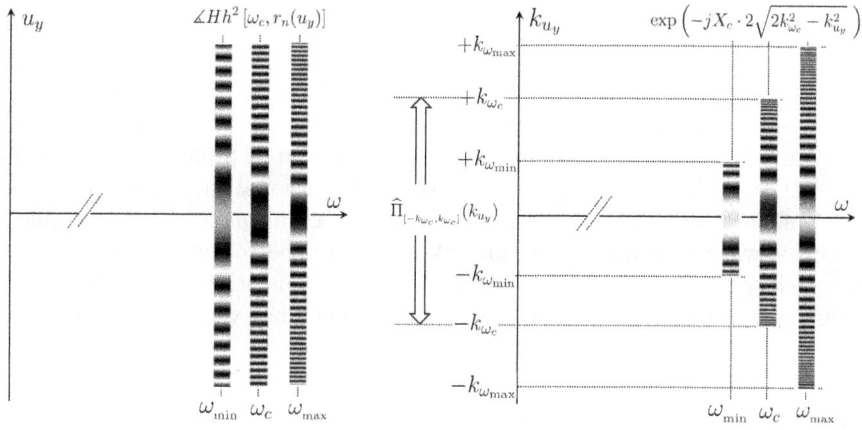

a. Space-time phase curtains. b. Fourier transforms.

Figure 4.11: Frequency-space and frequency-wavenumber phase histories.

The slow-time phase curtains of Fig. 4.8b are shown again in Fig. 4.11a. These phase maps are the frequency-space phase histories for a single scatterer gathered by SAR systems using continuous wave signals with three different ω_c values. Shown in Fig. 4.11b are the spatial Fourier transforms of the three slow-time phase curtains shown in Fig. 4.11a. The vertical axis in Fig. 4.11a corresponds to u_y. The vertical axis in Fig. 4.11b corresponds to k_{u_y}. The horizontal axis in both figures is signal frequency ω.

Frequency-Wavenumber Weight Function

Part of the amplitude function of (4.27) is the x *dimensional wavelength,*

$$\lambda_x(\omega_c, k_{u_y}) = \frac{2\pi}{\sqrt{k_{\omega_c}^2 - k_{u_y}^2}}. \tag{4.28}$$

We next explain how the dimensional wavelength $\lambda_x(\omega_c, k_{u_y})$ can be described as the *angled cut* or *miter cut* of the temporal signal wavelength λ_{ω_c} along the down-range x axis. The wavelength related scale factor $\lambda_x(\omega_c, k_{u_y})$ of (4.27) is noted in Fresnel's formulation of Huygens's principle [35]. We also show how (4.28) is determined by the Pythagorean constraint that exists among the various wavenumber parameters of (4.26).

4.6 Pythagorean Projections and Miter Cuts

We pause to explore the notion of miter cuts of wavelength and k-space projections of temporal wavenumber vectors. Miter cuts and projections provide a foundation for a clear understanding of the geometry associated with the phase factor

$$\exp\left(-jx_n \cdot 2\sqrt{k_{\omega_c}^2 - k_{u_y}^2}\right)$$

and the weight function

$$\lambda_x\left(\omega_c, k_{u_y}\right)$$

within the frequency-wavenumber phase history (4.26).

As background material, we first present a description of harmonic (sinusoidal) plane wave fields. We then define the decomposition of a harmonic (sinusoidal) spherical wave field in terms of a full (three dimensional) angular sweep of harmonic plane waves. Then for each directed sinusoidal plane wave of a spherical sinusoidal wave field decomposition, we illustrate the miter cuts of the temporal wavelength vector along the cardinal spatial dimensions. We also describe the associated projections of the temporal wavenumber vector onto the cardinal dimensions of spatial wavenumbers.

Directed Sinusoidal Plane Waves

Consider a three dimensional, sinusoidal plane wave traveling in the direction of the unit vector

$$\begin{aligned} \hat{\boldsymbol{r}} &= [|\hat{\boldsymbol{r}}|\cos\theta_{r_x} \ |\hat{\boldsymbol{r}}|\cos\theta_{r_y} \ |\hat{\boldsymbol{r}}|\cos\theta_{r_z}], \\ &= [\hat{r}_x \ \hat{r}_y \ \hat{r}_z], \end{aligned}$$

where θ_{r_x}, θ_{r_y} and θ_{r_z} are direction angles of $\hat{\boldsymbol{r}}$ and where $|\hat{\boldsymbol{r}}| = 1$. The harmonic plane wave traveling in the direction of $\hat{\boldsymbol{r}}$ is

$$\begin{aligned} pp(t, \boldsymbol{x})_{\hat{\boldsymbol{r}}} &= \cos(\omega_c t - k_{\omega_c}\boldsymbol{x} \cdot \hat{\boldsymbol{r}}), \\ &= \cos(\omega_c t - \boldsymbol{x} \cdot \boldsymbol{k}_{\hat{\boldsymbol{r}}}). \end{aligned} \tag{4.29}$$

The tonal or harmonic wavenumber vector $\boldsymbol{k}_{\hat{\boldsymbol{r}}}$ of (4.29) is determined by

$$\begin{aligned}\boldsymbol{k}_{\hat{\boldsymbol{r}}} &= k_{\omega_c}\hat{\boldsymbol{r}},\\ &= [k_{\omega_c}\cos\theta_{r_x} \;\; k_{\omega_c}\cos\theta_{r_y} \;\; k_{\omega_c}\cos\theta_{r_z}],\\ &= [k_{\omega_c}\hat{r}_x \;\; k_{\omega_c}\hat{r}_y \;\; k_{\omega_c}\hat{r}_z],\\ &= [k_{\hat{r}_x} \;\; k_{\hat{r}_y} \;\; k_{\hat{r}_z}],\end{aligned}$$

where $k_{\hat{r}_i} = k_{\omega_c}\hat{r}_i$ for $i \in \{x, y, z\}$. The directed sinusoidal plane wave function (4.29) can be expressed as the real part of the product of a pair of complex phase factors,

$$\cos(\omega_c t - \boldsymbol{x}\cdot\boldsymbol{k}_{\hat{\boldsymbol{r}}}) = \mathsf{Real}\left\{\exp(j\omega_c t)\cdot\exp(-j\boldsymbol{x}\cdot\boldsymbol{k}_{\hat{\boldsymbol{r}}})\right\}. \tag{4.30}$$

The phase factor $\exp(j\omega_c t)$ of (4.30) is the temporal element and $\exp(-j\boldsymbol{x}\cdot\boldsymbol{k}_{\hat{\boldsymbol{r}}})$ is the static spatial element of a complex and dynamic harmonic plane wave field,

$$\exp[j(\omega_c t - \boldsymbol{x}\cdot\boldsymbol{k}_{\hat{\boldsymbol{r}}})].$$

Plane Wave Decomposition of a Harmonic Spherical Phase Field

A key tool in directional analysis for SAR is the ability to describe a static harmonic spherical phase field,

$$\exp(-jk_{\omega_c}|\boldsymbol{x}|) \tag{4.31}$$

in terms of a full spherical sweep of static harmonic plane wave fields,

$$\exp(-j\boldsymbol{x}\cdot\boldsymbol{k}_{\hat{\boldsymbol{r}}}) = \exp(-jk_{\omega_c}\boldsymbol{x}\cdot\hat{\boldsymbol{r}}). \tag{4.32}$$

Synthesis of the monochromatic spherical phase field (4.31) from a full sweep of static harmonic plane wave field elements (4.32) is given by [84, 195, 988]

$$\begin{aligned}\exp(-jk_{\omega_c}|\boldsymbol{x}|) &= \frac{1}{(2\pi)^3}\int \frac{1}{\sqrt{k_{\omega_c}^2 - \boldsymbol{k}_{\hat{r}}^2}}\cdot\exp(-j\boldsymbol{x}\cdot\boldsymbol{k}_{\hat{r}})\cdot d\boldsymbol{k}_{\hat{r}}, \qquad (4.33)\\ &= \frac{1}{(2\pi)^3}\int_{-\infty}^{+\infty}\int_{-\infty}^{+\infty}\int_{-\infty}^{+\infty}\frac{1}{\sqrt{k_{\omega_c}^2 - k_{\hat{r}_x}^2 - k_{\hat{r}_y}^2 - k_{\hat{r}_z}^2}}\\ &\qquad\cdot\exp\left[-j(xk_{\hat{r}_x} + yk_{\hat{r}_y} + zk_{\hat{r}_z})\right]\cdot dk_{\hat{r}_x}dk_{\hat{r}_y}dk_{\hat{r}_z}.\end{aligned}$$

The triple integral over the dimensional elements of $\boldsymbol{k}_{\hat{r}} = [k_{\hat{r}_x} \;\; k_{\hat{r}_y} \;\; k_{\hat{r}_z}]$ induces a full sweep over the unit sphere with radius $\hat{\boldsymbol{r}}$.

Solutions of inverse wave field problems using an *angular spectrum* of directed harmonic plane wave fields were suggested by many researchers during the 1940s and 1950s [84, 91, 195, 289, 354, 466, 579, 810, 835].

Directed Plane Wave Elements within a Spherical Wave Field

We can use a plane wave decomposition of a spherical wave field to model the plane wave element of the spherical wave field sensed along an ever changing line of sight between sensor and scatterer as the sensor moves down the flight line.

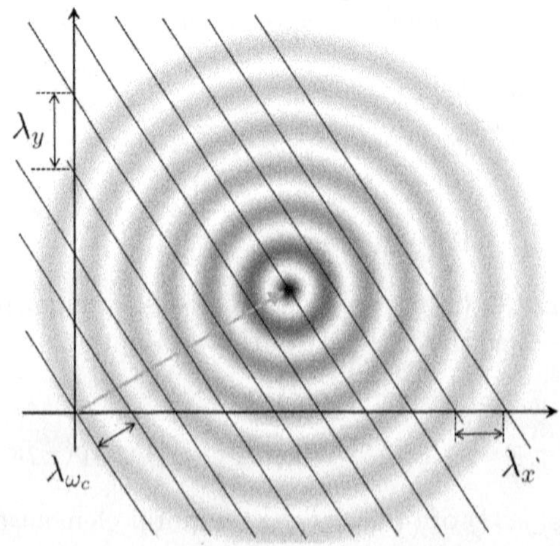

Figure 4.12: Line of sight plane wave element of spherical wave field.

Consider the illustration of Fig. 4.12. Shown is a line of sight vector (gray) between sensor and scatterer. Consider the harmonic plane wave of the same frequency along this line of sight vector. This particular harmonic plane wave is viewed as one of the constitute harmonic plane waves that synthesizes the scatterer's spherical phase field as expressed in (4.33).

The spatially projected temporal wavelength λ_{ω_c} of the harmonic plane wave is indicated by the slanted (black) lines in Fig. 4.12. As the view angle $\theta_x(u_y)_n$ changes, the particular plane wave element of the spherical wave field decomposition that is aligned with the line of sight vector changes. Hence, we obtain a direct coupling of the ever changing line of sight vector between sensor and scatterer and a similarly directed sweep of harmonic plane waves between sensor and scatterer.

Pythagorean Triplets of Temporal/Spatial Wavenumbers

An important aspect of the plane wave decomposition of the scattered spherical wave field is the Pythagorean relationship that exists between temporal and spatial wavenumber parameters as contained in (4.33). If we test the complex expression of the dynamic harmonic plane wave

$$\exp(j\omega_c t - \boldsymbol{x} \cdot \boldsymbol{k}_{\hat{\boldsymbol{r}}}) = \exp(j\omega_c t) \cdot \exp(-jk_{\omega_c} \boldsymbol{x} \cdot \hat{\boldsymbol{r}}) \tag{4.34}$$

as a solution of the free space Helmholtz wave motion equation, then we obtain a relationship between temporal and spatial wavenumbers,

$$\begin{aligned} k_{\omega_c}^2 &= |\boldsymbol{k}_{\hat{\boldsymbol{r}}}|^2, \\ &= k_{\hat{r}_x}^2 + k_{\hat{r}_y}^2 + k_{\hat{r}_z}^2. \end{aligned} \tag{4.35}$$

Each dimensional spatial wavenumber has an associated dimensional spatial wavelength,

$$\begin{aligned} \lambda_x &= \frac{2\pi}{k_{\hat{r}_x}}, \\ \lambda_y &= \frac{2\pi}{k_{\hat{r}_y}}, \\ \lambda_z &= \frac{2\pi}{k_{\hat{r}_z}}. \end{aligned}$$

We will explain how λ_x, λ_y and λ_z are all geometrically related to λ_{ω_c}; but first, we focus on the dimensional wavenumbers $k_{\hat{r}_x}$, $k_{\hat{r}_y}$ and $k_{\hat{r}_z}$.

Pythagorean Constraint Among Temporal-Spatial Wavenumbers

Equation (4.35) establishes a Pythagorean constraint among the wavenumbers of (4.33). We say the set of wavenumbers k_{ω_c}, $k_{\hat{r}_x}$, $k_{\hat{r}_y}$ and $k_{\hat{r}_z}$ that satisfy (4.35) comprise a *Pythagorean quadruple.* The squared spatial wavenumbers must sum to equal the square of the temporal wavenumber. As a result, the wavenumber vector

$$\boldsymbol{k}_{\hat{\boldsymbol{r}}} = [k_{\hat{r}_x} \; k_{\hat{r}_y} \; k_{\hat{r}_z}]$$

of a harmonic plane wave of temporal frequency ω_c traveling in the direction of $\hat{\boldsymbol{r}}$ must be such that

$$k_{\omega_c} = \sqrt{k_{\hat{r}_x}^2 + k_{\hat{r}_y}^2 + k_{\hat{r}_z}^2} \tag{4.36}$$

is always satisfied for the fixed k_{ω_c} as $\boldsymbol{k}_{\hat{\boldsymbol{r}}}$ varies within the plane wave decomposition of a spherical wave field.

Two Dimensional Pythagorean Constraint of Slant-Plane SAR

Consider the two dimensional version of (4.36),

$$k_{\omega_c} = \sqrt{k_{\hat{r}_x}^2 + k_{\hat{r}_y}^2}. \tag{4.37}$$

This two dimensional Pythagorean constraint limits the set of possible dimensional wavenumber combinations $(k_{\hat{r}_x}, k_{\hat{r}_y})$ of a sinusoidal plane wave of temporal wavenumber k_{ω_c}. The wavenumber parameters k_{ω_c}, $k_{\hat{r}_x}$ and $k_{\hat{r}_y}$ that satisfy (4.37) form a *Pythagorean triplet.*

If the temporal frequency ω_c of the sinusoidal plane wave is known and if it is possible to measure the along-track wavenumber $k_{\hat{r}_y}$ of the harmonic plane wave, then the down-range wavenumber $k_{\hat{r}_x}$ must be related to k_{ω_c} and $k_{\hat{r}_y}$ via

$$k_{\hat{r}_x} = \sqrt{k_{\omega_c}^2 - k_{\hat{r}_y}^2}. \tag{4.38}$$

If we equate $k_{u_y} = k_{\hat{r}_y}$ in the frequency-wavenumber phase history (4.26)

$$\sqrt{k_{\omega_c}^2 - k_{\hat{r}_y}^2} \mapsto \sqrt{k_{\omega_c}^2 - k_{u_y}^2},$$

then

$$\breve{k}_x = \sqrt{k_{\omega_c}^2 - k_{u_y}^2} \tag{4.39}$$

must be descriptive of an x dimensional wavenumber. Hence, if we equate an x dimensional phase factor with the first phase factor (4.26) of the spatial Fourier transform of the slow-time phase curtain,

$$\exp(-jx_n\breve{k}_x) \equiv \exp\left(-jx_n \cdot 2\sqrt{k_{\omega_c}^2 - k_{u_y}^2}\right), \tag{4.40}$$

then

$$2\sqrt{k_{\omega_c}^2 - k_{u_y}^2}$$

must be descriptive of an x dimensional wavenumber (4.39) of a sinusoidal plane wave (from the spherical wave field decomposition) directed along the line of sight between the sensor and a scatterer with down-range position x_n. Additionally, the phase factor

$$\exp(-jy_n \cdot 2k_{u_y}) \tag{4.41}$$

of the frequency-wavenumber phase history (4.26) is such that $2k_{u_y}$ is descriptive of a y dimensional wavenumber of the same scatterer with along-track position y_n,

$$\exp(-jy_nk_y) \equiv \exp(-jy_n \cdot 2k_{u_y}).$$

Using the newly defined phase factors (4.40) and (4.41) descriptive of plane wave motion, the core of the spatial Fourier transform of a slant-plane frequency-space phase curtain can be expressed as

$$SS(\omega_c, k_{u_y})_n \approx g_n \cdot a(\omega_c, k_{u_y})_{\text{psp}} \cdot \exp(-jx_n\breve{k}_x) \cdot \exp(-jy_nk_y). \tag{4.42}$$

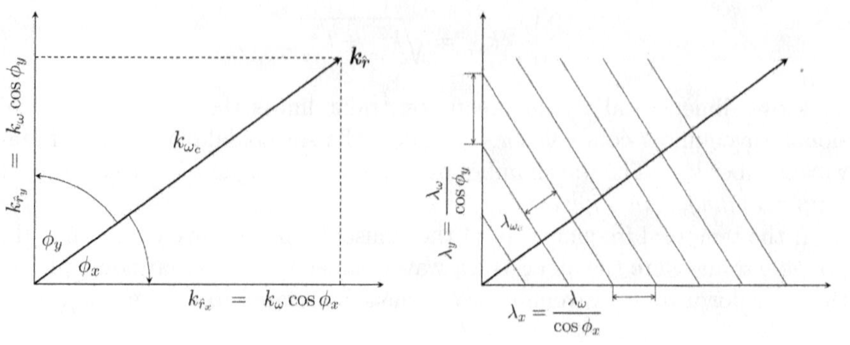

a. Dimensional projections. b. Miter cuts of temporal wavelength.

Figure 4.13: Wavenumber projections and miter cuts of wavelengths.

Pythagorean Triplet Wavenumber Projections

The relationship (4.38) between a sinusoid signal's temporal wavenumber k_{ω_c} and the along-track dimensional wavenumbers $k_{\hat{r}_y}$ and $k_{\hat{r}_x}$ can be explained in terms of k-space projections as illustrated in Fig. 4.13a. The three wavenumbers $(k_{\omega_c}, k_{\hat{r}_x}, k_{\hat{r}_y})$ form a right triangle in the k-space plane. The temporal wavenumber k_{ω_c} is the hypotenuse of this right triangle. The spatial wavenumbers $k_{\hat{r}_x}$ and $k_{\hat{r}_y}$ comprise the sides of this right triangle.

Wavenumber Domain Direction Angles

As shown in Fig. 4.13a, direction angles ϕ_x and ϕ_y exist in the k-space plane such that

$$\begin{aligned} k_{\hat{r}_x} &= k_{\omega_c} \cos \phi_x, \\ k_{\hat{r}_y} &= k_{\omega_c} \cos \phi_y. \end{aligned}$$

Hence, the relationship (4.38) establishes a geometric coupling between the directed plane wave temporal wavenumber k_{ω_c} and the dimensional spatial wavenumbers

$$(k_{\hat{r}_x}, k_{\hat{r}_y}) \mapsto (k_x, k_y)$$

of the static element of this harmonic plane wave field and as used in the dimensional phase factors contained in (4.42).

Spatial Miter Cuts of Temporal Wavelength

The Pythagorean constraint of temporal and spatial wavenumbers also determines relationships among the associated temporal wavelength λ_ω and dimensional wavelengths (λ_x, λ_y) [218]. The geometry of these relationships can be described as *angled cuts* or *miter cuts* of temporal wavelength. The "wavelength" of the angled cut of the temporal wavelength λ_ω in the x dimension, as shown in Fig. 4.13b, is given by

$$\lambda_x = \frac{\lambda_\omega}{\cos \phi_x}.$$

Likewise, the angled cut of the temporal frequency wavelength in the y dimension is given by

$$\lambda_y = \frac{\lambda_\omega}{\cos \phi_y}.$$

Derived Dimensional Wavenumbers

If the temporal wavelength of the sinusoidal wave field is known and if the dimensional wavelength λ_y can be measured along the y axis to determine the y dimensional wavenumber

$$k_y = \frac{2\pi}{\lambda_y},$$

then the x dimensional miter cut wavelength must be such that the corresponding wavenumber $\breve{k}_x$ satisfies,

$$\begin{aligned}\breve{k}_x(\omega_c, k_y) &= \sqrt{k_{\omega_c}^2 - k_y^2}, \\ &= \frac{2\pi}{\lambda_x}.\end{aligned}$$

Together, the wavelength parameters (λ_x, λ_y) define the wavenumber parameters $(\breve{k}_x, k_y)$ of the directed sinusoidal plane wave field (4.30) that exists along the line of sight vector between sensor and scatterer, and which exist in the dimensional phase factors of the frequency-wavenumber phase history (4.42).

4.7 Frequency-Wavenumber Windows

Evaluation of the spatial Fourier transform (4.26) and (4.42) of the frequency-space phase curtain implies the existence of a slow-time wavenumber window function $\Pi(k_{u_y})$ over which the k_{u_y}-domain spectrum is non-zero. This window indicates the bandwidth of the aperture wavenumber k_{u_y} over which (4.26) holds true. This region of support, specific to the case of the infinite synthetic aperture with a continuous wave signal, is

$$k_{u_y} \in [-k_{\omega_c}, +k_{\omega_c}]. \tag{4.43}$$

Outside of this region of support, the SAR signal spectrum is exactly null,

$$SS(\omega_c, k_{u_y})_n = 0.$$

We can express this manifold for the aperture wavenumber k_{u_y} as a window function,

$$\Pi_{[-k_{\omega_c}, k_{\omega_c}]}(k_{u_y}) = \begin{cases} 1 & \text{for} \quad |k_{u_y}| \le k_{\omega_c}, \\ 0 & \text{otherwise.} \end{cases} \tag{4.44}$$

Infinite Aperture Frequency-Wavenumber Rectilinear Spectrum

We use the $\Pi(\cdot)$ notation to more fully express the (ω_c, k_{u_y}) domain SAR signal spectrum as

$$\begin{aligned} &S_{\Pi(\omega_c)} S_{\Pi[-k_{\omega_c}, k_{\omega_c}]}(\omega_c, k_{u_y})_n \\ &= g_n \cdot \underbrace{a(\omega_c, k_{u_y})_{\text{psp}}}_{\substack{\text{amplitude} \\ \text{function}}} \cdot \underbrace{\Pi_{[-k_{\omega_c}, k_{\omega_c}]}(k_{u_y})}_{\substack{\text{wavenumber} \\ \text{window}}} \\ &\quad \cdot \underbrace{\exp(-jX_c \cdot 2\sqrt{k_{\omega_c}^2 - k_{u_y}^2}\,)}_{\substack{\text{down-range} \\ \text{information}}} \cdot \underbrace{\exp(-jy_n \cdot 2k_{u_y})}_{\substack{\text{along-track} \\ \text{information}}}. \end{aligned} \tag{4.45}$$

(Remember that x_n is set to X_c for all scatterers in the cross-range profiling analysis of this chapter.)

Frequency-Wavenumber Weight Function

We combine the window of the frequency-wavenumber phase history

$$\Pi_{[-k_{\omega_C}, k_{\omega_C}]}(k_{u_y})$$

with the principle of stationary phase amplitude function

$$a(\omega_c, k_{u_y})_{\text{psp}} = e^{j\frac{\pi}{4}} \cdot \lambda_x(\omega_c, u_y)$$

to create the aperture wavenumber weight function

$$\widehat{\Pi}_{[-k_{\omega_C}, k_{\omega_C}]}(k_{u_y}) = a(\omega_c, k_{u_y})_{\text{psp}} \cdot \Pi_{[-k_{\omega_C}, k_{\omega_C}]}(k_{u_y}). \tag{4.46}$$

Using the weight function (4.46), the frequency-wavenumber spectrum of a cross-range line of scatterers at X_c is

$$\begin{aligned} & S_{\Pi(\omega_c)} S_{\widehat{\Pi}[-k_{\omega_C}, k_{\omega_C}]} \\ &= \sum_n g_n \cdot \widehat{\Pi}_{[-k_{\omega_C}, k_{\omega_C}]}(k_{u_y}) \cdot \exp(-jX_c \cdot 2\sqrt{2k_{\omega_c}^2 - k_{u_y}^2}\ ) \cdot \exp(-jy_n \cdot 2k_{u_y}). \end{aligned} \tag{4.47}$$

The phase factor

$$\exp\left(-jX_c \cdot 2\sqrt{k_{\omega_c}^2 - k_{u_y}^2}\ \right)$$

of (4.47) depends on the frequency-wavenumber pair (ω_c, k_{u_y}) and provides information about scatterer position X_c in the down-range dimension. The phase factor,

$$\exp\left(-jy_n \cdot 2k_{u_y}\right)$$

depends only on the slow-time wavenumber k_{u_y} and is indicative of the along-track scatterer position y_n. For reasons that will become increasingly evident, we call a signal spectrum defined in the (ω_c, k_{u_y}) domain a *rectilinear spectrum.* We shall call another form of the SAR signal spectrum, obtained after Stolt formatting, the *angular spectrum.*

4.8 Cross-Range Profile Angular Spectrum

Key to SAR signal analysis is the mapping of measures of angular dwell as subtended by change of sensor position to phase information resident on circular *traces* on a slant-plane slice of the Ewald sphere. Each trace can be viewed as a scatterer specific data support manifold. To be fully descriptive in our signal analysis, the angular dwell specific to each scatterer and the associated trace on a common Ewald sphere must each be explicitly expressed.

Fourier Migration and Tonal Wavenumber Fields

The Fourier transforms of the slow-time phase curtains of Fig. 4.11b appear to be chirp like in that phase centers can be seen. Actually, these frequency-wavenumber phase histories are not chirps, but are tonal. The tonal aspect is not revealed until these rectilinear frequency-wavenumber phase histories are migrated onto semicircular traces on the Ewald sphere. This is accomplished with a process called *Fourier migration.* The Fourier migrated format is obtain by Stolt transformation [913, 914].

Fourier Transforms and Fourier Migrations

We have seen that

- *Fourier transformation* of the measured SAR frequency-space phase history yields the *SAR rectilinear spectrum,* i.e., the frequency-wavenumber phase history expressed in the (k_{ω_c}, k_{u_y}) domain.

We shall see that

- *Fourier migration* of the rectilinear SAR signal spectrum yields the *SAR angular spectrum,* i.e., the frequency-wavenumber phase history expressed in a wavenumber domain we indicate by $(\breve{k}_x, k_y)$.

We write a down-range wavenumber that has been obtained by Fourier migration as $\breve{k}_x$. The inverted curved caret annotation of $\breve{k}_x$ is a notational aid that helps us remember that the data of this down-range wavenumber is resident on an Ewald sphere and is obtained through Fourier migration.

Fourier Migration

The down-range and along-track position informative phase factors for the nth scatterer,

$$\begin{aligned} \exp(-jX_c\breve{k}_x) &= \exp\left(-jX_c \cdot 2\sqrt{k_{\omega_c}^2 - k_{u_y}^2}\right) \\ \exp(-jy_n k_y) &= \exp\left(-jy_n \cdot 2k_{u_y}\right) \end{aligned} \tag{4.48}$$

of (4.47) are similar in form to the phase factors

$$\begin{aligned} &\exp(-jX_c \cdot k_x) \\ &\exp(-jy_n \cdot k_y) \end{aligned}$$

contained in the along-track terrain spectrum,

$$G_{\Pi(\infty)}A_{\Pi(\infty)}(k_x, k_y) = \sum_n \exp(-jX_c \cdot k_x) \cdot \exp(-jy_n \cdot k_y).$$

This similarity, along with our discussion in Section 4.6, suggests that we define a new down-range wavenumber as

$$\breve{k}_x \equiv 2\sqrt{k_{\omega_c}^2 - k_{u_y}^2}. \tag{4.49}$$

This similarity also suggests that we consider values of the along-track wavenumber k_y that are double the values of the aperture wavenumber k_{u_y},

$$k_y \equiv 2k_{u_y}. \tag{4.50}$$

Down-Range Stolt Mapping

The down-range wavenumber $\breve{k}_x$ of (4.49) is the result of a transformation known as the *Stolt mapping* [189, 913, 914]. The Stolt transform (4.49) is functionally dependent on both the fast-time signal's angular frequency ω_c and the aperture wavenumber k_{u_y}.

Along-Track Stolt Mapping

The mapping (4.50) of the aperture wavenumber k_{u_y} to k_y changes the phase history of the SAR signal spectrum from dependency on the aperture wavenumber k_{u_y} in $\exp(-jy_n \cdot 2k_{u_y})$ to dependence on a new wavenumber equivalent to the along-track ground scene wavenumber k_y.

Fourier Migration and Stolt Formatting

Together, the two new wavenumbers $\breve{k}_x$ and k_y perform a mapping (4.48) of phase factors defined in the rectilinear spectrum domain (ω_c, k_{u_y}) to phase factors defined in the angular spectrum domain $(\breve{k}_x, k_y)$. This operation is called *Fourier migration* or *Stolt formatting*.

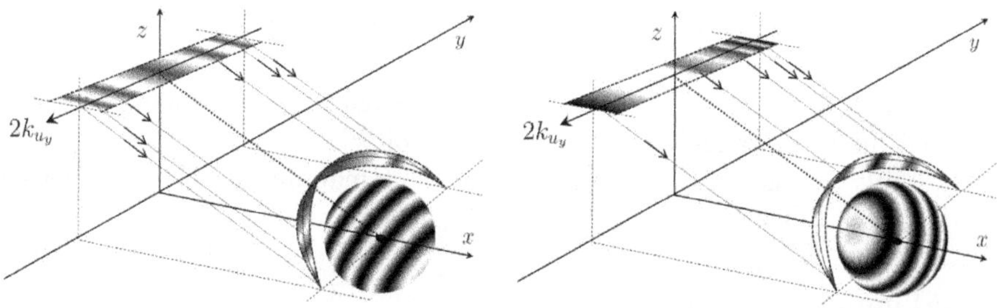

a. Boresight scatterer migration. b. Off boresight scatterer migration.

Figure 4.14: Scatterer specific Fourier migrations.

Example Stolt Mappings

The Fourier migration process maps a straight linear trace in the rectilinear frequency-wavenumber domain (ω_c, k_{u_y}) to a semicircular trace $(\breve{k}_x, k_y)$ on the surface of the Ewald sphere as illustrated by the two examples of Figs. 4.14a-b.

The semicircular traces in Figs. 4.14a-b are expanded into circular wedges or "lemon slices" to better show in phase map form the phase function resident on semicircle traces on the Ewald sphere. Note that this mapping does not change the value of the resident phase factors, only the nature of the supporting manifold of the data of these phase factors.

Infinite Aperture Angular Spectrum Phase Histories

We now proceed to more fully express the process of Stolt transformation. Using the new wavenumbers $(\breve{k}_x, k_y)$ as substitute variables in

$$\begin{aligned} & S_{\Pi(\omega_c)} S_{\Pi[-k_{\omega_c},k_{\omega_c}]}(\omega_c, k_{u_y})_n \\ &= \underbrace{g_n \cdot \widehat{\Pi}_{[-k_{\omega_c},k_{\omega_c}]}(k_{u_y})}_{\text{1D weight function}} \cdot \exp(-jX_c \cdot 2\sqrt{k_{\omega_c}^2 - k_{u_y}^2}\,) \cdot \exp(-jy_n \cdot 2k_{u_y}), \end{aligned}$$

we obtain

$$\begin{aligned} & \overset{\text{cw}}{M}_{\Pi(\oplus)} S_{\Pi[-2k_{\omega_c},2k_{\omega_c}]}(\breve{k}_x, k_y)_n \\ &= \underbrace{g_n \cdot \overset{\text{cw}}{\widehat{\Pi}}_{\oplus}(\breve{k}_x, k_y) \cdot \widehat{\Pi}_{[-2k_{\omega_c},2k_{\omega_c}]}(k_y)}_{\text{2D weight function}} \cdot \exp(-jX_c\breve{k}_x) \cdot \exp(-jy_nk_y). \end{aligned} \tag{4.51}$$

Fourier Migrated Spectrum Notation

Our notation for the phase history after Fourier migration changes $SS(\omega_c, k_{u_y})$ of the rectilinear spectrum to $MS(\breve{k}_x, k_y)$ of the Fourier migrated angular spectrum. The symbol "M" in the (ordered) multiple scripted notation of $MS_n(\breve{k}_x, k_y)$ indicates a Fourier migration was performed from the (ω_c, k_{u_y}) domain into the phase factor $\exp(jX_c\breve{k}_x)$ corresponding to the first wavenumber parameter, i.e., Stolt formatted the down-range wavenumber dimension $\breve{k}_x$.

The Stolt transformation is an ω_c dependent mapping from one dimension (k_{u_y}) into two dimensions $(\breve{k}_x, k_y)$.

Stolt Mapping Operator

The conversion of $SS(\omega_c, k_{u_y})_n$ to $MS(\breve{k}_x, k_y)_n$,

$$\overset{\text{cw}}{M}_{\Pi(\oplus)} S_{\Pi(4k_{\omega_c})}(\breve{k}_x, k_y)_n = \mathsf{ST}_{1\text{D}}^{\breve{k}_x}\left\{S_{\Pi(\omega_c)} S_{\Pi(2k_{\omega_c})}(\omega_c, k_{u_y})_n\right\},$$

is performed by the *Stolt operator*

$$\mathsf{ST}_{1\text{D}}^{\breve{k}_x}\{\cdot\}.$$

Angular Spectrum Weight Functions

The pair of angular spectrum weight functions

$$\overset{\text{cw}}{\widehat{\Pi}}_{\oplus}(\breve{k}_x, k_y) \ \text{ and } \ \widehat{\Pi}_{[-2k_{\omega_c},2k_{\omega_c}]}(k_y) \tag{4.52}$$

of (4.51) are the weight functions obtained by applying the Stolt transformation to the aperture wavenumber window function,

$$\widehat{\Pi}_{\oplus}^{\mathrm{cw}}(\breve{k}_x, k_y) \cdot \widehat{\Pi}_{[-2k_{\omega_c}, 2k_{\omega_c}]}(k_y) = \mathsf{ST}_{\mathrm{1D}}^{\breve{k}_x}\left\{\widehat{\Pi}_{[-k_{\omega_c}, k_{\omega_c}]}(k_{u_y})\right\}.$$

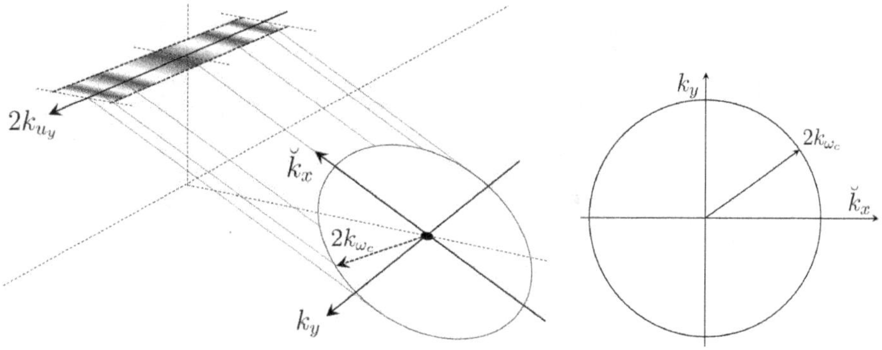

a. Slant-plane Stolt mapping. b. Semicircular manifold.

Figure 4.15: Angular spectrum slant-plane semicircle manifold geometry.

Angular Spectrum Semicircle Manifold

In cross-range profiling with linear flight SAR, the Stolt mapping from the rectilinear to the angular spectrum creates a mapping from a line of k_{u_y} data to a semicircle trace of $(\breve{k}_x, k_y)$ data as illustrated in Fig. 4.15a. The phase history data resident on this semicircle trace was previously illustrated as a phase map on the lemon slice wedge of Fig. 4.14a. The semicircle trace is a submanifold of the Ewald sphere. The semicircle's slant angle on the Ewald sphere is determined by the depression angle of the SAR's slant-plane data collection geometry. When plotted as a two dimensional angular spectrum, the slant-plane semicircle manifold is extracted from a slant-plane slice of the Ewald sphere and drawn as illustrated in Fig. 4.15b. The mapping of chirp like phase data of the rectilinear spectrum to tonal phase data on the Ewald sphere is illustrated in Figs. 4.16a-b for boresight and off boresight scatterer positions.

Separable Phase Factors on Nonseparable Manifolds

A key characteristic of SAR frequency-wavenumber phase histories migrated into the angular spectrum domain is the separability of tonal Stolt transformed phase factors resident on the Ewald sphere. One phase factor depends only on the down-range ground scene parameter x_n (here, equal to X_c). The other phase factor depends only on the along-track scatterer position parameter y_n. As shown in Figs. 4.16a-b, the wavenumber phase signals $\exp(-jX_c\breve{k}_x)$ and $\exp(-jy_nk_y)$ are tonal phase oscillations that are co-resident on the angular spectrum's semicircle manifold. Together, the two phase factors determine the scatterer specific "direction" of the tonal wave field resident on the Ewald sphere.

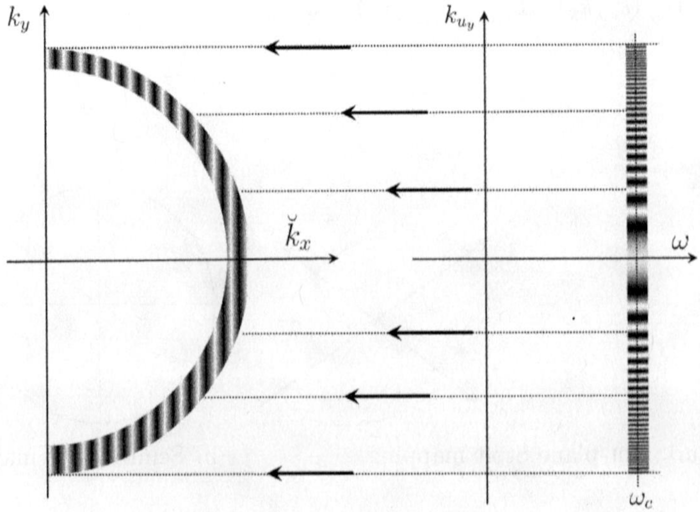

a. Boresight Stolt mapping.

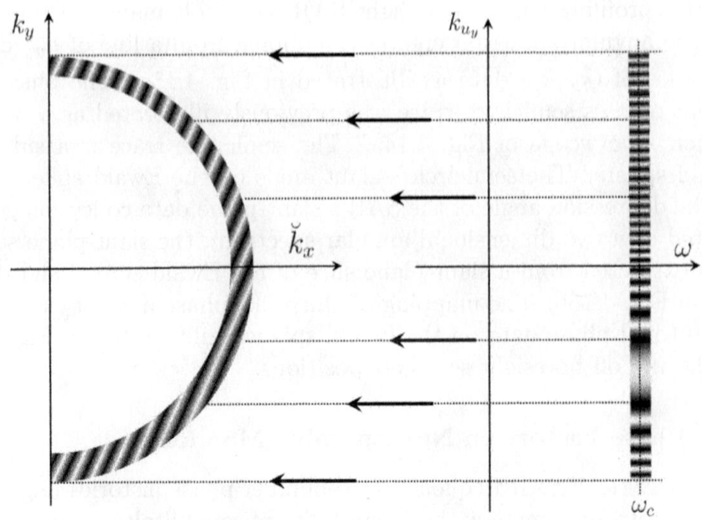

b. Off boresight Stolt mapping.

Figure 4.16: Rectilinear to angular spectrum mapping.

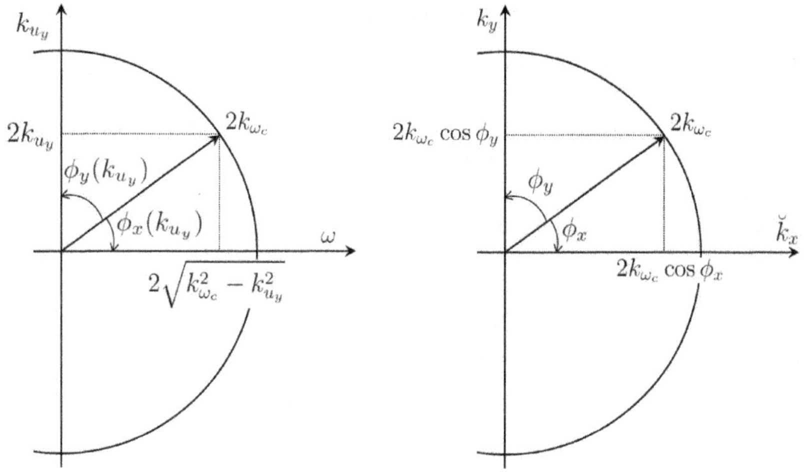

a. Rectilinear spectrum parameters. b. Angular spectrum parameters.

Figure 4.17: Angular spectrum angles, parameters and dimensional projections.

4.9 Angular Spectrum Geometry

The radius vector of the angular spectrum semicircle manifold defines wavenumber domain direction cosine angles ϕ_x and ϕ_y as shown in Figs. 4.17a-b. We call these direction cosine angles the *angular spectrum angles.*

Angular Spectrum Radius Vector Projections

We now define the Fourier migrated wavenumbers in such a way that there is explicit functional dependency on the angular spectrum angles ϕ_x and ϕ_y. As shown in Figs. 4.17a-b, a point on the Stolt formatted semicircle manifold has projections onto the $\breve{k}_x$ and k_y axis given by

$$\begin{aligned}\breve{k}_x(\phi_x) &= 2k_{\omega_c}\cos\phi_x,\\ k_y(\phi_y) &= 2k_{\omega_c}\cos\phi_y.\end{aligned} \tag{4.53}$$

The projections onto the $\breve{k}_x$ and k_y axis form right triangles with the direction cosines as expressed in (4.53). The length $2k_{\omega_c}$ of the hypotenuse of each of the angular spectrum triangles is determined by the circular manifold's radius vector. The length of the radius vector is determined by twice the transmitted signal's temporal frequency wavenumber k_{ω_c}. The factor of two is due to two way signal travel. The length of the angular spectrum radius vector remains fixed as the values of the angular spectrum angles ϕ_x and ϕ_y are allowed to vary.

Angular Spectrum Angles and Aperture Wavenumber

As discussed in Section 4.6, there is a Pythagorean relationship that exists among the angular spectrum wavenumbers $(k_{\omega_c}, \breve{k}_x, k_y)$,

$$(2k_{\omega_c})^2 = \breve{k}_x^2 + k_y^2. \tag{4.54}$$

An important implication of the angular spectrum's triangle geometry is that with the definition of the Stolt mapping from the rectilinear (k_{ω_c}, k_{u_y}) to the angular spectrum $(\breve{k}_x, k_y)$ wavenumbers,

$$\begin{aligned} 2\sqrt{k_{\omega_c}^2 - k_{u_y}^2} &\mapsto \breve{k}_x, \\ 2k_{u_y} &\mapsto k_y, \end{aligned}$$

we have expressions that define relationships between the aperture wavenumber k_{u_y} and the angular spectrum angles,

$$\begin{aligned} 2k_{\omega_c} \cos\phi_x(k_{u_y}) &= 2\sqrt{k_{\omega_c}^2 - k_{u_y}^2}, \\ 2k_{\omega_c} \cos\phi_y(k_{u_y}) &= 2k_{u_y}. \end{aligned}$$

These relationships are illustrated in Figs. 4.17a-b and will prove useful when we find additional relationships that couple angular spectrum angles $\phi_x(k_{u_y})$ and $\phi_y(k_{u_y})$ to spatial domain sensor-scatterer directional view angles $\theta_x(u_y)$ and $\theta_y(u_y)$.

4.10 Angular Spectrum Huygens-Fresnel Analysis

We next present a Huygens-Fresnel analysis of the angular spectrum of the cross-range profile of the infinitely long synthetic aperture. This decomposition is doable when the SAR data measurements are described in the aperture coordinate system. This analysis is not possible with a SAR signal model referenced to a scene centered coordinate system.

Angular Spectrum of a Squint Scene

If the center of the ground scene is squinted ($Y_c \neq 0$), then using both the aperture centered and scene centered coordinate systems, the location of the nth scatterer is expressible as

$$(x_n, y_n) = (X_c + \hat{x}_n, Y_c + \hat{y}_n)$$

and the frequency-space phase history becomes

$$\begin{aligned} &S_{\Pi(\omega_c)} s_{\Pi(\infty)}(\omega_c, u_y)_{\text{ACS}} \\ &= \sum_n g_n \cdot \exp\left\{-jk_{\omega_c} \cdot 2\sqrt{(X_c + \hat{x}_n)^2 + [u_y - (Y_c + \hat{y}_n)]^2}\right\}. \end{aligned} \tag{4.55}$$

The slow-time Fourier transform gives

$$\begin{aligned}
&S_{\Pi(\omega_c)} S_{\Pi(2k_{\omega_c})}(\omega_c, k_{u_y})_{\text{ACS}} \\
&= \exp(-jX_c \cdot 2\sqrt{k_{\omega_c}^2 - k_{u_y}^2}\,) \cdot \exp(-jY_c \cdot 2k_{u_y}) \\
&\quad \cdot a(\omega_c, k_{u_y})_{\text{psp}} \cdot \Pi_{[-k_{\omega_c}, k_{\omega_c}]}(k_{u_y}) \\
&\quad\quad \cdot \sum_n g_n \cdot \exp(-j\hat{x}_n \cdot 2\sqrt{k_{\omega_c}^2 - k_{u_y}^2}\,) \cdot \exp(-j\hat{y}_n \cdot 2k_{u_y}).
\end{aligned}$$

Fourier migration of (4.55) via the Stolt mapping produces the angular spectrum phase history,

$$\begin{aligned}
&M^{\text{cw}}_{\widehat{\Pi}(\oplus)} S_{\Pi(4k_{\omega_c})}(\breve{k}_x, k_y)_{\text{ACS}} \\
&= \mathsf{ST}^{\breve{k}_x}_{\text{1D}} \left\{ S_{\Pi(\omega_c)} S_{\Pi(2k_{\omega_c})}(\omega_c, k_{u_y})_{\text{ACS}} \right\}, \\
&= \exp(-jX_c \breve{k}_{\hat{x}}) \cdot \exp(-jY_c k_{\hat{y}}) \\
&\quad \cdot \widehat{\Pi}^{\text{cw}}_{\oplus}(\breve{k}_{\hat{x}}, k_{\hat{y}}) \cdot \widehat{\Pi}_{[-2k_{\omega_c}, 2k_{\omega_c}]}(k_{\hat{y}}) \\
&\quad \cdot \sum_n g_n \cdot \exp(-j \cdot 0 \cdot \breve{k}_{\hat{x}}) \cdot \exp(-j\hat{y}_n k_{\hat{y}}), \\
&= H_{X_c}(\breve{k}_{\hat{x}}) \cdot H_{Y_c}(k_{\hat{y}}) \\
&\quad \cdot \widehat{\Pi}^{\text{cw}}_{\oplus}(\breve{k}_{\hat{x}}, k_{\hat{y}}) \cdot \widehat{\Pi}_{[-2k_{\omega_c}, 2k_{\omega_c}]}(k_{\hat{y}}) \cdot G_{\Pi(\infty)} A_{\Pi(\infty)}(\breve{k}_{\hat{x}}, k_{\hat{y}})_{\text{SCS}},
\end{aligned}$$

where

$$\begin{aligned}
\widehat{\Pi}^{\text{cw}}_{\oplus}(\breve{k}_{\hat{x}}, k_{\hat{y}}) \cdot \widehat{\Pi}_{[-2k_{\omega_c}, 2k_{\omega_c}]}(k_{\hat{y}}) &= \mathsf{ST}^{\breve{k}_x}_{\text{1D}} \left\{ a(\omega_c, k_{u_y})_{\text{psp}} \cdot \Pi_{[-k_{\omega_c}, k_{\omega_c}]}(k_{u_y}) \right\}, \\
H_{X_c}(\breve{k}_{\hat{x}}) &= \exp(-jX_c \breve{k}_{\hat{x}}), \\
H_{Y_c}(k_{\hat{y}}) &= \exp(-jY_c k_{\hat{y}}), \\
G_{\Pi(\infty)} A_{\Pi(\infty)}(\breve{k}_{\hat{x}}, k_{\hat{y}})_{\text{SCS}} &= \sum_n g_n \cdot \exp(-j \cdot 0 \cdot \breve{k}_{\hat{x}}) \cdot \exp(-j\hat{y}_n k_{\hat{y}}),
\end{aligned}$$

where we have $\hat{x}_n = 0$ in the cross-range profiling problem. The phase factor

$$\begin{aligned}
H_{X_c}(\breve{k}_{\hat{x}}) &= \exp(-jX_c \cdot \breve{k}_{\hat{x}}), \\
&= \exp(-jX_c \cdot 2\sqrt{k_{\omega_c}^2 - k_{u_y}^2}\,)
\end{aligned}$$

is a Huygens-Fresnel transfer in the Fourier migrated down-range dimension. Likewise, the phase factor

$$\begin{aligned}
H_{Y_c}(k_{\hat{y}}) &= \exp(-jY_c \cdot k_{\hat{y}}), \\
&= \exp(-jY_c \cdot 2k_{u_y})
\end{aligned}$$

is a Huygens-Fresnel transfer in the along-track dimension. Hence, the angular spectrum for a squinted along-track ground scene has two embedded Huygens-

Fresnel transfers,

$$\begin{aligned}
&M^{\mathrm{cw}}_{\widehat{\Pi}(\oplus)} S_{\Pi(4k_{\omega_c})}(\breve{k}_x, k_y)_{\mathrm{ACS}} \\
&= H_{X_c}(\breve{k}_{\hat{x}}) \cdot H_{Y_c}(k_{\hat{y}}) \cdot \underbrace{\widehat{\Pi}^{\mathrm{cw}}_{\oplus}(\breve{k}_{\hat{x}}, k_{\hat{y}}) \cdot \widehat{\Pi}_{[-2k_{\omega_c}, 2k_{\omega_c}]}(k_{\hat{y}}) \cdot G_{\Pi(\infty)} A_{\Pi(\infty)}(\breve{k}_{\hat{x}}, k_{\hat{y}})_{\mathrm{SCS}}}_{G^{\mathrm{cw}}_{\widehat{\Pi}(\oplus)} A_{\widehat{\Pi}(4k_{\omega_c})}(\breve{k}_{\hat{x}}, k_{\hat{y}})_{\mathrm{SCS}}}, \\
&= H_{X_c}(\breve{k}_{\hat{x}}) \cdot H_{Y_c}(k_{\hat{y}}) \cdot G^{\mathrm{cw}}_{\widehat{\Pi}(\oplus)} A_{\widehat{\Pi}(4k_{\omega_c})}(\breve{k}_{\hat{x}}, k_{\hat{y}})_{\mathrm{SCS}}, \\
&= G^{\mathrm{cw}}_{\widehat{\Pi}(\oplus)} A_{\widehat{\Pi}(4k_{\omega_c})}(\breve{k}_x, k_y)_{\mathrm{ACS}},
\end{aligned}$$

where we applied the weight functions of the angular spectrum domain to the along-track terrain spectrum,

$$\begin{aligned}
&G^{\mathrm{cw}}_{\widehat{\Pi}(\oplus)} A_{\widehat{\Pi}(4k_{\omega_c})}(\breve{k}_{\hat{x}}, k_{\hat{y}})_{\mathrm{SCS}} \\
&\quad = \widehat{\Pi}^{\mathrm{cw}}_{\oplus}(\breve{k}_{\hat{x}}, k_{\hat{y}}) \cdot \widehat{\Pi}_{[-2k_{\omega_c}, 2k_{\omega_c}]}(k_{\hat{y}}) \cdot G_{\Pi(\infty)} A_{\Pi(\infty)}(\breve{k}_{\hat{x}}, k_{\hat{y}})_{\mathrm{SCS}}.
\end{aligned}$$

The angular spectrum is decomposed into a down-range Huygens-Fresnel transfer $H_{X_c}(\breve{k}_{\hat{x}})$, an along-track Huygens-Fresnel transfer $H_{Y_c}(k_{\hat{y}})$ and the along-track terrain spectrum

$$G_{\Pi(\infty)} A_{\Pi(\infty)}(\breve{k}_{\hat{x}}, k_{\hat{y}})_{\mathrm{SCS}}$$

of the ground scene described in the scene coordinate system. The semicircle angular spectrum manifold is

$$\widehat{\Pi}^{\mathrm{cw}}_{\oplus}(\breve{k}_{\hat{x}}, k_{\hat{y}}).$$

The two way diffraction limit imposed by the along-track wavenumber manifold is

$$\widehat{\Pi}_{[-2k_{\omega_c}, 2k_{\omega_c}]}(k_{\hat{y}}).$$

Angular Spectrum of a Broadside Scene

The Huygens-Fresnel decomposition of the angular spectrum of a broadside scene simplifies to an expression that contains only the down-range Huygens-Fresnel transfer function,

$$\begin{aligned}
M^{\mathrm{cw}}_{\widehat{\Pi}(\oplus)} S_{\widehat{\Pi}(4k_{\omega_c})}(\breve{k}_x, k_y)_{\mathrm{ACS}} &= H_{X_c}(\breve{k}_{\hat{x}}) \cdot G^{\mathrm{cw}}_{\widehat{\Pi}(\oplus)} A_{\widehat{\Pi}(4k_{\omega_c})}(\breve{k}_{\hat{x}}, k_{\hat{y}})_{\mathrm{SCS}}, \\
&= G^{\mathrm{cw}}_{\widehat{\Pi}(\oplus)} A_{\widehat{\Pi}(4k_{\omega_c})}(\breve{k}_x, k_y)_{\mathrm{ACS}}.
\end{aligned}$$

The decomposition of the angular spectrum of a broadside scene does not contain the along-track Huygens-Fresnel transfer function.

4.11 Angular Spectrum Huygens-Fresnel Inversion

Once a SAR phase history has been formatted as an angular spectrum through Fourier migration by Stolt transformation (and uniformly resampled), a SAR image (cross-range profile) can be obtained with inverse Fourier transforms.

Forming the cross-range profile via Fourier inversion in the aperture coordinate system can be problematic because of the large number of samples required to represent the scene's offset from the flight line. The size of such an image would span the imaging geometry from the flight path to the far edge of the scene.

The required fast Fourier transform (FFT) [204] sizes can be reduced by application of inverse Huygens-Fresnel transfers before the inverse Fourier transforms. These wavenumber domain inversions remove the down-range and along-track (if squinted) offsets between sensor and scene. These inversions place the SAR image at the center of the scene coordinate system and thereby reduce the FFT sizes required for Fourier inversion.

Inverse Huygens-Fresnel Transfers of Broadside Scene

For the broadside scene, the Huygens-Fresnel transfer $H_{X_c}(\breve{k}_{\hat{x}})$ embedded in the angular spectrum is not scatterer informative. Applying the inverse Huygens-Fresnel transfer

$$H^*_{X_c}(\breve{k}_{\hat{x}}) = \exp(jX_c\breve{k}_{\hat{x}})$$

to

$$\overset{\text{cw}}{M}_{\text{ñ}(\oplus)} S_{\text{ñ}(4k_{\omega_c})}(\breve{k}_x, k_y)_{\text{ACS}}$$

removes the embedded down-range coordinate shift phase factor $H_{X_c}(\breve{k}_{\hat{x}})$ from the angular spectrum's decomposition,

$$\begin{aligned}
\overset{\text{cw}}{I}_{\text{ñ}(\oplus)} A_{\text{ñ}(4k_{\omega_c})}(\breve{k}_{\hat{x}}, k_{\hat{y}})_{\text{SCS}} &= H^*_{X_c}(\breve{k}_{\hat{x}}) \times \left\{ \overset{\text{cw}}{M}_{\text{ñ}(\oplus)} S_{\text{ñ}(4k_{\omega_c})}(\breve{k}_x, k_y)_{\text{ACS}} \right\}, \\
&= \underbrace{H^*_{X_c}(\breve{k}_{\hat{x}}) \cdot H_{X_c}(\breve{k}_{\hat{x}})}_{\text{cancels}} \cdot \overset{\text{cw}}{G}_{\text{ñ}(\oplus)} A_{\text{ñ}(4k_{\omega_c})}(\breve{k}_{\hat{x}}, k_{\hat{y}})_{\text{SCS}}, \\
&= \overset{\text{cw}}{G}_{\text{ñ}(\oplus)} A_{\text{ñ}(4k_{\omega_c})}(\breve{k}_{\hat{x}}, k_{\hat{y}})_{\text{SCS}}
\end{aligned}$$

and leaves the desired SAR image spectrum $\overset{\text{cw}}{I}_{\text{ñ}(\oplus)} A_{\text{ñ}(4k_{\omega_c})}(\breve{k}_{\hat{x}}, k_{\hat{y}})_{\text{SCS}}$ of the along-track profile in the scene coordinate system.

Inverse Huygens-Fresnel Transfers of Squint Scene

Likewise, for the squint scene, we apply two inverse Huygens-Fresnel transfer functions,

$$\begin{aligned}
H^*_{X_c}(\breve{k}_{\hat{x}}) &= \exp(jX_c\breve{k}_{\hat{x}}), \\
H^*_{Y_c}(k_{\hat{y}}) &= \exp(jY_c k_{\hat{y}}),
\end{aligned}$$

to the angular spectrum to obtain,

$$\begin{aligned}
\overset{\text{cw}}{I}_{\text{ñ}(\oplus)} A_{\text{ñ}(4k_{\omega_c})}(\breve{k}_{\hat{x}}, k_{\hat{y}})_{\text{SCS}} &= H^*_{X_c}(\breve{k}_{\hat{x}}) \cdot H^*_{Y_c}(k_{\hat{y}}) \times \left\{ \overset{\text{cw}}{M}_{\text{ñ}(\oplus)} S_{\text{ñ}(4k_{\omega_c})}(\breve{k}_x, k_y)_{\text{ACS}} \right\}, \\
&= \underbrace{H^*_{X_c}(\breve{k}_{\hat{x}}) \cdot H_{X_c}(\breve{k}_{\hat{x}})}_{\text{cancels}} \cdot \underbrace{H^*_{Y_c}(k_{\hat{y}}) \cdot H_{Y_c}(k_{\hat{y}})}_{\text{cancels}} \\
&\quad \cdot \overset{\text{cw}}{G}_{\text{ñ}(\oplus)} A_{\text{ñ}(4k_{\omega_c})}(k_{\hat{x}}, k_{\hat{y}})_{\text{SCS}}, \\
&= \overset{\text{cw}}{G}_{\text{ñ}(\oplus)} A_{\text{ñ}(4k_{\omega_c})}(k_{\hat{x}}, k_{\hat{y}})_{\text{SCS}}.
\end{aligned}$$

The resulting along-track profile spectrum $\overset{\text{cw}}{I}_{\widehat{\sqcap}(\oplus)} A_{\widehat{\sqcap}(4k_{\omega_c})}(\breve{k}_{\hat{x}}, k_{\hat{y}})_{\text{SCS}}$ is in the scene coordinate system.

Huygens-Fresnel Transfers of Ewald Spheres

The operation of inverse Huygens-Fresnel transfers in the wavenumber domain, as illustrated in Figs. 4.18a-b, has a rather pleasing visual interpretation in terms of single scatterer Ewald spheres. Fig. 4.18a shows the full angular spectrum (not just slant-plane) phase signals on surfaces of the Ewald spheres of three scatterers. One is at boresight. Two are off boresight. Each scatterer has been given its own Ewald sphere, where in reality, the phase histories of all three scatterers reside on a single shared Ewald sphere.

The tonal phase signals on the Ewald spheres describe scatterer location within the aperture coordinate system as measured with a passband sensor that transmits a continuous wave signal of frequency ω_c.

This illustration with angular spectrums described over full Ewald spheres requires data collections from full spherical sweeps of possible view angles, which of course, is impossible in linear flight operations. The angular spectrum in slant-plane imaging reduces to a semicircle slice submanifold of the Ewald sphere. In monochromatic passband sensing, this semicircle submanifold has a radius determined by $2k_{\omega_c}$.

The wavenumber domain phase signal on each Ewald sphere is dimensionally decomposable into three tonal phase factors that describe the down-range, along-track and above-ground positions of the scatterers as expressed in the aperture coordinate system.

Fig. 4.18b illustrates the result of applying a single down-range inverse Huygens-Fresnel transfer that effectively shifts the center of the aperture coordinate system to a point closer to the center scatterer's position. The three scatterer's Ewald sphere phase signals show reduced tonal phase history frequencies because of the shorter offsets from the center of the new coordinate system's origin.

The phase modulation of the Huygens-Fresnel transfer function itself can be illustrated on the surface of an Ewald sphere. This is shown as the introduced fourth sphere in Fig. 4.18b. The tonal phase pattern on that Ewald sphere represents the wavenumber domain conjugated phase adjustment required to move from the aperture coordinate system to a position closer to the center of the scene coordinate system. A partial Huygens-Fresnel transfer is demonstrated in this example for the sake of illustration.

If a full Huygens-Fresnel transfer were performed, the boresight scatterer would be at scene center in the scene coordinate system and its Ewald sphere phase history would be free of tonal phase oscillations over the sphere's surface. The other two off boresight scatterers would have phase history tonal variations in only the along-track wavenumber dimension after Huygens-Fresnel conversion to the scene coordinate system.

Presumming

SAR image formation processing with Stolt formatting often involves an along-track wavenumber subsampling operation called "presumming" or "preintegra-

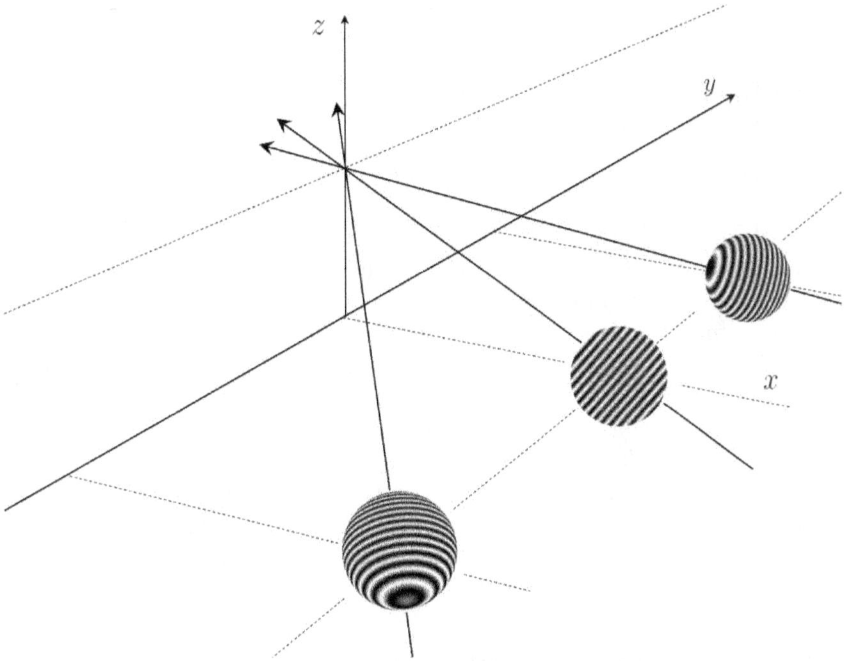

a. Ewald sphere phase histories of aperture coordinate system.

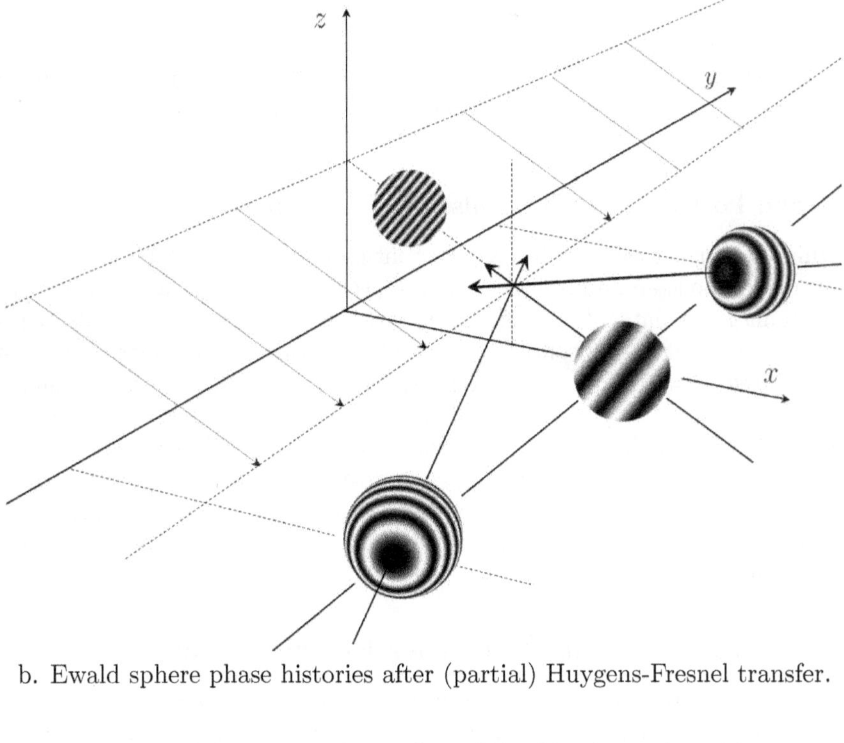

b. Ewald sphere phase histories after (partial) Huygens-Fresnel transfer.

Figure 4.18: Huygens-Fresnel transfers of Ewald sphere phase histories.

tion" [142, 250]. We introduce the process of presumming here and explain it in more detail in Section 7.12.

Pulsed Radar Data Collections

At certain stages of the signal analysis, we need to deal with a slow-time signal model where the SAR measurements are taken at discrete points $u_y[m]$ along the synthetic aperture. The operation of a pulsed radar can be viewed as defining an along-track spatial sampling function corresponding to pulse transmission/reception points along the SAR flight line [137].

The radar sensor can be imagined to "hop" along the synthetic aperture from $u_y[m]$ to $u_y[m+1]$. At each hop position the radar transmits a pulse and waits to record the echo data. This is sometimes called the "stop-and-hop" assumption for modeling SAR data collection across slow-time [683]. The stop-and-hop model may be viewed, from a signal processing perspective, as a slow-time Dirac impulse train that samples the slow-time phase curtain.

The continuous version of the slow-time phase curtain is not physically measurable with a pulsed radar system [407]. The continuous version can be obtained with a non-pulsed sensor using *frequency modulated continuous wave* signals [46, 86, 871, 1026, 1027].

Along-Track Sample Rate Adjustment

Presumming is a sample rate adjustment in the along-track wavenumber dimension that changes the slow-time sample rate from that required to support the phase curtain to a sample rate that supports the cross-range profile's along-track image size (recall that sample density in a wavenumber domain determines the supportable size of the scene to be imaged).

Pre and Post Presumming Pulse Data Indexing

In this chapter, we assume the use of a continuous wave signal with stop-and-hop sensor movement and an infinitely long synthetic aperture. Under this scenario, the manifold of data of the aperture wavenumber domain has finite support. The Nyquist issues that control the sampling of this wavenumber data must be investigated. However, we choose to postpone a detailed discussion of presumming until Section 7.12 and we assume here that by some appropriate method the sample density in the along-track dimension has been adjusted to the Nyquist rate required to support the along-track extent of the cross-range profile. The index m is used to reference aperture wavenumber pulse data before presumming. The index $n_{\hat{y}}$ is used to reference along-track image wavenumber samples after presumming.

Uniform Resampling of Fourier Migrated Data

The Stolt transform and Huygens-Fresnel transfers that convert the signal spectrum from $SS(\omega_c, k_{u_y})$ to the Fourier migrated angular spectrum $MS(\breve{k}_{\hat{x}}, k_{\hat{y}})$ create a non-uniformly spaced set of samples in the $\breve{k}_{\hat{x}}$ dimension of the $(\breve{k}_{\hat{x}}, k_{\hat{y}})$ domain. This migration from uniformly spaced slow-time wavenumber samples to

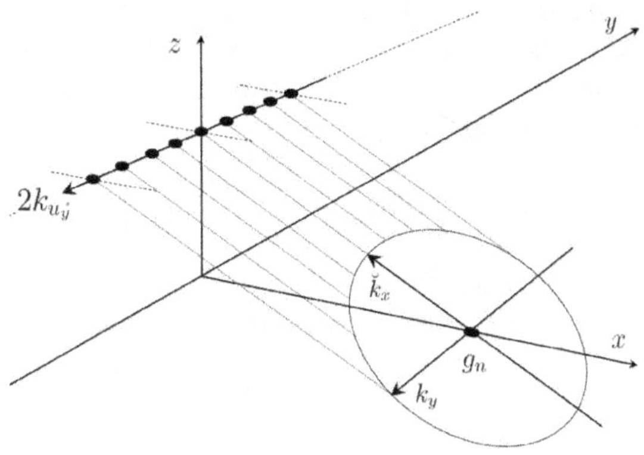

a. Rectilinear aperture wavenumber data samples.

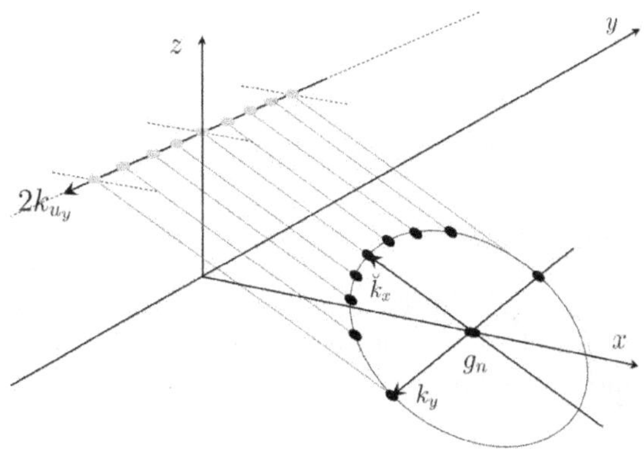

b. Rectilinear to angular spectrum migration.

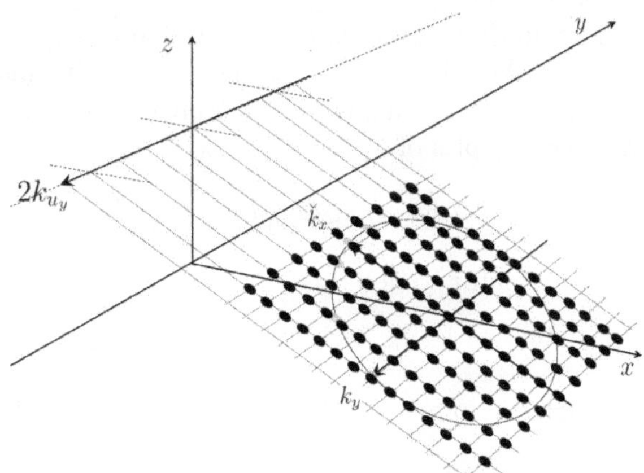

c. Uniform resampling of the angular spectrum data.

Figure 4.19: Uniform resampling of Fourier migrated wavenumber data.

non-uniformly spaced samples in the $\breve{k}_{\hat{x}}$ domain is shown in Figs. 4.19a-b. Non-uniformly spaced samples are not compatible with the uniform sample spacing requirement of the (inverse) discrete Fourier transforms required to form the SAR image. Hence, uniform resampling of

$$IA\left[\breve{k}_{\hat{x}}\left(\omega_c[n_{\omega_c}], k_{\hat{y}}[n_{\hat{y}}]\right), k_{\hat{y}}[n_{\hat{y}}]\right]$$

is required to produce the data set of an invertible cross-range profile spectrum. We indicate the uniformly resampled spectrum of the cross-range profile by

$$IA(\ddot{k}_{\hat{x}}[n_{\hat{x}}], k_y[n_{\hat{y}}]).$$

The index $n_{\hat{x}}$ identifies uniformly spaced output samples of the $\ddot{k}_{\hat{x}}$ domain obtained by resampling the phase history data of the $\breve{k}_{\hat{x}}$ domain.

Uniform Outputs from Non-Uniform Input Samples

We indicate the uniform resampling operator as

$$IA(\ddot{k}_{\hat{x}}[n_{\hat{x}}], k_y[n_{\hat{y}}]) = \mathsf{Uniform}_{1\mathrm{D}}^{\ddot{k}_{\hat{x}}}\left\{MS\left[\breve{k}_{\hat{x}}\left(\omega_c[n_{\omega_c}], k_{\hat{y}}[n_{\hat{y}}]\right), k_{\hat{y}}[n_{\hat{y}}]\right]\right\}.$$

The resample operator is subscripted by "1D" to emphasize that uniform resampling after slant-plane Stolt formatting is a one dimensional operation. Although Fig. 4.19c shows many samples in the uniform two dimensional output grid of this resampling process, only those samples near the semicircle manifold of the Ewald sphere are significantly non-zero.

Interpolation with Non-Uniform Input Sample Grids

After Fourier migration of $SS(\omega_c, k_{u_y})_{\mathrm{ACS}}$ to $MS(\breve{k}_x, k_y)_{\mathrm{ACS}}$ and after inverse Huygens transfers to obtain $IA(\breve{k}_{\hat{x}}, k_{\hat{y}})_{\mathrm{SCS}}$ from $MS(\breve{k}_x, k_y)_{\mathrm{ACS}}$, interpolation is performed to achieve uniform resampling in the $IA(\ddot{k}_{\hat{x}}, k_{\hat{y}})$ angular spectrum domain. Interpolation should be based on (non-uniformly spaced) samples of $IA(\breve{k}_{\hat{x}}, k_{\hat{y}})_{\mathrm{SCS}}$. Neighborhoods within the wavenumber domain of (4.49) and (4.50) determine which phase samples of

$$IA[\breve{k}_{\hat{x}}(\omega_c, k_{u_y}), k_{\hat{y}}(k_{u_y})]_{\mathrm{SCS}}$$

are input into an interpolator to determine the value of the interpolated output sample at

$$IA(\ddot{k}_{\hat{x}}, k_{\hat{y}})_{\mathrm{SCS}}.$$

Details of this interpolation process will be covered in Section 7.14.

"Stolt Interpolation" and "Fourier Interpolation"

There is a misstep common to new practitioners of angular spectrum inversion methods, namely the confusion between the Stolt transform and subsequent interpolation and resampling processing requirements. Interpolation should not be based on the uniformly spaced sample grid in the (ω_c, k_{u_y}) domain of $SS(\omega_c, k_{u_y})$ phase history data (see Section 7.14 for more detail).

Occasionally, novice practitioners define an errant algorithm that seeks to combine Stolt transformation and uniform resampling into a single process. This combined process is sometimes called "Stolt interpolation" [136, 137] or "Fourier interpolation."

To avoid confusion, terminology such as "Stolt transformation" or "Fourier migration" should be used for the mapping of (ω_c, k_{u_y}) to $\breve{k}_x$. The terminology of "Stolt interpolation" or "Fourier interpolation" should be avoided.

Simply referring to the uniform resampling process (without reference to Stolt processing) in the Fourier migrated wavenumber (angular spectrum) domain reduces chances for confusion. Interpolation is a separate, subsequent process performed over the non-uniformly spaced sample grid of $IA(\breve{k}_{\hat{x}}, k_{\hat{y}})_{\text{SCS}}$ to obtain $IA(\ddot{k}_{\hat{x}}, k_{\hat{y}})_{\text{SCS}}$.

4.12 Infinite Aperture Cross-Range Profile

We describe the result of Fourier inversion without and with Huygens-Fresnel transfers.

Fourier Inversion without Huygens-Fresnel Transfers

Let

$$IA[n_x, n_y]_{\text{ACS}} = IA(\ddot{k}_x[n_x], k_y[n_y])_{\text{ACS}}$$

be the uniformly resampled Stolt formatted angular spectrum data in the aperture coordinate system. An inverse two dimensional Fourier transform of

$$IA[n_x, n_y]_{\text{ACS}}$$

yields the SAR pixel array

$$ia[n_x, n_y]_{\text{ACS}}$$

that contains the along-track profile. For the broadside case, the SAR image obtained from the angular spectrum data described in the aperture coordinate system is

$$i_{\Pi[X_{\min}, X_{\max}]} a_{\Pi[-Y_o, +Y_o]}(x, y)_{\text{ACS}} = \mathsf{iFT}_{\text{2D}} \left\{ IA(\ddot{k}_x, k_y)_{\text{ACS}} \right\}.$$

For the squint scene case, the SAR image containing the along-track profile is

$$i_{\Pi[X_{\min}, X_{\max}]} a_{\Pi[Y_{\min}, Y_{\max}]}(x, y)_{\text{ACS}} = \mathsf{iFT}_{\text{2D}} \left\{ IA(\ddot{k}_x, k_y)_{\text{ACS}} \right\}.$$

This type of angular spectrum processing, where compensation of the resident Huygens-Fresnel transfers has not been performed, produces a relatively large image (which contains the desired cross-range profile) in the aperture coordinate system. This large image spans from $X_{\min}$ to $X_{\max}$, and from $-Y_o$ to $+Y_o$ (broadside) or from $Y_{\min}$ to $Y_{\max}$ (squint). This (mostly dark) image covers the region from flight line $X_{\min}$ to the far edge $X_{\max}$ of the ground patch.

Fourier Inversion with Huygens-Fresnel Transfers

The along-track image in the scene coordinate system is also formed with an inverse Fourier transform,

$$i_{\Pi[-X_o,X_o]} a_{\Pi[-Y_o,Y_o]}(\hat{x},\hat{y})_{\text{SCS}} = \mathsf{iFT}_{2D}\left\{IA(\ddot{k}_{\hat{x}}, k_{\hat{y}})_{\text{SCS}}\right\}.$$

This smaller along-track image of size $[-X_o, X_o]$ and $[-Y_o, Y_o]$ is centered at zero in the output of the inverse Fourier transform and contains the cross-range profile (located at $\hat{x} = 0$ in the scene coordinate system). The cross-range profile is given by

$$a_{\Pi[-Y_o,Y_o]}(\hat{y})_{\text{SCS}} = \sum_n g_n \cdot \mathsf{sinc}\left(\frac{\hat{y} - \hat{y}_n}{\Delta_{\hat{y}}}\right). \tag{4.56}$$

Along-Track Resolution with CW Tone Signals

Since the bandwidth of an infinite duration sinusoid is infinitely narrow, one may wonder why a large bandwidth signal is not required to achieve high resolution in along-track imaging. Understanding the answer to this question is a key to understanding SAR. High resolution aperture synthesis can be achieved without frequency diversity, but must rely on the angular diversity afforded by the changing sensor-scatterer imaging geometry [713].

Narrow Band Waveforms and Broadband Phase Curtains

Along-track imaging is based on the phase difference between the transmitted and received echo signal as a function of sensor position. This phase difference forms the slow-time phase curtain. This collected (spatial) signal can be broadband, even if the transmitted (temporal) signal is infinitely narrow band.

Infinite Aperture Slow-Time Bandwidth and Round Trip Along-Track Resolution

The aperture wavenumber window function $\Pi_{(2k_{\omega_c})}(k_{u_y})$ determines the ultimate resolution possible in along-track profiling. The aperture wavenumber bandwidth of

$$2k_{\omega_c} = k_{\omega_c} - (-k_{\omega_c})$$

of (4.44) is the support of the aperture wavenumber k_{u_y} obtained by an infinite linear synthetic aperture. Additionally, the round trip signal propagation path doubles the size of the wavenumber domain manifold by doubling the number of round trip phase cycles through the mapping $k_y = 2k_{u_y}$. Hence, the along-track image resolution achieved by an infinite synthetic aperture is

$$\begin{aligned}
\Delta_y &= \frac{2\pi}{2 \times \Pi_{[-k_{\omega_c},k_{\omega_c}]}(k_{u_y})}, \\
&= \frac{2\pi}{4k_{\omega_c}}, \\
&= \frac{\lambda_c}{4}.
\end{aligned} \tag{4.57}$$

Ultimate Along-Track Resolution Limits

The resolution limit (4.57) is approached in linear flight SAR as the length of the synthetic aperture approaches infinity. This is equivalent to the view angle span approaching the angular dwell of $180^o = [-90^o, +90^o]$. There is nothing that fundamentally precludes a SAR from gathering an objective angular dwell by non-linear flight operations (see Chapter 13).

We next give a preview of how the resolution limit (4.57) is modified by specific SAR operational modes and various choices of SAR image formation algorithms.

Resolution Limit of Spotlight Mode with Stolt Formatting

A quarter wavelength lower bound

$$\Delta_y^{\text{spot:line}} \geq \frac{\lambda_c}{4}$$

is applicable to SAR systems operated in spotlight mode that employ Stolt formatting algorithms (see Section 12.4). Linear flight Stolt formatting systems require phase stabilization to a reference down-range value, i.e., these systems[1] require motion compensation to a line [389].

Resolution Limit of Spotlight Mode with Polar Formatting

A lower bound of one fourth of the antenna width in the along-track direction

$$\Delta_y^{\text{spot:point}} \geq \frac{D_y}{4}$$

applies in spotlight mode SAR systems where polar formatting is used for image formation processing (see Section 12.5). Polar formatting requires phase stabilization to a ground reference point, i.e., these systems require motion compensation to a point.

Resolution Limit of Stripmap Mode with Stolt Formatting

The lower bound of one fourth of the antenna width in the along-track direction also applies in stripmap mode SAR with Stolt formatting (see Section 12.7), range-Doppler (see Section 12.9) and chirp scaling algorithms (see Section 12.10),

$$\Delta_y^{\text{strip:line}} \geq \frac{D_y}{4}.$$

Hence, the resolution of polar formatting in spotlight mode is similar to the resolution of Stolt formatting in stripmap mode [885].

[1]Stolt formatting in three dimensional imaging can utilize phase stabilization to a point, but with motion compensation adjustments made relative to a grid structure within an aperture volume that encloses the non-linear flight profile (see Chapter 13).

Spatially Variant and Invariant Resolutions

The point-spread function (4.57) of the along-track profile obtained with an infinite synthetic aperture is independent of the along-track position of the scatterer. This is not the case for cross-range profiling with finite synthetic apertures (see Section 8.10).

Chapter 5

Down-Range Profiling with Pulse Compression

Down-range profiling is a one dimensional subset of the SAR image formation process [638, 717, 788]. To simplify our down-range profile model, we continue to use an isotropic point antenna and assume each integrative range bin echo contains no more than the echo of one scatterer.

Pulsed radars that transmit waveforms with phase modulations can improve down-range resolution through the signal processing operations of *pulse compression* [202, 840, 864, 925]. Pulse compression is also called *range compression* [23]. Conventionally, but not necessarily, waveform pulse compression is the primary way of forming the SAR image in the down-range dimension [639, 679].

This chapter explores waveform pulse compression by matched filtering. Section 12.5 will explain the use of stretch processing as an alternate method for pulse compression of LFM chirp signals.

This chapter is complementary to Chapter 4, which dealt with one dimensional along-track image formation processing via aperture synthesis. Chapter 7 will combine one dimensional down-range profiling via pulse compression with one dimensional along-track profiling in a study of two dimensional SAR image formation processing.

This chapter is outlined as follows: Section 5.1 presents a model of the down-range sensor-scatterer imaging geometry. Section 5.2 gives a model of the received echo signal and presents an analysis of the received pulse echo spectrum. Section 5.3 overviews the process of waveform pulse compression. Section 5.4 discusses differences between various forms of coherent (complex) and detected (real) versions of down-range scatterer profiles. Finally, Section 5.5 describes the Huygens-Fresnel transfer decomposition of the range profile spectrum.

5.1 Down-Range Profiling

We establish spatial and spectral descriptions of a set of scatterers oriented along a line parallel to the down-range direction. We shall use an impulse function as an idealized point scatterer model and shall describe our imaging geometry in the aperture coordinate system.

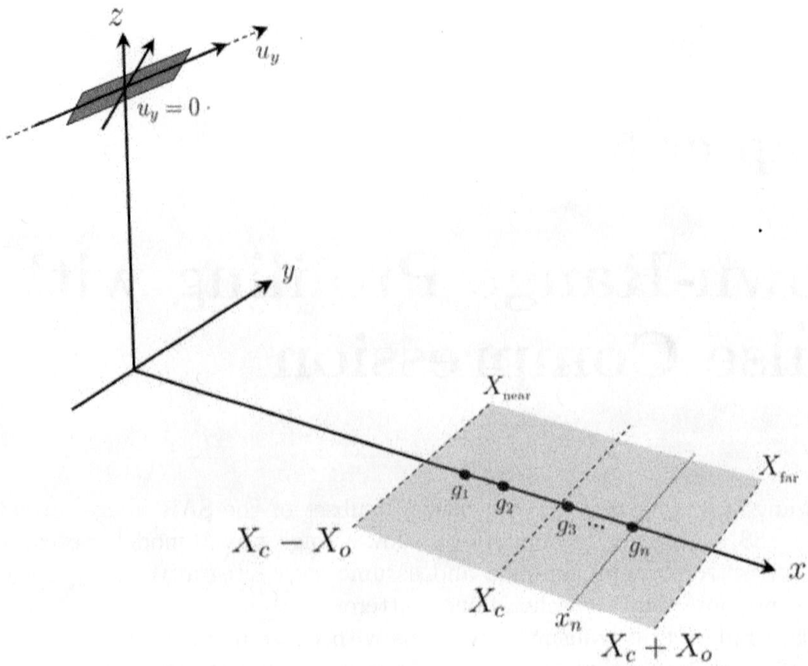

Figure 5.1: Imaging geometry for broadside down-range profiling.

One Dimensional Down-Range Profile

Assume the existence of a collection of discrete scatterers dispersed along a range line as shown in Fig. 5.1. This range line comprises a one dimensional broadside down-range terrain function [683],

$$r_{\Pi(2X_o@X_c)}(x)_{\text{ACS}} = \sum_n g_n \cdot \delta(x - x_n), \tag{5.1}$$

where $\delta(\cdot)$ is the Dirac impulse function. The subscript $\Pi(2X_o@X_c)$ of (5.1) indicates the scatterers are constrained to be within a range swath of width $2X_o$ centered at X_c. The index n identifies individual point scatterers with radar cross sections g_n and with down-range positions x_n.

Down-Range Profile Spectrum

The Fourier transform of the one dimensional down-range profile (5.1) is

$$R_{\Pi(\infty)}(k_x)_{\text{ACS}} = \sum_n g_n \cdot \exp(-jx_n k_x). \tag{5.2}$$

The ground scene wavenumber function (5.2) is a collection of g_n weighted complex sinusoids $\exp(-jx_n k_x)$ in the k_x domain, each with a frequency determined by the corresponding scatterer's down-range position x_n. The subscript $\Pi(\infty)$ of $R_{\Pi(\infty)}(k_x)$ indicates infinite bandwidth in the k_x wavenumber domain. Infinite spatial bandwidth is a result of the idealized infinitesimal width of the impulse function $\delta(\cdot)$ used as the scatterer model.

Broadside Down-Range Profile

To describe a range line of scatterers located in a nominal two dimensional area of a ground patch with width $2X_o$ and length $2Y_o$, we write for a broadside scene,

$$r_{\Pi(2X_o@X_c)}g_{\Pi(2Y_o@0)}(x,y)_{\text{ACS}} = \sum_n g_n \cdot \delta(x-x_n)\cdot\delta(y). \tag{5.3}$$

Although the terrain patch is two dimensional over the nominal area $2X_o \times 2Y_o$, the scene only contains one range line of scatterers. The subscript $\Pi(2Y_o@0)$ of (5.3) indicates a broadside range line of scatterers along $Y_c = 0$.

Broadside Down-Range Profile Spectrum

The two dimensional Fourier transform of (5.3) with respect to x and y gives

$$\begin{aligned} R_{\Pi(\infty)}G_{\Pi(\infty)}(k_x,k_y)_{\text{ACS}} &= \sum_n g_n \cdot \exp(-jx_nk_x)\cdot\exp(-j\cdot 0\cdot k_y), \quad (5.4)\\ &= \sum_n g_n \cdot \exp(-jx_nk_x)\cdot 1. \end{aligned}$$

Since $y_n = 0$, the Fourier transform (5.4) in the along-track dimension of a broadside scene has an along-track phase factor with a constant value of unity in the k_y wavenumber dimension,

$$\exp(-j\cdot 0\cdot k_y) = 1.$$

Squint Down-Range Profile

The two dimensional model of a squint ($Y_c \neq 0$) range line of scatterers is

$$r_{\Pi(2X_o@X_c)}g_{\Pi(2Y_o@Y_c)}(x,y)_{\text{ACS}} = \sum_n g_n \cdot \delta(x-x_n)\cdot\delta(y-Y_c) \tag{5.5}$$

and is configured as shown in Fig. 5.2. The subscript $\Pi(2Y_o@Y_c)$ of (5.5) indicates a line of scatterers aligned with $y = Y_c$ within a ground patch with an along-track length of $2Y_o$.

Squint Down-Range Profile Spectrum

The Fourier transform of the squint down-range ground scene is

$$R_{\Pi(\infty)}G_{\Pi(\infty)}(k_x,k_y)_{\text{ACS}} = \exp(-jY_ck_y)\cdot\sum_n g_n\cdot\exp(-jx_nk_x). \tag{5.6}$$

A non-trivial along-track phase factor $\exp(-jY_ck_y)$ that is independent of scatterer index n exists in (5.6). This phase factor has separated out of the scatterer

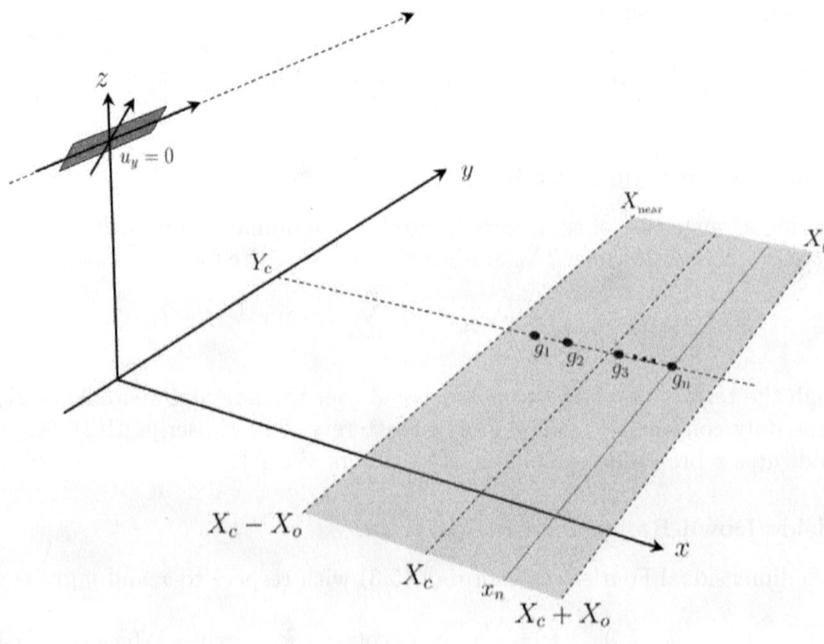

Figure 5.2: Imaging geometry for squint down-range profiling.

dependent summation over n. The phase factor $\exp(-jY_c k_y)$ is a Huygens-Fresnel transfer in the along-track dimension. Equating the wavenumber k_x of the aperture coordinate system with the wavenumber $k_{\hat{x}}$ of the scene coordinate system,

$$\begin{aligned} H_{Y_c}(k_y) &= \exp(-jY_c k_y), \\ &= \exp(-jY_c k_{\hat{y}}), \\ &= H_{Y_c}(k_{\hat{y}}) \end{aligned}$$

and using

$$R_{\Pi(\infty)}(k_{\hat{x}})_{\mathrm{SCS}} = \sum_n g_n \cdot \exp(-jx_n k_{\hat{x}}),$$

we have

$$R_{\Pi(\infty)} G_{\Pi(\infty)}(k_x, k_y)_{\mathrm{ACS}} = H_{Y_c}(k_{\hat{y}}) \cdot R_{\Pi(\infty)}(k_{\hat{x}})_{\mathrm{SCS}}.$$

A squint scene range profile spectrum in the aperture coordinate system is expressible as the range profile spectrum in the scene coordinate system subject to an along-track Huygens-Fresnel transfer.

5.2 Pulse Radar Echo Signal

We next examine several defining characteristics of the echo signals of pulsed radars useful in developing an approach to range profiling.

Range Swath

If a radar transmits a pulse beginning at the time instant $t = 0$ and if the echo from the nth scatterer at x_n is received beginning at the time instant t_n seconds later, then (assuming a direct path for transmitted signal and echo) the sensor-scatterer distance and the round trip echo delay are related by

$$t_n = \frac{1}{c} \times 2x_n.$$

If the receiver starts recording echoes immediately after transmission of a pulse with duration T_p, the maximum (but impractical) range over which an unambiguous range profile could be generated is

$$x \in \left[\frac{1}{2} \times cT_p, \infty\right). \tag{5.7}$$

There is a finite interval of interest,

$$\mathbb{X}_{\text{swath}} = [X_{\text{near}}, X_{\text{far}}], \tag{5.8}$$

for down-range profiling within the extrema of (5.7). This is the *range swath* over which the radar will be tasked to create a range profile.

Range Swath Size Limitations

Factors that limit the achievable size of the range swath include the unambiguous range imposed by the system's pulse repetition frequency and restrictions due to the sensitivity limitations of the radar's antenna and receiver. The range swath size can be further restricted during system design by limitations of the type of receiver subsystem used in the radar's front end, e.g., by use of a stretch receiver (see Section 12.5) and the type of phase stabilization implemented in the motion compensation system, e.g., motion compensation to a point (i.e., scene center) imposes restrictions on scene diameter when used with polar formatting (see Section 12.5).

Range Swath Boundaries

The limits of the range swath $\mathbb{X}_{\text{swath}}$ are indicated by the near edge distance X_{near} and the far edge distance X_{far}. The radar's range swath can be indicated with a window function,

$$\Pi_{[X_{\text{near}}, X_{\text{far}}]}(x).$$

The range swath half-width is denoted by X_o. In terms of this parameter, the full range swath width is

$$X_{\text{far}} - X_{\text{near}} = 2X_o.$$

The range swath can also be represented by the window function

$$\Pi_{(2X_o @ X_c)}(x) = [X_c - X_o, X_c + X_o],$$

which is often more simply written as $\Pi_{(2X_o)}(x)$.

Receive Window

Let the points in time T_s and T_f refer to instances after pulse transmission at which the receiver starts and finishes sampling of echo signals. The window function

$$\Pi_{[T_s,T_f]}(t) = [T_s, T_f]$$

represents the *fast-time receive window* of the radar. The receive window duration[1] or temporal manifold is

$$\mathbb{T}_{\text{receive}} = T_f - T_s.$$

Manipulation of the start and finish receive window parameters $[T_s, T_f]$ controls the range swath

$$\mathbb{X}_{\text{swath}} = [X_{\text{near}}, X_{\text{far}}]$$

over which returned pulse echoes are recorded. The *range swath echo duration* or manifold is

$$\mathbb{T}_{\text{swath}} = \frac{2}{c} \cdot \mathbb{X}_{\text{swath}}.$$

The time at which the receiver begins sampling is

$$T_s = \frac{1}{c} \cdot 2X_{\text{near}}.$$

To avoid range ambiguities, X_{near} must satisfy

$$X_{\text{near}} \geq \frac{1}{2} \cdot cT_p.$$

The receiver ceases data sampling after the pulse echo from a scatterer at X_{far} returns to the receiver. The receiver must allow time for the entire pulse duration T_p to be recorded. Hence, the end time of data recording is no sooner than

$$T_f = \frac{2}{c} X_{\text{far}} + T_p.$$

Range Swath Center

The spatial center of the range swath is

$$X_c = \frac{X_{\text{near}} + X_{\text{far}}}{2}.$$

Range Gating

Because of possible range gate ambiguities due to the long slant-range distance with multiple pulses in transit, spaceborne SAR systems have receive windows with fast-time gating designed to receive echoes only from the range swath of interest [221, 301].

[1] Whether notation like $\mathbb{T}_{\text{receive}}$ indicates a manifold or the size of the manifold will be clear from context.

Receive Window Center Time Reference

The receive window is offset from time zero ($t = 0$) by a nominal receive window center time value T_c. The center time reference T_c must be selected to satisfy

$$T_c \in \left[\frac{2}{c}(X_c - X_o), \frac{2}{c}(X_c + X_o) + T_p\right].$$

One possible choice of the receive window center time reference is

$$T_c = \frac{2}{c}X_c.$$

This choice is not the only possibility and T_c can be arbitrarily assigned anywhere within the limits of the receive window.

Time Shifted Pulses

A baseband pulse with a right shift of T_c is algebraically expressed as $p_{\Pi(T_p)}(t-T_c)$. A baseband pulse with a left shift of T_c is $p_{\Pi(T_p)}(t+T_c)$. The scripted notations for right and left shifted baseband pulse signals are

$$\begin{aligned} p_{\Pi(T_p)}^{\mathrm{R.shift}(T_c)}(t) &= p_{\Pi(T_p)}(t - T_c), \\ p_{\Pi(T_p)}^{\mathrm{L.shift}(T_c)}(t) &= p_{\Pi(T_p)}(t + T_c), \end{aligned}$$

where "R.shift(T_c)" indicates a temporal fast forward of T_c seconds and where "L.shift(T_c)" indicates a temporal rewind of T_c seconds. We sometimes use an even more abusive but compact notation for shifted pulses,

$$p_{\Pi(T_p@T_c)}(t) = p_{\Pi(T_p)}(t - T_c).$$

The $\Pi(T_p@T_c)$ subscript of $p_{\Pi(T_p@T_c)}(t)$ indicates the pulse is shifted to time T_c. Whether the pulse is centered at T_c or starts at T_c depends on the specific definition of the window function. We find occasion to use pulse signals defined in each of these ways.

Passband Echo Signal

Consider a radar that transmits a passband pulse waveform

$$w_{\Pi(T_p)}(t) = \exp(j\omega_c t) \times p_{\Pi(T_p)}(t)$$

with a pulse duration of T_p. The passband echo signal

$$s_{\Pi[T_s,T_f]}(t)$$

recorded during the radar's receive window $[T_s, T_f]$ is a collection of delayed waveform echoes $w_{\Pi(T_p)}(t - t_n)$ scattered by the set of down-range scatterers. The echo signal created by all illuminated scatterers is

$$s_{\Pi[T_s,T_f]}(t) = \sum_n g_n \cdot w_{\Pi(T_p)}(t - t_n). \tag{5.9}$$

Time-Space Decomposition of Echo Signal

Insights into the structure of the recorded passband signal can be obtained by decomposing the received echo signal (5.9) into elements related to the transmitted pulse waveform and elements related to scene characteristics. The shifting property of functions convolved with impulse functions is useful for achieving this decomposition. The received echo signal can be expressed as the convolution of the transmitted pulse waveform $w_{\Pi(T_p)}(t)$ with the collection of time shifted impulses $\delta(t - t_n)$ that represent echo delay instances induced by down-range scatterer locations $\delta(x - x_n)$,

$$\begin{aligned} s_{\Pi[T_s,T_f]}(t) &= \sum_n g_n \cdot w_{\Pi(T_p)}\left(t - t_n\right), \\ &= w_{\Pi(T_p)}(t) * \left[\sum_n g_n \cdot \delta\left(t - t_n\right)\right], \\ &= w_{\Pi(T_p)}(t) * \left\{\sum_n g_n \cdot \delta\left[\frac{2}{c}(x - x_n)\right]\right\}, \end{aligned} \tag{5.10}$$

where $*$ is the convolution operator. By expressing the down-range terrain function as

$$r_{\Pi[X_{\text{near}},X_{\text{far}}]}(x) = \sum_n g_n \cdot \delta(x - x_n)$$

and by considering

$$\begin{aligned} r_{\Pi[X_{\text{near}},X_{\text{far}}]}\left(\frac{2}{c}x\right) &= \sum_n g_n \cdot \delta\left[\frac{2}{c}(x - x_n)\right], \\ &= \sum_n g_n \cdot \delta\left(t - t_n\right), \\ &= r_{\Pi[T_s,T_f]}\left(t\right), \end{aligned} \tag{5.11}$$

we can write (5.10) as

$$\begin{aligned} s_{\Pi[T_s,T_f]}(t) &= w_{\Pi(T_p)}(t) * r_{\Pi[T_s,T_f]}\left(t\right), \\ &= w_{\Pi(T_p)}(t) * r_{\Pi[X_{\text{near}},X_{\text{far}}]}\left(\frac{2}{c}x\right). \end{aligned} \tag{5.12}$$

Equation (5.12) shows that the received echo signal is separable or decomposable into a temporal element $w_{\Pi(T_p)}(t)$ and a spatial element $r_{\Pi[X_{\text{near}},X_{\text{far}}]}\left[(2/c)\,x\right]$.

Pulse Radar Echo Spectrum

With the convolutional formulation (5.12) of the received echo signal, we are now in a position to more easily determine the Fourier transform of the received passband signal

$$s_{\Pi[T_s,T_f]}(t).$$

This Fourier transform yields the *echo signal spectrum*

$$S_{\Pi(2\omega_o@\omega_c)}(\omega_p).$$

We shall decompose $S(\cdot)$ to find the range profile's Fourier transform $R(\cdot)$ embedded in the received signal spectrum $S(\cdot)$. We then look for a way to process $S(\cdot)$ to isolate $R(\cdot)$ in preparation for recovering the range profile $r(\cdot)$ through an inverse Fourier transform.

Fourier Analysis Overview

Descriptions of baseband and passband pulse signals subject to shifts in time or frequency, or subject to carrier modulations (multiplication by complex sinusoids) are common in SAR signal analysis. Before we develop a model of the spectrum of the pulse echo signal, we briefly review background material related to Fourier analysis of such signals.

Fourier Transform Time Shift Property

Signal analysis of time shifted pulses engages a useful property of the Fourier transform called the *time shift property,*

$$p(t-T_c) \quad {}_t\!\Leftrightarrow_{\omega_b} \quad P(\omega_b)\cdot\exp(-jT_c\omega_b), \tag{5.13}$$
$$p(t+T_c) \quad {}_t\!\Leftrightarrow_{\omega_b} \quad P(\omega_b)\cdot\exp(+jT_c\omega_b), \tag{5.14}$$

or as expressed with scripted notations,

$$p^{\text{R.shift}(T_c)}(t) \quad {}_t\!\Leftrightarrow_{\omega_b} \quad P^{\text{carrier}(-T_c)}(\omega_b),$$
$$p^{\text{L.shift}(T_c)}(t) \quad {}_t\!\Leftrightarrow_{\omega_b} \quad P^{\text{carrier}(+T_c)}(\omega_b).$$

The ω_b domain complex sinusoid $\exp(-j\ T_c\ \omega_b)$ of (5.13) with negative frequency $-T_c$ is descriptive of a right time shift of $p(t)$ by T_c. Likewise, an ω_b domain sinusoid $\exp(+j\ T_c\ \omega_b)$ of (5.14) with positive frequency $+T_c$ is descriptive of a left time shift of $p(t)$ by T_c.

Fourier Transform Frequency Shift Property

Complementary to the time shift property is the *frequency shift property* of the Fourier transform,

$$p(t)\cdot\exp(+j\omega_c t) \quad {}_t\!\Leftrightarrow_{\omega_b} \quad P(\omega_b-\omega_c), \tag{5.15}$$
$$p(t)\cdot\exp(-j\omega_c t) \quad {}_t\!\Leftrightarrow_{\omega_b} \quad P(\omega_b+\omega_c),$$

or as expressed with scripted notation,

$$p^{\text{carrier}(+\omega_c)}(t) \quad {}_t\!\Leftrightarrow_{\omega_b} \quad P^{\text{R.shift}(\omega_c)}(\omega_b),$$
$$p^{\text{carrier}(-\omega_c)}(t) \quad {}_t\!\Leftrightarrow_{\omega_b} \quad P^{\text{L.shift}(\omega_c)}(\omega_b).$$

Up Conversion and Down Conversion

The carriers $\exp(+j\omega_c t)$ and $\exp(-j\omega_c t)$ of (5.15) are complex sinusoids in the time domain. The t domain carrier $\exp(+j\omega_c t)$ with positive (carrier) frequency $+\omega_c$ induces a right shift of $P(\omega_b)$ by ω_c. A t domain carrier $\exp(-j\omega_c t)$ with a negative frequency $-\omega_c$ induces a left frequency shift of $P(\omega_b)$ by ω_c. The equations of (5.15) are descriptions of the processes of *up conversion* and *down conversion.*

Baseband and Passband Pulse Waveforms and Spectra

Note that (5.15) can be more concisely written as the Fourier transform pair of a passband waveform,

$$w(t) \;{}_t\!\Leftrightarrow_{\omega_p}\; W(\omega_p),$$

where the relationship between passband and baseband pulse waveforms is implicit in our notation, (i.e., we always use $p(\cdot)$ for baseband signals, $w(\cdot)$ for transmitted passband waveforms and $s(\cdot)$ for received passband echo signals). We can also write

$$w(t) \quad = \quad p(t) \cdot \exp(+j\omega_c t).$$

Ways to express the passband pulse spectrum $W(\omega_p)$ for $\omega_p \in [\omega_c - \omega_o, \omega_c + \omega_o]$ in terms of the up converted baseband pulse spectrum include the following:

$$\begin{aligned} W(\omega_p) &= P^{\texttt{R.shift}(\omega_c)}(\omega_b) && \text{for} \quad \omega_b \in [\omega_c - \omega_o, \omega_c + \omega_o], \\ &= P(\omega_p - \omega_c) && \text{for} \quad \omega_p \in [\omega_c - \omega_o, \omega_c + \omega_o], \\ &= W(\omega_b + \omega_c) && \text{for} \quad \omega_b \in [-\omega_o, +\omega_o]. \end{aligned}$$

Likewise, ways to express the baseband pulse spectrum $P(\omega_b)$ for $\omega_b \in [-\omega_o, +\omega_o]$ in terms of the down converted passband pulse spectrum include the following:

$$\begin{aligned} P(\omega_b) &= W^{\texttt{L.shift}(\omega_c)}(\omega_p) && \text{for} \quad \omega_p \in [\omega_c - \omega_o, \omega_c + \omega_o], \\ &= W(\omega_p + \omega_c) && \text{for} \quad \omega_p \in [\omega_c - \omega_o, \omega_c + \omega_o], \\ &= P(\omega_p - \omega_c) && \text{for} \quad \omega_p \in [\omega_c - \omega_o, \omega_c + \omega_o]. \end{aligned}$$

Time Shifted Passband Pulse Echo Signals

We now use shift and carrier related properties of the Fourier transform to analyze time shifted passband pulse signals [23]. A passband pulse waveform with a right time shift of T_c is algebraically expressed as $w_{\Pi(T_p)}(t - T_c)$. The scripted notation for a right shifted passband pulse waveform is

$$w_{\Pi(T_p)}^{\texttt{R.shift}(T_c)}(t) \quad = \quad w_{\Pi(T_p)}(t - T_c). \tag{5.16}$$

The Fourier transform pair of a time shifted passband pulse signal is

$$w_{\Pi(T_p)}^{\texttt{R.shift}(T_c)}(t) \;{}_t\!\Leftrightarrow_{\omega_p}\; W_{\Pi(2\omega_o @\omega_c)}^{\texttt{carrier}(-T_c)}(\omega_p),$$

where

$$W_{\Pi(2\omega_o @\omega_c)}^{\texttt{carrier}(-T_c)}(\omega_p) = W_{\Pi(2\omega_o @\omega_c)}(\omega_p) \cdot \exp(-jT_c\omega_p). \tag{5.17}$$

The shift in time (5.16) injects a complex sinusoid (5.17) into the passband phase history. Hence, a time shifted passband signal has a carrier resident in both the time domain (to take the baseband pulse to passband) and a carrier in the frequency domain (induced by the shift in time).

Furthermore, the spectrum of a time shifted passband pulse is described in terms of the up shifted baseband pulse spectrum by

$$\begin{aligned} W^{\text{carrier}(-T_c)}_{\Pi(2\omega_o @ \omega_c)}(\omega_p) &= W_{\Pi(2\omega_o @ \omega_c)}(\omega_p) \cdot \exp(-jT_c\omega_p), \\ &= P^{\text{R.shift}(\omega_c)}_{\Pi(2\omega_o)}(\omega_b) \cdot \exp(-jT_c\omega_p), \\ &= P_{\Pi(2\omega_o)}(\omega_p - \omega_c) \cdot \exp(-jT_c\omega_p), \end{aligned}$$

for $\omega_p \in [\omega_c - \omega_o, \omega_c + \omega_o]$.

The time delayed passband signal $w^{\text{R.shift}(T_c)}_{\Pi(T_p)}(t)$ can likewise be expressed in terms of the baseband pulse signal,

$$\begin{aligned} w^{\text{R.shift}(T_c)}_{\Pi(T_p)}(t) &= w_{\Pi(T_p)}(t - T_c), \\ &= \exp[j\omega_c(t - T_c)] \times p_{\Pi(T_p)}(t - T_c). \end{aligned} \tag{5.18}$$

Both the up conversion carrier $\exp[j\omega_c(t-T_c)]$ and baseband pulse $p_{\Pi(T_p)}(t-T_c)$ are time shifted to the right by T_c in (5.18).

Armed with these Fourier analysis tools descriptive of shifts and carriers applied to pulse waveforms, we return to our analysis of the spectrum of the received SAR echo signal.

Pulse Echo Spectrum Decomposition

The Fourier transform of (5.12) gives the spectrum of the passband echo signal,

$$\begin{aligned} S_{\Pi(2\omega_o @ \omega_c)}(\omega_p) &= \mathsf{FT}^t_{1\text{D}}\left\{ s_{\Pi[T_s,T_f]}(t) \right\}, \\ &= \mathsf{FT}^t_{1\text{D}}\left\{ w_{\Pi(T_p)}(t) * \left[\sum_n g_n \cdot \delta(t - t_n) \right] \right\}, \\ &= W_{\Pi(2\omega_o @ \omega_c)}(\omega_p) \cdot \sum_n g_n \cdot \exp\left(-j\; t_n\; \omega_p\right), \\ &= W_{\Pi(2\omega_o @ \omega_c)}(\omega_p) \cdot \sum_n g_n \cdot \exp\left(-j\; \frac{2}{c} x_n\; \omega_p\right). \end{aligned} \tag{5.19}$$

Next, recall the model (5.11) of the one dimensional down-range ground scene,

$$r_{\Pi(2X_o @ X_c)}(x) = \sum_n g_n \cdot \delta(x - x_n).$$

The spectrum of the ground scene in k_x wavenumber space is

$$R_{\Pi(\infty)}(k_x) = \sum_n g_n \cdot \exp(-j \cdot x_n \cdot k_x). \tag{5.20}$$

Note the similarity of the phase element in (5.20) to the phase element

$$\sum_n g_n \cdot \exp\left(-j \cdot x_n \cdot \frac{2}{c}\omega_p\right)$$

contained within the signal spectrum decomposition of (5.19). If we define

$$k_x(\omega_p) = 2k_{\omega_p},$$

where

$$k_{\omega_p} = \frac{\omega_p}{c},$$

and substitute $k_x(\omega_p)$ for k_x in (5.20), we can write

$$R_{\Pi(\infty)}\left[k_x(\omega_p)\right] = \sum_n g_n \cdot \exp\left[-j \cdot x_n \cdot k_x(\omega_p)\right].$$

Hence, we can express (5.19) as a decomposition in the passband ω_p domain,

$$S_{\Pi(2\omega_o @ \omega_c)}(\omega_p) = W_{\Pi(2\omega_o @ \omega_c)}(\omega_p) \cdot R_{\Pi(\infty)}\left[k_x(\omega_p)\right]. \tag{5.21}$$

The temporal angular frequency spectrum (5.21) of the received signal is the terrain spectrum $R_{\Pi(\infty)}\left[k_x(\omega_p)\right]$ windowed and weighted by the passband spectrum $W_{\Pi(2\omega_o @ \omega_c)}(\omega_p)$ of the transmitted pulse waveform.

Passband Extract of the Remotely Sensed Terrain Spectrum

The received passband echo signal spectrum occupies the angular frequency passband,

$$[\omega_p^{\min}, \omega_p^{\max}] = [\omega_c - \omega_o,\ \omega_c + \omega_o], \tag{5.22}$$

where

$$\omega_p^{\min} = \omega_c - \omega_o,$$
$$\omega_p^{\max} = \omega_c + \omega_o.$$

We consider $W_{\Pi(2\omega_o @ \omega_c)}(\omega_p)$ of (5.21) to be a weight function centered at ω_c that extracts (windows) and amplitude modulates (weights) a portion of

$$R_{\Pi(\infty)}\left[k_x(\omega_p)\right]$$

over the passband (5.22).

Remotely Sensed Terrain Spectra

We state a somewhat obvious, important, but sometimes forgotten rule:

The portion of the terrain's (spatial wavenumber) spectrum sensed by the SAR system is only within a range defined by the radar signal's (temporal wavenumber) passband. There are defined relationships between the sensor's temporal (angular frequency) wavenumber and the scene's spatial wavenumber parameters sensed by the remote sensor.

This statement has certain implications that must be remembered when we deal with sensor systems that down convert passband data as part of the process of image formation, especially if complex pixel data are desired for computational exploitation (details to be covered in Section 5.4).

Passband Terrain Spectrum

The weight function defined by the passband pulse spectrum $W_{\Pi(2\omega_o @ \omega_c)}(\omega_p)$ applied to $R_{\Pi(\infty)}[k_x(\omega_p)]$ yields a segment of the passband ground scene spectrum,

$$\begin{aligned} R_{\widehat{\Pi}(2k_{\omega_o} @ k_{\omega_c})}[k_x(\omega_p)] &= W_{\Pi(2\omega_o @ \omega_c)}(\omega_p) \cdot R_{\Pi(\infty)}[k_x(\omega_p)], \\ &= W_{\Pi(2\omega_o @ \omega_c)}(\omega_c + \omega_b) \cdot R_{\Pi(\infty)}[k_x(\omega_c + \omega_b)], \quad (5.23) \end{aligned}$$

for

$$\omega_b \in [-\omega_o, +\omega_o].$$

The infinitely broadband range profile spectrum $R_{\Pi(\infty)}[k_x(\omega_p)]$ has been windowed (passband filtered) to a passband of $2k_{\omega_o}$ centered at k_{ω_c} and is weighted by $W_{\Pi(2\omega_o @ \omega_c)}(\omega_p)$, where

$$\begin{aligned} k_{\omega_o} &= \frac{\omega_o}{c}, \\ k_{\omega_c} &= \frac{\omega_c}{c}. \end{aligned}$$

We express all of these signal characteristics in the scripted notation of

$$R_{\widehat{\Pi}(2k_{\omega_o} @ k_{\omega_c})}[k_x(\omega_p)]. \quad (5.24)$$

The weighting of $R(\cdot)$ by $W(\cdot)$ is indicated in the subscript $\widehat{\Pi}(2k_{\omega_o} @ k_{\omega_c})$ of (5.24) with the addition of the caret placed on the window function $\Pi(\cdot)$, i.e., $\widehat{\Pi}(\cdot)$.

Down Converted Terrain Spectrum

Received signal down conversion is performed by time domain multiplication of the passband echo signal $s_{\Pi[T_s,T_f]}(t)$ with the complex sinusoid $\exp(-j\omega_c t)$,

$$s^{\text{carrier}(-\omega_c)}_{\Pi[T_s,T_f]}(t) = \exp(-j\omega_c t) \times \left\{ s_{\Pi[T_s,T_f]}(t) \right\}.$$

The induced result in the frequency domain is

$$\begin{aligned} & S^{\text{L.shift}(\omega_c)}_{\Pi(2\omega_o @ \omega_c)}(\omega_p) \\ &= \mathsf{FT}^t_{1\text{D}} \left\{ s^{\text{carrier}(-\omega_c)}_{\Pi[T_s,T_f]}(t) \right\}, \\ &= \mathsf{FT}^t_{1\text{D}} \left\{ \exp(-j\omega_c t) \times \left[w_{\Pi(T_p)}(t) * r_{\Pi[T_s,T_f]}(t) \right] \right\}, \\ &= \mathsf{FT}^t_{1\text{D}} \left\{ \exp(-j\omega_c t) \times \sum_n \exp[+j\omega_c(t - t_n)] \cdot p_{\Pi(T_p)}(t - t_n) \right\}, \\ &= \mathsf{FT}^t_{1\text{D}} \left\{ \sum_n \exp(-j\omega_c t_n) \cdot p_{\Pi(T_p)}(t - t_n) \right\}. \quad (5.25) \end{aligned}$$

Hence,

$$
\begin{aligned}
S^{\text{L.shift}(\omega_c)}_{\Pi(2\omega_o @ \omega_c)}(\omega_p) &= \sum_n \exp(-j\omega_c t_n) \cdot \exp(-j\omega_b t_n) \cdot P_{\Pi(2\omega_o @ 0)}(\omega_b), \\
&= P_{\Pi(2\omega_o @ 0)}(\omega_b) \cdot R_{\Pi(\infty)}\left[k_{\omega_c} + k_x(\omega_b)\right], \\
&= P_{\Pi(2\omega_o @ 0)}(\omega_b) \cdot R^{\text{L.shift}(k_{\omega_c})}_{\Pi(\infty)}\left[k_x(\omega_b)\right], \\
&= R^{\text{L.shift}(k_{\omega_c})}_{\hat{\Pi}(2k_{\omega_o} @ 0)}\left[k_x(\omega_b)\right],
\end{aligned}
\tag{5.26}
$$

for $\omega_b \in [-\omega_o, +\omega_o]$. Since

$$
\begin{aligned}
R^{\text{L.shift}(k_{\omega_c})}_{\Pi(\infty)}\left[k_x(\omega_b)\right] &= R^{\text{L.shift}(k_{\omega_c})}_{\Pi(\infty)}\left[k_x(\omega_p)\right], \\
P_{\Pi(2\omega_o @ 0)}(\omega_b) &= W^{\text{L.shift}(\omega_c)}_{\Pi(2\omega_o @ \omega_c)}(\omega_p),
\end{aligned}
$$

we can also write

$$
\begin{aligned}
S^{\text{L.shift}(\omega_c)}_{\Pi(2\omega_o @ \omega_c)}(\omega_p) &= W^{\text{L.shift}(\omega_c)}_{\Pi(2\omega_o @ \omega_c)}(\omega_p) \cdot R^{\text{L.shift}(k_{\omega_c})}_{\Pi(\infty)}\left[k_x(\omega_p)\right], \\
&= R^{\text{L.shift}(k_{\omega_c})}_{\hat{\Pi}(2k_{\omega_o} @ k_{\omega_c})}\left[k_x(\omega_p)\right],
\end{aligned}
\tag{5.27}
$$

for $\omega_p \in [\omega_c - \omega_o, \omega_c + \omega_o]$. The remotely sensed terrain spectrum of (5.27) is down converted from the original passband location centered at k_{ω_c} and resides at baseband centered about zero.

Baseband Extracted Ground Scene Spectrum at Baseband

We next compare (5.26) with yet another (conceptual) possibility, a baseband portion of the terrain spectrum described by

$$
R_{\hat{\Pi}(2k_{\omega_o} @ 0)}\left[k_x(\omega_b)\right] = P_{\Pi(2\omega_o @ 0)}(\omega_b) \cdot R_{\Pi(\infty)}\left[k_x(\omega_b)\right]. \tag{5.28}
$$

The pulse spectrum $P_{\Pi(2\omega_o @ 0)}(\omega_b)$ in (5.28) is a baseband window that extracts a baseband portion of the terrain spectrum $R_{\Pi(\infty)}\left[k_x(\omega_b)\right]$. Although the spectra of both (5.26) and (5.28) reside at baseband, the ground scene related spectral information extracted by these two windowed portions of the infinite bandwidth ground scene are different. The extracted baseband terrain spectrum of (5.28) is not obtainable by a SAR sensor. Interpreting a down converted range profile signal as containing the terrain spectral information of (5.28) is incorrect.

Pixel Domain Comparisons of Passband and Baseband Imagery

To further emphasize the differences among these three types of extracted range profile spectra, we examine the complex pixel imagery obtained by inverse Fourier transforms of (5.26), (5.27) and (5.28). We shall examine their similarities and differences in the pixel domain, with and without complex to real conversion of pixel data (see Section 5.4). But first, we describe the process of pulse compression.

5.3 Pulse Compression

Now that we have a model of the received passband signal spectrum with a decomposition that exposes a constituent passband element of the range profile spectrum $R(\cdot)$, we shall use matched filtering for pulse compression to compensate the phase of $P(\cdot)$ within the down converted signal spectrum $S(\cdot)$, leaving a windowed, weighted, down converted passband range profile spectrum

$$R^{\mathrm{L.shift}(k_{\omega_c})}_{\hat{\Pi}(2k_{\omega_o}@k_{\omega_c})}[k_x(\omega_p)], \tag{5.29}$$

extracted from the infinite bandwidth terrain spectrum,

$$R_{\Pi(\infty)}[k_x(\omega_p)].$$

Once extracted, the down-range ground scene spectrum (5.29) can be inverse Fourier transformed to yield a complex pixel range profile $r(x)$. We shall see, however, that received signal demodulation leaves an undesirable phase artifact in the complex pixel domain of the down converted range profile. Awareness of the presence of this systemic phase noise is the first step towards compensation that creates clean coherent range profile signatures.

Deconvolution Based Range Profiling

We desire to extract $R(\cdot)$ from the down converted signal spectrum,

$$S^{\mathrm{L.shift}(\omega_c)}_{\Pi(2\omega_o@\omega_c)}(\omega_p) = P_{\Pi(2\omega_o@0)}(\omega_b) \cdot R^{\mathrm{L.shift}(k_{\omega_c})}_{\Pi(\infty)}[k_x(\omega_b)]. \tag{5.30}$$

One way to do this is by division by the baseband pulse spectrum [493, 885],

$$R^{\mathrm{L.shift}(k_{\omega_c})}_{\Pi(2k_{\omega_o}@0)}[k_x(\omega_b)] = \frac{P_{\Pi(2\omega_o@0)}(\omega_b) \cdot R^{\mathrm{L.shift}(k_{\omega_c})}_{\Pi(\infty)}[k_x(\omega_b)]}{P_{\Pi(2\omega_o@0)}(\omega_b)}.$$

Although division based deconvolution in the frequency domain compensates both unwanted phase and amplitude modulations caused by the pulse spectrum weight function

$$P_{\Pi(2\omega_o@0)}(\omega_b),$$

this approach is computationally unstable because of the possibility of divide by zero faults at spectral nulls or near band edges. Division by the pulse spectrum is an example of idealized but unstable inverse filtering[2].

Matched Filter Based Range Profiling

A numerically stable approach [841] that removes the unwanted phase factor

$$\measuredangle P_{\Pi(2\omega_o@0)}(\omega_b) = \exp[j\angle P_{\Pi(2\omega_o@0)}(\omega_b)]$$

[2]Inverse filtering, if performed, must be modified to achieve stability. Weiner[3] filtering is one approach for dealing with the presence of noise and finite bandwidth effects in inverse filtering [713].

from (5.30) is provided by *matched filtering* [201, 203]. Matched filtering maximizes the signal to noise ratio in white noise environments [202], but leaves an embedded residual amplitude modulation element $|P_{\Pi(2\omega_o @0)}(\omega_b)|^2$.

To describe the process of matched filtering, (5.30) can be expanded as

$$S^{\text{L.shift}(\omega_c)}_{\Pi(2\omega_o @\omega_c)}(\omega_p) = |P_{\Pi(2\omega_o @0)}(\omega_b)| \cdot \exp[j\angle P_{\Pi(2\omega_o @0)}(\omega_b)] \cdot R^{\text{L.shift}(k_{\omega_c})}_{\Pi(\infty)}[k(\omega_b)]. \quad (5.31)$$

The phase factor $\exp[j\angle P_{\Pi(2\omega_o @0)}(\omega_b)]$ of the pulse signal spectrum provides no information about scatterer echo strength or scatterer down-range location x_n. Matched filtering performed by multiplication of (5.31) in the frequency domain with the conjugate phase function $P^*_{\Pi(2\omega_o @0)}(\omega_b)$ gives

$$\begin{aligned} R^{\text{L.shift}(\omega_c)}_{\widehat{\Pi}(2\omega_o @0)}[k_x(\omega_b)] &= P^*_{\Pi(2\omega_o @0)}(\omega_b) \times \left\{ S^{\text{L.shift}(\omega_c)}_{\Pi(2\omega_o @\omega_c)}(\omega_p) \right\}, \\ &= |P_{\Pi(2\omega_o @0)}(\omega_b)|^2 \cdot R^{\text{L.shift}(\omega_c)}_{\Pi(\infty)}[k_x(\omega_b)]. \quad (5.32) \end{aligned}$$

The weight function in the subscript of $\widehat{\Pi}(\cdot)$ of $R^{\text{L.shift}(\omega_c)}_{\widehat{\Pi}(2k_{\omega_o} @0)}(\cdot)$ is now determined by the modulus squared of the pulse spectrum,

$$|P_{\Pi(2\omega_o @0)}(\omega_b)|^2 \mapsto \widehat{\Pi}(2k_{\omega_o} @0).$$

The range compressed spectrum (5.32) contains a down converted portion of the passband range profile spectrum

$$R^{\text{L.shift}(\omega_c)}_{\Pi(\infty)}[k_x(\omega_b)] = R^{\text{L.shift}(\omega_c)}_{\Pi(\infty)}[k_x(\omega_p)]$$

windowed and weighted by

$$|P_{\Pi(2\omega_o @0)}(\omega_b)|^2.$$

The inverse Fourier transform of (5.32) yields a complex down converted down-range image (range profile).

5.4 Pulse Compressed Range Profile

We next explore properties of complex range profiles obtained from down converted data output by passband sensors. Passband sensors that perform down conversion of received signal data are sometimes called *frequency offset* [681, 681] or *offset Fourier* sensors [440].

Systemic phase noise can exist in the imagery produced from passband data that has been down converted to baseband. We identify a simple compensation method for such phase artifacts.

Three Forms of Range Profile Spectra

Various forms of complex down-range imagery can be obtained from the three different signal types discussed previously in Section 5.2:

Passband:

$$R_{\widehat{\Pi}(2k_{\omega_o} @k_{\omega_c})}[k_x(\omega_p)] = \left|W_{\Pi(2\omega_o @\omega_c)}(\omega_p)\right|^2 \cdot R_{\Pi(\infty)}[k_x(\omega_p)] \quad (5.33)$$

Down Converted:

$$
\begin{aligned}
R^{\text{L.shift}(k_{\omega_c})}_{\widehat{\Pi}(2k_{\omega_o}@k_{\omega_c})}[k_x(\omega_p)] &= \left|W^{\text{L.shift}(\omega_c)}_{\Pi(2\omega_o@\omega_c)}(\omega_p)\right|^2 \cdot R^{\text{L.shift}(\omega_c)}_{\Pi(\infty)}[k_x(\omega_p)] \qquad (5.34)\\
&= \left|P_{\Pi(2\omega_o@0)}(\omega_b)\right|^2 \cdot R^{\text{L.shift}(\omega_c)}_{\Pi(\infty)}[k_x(\omega_b)]
\end{aligned}
$$

Baseband:

$$
R_{\widehat{\Pi}(2k_{\omega_o}@0)}[k_x(\omega_b)] = \left|P_{\Pi(2\omega_o@0)}(\omega_b)\right|^2 \cdot R_{\Pi(\infty)}[k_x(\omega_b)] \qquad (5.35)
$$

These are all available to yield range compressed imagery by inverse Fourier transformation. The baseband spectrum (5.35), however, is not measurable by a realizable SAR sensor, but is included in the comparison for the sake of completeness.

Detected Imagery

Fourier inversion of range profile spectra followed by complex to real pixel conversion yields *detected* down-range imagery. "Detection processing" is terminology used in the radar community to describe the conversion of complex samples to real samples. We shall see that Fourier inversion and magnitude detection of the passband range spectrum (5.33) and the down converted spectrum (5.34) produce the same detected range profiles. The detected version of the range profile produced by the baseband spectrum (5.35) is different. The Fourier inverted complex versions of (5.33), (5.34) and (5.35) are all distinct. When algorithms (e.g., autofocus, interferometry, moving object detection, etc.) are developed to exploit SAR imagery in complex pixel format, awareness of these differences is important.

Scaling Property of Fourier Transform

Before proceeding, we review the scaling property of the Fourier transform. This is useful because of conversions required between the temporal and spatial domains. In particular, the $c/2$ scaling between time and space variables

$$
x = \frac{c}{2} \times t, \qquad (5.36)
$$

and the $2/c$ scaling between temporal angular frequency and spatial wavenumber

$$
k_x = \frac{2}{c} \times \omega,
$$

motivate our review of this topic.

Fourier Transforms of Time Scaled Expressions

A Fourier transform applied to a temporal signal creates a mapping between the time and angular frequency domains,

$$
s(t) \quad {}_t\Leftrightarrow_\omega \quad S(\omega).
$$

Scaling of the time parameter t by $c/2$ induces a mapping via (5.36) of the temporal variable t to the spatial domain range variable x. The Fourier transform pair of a temporal function with a $c/2$ scaled time variable conforms to the rule,

$$s\left(\frac{c}{2}t\right) \quad {}_t\Leftrightarrow_\omega \quad \left|\frac{2}{c}\right| \cdot S\left(\frac{2}{c}\omega\right). \tag{5.37}$$

Likewise, given the range profile Fourier transform pair,

$$r(x) \quad {}_x\Leftrightarrow_{k_x} \quad R(k_x),$$

scaling of x by $2/c$ produces a Fourier transform pair that conforms to the rule,

$$r\left(\frac{2}{c}x\right) \quad {}_x\Leftrightarrow_{k_x} \quad \left|\frac{c}{2}\right| \cdot R\left(\frac{c}{2}k_x\right). \tag{5.38}$$

The Fourier transform pairs (5.37) and (5.38), each with scaled parameters, allow us to relate the point spread function of range profiles computed in terms of temporal domain variables with those computed in terms of spatial domain variables.

Point Spread Function in Temporal and Spatial Domains

Consider the Fourier transform pairs of a point spread function defined in the temporal domain

$$\mathsf{psf}_t\,(t) \quad {}_t\Leftrightarrow_\omega \quad \mathsf{PSF}_\omega(\omega)$$

and the point spread function defined in the spatial domain

$$\mathsf{psf}_x\,(x) \quad {}_x\Leftrightarrow_{k_x} \quad \mathsf{PSF}_{k_x}(k_x).$$

$\mathsf{PSF}_\omega(\omega)$ and $\mathsf{PSF}_{k_x}(k_x)$ are the spectrums of the temporal and spatial domain pulse spread functions, respectively. Using $\mathsf{PSF}_\omega(\omega)$ and $\mathsf{PSF}_{k_x}(k_x)$, we seek relationships between the temporal and spatial point spread functions $\mathsf{psf}_t\,(t)$ and $\mathsf{psf}_x\,(x)$.

Relationships of Temporal and Spatial Point Spread Function Spectra

Invoking the principle of conservation of energy, the point spread function spectrums $\mathsf{PSF}_\omega(\omega)$ and $\mathsf{PSF}_{k_x}\,(k_x)$ are assumed equal (without scaling) with direct substitution of $\omega = \frac{c}{2}k_x$ or $k_x = \frac{2}{c}\omega$, as appropriate. Hence,

$$\mathsf{PSF}_\omega(\omega) = |P(\omega)|^2 = \mathsf{PSF}_{k_x}\left(\frac{c}{2}k_x\right) \tag{5.39}$$

holds true, and likewise,

$$\mathsf{PSF}_\omega\left(\frac{2}{c}\omega\right) = |P\,(k_x)|^2 = \mathsf{PSF}_{k_x}\,(k_x) \tag{5.40}$$

holds true.

Relationships of Temporal and Spatial Point Spread Functions

The Fourier relationships between the spatial point spread function $\mathsf{psf}_x(\cdot)$ with scaled spatial parameter $\frac{2}{c}x$ and the temporal point spread function $\mathsf{psf}_t(\cdot)$ with unscaled temporal parameter t are

$$\begin{array}{lclcl} \mathsf{psf}_t\,(t) & {}_{t} \Leftrightarrow {}_{\omega} & \mathsf{PSF}_{\omega}\,(\omega) & & \\ & & \| & & \\ & & \mathsf{PSF}_{k_x}\left(\frac{c}{2}k_x\right) & {}_{k_x} \Leftrightarrow {}_{x} & \left|\frac{2}{c}\right| \cdot \mathsf{psf}_x\left(\frac{2}{c}x\right), \end{array}$$

where $\|$ expresses the energy conservation equality (5.39). Likewise, the sequence of Fourier relationships between the temporal point spread function $\mathsf{psf}_t(\cdot)$ with scaled temporal parameter $\frac{c}{2}t$ and the spatial point spread function $\mathsf{psf}_x(\cdot)$ with unscaled spatial parameter x are

$$\begin{array}{lclcl} \mathsf{psf}_x\,(x) & {}_{x} \Leftrightarrow {}_{k_x} & \mathsf{PSF}_{k_x}\,(k_x) & & \\ & & \| & & \\ & & \mathsf{PSF}_{\omega}\left(\frac{2}{c}\omega\right) & {}_{\omega} \Leftrightarrow {}_{t} & \left|\frac{c}{2}\right| \cdot \mathsf{psf}_t\left(\frac{c}{2}t\right), \end{array}$$

where $\|$ expresses the energy conservation equality (5.40). Hence, given the point spread function in either the spatial or temporal domain, the point spread function expressed in the other domain is

$$\begin{aligned} \mathsf{psf}_t\,(t) &= \left|\frac{2}{c}\right| \cdot \mathsf{psf}_x\left(\frac{2}{c}x\right), \\ \mathsf{psf}_x\,(x) &= \left|\frac{c}{2}\right| \cdot \mathsf{psf}_t\left(\frac{c}{2}t\right). \end{aligned} \tag{5.41}$$

Coherent Range Profiles

We use the point spread function relationships (5.41) to explore the inverse Fourier transforms of (5.33), (5.34) and (5.35) in temporal and spatial domains.

Passband Receive Window Profile

By the temporal inverse Fourier transform of (5.33), the passband matched filtered range profile (without down conversion) is

$$\begin{aligned} & \mathsf{iFT}_{1\mathrm{D}}^{\omega_p}\left\{R_{\hat{\Pi}(2k_{\omega_o} @ k_{\omega_c})}\left[k_x(\omega_p)\right]\right\} \\ &= \mathsf{iFT}_{1\mathrm{D}}^{\omega_p}\left\{\left|W_{\Pi(2\omega_o @ \omega_c)}(\omega_p)\right|^2 \cdot R_{\Pi(\infty)}\left[k_x(\omega_p)\right]\right\} \\ &= \sum_n g_n \cdot \mathsf{psf}_t\,(t - t_n)^{\text{passband}}, \\ &= r_{\Pi[T_s, T_f]}(t)^{\text{passband}}, \end{aligned} \tag{5.42}$$

where

$$\mathsf{psf}_t(t)^{\text{passband}} = \mathsf{iFT}_{1\mathrm{D}}^{\omega_p}\left\{\left|W_{\Pi(2\omega_o @ \omega_c)}(\omega_p)\right|^2\right\}.$$

Passband Range Swath Profile

Next, to move from the time domain result (5.42) to a spatial domain range profile. We scale time and $\mathsf{psf}_t(t)$ values as appropriate in

$$\begin{aligned} r_{\Pi(2X_o @ X_c)}(x)^{\text{passband}} &= \left|\frac{c}{2}\right| \cdot r_{\Pi[T_s,T_f]}\left(\frac{c}{2}t\right)^{\text{passband}} , \\ &= \sum_n g_n \cdot \left|\frac{c}{2}\right| \cdot \mathsf{psf}_t\left[\frac{c}{2}(t-t_n)\right]^{\text{passband}} , \\ &= \sum_n g_n \cdot \mathsf{psf}_x(x-x_n)^{\text{passband}} . \end{aligned}$$

Down Converted Receive Window Profile

The temporal inverse Fourier transform of the down converted matched filtered range profile data (5.34) is

$$\begin{aligned} &\mathsf{iFT}_{1\text{D}}^{\omega_p}\left\{R_{\hat{\Pi}(2k_{\omega_o} @ k_{\omega_c})}^{\text{L.shift}(k_{\omega_c})}[k_x(\omega_p)]\right\} \\ &= \mathsf{iFT}_{1\text{D}}^{\omega_p}\left\{\left|W_{\Pi(2\omega_o @ \omega_c)}^{\text{L.shift}(\omega_c)}(\omega_p)\right|^2 \cdot R_{\Pi(\infty)}^{\text{L.shift}(k_{\omega_c})}[k_x(\omega_p)]\right\} \\ &= \sum_n g_n \cdot \mathsf{psf}_t^{\text{carrier}(-\omega_c)}(t-t_n)^{\text{passband}} , \\ &= \exp(-j\omega_c t) \cdot \sum_n g_n \cdot \mathsf{psf}_t(t-t_n)^{\text{passband}} , \\ &= r_{\Pi[T_s,T_f]}^{\text{carrier}(-\omega_c)}(t)^{\text{passband}} . \end{aligned}$$

The ω_p frequency domain left shift caused by baseband conversion has injected a carrier

$$\exp(-j\omega_c t) = \exp\left(-j \cdot \frac{2}{c}\omega_c \cdot \frac{c}{2}t\right) = \exp\left(-jk_{\omega_c}x\right) \tag{5.43}$$

into the complex range profile.

Down Converted Range Swath Profile

Again, to move from the time domain to the spatial domain,

$$\begin{aligned} r_{\Pi(2X_o @ X_c)}^{\text{carrier}(-k_{\omega_c})}(x)^{\text{passband}} &= \left|\frac{c}{2}\right| \cdot r_{\Pi[T_s,T_f]}^{\text{carrier}(-\omega_c)}\left(\frac{c}{2}t\right)^{\text{passband}} , \\ &= \exp(-j\omega_c t) \cdot \sum_n g_n \cdot \left|\frac{c}{2}\right| \cdot \mathsf{psf}_t\left[\frac{c}{2}(t-t_n)\right]^{\text{passband}} , \\ &= \exp(-jk_{\omega_c}x) \cdot \sum_n g_n \cdot \mathsf{psf}_x(x-x_n)^{\text{passband}} . \end{aligned}$$

Hence, the range profile formed from a down converted passband signal is

$$\begin{aligned} r_{\Pi(2X_o @ X_c)}(x)^{\text{down converted}} &= r_{\Pi(2X_o @ X_c)}^{\text{carrier}(-k_{\omega_c})}(x)^{\text{passband}} , \\ &= \exp(-jk_{\omega_c}x) \cdot r_{\Pi(2X_o @ X_c)}(x)^{\text{passband}} . \end{aligned}$$

The down converted range profile contains an extraneous carrier $\exp(-jk_{\omega_c}x)$ in the complex pixel range profile that is not present in the passband range profile.

Frequency Offset Carrier Removal

The down converted range profile has the embedded complex carrier (5.43) due to the left frequency shift that occurred when down conversion was performed on the passband echo signal. This systemic phase "noise" is not a problem if the complex range profile is detected (converted from complex to real) before visual exploitation [721]. But if a computer based exploitation algorithm expects clean coherent (complex data) range profiles, then leaving this phase noise uncompensated is problematic.

The fix is simple enough. To obtain clean coherent data, the down conversion carrier residue is removed by "up conversion" or "carrier reinsertion" in the range compressed pixel domain after pulse compression of down converted data is complete [424, 493]. This is done with the phase adjustment,

$$r_{\Pi(2X_o@X_c)}(x)^{\text{passband}} = \exp\left(+jk_{\omega_c}x\right) \times \left\{ r^{\text{carrier}(-k_{\omega_c})}_{\Pi(2X_o@X_c)}(x)^{\text{down converted}} \right\}.$$

This phase multiply in the complex pixel domain compensates for the frequency offset caused by baseband conversion.

Baseband Range Swath Profile

The inverse Fourier transform of a portion of terrain spectrum centered at baseband (5.35) is

$$\begin{aligned} \mathsf{iFT}^{\omega_b}_{1\text{D}}\left\{R_{\hat{\Pi}(2k_{\omega_o}@0)}\left[k_x(\omega_b)\right]\right\} &= \mathsf{iFT}^{\omega_b}_{1\text{D}}\left\{\left|P_{\hat{\Pi}(2\omega_o@0)}(\omega_b)\right|^2 \cdot R_{\Pi(\infty)}\left[k_x(\omega_b)\right]\right\}, \\ &= \sum_n g_n \cdot \mathsf{psf}_t\left(t - t_n\right)^{\text{baseband}}, \\ &= r_{\Pi[T_s,T_f]}(t)^{\text{baseband}}, \end{aligned}$$

where

$$\mathsf{psf}_t(t)^{\text{baseband}} = \mathsf{iFT}^{\omega_b}_{1\text{D}}\left\{\left|P_{\hat{\Pi}(2\omega_o@0)}(\omega_b)\right|^2\right\}.$$

Coherent and Non-Coherent Sums

If scatterers are separated by at least the resolution distance as determined by $\mathsf{psf}_x(\cdot)$, then the following expression holds for detected imagery[4],

$$\left|\sum_n g_n \cdot \mathsf{psf}_x(x - x_n)^{\text{passband}}\right| \approx \sum_n |g_n| \left|\mathsf{psf}_x(x - x_n)^{\text{passband}}\right|.$$

For discrete scatterers more closely spaced than the resolution of the sensor, the detected version of a coherent sum does not necessarily equal the sum of the detections [691].

[4]There are claims that taking the modulus of complex imagery increases the wavenumber bandwidth of the image and that an over sampling factor should be employed to prevent aliasing during data collections from such effects [42].

Range Resolution

Rayleigh Resolution

The null to null or Rayleigh resolution of the range point spread function $\mathsf{psf}_x(x)$ is determined by the Heisenberg relationship,

$$\begin{aligned}\Delta_x &= \frac{2\pi}{\Pi_{(2k_{x_o})}(k_x)},\\ &= \frac{2\pi}{\frac{2}{c}2\omega_o},\\ &= \frac{c}{2B}.\end{aligned}$$

The achievable resolution is inversely proportional to pulse bandwidth. Typically, for signals with time-bandwidth products T_pB greater than about twenty, the matched filter response is well approximated by a sinc response with a peak sidelobe level at approximately −13 dB below the mainlobe peak.

3 dB Resolution

An alternative measure of the range resolution Δ_x is the width of the point spread function at the half power point below the peak of the mainlobe. For a point spread function modeled as an unapodized sinc function, this value is about

$$\Delta_x \approx 0.886 \cdot \frac{c}{2B}. \tag{5.44}$$

3 dB Resolution with Pulse Weighting

A generalized form of (5.44) that accounts for the broadening of the point spread function with application of a weighting function to reduce the peak sidelobe level [435] is

$$\Delta_x = \kappa \cdot \frac{c}{2B}.$$

For example, if a Taylor weighting that achieves nearly constant sidelobes at a level of −35 dB is used, then $\kappa = 1.2$.

Resolution and Pulse Compression

We make an observation about the terminology of "pulse compression." The physical length of a propagating electromagnetic pulse in free space can be quite long. Each microsecond of pulse duration gives a propagating signal length of about 300 meters. The effectiveness of pulse compression can be measured as the ratio of the propagating signal's length in free space compared to the length of the point spread function in the range profile after matched filtering. This creates the pulse length "compression" measure.

With pulse compression, the achievable resolution is a function of the signal's bandwidth and not its time duration. The time duration of a phase or frequency modulated pulse may be increased without degrading range resolution and thereby provide advantages in achieving high levels of transmitted signal power without increasing the peak power level of the transmitted signal.

For example, an unmodulated pulse with a pulse duration of about $T_p = 2$ nanoseconds is required to have a range resolution of about 0.3 m without pulse compression. The ability to transmit such a short pulse with enough power to meet even the most modest signal to noise requirements would be difficult. On the other hand, a modestly powered system that transmits a linearly frequency modulated chirp pulse signal that spans 430 MHz over 10 microseconds (3 km free space pulse length) would yield a range resolution of about 0.3 m after pulse compression.

The pulse compression ratio is approximately determined by the time-bandwidth product T_pB of the transmitted pulse, which for this example is

$$430 \text{ MHz } \times 10 \text{ microseconds } = 4300 \mapsto 36.3 \text{ dB}.$$

This compares to the pulse length compression factor of

$$3 \text{ km } / \ 0.3 \text{ m } = 10,000 \mapsto 40.0 \text{ dB}.$$

Pulse Echo Sampling

Now that we have a handle on range profile spatial resolution, we next describe Nyquist data rates required for signal acquisition and count the numbers of samples obtained in various fast-time data collection scenarios for down-range profiling. We overview sampling of down converted passband pulse signals, direct sampling of passband signals (at baseband sampling rates with subsequent down conversion performed in software) and sampling of time shifted pulses [41, 111, 444, 671].

Nyquist Sampling of Baseband Pulses

With the spectral support of a baseband pulse described as $\Pi_{\omega_b} = 2\omega_o$, we know from the Heisenberg uncertainty principle that the fast-time Nyquist sampling interval Δ_{t_b} for a baseband pulse signal is

$$\Delta_{t_b} = \frac{2\pi}{2\omega_o}. \tag{5.45}$$

The "b" subscript of t_b is a reminder that sampling is occurring at Nyquist rates for a baseband pulse. Nyquist sampling over the pulse duration T_p with the sample density (5.45) produces a number of time domain baseband samples N_{t_b} given by

$$N_{t_b} = \left\lceil \frac{T_p}{\Delta_{t_b}} \right\rceil.$$

The Nyquist frequency bin spacing,

$$\begin{aligned} \Delta_{\omega_b} &= \frac{2\pi}{\Pi_{t_b}}, \\ &= \frac{2\pi}{N_{t_b}\Delta_{t_b}}, \end{aligned}$$

of sampled baseband pulse data converted to the Fourier domain by a discrete Fourier transform is determined by the number of time samples in the time domain window Π_{t_b}, where

$$\Pi_{t_b} = N_{t_b} \Delta_{t_b}.$$

The number of frequency samples N_{ω_b} produced by a discrete Fourier transform is equal to the number of time samples N_{t_b}. The number of frequency samples also satisfies,

$$N_{\omega_b} = \left\lceil \frac{2\omega_o}{\Delta_{\omega_b}} \right\rceil.$$

Hence, the number of time samples and frequency bins in the discrete Fourier transform obtained in Nyquist sampling of a baseband pulse has this summary characterization,

$$\left\lceil \frac{T_p}{\Delta_{t_b}} \right\rceil = N_{t_b} = N_{\omega_b} = \left\lceil \frac{2\omega_o}{\Delta_{\omega_b}} \right\rceil.$$

Nyquist Sampling of Passband Pulses

Consider a situation where direct sampling of a passband pulse signal (without down conversion) is desired. This is a (complex data) sampling problem that, as one approach, has a Nyquist rate governed by the highest frequency $\omega_{\max}$ of the passband signal. Unaliased sampling of this passband pulse signal with the "maximum frequency rule" requires a fast-time complex sampling interval of

$$\Delta_{t_p} = \frac{2\pi}{\omega_{\max}}.$$

The "p" subscript of t_p is a reminder that sampling is occurring at Nyquist rates for a passband signal. Using this sampling interval over the passband pulse signal duration T_p produces a number of complex samples given by

$$N_{t_p}^{\text{carrier}(\omega_c)} = \left\lceil \frac{T_p}{\Delta_{t_p}} \right\rceil.$$

The "carrier" superscript is a reminder that this sampling function must be sufficiently dense to represent the passband signal with its embedded carrier signal $\exp(j\omega_c t)$. The Nyquist frequency bin spacing Δ_{ω_p} of sampled passband pulse data converted to the Fourier domain by a discrete Fourier transform is

$$\Delta_{\omega_p} = \frac{2\pi}{N_{t_p}^{\text{carrier}(\omega_c)} \Delta_{t_p}}.$$

The number of frequency samples $N_{\omega_p}^{\text{R.shift}(\omega_c)}$ resulting from the discrete Fourier transform is equal to the number of time samples,

$$N_{\omega_p}^{\text{R.shift}(\omega_c)} = N_{t_p}^{\text{carrier}(\omega_c)},$$

which also satisfies

$$N_{\omega_p}^{\text{R.shift}(\omega_c)} = \left\lceil \frac{\omega_{\max}}{\Delta_{\omega_p}} \right\rceil.$$

The "R.shift" superscript is a reminder that the number of samples must be sufficient to represent the shift that defines the location of the passband in the frequency domain.

In summary, we have the following relationships among the parameters that characterize Nyquist sampling at passband frequencies,

$$\left\lceil \frac{T_p}{\Delta_{t_p}} \right\rceil = N_{t_p}^{\text{carrier}(\omega_c)} = N_{\omega_p}^{\text{R.shift}(\omega_c)} = \left\lceil \frac{\omega_{\max}}{\Delta_{\omega_p}} \right\rceil .$$

The increased passband sampling density (maximum frequency rule) is required to sample the resident carrier $\exp(j\omega_c t)$ without aliasing. The increased sample count in the frequency domain also provides enough samples to reveal the up shift of the baseband pulse spectrum to the passband.

Nyquist Sampling of Time Shifted Baseband Pulses

The Fourier transform pair of a right time shifted baseband pulse signal is

$$p_{\Pi(T_p)}^{\text{R.shift}(T_c)}(t_b) \;{}_{t_b}\!\Leftrightarrow_{\omega_b}\; P_{\Pi(2\omega_o)}^{\text{carrier}(-T_c)}(\omega_b),$$

where

$$P_{\Pi(2\omega_o)}^{\text{carrier}(-T_c)}(\omega_b) = \exp(-jT_c\omega_b) \cdot P_{\Pi(2\omega_o)}(\omega_b).$$

The Nyquist rate for a time shifted baseband pulse signal is the same as the Nyquist rate for the unshifted version of the pulse,

$$\Delta_{t_b} = \frac{2\pi}{2\omega_o}.$$

However, since the pulse has been time shifted, the number of fast-time samples required to expose the existence of the time shift within the radar's receive window is

$$N_{t_b}^{\text{R.shift}(T_c)} = \left\lceil \frac{T_f}{\Delta_{t_b}} \right\rceil ,$$

where T_f is the final time instant of the shifted pulse.

If these samples are processed with a discrete Fourier transform, then the number of frequency samples is equal to the number of time samples,

$$N_{\omega_b}^{\text{carrier}(-T_c)} = N_{t_b}^{\text{R.shift}(T_c)},$$

where $N_{\omega_b}^{\text{carrier}(-T_c)}$ is the number of finer grain frequency bins of width $\Delta_{\omega_b}^{\text{carrier}(-T_c)}$. The frequency bin spacing produced by the discrete Fourier transform is

$$\Delta_{\omega_b}^{\text{carrier}(-T_c)} = \frac{2\pi}{N_{t_b}^{\text{R.shift}(T_c)}\Delta_t}.$$

The increased sample density is required to effectively sample the injected carrier $\exp(-jT_c\omega_b)$ in the ω_b frequency domain caused by the time shift T_c in the time domain. In summary, the relationships,

$$\left\lceil \frac{T_f}{\Delta_{t_b}} \right\rceil = N_{t_b}^{\text{R.shift}(T_c)} = N_{\omega_b}^{\text{carrier}(-T_c)} = \left\lceil \frac{2\omega_o}{\Delta_{\omega_b}^{\text{carrier}(-T_c)}} \right\rceil$$

characterize the number and density of time shifted baseband signal sampling parameters.

Nyquist Sampling of Time Shifted Passband Pulses

Consider the situation of a time shifted passband pulse. The time sample density requirement is equal to that of the unshifted passband pulse, but the number of samples must increase to express the time shift. This increased number of samples in time provides sufficient sample density after the discrete Fourier transform to reveal the injected carrier $\exp(-jT_c\omega_p)$ in the passband spectrum.

Direct Sampling of Passband Signal at Baseband Rates

If direct sampling of a passband signal at the baseband Nyquist rate is used (this approach aliases the carrier), then the sample count and bin spacing for the time shifted baseband signal holds. Modern radar systems are increasingly using direct sampling of analog passband signals where the digital conversion is at an intermediate or baseband sampling rate. These systems perform down conversion (with a similarly aliased carrier) by digital processing in software [444,671] or rely on extraction of the baseband element of the replicated spectrum of the sampled passband signal [389, 748, 751].

5.5 Range Profile Huygens-Fresnel Inversion

The Fourier transform pair that describes the down converted passband range profile is

$$\overset{\text{carrier}(-k_{\omega_c})}{r}_{\Pi[X_{\min},X_{\max}]}(x)^{\text{passband}}_{\text{ACS}} \quad \hat{x} \Leftrightarrow k_{\hat{x}} \quad \overset{\text{L.shift}(k_{\omega_c})}{R}_{\hat{\Pi}(k_{\omega_o}@k_{\omega_c})}(k_x)^{\text{passband}}_{\text{ACS}} .$$

The down converted range profile

$$\overset{\text{carrier}(-\omega_c)}{r}_{\Pi[X_{\min},X_{\max}]}(x)^{\text{passband}}_{\text{ACS}} \tag{5.46}$$

is nominally centered at a down-range distance of $X_c \in [X_{\min}, X_{\max}]$. A spatial inverse Fourier transform of the spectrum of a range profile that is offset to a range swath center position X_c is not computationally efficient. The number of samples required by the range profile (5.46) must accommodate the maximum range offset $X_{\max}$.

Shifting the data to a scene center value of zero (i.e., the center of the scene coordinate system) reduces the number of samples required to represent the range profile. This shift increases the computational efficiency of the inverse discrete Fourier transform operation. This shift to the origin of the scene coordinate system can be implemented through inverse Huygens-Fresnel transfers executed in the wavenumber domain.

Huygens-Fresnel Decomposition of Range Profile Spectrum

Using parameters from both the aperture and scene coordinate systems,

$$\begin{aligned} \exp(-jx_n k_x) &= \exp\left[-j\left(X_c + \hat{x}_x\right)k_x\right], \\ &= \exp(-jX_c k_x) \cdot \exp(-j\hat{x}_n k_x), \end{aligned}$$

and since we can equate the down-range wavenumber k_x of the aperture coordinate system with the down-range wavenumber $k_{\hat{x}}$ of the scene coordinate system, we have

$$\exp(-jx_n k_x) = \exp(-jX_c k_{\hat{x}}) \cdot \exp(-j\hat{x}_n k_{\hat{x}}).$$

The relationship of the down converted range profile spectrum expressed in the aperture coordinate system to one expressed in the scene coordinate system is

$$\begin{aligned} R^{\text{L.shift}(k_{\omega_c})}_{\hat{\Pi}(2k_{\omega_o}@k_{\omega_c})}(k_x)_{\text{ACS}} &= \exp\left(-jX_c k_{\hat{x}}\right) \cdot R^{\text{L.shift}(k_{\omega_c})}_{\hat{\Pi}(2k_{\omega_o}@k_{\omega_c})}(k_{\hat{x}})_{\text{SCS}}, \\ &= H_{X_c}(k_{\hat{x}}) \cdot R^{\text{L.shift}(k_{\omega_c})}_{\hat{\Pi}(2k_{\omega_o}@k_{\omega_c})}(k_{\hat{x}})_{\text{SCS}}, \end{aligned}$$

where

$$H_{X_c}(k_{\hat{x}}) = \exp\left(-jX_c k_{\hat{x}}\right). \tag{5.47}$$

A higher sample density is required in the aperture coordinate system to support the sampling of the resident Huygens-Fresnel transfer function (5.47) that describes (in the down-range wavenumber domain of $k_{\hat{x}}$) the ground scene location with respect to the radar.

Range Profile Range Swath Offset Removal

The removal of the range offset X_c can be accomplished in the wavenumber domain before inverse discrete Fourier transforms by multiplication with an inverse Huygens-Fresnel transfer function,

$$\begin{aligned} R^{\text{L.shift}(k_{\omega_c})}_{\hat{\Pi}(2k_{\omega_o}@k_{\omega_c})}(k_{\hat{x}})_{\text{SCS}} &= H^*_{X_c}(k_x) \times \left\{R^{\text{L.shift}(k_{\omega_c})}_{\hat{\Pi}(2k_{\omega_o}@k_{\omega_c})}(k_x)_{\text{ACS}}\right\}, \\ &= \exp\left(+jX_c k_x\right) \times \left\{R^{\text{L.shift}(k_{\omega_c})}_{\hat{\Pi}(2k_{\omega_o}@k_{\omega_c})}(k_x)_{\text{ACS}}\right\}. \end{aligned} \tag{5.48}$$

The inverse Huygens-Fresnel transfer reduces the number of complex samples required in the inverse discrete Fourier transform from the number required to span the distance to the far edge of the range swath in the aperture coordinate system to a smaller number required to span the zero centered range swath $[-X_o, +X_o]$ in the scene coordinate system,

$$r^{\text{carrier}(-\omega_c)}_{\Pi[-X_o,+X_o]}(\hat{x})^{\text{passband}}_{\text{SCS}} = \mathsf{iFT}^{k_x}_{1\text{D}}\left\{R^{\text{L.shift}(k_{\omega_c})}_{\hat{\Pi}(2k_{\omega_o}@k_{\omega_c})}(k_{\hat{x}})_{\text{SCS}}\right\}.$$

Receive Window Time Offset Removal

The coordinate system conversion of (5.48) is identical to removing the time offset of the receive window, as can be seen by

$$\exp\left[j \cdot X_c \cdot k_{\hat{x}}(\omega_p)\right] = \exp\left(j \cdot \frac{c}{2}T_c \cdot \frac{2}{c}\omega_p\right) = \exp(j \cdot T_c \cdot \omega_p),$$

where we assume the receive window has been designed such that $T_c = \frac{2}{c}X_c$. Hence, the time offset of the receive window, or equivalently, the spatial offset to profile center can be removed by

$$\begin{aligned} & R_{\hat{\Pi}(2k_{\omega_o}@k_{\omega_c})}^{\text{L.shift}(k_{\omega_c})}\left[k_{\hat{x}}(\omega_p)\right]_{\text{SCS}} \\ &= H_{X_c}^*\left[k_x(\omega_p)\right] \times \left\{R_{\hat{\Pi}(2k_{\omega_o}@k_{\omega_c})}^{\text{L.shift}(k_{\omega_c})}\left[k_x(\omega_p)\right]_{\text{ACS}}\right\}, \\ &= \exp(+jT_c\omega_p) \times \left\{R_{\hat{\Pi}(2k_{\omega_o}@k_{\omega_c})}^{\text{L.shift}(k_{\omega_c})}\left[k_x(\omega_p)\right]_{\text{ACS}}\right\}. \end{aligned} \tag{5.49}$$

The inverse Fourier transform of the offset compensated phase history gives

$$r_{\Pi[-X_o,X_o]}^{\text{carrier}(-k_{\omega_c})}(\hat{x})_{\text{SCS}}^{\text{passband}} \quad \hat{x} \Leftrightarrow k_{\hat{x}} \quad R_{\hat{\Pi}(2k_{\omega_o}@k_{\omega_c})}^{\text{L.shift}(k_{\omega_c})}(k_{\hat{x}})_{\text{SCS}}.$$

The discrete Fourier transform need have only enough samples to span a down-range scene size $2X_o$ centered at $x = 0$ after the shift of (5.48) or (5.49) from the aperture coordinate system to the scene coordinate system by inverse Huygens-Fresnel transfers.

Coordinate Shifts: Important Concept or Insignificant Detail?

Although the shift achieved by $\exp(+jX_c k_{\hat{x}})$ or by $\exp(+jT_c\omega_p)$ might seem to involve a rather trivial signal processing operation of little theoretical significance (i.e., receive window offset removal), we have interpreted this coordinate shift as related to fundamental properties of spherical wave field inversions performed in wavenumber domains. This perspective is based in the wave theories of Huygens and Fresnel as embodied in the Huygens-Fresnel transfer function. This association shall become more explicit and more obvious as our SAR signal models become more advanced.

Summary Overview of Key Concepts

The signal analysis and signal processing development of the down-range profiling method of this chapter illustrates major themes in SAR image formation approaches performed in the Fourier domain. Such algorithms seek to isolate the scene's terrain spectrum by removing non-informative phase factors from the received signal spectrum. These algorithms then shift the data to scene centered coordinate representations before producing the image with inverse Fourier transforms. This coordinate shift reduces the required FFT sizes to more manageable levels. This is part of the role of inverse Huygens-Fresnel transfer functions.

Down Conversion Carrier Removal

The other major theme is that care must be taken to obtain clean coherent data without extraneous systemic phase artifacts, in this case, a phase residual caused by down conversion of passband signals. For the case of down-range profiling, down conversion carrier artifact removal gives

$$r_{\Pi[-X_o,X_o]}(\hat{x})_{\text{SCS}}^{\text{passband}} = \exp(jk_{\omega_c}\hat{x}) \times r_{\Pi[-X_o,X_o]}^{\text{carrier}(-k_{\omega_c})}(\hat{x})_{\text{SCS}}^{\text{passband}}.$$

Chapter 6

Chirp Signals

Sinusoidal signals characterized by time or space dependent frequency modulation are *chirp signals* [269, 534]. Temporal chirp signals exist in SAR when the selected waveform is a chirp. Linear frequency modulated (LFM) chirp signals are a common choice. SAR phase histories always contain spatial chirp signals. Scatterer specific spatial frequency modulations are induced as the sensor changes position within the synthetic aperture. Scatterer specific phase curtains comprise the slow-time (spatial) chirp signals of SAR.

We introduce in this chapter a general description of chirp signals and we explore their Fourier transforms. The concept of a *chirp couplet* is used to describe the time-frequency coupling of temporal chirp waveforms. Chirp couplets will also be used in subsequent chapters to describe space-wavenumber couplings of spatial chirp waveforms (see Sections 12.5 and 13.5).

We study the LFM chirp signal in detail. We make particular note of attributes of the Fourier transform of *windowed* LFM chirp signals, i.e., LFM chirp *pulses*. The time-frequency coupling of LFM chirp pulses makes description of the Fourier transform of windowed LFM chirp signals curiously interesting in various aspects.

This chapter is outlined as follows: Section 6.1 introduces the *time-frequency chirp couplet* and demonstrates its use in describing a general class of chirp waveforms. Section 6.2 presents unwindowed LFM chirp signals described with chirp couplets. Section 6.3 analyzes the Fourier transform of the unwindowed LFM chirp signal. The dual *frequency-time couplet* of the chirp spectrum is introduced. Finally, Section 6.4 examines LFM chirp pulse signals obtained by applying a time domain window function to an infinite time duration LFM chirp phase signal. Determining attributes of the spectrum of a windowed chirp signal is given special attention.

6.1 General Chirp Waveforms

We define and set up notation for a general class of temporal chirp signals in both baseband and passband forms.

Baseband Chirp Pulses

A *baseband chirp pulse signal* can be expressed as

$$\begin{aligned} & p_{\Pi(T_p @ t_o)}(t-t_o)^{\text{chirp}} \\ &= \overline{\Pi_{[t_o-T_p/2,t_o+T_p/2]}(t-t_o)} \cdot \exp\left[j \cdot \overline{\omega_b(t-t_o)} \cdot (t-t_o)\right], \end{aligned} \tag{6.1}$$

where t_o is the chirp pulse time reference and T_p is the pulse duration. Signal elements related specifically to the defining attributes of a chirp waveform are indicated with over lines. These include the *chirp window function* centered at t_o,

$$\overline{\Pi_{[t_o-T_p/2,t_o+T_p/2]}(t-t_o)}$$

and a *chirp phase signal* with a time reference of t_o,

$$\exp\left[j \cdot \overline{\omega_b(t-t_o)} \cdot (t-t_o)\right]. \tag{6.2}$$

The chirp phase signal (6.2) of the chirp pulse signal (6.1) contains the *chirp couplet*

$$\overline{\omega_b(t-t_o)} \in \mathbb{C}_b,$$

which describes the time dependent frequency coupling of the chirp signal. The chirp couplet assumes angular frequency values within the *chirp couplet frequency span,*

$$\mathbb{C}_b = \left[\omega_b^{\text{min}}, \omega_b^{\text{max}}\right]$$

as a function of shifted time values $t-t_o$ that occur within the shifted intra-pulse period,

$$t-t_o \in \left[-\frac{T_p}{2}, \frac{T_p}{2}\right].$$

We sometimes suppress the time reference by setting $t_o = 0$ for notational simplicity or when we seek the spectral analysis of a signal with no time shift.

Time-Frequency Chirp Couplets

The time-frequency (i.e., time-maps-to-frequency) chirp couplet $\overline{\omega_b(t)}$ is a useful formalism that expresses the time dependent angular frequency element of a general baseband chirp phase signal,

$$\exp[j\phi(t)] = \exp[j \cdot \overline{\omega_b(t)} \cdot t],$$

where we require the functional form,

$$\phi(t) = \overline{\omega_b(t)} \cdot t. \tag{6.3}$$

Baseband Chirp Pulse Instantaneous Frequency

The baseband chirp pulse has an instantaneous angular frequency of

$$\omega_i(t) = \partial_t \{\phi(t)\}.$$

The instantaneous frequency sweeps through the baseband

$$\beta_b = [-\omega_o, \omega_o]$$

in some manner as determined by the chirp couplet $\overline{\omega_b(t)}$.

Instantaneous Chirp Couplet Frequency and Instantaneous Chirp Frequency

The instantaneous chirp couplet frequency $\overline{\omega_b(t)}$ is not the same as the chirp's instantaneous frequency $\omega_i(t)$. This can be seen by taking the time derivative of (6.3) to determine the chirp signal's instantaneous frequency in terms of the chirp couplet,

$$\begin{aligned} \omega_i(t) &= \partial_t \left\{ \overline{\omega_b(t)} \cdot t \right\}, \\ &= \overline{\omega_b(t)} + \partial_t \left\{ \overline{\omega_b(t)} \right\} \cdot t. \end{aligned}$$

The difference between the instantaneous frequency and the instantaneous chirp couplet frequency is $\partial_t \left\{ \overline{\omega_b(t)} \right\} \cdot t$.

Fourier Transforms of Baseband Chirp Pulse Signals

The spectrum of the baseband chirp pulse waveform $p_{\Pi(T_p)}(t)^{\text{chirp}}$ is determined by the Fourier transform,

$$P_{\Pi(2\omega_o @0)}(\omega_b)^{\text{chirp}} = \mathsf{FT}\left\{ p_{\Pi(T_p @0)}(t)^{\text{chirp}} \right\}.$$

We express the spectrum of a general chirp pulse as

$$P_{\Pi(2\omega_o @0)}(\omega_b)^{\text{chirp}} = \overline{\Pi_{(2\omega_o @0)}(\omega_b)} \cdot \exp[j \cdot \overline{t(\omega_b)} \cdot \omega_b].$$

This notation allows us to write the Fourier transform pair of a general baseband chirp pulse as

$$\overline{\Pi_{[-T_p/2,+T_p/2]}(t)} \cdot \exp[j \cdot \overline{\omega_b(t)} \cdot t] \;\; {}_t\Leftrightarrow_{\omega_b} \;\; \overline{\Pi_{(2\omega_o @0)}(\omega_b)} \cdot \exp[j \cdot \overline{t(\omega_b)} \cdot \omega_b].$$

Time-Frequency and Frequency-Time Chirp Couplets

The Fourier transform of $p_{\Pi(T_p @0)}(t)^{\text{chirp}}$ maps the baseband time-frequency chirp couplet

$$\overline{\omega_b(t)}$$

to a baseband frequency-time (i.e., frequency-maps-to-time) chirp couplet

$$\overline{t(\omega_b)}.$$

The *frequency-time chirp couplet* $\overline{t(\omega_b)}$ of the chirp pulse baseband spectrum is such that

$$\overline{t(\omega_b)} \in \mathbb{T} \text{ for } \omega_b \in \mathbb{C}_b,$$

with $\mathbb{T}$ defined as the *couplet time span.* The couplet time span $\mathbb{T}$ is the dual of the couplet frequency span $\mathbb{C}_b$.

Passband Chirp Pulse Waveforms

A *passband chirp pulse waveform* is obtained, in terms of a general time reference t_o, by up conversion of a baseband chirp pulse,

$$\begin{aligned}
& w_{\Pi(T_p@t_o)}(t-t_o)^{\text{chirp}} \\
&= \exp[j\omega_c(t-t_o)] \times \left\{p_{\Pi(T_p@t_o)}(t-t_o)^{\text{chirp}}\right\}, \\
&= \overline{\Pi_{[t_o-T_p/2,t_o+T_p/2]}(t-t_o)} \cdot \exp\left\{j \cdot [\omega_c + \overline{\omega_b(t-t_o)}] \cdot (t-t_o)\right\}, \qquad (6.4) \\
&= \overline{\Pi_{[t_o-T_p/2,t_o+T_p/2]}(t-t_o)} \cdot \exp[j \cdot \overline{\omega_p(t-t_o)} \cdot (t-t_o)].
\end{aligned}$$

The *passband chirp couplet*

$$\overline{\omega_p(t-t_o)} = \omega_c + \overline{\omega_b(t-t_o)}$$

assumes time dependent angular frequency values within the *passband couplet span,*

$$\begin{aligned}
\overline{\omega_p(t-t_o)} &\in \omega_c + \mathbb{C}_b, \\
&= \mathbb{C}_p,
\end{aligned}$$

where ω_c is the passband signal's *carrier frequency* and where

$$\begin{aligned}
t-t_o &\in \left[-\frac{T_p}{2}, \frac{T_p}{2}\right], \\
&= \mathbb{T}_o,
\end{aligned}$$

with $\mathbb{T}_o$ defined as the couplet time span centered on the time instant t_o.

Fourier Transforms of Passband Chirp Pulse Signals

Next, assume that $t_o = 0$. The spectrum of the passband chirp pulse waveform

$$w_{\Pi(T_p@0)}(t)^{\text{chirp}}$$

is determined by the Fourier transform,

$$W_{\Pi(2\omega_o@\omega_c)}(\omega_p)^{\text{chirp}} = \mathsf{FT}\left\{w_{\Pi(T_p@0)}(t)^{\text{chirp}}\right\},$$

which we express in a general chirp couplet form as

$$W_{\Pi(2\omega_o@\omega_c)}(\omega_p)^{\text{chirp}} = \overline{\Pi_{(2\omega_o@\omega_c)}(\omega_p)} \cdot \exp[j \cdot \overline{t(\omega_p)} \cdot \omega_p].$$

The passband chirp spectrum has *frequency dependent time modulation* expressed by

$$\exp[j \cdot \overline{t(\omega_p)} \cdot \omega_p],$$

where the chirp couplet's frequency dependent time instances

$$\overline{t(\omega_p)} \in \mathbb{T}$$

are determined by passband frequencies ω_p contained within

$$\omega_p \in \mathbb{C}_p.$$

6.2 LFM Chirp Waveform

We next examine linear frequency modulated chirp signals [46, 75, 534] as a specialization within the class of generalized chirp waveforms.

LFM and Quadratic Phase Modulation

Consider an LFM waveform with infinite time duration. This LFM waveform can be described in various ways,

$$\begin{aligned} p_{\Pi(\infty)}(t-t_o)^{\text{LFM}} &= \exp[j \cdot \overline{\omega_b(t-t_o)^{\text{LFM}}} \cdot (t-t_o)], \\ &= \exp[j \cdot \overline{\alpha \cdot (t-t_o)} \cdot (t-t_o)], \\ &= \exp\left[j\alpha \cdot (t-t_o)^2\right], \end{aligned} \tag{6.5}$$

where

$$\overline{\omega_b(t-t_o)^{\text{LFM}}} = \overline{\alpha \cdot (t-t_o)}.$$

We see in (6.5) that linear frequency modulation is synonymous with quadratic phase modulation[1].

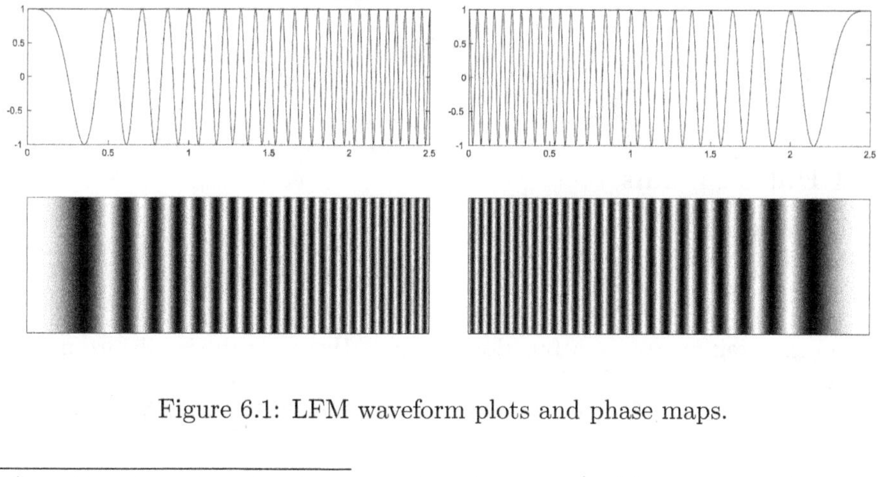

Figure 6.1: LFM waveform plots and phase maps.

[1]The LFM chirp phase function is sometimes defined as $\frac{1}{2}\alpha t^2$ [23, 349, 534, 976] instead of αt^2.

Up Chip and Down Chirp LFM Signals

If the LFM quadratic phase modulation parameter α is positive, then the LFM signal is an *up chirp LFM signal.* If negative, then a *down chirp LFM signal.*

Fig. 6.1 shows plots and phase maps of the real part

$$\cos(\omega_c + \alpha t^2)$$

of complex passband chirp signals for positive and negative α values and $t \in [0, T_p]$.

Quadratic Phase Modulation at Baseband

A plot and phase map of the quadratic phase modulation are shown in Fig. 6.2. The phase map of the quadratic phase modulated signal of Fig. 6.2 is the result of baseband conversion of the passband up chirp signal of Fig. 6.1.

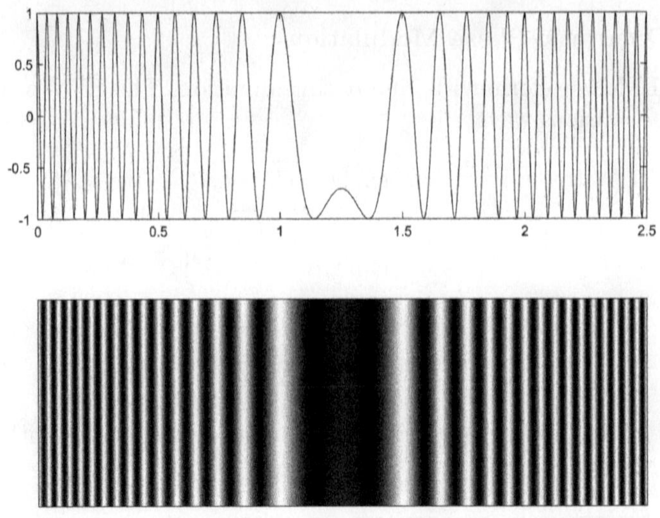

Figure 6.2: LFM baseband quadratic phase modulation.

LFM Pulse Signals

The unwindowed LFM chirp signal

$$p_{\Pi(\infty)}(t - t_o)^{\text{LFM}} \quad = \quad \exp[j \cdot \overline{\alpha \cdot (t - t_o)} \cdot (t - t_o)]$$

has no beginning or end as a function of time. Hence, such signals are not realizable and do not provide an adequate model for LFM *pulses.* The addition of a window or weight function provides a model for a pulsed LFM (baseband) waveform,

$$p_{\Pi(T_p @ t_o)}(t - t_o)^{\text{LFM}} \quad = \quad \overline{\Pi_{(T_p)}(t - t_o)} \cdot \exp[j \cdot \overline{\alpha \cdot (t - t_o)} \cdot (t - t_o)].$$

LFM Instantaneous Baseband Frequency

The instantaneous angular frequency of a baseband LFM pulse (6.5) is determined by the time derivative of the phase function [132, 133],

$$
\begin{aligned}
\omega_i(t-t_o)^{\text{LFM}} &= \partial_t\left\{\phi(t-t_o)^{\text{LFM}}\right\}, \\
&= \partial_t\left\{\overline{\alpha\cdot(t-t_o)}\cdot(t-t_o)\right\}, \\
&= \partial_t\left\{\alpha\cdot(t-t_o)^2\right\}, \\
&= 2\alpha\cdot(t-t_o). \qquad (6.6)
\end{aligned}
$$

If $t_o = 0$, then

$$
\begin{aligned}
\omega_i(t)^{\text{LFM}} &= \partial_t\left\{\overline{\omega_b(t)^{\text{LFM}}}\cdot t\right\}, \\
&= \overline{\omega_b(t)^{\text{LFM}}} + \partial_t\left\{\overline{\omega_b(t)^{\text{LFM}}}\right\}\cdot t, \\
&= \overline{\alpha\cdot t} + \partial_t\left\{\overline{\alpha\cdot t}\right\}\cdot t, \\
&= 2\alpha t.
\end{aligned}
$$

LFM Instantaneous Baseband Frequency and Couplet Frequency

Since, for LFM pulses, the baseband LFM time-frequency couplet is

$$
\overline{\omega_b(t-t_o)^{\text{LFM}}} = \overline{\alpha\cdot(t-t_o)},
$$

the instantaneous baseband frequency (6.6) of the LFM pulse is twice the instantaneous baseband couplet value,

$$
\begin{aligned}
\omega_i(t-t_o)^{\text{LFM}} &= 2\cdot\overline{\omega_b(t-t_o)^{\text{LFM}}}, \\
&= 2\alpha\cdot(t-t_o).
\end{aligned}
$$

The instantaneous baseband frequency of the LFM chirp sweeps through

$$
\omega_i(t-t_o)^{\text{LFM}} \in \left[-\alpha T_p, +\alpha T_p\right],
$$

while the LFM chirp couplet only sweeps through the half-band,

$$
\overline{\omega_b(t-t_o)^{\text{LFM}}} \in \left[-\alpha\frac{T_p}{2}, +\alpha\frac{T_p}{2}\right].
$$

Passband LFM Chirp Signals

The *passband chirp couplet* of a passband LFM chirp signal is

$$
\overline{\omega_p(t-t_o)^{\text{LFM}}} = \omega_{\text{mid}} + \overline{\alpha\cdot(t-t_o)}.
$$

The instantaneous LFM passband frequency is the time derivative of the passband phase function,

$$
\begin{aligned}
\partial_t\left\{\phi(t-t_o)^{\text{LFM}}\right\} &= \omega_{\text{mid}} + \omega_i(t-t_o), \\
&= \omega_{\text{mid}} + 2\alpha\cdot(t-t_o),
\end{aligned}
$$

where the carrier frequency is $\omega_c = \omega_{\text{mid}}$. In this instance, the center frequency of the passband pulse is viewed as the carrier frequency.

LFM Chirp Signal Bandwidth

A passband LFM pulse with time duration of

$$[t_o - T_p/2, t_o + T_p/2]$$

has a passband given by the frequency interval between $\omega_{\text{mid}} - \alpha T_p$ and $\omega_{\text{mid}} + \alpha T_p$. Hence, the passband LFM pulse has an angular frequency bandwidth of

$$\beta_{\text{LFM}} = 2\alpha T_p.$$

The angular frequency bandwidth of an LFM chirp signal is often characterized in terms of the half-bandwidth parameter ω_o [276],

$$\beta_{\text{LFM}} = 2\omega_o,$$

where

$$\omega_o = \alpha T_p.$$

The cyclic frequency bandwidth is

$$B_{\text{LFM}} = \frac{1}{2\pi}\beta_{\text{LFM}}.$$

LFM Chirp Rate

The slope of the instantaneous cyclic frequency of an LFM signal is determined by the parameter,

$$\gamma = \frac{B_{\text{LFM}}}{T_p},$$

which has units of hertz per second. The parameter γ is called the *chirp rate.*

LFM Chirp Rate and Instantaneous Frequency

The LFM chirp rate is the rate of change of the instantaneous cyclic frequency,

$$\begin{aligned}
\gamma &= \partial_t \left\{ f_i(t)^{\text{LFM}} \right\}, \\
&= \frac{1}{2\pi}\partial_t^2 \left\{ \phi(t)^{\text{LFM}} \right\}, \\
&= \frac{1}{2\pi}\partial_t^2 \left\{ \alpha t^2 \right\}, \\
&= \frac{\alpha}{\pi}.
\end{aligned} \tag{6.7}$$

An LFM signal has a fixed chirp rate. This is not case for non-linear frequency modulated waveforms [276, 278, 279].

LFM Cyclic and Angular Frequency Bandwidth

Since the angular frequency bandwidth is expressible by both

$$\beta_{\text{LFM}} = 2\alpha T_p \tag{6.8}$$

and

$$\beta_{\text{LFM}} = 2\pi B_{\text{LFM}}, \tag{6.9}$$

by equating (6.8) and (6.9),

$$\alpha = \pi \frac{B_{\text{LFM}}}{T_p}.$$

Linear Frequency Parameter and Quadratic Phase Parameter

From (6.7) the quadratic phase modulation parameter α and the linear cyclic frequency modulation slope parameter γ are related as

$$\alpha = \pi\gamma.$$

Instantaneous Frequency Expressions

Since the instantaneous angular frequency of an LFM chirp signal without a time delay is

$$\omega_i(t)^{\text{LFM}} = 2\alpha t, \tag{6.10}$$

the instantaneous cyclic frequency of the LFM chirp is

$$f_i(t)^{\text{LFM}} = \gamma t.$$

Dividing the instantaneous angular frequency function of (6.10) by 2π gives another expression of the instantaneous frequency of the LFM chirp signal,

$$f_i(t)^{\text{LFM}} = \frac{\alpha}{\pi} t. \tag{6.11}$$

With these LFM signal parameters, we can express an LFM phase signal in various parametric forms,

$$\begin{aligned} p(t)^{\text{LFM}} &= \exp\left(j\alpha t^2\right), \\ &= \exp\left(j\pi\gamma t^2\right), \\ &= \exp\left(j\pi \frac{B_{\text{LFM}}}{T_p} t^2\right). \end{aligned}$$

Chirp Rate Parameter Realizations

Any given value of the γ (or α) parameter has an infinite set of possible realizations in terms of combinations of pulse durations T_p and pulse bandwidths B_{LFM}. For example, a chirp signal that sweeps through 80 MHz of bandwidth in a 10 microsecond pulse produces an LFM frequency slope of

$$\begin{aligned} \gamma &= 80 \text{ MHz}/10\ \mu s, \\ &= 8 \times 10^{12} \text{ hertz/second}. \end{aligned}$$

The corresponding quadratic phase curvature is

$$\alpha \quad = \quad 8\pi \times 10^{12} \text{ radians/second/second.}$$

This pulse design shares the same γ (or α) parameter with a pulse that has a 160 MHz sweep over a 5 microsecond pulse, or with a pulse that has a 40 MHz sweep over a 20 microsecond pulse and so on. In all these cases, the time-bandwidth product $T_p B_{\text{LFM}} = 800$ remains fixed.

Transmitted LFM Pulse Signal

The transmitted LFM pulse *signal* is referenced over a time interval $[0, T_p]$ that starts at time zero[2]. For the up chirp case with $\alpha > 0$, the transmitted passband signal is

$$\begin{aligned} w(t)^{\text{LFM}} &= \overline{\Pi_{[0,T_p]}(t)} \cdot \exp(j\omega_{\min} t) \cdot \exp[j \cdot \overline{\omega_b(t)^{\text{LFM}}} \cdot t], \\ &= \overline{\Pi_{[0,T_p]}(t)} \cdot \exp[j \cdot \overline{\omega_p(t)^{\text{LFM}}} \cdot t]. \end{aligned}$$

where

$$\overline{\omega_p(t)^{\text{LFM}}} = \omega_{\min} + \overline{\omega_b(t)^{\text{LFM}}}$$

is the passband chirp couplet with $t \in [0, T_p]$. The instantaneous passband frequency sweeps through the LFM passband

$$\omega_i(t)^{\text{LFM}} \in [\omega_{\min}, \omega_{\min} + 2\alpha T_p]$$

with a positive quadratic phase modulation α parameter. The lowest instantaneous frequency is

$$\omega_{\min}.$$

The transmitted LFM pulse signal has a middle passband frequency of

$$\begin{aligned} \omega_c &= \omega_{\text{mid}}, \\ &= \omega_{\min} + \alpha T_p \end{aligned}$$

and has a highest instantaneous frequency of

$$\begin{aligned} \omega_i(T_p)^{\text{LFM}} &= \omega_{\min} + 2\alpha T_p, \\ &= \omega_{\max}. \end{aligned}$$

Received LFM Pulse Signals

Echo signals from LFM chirp pulses returned by the nth scatterer are defined over a time interval $[t_n, t_n + T_p]$ that starts at the echo delay time instant t_n.

---

[2]Hence, the transmitted signal can be dealt with as an analytic signal (see Section 14.5)

Mid-Frequency Demodulation of Passband LFM Signals

Demodulation of an LFM chirp can be performed with a local oscillator with a fixed reference *mid-frequency* of

$$\begin{aligned}\omega_{\text{LO}} &= \omega_c + \omega_o, \\ &= \omega_c + \alpha T_p, \\ &= \omega_{\text{mid}}.\end{aligned}$$

The frequency value $\omega_c + \omega_o$ is the middle passband frequency of the LFM chirp signal.

Min-Frequency and Max-Frequency Demodulation of Passband LFM Signals

Demodulation of an LFM chirp can also be performed with a local oscillator frequency equal to the minimum or maximum frequency of the passband signal.

Baseband LFM Signal's Mid-Frequency

A *min-frequency* demodulated chirp echo signal defined over $[t_n, t_n + T_p]$ is not a baseband signal as defined in the conventional sense of having a zero centered spectrum in the baseband frequency domain. The spectrum is a single sided spectrum that begins at zero frequency (see Section 14.5). This subtlety makes definition of a (non-zero) "mid-frequency" meaningful when dealing with *single-sided baseband* [691] or *analytic* [309] LFM signals. The LFM chirp signal's mid-frequency occurs in the middle ($t = T_p/2$) of the time window $[t_n, t_n + T_p]$. Hence, αT_p is the baseband signal's mid-frequency after carrier removal with a local oscillator signal frequency set to the chirp's minimum passband frequency.

6.3 Unwindowed LFM Waveform Spectrum

We next explore the Fourier transform of a linear frequency modulated chirp signal (see Section 15.1 for a more detailed derivation). We begin with infinite duration LFM chirp signals that have no window functions.

Quadratic Phase Modulation Signal Spectrum

A linear frequency modulation chirp signal with infinite time duration can be expressed in these notational forms,

$$\begin{aligned}p_{\Pi(\infty)}(t)^{\text{LFM}} &= \exp\left(j\alpha t^2\right), \qquad (6.12)\\ &= \exp\left(j \cdot \overline{\alpha t} \cdot t\right), \\ &= \exp[j \cdot \overline{\omega_b(t)^{\text{LFM}}} \cdot t].\end{aligned}$$

The LFM chirp signal (6.12) has exact (i.e., no approximations) Fourier transform expressions that can be written in forms that correspond to those of (6.12),

$$\begin{aligned}
P_{\Pi(?)}(\omega_b)^{\text{LFM}} &= \sqrt{\frac{\pi}{\alpha}} \cdot e^{j\frac{\pi}{4}} \cdot \exp\left(j\frac{\omega_b^2}{4\alpha}\right), \\
&= \sqrt{\frac{\pi}{\alpha}} \cdot e^{j\frac{\pi}{4}} \cdot \exp\left(j \cdot \overline{\frac{\omega_b}{4\alpha}} \cdot \omega_b\right), \\
&= \sqrt{\frac{\pi}{\alpha}} \cdot e^{j\frac{\pi}{4}} \cdot \exp[j \cdot \overline{t(\omega_b)^{\text{LFM}}} \cdot \omega_b]. \qquad (6.13)
\end{aligned}$$

The *LFM baseband frequency-time couplet,*

$$\overline{t(\omega_b)^{\text{LFM}}} = \overline{\frac{\omega_b}{4\alpha}}$$

of (6.13) is a description of LFM frequency-dependent chirp couplet time instances.

Chirp Couplet Frequency Dependent Time Instances

Analogous to instantaneous frequency, a frequency dependent time instant,

$$t_i(\omega_b),$$

dependent on ω_b within the spectrum of the chirp signal can be defined. For example, consider,

$$\phi(\omega_b) = \frac{\omega_b^2}{4\alpha}.$$

The derivative with respect to ω_b gives

$$\partial_{\omega_b}\{\phi(\omega_b)\} = \frac{\omega_b}{2\alpha}.$$

Hence, analogous to the time dependent instantaneous frequency

$$\omega_i(t) = 2\alpha t$$

of a time domain LFM chirp signal, a frequency dependent time instant

$$t_i(\omega_b) = \frac{1}{2\alpha}\omega_b$$

can be defined for each ω_b value within the chirp signal's spectrum.

Unwindowed LFM Signal Spectrum Infinite Bandwidth

Although approximation free, the expression of the Fourier transform given in (6.13) is incomplete. An ω_b domain window function has not been specified. The subscript of $P_{\Pi(?)}(\omega_b)^{\text{LFM}}$ has been temporally filled with "?" to indicate this unknown. When we consider this spectral domain window function, strict application of the Heisenberg uncertainty principle implies that infinite duration on one side of the Fourier transform should lead to precise localization on the other side of the transform. For example, the Fourier transform pair,

$$\delta(t) \quad \Leftrightarrow \quad 1$$

and the dual pair

$$1 \quad \Leftrightarrow \quad 2\pi\delta(\omega_b)$$

are examples of this infinite duration versus precise localization rule.

Since $p_{\Pi(\infty)}(t)^{\text{LFM}}$ has infinite time duration, we might expect $P_{\Pi(?)}(\omega_b)^{\text{LFM}}$ to be precisely localized, i.e., have zero bandwidth (i.e., a discrete frequency) in the frequency domain. We show through detailed derivation in Section 15.1 that this is not the case. *The spectrum of the unwindowed chirp signal is infinitely broadband.* This is one of the defining attributes of a chirp signal — increased duration of the time domain signal leads to increased bandwidth. Hence, the complete expression of the Fourier transform of the unwindowed LFM phase signal is

$$\Pi_{[-\infty,+\infty]}(t)\cdot\exp\left(j\alpha t^2\right) \quad {}_t\Leftrightarrow_{\omega_b} \quad \Pi_{[-\infty,+\infty]}(\omega_b)\cdot\sqrt{\frac{\pi}{\alpha}}\cdot e^{j\frac{\pi}{4}}\cdot\exp\left(-j\frac{\omega_b^2}{4\alpha}\right). \tag{6.14}$$

Which can be more compactly express in scripted notation as

$$p_{\Pi(\infty)}(t)^{\text{LFM}} \quad {}_t\Leftrightarrow_{\omega_b} \quad P_{\widehat{\Pi}(\infty)}(\omega_b)^{\text{LFM}},$$

where the weight function of the LFM spectrum is

$$\widehat{\Pi}_{(\infty)}(\omega_b) = \Pi_{[-\infty,+\infty]}(\omega_b)\cdot\sqrt{\frac{\pi}{\alpha}}\cdot e^{j\frac{\pi}{4}}.$$

In this result, we have a curious case where *support in time and frequency are both infinite.* This may lead one to wonder about the applicability of the Heisenberg uncertainty principle to chirp signals.

Uncertainty and the Phase Factor Fourier Transform Pair

We shall see that the phase element of LFM signals strictly adheres to the Heisenberg uncertainty principle. The phase signals in the various expressions of (6.12) and (6.13) are of key interest in SAR. The nominal Fourier transform pair for just the phase element of the LFM chirp signal, when written to expose the chirp couplet in both domains, is

$$\exp\left(j\cdot\overline{\alpha t}\cdot t\right) \quad \rightleftharpoons \quad \exp\left(j\cdot\overline{\frac{\omega_b}{4\alpha}}\cdot\omega_b\right).$$

This can be written in a form more conducive of insight relative to the Heisenberg uncertainty principle,

$$\exp\left(j\cdot 2\alpha\cdot\frac{t^2}{2}\right) \quad \rightleftharpoons \quad \exp\left(j\cdot\frac{1}{2\alpha}\cdot\frac{\omega_b^2}{2}\right). \tag{6.15}$$

The coefficients in the phase function arguments of exp(·) of (6.15) have been expanded to show the Heisenberg uncertainty relationship in regards to the spread of the quadratic phase centers at $t = 0$ and at $\omega_b = 0$. The time element $t^2/2$ of

$$\exp\left(j \cdot 2\alpha \cdot \frac{t^2}{2}\right)$$

is scaled by 2α. The corresponding phase element $\omega_b^2/2$ of

$$\exp\left(j \cdot \frac{1}{2\alpha} \cdot \frac{\omega_b^2}{2}\right)$$

is scaled by the reciprocal $1/(2\alpha)$. Hence, consistent with the Heisenberg uncertainty principle, localization *of the phase element* of the LFM chirp signal, as defined by the curvature parameter 2α of the time domain quadratic phase modulation characteristic, requires an inversely proportionate scaling $1/2\alpha$ of the frequency domain quadratic phase modulation.

6.4 Windowed LFM Waveform Spectrum

The Fourier transform of a *windowed* LFM chirp signal has characteristics that may seem curiously strange. To investigate, consider a windowed chirp signal with linear frequency modulation and finite time duration,

$$p_{\Pi(T_p)}(t)^{\text{LFM}} = \overline{\Pi_{(T_p)}(t) \cdot \exp\left(j \cdot \overline{\alpha t} \cdot t\right)}. \tag{6.16}$$

Methods for determining the Fourier transform of a time windowed LFM chirp signal include the approximate method of stationary phase [89, 307] and an exact analysis expressed in terms of *Fresnel integrals* [170, 445] (not the same as the Fresnel approximation of near field analysis, nor directly related to Fresnel wave fields). The method of stationary phase is more commonly used in SAR signal analysis, but provides less insight. The Fresnel integral approach explicitly indicates the bandwidth (frequency domain window) of the Fourier transform of an LFM chirp pulse signal.

In this section, we present and provide interpretations of the Fourier transform of the windowed LFM waveform. We provide in later sections detailed derivations using both the Fresnel integral approach (see Section 15.2) and stationary phase approach (see Section 15.5).

Exact Window Function of the LFM Pulse Spectrum

The windowed LFM signal (6.16) has an exact Fourier transform given by [445, 734]

$$P^{\text{exact}}_{\Pi(2\alpha T_p)}(\omega)^{\text{LFM}} = \overline{\widehat{\Pi}^{\text{exact}}_{(2\alpha T_p)}(\omega)^{\text{LFM}}} \cdot \exp\left(j \cdot \frac{\overline{\omega}}{4\alpha} \cdot \omega\right), \tag{6.17}$$

where the precise weight function of (6.17) is fully described by

$$\overline{\widehat{\Pi}^{\text{exact}}_{(2\alpha T_p)}(\omega)^{\text{LFM}}} = \frac{1}{\sqrt{2}} \cdot \left\{F\left[\frac{1}{2\sqrt{\alpha}} \cdot (\omega + \alpha T_p)\right] + F\left[\frac{1}{2\sqrt{\alpha}} \cdot (-\omega + \alpha T_p)\right]\right\}, \tag{6.18}$$

where $F(\cdot)$ is Fresnel's integral. See Section 15.2 for derivation.

Approximate Window Function of the LFM Pulse Spectrum

An approximate form of the Fourier transform of a windowed LFM signal is indicated by

$$P^{\text{approx}}_{\Pi(2\alpha T_p)}(\omega)^{\text{LFM}} \approx P^{\text{exact}}_{\Pi(2\alpha T_p)}(\omega)^{\text{LFM}},$$

where

$$P^{\text{approx}}_{\Pi(2\alpha T_p)}(\omega)^{\text{LFM}} = \overline{\widehat{\Pi}^{\text{approx}}_{(2\alpha T_p)}(\omega)^{\text{LFM}}} \cdot \exp\left(-j\frac{\omega^2}{4\alpha}\right) \tag{6.19}$$

and where we have defined an approximate LFM spectrum weight function,

$$\overline{\widehat{\Pi}^{\text{approx}}_{(2\alpha T_p)}(\omega)^{\text{LFM}}} = \overline{\Pi^{\text{approx}}_{(2\alpha T_p)}(\omega)^{\text{LFM}}} \cdot \sqrt{\frac{\pi}{\alpha}} \cdot e^{j\frac{\pi}{4}}. \tag{6.20}$$

The window function of (6.20) is defined as

$$\overline{\Pi^{\text{approx}}_{(2\alpha T_p)}(\omega)^{\text{LFM}}} = \begin{cases} 1 & \text{for} \quad |\omega| < \alpha T_p, \\ 0 & \text{otherwise.} \end{cases} \tag{6.21}$$

Exact (6.18) and approximation (6.21) window functions are shown in Fig. 6.3.

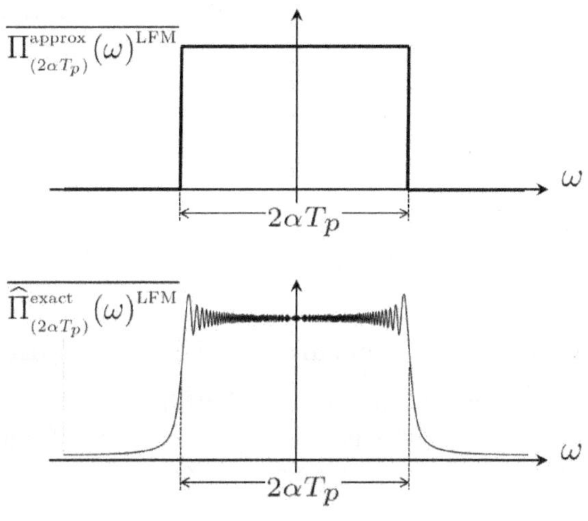

Figure 6.3: Approximate window and exact weight functions of LFM pulse spectrum.

Fourier Transform Pair of an LFM Pulse

Our expression for the Fourier transform of an LFM pulse signal is expressed in scripted notation as

$$p_{\Pi(T_p)}(t)^{\text{LFM}} \quad {}_t\!\Leftrightarrow_\omega \quad P_{\Pi(2\alpha T_p)}(\omega)^{\text{LFM}},$$

which is more fully expressed (in approximate form) as

$$\underbrace{\overline{\Pi_{(T_p)}(t)}^{\text{LFM}}}_{\text{time window}} \cdot \underbrace{\exp\left(j \cdot 2\alpha \cdot \frac{t^2}{2}\right)}_{\text{time quadratic phase}} \; {}_{t} \Leftrightarrow {}_{\omega} \; \underbrace{\overline{\widehat{\Pi}^{\text{approx}}_{(2\alpha T_p)}(\omega)}^{\text{LFM}}}_{\text{frequency window}} \cdot \underbrace{\exp\left(-j \cdot \frac{1}{2\alpha} \cdot \frac{\omega^2}{2}\right)}_{\text{frequency quadratic phase}}. \tag{6.22}$$

Chirp Signal Fourier Transforms and Nyquist Issues

The Fourier transform pair (6.22) is surprising in three ways. These curiosities relate to apparent inconsistencies that arise when this expression is compared to expectations that arise out of Heisenberg and Nyquist requirements, and when (6.22) is compared to the structure of more typical Fourier transform pair relationships.

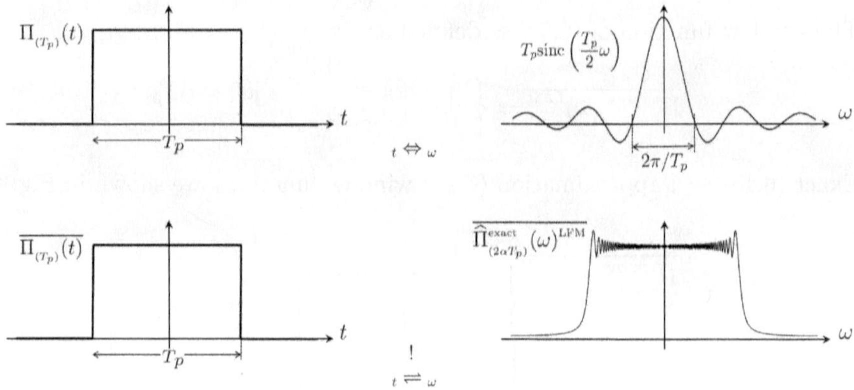

Figure 6.4: Square window functions of chirp pulse and spectrum.

Curiosity 1: Window Functions are Not Fourier Transforms Pairs

First, the LFM pulse window functions on both sides of the Fourier transform pair (6.22) are not themselves a Fourier transform pair! Contrary to common experience, an essentially square window function exists on both sides of the Fourier transform as shown in Fig. 6.4. One may have reasonably expected to see a rect(·) window function on one side of the Fourier relationship map to a sinc(·) window function on the other side of the Fourier transform. Instead of a window function relationship between time and frequency governed by

$$\Pi(t) \; {}_{t} \rightleftharpoons {}_{\omega} \; \text{sinc}(\omega),$$

we have the non-conforming (non-standard or unconventional Fourier) relationship,

$$\overline{\Pi_{(T_p)}(t)} \; \underset{t \rightleftharpoons \omega}{!} \; \overline{\widehat{\Pi}^{\text{exact}}_{(2\alpha T_p)}(\omega)}^{\text{LFM}}.$$

A square window function exists on both sides of the Fourier transform. The "!" mark is used above to indicate the exceptional nature of this relationship.

Curiosity 2: Multiplicative LFM Chirp Window Functions in Both Domains

The second curiosity is that the LFM chirp phase signals are *multiplied* by window functions in both time and frequency domains! Instead of

$$\Pi_{(\cdot)}(t) \times \exp(\cdot) \;\; {}_t\Leftrightarrow_\omega \;\; \Pi_{(\cdot)}(\omega) * \exp(\cdot),$$

we have the non-conforming relationship,

$$\overline{\Pi_{(\cdot)}(t)} \times \overline{\exp(\cdot)} \;\; {}_t\overset{!}{\rightleftharpoons}{}_\omega \;\; \overline{\Pi_{(\cdot)}(\omega)} \times \overline{\exp(\cdot)}.$$

One may have reasonably expected to see the multiplication-in-time window function map to a convolution-in-frequency window function. This multiplication (in time) mapped to multiplication[3] (in frequency) by the Fourier transform is sometimes characterized or justified with reference to the general Parseval[4] integral theorem as utilized in Fredholm[5] integral equations [883].

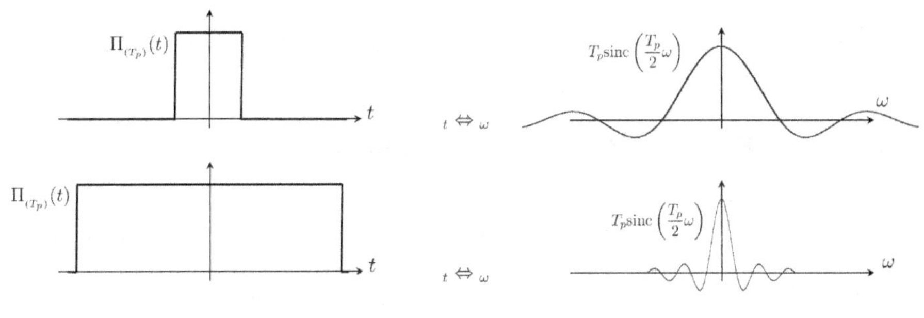

Figure 6.5: Pulse window functions subject to Heisenberg uncertainty.

Curiosity 3: Chirp Window Functions Escape Heisenberg Uncertainty

The third curiosity is that the window functions do not comprise a Heisenberg uncertainty pair with respect to observation duration versus localization preciseness! One may have reasonably expected to see a reciprocal relationship in the window function's durations, such as notionally represented by

$$\Pi_{\left(\frac{T_p}{1}\right)}(t) \;\; {}_t\Leftrightarrow_\omega \;\; \Pi_{\left(\frac{1}{T_p}\right)}(\omega)$$

[3]General inversion methods based on numerical solutions of the Fredholm integral equation of the first kind have also been investigated [15, 132, 257, 565]. The multiplicative relationships that exist in elements of these wavenumber domain models may sometimes be argued as justifiable because of the special properties of the Fourier transforms of temporal and spatial chirp signals explained here.

[4]Marc-Antoine Parseval des Chenes (1755–1836) was a French mathematician, most noted for Parseval's theorem, which was a precursor to the Fourier transform.

[5]Erik Ivar Fredholm (1866–1927) was a Swedish mathematician noted for his work on integral equations and operator theory.

and as illustrated in Fig. 6.5. Rather, a directly proportionate relationship exists between the time domain and frequency domain window sizes,

$$\overline{\Pi_{(T_p)}(t)} \quad \overset{!}{{}_t\rightleftharpoons_\omega} \quad \overline{\widehat{\Pi}_{(2\alpha T_p)}(\omega)}, \tag{6.23}$$

as illustrated in Fig. 6.6.

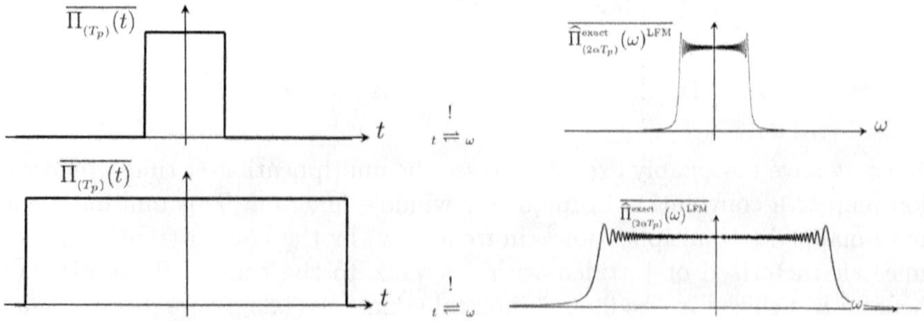

Figure 6.6: Chirp window functions not subject to Heisenberg uncertainty.

Chirp Phase Functions Conform to Heisenberg Uncertainty

We may be tempted to conclude that the Heisenberg uncertainty principle has somehow been violated by waveforms in the class of chirp signals. But before we jump to such a conclusion, note that the peculiar characteristics relate to only the pulsed LFM chirp signal's window element. The other signal element, the chirp quadratic phase factor, behaves as expected and conforms to the Heisenberg uncertainty principle,

$$\exp\left(j \cdot \frac{2\alpha}{1} \cdot \frac{t^2}{2}\right) \;{}_t\rightleftharpoons_\omega\; \exp\left(j \cdot \frac{1}{2\alpha} \cdot \frac{\omega^2}{2}\right).$$

The Fourier transform of a windowed chirp signal, with its defining characteristic of a time dependent, instantaneous frequency coupling (the chirp couplet), does not fit within the class of conventional non-chirp signals where the Heisenberg uncertainty principle is applicable to all signal elements without qualification.

Temporal and Spatial Chirp Signals and Heisenberg Uncertainty Principle

We conclude that the role of the uncertainty principle must be interpreted with care when Fourier transforms of windowed chirp signals are part of the signal analysis. This holds for chirp signals in both fast-time and slow-time (i.e., spatial phase curtains). An awareness and appreciation of these attributes of Fourier transforms of windowed chirp functions has important utility in understanding the nature of both temporal and spatial domain chirp signals in SAR analysis.

Chapter 7

SAR Image Formation by Pulse Compression and Stolt Formatting

We now combine one dimensional cross-range profiling by aperture synthesis with down-range profiling by pulse compression [159, 201, 202]. This provides our first look at SAR signal processing that results in a two dimensional slant-plane image [885].

We ignore in this chapter issues related to the antenna beam pattern and details related to operational modality (e.g., spotlight or stripmap modes). We continue to assume an isotropic point antenna. Antenna beam pattern related refinements will be developed in Chapter 9 and incorporated into mode specific SAR signal models in Chapter 12.

We seek to fully describe SAR phase histories when chirp pulse waveforms are transmitted. The resulting backscattered wave of each scatterer is modeled as an expanding spherical chirp wavelet.

We explore aperture synthesis through Stolt formatting with an infinite linear synthetic aperture. The assumption of an infinitely long synthetic aperture simplifies the analysis. The next chapter will revisit slant-plane imaging within the context of a finite synthetic aperture.

This chapter is outlined as follows: Section 7.1 develops a model for the slant-plane imaging geometry and presents the ground scene model. Section 7.2 introduces the use of chirp pulse signals in slant-plane imaging. Section 7.3 describes the phase history of time-space data measurements and presents example phase maps produced with LFM chirp waveforms. Section 7.4 describes the frequency-space domain phase history. Section 7.5 uses pulse compression to form the down-range image of the slant-plane system. Section 7.6 analyzes the range focused slant-plane image. Section 7.7 describes the slant-plane frequency-wavenumber domain phase history. Section 7.8 explores Stolt formatting for slant-plane imaging. Section 7.9 presents a Huygens-Fresnel analysis of the angular spectrum. Section 7.10 explains the option of performing Huygens-Fresnel transfers before or after Stolt formatting. Section 7.11 overviews the general system engineering work flow required to specify and derive requirements related to slant-plane SAR data collections. Section 7.12 explains the sample rate adjustment in the along-track dimension that can be performed before image formation. Section 7.13 examines all options related to the ordering of the Huygens-Fresnel transfer, Fourier migration and Fourier inversion processing steps. Section 7.14 describes uniform resampling of Fourier migrated angular spectrum data. Section 7.15 derives the Jacobian of the Stolt transformation. Section 7.16 describes the slant-plane image spectrum. Finally, Section 7.17 investigates characteristics of slant-plane SAR imagery formed via waveform pulse compression and Stolt formatting.

7.1 Slant-Plane Imaging Geometry

Up until now, we have assumed all scatterers exist on a single down-range line at an along-track position of Y_c or on a single cross-range line at the down-range distance of X_c. We now generalize our ground scene model to a more realistic case where scatterers have positions that are dispersed in both down-range and along-track dimensions.

Two Dimensional Slant-Plane Terrain Functions

The slant-plane terrain function is modeled as a collection of discrete scatterers with a two dimensional spatial distribution off to one side of the radar flight line as shown in Fig. 7.1. We reduce the problem from three dimensions to two dimensions by assuming the flight altitude is zero (although our illustrations are usually drawn otherwise). With this assumption, the scatterers are resident in a slant-plane that coincides with the ground plane [683, 834]. Scatterer reflectivities are assumed independent of transmitted signal frequency and independent of sensor-scatterer view angle.

Broadside and Squint Terrain Functions

We consider flight line to ground scene geometries for two cases. In one, the ground scene is in a broadside geometry ($Y_c = 0$) as shown in Fig. 7.1. In the other, the ground scene either leads ahead or lags behind (i.e., $Y_c \neq 0$) the "center" of the infinitely long synthetic aperture in a squint geometry as shown in Fig. 7.2.

Slant-Plane Terrain Functions in Aperture Coordinate System

A collection of point scatterers distributed in two dimensions forms the ground scene model [448],

$$\begin{aligned} g_{\Pi(2X_o@X_c)}g_{\Pi(2Y_o@Y_c)}(x,y)_{\text{ACS}} &= \sum_n g_n \cdot \delta(x-x_n, y-y_n), \\ &= \sum_n g_n \cdot \delta(x-x_n) \cdot \delta(y-y_n). \end{aligned} \tag{7.1}$$

The subscript $\Pi(2X_o@X_c)$ of $g_{\Pi(2X_o@X_c)}g_{\Pi(2Y_o@Y_c)}(x,y)_{\text{ACS}}$ indicates a bounded scene of width $2X_o$ centered at X_c in the down-range x dimension. The second subscript $\Pi(2Y_o@Y_c)$ indicates a bounded scene of length $2Y_o$ centered at Y_c in

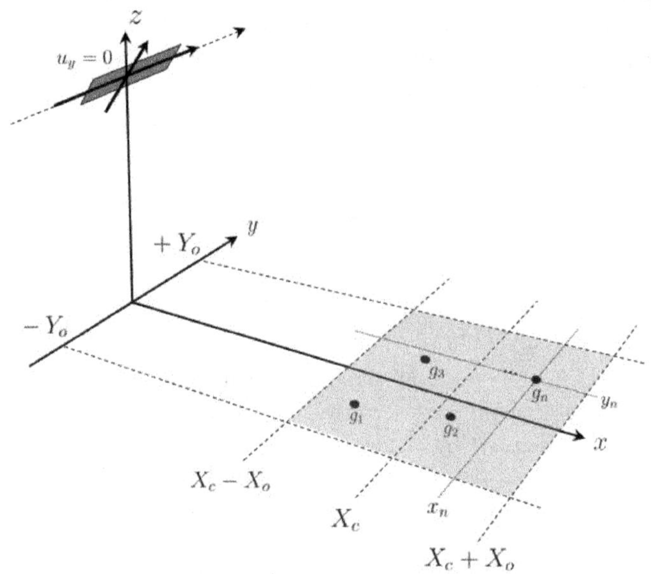

Figure 7.1: Slant-plane broadside imaging geometry for linear flight SAR.

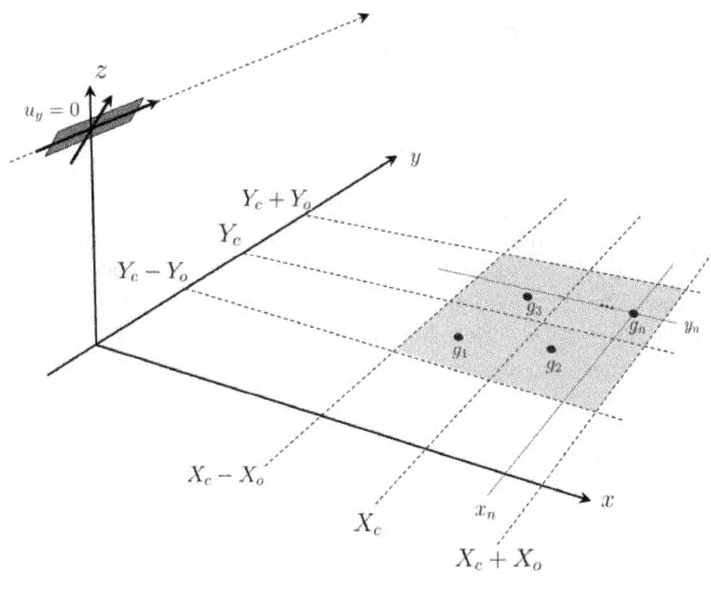

Figure 7.2: Slant-plane squint imaging geometry for linear flight SAR.

the along-track y dimension. Simple summing of the scatterer returns, as used in (7.1), is permitted if a weak scattering approximation is sufficiently accurate (i.e., multiple bounce scattering is deemed insignificant) [933].

Since the point scatterers are modeled as impulse functions, the wavenumber domain description of the terrain function is

$$\begin{aligned} G_{\Pi(\infty)}G_{\Pi(\infty)}(k_x,k_y)_{\mathrm{ACS}} &= \sum_n g_n \cdot \exp(-jx_nk_x)\cdot\exp(-jy_nk_y), \\ &= \sum_n G_{\Pi(\infty)}G_{\Pi(\infty)}(k_x,k_y)^n_{\mathrm{ACS}}, \end{aligned} \tag{7.2}$$

where

$$G_{\Pi(\infty)}G_{\Pi(\infty)}(k_x,k_y)^n_{\mathrm{ACS}} = g_n \cdot \exp(-jx_nk_x)\cdot\exp(-jy_nk_y).$$

The infinitely fine localization of the point scatterer model $\delta(x-x_n, y-y_n)$ requires infinite bandwidth in the terrain wavenumber domain (k_x, k_y). This infinite support is indicated by the subscripts $\Pi(\infty)$ of $G_{\Pi(\infty)}G_{\Pi(\infty)}(k_x,k_y)_{\mathrm{ACS}}$ in (7.2). The ground scene descriptions (7.1) and (7.2) form a Fourier transform pair in the aperture coordinate system,

$$g_{\Pi(2X_o@X_c)}g_{\Pi(2Y_o@Y_c)}(x,y)_{\mathrm{ACS}} \quad\Leftrightarrow\quad G_{\Pi(\infty)}G_{\Pi(\infty)}(k_x,k_y)_{\mathrm{ACS}}.$$

Slant-Plane Terrain Functions in Scene Coordinate System

The ground scene Fourier transform pair in the scene coordinate system is

$$g_{\Pi(2X_o@0)}g_{\Pi(2Y_o@0)}(\hat{x},\hat{y})_{\mathrm{SCS}} \quad\Leftrightarrow\quad G_{\Pi(\infty)}G_{\Pi(\infty)}(k_{\hat{x}},k_{\hat{y}})_{\mathrm{SCS}},$$

where $\Pi(2X_o@0)$ and $\Pi(2Y_o@0)$ indicate a scene of width $2X_o$ and length $2Y_o$ centered at $(0,0)$ in the scene coordinate system.

7.2 Slant-Plane Chirp Pulse Signal

We examine various phase histories (time-space, frequency-space, frequency-wavenumber, angular spectrum and image spectrum) for infinite aperture SAR systems that transmit chirp pulse signals. Our slant-plane signal analysis uses the general chirp pulse waveforms defined in Section 6.1. We are not specifically focused on the special case of the linear frequency modulated chirp waveform. LFM waveforms will be used or discussed in this chapter only to provide example phase map illustrations.

Baseband Chirp Pulse Signal

Consider a baseband chirp pulse

$$p_{\Pi(T_p@t_o)}(t-t_o)^{\text{chirp}}$$

with time reference t_o and a pulse duration of T_p defined by the chirp window function

$$\overline{\Pi_{(T_p)}(t-t_o)}$$

applied to a general baseband chirp phase factor

$$\exp[j \cdot \overline{\omega_b(t - t_o)} \cdot (t - t_o)].$$

The resulting baseband chirp pulse is

$$p_{\Pi(T_p @ t_o)}(t - t_o)^{\text{chirp}} = \overline{\Pi_{(T_p)}(t - t_o)} \cdot \exp[j \cdot \overline{\omega_b(t - t_o)} \cdot (t - t_o)]. \tag{7.3}$$

Baseband Chirp Pulse Spectrum

If $t_o = 0$, then the Fourier transform of a t domain windowed baseband chirp pulse

$$p_{\Pi(T_p @ 0)}(t)^{\text{chirp}}$$

yields the baseband ω_b domain windowed chirp pulse spectrum

$$P_{\Pi(2\omega_o @ 0)}(\omega_b)^{\text{chirp}} = \overline{\Pi_{(2\omega_o)}(\omega_b)} \cdot \exp[j \cdot \overline{t(\omega_b)} \cdot \omega_b], \tag{7.4}$$

with characteristics as previously described in Section 6.4. The Fourier transform of a general t_o time referenced baseband chirp pulse (7.3) is

$$p_{\Pi(T_p @ t_o)}(t - t_o)^{\text{chirp}} \quad {}_{t} \Leftrightarrow_{\omega_b} \quad P_{\Pi(2\omega_o @ 0)}(\omega_b)^{\text{chirp}} \cdot \exp(-jt_o\omega_b).$$

Passband Chirp Pulse Signal

With up conversion of the baseband chirp pulse by a carrier with frequency ω_c, the resulting passband chirp pulse signal is

$$w_{\Pi(T_p)}(t)^{\text{chirp}} = \exp(j\omega_c t) \times \left\{ p_{\Pi(T_p)}(t)^{\text{chirp}} \right\}.$$

The corresponding Fourier transform pair is

$$w_{\Pi(T_p)}(t)^{\text{chirp}} \quad {}_{t} \Leftrightarrow_{\omega_p} \quad W_{\Pi(2\omega_o @ \omega_c)}(\omega_p)^{\text{chirp}},$$

where

$$\omega_p = \omega_c + \omega_b$$

indicates passband frequencies. The instantaneous angular frequency of the transmitted chirp signal sweeps through values in the passband

$$\beta_p = [\omega_c - \omega_o, \omega_c + \omega_o]$$

during the intra-pulse time interval

$$t \in [0, T_p].$$

The transmitted passband chirp pulse signal is

$$\begin{aligned} w_{\Pi(T_p)}(t)^{\text{chirp}} &= \exp(j\omega_c t) \times \left\{ \overline{\Pi_{(T_p)}(t)} \cdot \exp[j \cdot \overline{\omega_b(t)} \cdot t] \right\}, \\ &= \overline{\Pi_{(T_p)}(t)} \cdot \exp\{j \cdot [\omega_c + \overline{\omega_b(t)}] \cdot t\}, \\ &= \overline{\Pi_{(T_p)}(t)} \cdot \exp[j \cdot \overline{\omega_p(t)} \cdot t]. \end{aligned} \tag{7.5}$$

The passband time-frequency chirp couplet $\overline{\omega_p(t)}$ is

$$\overline{\omega_p(t)} = \omega_c + \overline{\omega_b(t)}.$$

7.3 Time-Space Chirp Phase History

The chirp pulse echo signal returned by the nth scatterer g_n located at (x_n, y_n) is indicated in the time-space domain by $ss(t, u_y)_n^{\text{chirp}}$. The echo signal is assumed to be received without attenuation at all aperture positions $u_y \in [-\infty, \infty]$. The nth scatterer's echo signal has a fast-time delay of t_n. The echo delay is a function of slow-time u_y as determined by the round trip distance $r_n(u_y)$ between sensor and scatterer,

$$t_n = \frac{1}{c} \times 2r_n(u_y).$$

Full expression of a single scatterer's time-space phase history is

$$s_{\Pi(T_p)} s_{\Pi(\infty)} (t, u_y)_n^{\text{chirp}} = g_n \cdot \overline{\Pi_{(T_p)}(t - t_n)} \cdot \exp[j \cdot \overline{\omega_p(t - t_n)} \cdot (t - t_n)], \quad (7.6)$$

where the received echo time duration is over $[t_n, t_n + T_p]$.

Time Delayed Chirp Couplet

The time-frequency characteristic of the chirp echo signal (7.6) is indicated by the *time delayed passband chirp couplet,*

$$\overline{\omega_p(t - t_n)} \in \omega_c + \overline{\mathbb{C}}_b, \quad (7.7)$$

for

$$t - t_n \quad \in \quad [0, T_p] \, . \quad (7.8)$$

Since the time window (7.8) starts at time zero (instead of being centered about time zero), the echo signal can be dealt with as an analytic signal (see Section 14.5).

Time-Space Passband Phase History Decomposition

Consider a decomposition of the time-space phase history (7.6) given by

$$\begin{aligned} & s_{\Pi(T_p)} s_{\Pi(\infty)} (t, u_y)_n^{\text{chirp}} \quad (7.9) \\ & = g_n \cdot \overline{\Pi_{(T_p)}(t - t_n)} \cdot \exp[j \cdot \overline{\omega_p(t - t_n)} \cdot t] \cdot \exp[-j \cdot \overline{\omega_p(t - t_n)} \cdot t_n]. \end{aligned}$$

The window function

$$\overline{\Pi_{(T_p)}(t - t_n)} \quad (7.10)$$

is a shifted chirp window function that is "on" over the time interval

$$t \in [t_n, t_n + T_p] \, .$$

The first phase factor of (7.9),

$$\exp[j \cdot \overline{\omega_p(t - t_n)} \cdot t], \quad (7.11)$$

is a chirp phase signal with the time delayed chirp couplet of (7.7). There is a non-shifted time variable outside the chirp couplet in (7.11). Only the chirp couplet $\overline{\omega_p(t-t_n)}$ is time shifted. Both the chirp window function (7.10) and chirp couplet phase factor (7.11) of the echo signal are delayed by t_n. We combine these two to define a *delayed couplet chirp waveform,*

$$w_{\Pi(T_p @ t_n)}(t)_n^{\text{chirp}} = \overline{\Pi_{(T_p)}(t-t_n)} \cdot \exp[j \cdot \overline{\omega_p(t-t_n)} \cdot t], \tag{7.12}$$

which is a chirp pulse defined over the time window $[t_n, t_n + T_p]$. Equation (7.12) is a (non-delayed) time function with a time shifted chirp couplet and a time shifted chirp window function. Compare (7.12) with the transmitted chirp pulse waveform,

$$w_{\Pi(T_p @ 0)}(t)^{\text{chirp}} = \overline{\Pi_{(T_p)}(t)} \cdot \exp[j \cdot \overline{\omega_p(t)} \cdot t]. \tag{7.13}$$

The *delayed chirp couplet waveform* (7.12) is not the same as a *delayed chirp waveform,*

$$w_{\Pi(T_p @ t_n)}(t-t_n)_n^{\text{chirp}} = \overline{\Pi_{(T_p)}(t-t_n)} \cdot \exp[j \cdot \overline{\omega_p(t-t_n)} \cdot (t-t_n)].$$

Chirp Signal Time-Space Phase History Decomposition

Using the newly defined delayed couplet chirp waveform (7.12), we rewrite the echo signal of a single scatterer (7.6) as

$$\begin{aligned} s_{\Pi(T_p)} s_{\Pi(\infty)}(t, u_y)_n^{\text{chirp}} &= g_n \cdot w_{\Pi(T_p @ t_n)}(t)_n^{\text{chirp}} \cdot \exp[-j \cdot \overline{\omega_p(t-t_n)} \cdot t_n], \\ &= g_n \cdot w_{\Pi(T_p @ t_n)}(t)_n^{\text{chirp}} \cdot \exp[-j \cdot \overline{\omega_p(t-t_n)} \cdot \frac{2}{c} r_n(u_y)], \\ &= g_n \cdot \underbrace{w_{\Pi(T_p @ t_n)}(t)_n^{\text{chirp}}}_{\text{fast-time}} \underbrace{\cdot \exp[-jk_{\overline{\omega_p(t-t_n)}} \cdot 2r_n(u_y)]}_{\text{slow-time}}, \end{aligned} \tag{7.14}$$

where we have defined

$$\exp[-jk_{\overline{\omega_p(t-t_n)}} \cdot 2r_n(u_y)] = \exp[-j \cdot \overline{\omega_p(t-t_n)} \cdot t_n] \tag{7.15}$$

and where we have set

$$k_{\overline{\omega_p(t-t_n)}} = \frac{1}{c} \cdot \overline{\omega_p(t-t_n)}. \tag{7.16}$$

We view $k_{\overline{\omega_p(t-t_n)}}$ as the chirp form of the familiar temporal angular frequency wavenumber k_{ω_p}. Use of chirp pulse signals requires the generalization

$$k_{\omega_p} \mapsto k_{\overline{\omega_p(t-t_n)}}$$

to describe time-space domain chirp phase histories. The span of the *chirp passband temporal wavenumber* (7.16) is determined by the bandwidth of the transmitted chirp signal,

$$k_{\overline{\omega_p(t-t_n)}} \in \frac{1}{c} \cdot [\omega_{\text{mid}} - \omega_o, \omega_{\text{mid}} + \omega_o],$$

where ω_{mid} is the mid-frequency of the passband chirp pulse signal.

Chirp Slow-Time and Range Walk Phase Curtains

The phase factor (7.15) is the *chirp slow-time phase curtain.* This phase factor is a function of range walk $r_n(u_y)$ and is also called the *chirp range walk phase curtain.* This important phase factor is a remotely sensed linear flight sample of the spatial wave field of a spherical Huygens *chirp* wavelet[1].

Range Walk Phase Curtain Notation

We use the notation

$$\begin{aligned} \measuredangle hh^2[\overline{w_p(t-t_n), r_n(u_y)}] &= \exp[-j \cdot \overline{\omega_p(t-t_n)} \cdot t_n], \\ &= \exp[-jk_{\overline{\omega_p(t-t_n)}} \cdot 2r_n(u_y)] \end{aligned} \tag{7.17}$$

to represent the chirp phase curtain collected within the scattered field of a scatterer illuminated by a chirped Huygens wavelet. We employ the notation $hh(\cdot,\cdot)$ instead of $Hh(\cdot,\cdot)$ because the chirp couplet $\overline{\omega_p(t-t_n)}$ carries an explicit dependency on fast-time. Moreover, at this stage of analysis, we are concentrating on time-space attributes, not frequency-space attributes of the collected data.

Time-Space Attenuating Chirped Huygens Wavelet

The two way spherical Huygens chirp wavelet is

$$\begin{aligned} & hh^2[\overline{w_p(t-t_n), r_n(u_y)}] \\ &= \left(\frac{1}{4\pi}\right)^2 \cdot \left[\frac{1}{r_n(u_y)}\right]^2 \cdot \exp\left[-j \cdot \overline{\omega_p(t-t_n)} \cdot t_n\right], \\ &= \left(\frac{1}{4\pi}\right)^2 \cdot \left[\frac{1}{r_n(u_y)}\right]^2 \cdot \exp\left[-jk_{\overline{\omega_p(t-t_n)}} \cdot 2r_n(u_y)\right]. \end{aligned} \tag{7.18}$$

The expression (7.18) includes the signal's two way distance $2r_n(u_y)$ from the transmitter at $a_p(0,u_y)$ to the scattering ground point $g_n(x,y)$. The forward path distance of incident radiation that illuminates the ground scene is

$$r_n(u_y)^{\text{in}} = |g_n(x,y) - a_p(0,u_y)|. \tag{7.19}$$

The return distance of scattered radiation received back at the point antenna is the same, but expressed as

$$r_n(u_y)^{\text{sc}} = |a_p(0,u_y) - g_n(x,y)|. \tag{7.20}$$

Notational Rule: Field and Source Points

The spatial parameters on the right hand side of (7.19) and (7.20) are ordered to follow a notational convention of

$$\text{distance} = |\text{field point} - \text{source point}|.$$

The antenna is in transmit mode for (7.19) and in receive mode for (7.20). The "in" superscript indicates "incident." The "sc" superscript indicates "scattered."

[1] Rather than a linear flight sample of a Huygens *impulse* wavelet.

Round Trip and One Way Slow-Time Phase Shifts

The time dependent phase shift of the illuminating chirp signal transmitted from u_y and received at $g_n(x,y)$ is

$$\exp\left[-jk_{\overline{\omega_p(t-t_n)}} \cdot r_n(u_y)^{\text{in}}\right] = \measuredangle hh\left[\overline{w_p(t-t_n)}, r_n(u_y)^{\text{in}}\right].$$

Note that $hh(\cdot,\cdot)$ is no longer squared. The phase shift of the ground scattered return leg of this trip is

$$\exp\left[-jk_{\overline{\omega_p(t-t_n)}} \cdot r_n(u_y)^{\text{sc}}\right] = \measuredangle hh\left[\overline{w_p(t-t_n)}, r_n(u_y)^{\text{sc}}\right].$$

The combined two way trip and total phase shift is

$$\begin{aligned}
&\exp\left[-jk_{\overline{\omega_p(t-t_n)}} \cdot 2r_n(u_y)\right]^{\text{chirp}} \\
&= \measuredangle hh\left[\overline{w_p(t-t_n)}, r_n(u_y)^{\text{in}}\right] \cdot \measuredangle hh\left[\overline{w_p(t-t_n)}, r_n(u_y)^{\text{sc}}\right], \\
&= \measuredangle hh^2\left[\overline{w_p(t-t_n)}, r_n(u_y)\right].
\end{aligned}$$

The squared notation $hh^2(\cdot,\cdot)$ indicates a round trip migration of phase.

Time-Space Phase History: Chirp Time-Space Decomposition

We now have a descriptive and rather concise expression for the passband time-space phase history produced by a single scatterer illuminated by a chirp pulse signal,

$$\begin{aligned}
&s_{\Pi(T_p)}s_{\Pi(\infty)}(t,u_y)_n^{\text{chirp}} \\
&= g_n \cdot w_{\Pi(T_p@t_n)}(t)_n^{\text{chirp}} \cdot \measuredangle hh^2\left[\overline{w_p(t-t_n)}, r_n(u_y)\right]. \qquad (7.21)
\end{aligned}$$

The time-space phase history for a single scatterer includes a scatterer specific replica $w_{\Pi(T_p@t_n)}(t)_n^{\text{chirp}}$ of the transmitted chirp pulse signal $w_{\Pi(T_p@0)}(t)^{\text{chirp}}$, but with a chirp couplet time reference of t_n. The time-space phase history also includes a two way range walk chirp phase curtain $\measuredangle hh^2[\overline{w_p(t-t_n)}, r_n(u_y)]$.

Time-Space Phase History: Data Matrix Interpretation

A notional illustration of the measured data matrix associated with the received signal echoes in the time-space domain is shown in Figs. 7.3a-b. The measured fast-time samples contain a superposition of echoes from scatterers located under propagating spherical chirp wavefronts as illustrated in Fig. 7.3a. The measured fast-time data samples collected within the receive window $[T_s, T_f]$ of Fig. 7.3b provide a measure of range reflectivity from a superposition of scatterers under a spherical wavefront. The data also contain hyperbolic slow-time slant-range migrations of the echo signal delays from all scatterers. The measured spatial data samples capture each scatterer's range walk phase curtain as induced by changing sensor-scatterer geometry (both view angle direction and slant-range distance parameters vary as the radar moves through the synthetic aperture). The hyperbolic range migrations are expressed in the range walk chirp phase curtains of the time-space domain phase history.

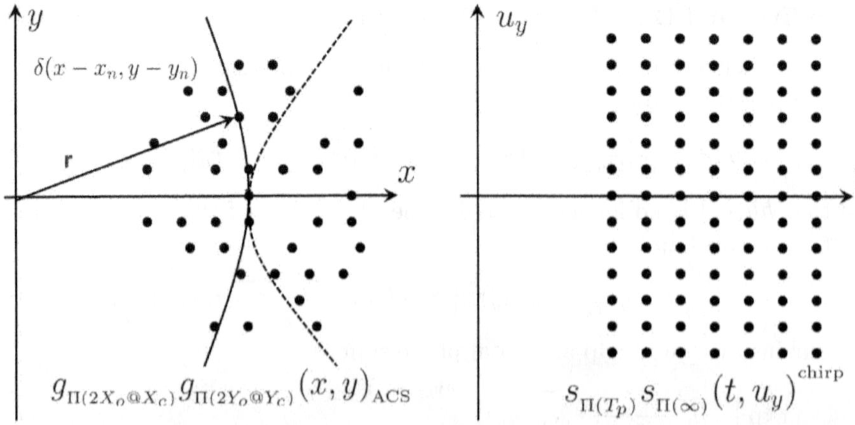

a. Wavefront and range walk. b. Time-space data samples.

Figure 7.3: Time-space phase history data matrix.

Time-Space Passband Phase Maps

We next examine phase maps of the time-space phase history of a single scatterer echo when LFM chirp pulse waveforms are transmitted. Fig. 7.4a shows an example phase map of the time-space chirp passband phase history before received signal down conversion. The imaging geometry involved with producing this phase map is such that the synthetic aperture is long enough and the distance to the scatterer is small enough that the hyperbolic range walk is clearly evident. The carrier frequency and bandwidth of the up chirp LFM pulses of this phase map are also chosen to be such that phase oscillations are clearly visible in printed format.

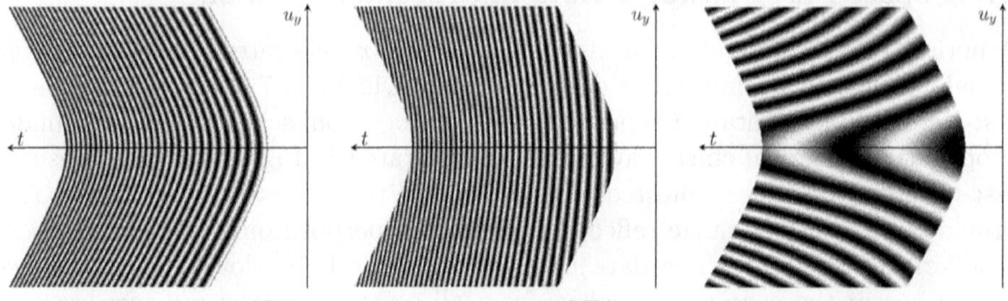

a. $s_{\Pi(T_p)} s_{\Pi(\infty)}(t, u_y)_n^{\text{LFM}}$ b. $w_{\Pi(T_p @ t_n)}(t)^{\text{LFM}}$ c. $hh^2[\overline{w_p(t - t_n), r_n(u_y)}]$

Figure 7.4: Up chirp LFM time-space phase history decomposition.

LFM Chirp Time-Space Phase History Decomposition

The phase map of Fig. 7.4a is the complete time-space phase history

$$
\begin{aligned}
& s_{\Pi(T_p)} s_{\Pi(\infty)}(t, u_y)_n^{\text{LFM}} \\
&= g_n \cdot \overline{\Pi_{(T_p)}(t - t_n)} \cdot \exp[j \cdot \overline{\omega_p(t - t_n)^{\text{LFM}}} \cdot (t - t_n)], \\
&= g_n \cdot \overline{\Pi_{(T_p)}(t - t_n)} \cdot \exp\{j \cdot [\omega_{\min} + \overline{\alpha \cdot (t - t_n)}] \cdot (t - t_n)\}
\end{aligned}
$$

collected from a single scatterer when an LFM up chirp pulse signal is transmitted. The horizontal axes of the phase maps of Fig. 7.4 represent fast-time. The vertical axes represent slow-time. The pulse transmit positions u_y are dense (closely spaced) and hence appear as a continuum in the vertical direction of this printed format. The time axes are flipped so the orientation of the range walk hyperbolae remain consistent with the hyperbolic range walk manifolds previously illustrated in Chapter 4. All displayed phase factors of Figs. 7.4a-c are windowed within the time-space domain by $\overline{\Pi_{(T_p)}[t - t_n(u_y)]}$ and hence all display the hyperbolic fast-time signal delay t_n caused by changing slant-range.

Minimum Chirp Frequency as Carrier Frequency

With an LFM up chirp pulse, it is usually more convenient to view the LFM carrier frequency as the lowest transmitted passband frequency within the intra-pulse period. The starting frequency of the up chirp frequency sweep is shown on the right edge of the phase map of Fig. 7.4a. The notation for the carrier frequency ω_c is usually replaced by the minimum chirp passband frequency $\omega_{\min}$ in expressions with up chirp waveforms.

Temporal Element Phase Map

The passband LFM chirp waveform with no time delay, but with a delayed chirp couplet determined by round trip signal delay is

$$
\begin{aligned}
& w_{\Pi(T_p @ t_n)}(t)^{\text{LFM}} \\
&= \overline{\Pi_{(T_p)}(t - t_n)} \cdot \exp[j \cdot \overline{\omega_p(t - t_n)^{\text{LFM}}} \cdot t], \qquad (7.22) \\
&= \overline{\Pi_{(T_p)}[t - t_n(u_y)]} \cdot \exp\{j \cdot [\omega_{\min} + \overline{\alpha \cdot [t - t_n(u_y)]}] \cdot t\}, \\
&= \overline{\Pi_{(T_p)}[t - t_n(u_y)]} \cdot \exp(j\omega_{\min} t) \cdot \exp(j\alpha t^2) \cdot \exp[-j \cdot \overline{\alpha t_n(u_y)} \cdot t].
\end{aligned}
$$

The phase map of the delayed chirp couplet LFM waveform (7.22) is shown in Fig. 7.4b.

Spatial Element Phase Map

The phase map of Fig. 7.4c is the LFM up chirp phase curtain,

$$
\begin{aligned}
& \measuredangle hh^2[\overline{w_p(t - t_n)}, r_n(u_y)] \\
&= \exp[-j \cdot \overline{\omega_p(t - t_n)^{\text{LFM}}} \cdot t_n], \qquad (7.23) \\
&= \exp\{-j \cdot [\omega_{\min} + \overline{\alpha \cdot (t - t_n)}] \cdot t_n\}, \\
&= \exp[-j\omega_{\min} t_n(u_y)] \cdot \exp[-j \cdot \overline{\alpha t} \cdot t_n(u_y)] \cdot \exp[j\alpha t_n^2(u_y)].
\end{aligned}
$$

The spatial phase curtain of Fig. 7.4c combines with the temporal phase map of Fig. 7.4b to produce the full phase history of a single scatterer,

$$s_{\Pi(T_p)}s_{\Pi(\infty)}(t,u_y)_n^{\text{LFM}} = g_n \cdot w_{\Pi(T_p@t_n)}(t)^{\text{LFM}} \cdot \measuredangle hh^2[\overline{w_p(t-t_n)}, r_n(u_y)],$$

as shown in Fig. 7.4a.

Couplet Tones in Both Temporal and Spatial Phase History Elements

It is important to note that both the temporal (7.22) and spatial (7.23) elements of the decomposed phase history contain phase factors comprised of couplet tones,

$$\exp[-j \cdot \overline{\alpha t_n(u_y)} \cdot t] = \exp[-j \cdot \overline{\alpha t} \cdot t_n(u_y)].$$

Hence, when the temporal and spatial phase elements are combined, the result is a locator tone with a frequency of $2\alpha t_n(u_y)$,

$$\exp[-j \cdot 2\alpha t_n(u_y) \cdot t] = \exp[-j \cdot \overline{\alpha t_n(u_y)} \cdot t] \cdot \exp[-j \cdot \overline{\alpha t} \cdot t_n(u_y)].$$

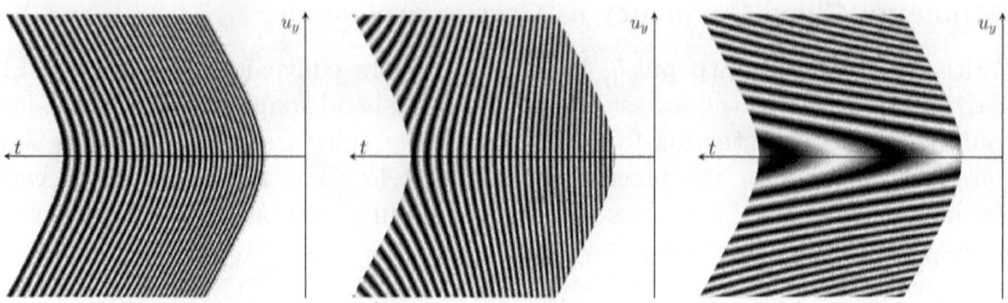

a. $s_{\Pi(T_p)}s_{\Pi(\infty)}(t,u_y)_n^{\text{LFM}}$ b. $w_{\Pi(T_p@t_n)}(t)^{\text{LFM}}$ c. $hh^2[\overline{w_p(t-t_n)}, r_n(u_y)]$

Figure 7.5: Down chirp LFM time-space phase history decomposition.

LFM Down Chirp

Figs. 7.5a-c show the same set of phase maps as Figs. 7.4a-c, but where the LFM signal is a down chirp waveform. With an LFM down chirp pulse, it is convenient to view the LFM carrier frequency as the highest transmitted passband frequency within the the intra-pulse period. Hence, ω_c is replaced by $\omega_{\max}$ in expressions with down chirp waveforms. This is the beginning frequency within the down chirp sweep as shown on the right edge of the phase map of Fig. 7.5a.

Phase Maps of Imaginary Parts of Baseband LFM Waveforms

When generating phase maps with both LFM up chirp and down chirp baseband examples, phase maps are more informative when produced from the imagery part of complex signals. The only difference between the phase factor of the baseband LFM up chirp phase signal,

$$\exp(+j\alpha t^2) = \cos(\alpha t^2) + j\sin(\alpha t^2)$$

and the baseband LFM down chirp phase signal,

$$\exp(-j\alpha t^2) = \cos(\alpha t^2) - j\sin(\alpha t^2),$$

is the sign of the imaginary part. The real parts are equal and would look the same if presented as phase maps.

Down Converted Time-Space Phase History

We next derive and examine the time-space phase history of a single scatterer after the received echo signal undergoes down conversion. We shall see that the slow-time phase curtain is unaffected by fast-time down conversion and remains dependent on passband frequencies.

Chirp Pulse Time-Space Phase History After Down Conversion

Consider the single scatterer passband time-space phase history when a generalized chirp waveform is transmitted,

$$\begin{aligned} & s_{\Pi(T_p)}\, s_{\Pi(\infty)}(t, u_y)_n^{\text{chirp}} \\ &= g_n \cdot \overline{\Pi_{(T_p)}(t-t_n)} \cdot \exp[j \cdot \overline{\omega_p(t-t_n)} \cdot (t-t_n)], \qquad (7.24) \\ &= g_n \cdot \overline{\Pi_{(T_p)}(t-t_n)} \cdot \exp[j \cdot \overline{\omega_b(t-t_n)} \cdot (t-t_n)] \cdot \exp[j\omega_c \cdot (t-t_n)], \end{aligned}$$

where we explicitly decompose the chirp passband couplet into a chirp baseband couplet added to the carrier frequency with a t_n shifted time reference,

$$\overline{\omega_p(t-t_n)} = \omega_c(t-t_n) + \overline{\omega_b(t-t_n)}. \qquad (7.25)$$

All temporal elements of (7.25) are subject to the delay echo time reference t_n.

Down conversion of the passband signal (7.24) gives

$$\begin{aligned} & \overset{\text{carrier}(-\omega_c)}{s_{\Pi(T_p)}}\, s_{\Pi(\infty)}(t, u_y)_n^{\text{chirp}} \\ &= \exp(-j\omega_c t) \\ &= \times \left\{ g_n \cdot \overline{\Pi_{(T_p)}(t-t_n)} \cdot \exp[j \cdot \overline{\omega_b(t-t_n)} \cdot (t-t_n)] \cdot \exp[j\omega_c(t-t_n)] \right\}, \\ &= g_n \cdot \overline{\Pi_{(T_p)}(t-t_n)} \cdot \exp[j \cdot \overline{\omega_b(t-t_n)} \cdot (t-t_n)] \cdot \exp(-j\omega_c t_n). \qquad (7.26) \end{aligned}$$

We rework this phase history decomposition in the following way,

$$\begin{aligned} & \overset{\text{carrier}(-\omega_c)}{s_{\Pi(T_p)}}\, s_{\Pi(\infty)}(t, u_y)_n^{\text{chirp}} \\ &= g_n \cdot \overline{\Pi_{(T_p)}(t-t_n)} \cdot \exp[j \cdot \overline{\omega_b(t-t_n)} \cdot t] \cdot \exp\{-j \cdot [\omega_c + \overline{\omega_b(t-t_n)}] \cdot t_n\}, \\ &= \underbrace{g_n \cdot \overline{\Pi_{(T_p)}(t-t_n)} \cdot \exp[j \cdot \overline{\omega_b(t-t_n)} \cdot t]}_{p_{\Pi(T_p @ t_n)}(t)^{\text{chirp}}} \cdot \underbrace{\exp[-j \cdot \overline{\omega_p(t-t_n)} \cdot t_n]}_{\measuredangle hh^2[\overline{w_p(t-t_n)}, r_n(u_y)]}. \qquad (7.27) \end{aligned}$$

Thus, the down converted time-space phase history can be expressed as

$$\begin{aligned} & \overset{\text{carrier}(-\omega_c)}{s_{\Pi(T_p)}}\, s_{\Pi(\infty)}(t, u_y)_n^{\text{chirp}} \\ &= g_n \cdot p_{\Pi(T_p @ t_n)}(t)^{\text{chirp}} \cdot \measuredangle hh^2[\overline{w_p(t-t_n)}, r_n(u_y)], \qquad (7.28) \end{aligned}$$

where

$$p_{\Pi(T_p@t_n)}(t)^{\text{chirp}} = \overline{\Pi_{(T_p)}(t-t_n)} \cdot \exp[j \cdot \overline{\omega_b(t-t_n)} \cdot t],$$

and

$$\begin{aligned}\measuredangle hh^2[\overline{w_p(t-t_n)}, r_n(u_y)] &= \exp[-j \cdot \overline{\omega_p(t-t_n)} \cdot t_n], \\ &= \exp[-j \cdot k_{\overline{\omega_p(t-t_n)}} \cdot 2r_n(u_y)]. \end{aligned} \quad (7.29)$$

The down converted time-space phase history of a point scatterer can be decomposed in a way that exposes a baseband chirp pulse $p_{\Pi(T_p@t_n)}(t)^{\text{chirp}}$ with a baseband chirp couplet $\overline{\omega_b[t-t_n(u_y)]}$ delayed by $t_n = (2/c) \cdot r_n(u_y)$ at aperture position u_y. The down converted chirp signal (7.28) also contains the passband chirp range walk phase curtain $\measuredangle hh^2[\overline{w_p(t-t_n)}, r_n(u_y)]$ of (7.29).

Received Echo Signal from Multiple Scatterers

The received time-space echo signal returned by all scatterers, measured within the radar's fast-time receive window $[T_s, T_f]$ and collected over the radar's (infinitely long) flight line, is

$$\begin{aligned} & {}^{\text{carrier}(-\omega_c)}_{\Pi[T_s,T_f]}s \quad s_{\Pi(\infty)}(t,u_y)^{\text{chirp}} \\ &= \sum_n {}^{\text{carrier}(-\omega_c)}_{\Pi(T_p)}s \quad s_{\Pi(\infty)}(t,u_y)_n^{\text{chirp}}, \\ &= \sum_n g_n \cdot p_{\Pi(T_p@t_n)}(t)^{\text{chirp}} \cdot \measuredangle hh^2[\overline{w_p(t-t_n)}, r_n(u_y)]. \end{aligned} \quad (7.30)$$

All signal elements in (7.30) are dependent on scatterer index n. We shall see that this is not the case in the frequency-space domain phase history obtained by the fast-time Fourier transform of (7.30).

Demodulated Time-Space Phase History Decomposition: Summary

The single scatterer's demodulated time-space domain phase history

$${}^{\text{carrier}(-\omega_c)}_{\Pi(T_p)}s \quad s_{\Pi(\infty)}(t,u_y)_n^{\text{chirp}} = g_n \cdot p_{\Pi(T_p@t_n)}(t)^{\text{chirp}} \cdot \measuredangle hh^2[\overline{w_p(t-t_n)}, r_n(u_y)]$$

has three factors that provide information about scatterer strength and scatterer location.

- Factor 1: g_n expresses the strength (complex radar cross section, etc.) of the point scatterer.
- Factor 2: $p_{\Pi(T_p@t_n)}(t)^{\text{chirp}}$ is a baseband phase history element that enables down-range imaging via chirp waveform pulse compression.
- Factor 3: $\measuredangle hh^2[\overline{w_p(t-t_n)}, r_n(u_y)]$ is a chirp passband range walk phase curtain that provides measurements along u_y of the scattered phase field of a Huygens chirp wavelet radiator (the illuminated scatterer) located at (x_n, y_n).

Factor 1 relates to scattering cross section and determines pixel brightness in the SAR image. If complex, the phase value of g_n represents, in part, the sub-wavelength positioning of the point scatterer (if only one scatterer exists within the resolution cell) or more generally, the collective reflective and dielectric properties of all scatterers within the resolution cell [290, 683]. A phase shift in the reflected signal may also occur because only a fraction of the incident radiation is reflected back to the radar [679].

Factor 2 is exploited by systems which rely on waveform pulse compression for the down-range part of the image formation process. This factor also describes the hyperbolic fast-time delay (within the chirp couplet) of baseband chirp pulse echoes as a function of slow-time. The fast-time delay of the chirp characteristic creates the slow-time hyperbolic range walk of the range only focused SAR image.

Factor 3 is the range walk phase curtain as determined by the chirp waveform's passband frequencies. The slow-time phase curtain is exploited for along-track imaging via aperture synthesis.

SAR signal models have been set up where the slow-time phase curtain phase factor has been essentially ignored [114, 679]. The phase curtain is absolutely required in near range, wide beamwidth, aperture synthesis. Although the phase curtain becomes approximately constant (independent of u_y) in long range, narrow beamwidth imaging scenarios, ignoring this important phase factor is not recommended in modern SAR systems.

Time-Space Baseband Phase Maps

The top row phase maps of Figs. 7.6a-c show the decomposition of a time-space passband phase history of a single scatterer,

$$s_{\Pi(T_p)} s_{\Pi(\infty)} (t, u_y)_n^{\text{chirp}}.$$

The bottom row phase maps of Figs. 7.6d-f show the corresponding baseband phase history elements of

$$\overset{\text{carrier}(-\omega_c)}{s_{\Pi(T_p)}} \; s_{\Pi(\infty)} (t, u_y)_n^{\text{chirp}}. \tag{7.31}$$

These phase maps are produced with signal simulations using LFM up chirp waveforms. The phase map of Fig. 7.6e shows the phase map of the fast-time baseband waveform (7.21) that has a delayed chirp couplet,

$$\begin{aligned} p_{\Pi(T_p @ t_n)}(t)^{\text{LFM}} &= \overline{\Pi_{(T_p)}(t - t_n)} \cdot \exp[j \cdot \overline{\omega_b(t - t_n)^{\text{LFM}}} \cdot t], \\ &= \overline{\Pi_{(T_p)}(t - t_n)} \cdot \exp[j \cdot \overline{\alpha \cdot (t - t_n)} \cdot t], \\ &= \overline{\Pi_{(T_p)}(t - t_n)} \cdot \exp(j \alpha t^2) \cdot \exp[-j \alpha t_n(u_y) t]. \end{aligned} \tag{7.32}$$

The phase map Fig. 7.6f shows the phase map of the slow-time passband range walk chirp phase curtain (7.17),

$$\begin{aligned} &\measuredangle hh^2 \overline{[w_p(t - t_n), r_n(u_y)]} \\ &= \exp[-j \cdot \overline{\omega_p(t - t_n)^{\text{LFM}}} \cdot t_n], \\ &= \exp\{-j \cdot [\omega_{\min} + \overline{\alpha \cdot (t - t_n)}] \cdot t_n\}, \\ &= \exp[-j \omega_{\min} t_n(u_y)] \cdot \exp[-j \alpha t_n(u_y) t] \cdot \exp[j \alpha t_n^2(u_y)]. \end{aligned} \tag{7.33}$$

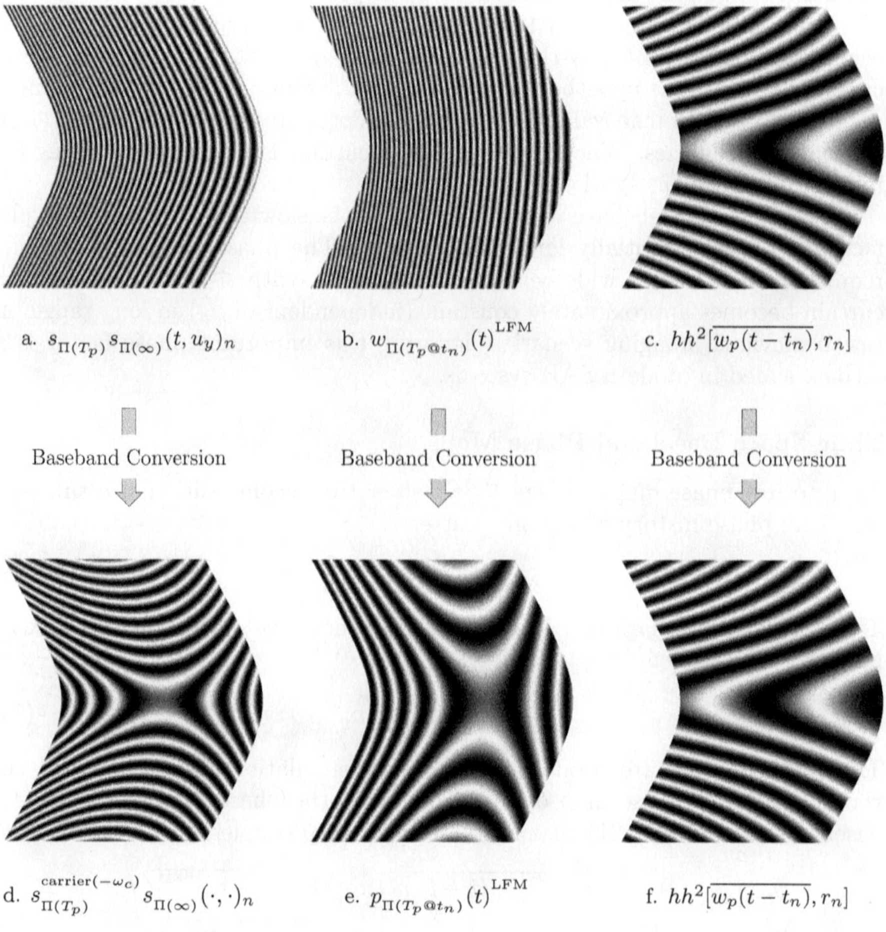

Figure 7.6: Baseband time-space up chirp LFM decomposition.

The phase history elements (7.32) and (7.33) combine to produce the down converted time-space phase history (7.31) shown in Fig. 7.6d. The down chirp version of these phase maps are shown in Figs. 7.7a-f.

Figure 7.7: Baseband time-space down chirp LFM decomposition.

7.4 Frequency-Space Phase History

We next determine the chirp frequency-space phase history $Ss(\omega_p, u_y)^{\text{chirp}}$ of the slant-plane SAR imaging system. We first examine the passband frequency-space phase history of a single scatterer. We then examine changes within the frequency-space phase history that occur with down conversion of the received signal.

Frequency-Space Phase History by Fourier Transform

The first step of SAR image formation processing with Stolt formatting is to convert the measured time-space phase history,

$$s_{\Pi[T_s,T_f]} s_{\Pi(\infty)}(t, u_y)^{\text{chirp}}$$

to the frequency-space domain,

$$S_{\Pi(2\omega_o @\omega_c)} s_{\Pi(\infty)}(\omega_p, u_y)^{\text{chirp}}.$$

Compared to the continuous wave case previously discussed in Chapter 4, down conversion alone of the received echo signal does not suffice for the transformation[2],

$$ss(t, u_y)^{\text{chirp}} \mapsto Ss(\omega_p, u_y)^{\text{chirp}},$$

when chirp pulse signals are transmitted. A marginal fast-time Fourier transform is required to convert $ss(t, u_y)^{\text{chirp}}$ to $Ss(\omega_p, u_y)^{\text{chirp}}$.

Frequency-Space Phase History with General Chirp Expressions

Our generalized expressions of time-space chirp signals, as developed in Section 6.1, facilitate evaluation and expression of the result of the fast-time Fourier transform applied to the time-space phase history. The fast-time Fourier transform of time-space data of a single scatterer is

$$\begin{aligned} & S_{\Pi(2\omega_o @\omega_c)} s_{\Pi(\infty)}(\omega_p, u_y)_n^{\text{chirp}} \\ &= \mathsf{FT}^t_{\text{1D}} \left\{ s_{\Pi(T_p)} s_{\Pi(\infty)}(t, u_y)_n^{\text{chirp}} \right\}, \\ &= \mathsf{FT}^t_{\text{1D}} \left\{ g_n \cdot \overline{\Pi_{(T_p)}(t - t_n)} \cdot \exp[j \cdot \overline{\omega_p(t - t_n)} \cdot (t - t_n)] \right\}, \\ &= g_n \cdot \underbrace{\overline{\Pi_{(2\omega_o @\omega_c)}(\omega_p)} \cdot \exp[j \cdot \overline{t(\omega_p)} \cdot \omega_p]}_{W_{\Pi(2\omega_o @\omega_c)}(\omega_p)^{\text{chirp}}} \cdot \exp(-j\omega_p t_n). \end{aligned} \tag{7.34}$$

Equation (7.34) is obtained by the time shift property of the Fourier transform and the Fourier transform of a general passband chirp pulse signal (see Section 6.1).

Frequency-Space Phase Curtains

In the phase history of (7.34), we find the spectrum of the passband chirp waveform,

$$W_{\Pi(2\omega_o @\omega_c)}(\omega_p)^{\text{chirp}} = \overline{\Pi_{(2\omega_o @\omega_c)}(\omega_p)} \cdot \exp[j \cdot \overline{t(\omega_p)} \cdot \omega_p]. \tag{7.35}$$

We also see a phase curtain

$$\exp(-j\omega_p t_n) = \exp\left[-jk_{\omega_p} \cdot 2r_n(u_y)\right],$$

[2]If a radar system employs stepped frequency waveforms, then the signal gathered in time and space directly resides in the frequency-space domain [389, 698].

which we choose to write as

$$\measuredangle Hh^2[\omega_p, r_n(u_y)] \;=\; \exp\left[-jk_{\omega_p} \cdot 2r_n(u_y)\right]. \qquad (7.36)$$

The phase factor (7.36) is similar to the time-space range walk phase curtain,

$$\measuredangle hh^2[\overline{\omega_p(t-t_n)}, r_n(u_y)] = \exp[-jk_{\overline{\omega_p(t-t_n)}} \cdot 2r_n(u_y)], \qquad (7.37)$$

the only difference being k_{ω_p} in (7.36) is replaced by $k_{\overline{\omega_p(t-t_n)}}$ in (7.37). We call (7.36) a *frequency-space phase curtain.*

The time-space phase curtain (7.37) is chirped, as indicated by the existence of over bars in the expression. The frequency-space phase curtain (7.36) is not chirped.

The frequency-space phase curtain (7.36) is independent of time and dependent on the passband angular frequency parameter ω_p that is emergent from the fast-time Fourier transform (7.34). The passband angular frequency parameter ω_p in the frequency-space phase curtain (7.36) does not originate from the chirp couplet of a time domain chirp signal.

Frequency-Space Phase History in Phase Modulated Spectral Form

With the decomposition provided by (7.35) and (7.36), the frequency-space phase history of a single scatterer is

$$S_{\Pi(2\omega_o @\omega_c)}s_{\Pi(\infty)}(\omega_p, u_y)_n^{\text{chirp}} \;=\; W_{\Pi(2\omega_o @\omega_c)}(\omega_p)^{\text{chirp}} \cdot g_n \cdot \measuredangle Hh^2[\omega_p, r_n(u_y)].$$

Frequency-Space Passband Phase History of Entire Ground Scene

By superposition, the fast-time Fourier transform of the passband echo signal from all scatterers is

$$\begin{aligned} &S_{\Pi(2\omega_o @\omega_c)}s_{\Pi(\infty)}(\omega_p, u_y)^{\text{chirp}} \\ &\quad= W_{\Pi(2\omega_o @\omega_c)}(\omega_p)^{\text{chirp}} \sum_n g_n \cdot \measuredangle Hh^2[\omega_p, r_n(u_y)]. \end{aligned} \qquad (7.38)$$

The chirp waveform spectrum $W_{\Pi(2\omega_o @\omega_c)}(\omega)^{\text{chirp}}$ is independent of the scatterer index n and can be moved outside the summation. Equation (7.38) provides a canonical form typical of frequency-space phase histories.

Passband Frequency-Space Phase History: Data Matrix Interpretation

As illustrated in Figs. 7.8a-b, the fast-time Fourier transform of the time-space phase history is computed for each row of the time-space data matrix. A discrete Fourier transform of the receive window's fast-time data samples along the rows of the measured data matrix yields fast-time frequency bins corresponding to the matrix columns. The frequency bin columns of the data matrix span the passband. Resident in the column data are the scatterer specific frequency-space phase curtains $\measuredangle Hh^2[\omega_p, r_n(u_y)]$. These phase curtains (one for each scatterer) are spatially induced phase differences of transmitted and received passband chirp pulse signals. These phase curtains are functions of sensor position and instantaneous chirp frequency within the passband.

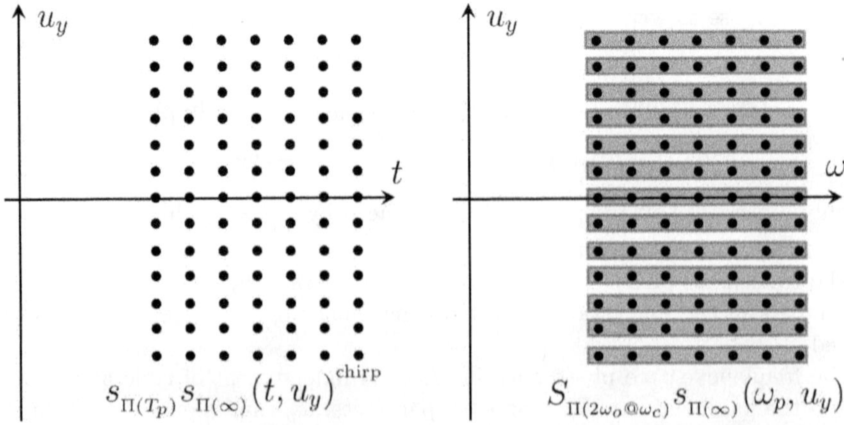

a. Time-space data. b. Frequency-space data.

Figure 7.8: Frequency-space phase history data matrix.

Down Converted Frequency-Space Phase History

We next explore the frequency-space phase history of the down converted chirp pulse echo of a single scatterer. We shall see that demodulation down converts some, but not all, passband frequency dependent phase factors of the frequency-space phase history.

Baseband Frequency-Space Phase History of Single Scatterer

Consider the fast-time Fourier transform of the down converted chirp pulse time-space phase history as described by (7.26),

$$
\begin{aligned}
&\mathsf{FT}^t_{1\text{D}}\left\{ s^{\text{carrier}(-\omega_c)}_{\Pi(T_p)} \; s_{\Pi(\infty)}(t, u_y)^{\text{chirp}}_n \right\} \\
&= \mathsf{FT}^t_{1\text{D}}\left\{ g_n \cdot \overline{\Pi_{(T_p)}(t-t_n)} \cdot \exp[j \cdot \overline{\omega_b(t-t_n)} \cdot (t-t_n)] \cdot \exp(-j\omega_c t_n) \right\}, \\
&= \exp(-j\omega_c t_n) \times \mathsf{FT}^t_{1\text{D}}\left\{ g_n \cdot \overline{\Pi_{(T_p)}(t-t_n)} \cdot \exp[j \cdot \overline{\omega_b(t-t_n)} \cdot (t-t_n)] \right\}.
\end{aligned}
$$

Using the time shift property of the Fourier transform and the expression of the Fourier transform of a general baseband chirp pulse waveform, evaluation of this fast-time Fourier transform gives, since $\exp\{-j\omega_c t_n\}$ is independent of fast-time,

$$
\begin{aligned}
&S^{\text{L.shift}(\omega_c)}_{\Pi(2\omega_o @ \omega_c)} \; s_{\Pi(\infty)}(\omega_p, u_y)^{\text{chirp}}_n \\
&= \mathsf{FT}^t_{1\text{D}}\left\{ s^{\text{carrier}(-\omega_c)}_{\Pi(T_p)} \; s_{\Pi(\infty)}(t, u_y)^{\text{chirp}}_n \right\} \\
&= \exp(-j\omega_c t_n) \times \left\{ g_n \cdot \overline{\Pi_{(2\omega_o @ 0)}(\omega_b)} \cdot \exp[j \cdot \overline{t(\omega_b)} \cdot \omega_b] \cdot \exp(-j\omega_b t_n) \right\}, \\
&= \underbrace{g_n \cdot \overline{\Pi_{(2\omega_o @ 0)}(\omega_b)} \cdot \exp[j \cdot \overline{t(\omega_b)} \cdot \omega_b]}_{P_{\Pi(2\omega_o @ 0)}(\omega_b)^{\text{chirp}}} \cdot \underbrace{\exp[-j(\omega_c + \omega_b) t_n]}_{\exp(-j\omega_p t_n)}.
\end{aligned}
$$

Using notations of the constituent baseband chirp pulse spectrum and the slow-time phase curtain elements appropriate for the frequency-space domain,

$$\begin{aligned} P_{\Pi(2\omega_o@0)}(\omega_b)^{\text{chirp}} &= \overline{\Pi_{(2\omega_o@0)}(\omega_b)} \cdot \exp[j \cdot \overline{t(\omega_b)} \cdot \omega_b], \\ \measuredangle Hh^2[\omega_p, r_n(u_y)] &= \exp(-j\omega_p t_n), \end{aligned}$$

we can concisely write the *down converted chirp frequency-space phase history* of a single scatterer as

$$S_{\Pi(2\omega_o@\omega_c)}^{\text{L.shift}(\omega_c)} s_{\Pi(\infty)}(\omega_p, u_y)_n^{\text{chirp}} = P_{\Pi(2\omega_o@0)}(\omega_b)^{\text{chirp}} \cdot g_n \cdot \measuredangle Hh^2[\omega_p, r_n(u_y)].$$

Even after down conversion, the slow-time phase curtain of the frequency-space phase history remains a function of passband frequencies.

Comparison of Time-Space and Frequency-Space Forms

We next summarize some of the results obtained so far in this chapter. This allows a side by side comparison of key equations related to time-space and frequency-space phase history decompositions.

Time-Space Phase Histories in Canonical Form

The passband chirp time-space phase history

$$s_{\Pi[T_s,T_f]} s_{\Pi(\infty)}(t, u_y)^{\text{chirp}} = \sum_n w_{\Pi(T_p@t_n)}(t)^{\text{chirp}} \cdot g_n \cdot \measuredangle hh^2[\overline{\omega_p(t-t_n)}, r_n(u_y)] \tag{7.39}$$

subjected to down conversion gives the baseband chirp time-space phase history,

$$s_{\Pi[T_s,T_f]}^{\text{carrier}(-\omega_c)} s_{\Pi(\infty)}(t, u_y)^{\text{chirp}} = \sum_n p_{\Pi(T_p@t_n)}(t)^{\text{chirp}} \cdot g_n \cdot \measuredangle hh^2[\overline{\omega_p(t-t_n)}, r_n(u_y)]. \tag{7.40}$$

The shifted time element $t - t_n(u_y)$ in the chirp couplet $\overline{\omega_p[t - t_n(u_y)]}$ of the slow-time range walk phase curtain

$$hh^2[\overline{\omega_p(t-t_n)}, r_n(u_y)]$$

is a parameter that specifies the coupling between chirp passband frequencies and fast-time echo delay as a function of sensor position within the synthetic aperture.

Frequency-Space Phase Histories in Canonical Form

The passband chirp frequency-space phase history of all ground scatterer echo signals, obtained by the temporal Fourier transform of (7.39), is

$$S_{\Pi(2\omega_o@\omega_c)} s_{\Pi(\infty)}(\omega_p, u_y)^{\text{chirp}} = W_{\Pi(2\omega_o@\omega_c)}(\omega_p)^{\text{chirp}} \cdot \sum_n g_n \cdot \measuredangle Hh^2[\omega_p, r_n(u_y)].$$

The baseband chirp frequency-space phase history,

$$S^{\text{L.shift}(\omega_c)}_{\Pi(2\omega_o @\omega_c)} s_{\Pi(\infty)}(\omega_p, u_y)^{\text{chirp}} = P_{\Pi(2\omega_o @0)}(\omega_b)^{\text{chirp}} \cdot \sum_n g_n \cdot \measuredangle Hh^2[\omega_p, r_n(u_y)],$$

is obtained by fast-time down conversion and temporal Fourier transformation of (7.40), where $\omega_b = \omega_p - \omega_c$.

Scatterer Independence in Frequency-Space Domain

An important difference between the time-space and frequency-space chirp phase histories is that the delayed chirp couplet waveform elements,

$$w_{\Pi(T_p @t_n)}(t)^{\text{chirp}}$$
$$p_{\Pi(T_p @t_n)}(t)^{\text{chirp}}$$

of the time-space domain are dependent on scatterer index, but the corresponding spectral elements in the frequency-space domain,

$$W_{\Pi(2\omega_o @\omega_c)}(\omega_p)^{\text{chirp}}$$
$$P_{\Pi(2\omega_o @0)}(\omega_b)^{\text{chirp}}$$

are independent of scatterer index. This observation is exploited in some SAR image formation algorithms that rely on waveform pulse compression as a form of polychromatic coherent processing gain (see Chapter 13).

Frequency-Space Baseband Phase Maps

We next present phase maps of the decomposition of the slant-plane frequency-space phase history. We examine phase maps before and after down conversion, and with up and down chirp LFM waveforms.

LFM Up Chirp

The phase maps of Figs. 7.6d-f are replicated in the top row of the phase maps of Figs. 7.9a-c. The phase map of Fig. 7.9d is the LFM frequency-space baseband phase history,

$$S^{\text{L.shift}(\omega_c)}_{\Pi(2\omega_o @\omega_c)} s_{\Pi(\infty)}(\omega_p, u_y)_n^{\text{LFM}}.$$

The horizontal axes of Figs. 7.9d and 7.9e represent baseband angular frequency,

$$\omega_b \in [-\omega_o, +\omega_o].$$

The horizontal axis of Fig. 7.9f indicates passband angular frequency,

$$\omega_p \in [\omega_c - \omega_o, \omega_c + \omega_o].$$

The vertical axes of all phase maps of Figs. 7.9a-f represent aperture position u_y.

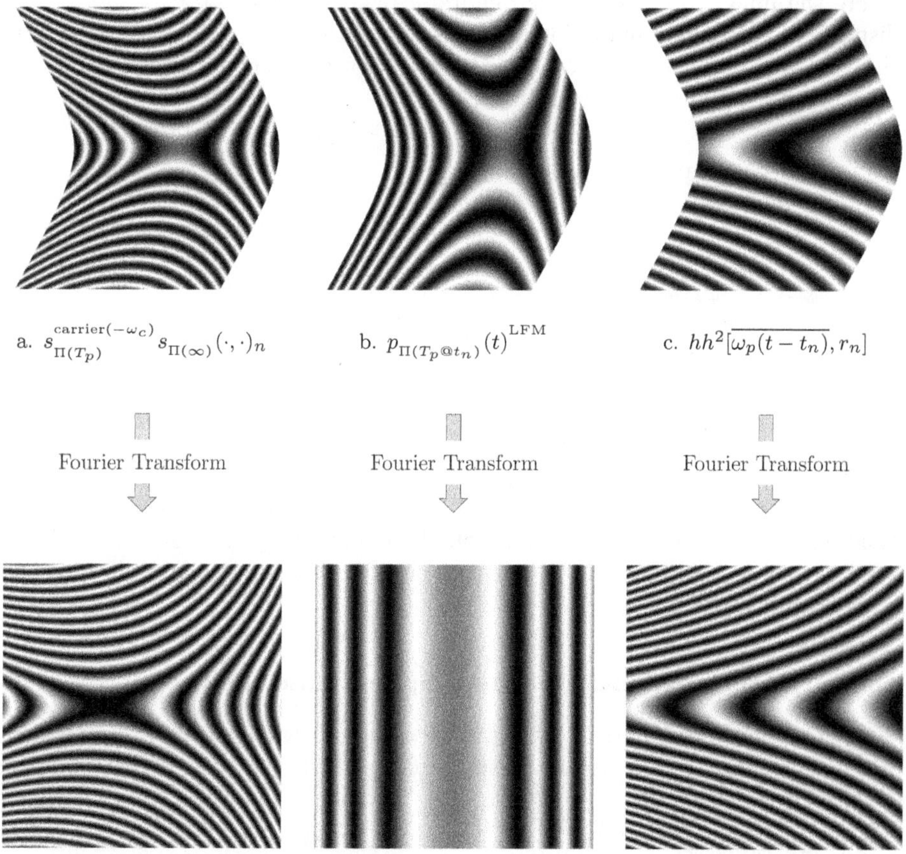

a. $s_{\Pi(T_p)}^{\text{carrier}(-\omega_c)} s_{\Pi(\infty)}(\cdot,\cdot)_n$ b. $p_{\Pi(T_p@t_n)}(t)^{\text{LFM}}$ c. $hh^2[\overline{\omega_p(t-t_n)}, r_n]$

d. $S_{\Pi(2\omega_o)}^{\text{L.shift}(\omega_c)} s_{\Pi(\infty)}(\cdot,\cdot)_n$ e. $P_{\Pi(2\omega_o)}(\omega_p-\omega_c)^{\text{LFM}}$ f. $\measuredangle Hh^2[\omega_p, r_n(u_y)]$

Figure 7.9: Frequency-space up chirp LFM phase history decomposition.

Baseband Pulse Spectrum

Fig. 7.9e shows the spectrum of an LFM baseband chirp pulse,

$$\begin{aligned} P_{\Pi(2\omega_o@0)}(\omega_b)^{\text{LFM}} &= \overline{\Pi_{(2\omega_o@0)}(\omega_b)} \cdot \exp[j \cdot \overline{t(\omega_b)^{\text{LFM}}} \cdot \omega_b], \\ &= \overline{\Pi_{(2\omega_o@0)}(\omega_b)} \cdot \exp\left(j\frac{\omega_b^2}{4\alpha}\right), \end{aligned}$$

where the imaginary part of the quadratic phase modulation $\exp\left[j\omega_b^2/(4\alpha)\right]$ is displayed in the phase map. The phase center of this quadratic phase function in the ω_b domain is clearly visible. This phase factor is independent of u_y and hence has no variation in the vertical dimension.

Passband Slow-Time Phase Curtain

The phase map of Fig. 7.9f is the frequency-space passband phase curtain,

$$\begin{aligned} \measuredangle Hh^2[\omega_p, r_n(u_y)] &= \exp\left(-j\omega_p t_n\right), \\ &= \exp\left[-j \cdot (\omega_{\text{mid}} + \omega_b) \cdot t_n\right], \end{aligned} \tag{7.41}$$

where

$$\omega_b \in 2\alpha \cdot \left[-\frac{T_p}{2}, +\frac{T_p}{2}\right]$$

for LFM waveforms (see Section 12.5). Since this is a phase map of the slow-time phase curtain plotted as a function of passband frequency bin value, there is no range walk. Range walk is evident in the time-space passband phase curtain displayed in Fig. 7.9c.

Down Converted Signal Frequency-Space Phase History

The slow-time phase curtain of Fig. 7.9f combines with the pulse spectrum of Fig. 7.9e to produce the frequency-space phase history of a single scatterer,

$$S_{\Pi(2\omega_o@\omega_c)}^{\text{L.shift}(\omega_c)} s_{\Pi(\infty)}(\omega_p, u_y)_n^{\text{LFM}} = g_n \cdot P_{\Pi(2\omega_o@0)}(\omega_b)^{\text{LFM}} \cdot \measuredangle Hh^2[\omega_p, r_n(u_y)],$$

as shown in Fig. 7.9d.

Time-Space and Frequency-Space Phase History Differences

A difference between phase maps of time-space phase histories and frequency-space phase histories is the absence of hyperbolic range walk in the frequency-space phase history. The Fourier transform that converts the fast-time dimension of the time-space phase history to the frequency domain changes the fast-time signal delay experienced at each pulse transmission/reception point to a u_y dependent complex sinusoidal carrier in the frequency-space phase history. We shall explore decompositions of the frequency-space phase history that isolate these delay specific tones.

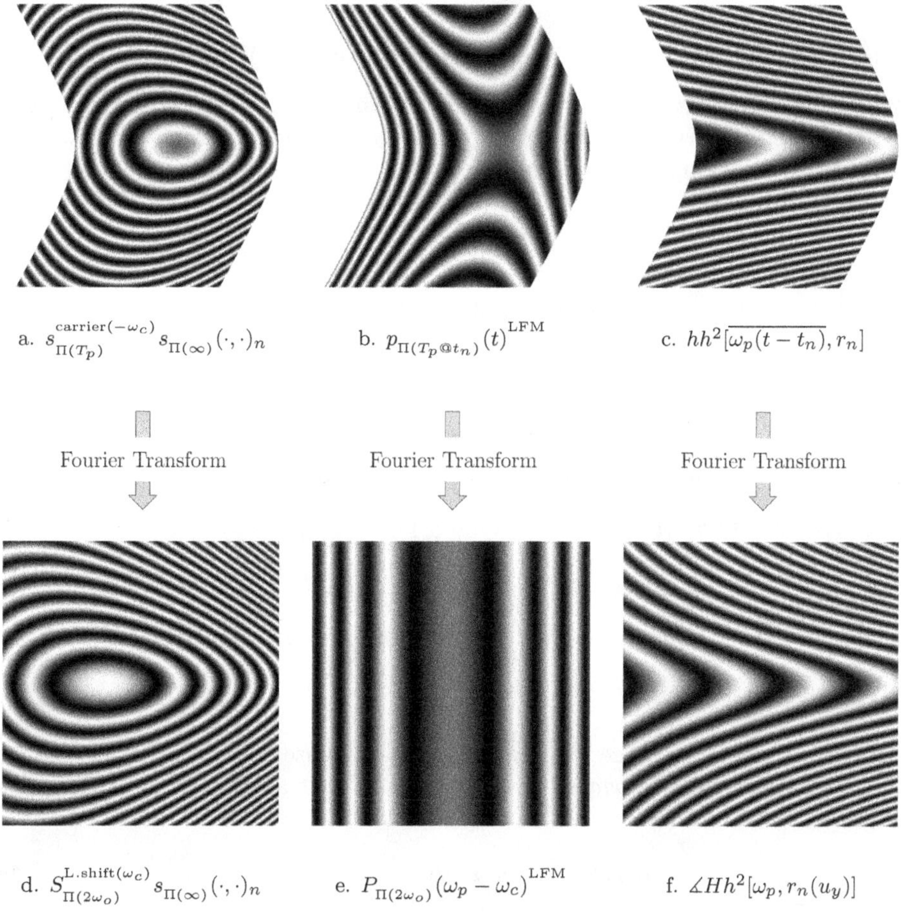

Figure 7.10: Frequency-space down chirp LFM phase history decomposition.

LFM Down Chirp

Figs. 7.10a-f show the same set of phase maps as Figs. 7.9a-f, but where the transmitted signal is an LFM down chirp waveform.

7.5 Pulse Compressed Phase History

We now examine in detail down-range imaging via chirp waveform two-pass (matched filter) pulse compression for slant-plane SAR image formation.

Slant-Plane Chirp Pulse Compression Matched Filtering

Waveform pulse compression removes the non-informative (about scatterer properties) phase factors associated with pulse signal attributes from the down converted frequency-space phase history,

$$S^{\text{L.shift}(\omega_c)}_{\Pi(2\omega_o @\omega_c)} s_{\Pi(\infty)}(\omega_p, u_y)^{\text{chirp}}. \tag{7.42}$$

As explained in Chapter 5, one approach to pulse compression involves canceling the phase modulation of the chirp pulse spectrum

$$\measuredangle P_{\Pi(2\omega_o)}(\omega_b)^{\text{chirp}} = \exp[j \cdot \overline{t(\omega_b)} \cdot \omega_b]$$

by matched filtering.

Chirp Waveform Matched Filter at Baseband

The matched filter for the baseband chirp pulse spectrum is

$$P^*_{\Pi(2\omega_o)}(\omega_b)^{\text{chirp}} = \overline{\Pi_{(2\omega_o)}(\omega_b)} \cdot \exp[-j \cdot \overline{t(\omega_b)} \cdot \omega_b], \tag{7.43}$$

where we have assumed the pulse spectrum window function $\overline{\Pi_{(2\omega_o)}(\omega_b)}$ is real.

Range Compression

Range compression changes the frequency-space phase history $Ss(\cdot,\cdot)$ to a *range focused frequency-space phase history* $Rs(\cdot,\cdot)$. This transformation is more fully express in scripted notation as

$$S^{\text{L.shift}(\omega_c)}_{\Pi(2\omega_o @\omega_c)} s_{\Pi(\infty)}(\omega_p, u_y)^{\text{chirp}} \mapsto R^{\text{L.shift}(\omega_c)}_{\hat{\Pi}(2\omega_o @\omega_c)} s_{\Pi(\infty)}(\omega_p, u_y)^{\text{chirp}}. \tag{7.44}$$

Multiplication of (7.42) with the matched filter (7.43) performs frequency domain range compression,

$$\begin{aligned} &R^{\text{L.shift}(\omega_c)}_{\hat{\Pi}(2\omega_o @\omega_c)} s_{\Pi(\infty)}(\omega_p, u_y)^{\text{chirp}} \\ &= P^*_{\Pi(2\omega_o)}(\omega_b)^{\text{chirp}} \times \left\{ S^{\text{L.shift}(\omega_c)}_{\Pi(2\omega_o @\omega_c)} s_{\Pi(\infty)}(\omega_p, u_y)^{\text{chirp}} \right\}, \\ &= P^*_{\Pi(2\omega_o)}(\omega_b)^{\text{chirp}} \times P_{\Pi(2\omega_o)}(\omega_b)^{\text{chirp}} \cdot \sum_n g_n \cdot \measuredangle Hh^2[\omega_p, r_n(u_y)], \\ &= \left| P_{\Pi(2\omega_o)}(\omega_b)^{\text{chirp}} \right|^2 \cdot \sum_n g_n \cdot \measuredangle Hh^2[\omega_p, r_n(u_y)]. \end{aligned} \tag{7.45}$$

Range Compression Weight Function

The weighting imposed on (7.45) by the modulus squared of the chirp pulse spectrum,

$$\left| P_{\Pi(2\omega_o)}(\omega_b)^{\text{chirp}} \right|^2$$

is indicated in (7.44) by the subscript weight function $\widehat{\Pi}(2\omega_o@\omega_c)$ contained in

$$R^{\text{L.shift}(\omega_c)}_{\widehat{\Pi}(2\omega_o@\omega_c)} s_{\Pi(\infty)}(\omega_p, u_y)^{\text{chirp}}.$$

Range Compressed Spectrum Phase Maps

Range compression through matched filtering is illustrated by the phase maps of (7.44) as shown in Figs. 7.11a-f. The phase maps of Fig. 7.11c and Fig. 7.11f are the frequency-space phase curtains, which are not changed by pulse compression. The phase map of Fig. 7.11b shows the chirp pulse spectrum. Multiplication of this phase factor by its conjugate leaves the modulus squared chirp pulse spectrum weight function (nearly a constant) as indicated in Fig. 7.11e. The phase map of Fig. 7.11a is the frequency-space phase history before pulse compression matched filtering. The phase map of Fig. 7.11d is the frequency-space phase history after pulse compression matched filtering. The only phase factor that remains after pulse compression is the weighted frequency-space slow-time phase curtain.

Pulse Compression: Data Matrix Interpretation

At this point, as illustrated in Figs. 7.12a-b, matched filtering for pulse compression involves multiplication of

$$S^{\text{L.shift}(\omega_c)}_{\Pi(2\omega_o@\omega_c)} s_{\Pi(\infty)}(\omega_p, u_y)^{\text{chirp}}$$

along the rows of the collected data matrix with the samples of

$$P^*_{\Pi(2\omega_o)}(\omega_b)^{\text{chirp}}.$$

This transforms the data matrix into the array represented by

$$R^{\text{L.shift}(\omega_c)}_{\widehat{\Pi}(2\omega_o@\omega_c)} s_{\Pi(\infty)}(\omega_p, u_y)^{\text{chirp}}.$$

The frequency-space phase curtain comprises the range focused frequency-space phase history.

7.6 Range Focused Slant-Plane Image

After pulse compression matched filtering, an inverse Fourier transform of the range compressed frequency-space phase history produces an image focused in the down-range dimension. Inverse Fourier transforms of the rows of the range compressed phase history data matrix can be formulated as

$$\begin{aligned}
&\mathsf{iFT}^{\omega_b}_{1\text{D}} \left\{ R^{\text{L.shift}(\omega_c)}_{\widehat{\Pi}(2\omega_o@\omega_c)} s_{\Pi(\infty)}(\omega_p, u_y)^{\text{chirp}} \right\} \\
&= \mathsf{iFT}^{\omega_b}_{1\text{D}} \left\{ \left| P_{\Pi(2\omega_o)}(\omega_b)^{\text{chirp}} \right|^2 \cdot \sum_n g_n \cdot \measuredangle Hh^2[\omega_p, r_n(u_y)] \right\}, \qquad (7.46)
\end{aligned}$$

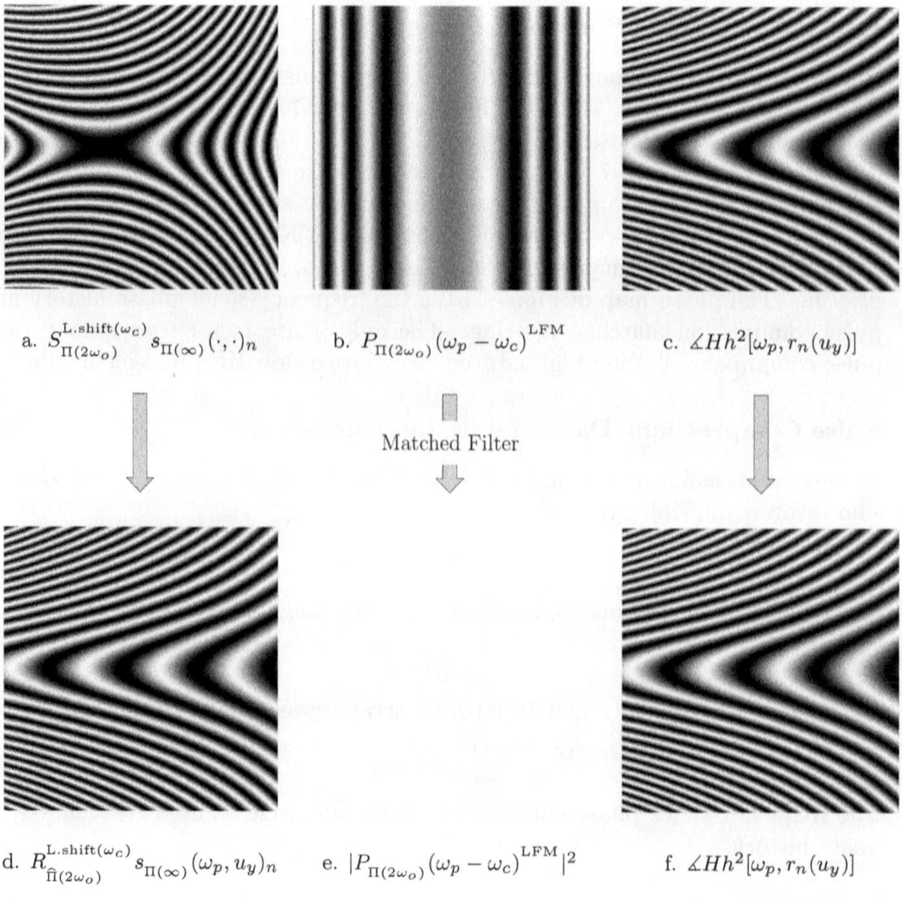

Figure 7.11: Frequency-space pulse compressed phase history decomposition.

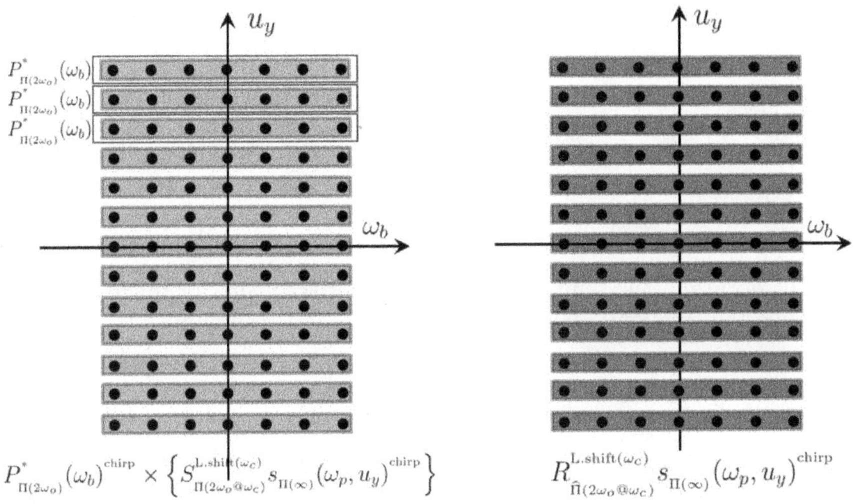

a. Pulse compression row multiplies. b. Range-only focused spectrum.

Figure 7.12: Pulse compressed frequency-space phase history data matrix.

where

$$\measuredangle Hh^2[\omega_p, r_n(u_y)] = \exp\left[-j \cdot (\omega_c + \omega_b) \cdot t_n(u_y)\right]$$

and where we have used $\omega_p = \omega_c + \omega_b$. The first step is to identify those elements of the range compressed down converted frequency-space phase history (7.46) that are dependent on baseband angular frequency. Only these factors are involved with the inverse Fourier transform (7.46). Hence, we have

$$\begin{aligned} &\mathsf{iFT}^{\omega_b}_{1D}\left\{ R^{\text{L.shift}(\omega_c)}_{\hat{\Pi}(2\omega_o @ \omega_c)} s_{\Pi(\infty)}(\omega_p, u_y)^{\text{chirp}} \right\} \\ &= \mathsf{iFT}^{\omega_b}_{1D}\left\{ \left| P_{\Pi(2\omega_o)}(\omega_b)^{\text{chirp}} \right|^2 \cdot \sum_n g_n \cdot \exp\left[-j \cdot (\omega_c + \omega_b) \cdot t_n(u_y)\right] \right\}, \\ &= \underbrace{\sum_n \mathsf{iFT}^{\omega_b}_{1D}\left\{ \left| P_{\Pi(2\omega_o)}(\omega_b)^{\text{chirp}} \right|^2 \cdot g_n \cdot \exp\left[-j\omega_b t_n(u_y)\right] \right\}}_{\text{temporal inverse Fourier transform}} \cdot \underbrace{\exp\left[-j\omega_c t_n(u_y)\right]}_{\text{not } \omega_b \text{ dependent}}. \end{aligned} \tag{7.47}$$

Scatterer Specific Range Walk

The point spread function $\mathsf{psf}(\cdot)$ in the down-range dimension is determined by the inverse Fourier transform of the modulus squared chirp baseband pulse spectrum $|P_{\Pi(2\omega_o)}(\omega_b)^{\text{chirp}}|^2$. If the inverse Fourier transform operates on data in the baseband angular frequency domain ω_b (instead of a spatial wavenumber domain), then the result of the inverse Fourier transform resides in the time domain. We indicate this temporal dependency by modifying our notation for a point spread function by adding a "t" subscript,

$$\mathsf{psf}(\cdot) \mapsto \mathsf{psf}_t(\cdot),$$

where

$$\mathsf{psf}_t(t)^{\text{chirp}} = \mathsf{iFT}_{\text{1D}}^{\omega_b}\left\{\left|P_{\Pi(2\omega_o)}(\omega_b)^{\text{chirp}}\right|^2\right\}.$$

Using this notation, the temporal inverse Fourier transform of range compressed down converted frequency-space phase history (7.47) results in

$$\begin{aligned}&\mathsf{iFT}_{\text{1D}}^{\omega_b}\left\{R_{\hat{\Pi}(2\omega_o @ \omega_c)}^{\text{L.shift}(\omega_c)} s_{\Pi(\infty)}(\omega_p, u_y)^{\text{chirp}}\right\}\\ &= \sum_n g_n \cdot \mathsf{psf}_t\left[t - t_n(u_y)\right]^{\text{chirp}} \cdot \exp\left[-j\omega_c t_n(u_y)\right].\end{aligned} \tag{7.48}$$

Inverse Fourier Transforms for Range Focusing

The time-space domain range focused image (7.48) has scatterer specific pulsed compressed range walk elements,

$$\mathsf{psf}_t\left[t - t_n(u_y)\right]^{\text{chirp}}.$$

This image remains unfocused in the along-track dimension and range walk is evident in the range (only) focused image. The degree of range walk for the nth scatterer at aperture position u_y is determined by the time difference $t - t_n(u_y)$. This time delay positions the point spread function $\mathsf{psf}_t(\cdot)$ at

$$t_n(u_y) = \frac{2}{c} r_n(u_y)$$

in the range dimension of the range compressed image.

Carrier Frequency Phase Curtain

The range focused image (7.48) has an embedded range walk phase curtain

$$\exp\left[-j\omega_c t_n(u_y)\right] \tag{7.49}$$

that exists along the hyperbolic trace of the range focused image. This phase curtain is dependent on the carrier frequency. This phase factor is called the *carrier range walk phase curtain.*

Down Converted Range Compressed Image

The range compressed image has an extraneous phase artifact (in the range focused pixel domain) related to down conversion. To see this, we begin by multiplying (7.48) with both $\exp(-j\omega_c t)$ and $\exp(+j\omega_c t)$,

$$\begin{aligned}
&\exp(-j\omega_c t)\cdot\exp(+j\omega_c t)\times\left\{\sum_n g_n\cdot\mathsf{psf}_t\left[t-t_n(u_y)\right]^{\text{chirp}}\cdot\exp\left[-j\omega_c t_n(u_y)\right]\right\}\\
&= \exp(-j\omega_c t)\cdot\sum_n \underbrace{g_n\cdot\mathsf{psf}_t\left[t-t_n(u_y)\right]^{\text{chirp}}\cdot\exp\left\{j\omega_c\left[t-t_n(u_y)\right]\right\}}_{r_{\Pi[T_s,T_f]}s_{\Pi(\infty)}(t,u_y)_n^{\text{chirp}}},\\
&= \exp(-j\omega_c t)\cdot\sum_n r_{\Pi[T_s,T_f]}s_{\Pi(\infty)}(t,u_y)_n^{\text{chirp}}.
\end{aligned}\tag{7.50}$$

In (7.50), the "clean" range compressed range walk profile of the nth scatterer is

$$r_{\Pi[T_s,T_f]}s_{\Pi(\infty)}(t,u_y)_n^{\text{chirp}} = g_n\cdot\mathsf{psf}_t\left[t-t_n(u_y)\right]^{\text{chirp}}\cdot\exp\left\{j\omega_c\left[t-t_n(u_y)\right]\right\}.$$

There exists an extraneous phase factor $\exp(-j\omega_c t)$ in (7.50) due to down conversion. The existence of this phase factor is rather covert, since it cancels the time dependent element $\exp(+j\omega_c t)$ of the carrier frequency phase curtain of the range focused image as expressed in (7.26) and (7.50), i.e.,

$$\exp(-j\omega_c t)\times\exp\left\{j\omega_c\left[t-t_n(u_y)\right]\right\} = \exp\left[-j\omega_c t_n(u_y)\right].$$

This cancelation should be avoided in working with expressions of the range focused image in the time-space domain. The full time shifted phase factor $\exp\left\{j\omega_c\left[t-t_n(u_y)\right]\right\}$, as contained in (7.50), is required to fully express scatterer specific, carrier frequency dependent, range walk phase curtains. The phase factor $\exp\left\{j\omega_c\left[t-t_n(u_y)\right]\right\}$ exists because no aperture synthesis has yet been performed in the along-track dimension.

Range Compressed Image of Ground Scene

Collectively, the range compressed range walk profiles of all scatterers form the range compressed image,

$$\begin{aligned}
&\overset{\text{carrier}(-\omega_c)}{r_{\Pi[T_s,T_f]}}s_{\Pi(\infty)}(t,u_y)^{\text{chirp}}\\
&= \exp(-j\omega_c t)\cdot\underbrace{\sum_n r_{\Pi[T_s,T_f]}s_{\Pi(\infty)}(t,u_y)_n^{\text{chirp}}}_{r_{\Pi[T_s,T_f]}s_{\Pi(\infty)}(t,u_y)^{\text{chirp}}},\\
&= \exp(-j\omega_c t)\cdot r_{\Pi[T_s,T_f]}s_{\Pi(\infty)}(t,u_y)^{\text{chirp}},
\end{aligned}\tag{7.51}$$

where

$$\begin{aligned}
&r_{\Pi[T_s,T_f]}s_{\Pi(\infty)}(t,u_y)^{\text{chirp}}\\
&= \sum_n g_n\cdot\mathsf{psf}_t\left[t-t_n(u_y)\right]_n^{\text{chirp}}\cdot\exp\left\{j\omega_c\left[t-t_n(u_y)\right]\right\}.
\end{aligned}\tag{7.52}$$

Fourier Transform Pair of Range Compressed SAR Imagery

The derivation above gives the frequency-space to time-space Fourier transform pair for the range compressed image. When expressed in scripted notation, this Fourier transform pair is compactly represented by

$$\overset{\text{carrier}(-\omega_c)}{r_{\Pi[T_s,T_f]}} \, s_{\Pi(\infty)}(t,u_y)^{\text{chirp}} \quad {}_t\Leftrightarrow_{\omega_b} \quad \overset{\text{L.shift}(\omega_c)}{R_{\hat{\Pi}(2\omega_o @ \omega_c)}} \, s_{\Pi(\infty)}(\omega_p,u_y)^{\text{chirp}}.$$

Phase Noise from Frequency Offset Imaging Down Conversion

The phase factor $\exp(-j\omega_c t)$ of (7.51) within the pulse compressed pixel domain (range only focused SAR image) is generated by down conversion of the received fast-time signal. This residual phase factor is extraneous and provides no information about the terrain function. A coherently "clean" range compressed image (7.52) does not include the embedded down conversion phase factor $\exp(-j\omega_c t)$. This phase factor should be compensated away by

$$\exp(+j\omega_c t) \times \{\cdot\} \tag{7.53}$$

applied to the range focused or fully focused SAR image. If done, artifact free complex pixel imagery can be exploited, and in the range only focused case, we are dealing with

$$r_{\Pi[T_s,T_f]} \, s_{\Pi(\infty)}(t,u_y)^{\text{chirp}} \quad {}_t\Leftrightarrow_{\omega_b} \quad R_{\hat{\Pi}(2\omega_o @ \omega_c)} \, s_{\Pi(\infty)}(\omega_p,u_y)^{\text{chirp}}.$$

Example Range Compressed Image and Phase Maps

An example range compressed image is shown in Fig. 7.13b. The phase map of Fig. 7.13a shows the phase history before range compression. The phase curtain of Fig. 7.13a changes from a description for all frequencies within the down converted passband to a single range walk phase curtain dependent on only the carrier frequency that resides along the range walk trace of Fig. 7.13b.

a. $R_{\hat{\Pi}(2\omega_o)} \, s_{\Pi(\infty)}(\omega_p,u_y)$ b. $r_{\Pi[T_s,T_f]} \, s_{\Pi(\infty)}(t,u_y)$

Figure 7.13: Pulse compressed range focused slant-plane image.

Pulse Compressed Huygens Wavelets

Since the range compressed image resides in the time-space domain, we revisit the Huygens-Fresnel wave theory framework to further identify wave theory related elements of (7.52). A little imagination allows us to view

$$\mathsf{psf}_t\left[t - t_n(u_y)\right]^{\text{chirp}}$$

as the waveform of a type of Huygens wavelet, where instead of an impulsive spherical wavefront or a chirp waveform wavefront, we are dealing with an expanding wavelet with a wavefront defined by the temporal point spread function $\mathsf{psf}_t(\cdot)^{\text{chirp}}$. This pulse compressed chirp wavelet expands with time in all directions from the scatterer located (x_n, y_n). The temporal radius is specified by $t - t_n(u_y)$ for all u_y within the synthesis aperture.

7.7 Frequency-Wavenumber Phase History

Now that we have explored the range compressed slant-plane image, we return to our study of the frequency-space phase history. The next step is to perform the spatial Fourier transform in the slow-time dimension to convert the frequency-space phase history into a frequency-wavenumber phase history.

Slant-Plane Slow-Time Fourier Transform

A slow-time spatial Fourier transform of the baseband range compressed chirp pulse frequency-space phase history

$$R^{\text{L.shift}(\omega_c)}_{\hat{\Pi}(2\omega_o @ \omega_c)} s_{\Pi(\infty)}(\omega_p, u_y)^{\text{chirp}}$$

yields the range compressed baseband frequency-wavenumber phase history,

$$R^{\text{L.shift}(\omega_c)}_{\hat{\Pi}(2\omega_o @ \omega_c)} S_{\hat{\Pi}[-k_{\omega_p}, k_{\omega_p}]}(\omega_p, k_{u_y})^{\text{chirp}}.$$

The details of this Fourier transform are

$$\begin{aligned}
&\mathsf{FT}^{u_y}_{1\text{D}}\left\{R^{\text{L.shift}(\omega_c)}_{\hat{\Pi}(2\omega_o @ \omega_c)} s_{\Pi(\infty)}(\omega_p, u_y)^{\text{chirp}}\right\} \\
&\quad = \mathsf{FT}^{u_y}_{1\text{D}}\left\{\left|P_{\Pi(2\omega_o @ 0)}(\omega_b)^{\text{chirp}}\right|^2 \cdot \sum_n g_n \cdot \measuredangle Hh^2\left[\omega_p, r_n(u_y)\right]\right\}, \\
&\quad = \left|P_{\Pi(2\omega_o @ 0)}(\omega_b)^{\text{chirp}}\right|^2 \cdot \mathsf{FT}^{u_y}_{1\text{D}}\left\{\sum_n g_n \cdot \exp\left[-jk_{\omega_p} \cdot 2r_n(u_y)\right]\right\}.
\end{aligned}$$

From our study of the infinite aperture case in Section 4.5,

$$\begin{aligned}
&\mathsf{FT}_{1\mathrm{D}}^{u_y}\left\{ R_{\widehat{\Pi}(2\omega_o @ \omega_c)}^{\mathrm{L.shift}(\omega_c)} s_{\Pi(\infty)}(\omega_p, u_y)^{\mathrm{chirp}} \right\} \\
&= \left| P_{\Pi(2\omega_o @ 0)}(\omega_p - \omega_c)^{\mathrm{chirp}} \right|^2 \cdot a(\omega_p, k_{u_y})_{\mathsf{FT}} \cdot \Pi_{[-k_{\omega_p}, k_{\omega_p}]}(k_{u_y}) \\
&\quad \cdot \sum_n g_n \cdot \exp(-jx_n \cdot 2\sqrt{k_{\omega_p}^2 - k_{u_y}^2}\,) \cdot \exp(-jy_n \cdot 2k_{u_y}), \\
&= \left| P_{\Pi(2\omega_o @ 0)}(\omega_p - \omega_c)^{\mathrm{chirp}} \right|^2 \cdot \widehat{\Pi}_{[-k_{\omega_p}, k_{\omega_p}]}(k_{u_y}) \\
&\quad \cdot \sum_n g_n \cdot \exp(-jx_n \cdot 2\sqrt{k_{\omega_p}^2 - k_{u_y}^2}\,) \cdot \exp(-jy_n \cdot 2k_{u_y}), \\
&= R_{\widehat{\Pi}(2\omega_o @ \omega_c)}^{\mathrm{L.shift}(\omega_c)} S_{\Pi[-k_{\omega_p}, k_{\omega_p}]}(\omega_p, k_{u_y})^{\mathrm{chirp}}.
\end{aligned}$$

As explained in Section 4.5, the support of the aperture wavenumber k_{u_y} is limited to

$$k_{u_y} \in \Pi_{[-k_{\omega_p}, k_{\omega_p}]}(k_{u_y}) \tag{7.54}$$

for each ω_p within the passband of the chirp waveform.

Frequency-Wavenumber Phase History Weight Function

The *aperture wavenumber weight function*

$$\widehat{\Pi}_{[-k_{\omega_p}, k_{\omega_p}]}(k_{u_y}) = a(\omega_p, k_{u_y})_{\mathsf{FT}} \cdot \Pi_{[-k_{\omega_p}, k_{\omega_p}]}(k_{u_y})$$

includes a complex amplitude factor

$$\begin{aligned}
a(\omega_p, k_{u_y})_{\mathsf{FT}} &= e^{-j\frac{\pi}{4}} \cdot \lambda_x(\omega_p, k_{u_y}), \\
&= e^{-j\frac{\pi}{4}} \cdot \frac{2\pi}{\sqrt{k_{\omega_p}^2 - k_{u_y}^2}}
\end{aligned}$$

that results from the slow-time Fourier transform [885].

Frequency-Wavenumber Slant-Plane Phase Maps

Example phase maps based on LFM waveforms before and after the along-track slow-time Fourier transform are shown in Figs. 7.14a-b. The frequency-wavenumber domain manifold shown in Fig. 7.14b has a trapezoidal shape as required by (7.54). Increasing aperture wavenumber bandwidth as a function of increasing passband frequencies is marked by the two lines that form the trapezoidal "wedge" in Fig. 7.14b. The expanding trapezoid indicates the dependency of the slow-time bandwidth of this phase history on the instantaneous passband frequency of the transmitted chirp signal. (Remember that the horizontal axis is reversed in these phase maps. The passband frequency increases to the left in these frequency-wavenumber domain phase maps.)

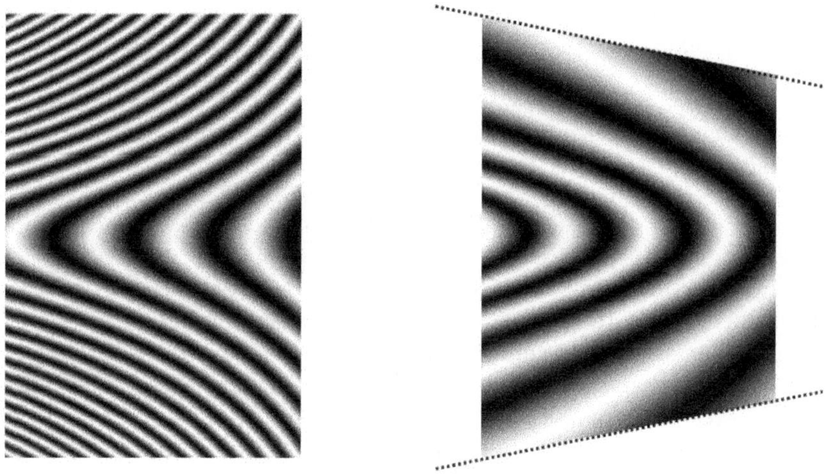

a. $R^{\text{L.shift}(\omega_c)}_{\widehat{\Pi}(2\omega_o @ \omega_c)} s_{\Pi(\infty)}(\omega_p, u_y)^{\text{chirp}}$ b. $R^{\text{L.shift}(\omega_c)}_{\widehat{\Pi}(2\omega_o @ \omega_c)} S_{\widehat{\Pi}[-k_{\omega_p}, k_{\omega_p}]}(\omega_p, k_{u_y})^{\text{chirp}}$

Figure 7.14: Frequency-space to frequency-wavenumber transformation.

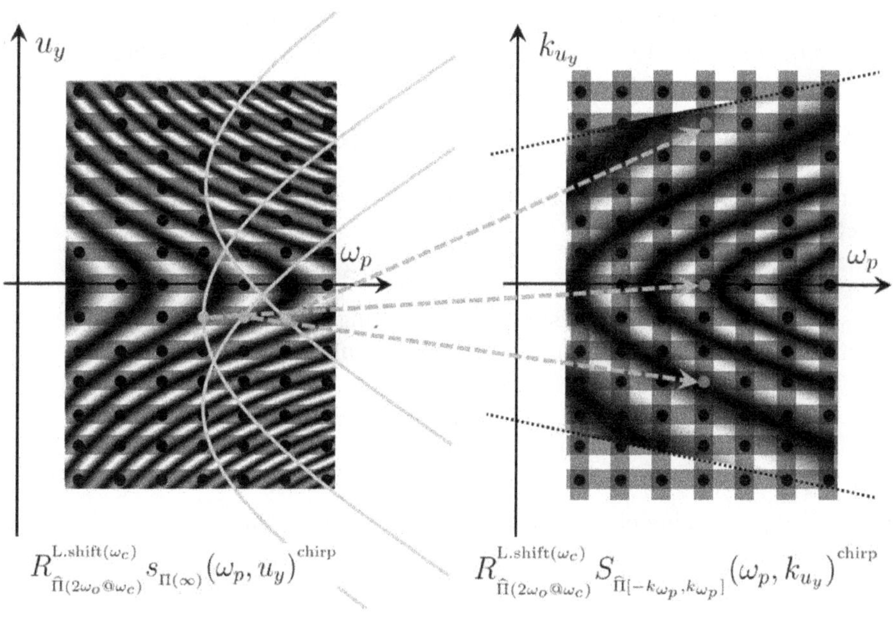

Figure 7.15: Frequency-wavenumber phase history data matrix.

Slow-Time Fourier Transform: Data Matrix Interpretation

Beginning with the data matrix in the frequency-space domain

$$R^{\text{L.shift}(\omega_c)}_{\hat{\Pi}(2\omega_o @\omega_c)} s_{\Pi(\infty)}(\omega_p, u_y)^{\text{chirp}}, \tag{7.55}$$

a slow-time Fourier transform is performed along the columns of the data matrix to obtain samples in the rectilinear frequency-wavenumber domain

$$R^{\text{L.shift}(\omega_c)}_{\hat{\Pi}(2\omega_o @\omega_c)} S_{\hat{\Pi}[-k_{\omega_p}, k_{\omega_p}]}(\omega_p, k_{u_y})^{\text{chirp}}. \tag{7.56}$$

Shown in Fig. 7.15 is a notional illustration of the effect of the slow-time Fourier transform as applied to the columns of the frequency-space data matrix (7.55). This Fourier transform converts frequency-space column data (left phase map) into frequency-wavenumber column data (right phase map) (7.56). The horizontal frequency axes of the phase maps of Fig. 7.15 have been flipped back to follow the convention that passband frequency increases to the right.

To examine the mapping of a scatterer's aperture position to specific aperture wavenumbers within the received (slow-time) signal spectrum, consider the data of a single pulse (row) near the center of the synthetic aperture as shown on the left side of Fig. 7.15. The superposition of range walk hyperbolas illustrate in a notional way how different points along the range walk profile, corresponding to specific sensor-scatterer view angles and slant-ranges as captured within the one pulse of interest, are mapped to different aperture wavenumber k_{u_y} values. Echoes from scatterers at up-track positions (forward of sensor) collected within this pulse are mapped (through the slow-time Fourier transform) to positive along-track (aperture) wavenumber values. Echoes from down-track (behind sensor) scatterers are mapped to negative along-track (aperture) wavenumber values.

7.8 Slant-Plane Angular Spectrum

We next develop an expression of the down converted Fourier migrated angular spectrum,

$$M^{\text{L.shift}(\omega_c)}_{\hat{\Pi}(\oplus)} S_{\hat{\Pi}[-2k_{\omega_p}, 2k_{\omega_p}]}(\breve{k}_x, k_y)^{\text{chirp}}_{\text{ACS}}$$

for slant-plane chirp pulse systems. Fourier migration by Stolt transformation from the rectilinear (ω_p, k_{u_y}) domain to the angular $(\breve{k}_x, k_y)$ domain,

$$\begin{aligned} 2\sqrt{k_{\omega_p}^2 - k_{u_y}^2} &\mapsto \breve{k}_x, \\ 2k_{u_y} &\mapsto k_y, \end{aligned} \tag{7.57}$$

causes a repositioning of resident phase factor data samples

$$\begin{aligned} \exp(-jx_n \cdot 2\sqrt{k_{\omega_p}^2 - k_{u_y}^2}\,) &\mapsto \exp(-jx_n \breve{k}_x) \\ \exp(-jy_n \cdot 2k_{u_y}) &\mapsto \exp(-jy_n k_y) \end{aligned} \tag{7.58}$$

from the rectilinear frequency-wavenumber domain (ω_p, k_{u_y}) to non-uniformly spaced positions in the $(\breve{k}_x, k_y)$ angular spectrum domain.

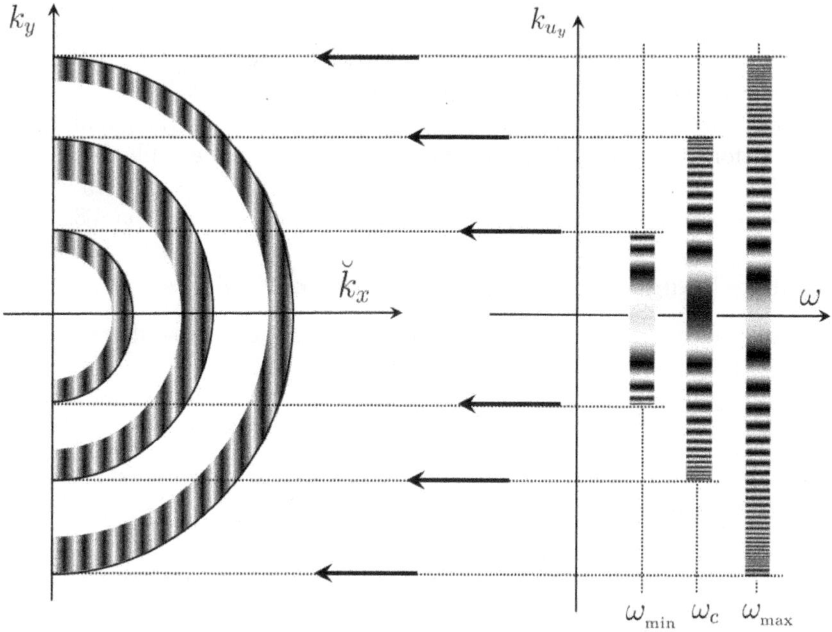

a. Boresight scatterer.

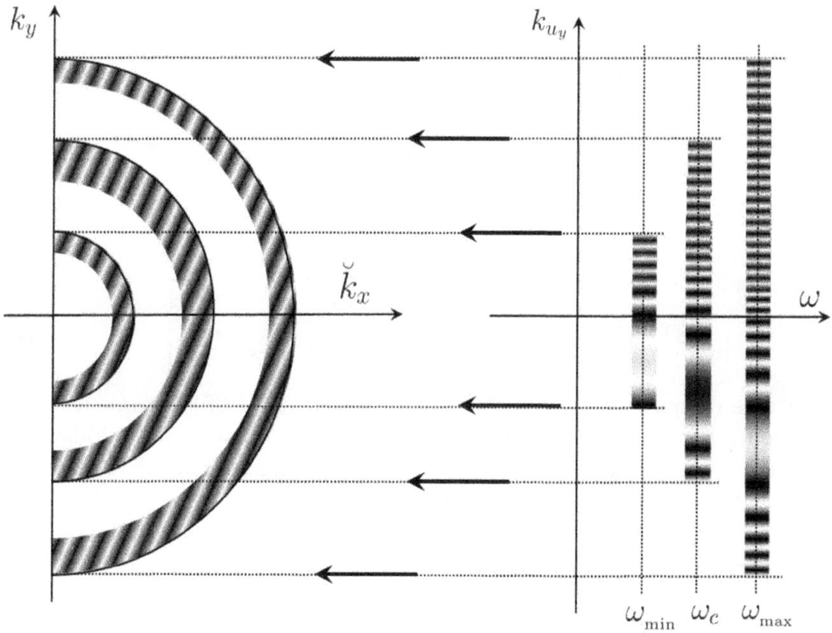

b. Off boresight scatterer.

Figure 7.16: Phase history before and after Stolt formatting.

Frequency-Wavenumber Phase History Mapping to Ewald Sphere

Fourier migration of data acquired within the passband of chirp pulse waveforms can be considered a polychromatic extension of the monochromatic continuous wave case of Section 4.8. This mapping is illustrated in Figs. 7.16a-b for the rectilinear spectrum of three discrete passband frequencies: $\omega_{\min}$, ω_c and $\omega_{\max}$. The Stolt transforms for the (infinite aperture) polychromatic slant-plane imaging problem assumes all the attributes explained for the (infinite aperture) monochromatic along-track profiling problem, as described previously in Section 4.8.

Rectilinear and Angular Spectrum Windows and Weights

The aperture wavenumber k_{u_y} dimension of the range compressed frequency-wavenumber phase history data is weighted by

$$\widehat{\Pi}_{[-k_{\omega_p},k_{\omega_p}]}(k_{u_y}) = a(\omega_p, k_{u_y})_{\mathsf{FT}} \cdot \Pi_{[-k_{\omega_p},k_{\omega_p}]}(k_{u_y}) \tag{7.59}$$

and the temporal passband wavenumber k_{ω_p} dimension is weighted by the compressed chirp pulse spectrum

$$\left| P_{\Pi(2\omega_o)}(\omega_p - \omega_c)^{\text{chirp}} \right|^2 . \tag{7.60}$$

Both of these weights of the rectilinear spectrum are jointly subject to the Stolt transformation. We indicate the weight functions of the angular spectrum of an infinite synthetic aperture induced by the Stolt mapping of the product of (7.59) and (7.60) by

$$\begin{aligned} &\widehat{\Pi}_{\oplus}^{\text{chirp}}(\breve{k}_x, k_y) \cdot \widehat{\Pi}_{[-2k_{\omega_p},2k_{\omega_p}]}(k_y) \\ &= \mathsf{ST}_{1\text{D}}^{\breve{k}_x} \left\{ \left| P_{\Pi(2\omega_o @ 0)}(\omega_p - \omega_c)^{\text{chirp}} \right|^2 \cdot \widehat{\Pi}_{[-k_{\omega_p},k_{\omega_p}]}(k_{u_y}) \right\}, \end{aligned}$$

or equivalently by

$$\begin{aligned} &\widehat{\Pi}_{\oplus}^{\text{chirp}}(\breve{k}_x, k_y) \cdot \widehat{\Pi}_{[-2k_{\omega_p},2k_{\omega_p}]}(k_y) \\ &= \mathsf{ST}_{1\text{D}}^{\breve{k}_x} \left\{ \left| W_{\Pi(2\omega_o @ \omega_c)}(\omega_p)^{\text{chirp}} \right|^2 \cdot \widehat{\Pi}_{[-k_{\omega_p},k_{\omega_p}]}(k_{u_y}) \right\}. \end{aligned} \tag{7.61}$$

The *pair* of one dimensional weight functions (7.59) and (7.60) become a *single* two dimensional (even though the notation seems to indicate otherwise) weight function given by (7.61). Hence, applying the Stolt transform to the pulse compressed frequency-wavenumber phase history gives

$$\begin{aligned} &M_{\widehat{\Pi}(\oplus)}^{\text{L.shift}(\omega_c)} S_{\widehat{\Pi}[-2k_{\omega_p},2k_{\omega_p}]}(\breve{k}_x, k_y)_{\text{ACS}}^{\text{chirp}} \\ &= \mathsf{ST}_{1\text{D}}^{\breve{k}_x} \left\{ R_{\widehat{\Pi}(2\omega_o @ \omega_c)}^{\text{L.shift}(\omega_c)} S_{\widehat{\Pi}[-k_{\omega_p},k_{\omega_p}]}(\omega_p, k_{u_y})_{\text{ACS}}^{\text{chirp}} \right\}, \\ &= \widehat{\Pi}_{\oplus}^{\text{chirp}}(\breve{k}_x, k_y) \cdot \widehat{\Pi}_{[-2k_{\omega_p},2k_{\omega_p}]}(k_y) \cdot \sum_n g_n \cdot \exp(-jx_n\breve{k}_x) \cdot \exp(-jy_n k_y). \end{aligned}$$

From this point on, we assume compensation of down conversion will eventually be performed with the pixel domain operations of (7.53). In which case, the passband weighting of (7.61) is allowed. Hence the "L.shift(ω_c)" superscript is no longer required and in our frequency-wavenumber domain analysis we can write

$$\begin{aligned}
&M^{\text{chirp}}_{\widehat{\Pi}(\oplus)} S_{\widehat{\Pi}[-2k_{\omega_p},2k_{\omega_p}]}(\breve{k}_x, k_y)_{\text{ACS}} \\
&= \widehat{\Pi}^{\text{chirp}}_{\oplus}(\breve{k}_x, k_y) \cdot \widehat{\Pi}_{[-2k_{\omega_p},2k_{\omega_p}]}(k_y) \cdot \sum_n g_n \cdot \exp(-jx_n\breve{k}_x) \cdot \exp(-jy_n k_y).
\end{aligned}$$

7.9 Slant-Plane Huygens-Fresnel Analysis

We next perform the Huygens-Fresnel decomposition of the slant-plane angular spectrum phase history. Start with the Fourier migrated spectrum in the aperture coordinate system,

$$\begin{aligned}
&M^{\text{chirp}}_{\widehat{\Pi}(\oplus)} S_{\widehat{\Pi}[-2k_{\omega_p},2k_{\omega_p}]}(\breve{k}_x, k_y)_{\text{ACS}} \\
&= \widehat{\Pi}^{\text{chirp}}_{\oplus}(\breve{k}_x, k_y) \cdot \widehat{\Pi}_{[-2k_{\omega_p},2k_{\omega_p}]}(k_y) \cdot \sum_n g_n \cdot \exp(-jx_n\breve{k}_x) \cdot \exp\left(-jy_n k_y\right).
\end{aligned}$$

Express scatterer positions in terms of both aperture and scene coordinate system parameters,

$$\begin{aligned}
&M^{\text{chirp}}_{\widehat{\Pi}(\oplus)} S_{\widehat{\Pi}[-2k_{\omega_p},2k_{\omega_p}]}(\breve{k}_x, k_y)_{\text{ACS}} \\
&= \widehat{\Pi}^{\text{chirp}}_{\oplus}(\breve{k}_{\hat{x}}, k_{\hat{y}}) \cdot \widehat{\Pi}_{[-2k_{\omega_p},2k_{\omega_p}]}(k_{\hat{y}}) \\
&\quad \cdot \sum_n g_n \cdot \exp[-j\left(X_c + \hat{x}_n\right)\breve{k}_{\hat{x}}] \cdot \exp[-j\left(Y_c + \hat{y}_n\right)k_{\hat{y}}], \\
&= \exp(-jX_c\breve{k}_{\hat{x}}) \cdot \exp\left(-jY_c k_{\hat{y}}\right) \cdot \widehat{\Pi}^{\text{chirp}}_{\oplus}(\breve{k}_{\hat{x}}, k_{\hat{y}}) \cdot \widehat{\Pi}_{[-2k_{\omega_p},2k_{\omega_p}]}(k_{\hat{y}}) \\
&\quad \cdot \sum_n g_n \cdot \exp(-j\hat{x}_n\breve{k}_{\hat{x}}) \cdot \exp\left(-j\hat{y}_n k_{\hat{y}}\right),
\end{aligned} \tag{7.62}$$

where aperture and scene wavenumbers are considered equivalent,

$$\begin{aligned}
\breve{k}_x &= \breve{k}_{\hat{x}}, \\
k_y &= k_{\hat{y}}.
\end{aligned}$$

At this point, the forward Huygens-Fresnel transfer elements,

$$\begin{aligned}
H_{X_c}(\breve{k}_{\hat{x}}) &= \exp(-jX_c\breve{k}_{\hat{x}}), \\
H_{Y_c}(k_{\hat{y}}) &= \exp\left(-jY_c k_{\hat{y}}\right),
\end{aligned} \tag{7.63}$$

are evident in (7.62). We make explicit the terrain spectrum that is embedded within the Stolt formatted phase history by writing,

$$
\begin{aligned}
&\underset{\widehat{\Pi}(\oplus)}{M}^{\text{chirp}} S_{\widehat{\Pi}[-2k_{\omega_p},2k_{\omega_p}]}(\breve{k}_x, k_y)_{\text{ACS}} \\
&= H_{X_c}(\breve{k}_{\hat{x}}) \cdot H_{Y_c}(k_{\hat{y}}) \cdot \widehat{\Pi}_{\oplus}^{\text{chirp}}(\breve{k}_{\hat{x}}, k_{\hat{y}}) \cdot \widehat{\Pi}_{[-2k_{\omega_p},2k_{\omega_p}]}(k_{\hat{y}}) \\
&\quad \underbrace{\cdot \sum_n g_n \cdot \exp(-j\hat{x}_n \breve{k}_{\hat{x}}) \cdot \exp(-j\hat{y}_n k_{\hat{y}}),}_{G_{\Pi(\infty)} G_{\Pi(\infty)}(\breve{k}_{\hat{x}}, k_{\hat{y}})_{\text{SCS}}} \\
&= H_{X_c}(\breve{k}_{\hat{x}}) \cdot H_{Y_c}(k_{\hat{y}}) \cdot \underset{\widehat{\Pi}(\oplus)}{G}^{\text{chirp}} G_{\widehat{\Pi}[-2k_{\omega_p},2k_{\omega_p}]}(\breve{k}_{\hat{x}}, k_{\hat{y}})_{\text{SCS}}, \\
&= \underset{\widehat{\Pi}(\oplus)}{G}^{\text{chirp}} G_{\widehat{\Pi}[-2k_{\omega_p},2k_{\omega_p}]}(\breve{k}_x, k_y)_{\text{ACS}}.
\end{aligned}
$$

Huygens-Fresnel analysis has exposed a windowed sample of the terrain spectrum expressed in the angular spectrum of the scene coordinate system,

$$
\begin{aligned}
&\underset{\widehat{\Pi}(\oplus)}{G}^{\text{chirp}} G_{\widehat{\Pi}[-2k_{\omega_p},2k_{\omega_p}]}(\breve{k}_x, k_y)_{\text{ACS}} \\
&= \widehat{\Pi}_{\oplus}^{\text{chirp}}(\breve{k}_{\hat{x}}, k_{\hat{y}}) \cdot \widehat{\Pi}_{[-2k_{\omega_p},2k_{\omega_p}]}(k_{\hat{y}}) \cdot G_{\Pi(\infty)} G_{\Pi(\infty)}(\breve{k}_{\hat{x}}, k_{\hat{y}})_{\text{SCS}},
\end{aligned}
$$

subject to the forward Huygens-Fresnel transfer functions (7.63).

7.10 Slant-Plane Huygens-Fresnel Inversion

SAR image formation processing via Fourier migration requires a Stolt transformation of the rectilinear SAR signal spectrum. Optionally, one or more (depending on whether the scene is broadside or squint) inverse Huygens-Fresnel transfers can be performed to better prepare the collected phase history data for Fourier inversion by reducing sampling rate requirements and data set sizes. Application of inverse Huygens-Fresnel transforms reduces computational and memory loading of the SAR processing system.

Huygens-Fresnel Inversion Options

If processed without the inverse Huygens-Fresnel transfers, then the SAR image will be formed in a two dimensional image space that spans the region from flight line to the most remote point in the ground scene. This two dimensional image is mostly empty except in the vicinity of the illuminated ground scene (if we were to model the field of view induced by an antenna beam pattern). With application of the inverse Huygens-Fresnel transfers, the span of the image array reduces to a size sufficient to support a smaller image centered at the origin of the scene coordinate system. The number of samples and required wavenumber sample density is reduced by application of the inverse Huygens-Fresnel transfers.

Inverse Huygens-Fresnel Transfers are Optional

There is no set requirement for Huygens-Fresnel transfers to be performed in angular spectrum processing. Application of inverse Huygens-Fresnel transfers is optional. However, use of Huygens-Fresnel transfers is recommended for the reasons stated above.

Huygens-Fresnel Inversion Before or After Stolt Formatting

If used, the Huygens-Fresnel transfers can be applied either before or after the Stolt transform.

In some of the early 1990s versions of the omega-k algorithm, the Huygens-Fresnel transfers were applied after Stolt formatting [137, 790, 791]. Other early implementations applied Huygens-Fresnel transfers before Stolt formatting [352, 771].

Although there are examples of both possibilities, the option of performing inverse Huygens-Fresnel transfers before or after the Stolt transform is not widely recognized [40, 389, 885].

Huygens-Fresnel Inversion Without Stolt Formatting

It is also possible to formulate SAR image formation processing in the frequency-wavenumber domain without Stolt formatting [885]. This possibility involves a set of multiple Huygens-Fresnel transfers. Each Huygens-Fresnel transfer within this set is designed specifically for a single cross-range line of the SAR image being formed (see Section 12.8).

Huygens-Fresnel Inversion Options

Given these options, SAR image formation processing can be performed in the rectilinear or angular spectrum domains with four major approaches:

1. Huygens-Fresnel transfers before Stolt formatting,
2. Huygens-Fresnel transfers after Stolt formatting,
3. Stolt formatting without Huygens-Fresnel transfers and
4. Huygens-Fresnel transfers without Stolt formatting.

The first two options are described in this section. The third option is explained in Section 7.16. The fourth option is described in Section 12.8.

Huygens-Fresnel Inversion After Stolt Transformation

Given a Huygens-Fresnel decomposition of the pulse compressed and Fourier migrated SAR angular spectrum,

$$\begin{aligned} & M^{\text{chirp}}_{\text{ñ}(\oplus)} S_{\text{ñ}[-2k_{\omega_p}, 2k_{\omega_p}]}(\breve{k}_x, k_y)_{\text{ACS}} \\ &= H_{X_c}(\breve{k}_{\hat{x}}) \cdot H_{Y_c}(k_{\hat{y}}) \cdot G^{\text{chirp}}_{\text{ñ}(\oplus)} G_{\text{ñ}[-2k_{\omega_p}, 2k_{\omega_p}]}(\breve{k}_{\hat{x}}, k_{\hat{y}})_{\text{SCS}}, \end{aligned}$$

application of inverse Huygens-Fresnel transfers gives

$$
\begin{aligned}
& \overset{\text{chirp}}{I}_{\widehat{\sqcap}(\oplus)} I_{\widehat{\sqcap}[-2k_{\omega_p}, 2k_{\omega_p}]}(\breve{k}_{\hat{x}}, k_{\hat{y}})_{\text{SCS}} \\
&= H^*_{X_c}(\breve{k}_x) \cdot H^*_{Y_c}(k_y) \times \left\{ \overset{\text{chirp}}{M}_{\widehat{\sqcap}(\oplus)} S_{\widehat{\sqcap}[-2k_{\omega_p}, 2k_{\omega_p}]}(\breve{k}_x, k_y)_{\text{ACS}} \right\}, \\
&= \overset{\text{chirp}}{G}_{\widehat{\sqcap}(\oplus)} G_{\widehat{\sqcap}[-2k_{\omega_p}, 2k_{\omega_p}]}(\breve{k}_{\hat{x}}, k_{\hat{y}})_{\text{SCS}}.
\end{aligned}
$$

The resulting SAR image spectrum

$$
\overset{\text{chirp}}{I}_{\widehat{\sqcap}(\oplus)} I_{\widehat{\sqcap}[-2k_{\omega_p}, 2k_{\omega_p}]}(\breve{k}_{\hat{x}}, k_{\hat{y}})_{\text{SCS}}
$$

is comprised of a windowed and weighted sample of the terrain function

$$
\begin{aligned}
& \overset{\text{chirp}}{G}_{\widehat{\sqcap}(\oplus)} G_{\widehat{\sqcap}[-2k_{\omega_p}, 2k_{\omega_p}]}(\breve{k}_{\hat{x}}, k_{\hat{y}})_{\text{SCS}} \\
&= \overset{\text{chirp}}{\widehat{\Pi}}_{\oplus}(\breve{k}_{\hat{x}}, k_{\hat{y}}) \cdot \widehat{\Pi}_{[-2k_{\omega_p}, 2k_{\omega_p}]}(k_{\hat{y}}) \cdot G_{\Pi(\infty)} G_{\Pi(\infty)}(\breve{k}_{\hat{x}}, k_{\hat{y}})_{\text{SCS}}
\end{aligned}
$$

expressed in the scene coordinate system.

Slant-Plane Huygens-Fresnel Transfers of Ewald Spheres

The operation of slant-plane inverse Huygens-Fresnel transfers can be interpreted in terms of single scatterer Ewald spheres as shown in Figs. 7.17a-b. In slant-plane passband imaging, data on only slant-plane slices of the Ewald sphere are resident in the angular spectrum phase history. The slant-plane slice of the Ewald sphere maps out a semicircular manifold for each frequency bin within the transmitted signal's passband [40, 638, 722]. The linear flight path must be infinitely long to capture a complete semicircle trace of data [258].

Fig. 7.17b shows the slant-plane slices of the Ewald sphere phase signals of Fig. 7.17a. One is at boresight and two are off boresight. The slant-plane phase signals resident on the sliced Ewald sphere manifolds describe scatterer location. The slant-plane wavenumber domain phase signals are dimensionally decomposable into three tonal phase factors that provide measures of information about the down-range, along-track and above-ground elements of scatterer positions.

Fig. 7.17b also shows application of a slant-plane inverse Huygens-Fresnel transfer that shifts the aperture referenced coordinate system to a coordinate system nearer scene center. The tonal phase pattern on that sliced Ewald sphere represents the wavenumber domain's conjugated phase adjustment required to move the expression of data from the aperture coordinate system to a point reference closer to the center of the scene coordinate system.

Huygens-Fresnel Inversion Before Stolt Transformation

Inverse Huygens-Fresnel transfers and Fourier migration via Stolt transform are commutative processing steps and can be reordered. Processing with Huygens-Fresnel inversion before Stolt formatting must be described in terms of the rectilinear spectrum.

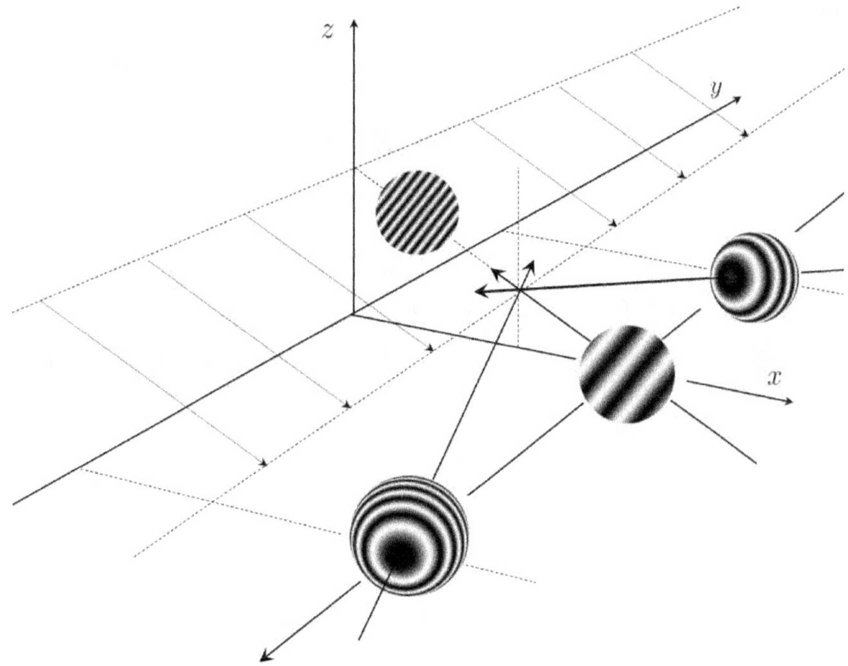

a. Ewald sphere phase histories after (partial) Huygens-Fresnel transfers.

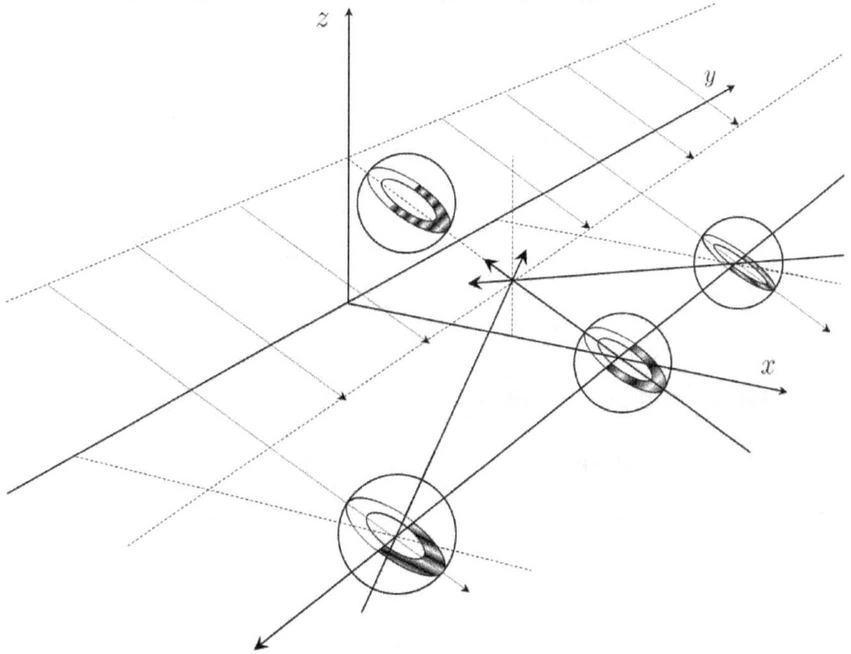

b. Slant-plane slice manifolds of Ewald spheres.

Figure 7.17: Ewald sphere slant-plane slice manifolds and Huygens-Fresnel transfers.

Rectilinear Spectrum Huygens-Fresnel Transfer Functions

The down-range Huygens-Fresnel transfer $H_{X_c}(\breve{k}_x)$ can be described in the rectilinear (ω_p, k_{u_y}) frequency-wavenumber domain as

$$\begin{aligned} H_{X_c}(\breve{k}_x) &= \exp(-jX_c\breve{k}_x), \\ &= \exp(-jX_c \cdot 2\sqrt{k_{\omega_p}^2 - k_{u_y}^2}\ ), \\ &= H_{X_c}(\omega_p, k_{u_y}). \end{aligned} \tag{7.64}$$

Huygens-Fresnel Transfers of the Rectilinear Spectrum

Applying the (ω_p, k_{u_y}) domain version of the inverse Huygens-Fresnel transfer function

$$H_{X_c}^*(\omega_p, k_{u_y})$$

to the (un-migrated) pulse compressed SAR signal spectrum

$$R_{\hat{\Pi}(2\omega_o @ \omega_c)} S_{\Pi[-k_{\omega_p}, k_{\omega_p}]}(\omega_p, k_{u_y})_{\text{ACS}}$$

of a broadside scene gives

$$\begin{aligned} & H_{X_c}^*(\omega_p, k_{u_y}) \times \left\{ R_{\hat{\Pi}(2\omega_o @ \omega_c)} S_{\Pi[-k_{\omega_p}, k_{\omega_p}]}(\omega_p, k_{u_y})_{\text{ACS}} \right\} \\ &= R_{\hat{\Pi}(2\omega_o @ \omega_c)} S_{\Pi[-k_{\omega_p}, k_{\omega_p}]}(\omega_p, k_{u_y})_{\text{SCS}}. \end{aligned}$$

This inverse Huygens-Fresnel transfer has moved the rectilinear frequency-wavenumber SAR phase history from the aperture coordinate system

$$R_{\hat{\Pi}(2\omega_o @ \omega_c)} S_{\Pi[-k_{\omega_p}, k_{\omega_p}]}(\omega_p, k_{u_y})_{\text{ACS}}$$

to the scene coordinate system

$$R_{\hat{\Pi}(2\omega_o @ \omega_c)} S_{\Pi[-k_{\omega_p}, k_{\omega_p}]}(\omega_p, k_{u_y})_{\text{SCS}}.$$

Rectilinear Spectrum Transfer and Squinted Scene

If the scene is squinted, a Huygens-Fresnel inversion with

$$H_{X_c}^*(k_{u_y}) = \exp(jY_c k_{u_y}) \tag{7.65}$$

in the along-track dimension should also be applied.

Stolt Transform After Huygens-Fresnel Inversion

Stolt transformation of the rectilinear spectrum after Huygens-Fresnel inversion,

$$\begin{aligned} & M_{\hat{\Pi}(\oplus)}^{\text{chirp}} S_{\Pi[-2k_{\omega_p}, 2k_{\omega_p}]}(\breve{k}_{\hat{x}}, k_{\hat{y}})_{\text{SCS}} \\ &= \mathsf{ST}_{1\text{D}}^{\breve{k}_{\hat{x}}} \left\{ R_{\hat{\Pi}(2\omega_o @ \omega_c)} S_{\Pi[-k_{\omega_p}, k_{\omega_p}]}(\omega_p, k_{u_y})_{\text{SCS}} \right\}, \\ &= \hat{\Pi}_{\oplus}^{\text{chirp}}(\breve{k}_{\hat{x}}, k_{\hat{y}}) \cdot \hat{\Pi}_{[-2k_{\omega_p}, 2k_{\omega_p}]}(k_{\hat{y}}) \cdot G_{\Pi(\infty)} G_{\Pi(\infty)} [\breve{k}_{\hat{x}}(\omega, k_{u_y}), k_{\hat{y}}(k_{u_y})]_{\text{SCS}}, \\ &= G_{\Pi(\oplus)}^{\text{chirp}} G_{\Pi(4k_{\omega_p})}(\breve{k}_{\hat{x}}, k_{\hat{y}})_{\text{SCS}} \end{aligned}$$

obtains the same result where Fourier migration is performed before Huygens-Fresnel transfer.

Huygens-Fresnel Transfers as Slow-Time Matched Filters

An along-track matched filter analogous to the down-range pulse compression matched filter $P^*(\omega)$ has not been identified. It is tempting to call the conjugates of the phase signals (7.64) and (7.65) "matched filters," especially the along-track phase factor $H_{Y_c}(k_y) = \exp(-jY_c k_y)$. Although the conjugate of $H_{Y_c}(k_y)$ is sometimes interpreted as the slow-time matched filter [397], this transfer function is not a matched filter [648, 649]. We shall see in Section 7.16 that the inverse Huygens-Fresnel transfer function $H_{Y_c}(k_y)$ does not determine in any way the characteristics of the point spread function in the along-track dimension.

7.11 SAR System Requirements Engineering

We next identify the system engineering steps required to properly specify requirements for along-track SAR signal acquisition.

Slant-Plane Image Nyquist Specifications

We begin with the Nyquist requirements that are derived from free parameters associated with desired slant-plane image characteristics.

Step 1: Specify the Desired Along-Track Image Resolution: The desired along-track image resolution parameter

$$\Delta_{\hat{y}} \tag{7.66}$$

is a free parameter provided to the SAR system engineer as a design requirement related to image quality objectives. Specifications related to Nyquist constraints in the image domain are determined in the scene coordinate system $(\hat{x}, \hat{y})$ and corresponding wavenumber domain $(k_{\hat{x}}, k_{\hat{y}})$.

Step 2: Determine Along-Track Image Bandwidth Requirement: The requirement for the along-track image wavenumber bandwidth $\Pi_{k_{\hat{y}}}$ is derived from the along-track image resolution specification (7.66) and by the Heisenberg relationship,

$$\Pi_{k_{\hat{y}}} \geq \frac{2\pi}{\Delta_{\hat{y}}}. \tag{7.67}$$

The wavenumber bandwidth requirement $\Pi_{k_{\hat{y}}}$ supports unaliased sample spacing $\Delta_{\hat{y}}$ for the along-track image (pixel density in the $\hat{y}$ domain). If (7.67) is satisfied, at least one (complex) SAR image pixel is dedicated to each along-track resolution cell.

Over Sampled Pixel Representations: An over sampled pixel representation in the along-track dimension of the image provides multiple pixels per resolution cell and allows the along-track point spread function to be better observed. If desired, this over sampling is achieved by zero padding the spectral window $\Pi_{k_{\hat{y}}}$ before Fourier inversion of $II(\cdot,\cdot)$ to obtain $ii(\cdot,\cdot)$.

Step 3: Specify the Along-Track Image Size: The along-track image size

$$\Pi_{\hat{y}}$$

is a free parameter provided to the SAR system engineer as a requirement.

Step 4: Determine the Along-Track Image Size Pixel Count: The number of pixels $N_{\hat{y}}$ in the along-track dimension of a critically sampled SAR image is determined by the desired resolution Δ_y and desired scene size $\Pi_{\hat{y}}$,

$$N_{\hat{y}} = \left\lceil \frac{\Pi_{\hat{y}}}{\Delta_{\hat{y}}} \right\rceil .$$

Step 5: Determine Along-Track Image Wavenumber Sample Density: The required Nyquist sample density $\Delta_{k_{\hat{y}}}$ for the along-track image wavenumber $k_{\hat{y}}$ is determined by the image size $\Pi_{\hat{y}}$, which in turn, is determined by the number of pixels and pixel spacing in the along-track dimension of the critically sampled image,

$$\begin{aligned} \Delta_{k_{\hat{y}}} &= \frac{2\pi}{\Pi_{\hat{y}}}, \\ &\approx \frac{2\pi}{N_{\hat{y}}\Delta_{\hat{y}}}. \end{aligned}$$

Slow-Time Domain Nyquist Specifications

After completion of Steps 1 though 5, the independent and dependent parameters associated with desired image characteristics have been specified or derived. The next steps are to determine parameters of data acquisition tasks, i.e., specify or derive parameters related to the synthetic aperture in the data collection domain. The primary parameters are the synthetic aperture length and spatial (not temporal) pulse repetition interval (PRI).

Step 6: Determine Synthetic Aperture Slow-Time Bandwidth Requirement: The along-track image bandwidth requirement $\Pi_{k_{\hat{y}}}$ determined in Step 2 flows directly into a requirement for the bandwidth $\Pi_{k_{u_y}}$ of the aperture wavenumber, i.e., sets a lower bound to the bandwidth requirement of the slow-time phase curtain of the synthetic aperture. The mapping between the required aperture wavenumber k_{u_y} and the along-track image wavenumber $k_{\hat{y}}$ is

$$k_{\hat{y}} = 2k_{u_y},$$

hence, we must have

$$\Pi_{k_{u_y}} \geq \frac{1}{2}\Pi_{k_{\hat{y}}}. \tag{7.68}$$

One must deal with different cases for specification of $\Pi_{k_{u_y}}$ for infinite and finite apertures, for spotlight and stripmap operations, and for Stolt and polar formatting approaches to SAR image formation processing. At this point, we only address one case, the infinitely long synthetic aperture.

Required Slow-Time Bandwidth of the Infinite Aperture: The aperture wavenumber manifold generated by the infinitely long synthetic aperture is

$$\Pi_{k_{u_y}} = \left[-k_{\omega_{\max}}, +k_{\omega_{\max}}\right],$$

where

$$\Pi_{k_{\hat{y}}} = \left[-2k_{\omega_{\max}}, +2k_{\omega_{\max}}\right].$$

Fast-Time Domain Nyquist Specifications: For the extreme case of the infinite synthetic aperture, the aperture wavenumber k_{u_y} at least spans the slow-time baseband interval determined by the lowest instantaneous chirp frequency $\omega_{\min}$ within the chirp passband,

$$\begin{aligned} \Pi_{k_{u_y}} &\geq \left[-k_{\omega_{\min}}, +k_{\omega_{\min}}\right], \\ &\geq \left[-\frac{2\pi}{\lambda_{\max}}, +\frac{2\pi}{\lambda_{\max}}\right]. \end{aligned}$$

The acceptable maximum value of the temporal wavelength parameter λ is related to the requirement on the bandwidth of the spatial wavenumber parameter k_{u_y}.

Lower Bound to Passband Signal Frequencies: For infinite aperture SAR and a specified desired resolution, there is a lower bound to the acceptable *minimum* chirp frequency (or *maximum* chirp wavelength) within the passband of the transmitted chirp signal. To see this, consider

$$\begin{aligned} \Pi_{k_{u_y}} &\geq 2 \cdot k_{\omega_{\min}}, \\ &\geq \frac{4\pi}{\lambda_{\max}}, \end{aligned}$$

which leads to

$$\begin{aligned} \lambda_{\max} &\leq \frac{4\pi}{\Pi_{k_{u_y}}}, \\ &\leq \frac{4\pi}{\frac{1}{2}\Pi_{k_{\hat{y}}}}, \end{aligned}$$

and since

$$\Pi_{k_{\hat{y}}} \geq \frac{2\pi}{\Delta_{\hat{y}}},$$

we have the requirement

$$\lambda_{\max} \leq 4 \cdot \Delta_{\hat{y}}.$$

A SAR imaging system must utilize fast-time signals with maximum wavelength no larger than four times the objective slow-time resolution $\Delta_{\hat{y}}$. This places a lower bound on the acceptable minimum chirp frequency for a given objective along-track resolution $\Delta_{\hat{y}}$.

Step 7 Determine Synthetic Aperture Sample Density: Given the requirement (7.68) for $\Pi_{k_{u_y}}$, the derived slow-time sampling interval (spatial pulse repetition interval) must satisfy the Nyquist constraint,

$$\Delta_{u_y} \leq \frac{2\pi}{\Pi_{k_{u_y}}}.$$

Since the size of $\Pi_{k_{u_y}}$ covers up to a span of $2k_{\omega_{\max}}$, the Nyquist sample density requirement for alias free sampling along the synthetic aperture flight line must satisfy

$$\Delta_{u_y} \leq \frac{\lambda_{\min}}{2},$$

where complex data are collected.

Over Sampled Pixel Representations: If there is a desire for an over sampled pixel density $\Delta_{\hat{y}}$ in the along-track dimension, then it does not follow that there must be an over sampled pulse repetition interval Δ_{u_y}.

Step 8 Determine Required Synthetic Aperture Length Requirement: By definition, the synthetic aperture length is $[-\infty, +\infty]$ for the case under study in this chapter.

Step 9 Determine Number of Synthetic Aperture Samples: The discrete, along-track sample positions,

$$u_y[m_y] = m_y \cdot \Delta_{u_y} \text{ with } m_y = \ldots, -2, -1, 0, +1, +2, \ldots$$

are the points along the synthetic aperture where pulses are transmitted and the echo signals are received and recorded. The number of pulses in a linear synthetic aperture along u_y is indicated by

$$M_{u_y}.$$

The number of pulses required in an infinite synthetic aperture is, of course,

$$M_{u_y} = \infty.$$

Simulation of Infinite Aperture Imagery: Although direct (spatial domain) simulation of an infinitely long synthetic aperture is impossible, simulations of the imagery of the infinite aperture case are possible if the data of the frequency-wavenumber domain phase history are directly simulated. Although $M_{u_y} = \infty$ pulses are required for infinite aperture data collections, the image need only contain at least $N_{\hat{y}}$ pixels in the along-track direction. In which case, the number of simulated frequency-wavenumber domain samples $N_{\hat{y}}$ can be set to $\Pi_{k_{\hat{y}}}/\Delta_{k_{\hat{y}}}$ without loss of information because of the non-zero lower bound to the two way diffraction limited resolution.

7.12 Along-Track Sample Rate Adjustments

The Nyquist constraint that governs sample density $\Delta_{k_{u_y}}$ in the aperture wavenumber domain for slow-time signal acquisition is different than the Nyquist rate that governs the required sample density $\Delta_{k_{\hat{y}}}$ in the along-track SAR image wavenumber domain. The along-track image wavenumber sample density $\Delta_{k_{\hat{y}}}$ is governed by the desired image size in the along-track dimension. The aperture wavenumber sample density $\Delta_{k_{u_y}}$ is determined indirectly by the desired resolution, which determines the required aperture size [885]. Hence, there is often a need to adjust along-track wavenumber sample densities after k_{u_y} is mapped to $k_{\hat{y}}$."

Presumming

Preparatory to image formation in the along-track dimension, we adjust by some means the sample density of the along-track wavenumber phase history from $\Delta_{k_{u_y}}$ to $\Delta_{k_{\hat{y}}}$ and we adjust the number of along-track samples from M_{u_y} to N_y. The radar community uses so called "presumming" or "preintegration" methods to reduce the typically more stringent wavenumber sample density $\Delta_{k_{u_y}}$ of the slow-time signal spectrum to the typically less dense wavenumber sample spacing $\Delta_{k_{\hat{y}}}$ of the image spectrum [26, 142, 530, 664].

There are various ways to achieve a lower sample density [118, 624]. One may choose to simply subsample the data or less preferably, one may choose to coherently sum adjacent along-track wavenumber samples (hence the name of "presumming") [530]. Coherent presumming has the advantage that the signal-to-noise ratio of the along-track phase history data is increased.

Presumming and Computational Loading

Processing of over sampled data should be avoided to minimize computational and memory loading. Presumming can be done at a variety of points within the signal processing work flow to reduce sample counts. Placement of the presumming operation should be as early as possible in the work flow to reduce computational complexity.

Poor Processing Order Selections

If a computationally efficient permutation of operational steps is not selected, then the computational complexity of the Stolt formatting method can appear much more demanding than actually required. Poor choices of processing order are not so uncommon in attempts to assess omega-k processing complexity [495, 887].

7.13 Stolt Formatting Processing Order

The steps of the processing sequence involving the Huygens-Fresnel transfer operation, the Fourier migration operation (i.e, Stolt formatting) and the presumming operation are all commutative. These processing steps can be arbitrarily ordered. The selection of processing order can significantly impact the complexity of an implementation of a particular instance of a Stolt formatting algorithm.

We now describe the options for process order. To make the next few paragraphs easier to follow, we shall label the Stolt formatting process as the "A" process, Huygens-Fresnel transfers as the "B" process and presumming as the "C" process. We next examine each possible permutation of A-B-C.

Option A-B-C: Stolt Formatting, Huygens-Fresnel, Presumming

The A-B-C option is to perform the ordered steps of

1. A: Stolt formatting,
2. B: Huygens-Fresnel transfer,
3. C: Presumming.

This option is less desirable because Stolt formatting and Huygens-Fresnel transfers are performed on data that are subsequently down sampled by presumming. Hence, processing effort is wasted with this ordering because processed (Stolt formatted and Huygens-Fresnel transfered) data samples are discarded in the last presumming step. In spite of its inefficiency, this processing order has been used in a computational complexity comparison of polar formatting, overlapped subapertures and Stolt formatting [495].

Option A-C-B: Stolt Formatting, Presumming, Huygens-Fresnel

The A-C-B option is to perform the ordered steps of

1. A: Stolt formatting,
2. C: Presumming,
3. B: Huygens-Fresnel transfer.

This option also performs operations that are wasted by waiting to down sample after the processing of Stolt formatting.

Option B-A-C: Huygens-Fresnel, Stolt Formatting, Presumming

The B-A-C option is to perform the ordered steps of

1. B: Huygens-Fresnel transfer,
2. A: Stolt formatting,
3. C: Presumming.

This processing also wastes computational effort by waiting to down sample after Stolt formatting.

Option B-C-A: Huygens-Fresnel, Presumming, Stolt Formatting

The B-C-A option is to perform the ordered steps of

1. B: Huygens-Fresnel transfer,
2. C: Presumming,
3. A: Stolt formatting.

Again, computational effort is wasted by down sampling after Huygens-Fresnel transfers.

Option C-A-B: Presumming, Stolt Formatting, Huygens-Fresnel

The C-A-B option is to perform the ordered steps of

1. C: Presumming,
2. A: Stolt formatting,
3. B: Huygens-Fresnel transfer.

This option is more desirable, but has the disadvantage that the Huygens-Fresnel transfer functions must also be Stolt formatted for application in the angular spectrum domain. Stolt formatted Huygens-Fresnel transfers are applied to Stolt formatted phase history data. Stolt formatting operations must be performed on two different data arrays before these arrays are multiplied sample by sample.

Option C-B-A: Presumming, Huygens-Fresnel, Stolt Formatting

The most desirable option is to perform the ordered steps of

1. C: Presumming,
2. B: Huygens-Fresnel transfer,
3. A: Stolt formatting.

Performing down sampling first reduces the data set size that must be processed in subsequent steps. Performing Huygens-Fresnel transfers next allows processing in the rectilinear spectrum without requiring Stolt formatting of the Huygens-Fresnel transfer functions. Performing the Stolt transform after Huygens-Fresnel transfers also has the advantage that the conversion from the aperture centered to the scene centered coordinate system makes the phase oscillations of the phase history data vary less rapidly, which makes the interpolation operations of resampling more accurate with finite order interpolators.

We next discuss some of the signal processing details of the C-B-A processing order option.

Huygens-Fresnel Inversion in Presummed Data Spaces

A single scatterer's frequency-wavenumber phase history before presumming is shown in Fig. 7.18a. Recall that m_y is used to index the along-track samples (pulses) collected along the linear flight path. The rectilinear phase history after presumming is shown in Fig. 7.18c. Let n_y be used to index the lower sample density data obtained after presumming. Presumming of Fig. 7.18a results in aliasing of the embedded forward Huygens-Fresnel transfer phase factors, which are not scatterer informative. The phase curtain of Fig. 7.18c has a phase center that has been replicated in false phase centers above and below the true phase center.

Although the along-track data of Fig. 7.18c appears to be aliased, scatterer informative position data within the scene coordinate system has not been corrupted. If presumming is done correctly, the along-track phase factor $\exp(j\hat{y}_n k_{\hat{y}})$ that is embedded in the presummed

$$R_{\widehat{\Pi}(2\omega_o @ \omega_c)} S_{\Pi[-k_{\omega_p}, k_{\omega_p}]}(\omega_p, k_{u_y}[n_{\hat{y}}])_{\text{ACS}}$$

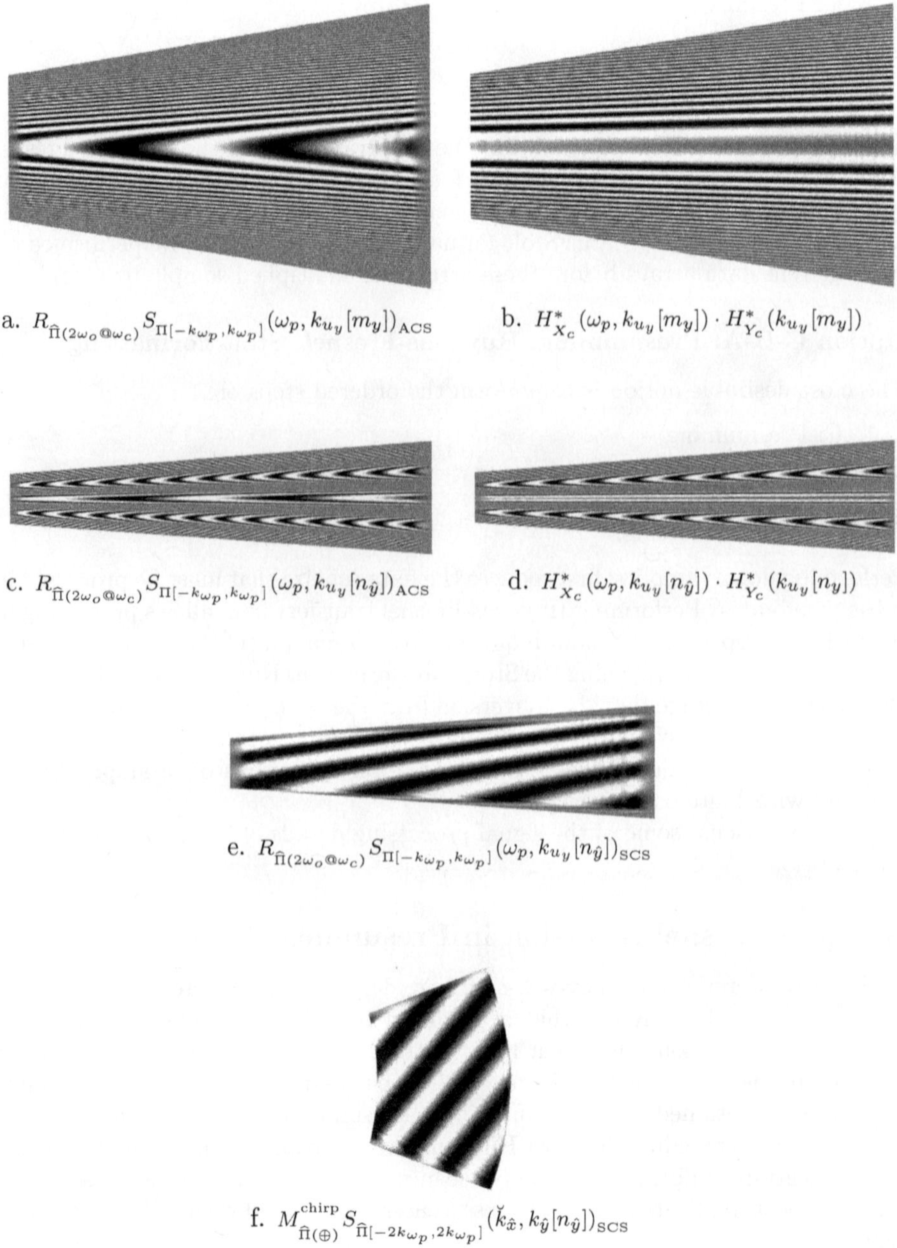

a. $R_{\widehat{\Pi}(2\omega_o @ \omega_c)} S_{\Pi[-k_{\omega_p}, k_{\omega_p}]}(\omega_p, k_{u_y}[m_y])_{\text{ACS}}$

b. $H^*_{X_c}(\omega_p, k_{u_y}[m_y]) \cdot H^*_{Y_c}(k_{u_y}[m_y])$

c. $R_{\widehat{\Pi}(2\omega_o @ \omega_c)} S_{\Pi[-k_{\omega_p}, k_{\omega_p}]}(\omega_p, k_{u_y}[n_{\hat{y}}])_{\text{ACS}}$

d. $H^*_{X_c}(\omega_p, k_{u_y}[n_{\hat{y}}]) \cdot H^*_{Y_c}(k_{u_y}[n_y])$

e. $R_{\widehat{\Pi}(2\omega_o @ \omega_c)} S_{\Pi[-k_{\omega_p}, k_{\omega_p}]}(\omega_p, k_{u_y}[n_{\hat{y}}])_{\text{SCS}}$

f. $M^{\text{chirp}}_{\widehat{\Pi}(\oplus)} S_{\widehat{\Pi}[-2k_{\omega_p}, 2k_{\omega_p}]}(\breve{k}_{\hat{x}}, k_{\hat{y}}[n_{\hat{y}}])_{\text{SCS}}$

Figure 7.18: Phase maps of Huygens-Fresnel transfers before Stolt formatting.

phase history is not aliased. All is fine if the reduced sample rate is such that the along-track locator sinusoid $\exp(j\hat{y}_n k_{\hat{y}})$ with positional frequency $\hat{y}_n$, which is embedded in the apparently aliased phase history of Fig. 7.18c, is sampled at or above the Nyquist rate for this along-track locator tone signal.

The design of inverse Huygens-Fresnel transfer functions can be tailored to operate on previously presummed data. As demonstrated in [885], this requires that the inverse Huygens-Fresnel transfer function be subsampled down to the presummed data rate, or initially designed at the reduced sample rate. The phase map of the inverse Huygens-Fresnel transfer function before presumming is shown in Fig. 7.18b. The phase map of the inverse Huygens-Fresnel transfer function after presumming is shown in Fig. 7.18d. The presummed inverse Huygens-Fresnel transfer function is aliased.

The result of applying the presummed inverse Huygens-Fresnel transfer function to the presummed rectilinear spectrum is shown in Fig. 7.18e, where

$$
\begin{aligned}
& R_{\hat{\Pi}(2\omega_o @ \omega_c)} S_{\Pi[-k_{\omega_p}, k_{\omega_p}]}(\omega_p, k_{u_y}[n_{\hat{y}}])_{\mathrm{SCS}} \\
&= \underbrace{H^*_{X_c}(\omega_p, k_{u_y}[n_{\hat{y}}])}_{\text{aliased}} \cdot \underbrace{H^*_{Y_c}(k_{u_y}[n_{\hat{y}}])}_{\text{aliased}} \times \underbrace{\left\{ R_{\hat{\Pi}(2\omega_o @ \omega_c)} S_{\Pi[-k_{\omega_p}, k_{\omega_p}]}(\omega_p, k_{u_y}[n_{\hat{y}}])_{\mathrm{ACS}} \right\}}_{\text{aliased}}, \\
&= \underbrace{H^*_{X_c}(\omega_p, k_{u_y}[n_{\hat{y}}]) \times H_{X_c}(\omega_p, k_{u_y}[n_{\hat{y}}])}_{\text{cancels}} \cdot \underbrace{H^*_{Y_c}(k_y[n_{\hat{y}}]) \times H_{Y_c}(k_y[n_{\hat{y}}])}_{\text{cancels}} \\
&\qquad \times \underbrace{M^{\mathrm{chirp}}_{\hat{\Pi}(\oplus)} S_{\hat{\Pi}[-2k_{\omega_p}, 2k_{\omega_p}]}(\breve{k}_x, k_{\hat{y}}[n_{\hat{y}}])_{\mathrm{SCS}}}_{\text{unaliased}} \, .
\end{aligned}
$$

The inverse Huygens-Fresnel transfers in effect translate the rectilinear spectrum from the aperture coordinate system to the scene coordinate system. The tonal wave field that locates the single scatterer within the rectilinear spectrum of the scene coordinate system is clearly unaliased in Fig. 7.18e. This tonal spectrum of Fig. 7.18e is latently resident in the phase history of Fig. 7.18c.

It is clever [40, 389, 885] to apply an intentionally aliased inverse Huygens-Fresnel transfer function to a phase history sampled in a way that the non-informative phase factors are also intentionally aliased, but where the embedded scatterer informative phase factors are not aliased. Such an approach will eliminate unwanted (and aliased) phase elements. Only the scatterer informative (and unaliased) phase elements remain. One advantage of this approach to Huygens-Fresnel inversion is computational and memory efficiency.

Stolt Transforms in Presummed Data Spaces

The complexity of the Stolt transform is likewise reduced, without any loss in image quality, if applied to appropriately presummed data. The only difference is that there are fewer rows of the Fourier migrated data matrix to process with the Stolt mapping after along-track presumming. For example, the Fourier migrated spectrum after presumming and Stolt transformation is shown in Fig. 7.18f. The slight warping of the tonal wave field of Fig. 7.18e has been corrected by Stolt formatting. The spectrum of Fig. 7.18f can undergo Fourier inversion to yield the SAR image.

Note the significant reduction in data volume from Fig. 7.18a to Fig. 7.18e. A Stolt mapping applied to Fig. 7.18a would be a significant waste of computational resources.

7.14 Angular Spectrum Uniform Resampling

After Stolt formatting, there is a need to uniformly resample the non-uniformly spaced data of the Fourier migrated down-range $\breve{k}_{\hat{x}}$ wavenumber dimension. The required uniform sample density and desired output sample locations of the uniformly resampled wavenumber data must be determined before interpolation from non-uniformly spaced migrated input data proceeds.

Furthermore, with finite kernel interpolation, a neighborhood around each of the desired uniformly spaced output sample points must be specified. These neighborhoods identify which subset of non-uniformly spaced Fourier migrated input data samples to include in calculating the value of the interpolated output sample [194, 890].

Indexed Discrete Data Representations

Describing the uniform resampling processing in detail is rather tedious, but important. We fully describe this process and include a description of the various indices used to index all the involved discrete data.

Index of Passband Angular Frequency Bins

Let n_{ω_p} index the passband angular frequency bins $\omega_p[n_{\omega_p}]$. The number of passband angular frequency bins is N_{ω_p}. With use of a monochromatic waveform, $N_{\omega_p} = 1$; otherwise, $N_{\omega_p} > 1$.

Passband Angular Frequency Bin Values

A passband angular frequency value is returned by the index operator

$$\omega_p[n_{\omega_p}] = \omega_c + n_{\omega_p} \cdot \Delta_{\omega_p},$$

where

$$n_{\omega_p} \in \left[-\frac{N_{\omega_p}}{2}, \cdots, 0, \cdots, +\frac{N_{\omega_p}}{2} - 1\right]$$

and

$$\Delta_{\omega_p} = \frac{2\pi}{\Pi_{k_{\omega_p}}},$$

with $\Pi_{k_{\omega_p}}$ being the fast-time signal's passband bandwidth.

Indexed Temporal Angular Frequency Wavenumber

The temporal angular frequency wavenumber values are determined by the passband angular frequency values,

$$k_{\omega_p}[n_{\omega_p}] = \frac{1}{c} \cdot \omega_p[n_{\omega_p}].$$

Index of Presummed Aperture Wavenumber Bins

Let $n_{\hat{y}}$ index the presummed slow-time aperture wavenumber bins $k_{u_y}[n_{\hat{y}}]$ in the scene coordinate system, with $N_{\hat{y}}$ being the number of presummed aperture or along-track image wavenumber bins after presumming.

Presummed Aperture Wavenumber Values

A value of the presummed aperture wavenumber for a broadside imaging geometry is returned by the index operator

$$k_{u_y}[n_{\hat{y}}] = n_{\hat{y}} \cdot \Delta_{k_{\hat{y}}},$$

where

$$n_{\hat{y}} \in \left[-\frac{N_{\hat{y}}}{2}, \cdots, 0, \cdots, +\frac{N_{\hat{y}}}{2} - 1\right],$$

with

$$\Delta_{k_{\hat{y}}} = \frac{2\pi}{\Pi_{\hat{y}}}.$$

Note that the sample density before presumming,

$$\Delta_{k_{u_y}} = \frac{2\pi}{\Pi_{u_y}},$$

is not being used, with Π_{u_y} being the synthetic aperture length.

Index of Along-Track Image Wavenumber Bins

Let $n_{\hat{y}}$ also index the along-track image wavenumber bins $k_{\hat{y}}[n_{\hat{y}}]$.

Along-Track Image Wavenumber Values

A value of the along-track image wavenumber is determined by the aperture wavenumber and returned by the operator,

$$k_{\hat{y}}[n_{\hat{y}}] = 2k_{u_y}[n_{\hat{y}}].$$

Indices of Uniformly Spaced Fourier Migrated Wavenumber Samples

Let $n_{\hat{x}}$ index the desired uniformly spaced $\ddot{k}_{\hat{x}}$ Fourier migrated down-range wavenumber samples,

$$\ddot{k}_{\hat{x}}[n_{\hat{x}}].$$

Uniformly Spaced Fourier Migrated Wavenumber Samples Points

The desired uniformly spaced Fourier migrated down-range wavenumber number sample points are returned by the index operator

$$\ddot{k}_{\hat{x}}[n_{\hat{x}}] = \breve{k}_{\hat{x}_c} + n_{\hat{x}} \cdot \Delta_{\ddot{k}_{\hat{x}}}, \tag{7.69}$$

where

$$n_{\hat{x}} \in \left[-\frac{N_{\hat{x}}}{2}, \cdots, 0, \cdots, +\frac{N_{\hat{x}}}{2} - 1\right]$$

with $N_{\ddot{k}_{\hat{x}}}$ being the number of uniformly spaced samples and where

$$\Delta_{\ddot{k}_{\hat{x}}} = \frac{2\pi}{\Pi_{(2X_o)}(\hat{x})},$$

with $2X_o$ being the image size in the down-range dimension. The center wave-number $\breve{k}_{\hat{x}_c}$ of the down-range Fourier migrated data (7.69) is

$$\breve{k}_{\hat{x}_c} = \frac{2}{c} \times \omega_c.$$

Indices of Image Pixels

Let $n_{\hat{x}}$ and $n_{\hat{y}}$ also index image pixels,

$$ii[n_{\hat{x}}, n_{\hat{y}}].$$

Indexed Non-Uniform Fourier Migrated Down-Range Wavenumber

The non-uniformly spaced Fourier migrated down-range wavenumber samples on the $n_{\hat{y}}^{\text{th}}$ row of the Stolt transformed SAR data matrix are located at

$$\breve{k}_{\hat{x}}[n_{\omega_p}, n_{\hat{y}}] = 2\sqrt{k^2_{\omega_p}[n_{\omega_p}] - k^2_{u_y}[n_{\hat{y}}]}.$$

Indexed Non-Uniform Fourier Migrated Down-Range Phase Factor

Note that the complex values of the non-uniformly spaced migrated down-range terrain phase factor samples,

$$\exp(-j\hat{x}_n \cdot \breve{k}_{\hat{x}}[n_{\omega_p}, n_{\hat{y}}])$$

located at $\breve{k}_{\hat{x}}[n_{\omega_p}, n_{\hat{y}}]$ are equal to the complex values of the rectilinear spectrum phase factor samples,

$$\exp(-j\hat{x}_n \cdot 2\sqrt{k^2_{\omega_p}[n_{\omega_p}] - k^2_{u_y}[n_{\hat{y}}]}\ )$$

indexed by $[n_{\omega_p}, n_{\hat{y}}]$ in the rectilinear (k_{ω_p}, k_{u_y}) data array. That is,

$$\exp(-j\hat{x}_n \cdot \breve{k}_{\hat{x}}[n_{\omega_p}, n_{\hat{y}}]) = \exp(-j\hat{x}_n \cdot 2\sqrt{k^2_{\omega_p}[n_{\omega_p}] - k^2_{u_y}[n_{\hat{y}}]}\ ).$$

Indexed Image Spectrum Along-Track Phase Factor

The complex values of the phase factor related to along-track terrain wavenumber

$$\exp(-j\hat{y}_n \cdot k_{\hat{y}}[n_{\hat{y}}])$$

and the phase factor expressed in terms of the aperture wavenumber

$$\exp(-j\hat{y}_n \cdot 2k_{u_y}[n_{\hat{y}}])$$

are also equal,

$$\exp(-j\hat{y}_n \cdot k_{\hat{y}}[n_{\hat{y}}]) = \exp(-j\hat{y}_n \cdot 2k_{u_y}[n_{\hat{y}}]).$$

Indexed Non-Uniformly Sampled Fourier Migrated Phase History

Hence, the indexed non-uniformly spaced (in the $\breve{k}_{\hat{x}}$ dimension) Fourier migrated (angular spectrum) phase history data samples

$$MS(\breve{k}_{\hat{x}}[n_{\omega_p}, n_{\hat{y}}], k_{\hat{y}}[n_{\hat{y}}])_{\text{SCS}} = \sum_n g_n \cdot \exp(-j\hat{x}_n \cdot \breve{k}_{\hat{x}}[n_{\omega_p}, n_{\hat{y}}]) \cdot \exp(-j\hat{y}_n \cdot k_{\hat{y}}[n_{\hat{y}}]) \tag{7.70}$$

are equal to the un-migrated rectilinear spectrum phase history data samples

$$SS[n_{\omega_p}, n_{\hat{y}}]_{\text{SCS}} = \sum_n g_n \cdot \exp(j\hat{x}_n \cdot 2\sqrt{k^2_{\omega_p}[n_{\omega_p}] - k^2_{u_y}[n_{\hat{y}}]}\ ) \cdot \exp(j\hat{y}_n \cdot 2k_{u_y}[n_{\hat{y}}]). \tag{7.71}$$

The only difference between (7.70) and (7.71) is that the data samples have been migrated from their positions in the uniformly spaced data array of the rectilinear spectrum to positions in a non-uniformly spaced data array of the angular spectrum. Our next step is to create a uniformly spaced array of resampled data within the Fourier migrated angular spectrum.

Indexed Uniformly Sampled Fourier Migrated Down-Range Phase Factor

The down-range phase factor obtained from uniformly resampled Fourier migrated wavenumber data is

$$\exp(-j\hat{x}_n \cdot \ddot{k}_{\hat{x}}[n_{\hat{x}}]).$$

Uniformly Sampled Image Spectrum

An interpolation operator applied to the non-uniformly spaced Fourier migrated spectrum (7.70) returns the uniformly sampled image spectrum

$$\begin{aligned} II[n_{\hat{x}}, n_{\hat{y}}]_{\text{SCS}} &= II(\ddot{k}_{\hat{x}}[n_{\hat{x}}], k_{\hat{y}}[n_{\hat{y}}])_{\text{SCS}}, \\ &= \sum_n g_n \cdot \exp(-j\hat{x}_n \cdot \ddot{k}_{\hat{x}}[n_{\hat{x}}]) \cdot \exp(-j\hat{y}_n \cdot k_{\hat{y}}[n_{\hat{y}}]). \end{aligned}$$

Interpolation Process

The operations of the interpolation from non-uniformly spaced Fourier migrated phase history data to uniformly spaced angular spectrum phase history data are

$$\begin{aligned} & II(\ddot{k}_{\hat{x}}[n_{\hat{x}}], k_{\hat{y}}[n_{\hat{y}}])_{\text{SCS}} \\ &\quad = \mathsf{Uniform}^{\breve{k}_{\hat{x}}}_{\text{1D}} \left\{ MS(\breve{k}_{\hat{x}}[n_{\omega_p}, n_{\hat{y}}], k_{\hat{y}}[n_{\hat{y}}])_{\text{SCS}} \right\}, \\ &\quad = \sum_{n_{\omega_p}: |\breve{k}_{\hat{x}}[n_{\omega_p}, n_{\hat{y}}] - \ddot{k}_{\hat{x}}[n_{\hat{x}}]| \in N_s \Delta_{\ddot{k}_{\hat{x}}}} MS(\breve{k}_{\hat{x}}[n_{\omega_p}, n_{\hat{y}}], k_{\hat{y}}[n_{\hat{y}}])_{\text{SCS}} \\ &\qquad\qquad \times \left\{ h(\breve{k}_{\hat{x}}[n_{\omega_p}, n_{\hat{y}}] - \ddot{k}_{\hat{x}}[n_{\hat{x}}]) \cdot J[n_{\omega_p}, n_{\hat{y}}] \right\}, \end{aligned} \tag{7.72}$$

where the sum over n_{ω_p} for each $n_{\hat{y}}$ are such that the interpolator's ingested neighborhood of non-uniformly spaced input samples of

$$MS(\breve{k}_{\hat{x}}[n_{\omega_p}, n_{\hat{y}}], k_{\hat{y}}[n_{\hat{y}}])_{\text{SCS}}$$

satisfies

$$|\breve{k}_{\hat{x}}[n_{\omega_p}, n_{\hat{y}}] - \ddot{k}_{\hat{x}}[n_{\hat{x}}]| \in N_s \Delta_{\ddot{k}_{\hat{x}}}$$

and where N_s defines the number of uniformly spaced data bins $\Delta_{\ddot{k}_{\hat{x}}}$ used to size the neighborhood of ingestion of the finite interpolator.

Interpolation and Uniform Resampling

Various interpolators are presented and analyzed in [146, 488, 583]. If the interpolation function is a sinc function, then $h(\cdot)$ in (7.72) is given by

$$h(\breve{k}_{\hat{x}}[n_{\omega_p}, n_{\hat{y}}] - \ddot{k}_{\hat{x}}[n_{\hat{x}}]) = 2X_o \cdot \mathsf{sinc}\left[2X_o \cdot (\breve{k}_{\hat{x}}[n_{\omega_p}, n_{\hat{y}}] - \ddot{k}_{\hat{x}}[n_{\hat{x}}])\right].$$

Sometimes the sinc kernel is broadened by about five percent and weighted,

$$\begin{aligned} & h(\breve{k}_{\hat{x}}[n_{\omega_p}, n_{\hat{y}}] - \ddot{k}_{\hat{x}}[n_{\hat{x}}]) \\ &= \mathsf{weight}(\breve{k}_{\hat{x}}[n_{\omega_p}, n_{\hat{y}}] - \ddot{k}_{\hat{x}}[n_{\hat{x}}]) \cdot 2X_o \cdot \mathsf{sinc}\left[2X_o \cdot (\breve{k}_{\hat{x}}[n_{\omega_p}, n_{\hat{y}}] - \ddot{k}_{\hat{x}}[n_{\hat{x}}])\right]. \end{aligned}$$

For polychromatic sensing, the weighting controls signal processing artifacts such as filter ripple in the passband, sidelobes in the stop-band and non-ideal transition bands [440]. A Hanning window is a common choice for the weight function $\mathsf{weight}(\cdot)$.

Interpolation for Monochromatic Sensing

Note that if a continuous wave signal is transmitted, then the interpolation operator (7.72) only has one input sample to interpolate to produce each output sample. For monochromatic sensing, the use of the weight function is not recommended (see Chapter 13).

Common Implementation Errors

Soumekh notes [887, 889, 899] that descriptions of Stolt formatting found in early omega-k literature "suffer from various misunderstandings, misconceptions and incorrect implementation of the algorithm." There is a misstep common to new practitioners of angular spectrum inversion methods. The processing required for Stolt transformation and the processing required for uniform resampling of Fourier migrated phase history data were sometimes inappropriately merged. We now address this potential error in more detail (beyond the discussion started in Section 4.8).

Clearly, uniform resampling is required after the non-linear Stolt transformation of the SAR signal spectrum to produce

$$II[n_{\hat{x}}, n_{\hat{y}}]_{\text{SCS}}.$$

But what is not clear is that interpolation and resampling should *follow* Fourier migration. That is, uniform resampling should occur over neighborhoods of non-uniformly spaced input samples of

$$MS[n_{\omega_p}, n_{\hat{y}}]_{\mathrm{SCS}}$$

defined within the Stolt transformed angular spectrum domain

$$(\breve{k}_{\hat{x}}[n_{\omega_p}, n_{\hat{y}}], k_{\hat{y}}[n_{\hat{y}}])$$

and should not be performed over uniformly spaced input samples of

$$SS[n_{\omega_p}, n_{\hat{y}}]_{\mathrm{SCS}}$$

defined within the un-migrated rectilinear spectrum

$$(\omega_p[n_{\omega_p}], k_{u_y}[n_{\hat{y}}]).$$

For each pair of index values $[n_{\omega_p}, n_{\hat{y}}]$, the phase values of $MS[n_{\omega_p}, n_{\hat{y}}]$ and $SS[n_{\omega_p}, n_{\hat{y}}]$ are equal,

$$MS[n_{\omega_p}, n_{\hat{y}}]_{\mathrm{SCS}} = SS[n_{\omega_p}, n_{\hat{y}}]_{\mathrm{SCS}},$$

but interpolator input neighborhoods within the angular domain of $MS[n_{\omega_p}, n_{\hat{y}}]$, not interpolator input neighborhoods within the rectilinear domain $SS[n_{\omega_p}, n_{\hat{y}}]$, must be used to determine which phase values of $MS[n_{\omega_p}, n_{\hat{y}}]$ (or equal values specified by $SS[n_{\omega_p}, n_{\hat{y}}]$) to ingest into the interpolator [898].

"Stolt Interpolation"

We associate the terminology of "Stolt interpolation" with erroneous implementations that inappropriately combine the processing steps of Stolt transformation and uniform resampling.

When "Stolt interpolation" is tried, one of the typical conclusions is that excessively "high quality" interpolators are required to achieve satisfactory results [39, 40, 62, 63, 135, 241]. Evaluation of this errant approach to Stolt formatting sometimes results in warnings against "simple and cheap" interpolators [137]. Expressions like "high quality," "critical" or "precise" [406, 407] are euphemisms where "excessively complex" is part of the intended accusation. The misguided use of "Stolt interpolation" has led some to claim angular spectrum processing methods to be unacceptably complex to achieve satisfactory performance levels.

In reality, the problems with such implementations extend beyond just the complexity of the interpolator.

Further Complications of Stolt Interpolation

Beyond the issues just discussed, there have been approaches formulated where, in addition to the interpolation for Fourier migration, an interpolation from the slant-plane to ground-plane is also integrated into the "Stolt interpolation" operation [440]. Moreover, approaches have also been proposed to speed up the processing by employing non-uniform FFT [412, 923] to intentionally blend the Stolt formatting and Fourier inversion, or to otherwise improve performance by modifying the FFT [350].

Interpolation After Huygens-Fresnel Transfers

Interpolation processes should be implemented after Huygens-Fresnel transfers. This makes the job of the interpolator easier because of the smoother phase functions associated with in situ phase histories described within the scene coordinate system, instead of the more rapidly varying phase histories as remotely sensed and described within the aperture coordinate system [218].

Interpolation and the Jacobian Determinant

Additionally, one of the elements sometimes missing in implementations of the signal processing of "Stolt interpolation" is the failure to account for the Jacobian of the Stolt transformation, as included in (7.72). The requirement for inclusion of the Jacobian in Fourier migration is one part of a more general concept we choose to call, *"Fourier reconstruction under variable transformation."* We will discuss principles of Fourier reconstruction under a change of variables transformation in Section 13.9.

7.15 Jacobian of Stolt Transformation

We next give overviews of the Jacobian matrix and the Jacobian determinant of slant-plane SAR Fourier migrations.

Jacobian Matrix

Consider an m dimensional function $\boldsymbol{y}(\boldsymbol{x})$ of an n dimensional vector $\boldsymbol{x}$. The matrix formed from the ordered first partial derivatives,

$$J = \begin{bmatrix} \dfrac{\partial y_1}{\partial x_1} & \cdots & \dfrac{\partial y_1}{\partial x_n} \\ \vdots & \ddots & \vdots \\ \dfrac{\partial y_m}{\partial x_1} & \cdots & \dfrac{\partial y_m}{\partial x_n} \end{bmatrix}$$

is the *Jacobian matrix.* Each row is the gradient $\nabla y_i(\boldsymbol{x})$ of the scalar component function $y_i(\boldsymbol{x})$ of $\boldsymbol{y}(\boldsymbol{x})$. Each row's gradient function describes a surface that characterizes the rate of change of y_i with respect to the dimensions of $\boldsymbol{x}$. Collectively, the rows of the Jacobian describe the orientation of a plane tangent to the surface $\boldsymbol{y}(\boldsymbol{x})$ at the point $\boldsymbol{x}$.

Jacobian Determinant

The absolute value of the determinant of the Jacobian matrix at $\boldsymbol{x}$ gives the factor by which a function $\boldsymbol{y}(\boldsymbol{x})$ expands or shrinks a differential volume at $\boldsymbol{x}$. For example, if

$$J\left[y_1(x_1, x_2), y_2(x_1, x_2)\right]$$

is the Jacobian matrix of a mapping from (x_1, x_2) to (y_1, y_2), then the Jacobian determinant, also called by some the Beylkin determinant in computed imaging

[69, 706], is

$$\begin{aligned} |J[y_1(x_1,x_2), y_2(x_1,x_2)]| &= \left| \frac{\partial(y_1,y_2)}{\partial(x_1,x_2)} \right|, \\ &= \begin{vmatrix} \frac{\partial y_1(x_1,x_2)}{\partial x_1} & \frac{\partial y_1(x_1,x_2)}{\partial x_2} \\ \frac{\partial y_2(x_1,x_2)}{\partial x_1} & \frac{\partial y_2(x_1,x_2)}{\partial x_2} \end{vmatrix}. \end{aligned}$$

7.16 Slant-Plane Image Spectrum

We next examine the image spectrum obtained through Fourier migration and the role of the Jacobian determinant of the Stolt transformation.

The Fourier migration of $SS(\omega_p, k_{u_y})_{\text{SCS}}$ to $MS(\breve{k}_{\hat{x}}, k_{\hat{y}})_{\text{SCS}}$ engages the change of variables theorem of calculus. The change of variables theorem also applies to Fourier migration of $H_{X_c}(\omega_p, k_{u_y})$ to $H_{X_c}(\breve{k}_{\hat{x}}, k_{\hat{y}})_{\text{SCS}}$.

To determine the Jacobian involved with this change of variables, consider the continuous version of the inverse Fourier transform that converts the image spectrum to the SAR image,

$$ii(\hat{x}, \hat{y})_{\text{SCS}} = \int\int II(\breve{k}_{\hat{x}}, k_{\hat{y}})_{\text{SCS}} \cdot \exp(j\hat{x}\breve{k}_{\hat{x}}) \exp(j\hat{y}k_{\hat{y}}) \cdot d\breve{k}_{\hat{x}} dk_{\hat{y}}.$$

We explicitly indicate the functional dependence of the output wavenumbers $(\breve{k}_{\hat{x}}, k_{\hat{y}})$ of Stolt formatting on the input wavenumbers of the rectilinear spectrum (ω_p, k_{u_y}) and the dependence of $II(\cdot,\cdot)$ on $MS(\cdot,\cdot)$ via the change of variables theorem,

$$\begin{aligned} ii(\hat{x}, \hat{y})_{\text{SCS}} = \int\int & MS[\breve{k}_{\hat{x}}(\omega_p, k_{u_y}), k_{\hat{y}}(\omega_p, k_{u_y})]_{\text{SCS}} \\ & \cdot \exp[j\hat{x} \cdot \breve{k}_{\hat{x}}(\omega_p, k_{u_y})] \exp[j\hat{y} \cdot k_{\hat{y}}(\omega_p, k_{u_y})] \\ & \cdot \begin{vmatrix} \frac{\partial \breve{k}_{\hat{x}}}{\partial \omega_p} & \frac{\partial \breve{k}_{\hat{x}}}{\partial k_{u_y}} \\ \frac{\partial k_{\hat{y}}}{\partial \omega_p} & \frac{\partial k_{\hat{y}}}{\partial k_{u_y}} \end{vmatrix} \cdot \partial\omega \partial k_{u_y}. \end{aligned}$$

Clearly, we must have

$$\begin{aligned} & II[\breve{k}_{\hat{x}}(\omega_p, k_{u_y}), k_{\hat{y}}(\omega_p, k_{u_y})]_{\text{SCS}} \\ &= MS[\breve{k}_{\hat{x}}(\omega_p, k_{u_y}), k_{\hat{y}}(\omega_p, k_{u_y})]_{\text{SCS}} \times \left| J[\breve{k}_{\hat{x}}(\omega_p, k_{u_y}), k_{\hat{y}}(\omega_p, k_{u_y})] \right|, \end{aligned}$$

with the Jacobian determinant given by

$$\left| J[\breve{k}_{\hat{x}}(\omega_p, k_{u_y}), k_{\hat{y}}(\omega_p, k_{u_y})] \right| = \begin{vmatrix} \frac{\partial \breve{k}_{\hat{x}}(\omega_p,k_{u_y})}{\partial \omega_p} & \frac{\partial \breve{k}_{\hat{x}}(\omega_p,k_{u_y})}{\partial k_{u_y}} \\ \frac{\partial k_{\hat{y}}(\omega_p,k_{u_y})}{\partial \omega_p} & \frac{\partial k_{\hat{y}}(\omega_p,k_{u_y})}{\partial k_{u_y}} \end{vmatrix}.$$

Jacobian of Two Dimensional Stolt Transformation

The Jacobian determinant describes the local expansions or contractions of the SAR signal spectrum $SS(\omega_p, k_{u_y})_{\text{SCS}}$ in the mapping to $MS(\breve{k}_{\hat{x}}, k_{\hat{y}})_{\text{SCS}}$ caused by the Stolt transformation,

$$MS(\breve{k}_{\hat{x}}, k_{\hat{y}})_{\text{SCS}} = \left| J\left[\breve{k}_x(\omega_p, k_{u_y}), k_y(\omega_p, k_{u_y})\right]\right| \times SS(\omega_p, k_{u_y})_{\text{SCS}}.$$

With

$$\begin{aligned} \breve{k}_{\hat{x}}(\omega_p, k_{u_y}) &= 2\sqrt{k_{\omega_p}^2 - k_{u_y}^2}, \\ k_{\hat{y}}(\omega_p, k_{u_y}) &= 2k_{u_y}, \end{aligned}$$

of the Stolt transformation, evaluation of the associated Jacobian determinant gives

$$\begin{aligned} \left| J\left[\breve{k}_x(\omega_p, k_{u_y}), k_y(\omega_p, k_{u_y})\right]\right| &= \begin{vmatrix} \frac{\partial \breve{k}_x(\omega_p, k_{u_y})}{\partial \omega_p} & \frac{\partial \breve{k}_x(\omega_p, k_{u_y})}{\partial k_{u_y}} \\ \frac{\partial k_y(\omega_p, k_{u_y})}{\partial \omega_p} & \frac{\partial k_y(\omega_p, k_{u_y})}{\partial k_{u_y}} \end{vmatrix}, \\ &= \frac{k_{\omega_p}}{\sqrt{k_{\omega_p}^2 - k_{u_y}^2}}, \\ &= \frac{1}{\cos\phi_x(k_{u_y})}, \end{aligned}$$

where ϕ_x is the angular spectrum angle defined with respect to the $\breve{k}_x$ axis and where $\phi_x(k_{u_y})$ is expressed as a function of k_{u_y}. The discrete version of the Jacobian of the Stolt mapping is

$$\left| J[n_{\omega_p}, n_{\hat{y}}]\right| = \frac{k_{\omega_p}[n_{\omega_p}]}{\sqrt{k_{\omega_p}[n_{\omega_p}]^2 - k_{u_y}[n_{\hat{y}}]^2}}.$$

The Jacobian determinant of the Stolt transformation captures the projection of a uniform distribution of mass spread over a semicircular manifold onto the k_{u_y} axis. This projection and change of variables allows the curvilinear (semicircle) integral traced through the $(\breve{k}_{\hat{x}}, k_{\hat{y}})$ plane to be replaced with the ordinary (straight) line integral along k_{u_y} [722].

7.17 Slant-Plane Image

The last step for slant-plane SAR image formation is to perform inverse Fourier transforms to create a complex pixel image array. A two dimensional inverse Fourier transform of $II(\ddot{k}_{\hat{x}}, k_{\hat{y}})_{\text{SCS}}$ produces the two dimensional SAR image $ii(\hat{x}, \hat{y})_{\text{SCS}}$.

Broadside Scene Image in Aperture Coordinate System

For the broadside case, the SAR image obtained from the uniformly resampled angular spectrum data described in the aperture coordinate system is

$$\mathsf{iFT}_{2D}\left\{ I^{\text{chirp}}_{\hat{\Pi}(\oplus)} I_{\hat{\Pi}[-2k_{\omega_p}, 2k_{\omega_p}]}(\ddot{k}_x, k_y)_{\text{ACS}} \right\} = i_{\Pi[X_{\min}, X_{\max}]}\, i_{(2Y_o @ 0)}(x, y)_{\text{ACS}}.$$

This image has a range extent

$$[X_{\min}, X_{\max}]$$

that spans the down-range distance from the flight line $X_{\min}$ to the far down-range edge of the scene $X_{\max}$ and is located at broadside with an along-track length of $2Y_o$ centered at 0.

Squint Scene Image in Aperture Coordinate System

For the squint scene case,

$$\mathsf{iFT}_{2D}\left\{I^{\text{chirp}}_{\hat{\Pi}(\oplus)} I_{\hat{\Pi}[-2k_{\omega_p},2k_{\omega_p}]}(\ddot{k}_x, k_y)_{\text{ACS}}\right\} = i_{\Pi[X_{\min},X_{\max}]} i_{\Pi[Y_{\min},Y_{\max}]}(x,y)_{\text{ACS}}.$$

This image has cross-range extent

$$[Y_{\min}, Y_{\max}]$$

that spans the minimum and maximum along-track offset positions of scatterers within the squint ground scene.

Slant-Plane Image in Scene Coordinate System

The along-track image in the scene coordinate system is also formed with an inverse Fourier transform,

$$\mathsf{iFT}_{2D}\left\{I^{\text{chirp}}_{\hat{\Pi}(\oplus)} I_{\hat{\Pi}[-2k_{\omega_p},2k_{\omega_p}]}(\ddot{k}_{\hat{x}}, k_{\hat{y}})_{\text{SCS}}\right\} = i_{\Pi[-X_o,X_o]} i_{\Pi[-Y_o,Y_o]}(\hat{x},\hat{y})_{\text{SCS}}.$$

The image phase history $II(k_{\hat{x}}, k_{\hat{y}})_{\text{SCS}}$ that is input into the inverse Fourier transform operator is the same for both the broadside and squint cases if the appropriate inverse Huygens-Fresnel transfers have been applied in conversions to scene centered coordinates. This along-track image is centered at $(0,0)$ with range extent $[-X_o, X_o]$ and along-track width of $[-Y_o, Y_o]$.

Frequency Offset Carrier Removal from Pixel Domain Data

As explained in Section 5.4, the down conversion artifact within

$$i^{\text{carrier}(-\omega_c)}_{\Pi[-X_o,X_o]} i_{\Pi[-Y_o,Y_o]}(\hat{x},\hat{y})^{\text{chirp}}_{\text{SCS}}$$

should be compensated to produce clean coherent slant-plane pixel imagery

$$i_{\Pi[-X_o,X_o]} i_{\Pi[-Y_o,Y_o]}(\hat{x},\hat{y})^{\text{chirp}}_{\text{SCS}},$$

if complex pixel imagery is to be exploited.

Chapter 8

Finite Aperture Stolt Formatting

We advance the SAR signal model by adding a spatial domain window element that describes a synthetic aperture of finite length. The primary impact of truncation to a finite aperture is that scatterer specific *arc window functions* must be incorporated onto the semicircle manifold of the Ewald sphere angular spectrum.

This chapter is outlined as follows: Section 8.1 presents changes to the slant-plane time-space phase history caused by introduction of a finite aperture into the SAR signal model. Section 8.2 repeats this for the frequency-space phase history. Section 8.3 shows the impact of a finite aperture on the along-track signal spectrum. Section 8.4 describes how the finite aperture effect is scatterer specific in the spatial domain. Section 8.5 illustrates how scatterer specific angular dwells are induced by truncation of the synthetic aperture to finite length. Section 8.6 explains how scatterer specific angular dwells "shatter" the remotely sensed terrain spectrum. Section 8.7 presents the angular spectrum of the finite aperture case. Section 8.8 gives the Huygens-Fresnel decomposition of the finite aperture angular spectrum. Section 8.9 describes Huygens-Fresnel inversion of the finite aperture angular spectrum. Finally, Section 8.10 explains how arc windows of the angular spectrum impact the image spectrum and describes characteristics of SAR imagery produced by a finite aperture.

8.1 Time-Space Phase History

In the prior chapter on slant-plane systems, we combined an infinite synthetic aperture element with a chirp pulse signal element. In this chapter, we introduce a finite synthetic aperture element into our signal model.

We return to a simpler fast-time waveform and assume the use of a non-pulsed, monochromatic, continuous wave tone signal,

$$w(t)^{\text{cw}} = \exp(j\omega_c t).$$

This approach allows us to focus on a simple waveform and its single frequency spectrum. As we shall outline in Section 8.10, the monochromatic development extends in a straight forward way to polychromatic chirp pulse signals.

Finite Aperture Imaging Geometry

The two dimensional terrain function of this chapter does not vary from the slant-plane model of the previous chapter. We assume the continuous wave signal is transmitted in all directions by a point antenna.

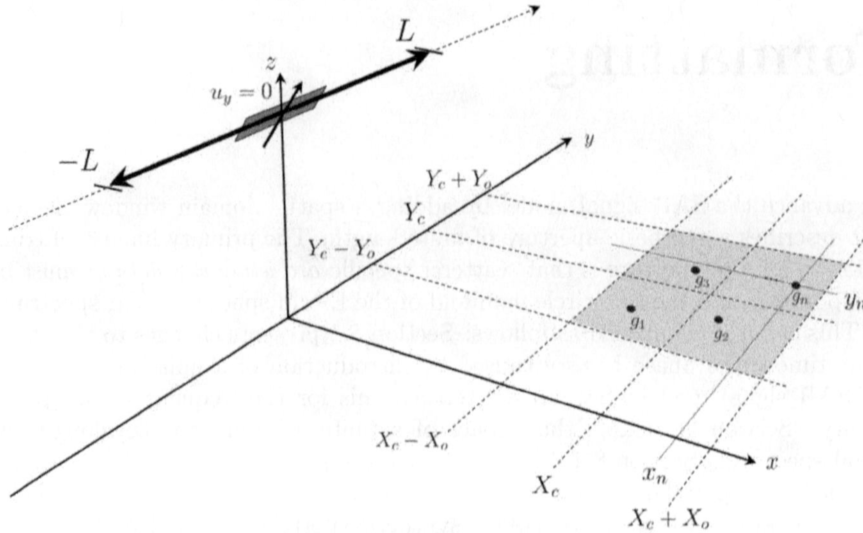

Figure 8.1: Finite synthetic aperture imaging geometry.

Finite Aperture Window Function

A description of a finite synthetic aperture is incorporated into the received signal model by adding the *synthetic aperture window function*,

$$\Pi_{[-L,L]}(u_y) = \begin{cases} 1 & \text{for} \qquad |u_y| \leq L, \\ 0 & \text{otherwise.} \end{cases}$$

The slow-time window function $\Pi_{[-L,L]}(u_y)$ indicates the synthetic aperture is bounded to an along-track interval of $[-L, L]$. This window function is centered at $u_y = 0$ as illustrated for a squint scene in Fig. 8.1.

Continuous Wave Echo Signal

The continuous wave echo signal collected across the finite synthetic aperture is

$$\begin{aligned} & s_{\Pi(\infty)} s_{\Pi[-L,L]}(t,u_y)^{\text{cw}} \\ &= \Pi_{[-L,L]}(u_y) \cdot \sum_n g_n \cdot w\left[t - t_n(u_y)\right]^{\text{cw}}, \\ &= \Pi_{[-L,L]}(u_y) \cdot \sum_n g_n \cdot \exp\left\{j\omega_c\left[t - t_n(u_y)\right]\right\}, \\ &= \Pi_{[-L,L]}(u_y) \cdot \sum_n g_n \cdot \exp(j\omega_c t) \cdot \exp\left[-j\omega_c t_n(u_y)\right]. \end{aligned} \tag{8.1}$$

The spatial slow-time window in (8.1) is indicated by the subscript $\Pi[-L, L]$ of

$$s_{\Pi(\infty)} s_{\Pi[-L,L]}(t,u_y)^{\text{cw}}, \tag{8.2}$$

where the span of positions of the point antenna within the synthetic aperture is

$$a_p(u_y) \in \Pi_{[-L,L]}(u_y).$$

With a continuous wave signal, the signal duration is infinite in time, hence the subscript $\Pi(\infty)$ of (8.2).

Along-Track Profile Time-Space Phase History

With substitution of the complex tone waveform

$$w(t)^{\text{cw}} = \exp(j\omega_c t) \tag{8.3}$$

into (8.1), we obtain the *finite aperture continuous wave time-space phase history,*

$$s_{\Pi(\infty)} s_{\Pi[-L,L]}(t,u_y)^{\text{cw}} = \Pi_{[-L,L]}(u_y) \cdot \sum_n g_n \cdot w(t)^{\text{cw}} \cdot \exp\left[-jk_{\omega_c} \cdot 2r_n(u_y)\right]. \tag{8.4}$$

8.2 Frequency-Space Phase History

With the transmission of a continuous wave signal, demodulation of the received echo signal removes the temporal phase factor (8.3) from the time-space phase history (8.4).

$$\begin{aligned} & \exp(-j\omega_c t) \times \left\{ s_{\Pi(\infty)} s_{\Pi[-L,L]}(t,u_y)^{\text{cw}} \right\} \\ &= \exp(-j\omega_c t) \times \left\{ \Pi_{[-L,L]}(u_y) \cdot \sum_n g_n \cdot w(t)^{\text{cw}} \cdot \exp\left[-jk_{\omega_c} \cdot 2r_n(u_y)\right] \right\}. \end{aligned}$$

This leaves the frequency-space phase history,

$$S_{\Pi(\omega_c)} s_{\Pi[-L,L]}(\omega_c,u_y) = \Pi_{[-L,L]}(u_y) \cdot \sum_n g_n \cdot \exp\left[-jk_{\omega_c} \cdot 2r_n(u_y)\right], \tag{8.5}$$

where the mapping (without Fourier transformation) from the time-space to the frequency-space phase history for a continuous wave echo signal is described by double scripted notation,

$$s_{\Pi(\infty)}s_{\Pi[-L,L]}(t,u_y)^{\text{cw}} \quad \mapsto \quad S_{\Pi(\omega_c)}s_{\Pi[-L,L]}(\omega_c,u_y).$$

There is no need to add the "CW" superscript to $S_{\Pi(\omega_c)}s_{\Pi[-L,L]}(\omega_c,u_y)$, since it is explicitly dependent on the carrier frequency ω_c.

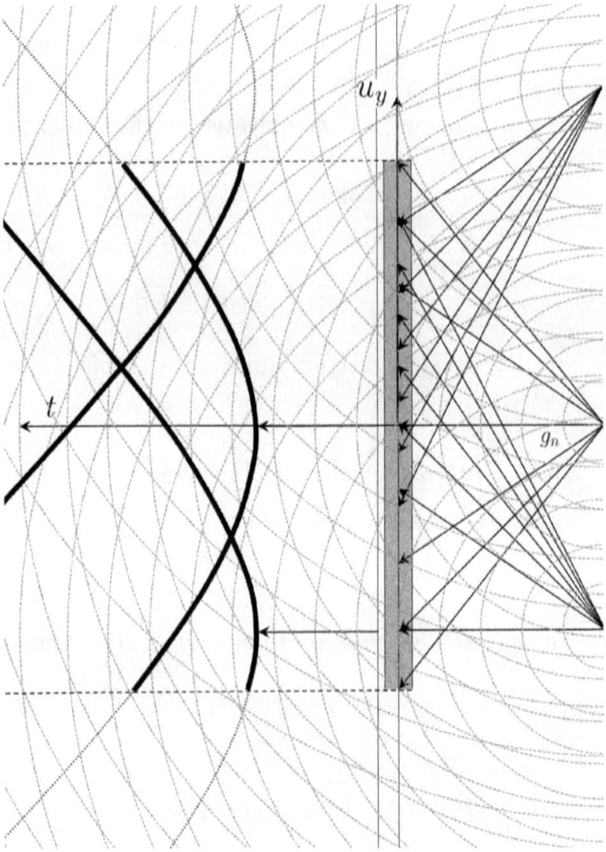

Figure 8.2: Truncated hyperbolic range walks.

Truncated Hyperbolic Range Walk in Frequency-Space Phase History

A key attribute of a phase history collected within a finite synthetic aperture is the different truncations of signal delay hyperbolas of scatterers located at different along-track positions. As the radar's position progresses through a finite synthetic aperture as illustrated in Fig. 8.2, different portions of scatterer specific range walk hyperbolas are sensed by the SAR system. This attribute is directly captured in the range walk phase curtains of (8.5).

Finite Aperture Collections within Fresnel Wave Fields

Another view of the impact of the finite aperture on data collections is that a finite synthetic aperture only permits a restricted measurement of a linear flight spatial sample of the frequency-space Fresnel wave field of each scatterer. These windowed phase signal measurements are

$$\begin{aligned} S_{\Pi(\omega_c)} s_{\Pi[-L,L]}(\omega_c, u_y)^{\text{cw}} &= \Pi_{[-L,L]}(u_y) \cdot \sum_n g_n \cdot \exp\left[-jk_{\omega_c} \cdot 2r_n(u_y)\right], \\ &= \Pi_{[-L,L]}(u_y) \cdot \sum_n g_n \cdot \measuredangle Hh\left[\omega_c, r_n(u_y)\right], \qquad (8.6) \end{aligned}$$

where

$$\measuredangle Hh\left[\omega_c, r_n(u_y)\right] = \exp\left[-jk_{\omega_c} \cdot 2r_n(u_y)\right]$$

is the passband (single frequency) dependent slow-time phase curtain of one scatterer.

8.3 Frequency-Wavenumber Phase History

The finite aperture window functions of the time-space domain and frequency-space domain are not scatterer specific. However, we shall now show that the induced window functions in the frequency-wavenumber domain are uniquely determined by the position of each scatterer for the finite synthetic aperture case.

Spatial Fourier Transform of the Finite Aperture Along-Track Signal

The finite aperture along-track signal spectrum of a single scatterer is obtained with the slow-time Fourier transform in the along-track dimension,

$$S_{\Pi(\omega_c)} s_{\Pi[-L,L]}(\omega_c, u_y)_n \;\mapsto\; S_{\Pi(\omega_c)} S_{\Pi(?)}(k_{u_y})(\omega_c, k_{u_y})_n.$$

The slow-time Fourier transform for one way signal travel is formulated as

$$\begin{aligned} &S_{\Pi(\omega_c)} S_{\Pi(?)}(k_{u_y})(\omega_c, k_{u_y})_n \\ &\quad = \mathsf{FT}_{1\text{D}}^{u_y}\left\{S_{\Pi(\omega_c)} s_{\Pi[-L,L]}(\omega_c, u_y)_n\right\}, \qquad (8.7) \\ &\quad = \int_{-\infty}^{\infty} \Pi_{[-L,L]}(u_y) \cdot g_n \cdot \exp[-jk_{\omega_c} \cdot r_n(u_y)] \cdot \exp(-jk_{u_y} u_y) \cdot du_y, \\ &\quad = \int_{-L}^{L} g_n \cdot \exp\left[-jk_{\omega_c} \cdot \sqrt{x_n^2 + (u_y - y_n^2)}\right] \cdot \exp(-jk_{u_y} u_y) \cdot du_y. \end{aligned}$$

Evaluation of this expression can be achieved by various means. One approach is with a method based on the principle of stationary phase [885] (see Section 15.4). Another approach is based on the use of expansions provided by Hankel functions (see Section 11.7).

With a method based on the principle of stationary phase, we obtain the signal spectrum of a single scatterer in the frequency-wavenumber (ω_c, k_{u_y}) domain, nominally expressed for two way signal travel by

$$\begin{aligned} & S_{\Pi(\omega_c)} S_{\Pi(?)}(k_{u_y})(\omega_c, k_{u_y})_n \\ &= g_n \cdot a(\omega_c, k_{u_y})_{\text{psp}} \cdot \Pi_{\Pi(?)}(k_{u_y}) \\ &\quad \cdot \exp(-jx_n \cdot 2\sqrt{k_{\omega_c}^2 - k_{u_y}^2}\,) \cdot \exp(-jy_n \cdot 2k_{u_y}), \end{aligned} \tag{8.8}$$

where ancillary amplitude weight functions and constant scale factors obtained by application of the principle of stationary phase (PSP) have been indicated by

$$a(\omega_c, k_{u_y})_{\text{psp}}.$$

The bandwidth and band location of the aperture wavenumber window function is temporarily indicated in (8.7) and (8.8) as unknown by the "?" placed in the relevant subscripts. We shall determine this aperture wavenumber window function parameter in the following sections.

8.4 Scatterer Specific Apertures

Equation (8.8) is identical to the result obtained for the infinite aperture case, except for one important caveat related to the supporting aperture wavenumber manifold for each scatterer. The frequency-wavenumber phase history (8.8) is significantly non-zero only over a scatterer specific aperture wavenumber interval given by [891, 970]

$$k_{u_y} \in [k_{\omega_c} \cos\theta_y(+L)_n, k_{\omega_c} \cos\theta_y(-L)_n]. \tag{8.9}$$

Outside of this region of support the frequency-wavenumber phase history

$$S_{\Pi(\omega_c)} s_{\Pi(?)}(\omega_c, k_{u_y})_n \approx 0$$

is equal to zero, or nearly so. The finite aperture's scatterer specific window function (8.9) is a subset of the wavenumber window function of the infinite aperture case,

$$\begin{aligned} [k_{\omega_c} \cos\theta_y(+L)_n, k_{\omega_c} \cos\theta_y(-L)_n] &\subset [k_{\omega_c} \cos\theta_y(+\infty)_n, k_{\omega_c} \cos\theta_y(-\infty)_n], \\ &= [k_{\omega_c} \cos(\pi), k_{\omega_c} \cos(0)], \\ &= [-k_{\omega_c}, +k_{\omega_c}]. \end{aligned}$$

The size and position parameters of the frequency-wavenumber domain window functions (8.9) are determined in part by the position of each scatterer within the ground scene. More specifically, it is the extrema of the view angle $\theta_y(u_y)_n$, as determined by each scatterer's position relative to the limits (beginning and end) of the finite synthetic aperture, that defines the extent of these scatterer specific window functions.

Scatterer Specific Aperture Wavenumber Manifold Limits

Among the parameters that define the scatterer dependent window function (8.9) is a relationship between the signal frequency temporal wavenumber k_{ω_c}, the nth

scatterer specific instantaneous view angle $\theta_y(u_y)_n$ and the aperture wavenumber k_{u_y}. This important relationship, implied in (8.9), is made explicit by defining

$$k_{u_y} \equiv k_{\omega_c} \cos\theta_y(u_y)_n. \tag{8.10}$$

Aperture Wavenumber and Direction Sines

The view angle $\theta_y(u_y)_n$ of (8.10) is a direction cosine angle defined with respect to the along-track y axis. If the view angle is defined with respect to the down-range x axis, then the aperture wavenumber (8.10) would be specified by a directional sine angle [885],

$$k_{u_y} \equiv k_{\omega_c} \sin\theta_x(u_y)_n.$$

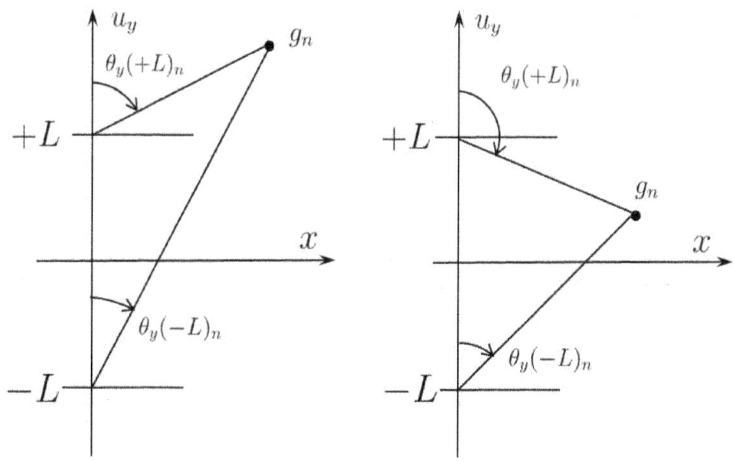

Figure 8.3: Scatterer specific extrema view angles of a finite aperture.

View Angle Extremes

The extent of the angular spectrum window function of the nth scatterer is determined by the values of $k_{u_y} = k_{\omega_c} \cos\theta_y(u_y)_n$ at the beginning ($u_y = -L$) and the end ($u_y = +L$) of the synthetic aperture as illustrated in Fig. 8.3. This scatterer specific window function is determined by the line of sight view angle that exists at each end of the synthetic aperture. The SAR signal spectrum's aperture wavenumber manifold (8.9) for the nth scatterer is determined by the sweep of sensor-scatterer view angle $\theta_y(u_y)_n$,

$$\theta_y(u_y)_n \in [\theta_y(-L)_n, \theta_y(+L)_n]. \tag{8.11}$$

8.5 Scatterer Specific Angular Dwells

The notion of a dwell in SAR can be associated with an observational interval in both fast-time and slow-time. Hence, "dwell" can be defined in both temporal and spatial domains. Furthermore, when dealing with measures of spatial observations over a sweep of view angles, the concept of dwell can be extended to that of an *angular dwell* [610].

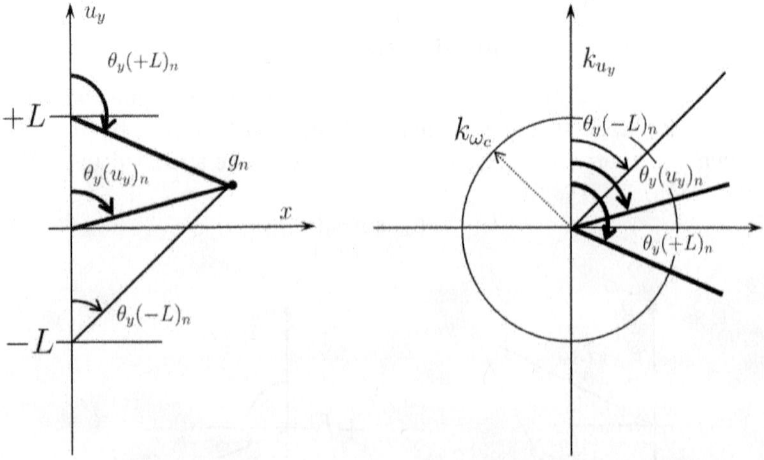

a. View angle dwell. b. Angular spectrum dwell.

Figure 8.4: View angle dwell to angular spectrum dwell mapping.

Angular Dwell

As the radar moves past a scatterer the instantaneous (scatterer dependent) view angle $\theta_y(u_y)_n$ sweeps out an angular dwell as shown in Fig. 8.4a. This angular dwell in the observation domain maps to a subset of the semicircle manifold of the angular spectrum as illustrated in Fig. 8.4b [610]. The angular dwell Θ_y of the nth scatterer is indicated by

$$\begin{aligned} \Theta_y(2L)_n &= \{\theta_y(u_y)_n\} \quad \text{for} \quad u_y \in [-L, +L], \\ &= [\theta_y(-L)_n, \theta_y(+L)_n]. \end{aligned} \tag{8.12}$$

If the length of the synthetic aperture is imagined to be infinite, then each scatterer has an angular dwell that extends from 0 radians (look ahead) to π radians (look behind). In the infinite aperture case, all scatterers share the same maximum angular dwell of

$$[\theta_y(-\infty)_n, \theta_y(+\infty)_n] = [0, \pi].$$

Aperture Wavenumber Window

Since,

$$k_{u_y} \equiv k_{\omega_c} \cos\theta_y(u_y)_n \quad \text{for} \quad u_y \in [-L, +L],$$

the aperture wavenumber window function of the nth scatterer can be expressed with a useful notation that indicates dependency on angular dwell,

$$\Pi_{\Theta_y}(k_{u_y})_n = [k_{\omega_c} \cos\theta_y(+L)_n, k_{\omega_c} \cos\theta_y(-L)_n]. \tag{8.13}$$

Scatterer Specific Aperture Wavenumber Windows

With the scatterer specific dwell (8.12) now established, and allowing ourselves a degree of freedom to (further) abuse the notation, we write (8.13) with the convenient short hand,

$$\Pi_{\Theta_y}(k_{u_y})_n \equiv k_{\omega_c} \cos\Theta_y(2L)_n,$$

which defines the aperture wavenumber window function in terms of the view angle sweep,

$$\Theta_y(2L)_n. \tag{8.14}$$

The scatterer specific dwell (8.14) is abbreviated as a subscript in $\Pi_{\Theta_y}(k_{u_y})_n$ to efficiently indicate (notationally) that the aperture wavenumber dwell of k_{u_y} is determined by the slow-time angular dwell Θ_y of the nth scatterer. The window function $\Pi_{\Theta_y}(k_{u_y})_n$ and angular dwell $\Theta_y(2L)_n$ are illustrated in Fig. 8.5.

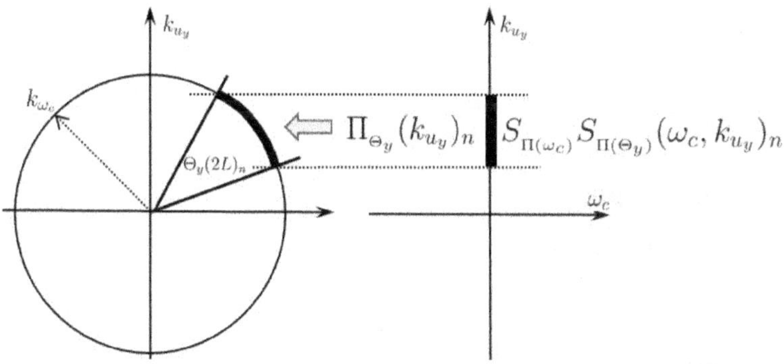

Figure 8.5: Scatterer specific angular dwell window function.

Finite Aperture Frequency-Wavenumber Phase History

We can now replace the "?" in the subscript of (8.8) with Θ_y, which indicates[1] the aperture wavenumber bandwidth is dependent on the angular dwell experienced by the nth scatterer,

$$\Pi_{(?)}(k_{u_y})_n \mapsto \Pi_{\Theta_y}(k_{u_y})_n.$$

This allows us to write the rectilinear frequency-wavenumber phase history of a single scatterer as

$$\begin{aligned}
&S_{\Pi(\omega_c)}S_{\Pi(\Theta_y)}(\omega_c, k_{u_y})_n \\
&\quad = a(\omega_c, k_{u_y})_{\text{psp}} \cdot \Pi_{\Theta_y}(k_{u_y})_n \\
&\qquad \cdot g_n \cdot \exp(-jx_n \cdot 2\sqrt{k_{\omega_c}^2 - k_{u_y}^2}\,) \cdot \exp(-jy_n \cdot 2k_{u_y}), \\
&\quad = \widehat{\Pi}_{\Theta_y}(\omega_c, k_{u_y})_n \cdot g_n \cdot \exp(-jx_n \cdot 2\sqrt{k_{\omega_c}^2 - k_{u_y}^2}\,) \cdot \exp(-jy_n \cdot 2k_{u_y}),
\end{aligned}$$

[1]The synthetic aperture length $2L$ is not indicated in $\Pi_{\Theta_y}(k_{u_y})_n$ to seek to simplify the notation.

where the amplitude modulation $a(\omega_c, k_{u_y})_{\text{psp}}$ that results from the slow-time Fourier transform (evaluated with a method based on the principle of stationary phase) is combined with the aperture wavenumber window function $\Pi_{\Theta_y}(k_{u_y})_n$,

$$\widehat{\Pi}_{\Theta_y}(\omega_c, k_{u_y})_n = a(\omega_c, k_{u_y})_{\text{psp}} \cdot \Pi_{\Theta_y}(k_{u_y})_n.$$

8.6 Terrain Shattered Spectrum

Ground Scene Angular Dwell

Consider the collection of scatterer specific angular dwells

$$\Theta_y(2L)_n$$

gathered from all scatterers within the ground scene, the overall *ground scene angular dwell*

$$\Theta_y(2L)$$

is the union of the angular dwells of the individual scatterers,

$$\Theta_y(2L) = \bigcup_n \Theta_y(2L)_n. \tag{8.15}$$

Shattered Signal Spectrum

Using (8.15), the frequency-wavenumber phase history for all scatterers can be written,

$$\begin{aligned} &S_{\Pi(\omega_c)} S_{\Pi(\Theta_y)}(\omega_c, k_{u_y}) \\ &\quad = \sum_n S_{\Pi(\omega_c)} S_{\Pi(\Theta_y)}(\omega_c, k_{u_y})_n, \\ &\quad = \sum_n g_n \cdot \widehat{\Pi}_{\Theta_y}(\omega_c, k_{u_y})_n \cdot \exp(-jx_n \cdot 2\sqrt{k_{\omega_c}^2 - k_{u_y}^2}\,) \cdot \exp(-jy_n \cdot 2k_{u_y}). \end{aligned} \tag{8.16}$$

When the n subscript is dropped from the frequency-wavenumber phase history,

$$S_{\Pi(\omega_c)} S_{\Pi(\Theta_y)}(\omega_c, k_{u_y})_n \quad \mapsto \quad S_{\Pi(\omega_c)} S_{\Pi(\Theta_y)}(\omega_c, k_{u_y}),$$

then the frequency-wavenumber phase history $S_{\Pi(\omega_c)} S_{\Pi(\Theta_y)}(\omega_c, k_{u_y})$ is descriptive of the spectrum of the echo signal received from the entire ground scene. Since the window function

$$\widehat{\Pi}_{\Theta_y}(\omega_c, k_{u_y})_n$$

remains within the summation of (8.16), we say the signal spectrum of the entire ground scene

$$S_{\Pi(\omega_c)} S_{\Pi(\Theta_y)}(\omega_c, k_{u_y})$$

has been *shattered* by the scatterer specific angular manifolds. This descriptive language is justified because the frequency-wavenumber phase history contains a collection "$\sum_n$" of pieces "$\Pi_{\Theta_y}(k_{u_y})_n$" of scatterer specific aperture wavenumber manifolds extracted from within the (unshattered) manifold $\Pi_{\oplus}(k_{u_y})$ of the infinite synthetic aperture [147].

Shattered Spectrum of the Remotely Sensed Terrain Function

Given the similarity of the phase factor (8.16) to an expression of the Fourier transform of the terrain function,

$$GG(\breve{k}_x, k_y) = \sum_n g_n \cdot \exp(-jx_n\breve{k}_x) \cdot \exp(-jy_n k_y),$$

where

$$\begin{aligned} \breve{k}_x &= 2\sqrt{k_{\omega_c}^2 - k_{u_y}^2}, \\ k_y &= 2k_{u_y}, \end{aligned}$$

we can write (8.16) as

$$\begin{aligned} &\widetilde{G}_{\Pi(\omega_c)}\widetilde{G}_{\Pi(\Theta_y)}(\omega_c, k_{u_y}) \qquad (8.17) \\ &= \sum_n \underbrace{g_n \cdot \widehat{\Pi}_{\Theta_y}(\omega_c, k_{u_y})_n \cdot \exp(-jx_n \cdot 2\sqrt{k_{\omega_c}^2 - k_{u_y}^2}\,) \cdot \exp(-j\hat{y}_n \cdot 2k_{u_y})}_{\widetilde{G}_{\Pi(\omega_c)}\widetilde{G}_{\Pi(\Theta_y)}(\omega_c, k_{u_y})_n}. \end{aligned}$$

We interpret (8.17) as a *shattered terrain spectrum.* The tildes of $\widetilde{G}\widetilde{G}(\cdot)$ are added to a terrain spectrum function $GG(\cdot)$ to indicate it has been shattered by finite aperture windowing effects. We write the individually windowed, scatterer specific, shattered terrain spectrum (submanifold) elements as

$$\begin{aligned} &\widetilde{G}_{\Pi(\omega_c)}\widetilde{G}_{\Pi(\Theta_y)}(\omega_c, k_{u_y})_n \\ &= g_n \cdot \widehat{\Pi}_{\Theta_y}(\omega_c, k_{u_y})_n \cdot \exp(-jx_n \cdot 2\sqrt{k_{\omega_c}^2 - k_{u_y}^2}\,) \cdot \exp(-j\hat{y}_n \cdot 2k_{u_y}) \end{aligned}$$

and hence,

$$\widetilde{G}_{\Pi(\omega_c)}\widetilde{G}_{\Pi(\Theta_y)}(\omega_c, k_{u_y}) = \sum_n \widetilde{G}_{\Pi(\omega_c)}\widetilde{G}_{\Pi(\Theta_y)}(\omega_c, k_{u_y})_n.$$

Notational Rule: Shattered Spectra of Scene and Scatterers

The angular dwell subscript $\Pi(\Theta_y)$ of

$$\widetilde{G}_{\Pi(\omega_c)}\widetilde{G}_{\Pi(\Theta_y)}(\omega_c, k_{u_y})$$

is descriptive of the angular dwell of the entire scene. When an n subscript is added,

$$\widetilde{G}_{\Pi(\omega_c)}\widetilde{G}_{\Pi(\Theta_y)}(\omega_c, k_{u_y})_n,$$

then the observation angular dwell subscript $\Pi(\Theta_y)$ of $\widetilde{G}_{\Pi(\omega_c)}\widetilde{G}_{\Pi(\Theta_y)}(\omega_c, k_{u_y})_n$ is descriptive of the angular dwell window of the nth scatterer within the scene.

8.7 Finite Aperture Angular Spectrum

We next describe steps of image formation processing based on Stolt formatting for a SAR signal model with a finite aperture. Many of the general concepts of Stolt formatting for the finite aperture case were covered in our discussion of the infinite aperture case.

Stolt Transformation of Finite Aperture Signal Spectrum

Fourier migration of the finite aperture's frequency-wavenumber phase history via Stolt transformation gives the angular spectrum of the finite aperture case,

$$\begin{aligned}
&\overset{\text{cw}}{M}_{\Pi(\otimes)} S_{\Pi(\Theta_y)}(\breve{k}_x, k_y)_{\text{ACS}} \\
&= \mathsf{ST}_{\text{1D}}^{\breve{k}_x} \left\{ S_{\Pi(\omega_c)} S_{\Pi(\Theta_y)}(\omega_c, k_{u_y})_{\text{ACS}} \right\}, \\
&= \mathsf{ST}_{\text{1D}}^{\breve{k}_x} \left\{ \sum_n g_n \cdot \widehat{\Pi}_{\Theta_y}(\omega_c, k_{u_y})_n \right. \\
&\qquad \left. \cdot \exp(-jx_n \cdot 2\sqrt{k_{\omega_c}^2 - k_{u_y}^2}\,) \cdot \exp(-jy_n \cdot 2k_{u_y}) \right\}, \\
&= \sum_n \overset{\text{cw}}{\widehat{\Pi}}_{\otimes}(\breve{k}_x, k_y)_n \cdot \widehat{\Pi}_{\Theta_y}(k_y)_n \cdot g_n \cdot \exp(-jx_n \breve{k}_x) \cdot \exp(-j\hat{y}_n k_y), \quad (8.18)
\end{aligned}$$

where the "cw" superscript of $\overset{\text{cw}}{M}_{\Pi(\otimes)} S_{\Pi(\Theta_y)}(\breve{k}_x, k_y)_{\text{ACS}}$ has been added to indicate the monochromatic continuous wave case.

Arc Weight Functions

The angular spectrum arc weight function of the nth scatterer contained in (8.18) is

$$\overset{\text{cw}}{\widehat{\Pi}}_{\otimes}(\breve{k}_x, k_y)_n \cdot \widehat{\Pi}_{\Theta_y}(k_y)_n = \mathsf{ST}_{\text{1D}}^{\breve{k}_x} \left\{ \widehat{\Pi}_{\Theta_y}(\omega_c, k_{u_y})_n \right\}. \quad (8.19)$$

The pair of weight functions on the left side of (8.5) represents the Stolt transform applied to the one dimensional finite aperture signal spectrum weight function

$$\widehat{\Pi}_{\Theta_y}(\omega_c, k_{u_y})_n.$$

The Stolt transform performs a mapping of the angular dwell $\Theta_y(2L)_n$ of k_{u_y}, as expressed in the rectilinear spectrum, to a pair of weight functions

$$\overset{\text{cw}}{\widehat{\Pi}}_{\otimes}(\breve{k}_x, k_y)_n \cdot \widehat{\Pi}_{\Theta_y}(k_y)_n,$$

that collectively represent a single arc window function in the angular spectrum domain.

Semicircle and Arc Window Icons

We previously used the icon $\oplus$ to characterize the scatterer independent semicircle wavenumber window function of the infinite aperture case. We use the icon $\otimes$ to characterize a scatterer specific wavenumber arc window function of the finite aperture case.

Semicircle Super Manifold

Recall that the infinite aperture angular spectrum window function induced by the Stolt transform in the wavenumber domain is a semicircle manifold

$$\widehat{\Pi}_{[-k_{\omega_c}, k_{\omega_c}]}(k_{u_y}) \quad \mapsto \quad \overset{\text{cw}}{\widehat{\Pi}}_{\oplus}(\breve{k}_x, k_y) \cdot \widehat{\Pi}_{[-2k_{\omega_c}, 2k_{\omega_c}]}(k_y).$$

This semicircular window is the same for all scatterers. For the finite aperture case, the wavenumber arc window functions that describe angular spectrum manifolds of individual scatterers differ in position and size, as indicated by the addition of the n subscripts in

$$\widehat{\Pi}_{\Theta_y}(\omega_c, k_{u_y})_n \quad \mapsto \quad \widehat{\Pi}_{\otimes}^{\text{cw}}(\breve{k}_x, k_y)_n \cdot \widehat{\Pi}_{\Theta_y}(k_y)_n.$$

Arc Windows on the Semicircle Manifold

With a finite synthetic aperture, the support of the angular spectrum phase history element induced by each scatterer is reduced from the semicircle manifold down to some arc window submanifold as shown in Fig. 8.5. The position of each arc on the semicircle is specific to the corresponding scatterer's position. The angular size of the arc is determined by the scatterer's angular dwell. The radius of the arc is set by the frequency ω_c of the transmitted continuous wave signal.

Scatterer View Angle and Angular Spectrum Angle

The view angle defined by the sensor-scatterer slant-plane vector is described by $\theta_y(u_y)_n$ for the nth scatterer. This view angle directly associates with aperture wavenumber values by

$$k_{u_y} = k_{\omega_c} \cos\theta_y(u_y)_n. \tag{8.20}$$

It is sometimes useful to express the dependence of the aperture wavenumber k_{u_y} on the scatterer specific view angle $\theta_y(u_y)_n$ by the functional notation,

$$k_{u_y}[\theta_y(u_y)_n] = k_{\omega_c} \cos\theta_y(u_y)_n.$$

Additionally, recall that the radius vector of the linear flight SAR's angular spectrum semicircle manifold defines wavenumber domain direction cosine angles ϕ_x and ϕ_y as shown in Fig. 8.6. A point on the semicircle manifold has a projection onto the k_y axis given by

$$k_y = 2k_{\omega_c} \cos\phi_y, \tag{8.21}$$

It is sometimes useful to express the dependence of the along-track wavenumber k_y on the angular spectrum angle ϕ_y by the functional notation,

$$k_y(\phi_y) = 2k_{\omega_c} \cos\phi_y.$$

Conversely, it is sometimes useful to express the association of ϕ_y with values of k_y by the inverse mapping,

$$\phi_y(k_y) = \arccos\left(\frac{k_y}{2k_{\omega_c}}\right).$$

Spatial View Angle and Angular Spectrum Angle Isomorphism

One way to view the functional impact of the Pythagorean constraint implied by the Stolt transform is to view the aperture wavenumber k_{u_y} or corresponding along-track wavenumber

$$k_y = 2k_{u_y} \tag{8.22}$$

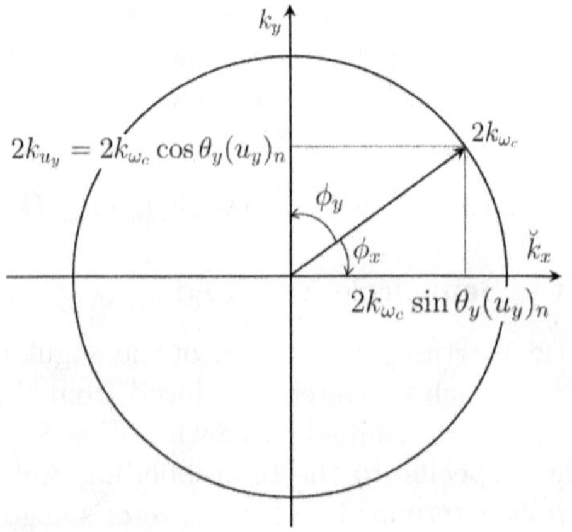

Figure 8.6: View angle and angular spectrum direction angles.

as free parameters. As k_{u_y} or k_y varies, the projected down-range wavenumber $\breve{k}_x$ must adjust to satisfy the Pythagorean constraint (4.54). Hence, different aperture wavenumber values k_{u_y} (associated with changes in along-track sensor position) induce migrations or traces of the down-range wavenumber $\breve{k}_x$ through changes in k_y as specified by (4.49) and (4.54), and as illustrated in Fig. 8.6.

Considering (8.20) and (8.21) together, and considering the along-track element (8.22) of the rectilinear to angular spectrum transformation, we obtain important relationships among wavenumbers, view angles and angular spectrum angles,

$$\begin{aligned} 2k_{\omega_c} \cos\theta_y(u_y)_n &= 2k_{\omega_c} \cos\phi_y(k_y), \\ \cos\theta_y(u_y)_n &= \cos\phi_y(k_y), \\ \theta_y(u_y)_n &= \phi_y(k_y). \end{aligned} \tag{8.23}$$

Hence, we have identified an important isomorphic mapping

$$\theta_y(u_y)_n \equiv \phi_y(k_y)$$

of sensor-scatterer view angle $\theta_y(u_y)_n$ to angular spectrum angle $\phi_y(k_y)$.

Angular Sweeps in Space and Wavenumber Domains

In (8.23) an association exists between view angles in the observation domain and angular spectrum angles in the wavenumber domain. Hence, *localized* phase history measurements are induced on the semicircle manifold of the angular spectrum domain from each sensor position within the synthetic aperture. The induced mapping of ϕ_y changes in lock step with changes in the view angle $\theta_y(u_y)_n$, as shown in Fig. 8.7.

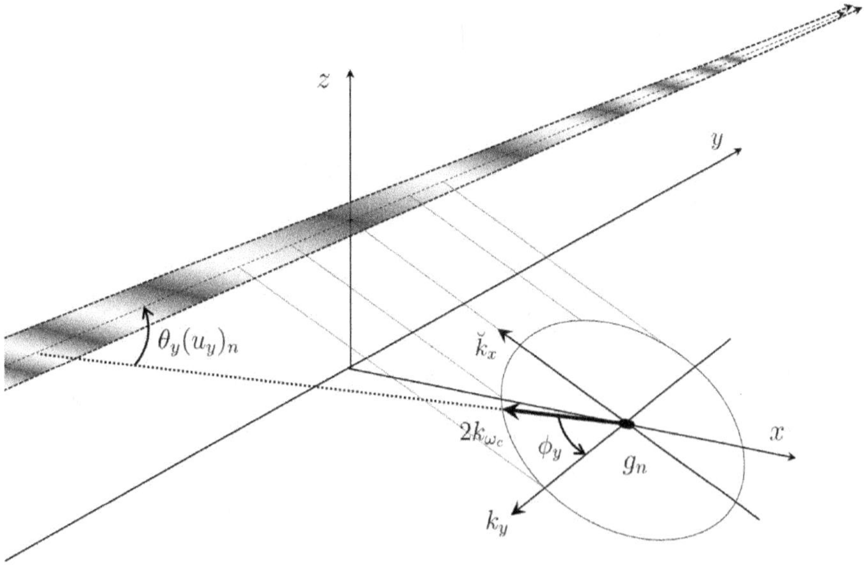

Figure 8.7: View angle mapping to angular spectrum semicircle manifold.

Ray Traced Manifolds

The arc manifolds for different scatterers located at different points (x_n, y_n) in the ground scene can be determined by the ray tracing method [132, 133] illustrated in Fig. 8.8. Lines are drawn from the start and end points of the linear synthetic aperture through each scatterer's location (x_n, y_n). The subtended angle on the other side of each scatterer defines that scatterer's angular dwell $\Theta_y(2L)_n$. The small circle about each scatterer represents the supporting circular manifold of radius $2k_{\omega_c}$. These circles and arc windows can be viewed as traces on a slant-plane slice of the Ewald sphere.

Ray Traced Manifolds on Ewald Spheres

The circle super-manifolds of Fig. 8.8 are shown as separated only for concept illustration. Scatterers of the ground scene all share the same supporting circle manifold within the frequency-wavenumber domain phase history. The three dimensional version of this ray tracing concept, as illustrated on scatterer specific Ewald spheres, is shown in Fig. 8.9.

8.8 Finite Aperture Huygens-Fresnel Analysis

Huygens-Fresnel decompositions can be performed in both the rectilinear and the angular spectrum domains. We start with the Huygens-Fresnel decomposition of the rectilinear frequency-wavenumber phase history for the finite aperture case.

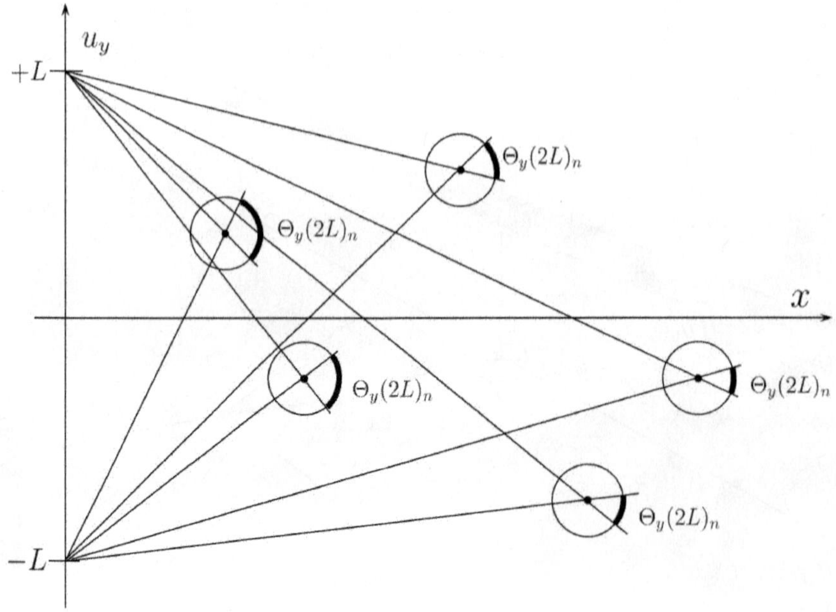

Figure 8.8: Angular manifolds from ray tracing.

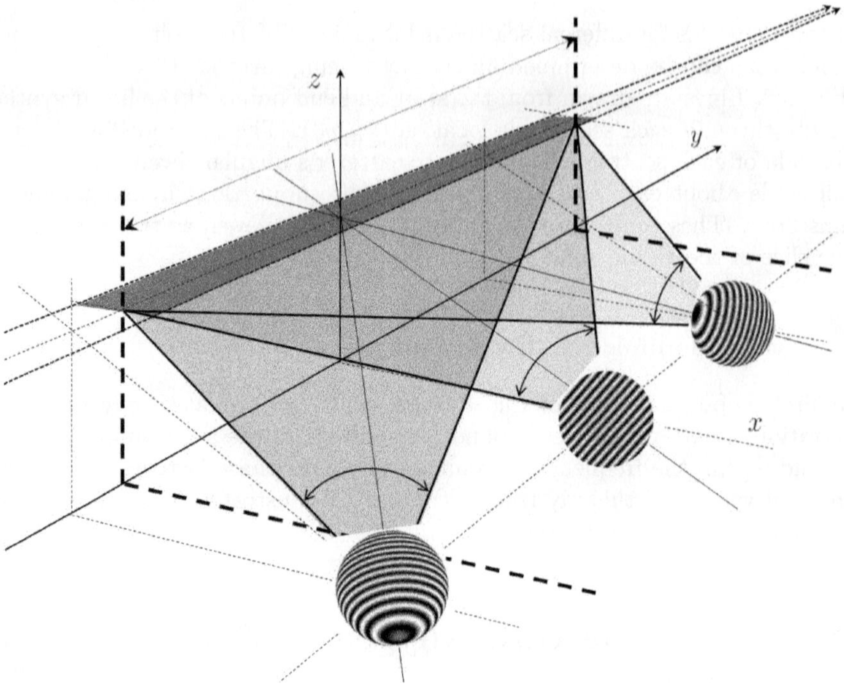

Figure 8.9: Ewald sphere traces of different scatterers.

Rectilinear Spectrum Huygens-Fresnel Analysis

Consider the superposition of all scatterer echoes from the ground scene expressed in the aperture coordinate system,

$$\begin{aligned}
&S_{\Pi(\omega_c)} S_{\Pi(\Theta_y)}(\omega_c, k_{u_y})_{\text{ACS}} \\
&= \underbrace{\exp(-jX_c \cdot 2\sqrt{k_{\omega_c}^2 - k_{u_y}^2}\,)}_{\substack{\text{down-range scene} \\ \text{center point information}}} \cdot \underbrace{\exp(-jY_c \cdot 2k_{u_y}\,)}_{\substack{\text{along-track scene} \\ \text{center point information}}} \\
&\quad \cdot \sum_n \underbrace{\widehat{\Pi}_{\Theta_y}(\omega_c, k_{u_y})_n}_{\substack{\text{scatterer angular} \\ \text{dwell windows}}} \cdot g_n \cdot \underbrace{\exp(-j\hat{x}_n \cdot 2\sqrt{k_{\omega_c}^2 - k_{u_y}^2}\,)}_{\substack{\text{down-range scatterer} \\ \text{information}}} \cdot \underbrace{\exp(-j\hat{y}_n \cdot 2k_{u_y})}_{\substack{\text{along-track scatterer} \\ \text{information}}}.
\end{aligned}$$

Using round trip Huygens-Fresnel transfers expressed in the rectilinear spectrum

$$\begin{aligned}
H_{X_c}(\omega_c, k_{u_y}) &= \exp(-jX_c \cdot 2\sqrt{k_{\omega_c}^2 - k_{u_y}^2}\,), \\
H_{Y_c}(k_{u_y}) &= \exp(-jY_c \cdot 2k_{u_y}),
\end{aligned}$$

the SAR signal spectrum can be decomposed into the shattered terrain spectrum expressed in the scene coordinate system and subject to Huygens-Fresnel transfers,

$$\begin{aligned}
&S_{\Pi(\omega_c)} S_{\Pi(\Theta_y)}(\omega_c, k_{u_y})_{\text{ACS}} \\
&= H_{X_c}(\omega_c, k_{u_y}) \cdot H_{Y_c}(k_{u_y}) \\
&\quad \cdot \sum_n \underbrace{\widehat{\Pi}_{\Theta_y}(\omega_c, k_{u_y})_n \cdot g_n \cdot \exp(-j\hat{x}_n \cdot 2\sqrt{k_{\omega_c}^2 - k_{u_y}^2}\,) \cdot \exp(-j\hat{y}_n \cdot 2k_{u_y})}_{\widetilde{G}_{\Pi(\omega_c)} \widetilde{G}_{\Pi(\Theta_y)}(\omega_c, k_{u_y})_{\text{SCS}}^n},
\end{aligned}$$

where a scatterer specific element of the shattered terrain spectrum is

$$\begin{aligned}
&\widetilde{G}_{\Pi(\omega_c)} \widetilde{G}_{\Pi(\Theta_y)}(\omega_c, k_{u_y})_{\text{SCS}}^n \\
&= \widehat{\Pi}_{\Theta_y}(\omega_c, k_{u_y})_n \cdot g_n \cdot \exp(-j\hat{x}_n \cdot 2\sqrt{k_{\omega_c}^2 - k_{u_y}^2}\,) \cdot \exp(-j\hat{y}_n \cdot 2k_{u_y}).
\end{aligned}$$

Hence, the model of the rectilinear SAR signal spectrum for the finite synthetic aperture is

$$\begin{aligned}
&S_{\Pi(\omega_c)} S_{\Pi(\Theta_y)}(\omega_c, k_{u_y})_{\text{ACS}} \\
&= H_{X_c}(\omega_c, k_{u_y}) \cdot H_{Y_c}(k_{u_y}) \cdot \underbrace{\sum_n \widetilde{G}_{\Pi(\omega_c)} \widetilde{G}_{\Pi(\Theta_y)}(\omega_c, k_{u_y})_{\text{SCS}}^n}_{\widetilde{G}_{\Pi(\omega_c)} \widetilde{G}_{\Pi(\Theta_y)}(\omega_c, k_{u_y})_{\text{SCS}}}, \\
&= H_{X_c}(\omega_c, k_{u_y}) \cdot H_{Y_c}(k_{u_y}) \cdot \widetilde{G}_{\Pi(\omega_c)} \widetilde{G}_{\Pi(\Theta_y)}(\omega_c, k_{u_y})_{\text{SCS}}, \qquad (8.24)
\end{aligned}$$

where the entire scene's shattered terrain spectrum is

$$\widetilde{G}_{\Pi(\omega_c)} \widetilde{G}_{\Pi(\Theta_y)}(\omega_c, k_{u_y})_{\text{SCS}} = \sum_n \widetilde{G}_{\Pi(\omega_c)} \widetilde{G}_{\Pi(\Theta_y)}(\omega_c, k_{u_y})_{\text{SCS}}^n.$$

The frequency-wavenumber phase history of the aperture coordinate system is comprised of the shattered ground scene spectrum described in the scene coordinate system and subject to rectilinearly formatted Huygens-Fresnel transfers.

Angular Spectrum Huygens-Fresnel Analysis

The Huygens-Fresnel decomposition of the SAR signal spectrum can also be performed in the angular spectrum domain. Recall the angular spectrum phase history of a finite aperture SAR system,

$$\begin{aligned} \overset{\text{cw}}{M}_{\widehat{\Pi}(\otimes)} S_{\widehat{\Pi}(\Theta_y)}(\breve{k}_x, k_y)_{\text{ACS}} \\ = \sum_n \overset{\text{cw}}{\widehat{\Pi}}_{\otimes}(\breve{k}_x, k_y)_n \cdot \widehat{\Pi}_{\Theta_y}(k_y)_n \cdot g_n \cdot \exp(-jx_n\breve{k}_x) \cdot \exp(-jy_n k_y). \quad (8.25) \end{aligned}$$

Shattered Terrain Spectrum Embedded in the Angular Spectrum

From (8.25), the following unwindowed element of the angular spectrum phase history,

$$G_{\Pi(\infty)} G_{\Pi(\infty)}(\breve{k}_x, k_y)^n_{\text{ACS}} = g_n \cdot \exp(-jx_n\breve{k}_x) \cdot \exp(-jy_n k_y)$$

can be associated with a single scatterer's portion of the (unshattered) terrain spectrum. Furthermore, by defining

$$\overset{\text{cw}}{\widetilde{G}}_{\widehat{\Pi}(\otimes)} \widetilde{G}_{\widehat{\Pi}(\Theta_y)}(\breve{k}_x, k_y)^n_{\text{ACS}} = \overset{\text{cw}}{\widehat{\Pi}}_{\otimes}(\breve{k}_x, k_y)^{\text{cw}}_n \cdot \widehat{\Pi}_{\Theta_y}(k_y)_n \cdot G_{\Pi(\infty)} G_{\Pi(\infty)}(\breve{k}_x, k_y)^n_{\text{ACS}},$$

the Fourier migrated angular spectrum can be expressed as

$$\overset{\text{cw}}{M}_{\widehat{\Pi}(\otimes)} S_{\widehat{\Pi}(\Theta_y)}(\breve{k}_x, k_y)_{\text{ACS}} = \sum_n \overset{\text{cw}}{\widetilde{G}}_{\widehat{\Pi}(\otimes)} \widetilde{G}_{\widehat{\Pi}(\Theta_y)}(\breve{k}_x, k_y)^n_{\text{ACS}}.$$

When decomposed by Huygens-Fresnel analysis,

$$\begin{aligned} \overset{\text{cw}}{M}_{\widehat{\Pi}(\otimes)} S_{\widehat{\Pi}(\Theta_y)}(\breve{k}_x, k_y)_{\text{ACS}} &= H_{X_c}(\breve{k}_{\hat{x}}) \cdot H_{Y_c}(k_{\hat{y}}) \cdot \sum_n \overset{\text{cw}}{\widetilde{G}}_{\widehat{\Pi}(\otimes)} \widetilde{G}_{\widehat{\Pi}(\Theta_y)}(\breve{k}_{\hat{x}}, k_{\hat{y}})^n_{\text{SCS}}, \\ &= H_{X_c}(\breve{k}_{\hat{x}}) \cdot H_{Y_c}(k_{\hat{y}}) \cdot \overset{\text{cw}}{\widetilde{G}}_{\widehat{\Pi}(\otimes)} \widetilde{G}_{\widehat{\Pi}(\Theta_y)}(\breve{k}_{\hat{x}}, k_{\hat{y}})_{\text{SCS}}, \end{aligned}$$

the finite aperture angular spectrum expressed in the aperture coordinate system is comprised of the shattered ground scene spectrum described in the scene coordinate system and subject to Stolt formatted Huygens-Fresnel transfers.

8.9 Finite Aperture Huygens-Fresnel Inversion

As explained in Section 7.13, Huygens-Fresnel transfers can be performed either before or after Stolt formatting, and can also be performed in either the rectilinear spectrum or angular spectrum domain.

Angular Spectrum Huygens-Fresnel Inversion

Application of the Stolt formatted inverse Huygens-Fresnel transfers

$$\begin{aligned} H^*_{X_c}(\breve{k}_x) &= \exp(jX_c\breve{k}_x), \\ H^*_{Y_c}(k_y) &= \exp(jY_c k_y) \end{aligned}$$

to

$$\overset{\text{cw}}{M}_{\hat{\Pi}(\otimes)} S_{\hat{\Pi}(\Theta_y)}(\breve{k}_x, k_y)_{\text{ACS}}$$

removes the embedded down-range phase factors $H_{X_c}(\breve{k}_{\hat{x}})$ and $H_{Y_c}(k_{\hat{y}})$ from the angular spectrum phase history. The desired ground scene spectrum shifts to the scene coordinate system,

$$\begin{aligned} &\overset{\text{cw}}{M}_{\hat{\Pi}(\otimes)} S_{\hat{\Pi}(\Theta_y)}(\breve{k}_{\hat{x}}, k_{\hat{y}})_{\text{SCS}} \\ &= H^*_{X_c}(\breve{k}_x) \cdot H^*_{Y_c}(k_y) \times \left\{ \overset{\text{cw}}{M}_{\hat{\Pi}(\otimes)} S_{\hat{\Pi}(\Theta_y)}(\breve{k}_x, k_y)_{\text{ACS}} \right\}, \\ &= \underbrace{H^*_{X_c}(\breve{k}_x) \cdot H_{X_c}(\breve{k}_{\hat{x}})}_{\text{cancels}} \times \underbrace{H^*_{Y_c}(k_y) \cdot H_{Y_c}(k_{\hat{y}})}_{\text{cancels}} \cdot \sum_n \overset{\text{cw}}{\widetilde{G}}_{\hat{\Pi}(\otimes)} \widetilde{G}_{\hat{\Pi}(\Theta_y)}(\breve{k}_{\hat{x}}, k_{\hat{y}})^n_{\text{SCS}}, \\ &= \sum_n \overset{\text{cw}}{\widetilde{G}}_{\hat{\Pi}(\otimes)} \widetilde{G}_{\hat{\Pi}(\Theta_y)}(\breve{k}_{\hat{x}}, k_{\hat{y}})^n_{\text{SCS}}, \\ &= \overset{\text{cw}}{\widetilde{G}}_{\hat{\Pi}(\otimes)} \widetilde{G}_{\hat{\Pi}(\Theta_y)}(\breve{k}_{\hat{x}}, k_{\hat{y}})_{\text{SCS}}. \end{aligned}$$

Rectilinear Spectrum Huygens-Fresnel Inversion

The inverse Huygens-Fresnel transfer operators can be described in the (ω_c, k_{u_y}) domain,

$$\begin{aligned} H^*_{X_c}(\omega_c, k_{u_y}) &= \exp(jX_c \cdot 2\sqrt{k^2_{\omega_c} - k^2_{u_y}}\ ), \\ H^*_{Y_c}(k_{u_y}) &= \exp(jY_c \cdot 2k_{u_y}\ ). \end{aligned}$$

Applying the (ω_c, k_{u_y}) domain version of the inverse Huygens-Fresnel transfer functions to the (un-migrated) SAR signal spectrum

$$S_{\Pi(\omega_c)} S_{\hat{\Pi}(\Theta_y)}(\omega_c, k_{u_y})_{\text{ACS}}$$

gives

$$\begin{aligned} &H_{\Pi(\omega_c)} S_{\hat{\Pi}(\Theta_y)}(\omega_c, k_{u_y})_{\text{SCS}} \\ &= H^*_{X_c}(\omega_c, k_{u_y}) \cdot H^*_{Y_c}(k_{u_y}) \times \left\{ S_{\Pi(\omega_c)} S_{\hat{\Pi}(\Theta_y)}(\omega_c, k_{u_y})_{\text{ACS}} \right\}, \qquad (8.26) \\ &= S_{\Pi(\omega_c)} S_{\hat{\Pi}(\Theta_y)}(\omega_c, k_{u_y})_{\text{SCS}}. \end{aligned}$$

Inverse Huygens-Fresnel transfers shift the rectilinear frequency-wavenumber SAR phase history from the aperture coordinate system to the scene coordinate system.

Stolt Transform After Huygens-Fresnel Inversion

Stolt transformation of (8.26) after the Huygens-Fresnel transfers gives

$$\underset{\hat{\Pi}(\otimes)}{\overset{\mathrm{cw}}{M}}\, S_{\hat{\Pi}(\Theta_y)}(\breve{k}_{\hat{x}}, k_{\hat{y}})_{\mathrm{SCS}} \;=\; \mathsf{ST}^{\breve{k}_{\hat{x}}}_{1\mathrm{D}}\left\{H_{\Pi(\omega_c)}\, S_{\hat{\Pi}(\Theta_y)}(\omega_c, k_{u_y})_{\mathrm{SCS}}\right\}$$

and the same result obtained by Fourier migration performed before Huygens-Fresnel transfer is achieved.

Uniform Resampling in Finite Aperture Case

Uniform resampling of $\breve{k}_{\hat{x}}$ is required to format the Fourier migrated spectrum preparatory to inverse Fourier transformation. Uniform resampling for the finite aperture case proceeds as described in Section 7.14 for the infinite synthetic aperture case. Uniform resampling of the Fourier migrated spectrum yields the image spectrum,

$$\underset{\hat{\Pi}(\otimes)}{\overset{\mathrm{cw}}{I}}\, I_{\hat{\Pi}(\Theta_y)}(\ddot{k}_{\hat{x}}, k_{\hat{y}})_{\mathrm{SCS}} = \mathsf{Uniform\ Resample}^{\breve{k}_{\hat{x}}}_{1\mathrm{D}}\left\{\underset{\hat{\Pi}(\otimes)}{\overset{\mathrm{cw}}{M}}\, S_{\hat{\Pi}(\Theta_y)}(\breve{k}_{\hat{x}}, k_{\hat{y}})_{\mathrm{SCS}}\right\}.$$

8.10 Finite Aperture Image Spectrum and Image

Fourier inversion for the finite aperture case proceeds without modification as described in Section 7.16 for the infinite synthetic aperture case. The two dimensional inverse discrete Fourier transform,

$$ii(\hat{x}, \hat{y})_{\mathrm{SCS}} = \mathsf{iFT}_{2\mathrm{D}}\left\{\underset{\hat{\Pi}(\otimes)}{\overset{\mathrm{cw}}{I}}\, I_{\hat{\Pi}(\Theta_y)}(\ddot{k}_{\hat{x}}, k_{\hat{y}})_{\mathrm{SCS}}\right\}$$

yields the SAR image $ii(\hat{x}, \hat{y}_{\mathrm{SCS}}$.

The key difference of the finite aperture case is the nature of the point spread function. Point spread function attributes are determined by scatterer specific wavenumber window functions induced by truncation of the infinite synthetic aperture to finite length. We next set up notation to identify, in general, whether an along-track point spread function is spatially variant or invariant.

Spatially Variant and Invariant SAR Images

If a SAR imaging system produces a spatially variant point spread function, then we write

$$\widetilde{ii}(\hat{x}, \hat{y})_{\mathrm{SCS}}.$$

If the point spread function is spatially invariant, then we write

$$\overline{ii}(\hat{x}, \hat{y})_{\mathrm{SCS}}.$$

In the infinitely long aperture case, all point scatterers have the same point spread function since they share the same semicircle angular spectrum manifold (the terrain spectrum is not shattered).

Along-track imagery of the finite aperture case is spatially variant (with a generic, modeless, signal analysis model). This is caused by the shattering of the terrain spectrum. In the finite aperture case, each scatterer has a unique arc manifold. Hence, the point spread function is different for each scatterer.

Stripmap Mode Exception to the Rule

We have not yet considered the impact of stripmap and spotlight operational modes on the point spread function of the finite aperture case. In stripmap mode, because of the interaction between the beamwidth and scatterer specific along-track dwells, Stolt formatted stripmap mode imagery is spatially invariant in the along-tack dimension (see Section 12.7).

Stolt Formatted Spotlight Mode Imagery is Spatially Variant

Spotlight mode imagery formed by Stolt formatting is spatially varying.

Spotlight Mode Long Range Small Spot Exception to the Rule

Long range, small spot, Stolt formatted spotlight mode imagery may have spatial variations too subtle to visually detect.

Polychromatic Extension of Monochromatic Results

We now generalize our transmitted waveform from the monochromatic continuous wave of frequency ω_c to a polychromatic chirp pulse waveform defined over the passband,

$$\omega_p \in [\omega_c - \omega_o, \omega_c + \omega_o].$$

We shall determine the point spread function of a pulsed SAR with a finite aperture by analyzing attributes of the polychromatic passband shattered spectrum.

Spatially Variant Point Spread Function Spectra

Consider the angular spectrum manifolds of a set of scatterers illustrated in Fig. 8.10a. The scatterers near boresight and closer to the sensor in range have larger k-space support. As the scatterer's along-track position moves off boresight or advances down-range, the angular dwell of the wavenumber manifold becomes smaller [883].

The angular spectra of all scatterers of Fig. 8.10a have increased radial wavenumber bandwidth compared to the case of monochromatic sensing (with continuous wave signals). The increased radial wavenumber bandwidth is obtained by the use of (polychromatic) pulse waveforms.

Bounding Box Approximation of an Arc Manifold

Also shown in Fig. 8.10a are square bounding boxes used to approximate the size of each scatterer's arc window manifold [827, 885]. Each bounding box has been rotated to align with its view angle to the center of the synthetic aperture.

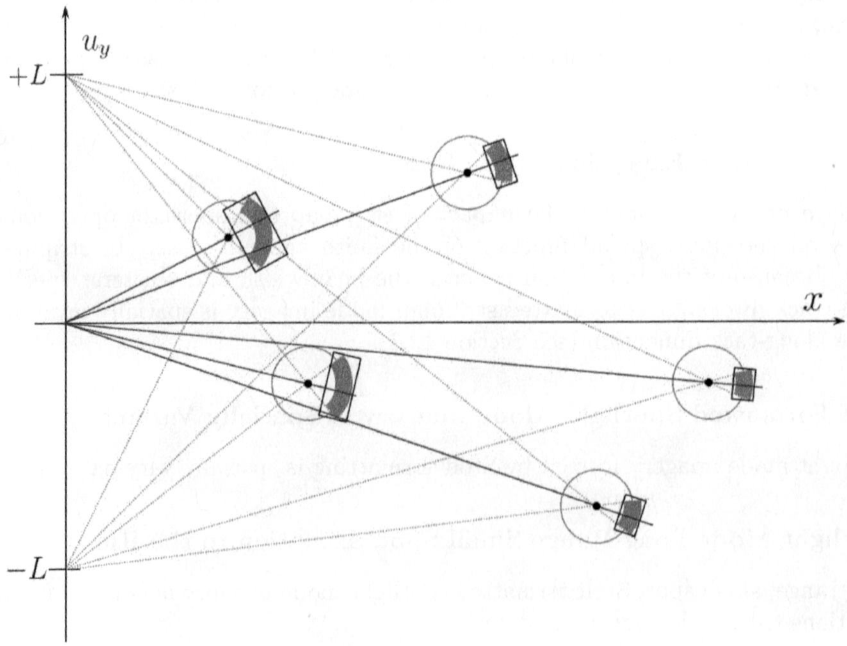

a. Bounding box approximations of polychromatic arc window functions.

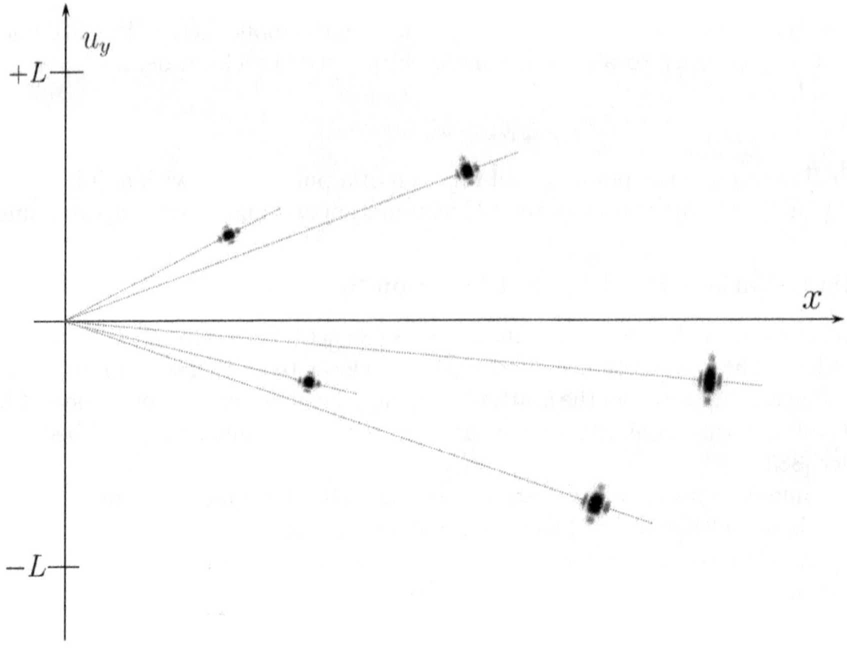

b. Spatially variant point spread functions.

Figure 8.10: Scatterer specific angular spectra and point spread functions.

This rotation allows the bounding box to provide a better approximation of the enclosed arc window manifold.

The rotated down-range wavenumber bandwidths of these rotated bounding boxes are given by

$$\begin{aligned}\beta_{x_n} &= 2k_{\omega_{\max}} - 2k_{\omega_{\min}}, \\ &= \frac{2}{c}\beta_\omega\end{aligned}$$

and the along-track heights of the rotated bounding boxes are given by

$$\beta_{y_n} = 2k_{\omega_c}\left[\cos\theta_y(-L)_n - \cos\theta_y(+L)_n\right].$$

Spatially Variant Point Spread Function Model

We can get an idea of the sizes of the SAR image point spread functions by considering the inverse Fourier transforms of table top functions with sizes determined by the scatterers' bounding boxes.

The rotated point spread function corresponding to a bounding box manifold is a two dimensional sinc function,

$$\mathsf{sinc}\left(\frac{x_n}{\Delta_{x_n}}\right)\cdot\mathsf{sinc}\left(\frac{y_n}{\Delta_{y_n}}\right),$$

as shown in Fig. 8.10b, where

$$\begin{aligned}\Delta_{x_n} &= \frac{2\pi}{\beta_{x_n}}, \\ \Delta_{y_n} &= \frac{2\pi}{\beta_{y_n}}.\end{aligned}$$

Slant-Range Resolution

The point spread function resolution of spotlight mode can be estimated by the down-range (slant-range) and along-track (azimuth) wavenumber bandwidths. Using the bounding box approximation, the resolution along the scatterer specific slant-range dimension is

$$\begin{aligned}\Delta_{x_n} &= \frac{2\pi}{2k_{\omega_{\max}} - 2k_{\omega_{\min}}}, \\ &= \frac{2\pi c}{2\beta_\omega}.\end{aligned}$$

Since the temporal cyclic frequency B and angular frequency β_ω bandwidths are related as

$$2\pi B = \beta_\omega,$$

the (rotated) down-range resolution is scatterer independent and we can write

$$\begin{aligned}\Delta_x &= \Delta_{x_n}, \\ &= \frac{c}{2B}.\end{aligned}$$

Azimuth Resolution

By changing measures of direction from directional cosines to directional sines, the scatterer specific rotated cross-range or azimuth resolution for the general case is

$$\Delta_{y_n} = \frac{2\pi}{2k_{\omega_c}[\sin\theta_x^{\max}(u_y)_n - \sin\theta_x^{\min}(u_y)_n]}.$$

Since, for a boresight scatterer in a broadside imaging geometry,

$$\sin\theta_x^{\min}(u_y)_n = -\sin\theta_x^{\max}(u_y)_n,$$

the along-track resolution is

$$\Delta_{y_n} = \frac{\lambda_c}{4}\cdot\frac{1}{\sin\theta_x^{\max}(u_y)_n}.$$

The azimuth support region of a scatterer's k-space manifold becomes smaller as the scatterer is positioned further away along-track from the ground scene's center point. Hence, the corresponding point spread function become less localized and the resolution degrades for off boresight scatterers. Furthermore, for large ground scenes, there will be an evident rotation of the point spread function if a scatterer is significantly off boresight as illustrated in Fig. 8.10b.

We see that in our generic (modeless) analysis, where we have not yet accounted for mode related specifics (stripmap and spotlight modes), that the along-track resolution of SAR imagery formed by Stolt formatting is spatially varying.

This behavior should not be confused with a similar behavior in SAR imagery formed by polar formatting methods. In polar formatting, the improvement in polar formatting (see Section 12.5) with decreasing (azimuth) distance from scene center is due to reduced approximation error in the plane wave Fraunhofer approximation.

Image Phase History Windowing

Because the semicircle manifold can be strongly non-square for low Q-factor systems, the point spread function can have non-rectangular sidelobes [397]. If there is a desire to minimize this effect, a square subset of the Stolt formatted manifold can be extracted, windowed and weighted [486]. The impact of the shattering of the terrain spectrum should be considered in the design of this window. A two dimensional Hamming weighting, for example, with a 1.30 mainlobe width expansion and a -43 dB peak sidelobe level can be used [435, 440]. Other submanifold extraction shapes, either enclosing or enclosed with respect to the arc window manifold of the entire scene, can be considered. The shape and orientation of this subaperture can have a major impact on the nature of the point spread function in the image domain, especially for ultra-wide bandwidth and ultra-wide beamwidth spotlight mode imagery produced by Stolt formatting.

Chapter 9

Surface Antenna Analysis

We now move beyond the simple point antenna to develop a model of a general *surface antenna.* The surface antenna is assumed stationary in the analysis of this chapter. A rectangular flat plate is used as the primary example of a surface antenna.

A surface antenna's beam pattern determines the spatial and angular field of view of the SAR sensor. This beam pattern can be described as a weight function in the time-space domain. The time-space domain weight function maps to weight functions induced in the frequency-space and frequency-wavenumber domains. We call such antenna induced weight functions, "beam functions."

This chapter is outlined as follows: Section 9.1 gives an overview of the characteristics of different types of physical apertures used in synthetic aperture sensor systems. Section 9.2 presents the use of Huygens-Fresnel methods to analyze antenna beam patterns in the time-space domain. Section 9.3 presents the use of Huygens-Fresnel methods to analyze antenna beam pattern weight functions induced in the frequency-space domain and Section 9.4 does the same for the frequency-wavenumber domain. Section 9.5 applies Huygens-Fresnel analysis methods to the particular case of a flat plate antenna. Section 9.6 presents time-space analysis for the flat plate antenna and Section 9.7 does the same in the frequency-space domain. Finally, Section 9.8 presents the frequency-wavenumber analysis of the flat plate antenna.

9.1 Antenna Analysis Overview

Types of real apertures or antennas used in synthetic aperture radars include rectangular arrays [243, 244, 955, 956], slotted wave guides [243, 1023], horns [164, 922], micro-strips [741], dipole line antennas [448, 922, 955, 956] and phased arrays. The antenna type of interest in this chapter is the class of surface antennas [358, 406, 407, 885].

Surface Antenna Excitation

Surface antennas are assumed to be energized from the "back-plane." Some surface antennas with "front-plane" horn feed excitation can also be modeled as back-plane energized surface antennas [885].

Far Field and Near Field Antenna Analysis

Far field Fraunhofer (plane wave) approximations and near field Fresnel (quadratic) approximations are conventionally used as tools in the analysis of antenna beam patterns. No such approximations are employed in Huygens-Fresnel based analysis [885].

Huygens-Fresnel Antenna Analysis

In the following development, Huygens-Fresnel wave theory is applied to produce a model of the radiated wave field of a general surface antenna. Surface antennas are analyzed by using a superposition principle to describe the composite Huygens-Fresnel wave field created by the collection of differential point radiators that comprise the antenna's surface.

Antenna Beam Patterns

Antenna analysis seeks to describe the amplitude and phase of the radiated field at points distant from the antenna. The remote field is produced by the aggregate of all radiating point elements on the antenna's surface. The spatial distribution of the totaled amplitude and phase field values constitute the antenna's *beam pattern*. The beam pattern defines the field of view in both transmit and receive modes of the radar. Antennas are reversible in their analysis. Developing a transmit or receive mode model also describes the other.

Beam Pattern and Surface Antenna Shape

Spatial attributes of the antenna beam pattern are determined by a surface antenna's physical shape. Surface antennas typically have beamwidths that decrease with increasing antenna size, but there are important exceptions to this rule [885]. An antenna with converse behavior allows the aperture gain to be increased by increasing the size of the antenna surface without reducing the size of the beam.

Data Support Manifolds of Stationary and Movable Antennas

An antenna beam pattern induces *beam function* weights within the various SAR phase history domains. We initially describe these beam functions in the time-space domain and then seek to determine the induced beam functions in the frequency-space and frequency-wavenumber domains. These beam functions define in part the data support manifolds afforded by the field of view of a stationary antenna. In the more general analysis of a moving SAR antenna, other portions of the data support manifold come from sweeping the position of the real aperture across the span of the synthetic aperture.

Spatially Variant Antenna Beam Patterns and Chirp Waveforms

When chirp waveforms are transmitted from a surface antenna, most surface antenna beam patterns vary with the instantaneous chirp frequency. The antenna beam pattern typically narrows as the transmitted chirp frequency increases. Hence, the field of view afforded by an antenna that transmits a chirp waveform changes during the waveform's intra-pulse time period. These antennas are called *chirp antennas.*

Spatially Invariant Antenna Beam Patterns and Chirp Waveforms

Some antennas' beam patterns remain independent of instantaneous chirp frequency. This type of antenna is useful when the bandwidth of the transmitted signal is a significant fraction of the carrier frequency [52,243,395,440,871]. These antennas are called *broadband antennas.*

Chirp and Broadband Antenna Beam Functions

When the induced beam functions of chirped and broadband antennas are examined and compared in the frequency-wavenumber domain, the characteristics of the beam functions induced by these two types of antennas are different [885]. Chirp and broadband antenna types can be considered complementary to each other in the following ways:

- Chirp antennas have beam patterns that vary with chirp frequency.
- Broadband antennas have beam patterns that are invariant to chirp frequency.
- Chirp antennas induce frequency-wavenumber domain beam functions that are invariant to chirp frequency.
- Broadband antennas induce frequency-wavenumber domain beam functions that vary with chirp frequency.

9.2 Surface Antenna Time-Space Analysis

We develop a model of the temporal and spatial characteristics of the radiated beam of a surface antenna. We characterize the antenna's beam pattern by describing the composite scalar electric field formed from the superposition of scaled expanding Huygens wavelets radiating from each point on the energized antenna's surface [174, 885].

Antenna Differential Point Elements

A surface antenna is modeled as consisting of differential elements a_e located at points

$$\boldsymbol{s} = [s_x \; s_y \; s_z]$$

on the antenna's surface $\mathbb{S}$. The spatial manifold $\mathbb{S}$ represents the three dimensional contour of the antenna surface. Points on the antenna surface are parametrically indicated by

$$a_e(\boldsymbol{s}) = \text{ for } \boldsymbol{s} \in \mathbb{S}.$$

The complex value of $a_e(\boldsymbol{s})$ is the excitation used to (time) impulsively energize the differential point element at $\boldsymbol{s}$. For the sake of notational simplicity, we choose not to add a "t" parameter to $a_e(\boldsymbol{s})$.

Antenna Field Points

We seek to describe the beam pattern at field points

$$\boldsymbol{r} = [r_x \; r_y \; r_z].$$

The distance between the point $\boldsymbol{s}$ on the antenna surface and the field point $\boldsymbol{r}$ is

$$|\boldsymbol{r} - \boldsymbol{s}| = \sqrt{(r_x - s_x)^2 + (r_y - s_y)^2 + (r_z - s_z)^2}.$$

Wave Field of Single Point Element on Antenna Surface

Consider the field at $\boldsymbol{r}$ produced by a single differential point element at $\boldsymbol{s}$ energized by a temporal impulse $\delta(t)$ with energy a_e. The resulting radiation is a scaled expanding Huygens wavelet,

$$a_e(\boldsymbol{s}) \cdot hh(t, \boldsymbol{r} - \boldsymbol{s}) = a_e(\boldsymbol{s}) \cdot \frac{1}{4\pi} \cdot \frac{1}{|\boldsymbol{r} - \boldsymbol{s}|} \cdot \delta\left[t - t_d(\boldsymbol{r} - \boldsymbol{s})\right],$$

where $t_d(\boldsymbol{r}-\boldsymbol{s})$ is the time delay required for the wavelet originating at the surface point $\boldsymbol{s}$ to reach the field point $\boldsymbol{r}$.

Time-Space Beam Pattern of Impulsively Energized Surface Antenna

Let

$$\Lambda(t, \boldsymbol{r})_{\mathbb{S}}^{\text{impulse}}$$

represent the composite antenna beam pattern produced by an impulsively energized surface antenna. By superposition, the field at $\boldsymbol{r}$ at the time instant t produced by all point elements on the antenna surface $\boldsymbol{s} \in \mathbb{S}$ with excitation $a_e(\boldsymbol{s})$ is

$$\Lambda(t, \boldsymbol{r})_{\mathbb{S}}^{\text{impulse}} = \int_{\mathbb{S}} \underbrace{a_e(\boldsymbol{s}) \cdot hh(t, \boldsymbol{r} - \boldsymbol{s})}_{\Lambda(t, \boldsymbol{r}-\boldsymbol{s})_{a_e}^{\text{impulse}}} \cdot d\boldsymbol{s}, \tag{9.1}$$

$$= \int_{\mathbb{S}} \Lambda(t, \boldsymbol{r} - \boldsymbol{s})_{a_e}^{\text{impulse}} \cdot d\boldsymbol{s},$$

where we express the field contribution of an impulsively energized point on the antenna surface as

$$\Lambda(t, \boldsymbol{r} - \boldsymbol{s})_{a_e}^{\text{impulse}} = a_e(\boldsymbol{s}) \cdot hh(t, \boldsymbol{r} - \boldsymbol{s}). \tag{9.2}$$

The scaled Huygens' wavelet of (9.2) propagates the impulsively generated wave field from points $\boldsymbol{s}$ on the antenna surface to distant points $\boldsymbol{r}$. The wavelets coherently interact to produce the field value $\Lambda(t, \boldsymbol{r})_{\mathbb{S}}^{\text{impulse}}$ [174].

Time-Space Beam Pattern of Pulse Energized Surface Antenna

The beam pattern produced by the surface antenna energized by a passband pulse waveform $w(t)$ is

$$\begin{aligned} \Lambda(t,\boldsymbol{r})_{\mathbb{S}}^{\text{pulse}} &= \int_{\mathbb{S}} a_e(\boldsymbol{s}) \cdot \frac{1}{4\pi} \cdot \frac{1}{|\boldsymbol{r}-\boldsymbol{s}|} \cdot w\left[t - t_d(\boldsymbol{r}-\boldsymbol{s})\right] \cdot d\boldsymbol{s}, \\ &= \int_{\mathbb{S}} \underbrace{\left[a_e(\boldsymbol{s}) \cdot hh(t,\boldsymbol{r}-\boldsymbol{s})\right]}_{\Lambda(t,\boldsymbol{r}-\boldsymbol{s})_{a_e}^{\text{impulse}}} *_t w(t) \cdot d\boldsymbol{s}, \\ &= \left[\int_{\mathbb{S}} \Lambda(t,\boldsymbol{r}-\boldsymbol{s})_{a_e}^{\text{impulse}} d\boldsymbol{s}\right] *_t w(t), \\ &= \Lambda(t,\boldsymbol{r})_{\mathbb{S}}^{\text{impulse}} *_t w(t). \end{aligned} \tag{9.3}$$

The time-space beam pattern of a pulsed surface antenna is the beam pattern of an impulsively energized surface antenna temporally convolved with the energizing waveform.

Center Referenced Beam Patterns in Time-Space Domain

At relatively large distances from the antenna, we describe spatial characteristics of the field by reference to a "center point" on the antenna's surface. This will allow us to identify the position of the surface antenna by reference to the location of a nominal center point.

Given the general expression,

$$\Lambda(t,\boldsymbol{r})_{\mathbb{S}}^{\text{impulse}} = \int_{\mathbb{S}} a_e(\boldsymbol{s}) \cdot \frac{1}{4\pi} \cdot \frac{1}{|\boldsymbol{r}-\boldsymbol{s}|} \cdot \delta\left[t - t_d(\boldsymbol{r}-\boldsymbol{s})\right] \cdot d\boldsymbol{s},$$

if we consider $\boldsymbol{r}$ sufficiently far from the point $\boldsymbol{a}_c$ on the antenna surface, then we have the approximation

$$\begin{aligned} \Lambda(t,\boldsymbol{r})_{\mathbb{S}}^{\text{impulse}} &\approx \frac{1}{4\pi} \cdot \frac{1}{|\boldsymbol{r}-\boldsymbol{a}_c|} \cdot \int_{\mathbb{S}} a_e(\boldsymbol{s}) \cdot \delta\left[t - t_d(\boldsymbol{r}-\boldsymbol{s})\right] \cdot d\boldsymbol{s}. \\ &= \Lambda(t,\boldsymbol{r}-\boldsymbol{a}_c)_{\mathbb{S}}^{\text{impulse}}. \end{aligned} \tag{9.4}$$

We call $\Lambda(t,\boldsymbol{r}-\boldsymbol{a}_c)_{\mathbb{S}}^{\text{impulse}}$ the *center referenced time-space antenna beam pattern*. The point $\boldsymbol{a}_c$ is identified as the nominal center point of the surface antenna. Note that (9.4) is not expressible in terms of a superposition integral involving the Huygens wavelet $hh(t,\boldsymbol{r}-\boldsymbol{s})$. The form of the transmitted beam pattern (9.4) has removed the surface dependent amplitude modulation $1/|\boldsymbol{r}-\boldsymbol{s}|$ for $\boldsymbol{s} \in \mathbb{S}$ and replaced it with a fixed distance parameter $1/|\boldsymbol{r}-\boldsymbol{a}_c|$. Although an approximation has been introduced in (9.4) for the antenna wave field, this wave field model has not invoked a far field Fraunhofer (plane wave) approximation typical of more conventional far field antenna analysis.

Center Referenced Beam Pattern of Pulse Signal

A surface antenna that transmits a passband waveform has a center referenced antenna beam pattern that is time modulated through temporal convolution with the transmitted pulse waveform,

$$\begin{aligned}\Lambda(t, \boldsymbol{r} - \boldsymbol{a}_c)_{\mathbb{S}}^{\text{pulse}} &= \Lambda(t, \boldsymbol{r} - \boldsymbol{a}_c)_{\mathbb{S}}^{\text{impulse}} \quad *_t \; w(t), \\ &= \frac{1}{4\pi} \cdot \frac{1}{|\boldsymbol{r} - \boldsymbol{a}_c|} \cdot \int_{\mathbb{S}} a_e(\boldsymbol{s}) \cdot w\left[t - t_d(\boldsymbol{r} - \boldsymbol{s})\right] \cdot d\boldsymbol{s}. \quad (9.5)\end{aligned}$$

9.3 Surface Antenna Frequency-Space Analysis

We next formulate a frequency-space domain description of a surface antenna's beam pattern.

Fourier Synthesis of Passband Pulse Signals

In preparation for our frequency-space analysis, recall that $w(t)$ is the passband pulse waveform transmitted by the SAR antenna. The waveform $w(t)$ has the passband spectrum $W(\omega_p)$. The relationship between $w(t)$ and $W(\omega_p)$ is the inverse Fourier transform,

$$w(t) = \frac{1}{2\pi} \int_{\beta_p} W(\omega_p) \cdot \exp(j\omega_p t) \cdot d\omega_p.$$

Frequency-Space Descriptions of Beam Patterns

Using monochromatic Fresnel spherical wave fields $Hh(\omega_p, \boldsymbol{r} - \boldsymbol{s})$, considered over all frequencies ω_p within the passband β_p, the surface antenna's time-space beam pattern can be expressed in terms of an inverse Fourier transform of a Huygens pulse wavelet defined by $w(t)$,

$$\begin{aligned}&\Lambda(t, \boldsymbol{r})_{\mathbb{S}}^{\text{pulse}} \\ &= \int_{\mathbb{S}} a_e(\boldsymbol{s}) \cdot \frac{1}{4\pi} \cdot \frac{1}{|\boldsymbol{r} - \boldsymbol{s}|} \cdot w\left[t - \underbrace{t_d(\boldsymbol{r} - \boldsymbol{s})}_{\text{time delay}}\right] \cdot d\boldsymbol{s}, \\ &= \int_{\mathbb{S}} \\ &\underbrace{\frac{1}{2\pi} \int_{\beta_p} a_e(\boldsymbol{s}) \cdot \underbrace{\frac{1}{4\pi} \cdot \frac{1}{|\boldsymbol{r} - \boldsymbol{s}|} \cdot \underbrace{\exp\left(-jk_{\omega_p}|\boldsymbol{r} - \boldsymbol{s}|\right)}_{\text{time delay phase factor}}}_{Hh(\omega_p, \boldsymbol{r} - \boldsymbol{s})} \cdot W(\omega_p) \cdot \exp\left(j\omega_p t\right) \cdot d\omega_p}_{\text{Inverse Fourier Transform}} \\ &\cdot d\boldsymbol{s},\end{aligned}$$

Hence,

$$\Lambda(t, \boldsymbol{r})_{\mathbb{S}}^{\text{pulse}} = \int_{\mathbb{S}} \frac{1}{2\pi} \int_{\beta_p} a_e(\boldsymbol{s}) \cdot Hh(\omega_p, \boldsymbol{r} - \boldsymbol{s}) \cdot W(\omega_p) \cdot \exp\left(j\omega_p t\right) \cdot d\omega_p \cdot d\boldsymbol{s}.$$

Changing the order of integration over the antenna surface with that of the inverse Fourier transform,

$$\Lambda(t,\boldsymbol{r})_{\mathbb{S}}^{\text{pulse}} = \frac{1}{2\pi}\int_{\beta_p} W(\omega_p)\cdot \underbrace{\int_{\mathbb{S}} a_e(\boldsymbol{s})\cdot Hh(\omega_p,\boldsymbol{r}-\boldsymbol{s})\cdot d\boldsymbol{s}}_{\Lambda(\omega_p,\boldsymbol{r})_{\mathbb{S}}^{\text{mono}}}\cdot \exp(j\omega_p t)\cdot d\omega_p, \tag{9.6}$$

$$= \frac{1}{2\pi}\int_{\beta_p} \underbrace{W(\omega_p)\cdot \Lambda(\omega_p,\boldsymbol{r})_{\mathbb{S}}^{\text{mono}}}_{\Lambda(\omega_p,\boldsymbol{r})_{\mathbb{S}}^{\text{pulse}}}\cdot \exp(j\omega_p t)\cdot d\omega_p, \tag{9.7}$$

$$= \frac{1}{2\pi}\int_{\beta_p} \Lambda(\omega_p,\boldsymbol{r})_{\mathbb{S}}^{\text{pulse}}\cdot \exp(j\omega_p t)\cdot d\omega_p. \tag{9.8}$$

We define in (9.6) a *monochromatic frequency-space surface antenna beam function,*

$$\Lambda(\omega_p,\boldsymbol{r})_{\mathbb{S}}^{\text{mono}} = \int_{\mathbb{S}} a_e(\boldsymbol{s})\cdot Hh(\omega_p,\boldsymbol{r}-\boldsymbol{s})\cdot d\boldsymbol{s}. \tag{9.9}$$

We define in (9.7) a *pulse (polychromatic) frequency-space surface antenna beam function.* This beam function is comprised of a passband pulse spectrum weighting of the monochromatic beam function,

$$\Lambda(\omega_p,\boldsymbol{r})_{\mathbb{S}}^{\text{pulse}} = W(\omega_p)\cdot \Lambda(\omega_p,\boldsymbol{r})_{\mathbb{S}}^{\text{mono}}.$$

We obtain in (9.8) a temporal Fourier transform pair of the beam pattern with pulse excitation,

$$\Lambda(t,\boldsymbol{r})_{\mathbb{S}}^{\text{pulse}} \quad {}_{t}\Leftrightarrow_{\omega_p} \quad \Lambda(\omega_p,\boldsymbol{r})_{\mathbb{S}}^{\text{pulse}}. \tag{9.10}$$

The left hand side of (9.10) is the beam function produced by the antenna beam pattern in the time-space domain. The right hand side of (9.10) is the beam function induced in the frequency-space domain.

Frequency-Space Domain Center Referenced Beam Patterns

The distant wave field with respect to a nominal center reference point on the surface antenna can also be expressed in terms of frequency-space parameters [173],

$$\Lambda(t,\boldsymbol{r}-\boldsymbol{a}_c)_{\mathbb{S}}^{\text{pulse}} = \frac{1}{2\pi}\int_{\beta_p} W(\omega_p)\cdot\frac{1}{4\pi}\cdot\frac{1}{|\boldsymbol{r}-\boldsymbol{a}_c|} \cdot\int_{\mathbb{S}} a_e(\boldsymbol{s})\cdot \exp\left(-jk_{\omega_p}|\boldsymbol{r}-\boldsymbol{s}|\right)\cdot d\boldsymbol{s}\cdot \exp(j\omega_p t)\cdot d\omega_p. \tag{9.11}$$

The far field model (9.11) does not rely on the Fraunhofer approximation to describe far field attributes[1].

[1]See Section 12.5 for an example of use of the Fraunhofer approximation to describe a plane wave approximation in the far field.

Because $1/|\boldsymbol{r}-\boldsymbol{s}|$ was replaced by $1/|\boldsymbol{r}-\boldsymbol{a}_c|$, (9.11) does not readily decompose into constituent monochromatic Fresnel wave fields like (9.6). However, as we show next, with an additional decomposition of the geometry of the antenna's surface with respect to a nominal center reference point, such a factorization is possible.

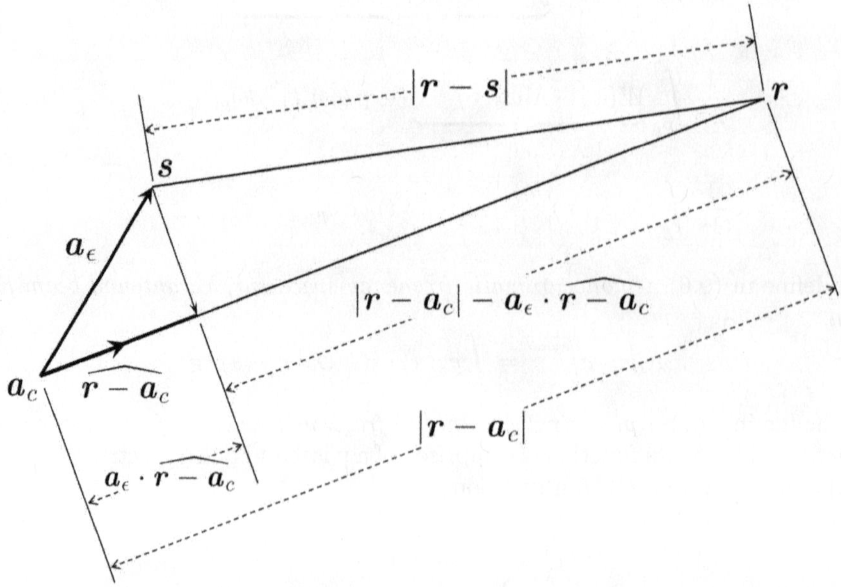

Figure 9.1: Center point referenced antenna surface model.

Center Point Referenced Antenna Surface Model

As shown in Fig. 9.1, points on the antenna surface $\boldsymbol{s}$ can be expressed in terms of a vector $\boldsymbol{a}_\epsilon$ that extends from the antenna center $\boldsymbol{a}_c$,

$$\boldsymbol{s} = \boldsymbol{a}_c + \boldsymbol{a}_\epsilon.$$

We are concerned with field distances from the antenna that are relatively large compared to the antenna size,

$$|\boldsymbol{a}_\epsilon| \ll |\boldsymbol{r}-\boldsymbol{a}_c|.$$

As illustrated in Fig. 9.1, an approximation of the distance between the field point $\boldsymbol{r}$ and each point $\boldsymbol{s}$ on the surface antenna is

$$|\boldsymbol{r}-\boldsymbol{s}| \approx |\boldsymbol{r}-\boldsymbol{a}_c| - \boldsymbol{a}_\epsilon \cdot \widehat{\boldsymbol{r}-\boldsymbol{a}_c}, \tag{9.12}$$

where the caret annotation of $\widehat{\boldsymbol{r}-\boldsymbol{a}_c}$ indicates a unit vector pointing in the direction of $\boldsymbol{r}-\boldsymbol{a}_c$ and where

$$\boldsymbol{a}_\epsilon \cdot \widehat{\boldsymbol{r}-\boldsymbol{a}_c} = (\boldsymbol{s}-\boldsymbol{a}_c) \cdot (\widehat{\boldsymbol{r}-\boldsymbol{a}_c})$$

is an inner product that defines a projection of $\boldsymbol{a}_\epsilon$ onto the unit direction vector $\widehat{\boldsymbol{r}-\boldsymbol{a}_c}$. The long range approximation (9.12) is used here to describe the antenna shape attributes. This long range approximation is different than using the Fraunhofer approximation to characterize the far field.

Center Referenced Frequency-Space Beam Model

By using the long range approximation (9.11), we obtain

$$\Lambda(t, \boldsymbol{r}-\boldsymbol{a}_c)_{\mathbb{S}}^{\text{pulse}} \approx \frac{1}{2\pi}\int_{\beta_p} W(\omega_p)\cdot \underbrace{\int_{\mathbb{S}} a_e(\boldsymbol{a}_c+\boldsymbol{a}_\epsilon)\cdot \exp\left[jk_{\omega_p}\left(\boldsymbol{a}_\epsilon\cdot\widehat{\boldsymbol{r}-\boldsymbol{a}_c}\right)\right]\cdot d\boldsymbol{a}_\epsilon}_{\mathbb{A}(\omega_p,\widehat{\boldsymbol{r}-\boldsymbol{a}_c})_{\mathbb{S}}} \cdot \underbrace{\frac{1}{4\pi}\cdot\frac{1}{|\boldsymbol{r}-\boldsymbol{a}_c|}\cdot\exp\left(-jk_{\omega_p}|\boldsymbol{r}-\boldsymbol{a}_c|\right)}_{Hh(\omega_p,\boldsymbol{r}-\boldsymbol{a}_c)}\cdot\exp\left(j\omega_p t\right)\cdot d\omega_p, \tag{9.13}$$

where $d\boldsymbol{a}_\epsilon = d\boldsymbol{s}$. Equation (9.13) contains a Fresnel wave field $Hh(\omega_p, \boldsymbol{r}-\boldsymbol{a}_c)$ referenced to the antenna surface center point.

In Situ Antenna Wave Field

We have identified in (9.13) a signal analysis element

$$\mathbb{A}(\omega_p, \widehat{\boldsymbol{r}-\boldsymbol{a}_c})_{\mathbb{S}} = \int_{\mathbb{S}} a_e(\boldsymbol{a}_c+\boldsymbol{a}_\epsilon)\cdot\exp\left[jk_{\omega_p}\times\left(\boldsymbol{a}_\epsilon\cdot\widehat{\boldsymbol{r}-\boldsymbol{a}_c}\right)\right]\cdot d\boldsymbol{a}_\epsilon \tag{9.14}$$

that is a frequency-space description of what we choose to call the *frequency-space in situ field* of the surface antenna. This in situ antenna field is defined within the frequency-space domain with respect to the direction $\widehat{\boldsymbol{r}-\boldsymbol{a}_c}$ from the antenna's center point $\boldsymbol{a}_c$ to the field point $\boldsymbol{r}$. We also call $\mathbb{A}(\omega_p, \widehat{\boldsymbol{r}-\boldsymbol{a}_c})_{\mathbb{S}}$ the *center referenced frequency-space antenna excitation function.*

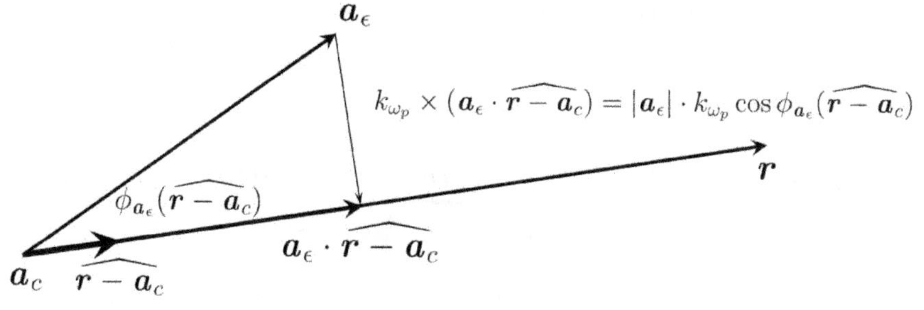

Figure 9.2: Wavenumber projection onto antenna beam angle.

Center Referenced Antenna Field Angle

The k_{ω_p} scaled projection $k_{\omega_p} \times (\boldsymbol{a}_\epsilon \cdot \widehat{\boldsymbol{r} - \boldsymbol{a}_c})$ contained in the phase factor

$$\exp[jk_{\omega_p} \times (\boldsymbol{a}_\epsilon \cdot \widehat{\boldsymbol{r} - \boldsymbol{a}_c})]$$

of (9.14) can be described in terms of a *center referenced antenna field angle*

$$\phi_{\boldsymbol{a}_\epsilon}(\widehat{\boldsymbol{r} - \boldsymbol{a}_c}) \tag{9.15}$$

as shown in Fig. 9.2. Hence, we use the equality

$$k_{\omega_p} \times (\boldsymbol{a}_\epsilon \cdot \widehat{\boldsymbol{r} - \boldsymbol{a}_c}) = |\boldsymbol{a}_\epsilon| \cdot k_{\omega_p} \cos\phi_{\boldsymbol{a}_\epsilon}(\widehat{\boldsymbol{r} - \boldsymbol{a}_c})$$

to express (9.14) as

$$\begin{aligned} &\mathbb{A}(\omega_p, \widehat{\boldsymbol{r} - \boldsymbol{a}_c})_{\mathbb{S}} \\ &= \int_{\mathbb{S}} a_e(\boldsymbol{a}_c + \boldsymbol{a}_\epsilon) \cdot \exp[j|\boldsymbol{a}_\epsilon| \cdot k_{\omega_p} \cos\phi_{\boldsymbol{a}_\epsilon}(\widehat{\boldsymbol{r} - \boldsymbol{a}_c})] \cdot d\boldsymbol{a}_\epsilon. \end{aligned} \tag{9.16}$$

The in situ field of the antenna is geometrically dependent upon the temporal wavenumber k_{ω_p} scaled (diminished) by the value of the cosine of the angle shown in Fig. 9.2 to distant field points (9.15).

Propagation of Monochromatic In Situ Antenna Wave Field

As expressed in (9.13),

$$Hh(\omega_p, \boldsymbol{r} - \boldsymbol{a}_c) = \frac{1}{4\pi} \cdot \frac{1}{|\boldsymbol{r} - \boldsymbol{a}_c|} \cdot \exp\left(-jk_{\omega_p}|\boldsymbol{r} - \boldsymbol{a}_c|\right), \tag{9.17}$$

depends on the single surface point $\boldsymbol{a}_c$ (not on the multiplicity of points $\boldsymbol{s} \in \mathbb{S}$), as a description of the propagation of the center referenced frequency-space in situ antenna wave field

$$\mathbb{A}(\omega_p, \widehat{\boldsymbol{r} - \boldsymbol{a}_c})_{\mathbb{S}}$$

to distant field points $\boldsymbol{r}$,

$$\Lambda(\omega_p, \boldsymbol{r} - \boldsymbol{a}_c)_{\mathbb{S}}^{\text{mono}} = \mathbb{A}(\omega_p, \widehat{\boldsymbol{r} - \boldsymbol{a}_c})_{\mathbb{S}} \cdot Hh(\omega_p, \boldsymbol{r} - \boldsymbol{a}_c). \tag{9.18}$$

We view (9.18) as the frequency-space monochromatic beam function induced by the antenna's beam pattern.

Antenna Wave Field Inverse Fourier Transform

Using (9.17) and (9.18),

$$\begin{aligned} &\Lambda(t, \boldsymbol{r} - \boldsymbol{a}_c)_{\mathbb{S}}^{\text{pulse}} \\ &\approx \frac{1}{2\pi} \int_{\beta_p} W(\omega_p) \cdot \underbrace{\mathbb{A}(\omega_p, \widehat{\boldsymbol{r} - \boldsymbol{a}_c})_{\mathbb{S}} \cdot Hh(\omega_p, \boldsymbol{r} - \boldsymbol{a}_c)}_{\Lambda(\omega_p, \boldsymbol{r} - \boldsymbol{a}_c)_{\mathbb{S}}^{\text{mono}}} \cdot \exp(j\omega_p t) \cdot d\omega_p, && (9.19) \\ &= \frac{1}{2\pi} \int_{\beta_p} \underbrace{W(\omega_p) \cdot \Lambda(\omega_p, \boldsymbol{r} - \boldsymbol{a}_c)_{\mathbb{S}}^{\text{mono}}}_{\Lambda(\omega_p, \boldsymbol{r} - \boldsymbol{a}_c)_{\mathbb{S}}^{\text{pulse}}} \cdot \exp(j\omega_p t) \cdot d\omega_p, && (9.20) \\ &= \frac{1}{2\pi} \int_{\beta_p} \Lambda(\omega_p, \boldsymbol{r} - \boldsymbol{a}_c)_{\mathbb{S}}^{\text{pulse}} \cdot \exp(j\omega_p t) \cdot d\omega_p, \end{aligned}$$

which determines a temporal Fourier transform pair relationship for the center referenced antenna beam pattern of a pulsed antenna,

$$\Lambda(t, \boldsymbol{r} - \boldsymbol{a}_c)_{\mathbb{S}}^{\text{pulse}} \quad {}_{t} \Leftrightarrow {}_{\omega_p} \quad \Lambda(\omega_p, \boldsymbol{r} - \boldsymbol{a}_c)_{\mathbb{S}}^{\text{pulse}} .$$

Wave Field Propagation of Pulsed Antenna In Situ Field

We view

$$\Lambda(\omega_p, \boldsymbol{r} - \boldsymbol{a}_c)_{\mathbb{S}}^{\text{pulse}} = W(\omega_p) \cdot \Lambda(\omega_p, \boldsymbol{r} - \boldsymbol{a}_c)_{\mathbb{S}}^{\text{mono}}$$

in (9.20) as the frequency-space antenna beam function produced by wave propagation per

$$Hh(\omega_p, \boldsymbol{r} - \boldsymbol{a}_c)$$

of the antenna's center referenced (direction dependent) in situ field

$$\mathbb{A}(\omega_p, \widehat{\boldsymbol{r} - \boldsymbol{a}_c})_{\mathbb{S}},$$

but where the pulse spectrum

$$W(\omega_p)$$

provides frequency dependent weighting of the polychromatic field over the passband β_p of the transmitted waveform.

Narrow Band SAR Antenna Time-Space Beam Pattern

We can use the frequency-space domain analysis to define a time-space characteristic of an antenna for the special case of transmission of a narrow band signal.

If the antenna waveform excitation is narrow band, then $\mathbb{A}(\omega_p, \widehat{\boldsymbol{r} - \boldsymbol{a}_c})_{\mathbb{S}}$ is nearly independent of the instantaneous frequency of the transmitted signal. For this special case, the passband frequency-space weight function $\mathbb{A}(\omega_p, \widehat{\boldsymbol{r} - \boldsymbol{a}_c})_{\mathbb{S}}$ is well approximated at the center frequency ω_c of the transmitted waveform,

$$\mathbb{A}(\omega_p, \widehat{\boldsymbol{r} - \boldsymbol{a}_c})_{\mathbb{S}} \approx \mathbb{A}(\omega_c, \widehat{\boldsymbol{r} - \boldsymbol{a}_c})_{\mathbb{S}}. \tag{9.21}$$

Because of the narrow bandwidth assumption (9.21), the beam pattern

$$\mathbb{A}(\omega_c, \widehat{\boldsymbol{r} - \boldsymbol{a}_c})_{\mathbb{S}}$$

does not significantly fluctuate with frequency. Hence, from (9.19),

$$\begin{aligned} \Lambda(t, \boldsymbol{r})_{\boldsymbol{a}_c}^{\text{narrow}} & \\ \approx \; & \mathbb{A}(\omega_c, \widehat{\boldsymbol{r} - \boldsymbol{a}_c})_{\mathbb{S}} \cdot \frac{1}{2\pi} \int_{\beta_p} W(\omega_p) \cdot Hh(\omega_p, \boldsymbol{r} - \boldsymbol{a}_c) \cdot \exp(j\omega_p t) \cdot d\omega_p, \\ = \; & \mathbb{A}(\omega_c, \widehat{\boldsymbol{r} - \boldsymbol{a}_c})_{\mathbb{S}} \times \frac{1}{4\pi} \cdot \frac{1}{|\boldsymbol{r} - \boldsymbol{a}_c|} \cdot w\left[t - t_d(\boldsymbol{r} - \boldsymbol{a}_c)\right]. \end{aligned} \tag{9.22}$$

The time-space domain description (9.22) of a narrow band antenna directly depends on the frequency-space (in situ) beam function $\mathbb{A}(\omega_c, \widehat{\boldsymbol{r} - \boldsymbol{a}_c})_{\mathbb{S}}$ defined by the center frequency of the transmitted narrow band waveform. The relationship (9.22) is multiplicative and not convolutional.

9.4 Surface Antenna Frequency-Wavenumber Analysis

An antenna beam pattern in the time-space domain induces a weight function in the frequency-wavenumber domain.

Angular Spectrum of Plane Waves

Consider the time-space antenna wave field expressed by synthesis through a temporal inverse Fourier transform of the frequency-space wave field,

$$\begin{aligned}\Lambda(t,\boldsymbol{r})^{\text{pulse}}_{\mathbb{S}} &= \frac{1}{2\pi}\int_{\beta_p} W(\omega_p)\cdot\int_{\mathbb{S}} a_e(\boldsymbol{s})\cdot Hh(\omega_p,\boldsymbol{r}-\boldsymbol{s})\cdot d\boldsymbol{s}\cdot\exp\left(j\omega_p t\right)\cdot d\omega_p,\\ &= \frac{1}{2\pi}\int_{\beta_p} W(\omega_p)\cdot\int_{\mathbb{S}} a_e(\boldsymbol{s})\\ &\quad\cdot\frac{1}{4\pi}\cdot\frac{1}{|\boldsymbol{r}-\boldsymbol{s}|}\cdot\exp\left(-jk_{\omega_p}|\boldsymbol{r}-\boldsymbol{s}|\right)\cdot d\boldsymbol{s}\cdot\exp\left(j\omega_p t\right)\cdot d\omega_p. \qquad (9.23)\end{aligned}$$

The ω_p dependent Fresnel phase field originating from the antenna surface point $\boldsymbol{s}$,

$$\measuredangle Hh(\omega_p,\boldsymbol{r}-\boldsymbol{s}) = \exp\left(-jk_{\omega_p}|\boldsymbol{r}-\boldsymbol{s}|\right)$$

can be synthesized in terms of a semicircle[2] angular spectrum of directed plane waves in the (ω_p, k_{s_y}) wavenumber domain [440, 871, 873]. This synthesis introduces an along-track real aperture wavenumber k_{s_y}. Since we assume the antenna is stationary, we choose to not have the (synthetic) aperture wavenumber k_{u_y} be the variable introduced by the spatial Fourier transform[3]. This frequency-wavenumber domain synthesis is

$$\begin{aligned}&\exp\left(-jk_{\omega_p}|\boldsymbol{r}-\boldsymbol{s}|\right)\\ &= \int_{-k_{\omega_p}}^{k_{\omega_p}} a(\omega_p,k_{s_y})_{\text{pwd}}\\ &\quad\cdot\exp[-j\,(r_x-s_x)\cdot\sqrt{k_{\omega_p}^2-k_{s_y}^2}\,]\cdot\exp[-j\,(r_y-s_y)\cdot k_{s_y}]\cdot dk_{s_y},\\ &= \int_{-k_{\omega_p}}^{k_{\omega_p}} a(\omega_p,k_{s_y})_{\text{pwd}}\cdot\exp[-jr_x\cdot\breve{k}_x(\omega_p,k_{s_y})]\cdot\exp[js_x\cdot\breve{k}_x(\omega_p,k_{s_y})]\\ &\quad\cdot\exp(-jr_y k_{s_y})\cdot\exp(js_y k_{s_y})\cdot dk_{s_y},\end{aligned} \qquad (9.24)$$

where

$$\breve{k}_x(\omega_p,k_{s_y}) = \sqrt{k_{\omega_p}^2-k_{s_y}^2}$$

---

[2]We model just the "forward scattering" for radiation emanating from the antenna surface, hence the synthesis is only over a semicircle, not the full circular angular spectrum.

[3]The *stationary antenna wavenumber* k_{s_y} is different, at least in notation, than the *moving antenna's aperture wavenumber* k_{u_y} that we have dealt with up to this point in our SAR analysis.

is a Fourier migrated wavenumber obtained through the Stolt mapping

$$(k_{\omega_p}, k_{s_y}) \mapsto \breve{k}_x,$$

and where

$$a(\omega_p, k_{s_y})_{\text{pwd}} \propto \frac{k_{\omega_p}}{\sqrt{k_{\omega_p}^2 - k_{s_y}^2}}$$

represents amplitude coefficients [440, 871, 873] of the plane wave decomposition (pwd) that corresponds to the semicircle angular spectrum. Proceed by substituting (9.24) into (9.23),

$$\begin{aligned}
&\Lambda(t, \boldsymbol{r})_{\mathbb{S}}^{\text{pulse}} \\
&= \frac{1}{2\pi} \int_{\beta_p} W(\omega_p) \int_{\mathbb{S}} a_e(\boldsymbol{s}) \cdot \frac{1}{4\pi} \cdot \frac{1}{|\boldsymbol{r} - \boldsymbol{s}|} \\
&\quad \cdot \int_{-k_{\omega_p}}^{k_{\omega_p}} a(\omega_p, k_{s_y})_{\text{pwd}} \cdot \exp[-jr_x \cdot \breve{k}_x(\omega_p, k_{s_y})] \cdot \exp[js_x \cdot \breve{k}_x(\omega_p, k_{s_y})] \\
&\quad \cdot \exp(-jr_y k_{s_y}) \cdot \exp(js_y k_{s_y}) \cdot dk_{s_y} \cdot d\boldsymbol{s} \cdot \exp(j\omega_p t) \cdot d\omega_p
\end{aligned}$$

and then changing the order of integration of the inverse Fourier transform with the antenna surface integral,

$$\begin{aligned}
&\Lambda(t, \boldsymbol{r})_{\mathbb{S}}^{\text{pulse}} \\
&= \frac{1}{2\pi} \int_{\beta_p} W(\omega) \int_{-k_{\omega_p}}^{k_{\omega_p}} a(\omega_p, k_{s_y})_{\text{pwd}} \\
&\quad \cdot \underbrace{\int_{\mathbb{S}} a_e(\boldsymbol{s}) \cdot \frac{1}{4\pi} \cdot \frac{1}{|\boldsymbol{r} - \boldsymbol{s}|} \cdot \exp[js_x \cdot \breve{k}_x(\omega_p, k_{s_y})] \cdot \exp(js_y k_{s_y}) \cdot d\boldsymbol{s}}_{\mathbb{A}(\omega_p, k_{s_y}, \boldsymbol{r})_{\mathbb{S}}} \\
&\quad \cdot \underbrace{\exp[-jr_x \cdot \breve{k}_x(\omega_p, k_{s_y})]}_{H_{r_x}(\omega_p, k_{s_y})} \cdot \underbrace{\exp(-jr_y k_{s_y})}_{H_{r_y}(k_{s_y})} \cdot dk_{s_y} \cdot \exp(j\omega_p t) \cdot d\omega_p. \quad (9.25)
\end{aligned}$$

We have defined in (9.25) an $\boldsymbol{r}$ dependent frequency-wavenumber domain antenna weight function,

$$\begin{aligned}
&\mathbb{A}(\omega_p, k_{s_y}, \boldsymbol{r})_{\mathbb{S}} \\
&= \int_{\mathbb{S}} a_e(\boldsymbol{s}) \cdot \frac{1}{4\pi} \cdot \frac{1}{|\boldsymbol{r} - \boldsymbol{s}|} \cdot \exp[js_x \cdot \breve{k}_x(\omega_p, k_{s_y})] \cdot \exp(js_y k_{s_y}) \cdot d\boldsymbol{s}.
\end{aligned} \quad (9.26)$$

The field parameter $\boldsymbol{r}$ in the frequency-wavenumber domain antenna weight function (9.26) determines, in part, the amplitude of $\mathbb{A}(\omega_p, k_{s_y}, \boldsymbol{r})_{\mathbb{S}}$ through the involvement of $1/|\boldsymbol{r} - \boldsymbol{s}|$ in the integral over $\boldsymbol{s} \in \mathbb{S}$. We also see in (9.25) the Huygens-Fresnel transfer functions,

$$\begin{aligned}
H_{r_x}(\omega_p, k_{s_y}) &= \exp[-jr_x \cdot \breve{k}_x(\omega_p, k_{s_y})], \\
H_{r_y}(k_{s_y}) &= \exp(-jr_y k_{s_y}).
\end{aligned} \quad (9.27)$$

With substitution of (9.26) and (9.27) into the decomposition of (9.25), we step through a synthesis sequence,

$$\begin{aligned}
\Lambda(t,\boldsymbol{r})_{\mathbb{S}}^{\text{pulse}} &= \frac{1}{2\pi}\int_{\Omega} W(\omega_p) \\
&\quad \cdot \int_{-k_{\omega_p}}^{k_{\omega_p}} \underbrace{a(\omega_p,k_{s_y})_{\text{pwd}} \cdot \mathbb{A}(\omega_p,k_{s_y},\boldsymbol{r})_{\mathbb{S}} \cdot H_{r_x}(\omega_p,k_{s_y}) \cdot H_{r_y}(k_{s_y})}_{\Lambda(\omega_p,k_{s_y},\boldsymbol{r})_{\mathbb{S}}^{\text{mono}}} \cdot dk_{s_y} \\
&\quad \cdot \exp(j\omega_p t)\cdot d\omega_p, \\
&= \frac{1}{2\pi}\int_{\Omega} W(\omega_p)\int_{-k_{\omega_p}}^{k_{\omega_p}} \Lambda(\omega_p,k_{s_y},\boldsymbol{r})_{\mathbb{S}}^{\text{mono}} \cdot dk_{s_y}\cdot \exp(j\omega_p t)\cdot d\omega_p. \qquad (9.28)
\end{aligned}$$

Antenna Monochromatic Frequency-Wavenumber Beam Function

We see in (9.28) the *monochromatic antenna frequency-wavenumber beam function,*

$$\begin{aligned}
&\Lambda(\omega_p,k_{s_y},\boldsymbol{r})_{\mathbb{S}}^{\text{mono}} \\
&\quad = a(\omega_p,k_{s_y})_{\text{pwd}} \cdot \mathbb{A}(\omega_p,k_{s_y},\boldsymbol{r})_{\mathbb{S}} \cdot H_{r_x}(\omega_p,k_{s_y}) \cdot H_{r_y}(k_{s_y}). \qquad (9.29)
\end{aligned}$$

This frequency-wavenumber domain beam function is defined for each field point $\boldsymbol{r}$. Along with being a part of the complex amplitude function $\mathbb{A}(\omega_p,k_{s_y},\boldsymbol{r})_{\mathbb{S}}$, the field parameter $\boldsymbol{r} = [r_x \ r_y]$ determines the dimensionally separated Huygens-Fresnel transfer functions $H_{r_x}(\omega_p,k_{s_y})$ and $H_{r_y}(k_{s_y})$. These Huygens-Fresnel transfers describe in the frequency-wavenumber domain the propagation of the in situ field from the antenna surface to field points identified by $\boldsymbol{r}$.

Frequency-Space Beam Functions from Frequency-Wavenumber Analysis

Similar to our prior use of frequency-space domain results to find novel time-space domain models, we now use results of the frequency-wavenumber analysis to find novel frequency-space and time-space domain descriptions of the antenna beam pattern. Synthesis from the frequency-wavenumber domain back to the time-space domain gives

$$\begin{aligned}
\Lambda(t,\boldsymbol{r})_{\mathbb{S}}^{\text{pulse}} &= \frac{1}{2\pi}\int_{\beta_p} W(\omega_p)\underbrace{\int_{-k_{\omega_p}}^{k_{\omega_p}} \Lambda(\omega_p,k_{s_y},\boldsymbol{r})_{\mathbb{S}}^{\text{mono}} \cdot dk_{s_y}}_{\Lambda(\omega_p,\boldsymbol{r})_{\mathbb{S}}^{\text{mono}}}\cdot \exp(j\omega_p t)\cdot d\omega_p \qquad (9.30) \\
&= \frac{1}{2\pi}\int_{\beta_p} \underbrace{W(\omega_p)\cdot \Lambda(\omega_p,\boldsymbol{r})_{\mathbb{S}}^{\text{mono}}}_{\Lambda(\omega_p,\boldsymbol{r})_{\mathbb{S}}^{\text{pulse}}} \cdot \exp(j\omega_p t)\cdot d\omega_p, \\
&= \frac{1}{2\pi}\int_{\beta_p} \Lambda(\omega_p,\boldsymbol{r})_{\mathbb{S}}^{\text{pulse}} \cdot \exp(j\omega_p t)\cdot d\omega_p. \qquad (9.31)
\end{aligned}$$

Monochromatic Frequency-Space Beam Function

We see in (9.30) that integrating (9.29) over the along-track real aperture wavenumber k_{s_y} gives the *monochromatic frequency-space beam function,*

$$\begin{aligned}\Lambda(\omega_p, \boldsymbol{r})_{\mathbb{S}}^{\text{mono}} &= \int_{-k_{\omega p}}^{k_{\omega p}} \Lambda(\omega_p, k_{s_y}, \boldsymbol{r})_{\mathbb{S}}^{\text{mono}} \cdot dk_{s_y}, \\ &= \int_{-k_{\omega p}}^{k_{\omega p}} a(\omega_p, k_{s_y})_{\text{pwd}} \cdot \mathbb{A}(\omega_p, k_{s_y}, \boldsymbol{r})_{\mathbb{S}} \cdot H_{r_x}(\omega_p, k_{s_y}) \cdot H_{r_y}(k_{s_y}) \cdot dk_{s_y}.\end{aligned} \tag{9.32}$$

The frequency-space domain description $\Lambda(\omega_p, \boldsymbol{r})_{\mathbb{S}}^{\text{mono}}$ of the antenna beam pattern is dependent on passband frequencies ω_p and field points $\boldsymbol{r}$. The frequency-space expression (9.32), which was obtained from the frequency-wavenumber domain (ω_p, k_{s_y}), complements the frequency-space result of (9.9),

$$\Lambda(\omega_p, \boldsymbol{r})_{\mathbb{S}}^{\text{mono}} = \int_{\mathbb{S}} a_e(\boldsymbol{s}) \cdot Hh(\omega_p, \boldsymbol{r} - \boldsymbol{s}) \cdot d\boldsymbol{s}, \tag{9.33}$$

obtained from direct frequency-space domain analysis. We can also compare (9.32) and (9.33) with the center referenced model (9.18),

$$\Lambda(\omega_p, \boldsymbol{r} - \boldsymbol{a}_c)^{\text{mono}} = \mathbb{A}(\omega_p, \widehat{\boldsymbol{r} - \boldsymbol{a}_c}) \cdot Hh(\omega_p, \boldsymbol{r} - \boldsymbol{a}_c). \tag{9.34}$$

So, in summary, we have three frequency-space descriptions of a monochromatic element of an antenna beam function:

- the beam function (9.32) derived from a Huygens-Fresnel decomposition in a frequency-wavenumber domain,
- the beam function (9.33) derived from a frequency-space domain superposition of Fresnel wave fields from all points of the antenna's surface and
- the beam function (9.34) based on a frequency-space description referenced to a nominal center point of the antenna's surface.

Antenna Pulse Frequency-Space Beam Function

Multiplying (9.32) by the passband pulse spectrum yields the *polychromatic frequency-space pulse beam function,*

$$\Lambda(\omega_p, \boldsymbol{r})_{\mathbb{S}}^{\text{pulse}} = W(\omega_p) \cdot \Lambda(\omega_p, \boldsymbol{r})_{\mathbb{S}}^{\text{mono}} \tag{9.35}$$

defined over the passband of the transmitted pulse.

Frequency-Space Domain Beam Pattern without Center Reference

Note that the descriptions (9.32) and (9.33) of the antenna beam $\Lambda(\omega_p, \boldsymbol{r})_{\mathbb{S}}$ for both monochromatic (9.32) and polychromatic (9.35) versions differ from the expression $\Lambda(\omega_p, \boldsymbol{r} - \boldsymbol{a}_c)_{\mathbb{S}}$ derived previously in (9.34) in reference to a nominal center point on the antenna surface. The definition of (9.26) removes the dependency on $\boldsymbol{a}_c$, but introduces a dependency on $\boldsymbol{r}$.

9.5 Flat Plate Surface Antenna

We have established approaches grounded in Huygens-Fresnel wave theory for descriptive analysis of an antenna's beam functions in time-space, frequency-space and frequency-wavenumber domains. We now apply these methods to a particular antenna shape, the flat plate surface antenna [693, 885].

Parametric Description of Flat Plate Antenna

The antenna diameter D_y is the along-track length of the flat plate antenna. The along-track dimension of the surface of a flat plate antenna is assumed to be aligned with the along-track dimension of the movable SAR platform. The flat plate surface is parametrically described by

$$\mathbb{S} = \{(0, s_y)\} \text{ for } s_y \in \left[-\frac{D_y}{2}, +\frac{D_y}{2}\right],$$

where the antenna surface has no down-range variability ($s_x = 0$) and where the continuous parameter s_y describes the position of each point radiator on the face of the flat plate antenna. We ignore the s_z dimension in slant-plane analysis. The antenna continues to be held stationary at this point of our analysis.

9.6 Flat Plate Antenna Time-Space Analysis

Consider a flat plate antenna energized by a time domain passband waveform $w(t)$ and by a uniform surface excitation function of unity,

$$a_e(s_y) = 1 \text{ for all } \boldsymbol{s} \in \mathbb{S}.$$

The beam pattern produced by the flat plate antenna can be expressed in the time-space domain by using (9.5), which gives

$$\begin{aligned} &\Lambda(t, \boldsymbol{r} - \boldsymbol{a}_c)_{\text{plate}}^{\text{pulse}} \\ &\quad \approx \frac{1}{4\pi} \cdot \frac{1}{|\boldsymbol{r} - \boldsymbol{a}_c|} \cdot \int_{\mathbb{S}} a_e(\boldsymbol{s}) \cdot w[t - t_d(\boldsymbol{r} - \boldsymbol{s})] \cdot d\boldsymbol{s}, \\ &\quad = \frac{1}{4\pi} \cdot \frac{1}{|\boldsymbol{r} - \boldsymbol{a}_c|} \cdot \int_{-D_y/2}^{+D_y/2} 1 \cdot w\left[t - \frac{1}{c}\sqrt{r_x^2 + (r_y - s_y)^2}\right] \cdot ds_y. \quad (9.36) \end{aligned}$$

This is the center referenced surface antenna beam pattern [395, 407, 885]. The waveform dependent time-space antenna analysis of (9.36) is rather difficult to analyze further. We move next to the expression of the flat plate antenna's beam pattern as formulated with frequency-space analysis. We hope to extend the flat plate antenna model from a more amenable frequency-space domain description.

9.7 Flat Plate Antenna Frequency-Space Analysis

Assume the flat plate antenna is located at the origin of the slant-plane aperture coordinate system,

$$\boldsymbol{a}_c = (0, 0).$$

Also, assume radiating points $\boldsymbol{a}_\epsilon$ on the flat plate antenna, with positions identified by

$$\boldsymbol{a}_\epsilon = s_y \hat{\boldsymbol{u}}_y,$$

having an excitation function of unity,

$$a_e(s_y) = 1.$$

From (9.14), the in situ field of a flat plate antenna described in terms of the cosine of the angle between the unit vector $\widehat{\boldsymbol{r} - \boldsymbol{a}_c}$ and the vector from $\boldsymbol{a}_c$ to $\boldsymbol{a}_\epsilon$ at

$$\boldsymbol{s} = s_y,$$

and described in the frequency-space domain is [175]

$$\begin{aligned}
&\mathbb{A}(\omega_p, \widehat{\boldsymbol{r} - \boldsymbol{a}_c})_{\text{plate}} \\
&\quad = \int_{\mathbb{S}} a_e(\boldsymbol{a}_c + \boldsymbol{a}_\epsilon) \cdot \exp\left[j k_{\omega_p} (\boldsymbol{a}_\epsilon \cdot \widehat{\boldsymbol{r} - \boldsymbol{a}_c}) \right] \cdot d\boldsymbol{a}_\epsilon, \\
&\quad = \int_{-D_y/2}^{+D_y/2} a_e(s_y) \cdot \exp\left[j k_{\omega_p} (s_y \hat{\boldsymbol{u}}_y \cdot \widehat{\boldsymbol{r} - \boldsymbol{a}_c}) \right] \cdot ds_y. \qquad (9.37)
\end{aligned}$$

We find it useful to rework the spatial and temporal wavenumber parameters of (9.37) in this manner,

$$\begin{aligned}
k_{\omega_p}(s_y \hat{\boldsymbol{u}}_y \cdot \widehat{\boldsymbol{r} - \boldsymbol{a}_c}) &= |s_y \hat{\boldsymbol{u}}_y| \cdot |k_{\omega_p}| \cdot \cos\phi_{\boldsymbol{a}_\epsilon}(\widehat{\boldsymbol{r} - \boldsymbol{a}_c}), \\
&= s_y \cdot k_{\omega_p} \cos\phi_{\boldsymbol{a}_\epsilon}(\widehat{\boldsymbol{r} - \boldsymbol{a}_c}), \qquad (9.38)
\end{aligned}$$

where, in this case, the angle $\phi_{\boldsymbol{a}_\epsilon}(\widehat{\boldsymbol{r} - \boldsymbol{a}_c})$ is the direction angle of the unit vector $\widehat{\boldsymbol{r} - \boldsymbol{a}_c}$ with respect to the y axis, which is aligned with the face of the flat plate antenna, hence

$$\cos\phi_{\boldsymbol{a}_\epsilon}(\widehat{\boldsymbol{r} - \boldsymbol{a}_c}) = \cos\phi_y(\widehat{\boldsymbol{r} - \boldsymbol{a}_c}). \qquad (9.39)$$

Thus (9.38) and (9.39) allow us to write

$$\begin{aligned}
&\mathbb{A}(\omega_p, \widehat{\boldsymbol{r} - \boldsymbol{a}_c})_{\text{plate}} \\
&\quad = \int_{-D_y/2}^{+D_y/2} a_e(s_y) \cdot \exp\left[j k_{\omega_p} (s_y \hat{\boldsymbol{u}}_y \cdot \widehat{\boldsymbol{r} - \boldsymbol{a}_c}) \right] \cdot ds_y, \\
&\quad = \int_{-\infty}^{+\infty} \Pi_{(D_y)}(s_y) \cdot \exp[j s_y \cdot \underbrace{k_{\omega_p} \cos\phi_y(\widehat{\boldsymbol{r} - \boldsymbol{a}_c})}_{\text{kernel}}] \cdot ds_y, \qquad (9.40)
\end{aligned}$$

which has the form of an inverse Fourier transform. If we introduce use of the real aperture wavenumber k_{s_y} and identify the kernel function as noted by the underscript in (9.40),

$$k_{s_y} \equiv k_{\omega_p} \cos\phi_y(\widehat{\boldsymbol{r} - \boldsymbol{a}_c}),$$

then by substituting the real aperture wavenumber k_{s_y} into (9.40),

$$\mathbb{A}(\omega_p, \widehat{\boldsymbol{r}-\boldsymbol{a}_c})_{\text{plate}} = \int_{-\infty}^{+\infty} \Pi_{(D_y)}(s_y) \cdot \exp\left(js_y k_{s_y}\right) \cdot ds_y. \tag{9.41}$$

Using the symmetry of $\Pi_{(D_y)}(s_y)$, which allows us to write $s_y = -s_y$ as required, we see that (9.41) is the Fourier transform of $\Pi_{(D_y)}(s_y)$. Hence,

$$\mathbb{A}(\omega_p, \widehat{\boldsymbol{r}-\boldsymbol{a}_c})_{\text{plate}} = D_y \cdot \operatorname{sinc}\left[\frac{D_y}{2} \cdot k_{\omega_p} \cos\phi_y(\widehat{\boldsymbol{r}-\boldsymbol{a}_c})\right] \tag{9.42}$$

which comes from the Fourier transform pair

$$\Pi_{(D_y)}(s_y) \Leftrightarrow D_y \cdot \operatorname{sinc}\left(\frac{D_y}{2} k_{s_y}\right). \tag{9.43}$$

Flat Plate Antenna's Monochromatic Frequency-Space Beam Function

From (9.42), we obtain the center referenced *flat plate antenna's monochromatic frequency-space beam function,*

$$\begin{aligned}
\Lambda(\omega_p, \boldsymbol{r}-\boldsymbol{a}_c)^{\text{mono}}_{\text{plate}} &= \mathbb{A}(\omega_p, \widehat{\boldsymbol{r}-\boldsymbol{a}_c})_{\text{plate}} \cdot Hh(\omega_p, \boldsymbol{r}-\boldsymbol{a}_c), \\
&= D_y \cdot \operatorname{sinc}\left[\frac{D_y}{2} \cdot k_{\omega_p} \cos\phi_y(\widehat{\boldsymbol{r}-\boldsymbol{a}_c})\right] \cdot Hh(\omega_p, \boldsymbol{r}-\boldsymbol{a}_c).
\end{aligned}$$

Flat Plate Antenna's Pulse Frequency-Space Beam Function

Furthermore, by adding the passband weight function of the pulse signal, we obtain the *flat plate antenna's frequency-space pulse beam function,*

$$\begin{aligned}
&\Lambda(\omega_p, \boldsymbol{r}-\boldsymbol{a}_c)^{\text{pulse}}_{\text{plate}} \\
&= W(\omega_p) \cdot \mathbb{A}(\omega_p, \widehat{\boldsymbol{r}-\boldsymbol{a}_c})_{\text{plate}} \cdot Hh(\omega_p, \boldsymbol{r}-\boldsymbol{a}_c), \\
&= W(\omega_p) \cdot D_y \cdot \operatorname{sinc}\left[\frac{D_y}{2} \cdot k_{\omega_p} \cos\phi_y(\widehat{\boldsymbol{r}-\boldsymbol{a}_c})\right] \cdot Hh(\omega_p, \boldsymbol{r}-\boldsymbol{a}_c).
\end{aligned} \tag{9.44}$$

In the frequency-space domain, the in situ field $\mathbb{A}(\omega_p, \widehat{\boldsymbol{r}-\boldsymbol{a}_c})_{\text{plate}}$ of the flat plate has a sinc weighting with respect to the temporal wavenumber k_{ω_p} projected onto the k_{s_y} dimension, as given by

$$k_{s_y} = k_{\omega_p} \cos\phi_y(\widehat{\boldsymbol{r}-\boldsymbol{a}_c}).$$

Each monochromatic element of this frequency-space domain wave field is weighted by the pulse spectrum characteristic $W(\omega_p)$. The propagation of the field is described by a Fresnel spherical wave field operator $Hh(\omega_p, \boldsymbol{r}-\boldsymbol{a}_c)$, referenced to the center of the flat plate.

Three Useful Results from Frequency-Space Analysis

Before moving on to the Huygens-Fresnel frequency-wavenumber analysis of the flat plate antenna, we utilize the frequency-space domain model of the antenna beam function to accomplish three additional tasks.

- First, we show how a temporal inverse Fourier transform of (9.44) can provide an alternate time-space domain beam function model.
- Second, we show how the frequency-space beam function can provide an alternate description of the frequency-wavenumber domain beam function.
- Third, we show how the frequency-space beam function can also be used to determine the angular beamwidth of the flat plate antenna.

Task 1: Temporal Inverse Fourier Transform of Frequency-Space Beam Function

We can insert (9.44) into an inverse Fourier transform of the frequency-space domain expression (9.19) of the flat plate's center referenced antenna beam pattern,

$$\Lambda(t, \boldsymbol{r} - \boldsymbol{a}_c)_{\text{plate}}^{\text{pulse}} \approx \frac{1}{2\pi} \int_{\beta_p} W(\omega_p) \cdot \mathbb{A}(\omega_p, \widehat{\boldsymbol{r} - \boldsymbol{a}_c})_{\text{plate}} \cdot Hh(\omega_p, \boldsymbol{r} - \boldsymbol{a}_c) \cdot \exp(j\omega_p t) \cdot d\omega_p,$$

to obtain

$$\begin{aligned}\Lambda(t, \boldsymbol{r} - \boldsymbol{a}_c)_{\text{plate}}^{\text{pulse}} \approx & \frac{1}{2\pi} \int_{\beta_p} W(\omega_p) \\ & \cdot D_y \cdot \operatorname{sinc}\left[\frac{D_y}{2} \cdot k_{\omega_p} \cos\phi_y(\widehat{\boldsymbol{r} - \boldsymbol{a}_c})\right] \\ & \cdot Hh(\omega_p, \boldsymbol{r} - \boldsymbol{a}_c) \cdot \exp(j\omega_p t) \cdot d\omega_p. \end{aligned} \quad (9.45)$$

This gives another time-space domain expression of the beam function of the flat plate antenna as an alternative to

$$\Lambda(t, \boldsymbol{r} - \boldsymbol{a}_c)_{\text{plate}}^{\text{pulse}} \approx \frac{1}{4\pi} \cdot \frac{1}{|\boldsymbol{r} - \boldsymbol{a}_c|} \cdot \int_{-D_y/2}^{+D_y/2} w\left[t - \frac{1}{c}\sqrt{r_x^2 + (r_y - s_y)^2}\right] \cdot ds_y.$$

Task 2: Frequency-Wavenumber Beam Function from Frequency-Space Beam Function

Expressing directional sensitivities of the flat plate antenna in terms of

$$\phi_y(\widehat{\boldsymbol{r} - \boldsymbol{a}_c}) = \hat{\boldsymbol{u}}_y \cdot \widehat{\boldsymbol{r} - \boldsymbol{a}_c},$$

we obtain

$$\begin{aligned}\mathbb{A}(\omega_p, \widehat{\boldsymbol{r} - \boldsymbol{a}_c})_{\text{plate}} &= \int_{\mathbb{S}} a_e(\boldsymbol{a}_c + \boldsymbol{a}_\epsilon) \cdot \exp[j|\boldsymbol{a}_\epsilon| \cdot k_{\omega_p} \cos\phi_y(\widehat{\boldsymbol{r} - \boldsymbol{a}_c})] \cdot d\boldsymbol{a}_\epsilon, \\ &= \int_{-D_y/2}^{+D_y/2} \exp[j s_y \times k_{\omega_p} \cos\phi_y(\widehat{\boldsymbol{r} - \boldsymbol{a}_c})] ds_y. \end{aligned} \quad (9.46)$$

Using the stationary antenna real aperture wavenumber defined as

$$k_{s_y} = k_{\omega_p} \cos \phi_y(\widehat{\boldsymbol{r} - \boldsymbol{a}_c}),$$

we recognize (9.46) as a rather simple expression of the Fourier transform previously expressed by (9.43),

$$\begin{aligned}
\mathbb{A}(\omega_p, k_{s_y})_{\text{plate}} &= \int_{-D_y/2}^{+D_y/2} \exp\left(j s_y k_{s_y}\right) ds_y, \\
&= \int_{-\infty}^{+\infty} \Pi_{(D_y)}(s_y) \cdot \exp\left(-j s_y k_{s_y}\right) ds_y, \\
&= D_y \cdot \operatorname{sinc}\left(\frac{D_y}{2} \cdot k_{s_y}\right),
\end{aligned} \tag{9.47}$$

where there is no longer dependencies on the parameters $\boldsymbol{a}_c$ and $\boldsymbol{r}$, but dependency on k_{s_y} as a free parameter within $\mathbb{A}(\omega_p, k_{s_y})_{\text{plate}}$.

Flat Plate Beam Pattern Spatial-Wavenumber Fourier Relationship

There is a Fourier transform pair relationship

$$\Pi_{(D_y)}(s_y) \quad {}_{s_y}\Leftrightarrow_{k_{s_y}} \quad D_y \cdot \operatorname{sinc}\left(\frac{D_y}{2} \cdot k_{s_y}\right)$$

between the antenna's physical shape

$$\mathbb{S}_{\text{plate}} = \Pi_{[-D_y/2, +D_y/2]}(s_y)$$

and the antenna frequency-wavenumber in situ field,

$$\mathbb{A}(\omega_p, k_{s_y})_{\text{plate}} = D_y \cdot \operatorname{sinc}\left(\frac{D_y}{2} \cdot k_{s_y}\right). \tag{9.48}$$

The flat plate antenna's in situ field, as described in the frequency-wavenumber domain by (9.48), is no longer a function of spatial parameters. It is only dependent upon the aperture wavenumber k_{s_y} of the stationary antenna.

Frequency-Space Beam Function from Frequency-Wavenumber Beam Function

Utilizing the fact that we can associate view angle with wavenumber angle (see Section 8.7), we have various expressions for the in situ field of the flat plate antenna. For example, we have a frequency-space description in terms of a y axis wavenumber angle,

$$\mathbb{A}[\omega_p, \phi_y(\widehat{\boldsymbol{r} - \boldsymbol{a}_c})]_{\text{plate}} = D_y \cdot \operatorname{sinc}\left[\frac{D_y}{2} \cdot k_{\omega_p} \cos \phi_y(\widehat{\boldsymbol{r} - \boldsymbol{a}_c})\right].$$

Using an equivalence between the spatial angle and wavenumber angle,

$$\theta_y(\widehat{\boldsymbol{r} - \boldsymbol{a}_c}) = \phi_y(\widehat{\boldsymbol{r} - \boldsymbol{a}_c}),$$

we obtain a frequency-space expression in terms of a spatial beam angle,

$$\mathbb{A}[\omega_p, \theta_y(\widehat{\boldsymbol{r} - \boldsymbol{a}_c})]_{\text{plate}} = D_y \cdot \text{sinc}\left[\frac{D_y}{2} \cdot k_{\omega_p} \cos\theta_y(\widehat{\boldsymbol{r} - \boldsymbol{a}_c})\right].$$

Representing spatial beam angle to some field point $\boldsymbol{r} = (r_x, r_y)$, by assuming $\boldsymbol{a}_c = (0,0,0)$ and by using

$$\sin\theta_x(\hat{\boldsymbol{r}}) = \cos\theta_y(\hat{\boldsymbol{r}})$$

for the unit vector $\hat{\boldsymbol{r}}$, we have an expression in terms of an antenna beam off boresight unit vector,

$$\begin{aligned}
\mathbb{A}(\omega_p, \hat{\boldsymbol{r}})_{\text{plate}} &= D_y \cdot \text{sinc}\left[\frac{D_y}{2} \cdot k_{\omega_p} \cos\theta_y(\hat{\boldsymbol{r}})\right], \\
&= D_y \cdot \text{sinc}\left[\frac{D_y}{2} \cdot k_{\omega_p} \sin\theta_x(\hat{\boldsymbol{r}})\right], \\
&= D_y \cdot \text{sinc}\left(\frac{D_y}{2} \cdot k_{\omega_p} \frac{r_y}{|\boldsymbol{r}|}\right),
\end{aligned}$$

and furthermore, by representing field points by $\boldsymbol{x} = (x, y)$, an additional frequency-space description of the antenna in situ field is obtained,

$$\mathbb{A}(\omega_p, \boldsymbol{x})_{\text{plate}} = D_y \cdot \text{sinc}\left(\frac{D_y}{2} \cdot k_{\omega_p} \frac{y}{\sqrt{x^2 + y^2}}\right).$$

Task 3: Stationary Flat Plate Antenna's Wavenumber Manifold

The unnormalized sinc function has nulls at multiples of π. Hence, the first nulls on each side of the sinc weight function (9.48) in the k_{s_y} domain occur at

$$\frac{D_y}{2} \cdot k_{s_y} = \pm\pi,$$

which gives nulls for the stationary antenna's cross-range wavenumber k_{s_y} at

$$k_{s_y} = \pm\frac{2\pi}{D_y}.$$

Thus, the flat plate antenna's wavenumber domain manifold has a nominal width definable as

$$k_{s_y} \quad \in \quad \left[-\frac{2\pi}{D_y}, +\frac{2\pi}{D_y}\right].$$

Flat Plate Antenna's Angular Beamwidth

Since

$$\cos\theta_y(\widehat{\boldsymbol{r} - \boldsymbol{a}_c}) = \sin\theta_x(\widehat{\boldsymbol{r} - \boldsymbol{a}_c}),$$

if we utilize the relationship

$$k_{s_y} = k_{\omega_p} \sin\theta_x(\widehat{\boldsymbol{r} - \boldsymbol{a}_c}),$$

then nulls can be identified in the angular spatial domain by find the directions $\widehat{\boldsymbol{r}-\boldsymbol{a}_c}$ where

$$\begin{aligned} k_{\omega_p} \sin\theta_x(\widehat{\boldsymbol{r}-\boldsymbol{a}_c}) &= \pm\frac{2\pi}{D_y}, \\ \sin\theta_x(\widehat{\boldsymbol{r}-\boldsymbol{a}_c}) &= \pm\frac{\lambda_\omega}{D_y}, \end{aligned}$$

which gives a null to null spatial angular beamwidth of

$$\begin{aligned} \Theta_{\text{ant}}^{\text{plate}} &= 2\times\theta_x(\widehat{\boldsymbol{r}-\boldsymbol{a}_c}), \\ &= 2\times\arcsin\left(\frac{\lambda_{\omega_p}}{D_y}\right). \end{aligned}$$

The angle of the first null or so called *divergence angle* [885] of the antenna beam pattern is

$$\phi_d(\omega_p) = \arcsin\left(\frac{\lambda_{\omega_p}}{D_y}\right).$$

The angular field of view or angular beamwidth of the antenna is

$$\Theta_{\text{ant}}^{\text{plate}}(\omega_p) = 2\times\phi_d(\omega_p).$$

This is the null to null angle of the mainlobe of the beam pattern. The angle $\phi_d(\omega_p)$ is the angle from the antenna's boresight to the first null of the beam pattern.

Flat Plate Antenna Instantaneous Beamwidth

The flat plate antenna has a beamwidth that decreases with increasing instantaneous chirp frequency ω_p (decreasing instantaneous chirp wavelength λ_{ω_p}). This property is a key to determining the nature of the window and weight functions induced in both frequency-space and frequency-wavenumber phase history domains by the flat plate antenna's beam pattern in spotlight mode SAR (see Section 12.4).

High Frequency / High-Q Approximation

When $\lambda_{\omega_c} << D_y$, then λ_{ω_c}/D_y is small and the angular beamwidth is well approximated by

$$\Theta_{\text{ant}}^{\text{plate}}(\omega_c)^{\text{high}-\text{Q}} \approx 2\times\frac{\lambda_{\omega_c}}{D_y}.$$

9.8 Flat Plate Antenna Frequency-Wavenumber Analysis

Although quite informative itself, the frequency-space domain expression (9.40) is also rather difficult to analyze further. We seek to gain additional insights by extending the analysis into the frequency-wavenumber domain.

Flat Plate Antenna Frequency-Wavenumber Beam Function

Assume $\boldsymbol{a}_c = (0,0,0)$ and consider the inverse Fourier transform described in (9.28),

$$\Lambda(t,\boldsymbol{r})_{\text{plate}}^{\text{pulse}} = \frac{1}{2\pi}\int_{\beta_p} W(\omega_p) \int_{-k_{\omega_p}}^{+k_{\omega_p}} \underbrace{\mathbb{A}(\omega_p,k_{s_y},\boldsymbol{r})_{\text{plate}} \cdot H_{r_x}(\omega_p,k_{s_y}) \cdot H_{r_y}(k_{s_y})}_{\Lambda(\omega_p,k_{s_y},\boldsymbol{r})_{\text{plate}}^{\text{mono}}} \cdot dk_{s_y} \cdot \exp(j\omega_p t)\cdot d\omega_p.$$

The $\boldsymbol{r}$ dependent frequency-wavenumber beam function $\Lambda(\omega_p,k_{s_y},\boldsymbol{r})_{\text{plate}}^{\text{mono}}$ of the flat plate surface antenna is determined by the in situ field $\mathbb{A}(\omega_p,k_{s_y},\boldsymbol{r})_{\text{plate}}$ subject to Huygens-Fresnel transfers,

$$\Lambda(\omega_p,k_{s_y},\boldsymbol{r})_{\text{plate}}^{\text{mono}} = \mathbb{A}(\omega_p,k_{s_y},\boldsymbol{r})_{\text{plate}} \cdot H_{r_x}(\omega_p,k_{s_y}) \cdot H_{r_y}(k_{s_y}).$$

The in situ field of the flat plate antenna, described in the frequency-wavenumber domain, with

$$\boldsymbol{s} = (s_x \cdot 0, s_y \hat{\boldsymbol{u}}_y, 0)$$

and $a_e(\boldsymbol{s}) = 1$ is

$$\begin{aligned}
\mathbb{A}(\omega_p,k_{s_y},\boldsymbol{r})_{\text{plate}} &= \int_{\mathbb{S}} a_e(\boldsymbol{s}) \cdot \frac{1}{4\pi} \cdot \frac{1}{|\boldsymbol{r}-\boldsymbol{s}|} \cdot \exp[j s_x \cdot \breve{k}_x(\omega_p,k_{s_y})] \cdot \exp(j s_y k_{s_y}) \cdot d\boldsymbol{s}, \\
&= \int_{-D_y/2}^{+D_y/2} 1 \cdot \frac{1}{4\pi} \cdot \frac{1}{\sqrt{r_x^2 + (r_y - s_y)^2}} \cdot \exp(j s_y k_{s_y}) \cdot ds_y.
\end{aligned}$$

If $\boldsymbol{r}$ is sufficiently distant such that

$$\begin{aligned}
\frac{1}{\sqrt{r_x^2 + (r_y - s_y)^2}} &\approx \frac{1}{\sqrt{r_x^2 + r_y^2}}, \\
&= \frac{1}{|\boldsymbol{r}|}
\end{aligned}$$

does not significantly vary as a function of s_y, then

$$\begin{aligned}
\mathbb{A}(\omega_p,k_{s_y},\boldsymbol{r})_{\text{plate}} &\approx \frac{1}{4\pi} \cdot \frac{1}{\sqrt{r_x^2 + r_y^2}} \cdot \int_{-\infty}^{+\infty} \Pi_{(D_y)}(s_y) \cdot \exp(j s_y k_{s_y}) \cdot ds_y, \\
&= \frac{1}{4\pi} \cdot \frac{1}{|\boldsymbol{r}|} \cdot D_y \cdot \operatorname{sinc}\left(\frac{D_y}{2} \cdot k_{s_y}\right), \\
&= \mathbb{A}(k_{s_y},\boldsymbol{r})_{\text{plate}}.
\end{aligned}$$

The $\boldsymbol{r}$ dependent frequency-wavenumber antenna weight function for the flat plate aperture is independent of temporal passband frequencies ω_p. A flat plate antenna's induced weight function in the frequency-wavenumber domain is invariant to instantaneous chirp frequency.

This behavior has an intuitive explanation. As the chirp frequency increases, the real aperture wavenumber

$$k_{s_y} = k_{\omega_p} \sin \theta_x(\widehat{\boldsymbol{r} - \boldsymbol{a}_s})$$

increases with increasing instantaneous chirp frequency k_{ω_p}, while the antenna's beamwidth

$$\begin{aligned}
\Theta_{\text{ant}}^{\text{plate}} &= \theta_x(\widehat{\boldsymbol{r} - \boldsymbol{a}_s}), \\
&= 2 \times \arcsin\left(\frac{\lambda_\omega}{D_y}\right), \\
&= 2 \times \phi_d^{\text{plate}}.
\end{aligned}$$

decreases with increasing instantaneous chirp frequency k_{ω_p}. These effects cancel one another and the flat plate's induced weight function $\mathbb{A}(k_{s_y}, \boldsymbol{r})_{\text{plate}}$ in frequency-wavenumber domain provides a constant support region in k_{s_y} space as ω_p is varied.

Frequency-Wavenumber Domain Antenna Rectangular Manifold

The null to null width of a flat plate's weight function $\mathbb{A}(k_{s_y}, \boldsymbol{r})_{\text{plate}}$ in the along-track real aperture wavenumber k_{s_y} is

$$k_{s_y} \in \widehat{\Pi}\left(-\frac{2\pi}{D_y}, +\frac{2\pi}{D_y}\right). \tag{9.49}$$

The beamwidth is dependent on instantaneous chirp frequency. The frequency-wavenumber weight function (9.49) is independent of instantaneous chirp frequency [873]. Other antenna types with frequency independent beamwidths have wavenumber weight functions $\mathbb{A}(\omega_p, k_{s_y}, \boldsymbol{r})_{\text{plate}}$ that are frequency dependent [143, 885].

Chapter 10

SAR Signal Channel Models

The signal pathway created by the transmission, scattering and reception of a SAR signal can be viewed as a *channel* [868,885]. Channel models for SAR can be formulated in time-space, frequency-space and frequency-wavenumber domains. The antenna analysis of Chapter 9 is an integral part of these channel models.

This chapter is outlined as follows: Section 10.1 describes the SAR channel model expressed in the time-space domain. Section 10.2 describes the SAR channel model in the frequency-space domain and Section 10.3 deals with the frequency-wavenumber domain. Finally, Section 10.4 presents integrative channel models based on a superposition assumption.

10.1 Channel Time-Space Analysis

The primary components of the time-space SAR channel model are the following:

- in situ field at the face of the transmitting antenna,
- propagation of the transmitted wave from the antenna to the ground scene,
- in situ terrain field produced by the incident wave at the ground scene,
- propagation of backscatter from the ground scene to the antenna and
- sensing of the backscattered field at the receiving antenna.

Simple Point Scattering Models

The illuminated ground scene is assumed to consist of a set of discrete scatterers. An empty scattering field, except for a finite collection of non-interacting point scatterers, simplifies channel modeling. Isotropic point scattering is assumed. The scatterer response may be frequency dependent. We will develop continuous models of electromagnetic scattering in Chapter 11.

Transmit Beam Time-Space Domain Analysis

The antenna's position $\boldsymbol{u}$ in the synthetic aperture is associated with the location of the nominal center point $\boldsymbol{a}_c$ of the surface antenna,

$$\boldsymbol{u} = \boldsymbol{a}_c.$$

The wave field produced by the antenna in transmit mode is determined by the beam function,

$$\begin{aligned} ss(t, \boldsymbol{r} - \boldsymbol{u})_{\text{S}}^{\text{transmit}} &= \Lambda(t, \boldsymbol{r} - \boldsymbol{u})_{\text{S}}^{\text{pulse}}, \\ &= w(t) \;*_t\; \Lambda(t, \boldsymbol{r} - \boldsymbol{u})_{\text{S}}^{\text{impulse}}, \end{aligned} \tag{10.1}$$

where, from (9.5),

$$\begin{aligned} &\Lambda(t, \boldsymbol{r} - \boldsymbol{u})_{\text{S}}^{\text{impulse}} \\ &= \frac{1}{4\pi} \cdot \frac{1}{|\boldsymbol{r} - \boldsymbol{u}|} \cdot \int_{\text{S}} a_e(\boldsymbol{s}) \cdot \delta\left\{t - t_d[\boldsymbol{r} - \boldsymbol{s}(\boldsymbol{u})]\right\} \cdot d\boldsymbol{s} \end{aligned} \tag{10.2}$$

and where $\boldsymbol{s}(\boldsymbol{u})$ describes surface points $\boldsymbol{s}$ of the surface antenna with the nominal center point located at $\boldsymbol{u}$. The model (10.1) includes the pulse waveform excitation $w(t)$ that determines time domain properties of the beam pattern through temporal convolution. The impulsive wavelet

$$\delta\left\{t - t_d[\boldsymbol{r} - \boldsymbol{s}(\boldsymbol{u})]\right\}$$

of (10.2) is the core element of Huygens' expanding spherical wavelet.

Incident Beam Time-Space Domain Analysis

An antenna operated in transmit mode creates a field over $\boldsymbol{r}$ and is assumed to produce a beam that illuminates ground points located at $\boldsymbol{x}$. The incident field at the ground scene is

$$ss(t, \boldsymbol{x} - \boldsymbol{u})_{\text{S}}^{\text{incident}} = \Lambda(t, \boldsymbol{x} - \boldsymbol{u})_{\text{S}}^{\text{pulse}}.$$

Time Dependent Point Scatterers

We are interested in the incident field at the location of each point scatterer, where the nth scatterer is described by

$$gg_n(t, \boldsymbol{x}) = g_n(t) \cdot \delta(\boldsymbol{x} - \boldsymbol{x}_n).$$

The scatterer at $\boldsymbol{x}_n$ has isotropic scattering cross section $g_n(t)$, which in general is complex and frequency dependent. When the terrain scattering function is sensitive to signal frequency, it is appropriate to consider a time dependent scattering model $gg_n(t, \boldsymbol{x})$ associated with $Gg_n(\omega, \boldsymbol{x})$. These are related by a temporal Fourier transform [181],

$$gg_n(t, \boldsymbol{x}) = \mathsf{iFT}_{\text{1D}}^{\omega}\left\{Gg_n(\omega, \boldsymbol{x})\right\}.$$

The time dependence of $gg_n(t, \boldsymbol{x})$ is an indirect description of the frequency dependence $Gg_n(\omega, \boldsymbol{x})$ of scattering behavior [176].

Ground Scene In Situ Field Time-Space Domain Analysis

By Huygens principles, the incident wave field $ss(t, \boldsymbol{x}_n - \boldsymbol{u})^{\text{incident}}$ can be viewed as energizing each point scatterer $gg_n(t, \boldsymbol{x} - \boldsymbol{x}_n)$. Hence, the in situ field that exists at the nth scatterer during scene illumination is

$$\begin{aligned} ss(t,\boldsymbol{x})_n^{\text{in situ}} &= gg_n(t,\boldsymbol{x}) \cdot ss(t,\boldsymbol{x}-\boldsymbol{u})_{\mathbb{S}}^{\text{incident}}, \\ &= gg_n(t,\boldsymbol{x}) \cdot \Lambda(t,\boldsymbol{x}-\boldsymbol{u})_{\mathbb{S}}^{\text{pulse}}, \\ &= g_n(t) \cdot \delta(\boldsymbol{x}-\boldsymbol{x}_n) \cdot \Lambda(t,\boldsymbol{x}-\boldsymbol{u})_{\mathbb{S}}^{\text{pulse}}, \end{aligned} \tag{10.3}$$

which is only spatially non-zero at $\boldsymbol{x} = \boldsymbol{x}_n$. The spatially multiplicative relationship of (10.3) can be expressed as a spatial convolution by using the property of convolution with shifted impulse functions,

$$g_n(t) \cdot \delta(\boldsymbol{x}-\boldsymbol{x}_n) \cdot \Lambda(t,\boldsymbol{x}-\boldsymbol{u})_{\mathbb{S}}^{\text{pulse}} = g_n(t) \cdot \delta(\boldsymbol{x}-\boldsymbol{x}_n) \ *_{\boldsymbol{x}} \ \Lambda(t,\boldsymbol{x}-\boldsymbol{u})_{\mathbb{S}}^{\text{pulse}}.$$

Using this rule, the in situ terrain field can be described in terms of a spatial convolution over the free variable $\boldsymbol{x}$,

$$ss(t,\boldsymbol{x})_n^{\text{in situ}} = gg_n(t,\boldsymbol{x}) \ *_{\boldsymbol{x}} \ \Lambda(t,\boldsymbol{x}-\boldsymbol{u})_{\mathbb{S}}^{\text{pulse}}.$$

Considering the transmitted pulse waveform $w(t)$ that defines the temporal characteristics of the incident field, the scattered in situ terrain field involves a model that includes convolutions in both time and space,

$$ss(t,\boldsymbol{x})_n^{\text{in situ}} = gg_n(t,\boldsymbol{x}) \ *_{\boldsymbol{x}} \ \left[w(t) \ *_t \ \Lambda(t,\boldsymbol{x}-\boldsymbol{u})_{\mathbb{S}}^{\text{impulse}} \right],$$

where

$$\Lambda(t,\boldsymbol{x}-\boldsymbol{u})_{\mathbb{S}}^{\text{pulse}} = w(t) \ *_t \ \Lambda(t,\boldsymbol{x}-\boldsymbol{u})_{\mathbb{S}}^{\text{impulse}}.$$

Scattered Field Time-Space Domain Analysis

With a point scatterer assumption, scattered spherical Huygens wavelets propagate over $\boldsymbol{r}$ and are returned back to the surface antenna, which is now passively operating in receive mode. By Huygens principles, each ground point scatterer produces a secondary Huygens wavelet $hh(t, \boldsymbol{r} - \boldsymbol{x}_n)$ scaled by the in situ reflectivity of the scatterer,

$$ss(t,\boldsymbol{r}-\boldsymbol{x})_n^{\text{scattered}} = \underbrace{ss(t,\boldsymbol{x})^{\text{in situ}}}_{\text{wavelet amplitude}} \cdot \underbrace{hh(t,\boldsymbol{r}-\boldsymbol{x})}_{\text{secondary wavelet}}.$$

By the property of convolution with shifted impulse functions, since $\delta(t, \boldsymbol{r} - \boldsymbol{x})$ is the core of $hh(t, \boldsymbol{r} - \boldsymbol{x})$, the scattered field can be described by convolutions,

$$\begin{aligned} &ss(t,\boldsymbol{r}-\boldsymbol{x})_n^{\text{scattered}} \\ &= \left\{ gg_n(t,\boldsymbol{x}) \ *_{\boldsymbol{x}} \ \left[w(t) \ *_t \ \Lambda(t,\boldsymbol{x}-\boldsymbol{u})_{\mathbb{S}}^{\text{impulse}} \right] \right\} \ *_{\boldsymbol{x}} \ hh(t,\boldsymbol{r}-\boldsymbol{x}). \end{aligned} \tag{10.4}$$

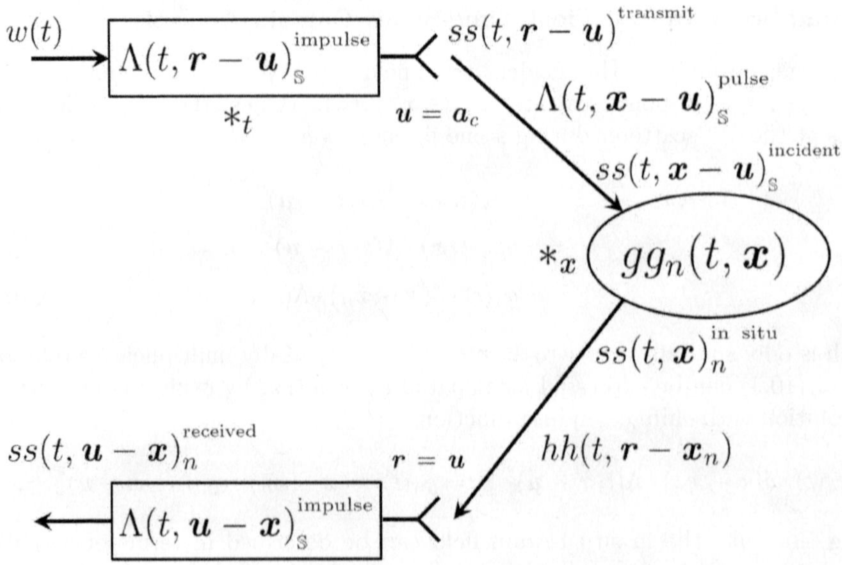

Figure 10.1: Time-space domain analysis of SAR channel model.

Receive Beam Time-Space Domain Analysis

Assuming that the scattered field is detected with the same sensitivity and directivity as when the antenna operates in transmit mode, we follow the propagation of Huygens wavelets $hh(t, \boldsymbol{r} - \boldsymbol{x})$ back to the receiving antenna located at $\boldsymbol{r} = \boldsymbol{u}$ and replace $hh(t, \boldsymbol{u} - \boldsymbol{x})$ with $\Lambda(t, \boldsymbol{u} - \boldsymbol{x})_{\mathrm{S}}^{\text{impulse}}$ in (10.4). This replacement utilizes the superposition principle to describe the signal collected by all points on the antenna's surface. The received field that expresses the conveyance of the in situ terrain field back to the antenna is

$$
\begin{aligned}
& ss(t, \boldsymbol{u} - \boldsymbol{x})_n^{\text{received}} \\
&\quad = \; ss(t, \boldsymbol{u} - \boldsymbol{x})_n^{\text{scattered}} , \\
&\quad = \; \left[gg_n(t, \boldsymbol{x}) \; *_{\boldsymbol{x}} \; \Lambda(t, \boldsymbol{x} - \boldsymbol{u})_{\mathrm{S}}^{\text{pulse}} \right] \; *_{\boldsymbol{x}} \; \Lambda(t, \boldsymbol{u} - \boldsymbol{x})_{\mathrm{S}}^{\text{impulse}} , \\
&\quad = \; \left\{ gg_n(t, \boldsymbol{x}) \; *_{\boldsymbol{x}} \; \left[w(t) \; *_t \; \Lambda(t, \boldsymbol{x} - \boldsymbol{u})_{\mathrm{S}}^{\text{impulse}} \right] \right\} \; *_{\boldsymbol{x}} \; \Lambda(t, \boldsymbol{u} - \boldsymbol{x})_{\mathrm{S}}^{\text{impulse}} .
\end{aligned}
$$

Time-Space Transmit/Receive Channel Model

By transmit and receive mode symmetry of the antenna beam pattern,

$$
\Lambda(t, \boldsymbol{x} - \boldsymbol{u})_{\mathrm{S}}^{\text{impulse}} = \Lambda(t, \boldsymbol{u} - \boldsymbol{x})_{\mathrm{S}}^{\text{impulse}} ,
$$

the time-space domain channel model is

$$
ss(t, \boldsymbol{u} - \boldsymbol{x})_n^{\text{received}} \; = \; w(t) \; *_t \; \left[gg_n(t, \boldsymbol{x}) \; *_{\boldsymbol{x}} \; \Lambda^2(t, \boldsymbol{u} - \boldsymbol{x})_{\mathrm{S}}^{\text{impulse}} \right] .
$$

The two way propagation path squares the weighting imposed by the antenna beam pattern as defined by impulsive excitation. This squared two way antenna weighting is convolved with both the transmitted waveform and the terrain's scattering function to define the two way time-space channel model, as conceptually illustrated in Fig. 10.1.

10.2 Channel Frequency-Space Analysis

We next perform the analysis of the SAR channel model as described in the frequency-space domain.

Transmit Beam Frequency-Space Domain Analysis

The transmit beam, as described in the frequency-space domain, can be determined by the temporal Fourier transform of the time-space domain pulse beam pattern,

$$\begin{aligned} Ss(\omega_p, \boldsymbol{r}-\boldsymbol{u})_{\mathrm{s}}^{\text{transmit}} &= \mathsf{FT}_{1\mathrm{D}}^{t}\left\{\Lambda(t, \boldsymbol{r}-\boldsymbol{u})_{\mathrm{s}}^{\text{pulse}}\right\}, \\ &= \mathsf{FT}_{1\mathrm{D}}^{t}\left\{w(t) \;*_t\; \Lambda(t, \boldsymbol{r}-\boldsymbol{u})_{\mathrm{s}}^{\text{impulse}}\right\}, \\ &= W(\omega_p)\cdot\Lambda(\omega_p, \boldsymbol{r}-\boldsymbol{u})_{\mathrm{s}}^{\text{mono}}, \end{aligned}$$

where

$$\begin{aligned} \Lambda(t, \boldsymbol{r}-\boldsymbol{u})_{\mathrm{s}}^{\text{impulse}} \;&{}_{t}\Leftrightarrow_{\omega_p}\; \Lambda(\omega_p, \boldsymbol{r}-\boldsymbol{u})_{\mathrm{s}}^{\text{mono}}, \\ w(t) \;&{}_{t}\Leftrightarrow_{\omega_p}\; W(\omega_p) \end{aligned}$$

holds true for all ω_p (the Fourier transform of an impulse is a constant that spans an infinite bandwidth). Using the results of the frequency-space antenna analysis of Section 9.3, we next decompose the frequency-space domain description of a monochromatic element of the beam pattern into the surface antenna's in situ field $\mathbb{A}(\omega_p, \widehat{\boldsymbol{r}-\boldsymbol{u}})_{\mathrm{s}}$ subject to a field transfer described by Fresnel's monochromatic wave field operator $Hh(\omega_p, \boldsymbol{r}-\boldsymbol{u})$, i.e.,

$$\begin{aligned} \Lambda(\omega_p, \boldsymbol{r}-\boldsymbol{u})_{\mathrm{s}}^{\text{mono}} &= \mathbb{A}(\omega_p, \widehat{\boldsymbol{r}-\boldsymbol{u}})_{\mathrm{s}}\cdot Hh(\omega_p, \boldsymbol{r}-\boldsymbol{u}), \\ &= \mathbb{A}(\omega_p, \widehat{\boldsymbol{r}-\boldsymbol{u}})_{\mathrm{s}}\cdot\frac{1}{4\pi}\cdot\frac{1}{|\boldsymbol{r}-\boldsymbol{u}|}\cdot\exp\left(-jk_{\omega_p}|\boldsymbol{r}-\boldsymbol{u}|\right). \end{aligned}$$

Hence, the frequency-space expression of the wave field over $\boldsymbol{r}$ of the antenna in transmit mode with a pulse waveform is

$$Ss(\omega_p, \boldsymbol{r}-\boldsymbol{u})_{\mathrm{s}}^{\text{transmit}} = W(\omega_p)\cdot\mathbb{A}(\omega_p, \widehat{\boldsymbol{r}-\boldsymbol{u}})_{\mathrm{s}}\cdot Hh(\omega_p, \boldsymbol{r}-\boldsymbol{u}). \quad (10.5)$$

Surface Antenna Location

The antenna's in situ wave field $\mathbb{A}(\omega_p, \widehat{\boldsymbol{r}-\boldsymbol{u}})_{\mathrm{s}}$ of (10.5) relies on the long range assumption relative to a nominal center point

$$\boldsymbol{u} = \boldsymbol{a}_c.$$

Incident Beam Frequency-Space Domain Analysis

The incident beam at the ground scene (where $\boldsymbol{r} = \boldsymbol{x}$) can be described in the frequency-space domain as

$$\begin{aligned} Ss(\omega_p, \boldsymbol{x} - \boldsymbol{u})_{\mathrm{s}}^{\text{incident}} &= W(\omega_p) \cdot \mathbb{A}(\omega_p, \widehat{\boldsymbol{x} - \boldsymbol{u}})_{\mathrm{s}} \cdot Hh(\omega_p, \boldsymbol{x} - \boldsymbol{u}), \\ &= W(\omega_p) \cdot \Lambda(\omega_p, \boldsymbol{x} - \boldsymbol{u})_{\mathrm{s}}^{\text{mono}}. \end{aligned}$$

Frequency Dependent Point Scatterers

The nth point scatterer on the ground is modeled in the frequency-space domain by

$$Gg_n(\omega_p, \boldsymbol{x}) = G_n(\omega_p) \cdot \delta(\boldsymbol{x} - \boldsymbol{x}_n).$$

This model includes the possibility of frequency dependent scattering.

Terrain In Situ Field Frequency-Space Domain Analysis

The terrain in situ field with frequency-space domain analysis retains the spatial convolution, but the time domain convolution becomes multiplicative in the frequency domain,

$$Ss(\omega_p, \boldsymbol{x})_n^{\text{in situ}} = Gg_n(\omega_p, \boldsymbol{x}) \ *_{\boldsymbol{x}} \left[W(\omega_p) \cdot \Lambda(\omega_p, \boldsymbol{x} - \boldsymbol{u})_{\mathrm{s}}^{\text{mono}} \right]. \quad (10.6)$$

Receive Beam Frequency-Space Domain Analysis

By adding another beam element $\Lambda(\omega_p, \boldsymbol{u} - \boldsymbol{x})_{\mathrm{s}}^{\text{mono}}$ to express conveyance of the in situ terrain field back to the receiving surface antenna, we obtain a model of the scattered field as received back at the antenna's surface,

$$\begin{aligned} &Ss(\omega_p, \boldsymbol{u} - \boldsymbol{x})_n^{\text{received}} \\ &\quad = \left[Gg_n(\omega_p, \boldsymbol{x}) \ *_{\boldsymbol{x}} \ W(\omega_p) \cdot \Lambda(\omega_p, \boldsymbol{x} - \boldsymbol{u})_{\mathrm{s}}^{\text{mono}} \right] \cdot \Lambda(\omega_p, \boldsymbol{u} - \boldsymbol{x})_{\mathrm{s}}^{\text{mono}}. \end{aligned}$$

Frequency-Space Transmit/Receive Channel Model

By transmit versus receive mode symmetry of the antenna beam pattern,

$$\Lambda(\omega_p, \boldsymbol{x} - \boldsymbol{u})_{\mathrm{s}}^{\text{mono}} = \Lambda(\omega_p, \boldsymbol{u} - \boldsymbol{x})_{\mathrm{s}}^{\text{mono}}.$$

The frequency-space domain description of the SAR channel model is given by the combined effect of transmit, scattered and receive channel model elements,

$$\begin{aligned} &Ss(\omega_p, \boldsymbol{u} - \boldsymbol{x}_n)_n^{\text{received}} \\ &\quad = W(\omega_p) \cdot \left[Gg_n(\omega_p, \boldsymbol{x}_n) \ *_{\boldsymbol{x}} \ \Lambda^2(\omega_p, \boldsymbol{u} - \boldsymbol{x})_{\mathrm{s}}^{\text{mono}} \right], \\ &\quad = W(\omega_p) \cdot \left\{ Gg_n(\omega_p, \boldsymbol{x}) \ *_{\boldsymbol{x}} \ \left[\mathbb{A}^2(\omega_p, \widehat{\boldsymbol{u} - \boldsymbol{x}})_{\mathrm{s}} \cdot Hh^2(\omega_p, \boldsymbol{u} - \boldsymbol{x}) \right] \right\}. \end{aligned}$$

Again, the two way propagation path squares the weighting imposed by the antenna in situ field $\mathbb{A}(\omega_p, \widehat{\boldsymbol{u} - \boldsymbol{x}})_{\mathrm{s}}$ and the Fresnel transference $Hh(\omega_p, \boldsymbol{u} - \boldsymbol{x})$ of

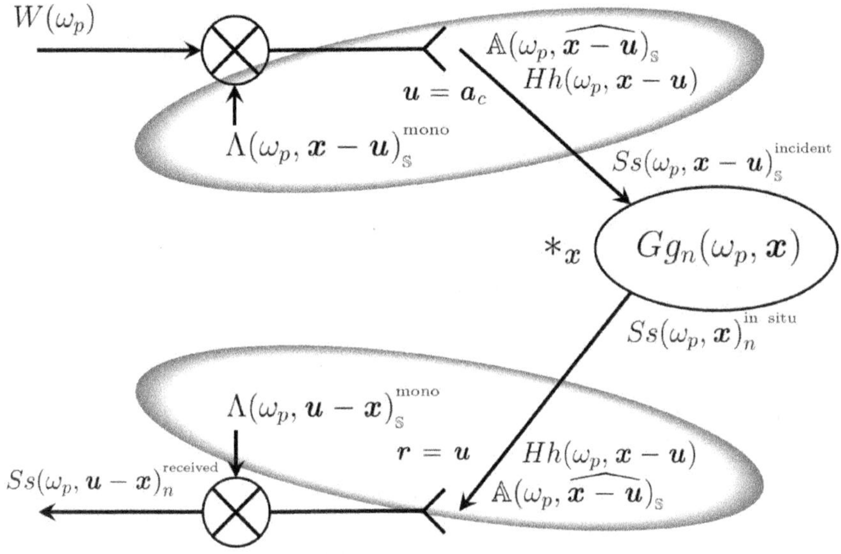

Figure 10.2: Frequency-space domain analysis of SAR channel model.

this wave field. This squared two way antenna weighting is convolved with the terrain's scattering function $Gg_n(\omega_p, \boldsymbol{x})$ to define the two way frequency-space channel model as shown in Fig. 10.2. The large ovals in Fig. 10.2 group related channel model elements.

10.3 Channel Frequency-Wavenumber Analysis

We next perform the analysis of the SAR channel model in the frequency-wavenumber domain. We employ Stolt mappings to bridge the analysis from the (ω_p, k_{s_y}) domain of the stationary antenna to the $(\breve{k}_x, k_y)$ domain of the angular spectrum.

Transmit Beam Frequency-Wavenumber Domain Analysis

Given our analysis in the frequency-wavenumber domain of the general antenna beam pattern (9.28), the transmit field described for the angular frequency value ω_p within the transmitted signal's passband is given by

$$SS(\omega_p, k_{s_y}, \boldsymbol{r}-\boldsymbol{u})_{\mathbb{S}}^{\text{transmit}} = W(\omega_p) \cdot \Lambda(\omega_p, k_{s_y}, \boldsymbol{r}-\boldsymbol{u})_{\mathbb{S}}^{\text{mono}}.$$

The monochromatic frequency-wavenumber beam pattern is described as the antenna's in situ wave field $\mathbb{A}(\omega_p, k_{s_y}, \boldsymbol{r}-\boldsymbol{u})_{\mathbb{S}}$ subject to dimensionally separable Huygens-Fresnel transfers,

$$\begin{aligned}&\Lambda(\omega_p, k_{s_y}, \boldsymbol{r}-\boldsymbol{u})_{\mathbb{S}}^{\text{mono}} \\ &\quad = a(\omega_p, k_{s_y})_{\text{pwd}} \cdot \mathbb{A}(\omega_p, k_{s_y}, \boldsymbol{r}-\boldsymbol{u})_{\mathbb{S}} \cdot H_{r_x(\boldsymbol{u})}(\omega_p, k_{s_y}) \cdot H_{r_y(\boldsymbol{u})}(k_{s_y}),\end{aligned}$$

where

$$\begin{aligned} H_{r_x(\boldsymbol{u})}(\omega_p, k_{s_y}) &= \exp[-jr_x(\boldsymbol{u}) \cdot \breve{k}_x(\omega_p, k_{s_y})], \\ H_{r_y(\boldsymbol{u})}(k_{s_y}) &= \exp\left[-jr_y(\boldsymbol{u}) \cdot k_{s_y}\right], \end{aligned}$$

with

$$\begin{aligned} r_x(\boldsymbol{u}) &= |r_x - u_x|, \\ r_y(\boldsymbol{u}) &= |r_y - u_y|. \end{aligned}$$

The antenna's in situ wave field $\mathbb{A}(\omega_p, k_{s_y}, \boldsymbol{r} - \boldsymbol{u})_{\mathbb{S}}$ at the passband frequency ω_p is given by

$$\begin{aligned} &\mathbb{A}(\omega_p, k_{s_y}, \boldsymbol{r} - \boldsymbol{u})_{\mathbb{S}} \\ &= \int_{\mathbb{S}} a_e\left[\boldsymbol{s}(\boldsymbol{u})\right] \cdot \frac{1}{4\pi} \cdot \frac{1}{|\boldsymbol{r} - \boldsymbol{s}(\boldsymbol{u})|} \\ &\qquad \cdot \exp[js_x(\boldsymbol{u}) \cdot \breve{k}_x(\omega_p, k_{s_y})] \cdot \exp\left[js_y(\boldsymbol{u}) \cdot k_{s_y}\right] \cdot d\boldsymbol{s}, \end{aligned}$$

with

$$\begin{aligned} |\boldsymbol{r} - \boldsymbol{s}(\boldsymbol{u})| &= \sqrt{[r_x - s_x(\boldsymbol{u})]^2 + [r_y - s_y(\boldsymbol{u})]^2}, \\ s_x(\boldsymbol{u}) &= |s_x - u_x|, \\ s_y(\boldsymbol{u}) &= |s_y - u_y|. \end{aligned}$$

In Situ Ground Scene Field Frequency-Wavenumber Domain Analysis

The in situ field with frequency-wavenumber domain analysis converts the spatial convolution of the frequency-space domain analysis (10.6) to multiplication in the wavenumber domain. It is necessary to consider the terrain in situ field at a specific ground point scatterer located at $\boldsymbol{x}_n$ as described in the frequency-wavenumber domain by $GG_n[\breve{k}_x(\omega_p, k_{s_y}), k_{s_y}]$,

$$\begin{aligned} &SS(\omega_p, k_{s_y}, \boldsymbol{x}_n)_n^{\text{in situ}} \\ &= W(\omega_p) \times_{\omega_p} \left\{GG_n[\breve{k}_x(\omega_p, k_{s_y}), k_{s_y}] \times_{k_{s_y}} \Lambda(\omega_p, k_{s_y}, \boldsymbol{x}_n - \boldsymbol{u})_{\mathbb{S}}^{\text{mono}}\right\}. \end{aligned}$$

Receive Beam Frequency-Wavenumber Domain Analysis

The frequency-wavenumber expression $\Lambda(\omega_p, k_{s_y}, \boldsymbol{u} - \boldsymbol{x}_n)_{\mathbb{S}}^{\text{mono}}$ of the beam for the antenna operating in receive mode is used to model the backscatter from $\boldsymbol{x}_n$ received back at the antenna located at $\boldsymbol{u}$,

$$\begin{aligned} &SS(\omega_p, k_{s_y}, \boldsymbol{u} - \boldsymbol{x}_n)_n^{\text{received}} \\ &= W(\omega_p) \times_{\omega_p} \left\{GG_n[\breve{k}_x(\omega_p, k_{s_y}), k_{s_y}] \times_{k_{s_y}} \Lambda(\omega_p, k_{s_y}, \boldsymbol{x}_n - \boldsymbol{u})_{\mathbb{S}}^{\text{mono}}\right\} \\ &\qquad \times_{k_{s_y}} \Lambda(\omega_p, k_{s_y}, \boldsymbol{u} - \boldsymbol{x}_n)_{\mathbb{S}}^{\text{mono}}. \end{aligned}$$

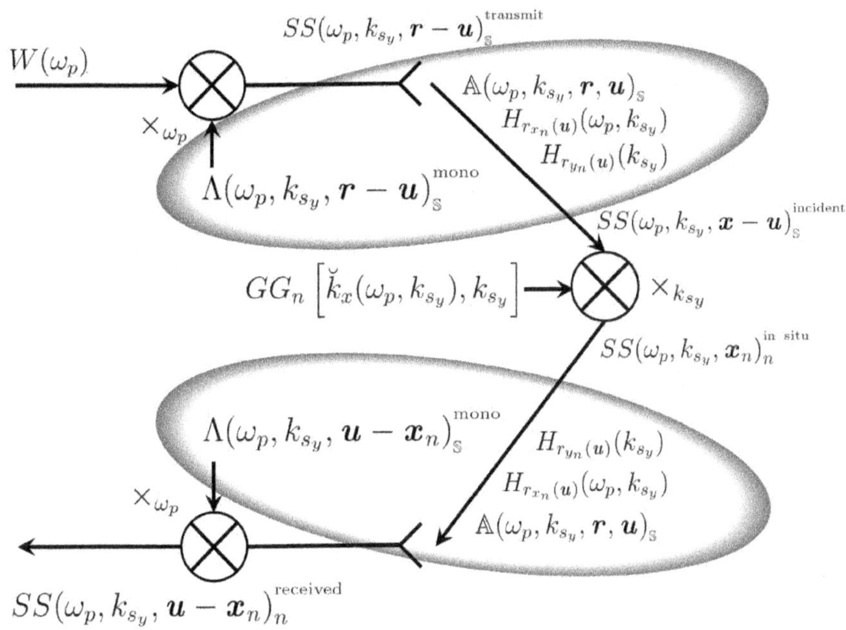

Figure 10.3: Frequency-wavenumber domain analysis of SAR channel model.

Frequency-Wavenumber Transmit/Receive Channel Model

It is reasonable to assume symmetry between the transmit and receive modes of the antenna's frequency-wavenumber domain characteristics,

$$\Lambda(\omega_p, k_{s_y}, \boldsymbol{x}_n - \boldsymbol{u})_{\mathbb{S}}^{\text{mono}} = \Lambda(\omega_p, k_{s_y}, \boldsymbol{u} - \boldsymbol{x}_n)_{\mathbb{S}}^{\text{mono}} . \tag{10.7}$$

Therefore, the frequency-wavenumber domain description of the SAR channel model is comprised of the concatenation of transmit, scattered and receive channel model elements,

$$\begin{aligned} &SS(\omega_p, k_{s_y}, \boldsymbol{u} - \boldsymbol{x}_n)_n^{\text{received}} \\ &= W(\omega_p) \times_{\omega_p} \left\{ GG_n[\breve{k}_x(\omega_p, k_{s_y}), k_{s_y}] \times_{k_{s_y}} \Lambda^2(\omega_p, k_{s_y}, \boldsymbol{u} - \boldsymbol{x}_n)_{\mathbb{S}}^{\text{mono}} \right\}. \end{aligned}$$

This gives more insight if the antenna beam pattern, expressed in the frequency-wavenumber domain, is decomposed into the in situ field at the antenna and the

Huygens-Fresnel transfers to and from the surface antenna,

$$
\begin{aligned}
SS(\omega_p, k_{s_y}, \boldsymbol{u} - \boldsymbol{x}_n)_n^{\text{received}} \\
= \; & W(\omega_p) \times_{\omega_p} \{ \\
& a(\omega_p, k_{s_y})_{\text{pwd}} \cdot \mathbb{A}(\omega_p, k_{s_y}, \boldsymbol{r} - \boldsymbol{u})_{\text{s}} \cdot H_{r_{xn}(\boldsymbol{u})}(\omega_p, k_{s_y}) \cdot H_{r_{yn}(\boldsymbol{u})}(k_{s_y}) \\
& \times_{k_{sy}} GG_n \left[\breve{k}_x(\omega_p, k_{s_y}), k_{s_y} \right] \times_{k_{sy}} \\
& a(\omega_p, k_{s_y})_{\text{pwd}} \cdot \mathbb{A}(\omega_p, k_{s_y}, \boldsymbol{r} - \boldsymbol{u})_{\text{s}} \cdot H_{r_{xn}(\boldsymbol{u})}(\omega_p, k_{s_y}) \cdot H_{r_{yn}(\boldsymbol{u})}(k_{s_y}) \\
& \}, \\
= \; & W(\omega_p) \\
& \times_{\omega_p} GG_n \left[\breve{k}_x(\omega_p, k_{s_y}), k_{s_y} \right] \\
& \times_{k_{sy}} \left[a^2(\omega_p, k_{s_y})_{\text{pwd}} \cdot \mathbb{A}^2(\omega_p, k_{s_y}, \boldsymbol{r} - \boldsymbol{u})_{\text{s}} \right. \\
& \left. \cdot H^2_{r_{xn}(\boldsymbol{u})}(\omega_p, k_{s_y}) \cdot H^2_{r_{yn}(\boldsymbol{u})}(k_{s_y}) \right] .
\end{aligned}
$$

The two way propagation path squares the weighting

$$a(\omega_p, k_{s_y})_{\text{pwd}} \cdot \mathbb{A}(\omega_p, k_{s_y}, \boldsymbol{r})_{\text{s}}$$

imposed by the in situ antenna beam pattern as mapped to the frequency-wavenumber domain.

The Huygens-Fresnel transfers

$$H_{r_x(\boldsymbol{u})}(\omega_p, k_{s_y}) \;\text{ and }\; H_{r_y(\boldsymbol{u})}(k_{s_y})$$

are also squared to reflect two way signal propagation.

The two way frequency-wavenumber channel model is diagramed in Fig. 10.3.

10.4 Scene Channel Models

Using a superposition principle and the weak scattering approximation among discrete scatterers (that is, assuming there is no wave field interactions among the discrete scatterers), we provide a summary of the various SAR channel models developed in this chapter. These summary channel models include returns from all discrete scatterers within the ground scene and are convolutional in nature [181, 865].

Time-Space Ground Scene Echo Signal

The composite echo signal of the entire ground scene as expressed in the time-space domain and received back at the surface antenna is

$$
\begin{aligned}
ss(t, \boldsymbol{u})_{\text{scene}}^{\text{received}} &= \sum_n ss(t, \boldsymbol{u} - \boldsymbol{x})_n^{\text{received}}, \\
&= \sum_n w(t) *_t \left[gg_n(t, \boldsymbol{x}) *_{\boldsymbol{x}} \Lambda^2(t, \boldsymbol{u} - \boldsymbol{x})_{\text{s}}^{\text{impulse}} \right] . \quad (10.8)
\end{aligned}
$$

In convolutional integral form,

$$
\begin{aligned}
ss(t,\boldsymbol{u})^{\text{received}}_{\text{scene}} \\
&= \sum_n \int_{\mathbb{T}} w(\tau) \int_{\mathbb{G}} gg_n(t-\tau,\boldsymbol{x}-\chi)\cdot \Lambda^2(t-\tau,\boldsymbol{u}-\chi)^{\text{impulse}}_{\mathbb{S}} \cdot d\chi d\tau, \\
&= \sum_n \int_{\mathbb{T}} w(\tau) \int_{\mathbb{G}} g_n(t-\tau)\cdot \delta(\boldsymbol{x}-\boldsymbol{x}_n-\chi)\cdot \Lambda^2(t-\tau,\boldsymbol{u}-\chi)^{\text{impulse}}_{\mathbb{S}} \cdot d\chi d\tau.
\end{aligned}
$$

Frequency-Space Ground Scene Echo Signal

The composite echo signal of the entire ground scene as expressed in the frequency-space domain and received back at the surface antenna is

$$
\begin{aligned}
Ss(\omega_p,\boldsymbol{u})^{\text{received}}_{\text{scene}} \\
&= \sum_n Ss(\omega_p,\boldsymbol{u}-\boldsymbol{x})^{\text{received}}_n, \\
&= \sum_n W(\omega_p)\cdot \Big\{ Gg_n(\omega_p,\boldsymbol{x}) *_{\boldsymbol{x}} \Big[\mathbb{A}^2(\omega_p,\widehat{\boldsymbol{u}-\boldsymbol{x}})_{\mathbb{S}}\cdot Hh^2(\omega_p,\boldsymbol{u}-\boldsymbol{x})\Big]\Big\}.
\end{aligned}
$$

In convolutional integral form,

$$
\begin{aligned}
Ss(\omega_p,\boldsymbol{u})^{\text{received}}_{\text{scene}} \\
&= \sum_n W(\omega_p)\cdot \int_{\mathbb{G}} Gg_n(\omega_p,\boldsymbol{x}-\chi)\cdot \mathbb{A}^2(\omega_p,\widehat{\boldsymbol{u}-\chi})_{\mathbb{S}}\cdot Hh^2(\omega_p,\boldsymbol{u}-\chi)\cdot d\chi, \\
&= \sum_n W(\omega_p)\cdot \int_{\mathbb{G}} G_n(\omega_p)\cdot \delta(\boldsymbol{x}-\boldsymbol{x}_n-\chi) \\
&\qquad \cdot \mathbb{A}^2(\omega_p,\widehat{\boldsymbol{u}-\chi})_{\mathbb{S}}\cdot Hh^2(\omega_p,\boldsymbol{u}-\chi)\cdot d\chi.
\end{aligned}
$$

Frequency-Wavenumber Ground Scene Echo Signal

The composite echo signal expressed in the frequency-wavenumber domain and received back at the antenna is

$$
\begin{aligned}
SS(\omega_p,k_{s_y},\boldsymbol{u})^{\text{received}}_{\text{scene}} \\
&= \sum_n SS(\omega_p,k_{s_y},\boldsymbol{u}-\boldsymbol{x}_n)^{\text{received}}_n, \\
&= \sum_n W(\omega_p) \\
&\quad \times_{\omega_p} GG_n[\breve{k}_x(\omega_p,k_{s_y}),k_{s_y}] \\
&\quad \times_{k_{s_y}} \Big[a^2(\omega_p,k_{u_y})_{\text{pwd}}\cdot \mathbb{A}^2(\omega_p,k_{s_y},\boldsymbol{r}-\boldsymbol{u})_{\mathbb{S}} \\
&\qquad \cdot H^2_{r_{xn}(\boldsymbol{u})}(\omega_p,k_{s_y})\cdot H^2_{r_{yn}(\boldsymbol{u})}(k_{s_y})\Big].
\end{aligned}
$$

There is no convolution to yield an integral form in the frequency-wavenumber domain, where multiplication is the equivalent operator.

Convolutional Models in SAR Analysis

A key assumption often, but not always, used in SAR signal model development is that the received radar echo signal can be accurately characterized by convolutions of the terrain scattering function with radar signal elements formulated in both fast-time and slow-time domains [868].

Deconvolution Methods in SAR Image Formation Processing

With convolutional signal models, the image formation problem can be set up as a deconvolution problem [250, 885]. Fourier transforms allow conversion of convolutional elements described in time and space domains into simpler multiplicative elements described in frequency and wavenumber domains. Solutions of this approach can be implemented in the Fourier domain with multiplicative system inversions.

Generalized versions of convolutional models lead to Fredholm integral expressions of spectral domain models [258].

Wave Field Inversion Models in SAR Analysis

The Huygens-Fresnel transfer functions are often interpreted as matched filter elements of the signal processing. However, SAR angular spectrum processing does not require matched filtering in the along-track dimension. Huygens-Fresnel transfers are more appropriately viewed as wave field inversion operators implemented in the wavenumber domain.

SAR image formation processing *is not* necessarily synonymous with matched filtering. SAR inversion can be accomplished in a more general context provided by wave theory tools, wave field modeling and wave field inversion methods.

The wave field inversion approach has advantages over the deconvolution approach in that algorithm developers are not constrained, due to an imposed convolutional modeling context, to search for deconvolution or matched filtering solutions for inversion.

Chapter 11

Wave Theory Revisited

We revisit the Huygens-Fresnel wave theory framework and elaborate on several key points. In addition to models based on partial differential equations and integral expressions based on convolutions, the scattering and propagation of electromagnetic waves can be expressed in terms of wave theory integral equations [89, 630]. For example, Helmholtz wave motion models can be described in both differential and integral forms [482,513,909,993]. Integral expressions are potentially more powerful in that they engage wave theory principles at macroscopic levels, but with connections to microscopic details.

We extend the wave theory framework by developing multiple approaches to integral formulations of Huygens-Fresnel wave theory principles. There are fundamentally four classes of integral expressions for wave motion that are of potential interest in SAR signal analysis.

1. Superposition integrals based on convolutional models of transmitted signal interactions with terrain functions.

2. Time-space domain integral expressions of scattered Huygens wavelet motion with impulsive wavefronts.

3. Frequency-space domain Helmholtz integral expressions of scattered harmonic Fresnel wave fields.

4. Time-space domain Kirchhoff integral expressions of scattered wavelet motion with expanding spherical wavefronts determined by general pulse waveforms.

Kirchhoff's formulation is the most general approach and is applicable to mechanical wave fields. Kirchhoff's formulation includes the time-space Huygens wavelet and frequency-space Helmholtz integral formulations as special cases.

However, Kirchhoff's formulation is only applicable to scalar wave fields. We must rely on formulation of Huygens principles in terms of Maxwell's equations for integral descriptions of electromagnetic vector wave fields.

This chapter is outlined as follows: Section 11.1 is a continuation of our historical overview of the study of light propagation (started in Section 3.1). Section 11.2 reexamines Huygens principles and provides interpretations and explanations of analytic subtleties. Section 11.3 shows how vector forms of Maxwell's electromagnetic equations rigorously satisfy Huygens principles. However, we set aside the vector form of electromagnetic wave analysis and turn to a simpler study of scalar acoustic waves. Section 11.4 demonstrates how the Huygens principles are satisfied by scalar acoustic fields in three dimensions, where the cause of scattering of acoustic waves is modeled as permutations of the propagation medium. This section presents an integral formulation of the Helmholtz wave motion equation and discusses a generalized version developed by Kirchhoff. Section 11.5 develops a model for terrain scattered electromagnetic signals based on perturbations of permittivity. Section 11.6 presents an integral form of the terrain scattered electromagnetic field. Finally, Section 11.7 examines complications of Huygens-Fresnel's wave theory that arise when a scattering problem is reduced from three dimensions to two dimensions. This section presents Hankel functions, which are descriptive of cylindrical wave fields in three dimensions and descriptive of circular wave fields in two dimensions.

11.1 Historical Overview of the Study of Light

We begin our revisit of the origins of electromagnetic wave theory by extending the overview that was started in Section 3.1 of the history of the study of light propagation.

Rays, Reflection and Refraction Observers in antiquity noted that light has ray like (non-spherical) propagation behaviors. The phenomena of reflection and refraction of straight light rays were also noted in antiquity [89].

Light Wave Longitudinal Motion Concepts René Descartes (1596–1650) postulated that a light wave is some form of a longitudinal pressure wavelet propagating through some unknown, invisible, pervasive, but perfectly elastic medium [239]. The notion of light as a longitudinal wave in an universal but unknown medium persisted for hundreds of years. This imagined covert medium was called "aether" [296].

Experimental Methods in Foundations of Optics Galileo Galilei[1] established rigorous experimental methods in studies of physics. The disciplined methods of Galileo led to empirical foundations in observational studies of light. Using Galileo's methods, Willebrord Snell[2] explored refraction in 1621. Snell passed away before publishing his discoveries. Descartes published in *Dioptrique*, without reference to Snell's work, what later came to be known as Snell's law [89].

Speed of Light Olaf Römer[3] used eclipses of the moons of Jupiter in 1675 to estimate longitudinal positions on Earth and was the first to make quantitative

[1]Galileo Galilei (1564–1642) was an Italian physicist, mathematician, astronomer and philosopher.

[2]Willebrord Snell (Snellius) (1580–1626) was a Dutch astronomer and mathematician.

[3]Olaf Römer (1644–1710) was a Danish astronomer.

measurements of the speed of light [795].

Light and Least Time Pierre de Fermat[4] postulated in 1657 that light will always traverse the path requiring the least time [236]. Einstein[5] later showed in his formulation of principles of general relativity that least time propagation is not concomitant with straight path propagation [298].

Light Diffraction The diffraction of light into regions of darkness was first noted in 1665 by Francesco Maria Grimaldi[6] [419]. This was the first evidence that light travels in ways other than straight paths, without a refractive medium involved in the change of direction.

White Light Contains Colors Isaac Newton (1642–1727) discovered that white light can be split into colors with a glass prism [694].

Optical Interference Some of the first observed indications of optical interference were bands of colors produced by reflections from thin floating films. These observations seem to be first noted by Robert Boyle[7] [94] and Robert Hooke[8] [467]. These color patterns came to be known as Newton's rings.

Spherical Impulse Wavelets of Light Huygens postulated in 1690 that the wave motion of light is founded in spherical impulsive wavelets. He hypothesized that primary wavelets caused secondary wavelets, leading to the propagation of the primary wavelet in a never ending chain reaction. The impulsive wavelets of Huygens' concepts were imagined to be longitudinal pressure waves traveling in aether. Huygens was the first to strongly champion a wave theory of light.

Spherical Wavelets of Light with Oscillations Hooke generalized Huygens' concept by postulating that propagating wavelets of light consist of rapid (non-impulsive) wavelet vibrations (not necessarily sinusoidal). He postulated that every vibration generated secondary expanding spherical wavelet vibrations.

Polarization of Light Huygens was studying the characteristics of light transmitted through Iceland spar when he discovered the property of light polarization. The polarization of light by reflection was first observed by Etinne Louis Malus[9] [618].

Huygens attempted to explain polarization with ellipsoidal wave concepts combined with notions of primary wavelets causing secondary wavelets.

Newton developed an explanation of polarization based on what he called the "transversality" of the polarization medium [89].

---

[4]Pierre de Fermat (1601–1665) was a French mathematician.
[5]Albert Einstein (1879-1955) was a German-born American theoretical physicist.
[6]Francesco Maria Grimaldi (1618–1663) was an Italian mathematician and physicist.
[7]Robert Boyle (1627–1691) was an Anglo-Irish philosopher, chemist and physicist.
[8]Robert Hooke (1635–1703) was an English philosopher, architect and mathematician.
[9]Etinne Louis Malus (1775–1812) was a French engineer, physicist and mathematician.

Transverse Light Wave Concepts Working with Arago[10], Fresnel found in 1816 that two beams of light with polarization at right angles to each other never interfere. The notion of a wave, as held at that time, was founded on the longitudinal wave concept. Fresnel's observation could not be conceptualized as consistent with the assumption of longitudinal wave motion.

Young, hearing of Fresnel's discovery, postulated in 1817 that vibrations of light are transverse.

Fresnel, building on Young's ideas, hypothesized that wave motion within the aether should be analyzed from a transverse wave perspective. Fresnel proceeded with a mechanical analysis that sought to explain the observed characteristics of wave interference. Fresnel was able to develop dynamic models of the "mechanisms of aether vibrations." However, Fresnel was not entirely successful in attempting to deduce optical properties from a mechanical analysis of transverse wave motion in an elastic medium.

About this time, in his study of sound waves of small amplitude, Fresnel introduced the key concept of acoustic longitudinal waves consisting of two parts, the compression wave (potential) element and the particle velocity (kinetic) element.

Particle versus Wave Theories of Light Difficulties faced by the postulated wave principles of Hooke and Huygens in explaining straight line propagation of light were viewed by Newton as so insurmountable that he persisted with postulates based in emission of corpuscular elements of light. The corpuscular theory of light was further developed during this time period by Pierre Simon de Laplace and Jean-Baptist Biot[11]. The corpuscular view held dominant for nearly a century, with only a few advocating the wave based theories. Leonhard Euler (1707–1783) was one who did.

Beginning in the nineteenth century, discoveries were made which renewed interest in wave theories of light. Foremost was the experimental work and ideas of Young [1020–1022], including his explanations in 1801 of interference of colors produced by thin floating films.

At the time of Young's and Fresnel's work, Newton's particle based theory of light was nearly universally accepted. Many were unwilling to supplant Newton's particle theory, even in the light of Young and Fresnel's empirical results that provided evidence in support of a wave theory framework.

Continuing Sceptism and Poisson's Spot Poisson, in an argument aligned with the mainstream corpuscular theories of the time, attempted to use Fresnel's own wave theories to conceptually argue against a wave theory for light. Poisson suggested that, if true, a careful consideration of Huygens-Fresnel wave theories would imply that a spot of light should appear along the center line of the shadow of an opaque circular disk. Poisson could imagine, based on the postulated theory he was seeking to prove wrong, that the perturbing edge of the circular disk should cause a diffracted wave field that creates a spot of light (a point of constructive interference) at points along the center line of the shadow. Since it should be easy

[10]Francois Jean Dominique Arago (1786–1753) was a French mathematician and physicist.
[11]Jean-Baptist Biot (1774–1862) was French physicist, astronomer and mathematician.

to show that the darkness of a shadow does not contain a spot of light, argued Poisson, the Huygens-Fresnel wave model would thereby be proved false.

A Grand Contest In 1818, the proponents of the particle theory of light proposed that the subject of diffraction be postulated as the prize question set forth by the French Academy of Sciences – with the expectation that the resulting treatment would lead to the crowning triumph of the emissive particle theory [89].

Fresnel's Submission: Sinusoidal Spherical Wave Field Concepts With confidence and courage to argue against consensus, Fresnel submitted an explanation of light diffraction that postulated that Poisson's spot should actually exist. Fresnel's treatment was based in a wave theory foundation formed by blending Young's principle of interference with one of Huygens' premises that the combined envelop of secondary wavelets explains the continued propagation of the primary wavelet. Fresnel extended the ideas of Huygens and Hooke by replacing the notion of a spherically expanding impulsive or oscillating wavelet with the concept of continuously radiating monochromatic sinusoidal spherical waves. Like Huygens' and Hooke's wavelets, Fresnel's spherical harmonic wave field "emerged" from a point "source" [35].

An Unexpected Winner Arago was the head of the French Academy of Sciences committee set up to conduct and judge the competition. Neither Poisson nor Fresnel had conducted experiments to probe the shadow of a disc to seek to detect the hypothesized spot of light. Arago was the first to venture to actually perform experiments. Arago's experiments, surprising to nearly all, proved Poisson's postulated spot of light actually exists! Fresnel was declared winner of this grand contest [603, 850, 851, 903].

Universal Nature of Light Fresnel also sought to determine in 1818 if there is any difference in the light from stellar or terrestrial sources. Arago was able to prove by experiment that there is no difference. Fresnel also investigated diffraction of light by straight edges, small apertures and screens. Given the wave (pun intended) of new empirical results, conventional wisdom seemed to swing away from Newton's corpuscular theories. A wave theory framework for light propagation enjoyed the lime light for most of the next hundred years.

Quantum Mechanics and Light Since the advent of quantum mechanics, it is no longer necessary to debate the relative merits of a particle theory versus a wave theory to explain light propagation. These apparent difficulties arise only when the study of light is approached from the perspective of a mechanical analysis. The inability of light to resolve into both transverse and longitudinal wave motion elements at inhomogeneous boundaries of an imagined propagation medium rules out analysis based in classical mechanics.

Modern science does not require mechanical explanations of all natural phenomena [89]. Maxwell's equations being the pinnacle example. Maxwell's equations are accepted at face value without recourse to decompose or physically motivate by reduction to something mechanical or simpler.

Emission of Light from Matter Maxwell's equations do not explain all aspects of light. Maxwell's equations fail to provide an explanation of the emission and absorption of light by matter. Such light-matter interactions occur, for example, in the photoelectric effect. Explanation of these properties requires modern physics.

Quantum theory postulates that a radiating body imparts energy into an electromagnetic field. This energy is imparted not in a continuous manner, but in small increments (quanta) given by Planck's[12] constant,

$$h = 6.626 \times 10^{-34} \text{ joule} \cdot \text{seconds},$$

where the imparted energy ϵ is determined by the Planck-Einstein relation,

$$\epsilon = h\nu, \tag{11.1}$$

where ν is the frequency of the emitted electromagnetic wave. It is the frequency of light, not the intensity, that determines the giving and taking of energy from matter. This quantum behavior is a key concept that separates the old and new worlds of physics.

Wave-Quantum Theory of Light

The foundation of an understanding of light in a modern context of relativity was established by Einstein. Einstein's willingness to accept seemingly unorthodox empirical results as a premise [296, 298], allowed him to return to Isaac's notion of the corpuscular theory of light in a new form coupled to the notion of the frequency of a wave as expressed by (11.1). Einstein's work opened to the world the blended wave-quantum theory of light.

Furthermore, in consideration of the behavior of light as observed by moving witnesses, Einstein's work in special relativity shattered forever a view that holds the rate of time flow as universally invariant among reference frames with relative motion.

11.2 Huygens Principles and Wave Properties

Two particular aspects of the Huygens principles are key in analytical wave theory. One is the sufficiency of wave state knowledge at one instant in time to describe the wave state at another instant. The other is the requirement that the interior field of an expanding impulsive wavelet must leave a null field (i.e., analytically equate to a null field). Analytic approaches for describing dynamic wave motion or static wave fields, based on either mechanical formulations of acoustic wave motion or on non-mechanical Maxwell equations, should satisfy these two requirements.

Huygens' Major Premise: Sufficiency

In addition to the key notions of primary and secondary expanding spherical wavelets, a somewhat implicit supposition required by Huygens principles is that

[12] Max Karl Ernst Ludwig Planck (1858–1947) was a German theoretical physicist who is viewed as the founding father of quantum theory.

knowledge of the state of a spherical wavelet at a particular instant in time is sufficient to determine the state of the wavelet at any prior or future instant. This sufficiency requirement allows one to characterize a system back in time to the instant at which impulsive wavelet energy is first introduced into the system through a point singularity [35].

More exactly, Huygens postulated that the wave motion state at a time instant $t = t_1$ caused by a disturbance at the initial instant $t = t_o$, can be calculated by determining the wave motion state at some intermediate instant $t = t'$ [35]. In a certain sense, from a signal processing perspective, this premise can be viewed as a sufficiency or Markovian[13] property. The utility of this premise is that information about the field on the surface of a sphere with radius

$$r = c \cdot (t' - t_o)$$

can be substituted without loss of information for analysis about the point (spatial) singularity at the prior impulsive time origin $t = t_o$. Hence, the difficulties of dealing with both temporal and spatial singularities are avoided.

Sufficiency allows macro-analysis over the Huygens spherical wavelet's surface to replace micro-analysis of the point singularity that caused the wavelet.

An important requirement of this aspect of Huygens' wave model is that knowledge of the wave field state must be over a fully *closed surface* in three dimensional space. This closed surface is often taken to be spherical, but application of generalized macro-analytic forms of the Huygens principles allows this closed surface to be arbitrary in shape.

Interior, Exterior and Induced Surface Sources

Wave field analysis based on Huygens' concepts can be formulated as a study of equivalent field sources induced on an arbitrarily shaped closed surface. These equivalent surface sources are induced when the closed surface is placed into an extant field caused by a given configuration of original field sources. These original field sources may be either inside or outside of the positioned closed surface. The key utility of this approach is an analysis that shows that the induced effect on the surface is sufficient to represent the original field sources that induced the effect.

We shall explore scenarios comprised of both interior and exterior sources relative to an arbitrarily shaped closed surface in our study of Huygens principles applied to both electromagnetic and acoustic wave fields.

Huygens' Major Premise: Interior Null Field

Consider first the case where the interior of a closed surface is free of sources and is placed into an existing field. Huygens principles require that the effect of all secondary (induced) sources over the entire closed surface collectively maintains an external field equivalent to the given field and equates to a null field within the closed surface. Once the induced surface sources are understood, these surface sources can then be viewed as causing an equivalent external field and the original field sources can be removed from the analysis of the given system.

[13] Andrei Andreyevich Markov (1856–1922) was a Russian mathematician.

Conversely, a closed surface can be selected that encloses the original field sources. The induced surface effect can be used to describe the original field as extant within the closed volume and the induced effect on the exterior of the surface must equate to a null field [35].

Analysis Based on the Huygens Principles

Beyond the intuitive arguments, these principles (sufficiency and equivalently induced extant and null fields) are strict analytic requirements for a candidate wave field model to conform to the Huygens principles. We shall seek analysis for both electromagnetic vectors wave fields and acoustic scalar wave fields that satisfy these two analytic requirements.

11.3 Huygens Principles and Vector Electromagnetic Fields

Huygens principles are proven to rigourously hold in three dimensional formulations of electromagnetic wave fields. To see this, consider time-space domain vector expressions of electric and magnetic field intensities,

$$\begin{aligned} \boldsymbol{E}(t,\boldsymbol{x}) &= E_x\hat{\mathbf{i}} + E_y\hat{\mathbf{j}} + E_z\hat{\mathbf{k}}, \\ \boldsymbol{H}(t,\boldsymbol{x}) &= H_x\hat{\mathbf{i}} + H_y\hat{\mathbf{j}} + H_z\hat{\mathbf{k}}. \end{aligned} \tag{11.2}$$

Each component of $\boldsymbol{E}$ and $\boldsymbol{H}$ independently satisfies its own scalar wave motion equation in the time-space domain. However, when $\boldsymbol{E}$ and $\boldsymbol{H}$ are analyzed separately within the context of a Huygens wave field problem, the secondary wavelets that arise out of these independent scalar descriptions of $\boldsymbol{E}$ and $\boldsymbol{H}$ do not collectively satisfy Maxwell's equations [91, 851]. Scalar and separate analysis of the electric and magnetic components of a vector wave field is not fully satisfactory in the context of a Huygens wave theory framework.

We now show, however, that Maxwell's equations do satisfy Huygens' wave field requirements when an analysis of $\boldsymbol{E}$ and $\boldsymbol{H}$ is performed jointly in vector form [35].

Analytic Expression of Huygens Principles for Vector Electromagnetic Fields

Start by recalling Maxwell equations,

$$\begin{aligned} \nabla \times \boldsymbol{H} - \frac{1}{c}\partial_t\{\boldsymbol{D}\} &= \frac{4\pi}{c}\boldsymbol{J}, \\ \nabla \times \boldsymbol{E} + \frac{1}{c}\partial_t\{\boldsymbol{B}\} &= 0, \end{aligned}$$

where

$$\begin{aligned}\nabla \cdot \boldsymbol{B} &= 0,\\ \nabla \cdot \boldsymbol{D} &= 4\pi\rho,\\ \boldsymbol{J} &= \sigma \boldsymbol{E},\\ \boldsymbol{D} &= \epsilon \boldsymbol{E},\\ \boldsymbol{B} &= \mu \boldsymbol{H}.\end{aligned}$$

Vector Fields Inside and Outside a Closed Surface

Using an approach [561] suggested by Larmor[14], consider a vector field analysis in and around a closed surface that is placed into an existing electromagnetic field. This closed surface is assumed empty of field sources. We must find a description of *induced equivalent surface sources* that can replace the external original field sources in the sense that they do not change the given vector field internal to the closed surface, but which equate to a null field exterior to the closed surface (once the original sources are removed).

We proceed to describe the induced equivalent surface sources with a joint formulation based on Maxwell equations.

Electromagnetic Equivalent Surface Sources

Consider an electromagnetic field in a source free volume V bounded by the closed surface S. The sources that cause the given field are outside of V. At a point $\boldsymbol{s}$ on S, we decompose the vector field into tangential components $\boldsymbol{E}_t$ and $\boldsymbol{H}_t$, and into inward normal components $\boldsymbol{E}_n$ and $\boldsymbol{H}_n$. Because of the analytic objective of an external null field and an internal equivalent field, the fields $\boldsymbol{E}$ and $\boldsymbol{H}$ must be discontinuous across S.

Induced Tangential Magnetic and Normal Electric Field Sources

There must be an electric surface current just on the inside surface of S at $\boldsymbol{s}$ of a certain density to produce the required tangential magnetic field $\boldsymbol{H}_t$. The required directed electric current sheet density in the differential element dS at $\boldsymbol{s}$ is

$$\boldsymbol{J} = c \cdot \frac{1}{4\pi} \hat{\mathbf{n}} \times \boldsymbol{H}.$$

To simplify vector notations, assume we are dealing with a point where the normal is aligned with z and $\boldsymbol{H}_t$ is aligned with y. In which case,

$$\begin{aligned}\boldsymbol{H}_t &= H_y\,\hat{\mathbf{j}},\\ \hat{\boldsymbol{n}} &= \hat{\mathbf{k}}\end{aligned}$$

and the current sheet that produces a surface charge distribution that gives the required discontinuity in $\boldsymbol{D}_n$ is

$$\boldsymbol{J} = -c \cdot \frac{1}{4\pi} H_y \hat{\mathbf{i}}.$$

[14] Sir Joseph Larmor (1857–1942) was an Irish mathematician.

By Maxwell's equations, the increase of charge density per unit of time

$$\partial_t\{\rho\}$$

on the differential (tangential planar) surface element

$$dS = dxdy$$

that touches S at $\boldsymbol{s}$ is

$$\partial_t\{\rho\} \cdot dxdy = \frac{1}{4\pi} \cdot \partial_t\{D_z\} \cdot dxdy$$

and the electric surface charge density is

$$\begin{aligned} \rho &= \frac{1}{4\pi} \cdot D_z, \\ &= \frac{1}{4\pi} \hat{\mathbf{n}} \cdot \boldsymbol{D}. \end{aligned}$$

This produces the required normal discontinuity $\boldsymbol{D}_n$ in the electric flux density.

Induced Normal Magnetic and Tangential Electric Field Sources

A similar analysis for the magnetic field shows that the magnetic surface current density is

$$\boldsymbol{K} = -c \cdot \frac{1}{4\pi} \hat{\mathbf{n}} \times \boldsymbol{D}$$

and the required magnetic surface charge density is

$$\tau = \frac{1}{4\pi} \hat{\mathbf{n}} \cdot \boldsymbol{B}.$$

These produce the discontinuities in the normal component $\boldsymbol{H}_n$ and the tangential component $\boldsymbol{E}_t$ required by Huygens principles.

Interior Field from Electric Equivalent Surface Sources

In free space, we have[15] $\boldsymbol{B} = \boldsymbol{H}$ and $\boldsymbol{D} = \boldsymbol{E}$ [89]. Hence, the electromagnetic vector field at the interior point $\boldsymbol{r}$ due to the surface current $\boldsymbol{J}$ and static charge density ρ over the entire interior surface is

$$\begin{aligned} \overrightarrow{\mathbb{D}}_+(t, \boldsymbol{r}) &= -\nabla \mathbb{D}_n(t, \boldsymbol{r}) - \frac{1}{c} \partial_t\{\overrightarrow{\mathbb{B}}_t(t, \boldsymbol{r})\}, \\ \overrightarrow{\mathbb{B}}_+(t, \boldsymbol{r}) &= \nabla \times \overrightarrow{\mathbb{B}_t}(t, \boldsymbol{r}), \end{aligned}$$

[15] Some have suggested that instead of describing the behavior of electromagnetic fields in matter via $\epsilon = D/E$ and $\mu = B/H$, that the differences $\boldsymbol{D} - \boldsymbol{E}$ and $\boldsymbol{B} - \boldsymbol{H}$ have simpler physical significance [89].

where $\mathbb{D}_n(t, \boldsymbol{r})$ is the electric scalar potential and $\overrightarrow{\mathbb{B}_t}(t, \boldsymbol{r})$ is the magnetic vector potential [88]. These are defined by

$$\begin{aligned}
\mathbb{D}_n(t, \boldsymbol{r}) &= \int_\circ \frac{1}{r} \cdot \rho(t - t_s, \boldsymbol{r} - \boldsymbol{s}) \cdot d\boldsymbol{s}, \\
&= \int_\circ \frac{1}{4\pi} \cdot \frac{1}{r} \cdot [\hat{\mathrm{n}} \cdot \boldsymbol{D}(t - t_s, \boldsymbol{r} - \boldsymbol{s})] \cdot d\boldsymbol{s}, \\
\overrightarrow{\mathbb{B}}_t(t, \boldsymbol{r}) &= \int_\circ \frac{1}{r} \cdot \boldsymbol{J}(t - t_s, \boldsymbol{r} - \boldsymbol{s}) \cdot d\boldsymbol{s}, \\
&= c \int_\circ \frac{1}{4\pi} \cdot \frac{1}{r} \cdot [\hat{\mathrm{n}} \times \boldsymbol{B}(t - t_s, \boldsymbol{r} - \boldsymbol{s})] \cdot d\boldsymbol{s},
\end{aligned}$$

where

$$r = |\boldsymbol{r} - \boldsymbol{s}|$$

is the distance from the interior field point $\boldsymbol{r}$ to the surface point $\boldsymbol{s}$ on S and where

$$t_s = \frac{1}{c}|\boldsymbol{r} - \boldsymbol{s}|$$

is the corresponding time delay of propagation [35, 89].

Interior Field from Magnetic Equivalent Surface Sources

The dual expressions derived for the associated magnetic fields are

$$\begin{aligned}
\overrightarrow{\mathbb{D}}_-(t, \boldsymbol{r}) &= -\nabla \times \overrightarrow{\mathbb{D}_t}(t, \boldsymbol{r}), \\
\overrightarrow{\mathbb{B}}_-(t, \boldsymbol{r}) &= -\nabla \mathbb{B}_n(t, \boldsymbol{r}) - \frac{1}{c}\partial_t\{\overrightarrow{\mathbb{D}_t}\}(t, \boldsymbol{r}),
\end{aligned}$$

where $\mathbb{B}_n(t, \boldsymbol{r})$ is the magnetic scalar anti-potential and $\overrightarrow{\mathbb{D}_t}(t, \boldsymbol{r})$ is the electric vector anti-potential [88]. These are defined by

$$\begin{aligned}
\overrightarrow{\mathbb{D}}_t(t, \boldsymbol{r}) &= \int_\circ \frac{1}{r} \cdot \boldsymbol{K}(t - t_s, \boldsymbol{r} - \boldsymbol{s}) \cdot d\boldsymbol{s}, \\
&= -c \int_\circ \frac{1}{4\pi} \cdot \frac{1}{r} \cdot [\hat{\mathrm{n}} \times \boldsymbol{D}(t - t_s, \boldsymbol{r} - \boldsymbol{s})] \cdot d\boldsymbol{s}, \\
\mathbb{B}_n(t, \boldsymbol{r}) &= \int_\circ \frac{1}{r} \cdot \tau(t - t_s, \boldsymbol{r} - \boldsymbol{s}) \cdot d\boldsymbol{s}, \\
&= \int_\circ \frac{1}{4\pi} \cdot \frac{1}{r} \cdot [\hat{\mathrm{n}} \cdot \boldsymbol{B}(t - t_s, \boldsymbol{r} - \boldsymbol{s})] \cdot d\boldsymbol{s}.
\end{aligned}$$

Total Interior Field from Equivalent Surface Sources

The total field at each time-space point $(t, \boldsymbol{r})$ is the combination of both potential and anti-potential elements,

$$\begin{aligned}
\boldsymbol{D}(t, \boldsymbol{r}) &= \overrightarrow{\mathbb{D}}_+(t, \boldsymbol{r}) + \overrightarrow{\mathbb{D}}_-(t, \boldsymbol{r}), \\
\boldsymbol{B}(t, \boldsymbol{r}) &= \overrightarrow{\mathbb{B}}_+(t, \boldsymbol{r}) + \overrightarrow{\mathbb{B}}_-(t, \boldsymbol{r}),
\end{aligned}$$

and hence,

$$\boldsymbol{D}(t,\boldsymbol{r}) = -\nabla\mathbb{D}_n(t,\boldsymbol{r}) - \frac{1}{c}\partial_t\{\overrightarrow{\mathbb{B}}_t(t,\boldsymbol{r})\} - \nabla\times\overrightarrow{\mathbb{D}}_t(t,\boldsymbol{r}), \tag{11.3}$$

$$\boldsymbol{B}(t,\boldsymbol{r}) = -\nabla\mathbb{B}_n(t,\boldsymbol{r}) - \frac{1}{c}\partial_t\{\overrightarrow{\mathbb{D}}_t(t,\boldsymbol{r})\} + \nabla\times\overrightarrow{\mathbb{B}}_t(t,\boldsymbol{r}). \tag{11.4}$$

Expressions (11.3) and (11.4), as derived in [561, 935], and similar expressions as derived in [915, 916] and in [355, 642, 930] (the latter only requires tangential field components), provide analytic descriptions of Huygens principles for vector electromagnetic fields consistent with Maxwell's equations. The required field interior to the closed surface is produced and all surface sources combine to equate to a null field on the exterior of the surface [35]. These equations establish that the required Huygens principles rigorously hold for joint formulations of electromagnetic vector fields based on Maxwell's equations.

We share another electromagnetic vector formulation of Huygens principles, but where the electric and magnetic parts are combined into a single complex expression. This formulation is wonderfully symmetric and concise [35, 929, 930].

Huygens Principles in Concise Maxwell Form

If we define [50]

$$\boldsymbol{M}(t,\boldsymbol{r}) = \boldsymbol{D}(t,\boldsymbol{r}) + j\boldsymbol{B}(t,\boldsymbol{r}),$$

then Maxwell's equations for free space take a simple notational form,

$$\begin{aligned}\nabla\times\boldsymbol{M}(t,\boldsymbol{r}) &= j\frac{1}{c}\partial_t\{\boldsymbol{M}(t,\boldsymbol{r})\},\\ \nabla\cdot\boldsymbol{M}(t,\boldsymbol{r}) &= 0.\end{aligned}$$

If we define,

$$\begin{aligned}\mathbb{M}_n(t,\boldsymbol{r}) &= \mathbb{D}_n(t,\boldsymbol{r}) + j\mathbb{B}_n(t,\boldsymbol{r}),\\ \overrightarrow{\mathbb{M}_t}(t,\boldsymbol{r}) &= \overrightarrow{\mathbb{B}}_t(t,\boldsymbol{r}) + j\overrightarrow{\mathbb{D}}_t(t,\boldsymbol{r}),\end{aligned}$$

where

$$\begin{aligned}\mathbb{M}_n(t,\boldsymbol{r}) &= \int_\circ \frac{1}{4\pi}\cdot\frac{1}{r}\cdot[\hat{\mathbf{n}}\cdot\boldsymbol{M}]\cdot d\boldsymbol{s},\\ \overrightarrow{\mathbb{M}_t}(t,\boldsymbol{r}) &= c\int_\circ \frac{1}{4\pi}\cdot\frac{1}{r}\cdot[\hat{\mathbf{n}}\times\boldsymbol{M}]\cdot d\boldsymbol{s},\end{aligned}$$

then

$$\boldsymbol{M}(t,\boldsymbol{r}) = -\nabla\mathbb{M}_n(t,\boldsymbol{r}) - \frac{1}{c}\cdot\partial_t\left\{\overrightarrow{\mathbb{M}_t}(t,\boldsymbol{r})\right\} + j\nabla\times\overrightarrow{\mathbb{M}_t}(t,\boldsymbol{r})$$

satisfies Huygens principles as required inside and outside[16] of S [35].

[16] Asymptotic bounding conditions are required when the analysis seeks to produce an interior null field and equate the exterior field to the given extant field [35].

Huygens-Maxwell Approaches in SAR Analysis

Although elegant, Huygens-Maxwell approaches present both opportunities and challenges to SAR analysis and image formation algorithm development. Maxwell equation based formulations remain an open opportunity in developing advanced models for SAR analysis and inversion methods [979].

Scalar Component Analysis of Electromagnetic Fields

We set aside the vector formulation of Maxwell's equations and continue next to investigate analysis of scalar components of electromagnetic fields. We turn to see what lessons can be learned from an analogous mechanical study of scalar acoustic wave fields. The analysis of acoustic waves shall prove enlightening. Mechanical scalar acoustic field analysis based on Helmholtz and Kirchhoff integral formulations provides analogous physical interpretations of scalar elements of electromagnetic wave fields (in terms of the potential and kinetic mechanics of acoustical wave fields).

11.4 Huygens Principles and Scalar Acoustic Fields

Huygens principles for any scalar wave phenomenon can be analyzed and expressed in terms of analogies with the physics of sound waves of small amplitude [35]. A study of acoustic waves [908] provides a mechanical tutorial on scalar wave motion that can be used as an analogy for scalar components of a (non-mechanical) electromagnetic vector wave field.

Acoustic Waves of Small Amplitude

Consider the sound (pressure) wave velocity vector

$$\boldsymbol{v}(t, \boldsymbol{r})$$

within a gaseous medium. An *acoustic field potential* $ss(t, \boldsymbol{r})$ exists where

$$\boldsymbol{v}(t, \boldsymbol{r}) = -\nabla \{ss(t, \boldsymbol{r})\}.$$

The pressure wave velocity vector is determined by the gradient of the acoustic field potential.

Perturbations of an Acoustic Wave Medium

Fresnel realized that an acoustic wave medium is not fully characterized by just the kinematic particle velocity attribute of longitudinal pressure wave motion – it is necessary to add a *compressibility* parameter descriptive of local density changes (the local potential energy reservoir) within the gaseous medium of propagation [70]. Hence, Fresnel modeled the state of the acoustic medium by a pair of parameters, the pressure

$$p(t, \boldsymbol{r})$$

and density

$$\rho(t, \boldsymbol{r}).$$

Variabilities of the density $\rho(t, \boldsymbol{r})$ are modeled as perturbations

$$\kappa(t, \boldsymbol{r})$$

described as a compression attribute. The variable density of the acoustic media is modeled by

$$\rho(t, \boldsymbol{r}) = \rho_o[1 + \kappa(t, \boldsymbol{r})],$$

where ρ_o is the nominal density of the medium at rest. Assuming time independence of the acoustic scattering response, local deviations or perturbations of an acoustic wave medium can be modeled by

$$\begin{aligned} g_\kappa(\boldsymbol{r}) &= \frac{\kappa(\boldsymbol{r}) - \kappa_o}{\kappa_o}, \\ g_\rho(\boldsymbol{r}) &= \frac{\rho(\boldsymbol{r}) - \rho_o}{\rho_o}, \end{aligned}$$

where the nominal speed of acoustic wave propagation is

$$v_o = \frac{1}{\sqrt{\rho_o \kappa_o}}$$

and where κ_o is the nominal compression state of the medium at rest. A model of acoustic waves of small amplitude requires $g_\kappa(\boldsymbol{r}) << 1$ and $g_\rho(\boldsymbol{r}) << 1$.

Acoustic Wave Motion Equation in Time and Space

Coupled mechanical equations for acoustic waves of small amplitude are

$$\begin{aligned} \nabla \cdot \{\boldsymbol{v}(t, \boldsymbol{r})\} + \partial_t \{\kappa(t, \boldsymbol{r})\} &= 0, \\ \nabla \{\kappa(t, \boldsymbol{r})\} + \frac{1}{v_o^2} \partial_t \{\boldsymbol{v}(t, \boldsymbol{r})\} &= 0. \end{aligned}$$

These equations of mechanical analysis lead to a familiar equation of wave motion for the (acoustic) field potential as expressed in time and space [35],

$$\nabla^2 \{ss(t, \boldsymbol{r})\} - \frac{1}{v_o^2} \partial_t^2 \{ss(t, \boldsymbol{r})\} = 0.$$

When compared to the scalar wave motion equation of just one component of the electric (or magnetic) field (3.5), application of models of acoustic wave motion to describe scalar elements of electromagnetic vector wave propagation is merely a matter of changing the terminology [35].

Wave State of Spherically Expanding Acoustic Wave Fields

Fresnel's work further characterized an acoustic field in terms of harmonic time-space domain wave fields given by (the real part of)

$$ss(t, \boldsymbol{r}) = \frac{1}{r} \cdot \exp[jk_\omega(ct - r)]. \tag{11.5}$$

The field described by (11.5) is a dynamic spherical harmonic wave field. If the acoustic spherical wave described by (11.5) reaches the shell S of radius r at time t, the wave continues to propagate outwardly and expands into a larger shell S' of radius r' at the time instant,

$$t' = t + \frac{1}{c}(r' - r).$$

Using Huygens principles, Fresnel reasoned that the harmonic wave field at S' and t' is fully (and only) dependent on the wave state of the inner shell S at time t. Issues related to the point source inside of S, which originally caused the wave at S at some prior time instant $t_o < t$, can be ignored in the analysis of the spherical wave S' at time t'.

By relating the pair of velocity and compression parameters, Fresnel strived to show that all equivalent sources on the spherical surface lead to an internal field that is null.

Fresnel's attempt was not completely successful. Fresnel had to make seemingly strange and arbitrary assumptions in his analysis that required the following:

- primary to secondary wave amplitudes are conveyed at the ratio of $1 : \lambda$ and
- secondary waves oscillate a quarter of a period ahead of the primary wave.

These unexpected mathematical oddities led many at that time to view Fresnel's analytical development as having no true physical basis.

Helmholtz Extension of Fresnel's Analysis

Fresnel's work eventually led to correct (i.e., consistent with the Huygens principles) analytical results in the work of others. An extension developed by Helmholtz was key. Helmholtz's approach was based on a surface and volume integral analysis over a closed surface that sought to describe the *static* harmonic spherical field with a distribution of equivalent surface sources. The wavelength related scaling and the phase advance, seemingly arbitrarily introduced in Fresnel's analysis, were correctly motivated by the physics based macro-analysis of Helmholtz.

Helmholtz's Formulation for Dynamic Harmonic Acoustic Wave Fields

If an acoustic wave field consists of a single tone harmonic signal, a solution of the time-space wave motion equation is decomposable into the form,

$$ss(t, \boldsymbol{r})^{\text{dynamic}}_{\text{harmonic}} = \exp(jk_{\omega_c}ct) \times Ss(\omega_c, \boldsymbol{r})^{\text{static}}.$$

Helmholtz showed in 1859 that the static element

$$Ss(\omega_c, \boldsymbol{r}) = Ss(\omega_c, \boldsymbol{r})^{\text{static}}$$

of the time harmonic wave field $ss(t, \boldsymbol{r})^{\text{dynamic}}_{\text{harmonic}}$ satisfies an important frequency-space domain equation of wave motion,

$$\nabla^2 \{Ss(\omega_c, \boldsymbol{r})\} + k_\omega^2 \cdot Ss(\omega_c, \boldsymbol{r}) = 0. \tag{11.6}$$

Using a static harmonic assumption, Helmholtz developed an integral based analysis of a distribution of acoustic harmonic field sources over a closed surface. This approach to Fresnel's problem proved to fully satisfy the requirements of the Huygens principles, i.e., it was shown that secondary wave sources propagate only outwardly and leave an interior null field.

Green's Formula – Relating Surface and Volume Integrals

To summarize Helmholtz's approach, we first describe a relationship between volume and surface integrals for wave fields called *Green's equivalent layer formula,* which is given by

$$\int_{\bullet} [Ss(\omega_c, \boldsymbol{v}) \cdot \nabla^2\{Ww(\omega_c, \boldsymbol{v})\} - Ww(\omega_c, \boldsymbol{v}) \cdot \nabla^2\{Ss(\omega_c, \boldsymbol{v})\}]\, dv$$
$$= \int_{\circ} [Ss(\omega_c, \boldsymbol{s}) \cdot \partial_{\hat{\mathrm{n}}}\{Ww(\omega_c, \boldsymbol{s})\} - Ww(\omega_c, \boldsymbol{s}) \cdot \partial_{\hat{\mathrm{n}}}\{Ss(\omega_c, \boldsymbol{s})\}]\, ds, \tag{11.7}$$

where $\circ$ represents a closed surface S bounding the volume V represented by $\bullet$ and where $\partial_{\hat{\mathrm{n}}}\{\cdot\}$ is differentiation along the outward normal to the surface S.

A surface integral in (11.7) is expressed as

$$\int_{\circ} \{\cdot\}\, ds$$

and a volume integral as

$$\int_{\bullet} \{\cdot\}\, dv.$$

Green's identity (11.7) is valid when

$$Ss(\omega_c, \boldsymbol{r})$$

and

$$Ww(\omega_c, \boldsymbol{v})$$

are regular in the sense that their first and second order partial derivatives are continuous within the volume V and on the surface S [35].

Helmholtz Integral Equation

When both $Ss(\omega_c, \boldsymbol{s})$ and $Ww(\omega_c, \boldsymbol{s})$ satisfy Helmholtz's frequency-space wave motion equation (11.6), then Green's equivalent layer formula (11.7) reduces to just a surface integral that expresses the field at some point $\boldsymbol{r}$ created by the distribution of sources on the volume's surface,

$$Ss(\omega_c, \boldsymbol{r})^{\text{induced}} \tag{11.8}$$
$$= \int_{\circ} [Ss(\omega_c, \boldsymbol{s}) \cdot \partial_{\hat{\mathrm{n}}}\{Ww(\omega_c, \boldsymbol{r} - \boldsymbol{s})\} - Ww(\omega_c, \boldsymbol{r} - \boldsymbol{s}) \cdot \partial_{\hat{\mathrm{n}}}\{Ss(\omega_c, \boldsymbol{s})\}]\, ds.$$

Helmholtz Integral Equation with Fresnel Wave Fields

In particular, consider the case where $Ww(\omega_c, \boldsymbol{r})$ is Fresnel's static harmonic spherical wave field (i.e., the Green's function of Helmholtz's partial differential wave motion equation),

$$\begin{aligned} Ww(\omega_c, \boldsymbol{r}-\boldsymbol{s}) &= Hh(\omega_c, \boldsymbol{r}-\boldsymbol{s}), \\ &= \frac{1}{4\pi}\cdot\frac{1}{r}\cdot\exp(-jk_\omega r). \end{aligned} \tag{11.9}$$

With substitution of Fresnel's harmonic wave field (11.9) into (11.8), we obtain

$$\begin{aligned} &Ss(\omega_c, \boldsymbol{r})^{\text{induced}} \\ &= \oint [Ss(\omega_c, \boldsymbol{s})\cdot\partial_{\hat{\mathrm{n}}}\{Hh(\omega_c, \boldsymbol{r}-\boldsymbol{s})\} - Hh(\omega_c, \boldsymbol{r}-\boldsymbol{s})\cdot\partial_{\hat{\mathrm{n}}}\{Ss(\omega_c, \boldsymbol{s})\}]\, d\boldsymbol{s}. \end{aligned} \tag{11.10}$$

This is *Helmholtz's integral wave equation.*

Fields Expressed by Helmholtz's Integral Equation

We consider four situations for evaluating (11.10). These four cases are determined by whether the point $\boldsymbol{r}$ of the induced field is inside or outside of S and whether the sources of a given "original" acoustic field

$$Ss(\omega_c, \boldsymbol{r})^{\text{original}}$$

are inside or outside of S.

Equivalent Surface Sources of Exterior Field Sources

Consider first the case where all original field sources are outside the closed surface S. It can be shown (see [35]) that the induced surface sources equate, as evaluated by (11.10), to a null acoustic field on the outside of the surface and preserve the original acoustic field on the interior of S:

$$\begin{aligned} Ss(\omega_c, \boldsymbol{r})^{\text{induced}} &= 0 \text{ for } \boldsymbol{r} \text{ outside } S, \\ Ss(\omega_c, \boldsymbol{r})^{\text{induced}} &= Ss(\omega_c, \boldsymbol{r})^{\text{original}} \text{ for } \boldsymbol{r} \text{ inside } S. \end{aligned}$$

The original field sources exterior to the closed surface have been replaced by the combined effect of all induced surface sources. The original exterior sources can now be assumed not to exist in further analysis within the closed surface. Knowledge of the field $Ss(\omega_c, \boldsymbol{s})$ over the entirety of the closed surface S is sufficient to say that the induced field in the interior of S is identical to the field produced by all original sources that were outside of S. This model allows the singularities external to S to be ignored and replaced with an integral analysis of the equivalent sources as distributed over the surface S.

Equivalent Surface Sources of Interior Field Sources

Next, consider the case where the original sources lie inside the closed surface. For this situation, additional regularity conditions must be satisfied outside the closed surface to ensure a system of expanding spherical waves exterior to the surface (see [35] for details). In this case, the induced field exterior to the closed surface is

$$\begin{aligned} Ss(\omega_c, \boldsymbol{r})^{\text{induced}} &= Ss(\omega_c, \boldsymbol{r})^{\text{original}} \text{ for } \boldsymbol{r} \text{ outside } S, \\ Ss(\omega_c, \boldsymbol{r})^{\text{induced}} &= 0 \text{ for } \boldsymbol{r} \text{ inside } S. \end{aligned}$$

Interior to the closed surface, the induced field is null and the exterior field is the same as that produced by the original field sources contained within the closed surface. Hence, the induced effect on the surface is sufficient to represent the original field outside the closed surface. The original field sources can be discarded and the closed surface can be assumed empty in subsequent analysis.

Kirchhoff's Integral Formula for Time-Space Domain Wave Field Models

Helmholtz's integral formula (11.10) can be shown to be a special case of a more general formula developed by Kirchhoff, which is applicable to scalar fields of any structure (not just static harmonic wave fields). In its most general form, the Kirchhoff integral formula expresses a general wave field at an arbitrary field point by integrative measures of the field in terms of boundary values (normal derivatives) over a closed surface and by integrative measures of the field within the volume [66, 665, 812]. Kirchhoff's integral formula for scalar wave fields has an easily understood interpretation in terms of the mechanics of acoustic field analysis (see [35]).

A solution (not necessarily harmonic) to the inhomogeneous partial differential wave motion equation

$$\nabla^2 \{ww(t, \boldsymbol{r})\} - \partial_t^2 \{ww(t, \boldsymbol{r})/c^2\} = q(t, \boldsymbol{r}).$$

within a volume V is given by a surface integral of boundary values on the surface S of V and by a volume integral within V involving both sources and initial values [665, 813]. With certain assumptions, a particular form of this solution in the time-space domain is given by

$$\begin{aligned} ww(t, \boldsymbol{r}) &= \int_{\mathbb{T}} \int_{\bigcirc} hh(t-\tau, \boldsymbol{r}-\boldsymbol{s}) \cdot \partial_{\hat{\mathrm{n}}} \{ww(\tau, \boldsymbol{s})\} \cdot d\tau ds \\ &\quad - \int_{\mathbb{T}} \int_{\bigcirc} ww(\tau, \boldsymbol{s}) \cdot \partial_{\hat{\mathrm{n}}} \{hh(t-\tau, \boldsymbol{r}-\boldsymbol{s})\} \cdot d\tau ds, \quad (11.11) \end{aligned}$$

with $\hat{\mathrm{n}}$ as the outward normal vector with respect to the surface, with $\mathbb{T}$ as the time manifold of integration and with

$$hh(t - t_o, \boldsymbol{r} - \boldsymbol{s}) = \frac{1}{4\pi} \cdot \frac{1}{r} \cdot \delta\left[t - t_o - t_d(\boldsymbol{r} - \boldsymbol{s})\right]$$

as the expanding Huygens spherical wavelet in the time-space domain. Equation (11.11) is *Kirchhoff's integral formula* and is a statement of Huygens principles for scalar fields and general waveforms described in the time-space domain.

Kirchhoff's integral expression can be viewed as a three dimensional convolution of the observed wave field or its normal derivative on the enclosing surface with Huygens' expanding spherical wavelet or its normal derivative [91].

One key value of Kirchhoff's formula is that it is not constrained by the harmonic wave field context of Helmholtz's frequency-space analysis [35].

Kirchhoff's formula allows the wave field on one side of a closed surface at a time instant to be determined by the values of the field and its first derivative at any previous time instant.

Actually, knowledge of a field and its derivative on this surface may be redundant. Knowledge of the field alone for all time is sufficient to determine the first derivative, or knowledge of the initial values of the field and its first derivative are sufficient to determine the field everywhere. This follows from a theorem that establishes that if the field and its first normal derivative are given initially and the field is given for all time on the closed surface, then the first normal derivative can be deduced for all time. The converse also holds (the field can be deduced from knowledge of the normal derivative) [35].

Kirchhoff's Integral Formula and Maxwell's Equations

Kirchhoff's integral formula can be formally applied to scalar components of an electric or magnetic field, but Kirchhoff's integral cannot be used for vector analysis of an electromagnetic vector field [35, 89]. If the six rectilinear components of an electromagnetic field are each described with Kirchhoff's formula, then the six wave motion equations so obtained do not, in general, collectively satisfy Maxwell's equations [91].

11.5 Terrain Scattered Differential Wave Motion Equations

We now set aside our discrete model of individual scatterers and give an overview of one approach for modeling a continuous distribution of electromagnetic scatterers. We present differential wave motion equations descriptive of electromagnetic wave fields scattered from a continuous terrain function [85]. Similar to the approach presented for modeling scattering in acoustics, these differential formulations model scattering phenomenology as perturbations of electromagnetic permittivity. We explore how the continuous electromagnetic scattering model fits into our expanding wave theory framework.

Perturbation Models in Scattered Wave Theory

The study of wave motion in computed imaging algorithm development can be framed as primarily an investigation of the effect of perturbations induced by a scene's or object's structure on an incident wave field [177, 325, 796, 837, 862].

There are more than twenty scattering models based on perturbation models identifiable in the literature [305]. The long list of published options comes from the importance of modeling wave scattering phenomena in many active sensor imaging applications such as medical imaging, optics, acoustics, geophysics, communications, terrestrial and extraterrestrial remote sensing, plasma physics [798] and so on [480, 501, 696].

Some models use single or doublet singularities within the scattering region [35,178,704,705,1009]. Other approaches use the tools of micro-local analysis [706] or Fourier integral operator methods to describe microwave scattering caused by weakly dispersive foliage layers [182, 705].

Matter Perturbations of Permittivity and Permeability

In substances that are essentially dielectrics, the electric and magnetic properties are completely determined by the permittivity ϵ and permeability μ of the matter encountered by a propagating electromagnetic wave. For most substances, the magnetic wave propagation permeability μ does not significantly deviate from μ_o, hence the scattered wave field is modeled as primarily caused by small perturbations ϵ of the otherwise free space homogenous permittivity ϵ_o [89].

Speed of Propagation in Space with Perturbed Permittivity

When electromagnetic waves propagate through matter, the speed of wave propagation changes as the permittivity varies from that of free space. For example, the speed of electromagnetic wave propagation at or within the ground scene at $\boldsymbol{x}$ is described by

$$c(\boldsymbol{x}) = \frac{1}{\sqrt{\epsilon(\boldsymbol{x})\mu_o}}.$$

Spatial variations in the speed $c(\boldsymbol{x})$ of in situ electromagnetic waves through the terrain function are modeled as caused by spatial variations of the terrain permittivity $\epsilon(\boldsymbol{x})$.

Incident Wave Motion Through the Ground Scene

The SAR transmitter at $\boldsymbol{u}$ illuminates the ground scene at $\boldsymbol{x}$ during signal transmit mode with an incident wave field indicated by

$$ss(t, \boldsymbol{x} - \boldsymbol{u})^{\text{in}}.$$

If we assume $\boldsymbol{u} = (0, 0, 0)$, then a general model of incident wave propagation through ground scene matter with a spatially dependent wave speed $c(\boldsymbol{x})$ is given by

$$\nabla^2 \left\{ ss(t, \boldsymbol{x})^{\text{in}} \right\} - \partial_t^2 \left\{ ss(t, \boldsymbol{x})^{\text{in}} / c(\boldsymbol{x})^2 \right\} = 0. \tag{11.12}$$

Total Wave Field within Scene

When the signal transmitted from the antenna impinges terrain matter, the composite electric field in the vicinity of each terrain point contains both an incident wave field element

$$ss(t, \boldsymbol{x})^{\text{in}}$$

and a scattered wave field element

$$ss(t,\boldsymbol{x})^{^{\text{sc}}}.$$

The resulting "total" or "in situ" wave field at the ground scene is modeled as the superposition of the incident and scattered fields,

$$ss(t,\boldsymbol{x})^{^{\text{to}}} = ss(t,\boldsymbol{x})^{^{\text{in}}} + ss(t,\boldsymbol{x})^{^{\text{sc}}}.$$

Clearly, in pulse echo imaging systems, the field that exists at the receiver during waveform transmit time intervals can be considered separately from the field that exists during echo receive time intervals. This is not the case with the in situ field scattered at or within the ground scene during the instances of electromagnetic wave interactions with matter. In that case, the incident and scattered fields are considered as coexistent for a period of time determined by the pulse duration [174, 257, 705, 707]. This model, based on the sum of the incident and scattered wave fields to describe the "total" field, assumes weak scattering and ignores possible multiple bounce interactions among the continuum of scatterers.

Total Wave Field Away from Scene

The total wave field outside or away from the scene is

$$\nabla^2\left\{ss(t,\boldsymbol{r})^{^{\text{to}}}\right\} - \partial_t^2\left\{ss(t,\boldsymbol{r})^{^{\text{to}}}/c^2\right\} = 0, \tag{11.13}$$

where $c(\boldsymbol{x})$ of (11.12) is replaced with c and $\boldsymbol{x}$ is replaced with $\boldsymbol{r}$ [173, 705]. The propagation speed of the total wave field away from the ground scene is the speed of light in free space.

Total and Incident Coupled Wave Motion Equations

We next view the incident wave field as a forcing function that drives the total wave field "response" to scene illumination. Hence, we rewrite the free space wave equation (11.13) by replacing the "0" with a forcing function given by (11.12),

$$\nabla^2\left\{ss(t,\boldsymbol{r})^{^{\text{to}}}\right\} - \partial_t^2\left\{ss(t,\boldsymbol{r})^{^{\text{to}}}/c^2\right\} = \nabla^2\left\{ss(t,\boldsymbol{r})^{^{\text{in}}}\right\} - \partial_t^2\left\{ss(t,\boldsymbol{r})^{^{\text{in}}}/c^2(\boldsymbol{x})\right\}. \tag{11.14}$$

Note the constant "c" on the left hand side and the spatially variant $c(\boldsymbol{x})$ on the right hand side of (11.14). A short exercise shows that

$$\nabla^2\left\{ss(t,\boldsymbol{r})^{^{\text{sc}}}\right\} - \partial_t^2\left\{ss(t,\boldsymbol{r})^{^{\text{sc}}}/c^2(\boldsymbol{x})\right\} = \underbrace{\left[\frac{1}{c^2} - \frac{1}{c^2(\boldsymbol{x})}\right]}_{g(\boldsymbol{x})} \cdot \partial_t^2\left\{ss(t,\boldsymbol{r})^{^{\text{to}}}\right\}. \tag{11.15}$$

We define in (11.15) the *continuous terrain scattering function* as

$$g(\boldsymbol{x}) = \frac{1}{c^2} - \frac{1}{c^2(\boldsymbol{x})},$$

where $\boldsymbol{x}$ represents a continuum of points within the terrain function. Hence, we obtain a differential wave motion equation

$$\nabla^2 \left\{ ss(t,\boldsymbol{r})^{\text{sc}} \right\} - \partial_t^2 \left\{ ss(t,\boldsymbol{r})^{\text{sc}} / c^2(\boldsymbol{x}) \right\} = g(\boldsymbol{x}) \cdot \partial_t^2 \left\{ ss(t,\boldsymbol{r})^{\text{to}} \right\} \tag{11.16}$$

driven by a continuous model of terrain scattering.

11.6 Terrain Scattered Integral Wave Motion Equations

An impinging electromagnetic wave is scattered by its interaction with terrain matter in a way that is well described by Huygens-Fresnel wave theory principles [797]. Using Huygens' wavelet, we can write (11.16) in the following time-space domain integral form [173, 199],

$$ss(t,\boldsymbol{r})^{\text{sc}} = \int_{\mathbb{G}} \int_{\mathbb{T}} g(\boldsymbol{x}) \cdot hh(t-\tau, \boldsymbol{r}-\boldsymbol{x}) \cdot \partial_\tau^2 \left\{ ss(\tau,\boldsymbol{x})^{\text{to}} \right\} \cdot d\tau d\boldsymbol{x}, \tag{11.17}$$

where $\mathbb{G}$ is the spatial manifold of integration (the ground scene) and where $\mathbb{T}$ is the temporal manifold of integration (fast-time dwell). The integral form (11.17) is obtained by multiplying (11.16) with $hh(t-\tau, \boldsymbol{r}-\boldsymbol{x})$ and by multiplying

$$\nabla^2 \left\{ hh(t-\tau, \boldsymbol{r}-\boldsymbol{x}) \right\} - \partial_t^2 \left\{ hh(t-\tau, \boldsymbol{r}-\boldsymbol{x}) / c^2 \right\} = \delta(t-\tau)\delta(\boldsymbol{r}-\boldsymbol{x})$$

with

$$ss(\tau,\boldsymbol{r})^{\text{in}},$$

and then by subtracting the two equations. The difference is temporally integrated over all time and spatially integrated over a large ball in space. Using a Green's identity as the radius of the ball goes to infinity, the integral over the asymptotic spatial derivatives vanishes and the result (11.17) remains [175].

Weak Scattering Approximation in Time-Space Integral Formulations

With a weak scattering or Born assumption, we can substitute the incident wave field $ss(\tau,\boldsymbol{x})^{\text{in}}$ for the total wave field $ss(\tau,\boldsymbol{x})^{\text{to}}$ on the right hand side of (11.17),

$$ss(t,\boldsymbol{r})^{\text{sc}} = \int_{\mathbb{G}} \int_{\mathbb{T}} g(\boldsymbol{x}) \cdot hh(t-\tau, \boldsymbol{r}-\boldsymbol{x}) \cdot \partial_\tau^2 \left\{ ss(\tau,\boldsymbol{x})^{\text{in}} \right\} \cdot d\tau d\boldsymbol{x}. \tag{11.18}$$

The Born approximation is valid when the average intensity of $ss(t,\boldsymbol{r})^{\text{sc}}$ is small compared to that of $ss(t,\boldsymbol{r})^{\text{in}}$. With this method, each point in the ground scene is viewed as a perturbing point radiator that produces a *scattered wave field* $ss(t,\boldsymbol{r})^{\text{sc}}$ energized only by $ss(t,\boldsymbol{r})^{\text{in}}$. The Born approximation linearizes the mapping between the perturbation profile and the scattered field by reducing the two unknowns of $ss(\tau,\boldsymbol{x})^{\text{to}}$ and $g(\boldsymbol{x})$ to an equation involving only one unknown, the terrain scattering function $g(\boldsymbol{x})$.

Frequency-Squared Parameter in Rayleigh and Bragg Scattering Models

If the time envelope (baseband pulse) $p(t)$ of the transmitted passband waveform

$$w(t) = \exp(j\omega_c t) \times p(t)$$

is slowly varying[17], then the time derivative $ss(\tau, \boldsymbol{x})^{\text{in}}$ of (11.18) becomes [175]

$$\partial_t^2 \left\{ ss(t, \boldsymbol{r})^{\text{in}} \right\} \approx \omega_c^2 \cdot ss(t, \boldsymbol{r})^{\text{in}}$$

and we have

$$ss(t, \boldsymbol{r})^{\text{sc}} \approx \int_{\mathbb{G}} \int_{\mathbb{T}} g(\boldsymbol{x}) \cdot hh(t - \tau, \boldsymbol{r} - \boldsymbol{x}) \cdot \omega_c^2 \cdot ss(t, \boldsymbol{r})^{\text{in}} \cdot d\tau d\boldsymbol{x}. \tag{11.19}$$

When a scatterer is small relative to transmitted signal wavelength, the integral expression (11.19) of wave motion derived from Helmholtz's integral equation (11.8) shows that this backscatter is proportional to ω^2. This frequency dependent modeling is called Rayleigh[18] scattering [11, 181]. When applied to atomic and electromagnetic wave interactions, it is sometimes called Bragg[19] scattering. Use of the ω^2 characteristic in describing backscatter is common in acoustic imaging, where the application of the small perturbation methods of Rayleigh and Bragg are typically volumetric approaches used to describe elastic acoustic wave-matter interactions.

Frequency-Squared Parameter in Helmholtz Equations

The genesis of the ω^2 element in the time-space domain expression (11.19) is different than the source of the

$$k_\omega^2 = \omega^2 / c^2$$

element in the frequency-space domain expression within the Helmholtz equation. There, ω^2 is the Fourier domain equivalent of a double derivative in the time domain, independent of any particular waveform choice as required in (11.19).

Frequency-Squared Parameter in Polar to Rectilinear Coordinate Conversions

An $|\omega|$ factor, or an approximation of $|\omega|$ that can be provided by ω^2, is sometimes used in SAR image formation processing based in filtered Fourier reconstruction algorithms that are formulated in polar coordinates (see Section 14.6). That instance of the $|\omega| \approx \omega^2$ factor has a different genesis than that of the equation that describes Rayleigh scattering (derived from Helmholtz's integral formula) or that of Helmholtz's wave motion equation in the frequency-space domain. The filtered Fourier reconstruction occurrence of the $|\omega|$ factor results from the Jacobian determinant of the transform between Cartesian and polar coordinate systems.

[17] This requirement is not satisfied by the class of chirp waveforms.

[18] John William Strutt, 3rd Baron Rayleigh (1842–1919) was an English physicist.

[19] Sir William Lawrence Bragg (1890–1971) was an Australian born British physicist.

Frequency-Squared Parameter in Convolutional Back-Projection Algorithms

There is yet another occurrence of ω^2 in some SAR algorithms. The inverse Radon transform has a factor of $|\omega|$, sometimes approximated by ω^2. In the inverse Radon transform case, the genesis of this $|\omega|$ factor is the frequency domain transfer function of the Hilbert transform-differentiation based spatial filter used in the inverse Radon transform. This is the filter of convolutional back-projection methods, which are sometimes used in SAR (see Section 14.5).

11.7 Huygens Principles in Two Dimensions

All formulae related to Huygens principles in three dimensions have their two dimensional analogues, but surprisingly, the proofs of the two dimensional versions are more difficult, if not impossible [35, 163].

The Huygens principles establish situations where full knowledge of wave state at any point in time allows determination of wave state at any other instant of time. "Full knowledge" requires wave state formulations in three dimensions. Reduction to two dimensions is problematic.

A consequence of the requirement for an analytically provable interior or exterior null field is that Huygens and Fresnel wave field models only hold rigorously in three dimensional problems [35, 134, 595, 790, 812]. Wave field models in two dimensions seem to only permit less elegant approaches that always end up falling short of strictly satisfying Huygens requirements. The lack of the third dimension prevents analytical developments where a wavelet expands only outwardly and leaves an interior null field. Alternative approaches must be formulated when dealing with two dimensional wave field inversion problems.

Circular Wave Motion Equations in Time-Space Domain

A wave theory foundation for problems described in two dimensions can be based on circular wave analysis or equations descriptive of cylindrical wave motion in three dimensions. The time-space domain equation of free space circular wave motion in two dimensions is

$$\partial_{r_x}^2\{cc(t,\boldsymbol{r})\} + \partial_{r_y}^2\{cc(t,\boldsymbol{r})\} - \partial_t^2\{cc(t,\boldsymbol{r})/c^2\} = 0. \tag{11.20}$$

The circular wave motion equation expressed in terms of polar coordinates is

$$\partial_r^2\{cc(t,r,\theta)\}+\frac{1}{r}\partial_r\{cc(t,r,\theta)\}+\frac{1}{r^2}\partial_\theta^2\{cc(t,r,\theta)\}-\frac{1}{c^2}\cdot\partial_t^2\{cc(t,r,\theta)\} = 0. \tag{11.21}$$

A wave function that satisfies (11.20) and (11.21), and which also satisfies initial conditions not involving the third dimension, must be a cylindrical wave. A cylindrical wave field can be envisioned as caused by a line of point sources orthogonal to the plane of two dimensional circular wave motion.

Circular Wave Motion Equations in Frequency-Space Domain

A general dynamic cylindrical wave function solution of (11.20) and (11.21) has the form

$$\overset{\text{dynamic}}{cc(t,r)} = \exp(jk_\omega ct) \cdot \overset{\text{static}}{Cc(\omega,r)} \tag{11.22}$$

where the frequency-space domain cylindrical waveform $\overset{\text{static}}{Cc(\omega,r)}$ is a circularly symmetric spatial function of $r = \sqrt{x^2+y^2}$ and is a function that does not depend on t.

Weber's Equation in Two Dimensions

The static element

$$Cc(\omega,r) = \overset{\text{static}}{Cc(\omega,r)}$$

of the general time-space solution $\overset{\text{dynamic}}{cc(t,r)}$ is a solution of the two dimensional analogue of Helmholtz's equation,

$$d_r^2\{Cc(\omega,r)\} + \frac{1}{r} \cdot d_r\{Cc(\omega,r)\} + k_\omega^2 \cdot Cc(\omega,r) = 0, \tag{11.23}$$

which is attributed to Weber [983].

Bessel Functions:
Not A Suitable Model for Waves from Point Singularities

A candidate solution for $Cc(\omega,r)$, which does not depend on ω and which satisfies (11.22) and (11.23), is the zeroth order Bessel[20] function of the first kind [980],

$$J_0(r) = \frac{1}{2\pi}\int_0^{2\pi} \exp[-jr\cos(\theta-\phi)] \cdot d\theta. \tag{11.24}$$

However, unlike the Green's function solution $Hh(\omega,r)$ of the three dimensional Helmholtz equation, the zeroth order Bessel function $J_0(r)$ solution of (11.23) does not have the required singularity (for describing point sources) at the origin of r. Hence,

$$J_0(\omega,r) = \exp(jk_\omega ct) \cdot J_0(r)$$

is not suitable for modeling Huygens-Fresnel type point sources in two dimensions.

Another well know solution of (11.22) is the zeroth order Bessel function of the second kind $Y_0(\omega,r)$. Although this function has the required point singularity at the origin, this solution is also deemed unsuitable with respect to required Huygens attributes [35].

Hankel Functions and Bessel Functions

Although the dynamic Bessel functions $J_0(\omega,r)$ and $Y_0(\omega,r)$ do not separately offer an acceptable solution, the Hankel[21] functions, which can be expressed in

[20] Friedrich Wilhelm Bessel (1784-1846) was a German mathematician and astronomer.
[21] Hermann Hankel (1839–1873) was a German mathematician.

terms of combined zeroth order Bessel functions of the first and second kinds,

$$\begin{aligned} H_0^{(1)}(\omega, r) &= J_0(\omega, r) + jY_0(\omega, r), \\ H_0^{(2)}(\omega, r) &= J_0(\omega, r) - jY_0(\omega, r), \end{aligned}$$

do offer useful solutions of circular wave motion equations with qualities more consistent with Huygens-Fresnel wave field principles adapted for two dimensional problems [91, 186].

Dynamic Hankel Functions as Circularly Symmetric Wave Fields

The first Hankel function, combined with the $\exp(jk_\omega ct)$ time dynamic phase factor, yields

$$Cc(\omega, r)_{\text{harmonic}}^{\text{dynamic}} = \exp(jk_\omega ct) \cdot H_0^{(1)}(\omega, r).$$

This dynamic wave field describes cylindrical waves diverging from the origin (a point source in two dimensions, a line source in three dimensions). The first Hankel function plays the role in two dimensions that the divergent dynamic harmonic Fresnel wave field function

$$\begin{aligned} \measuredangle Hh(\omega, r)_{\text{harmonic}}^{\text{dynamic}} &= \exp(jk_\omega ct) \cdot \measuredangle Hh(\omega, r), \\ &= \exp[jk_\omega(ct - r)] \end{aligned}$$

plays in three dimensions.

The second Hankel function, combined with a time dynamic phase factor with negated sign yields

$$Cc(\omega, r)_{\text{harmonic}}^{\text{dynamic}} = H_0^{(2)}(\omega, r) \cdot \exp(-jk_\omega ct).$$

This dynamic wave field describes cylindrical waves converging to the origin and hence is not of much use in describing the expanding waves of a scattered wave field.

Hankel Functions and Circularly Symmetric Harmonic Wave Fields

When the distance between the source singularity and a field point is significantly greater then the temporal signal's wavelength, the zeroth order Hankel function of the first kind can be approximated by [389, 434, 665]

$$H_0^{(1)}(\omega, r) \approx \frac{1}{\sqrt{k_\omega}} \cdot e^{j\frac{\pi}{4}} \cdot \sqrt{\frac{2}{\pi}} \cdot \frac{1}{|\boldsymbol{r}|} \cdot \exp(-jk_\omega|\boldsymbol{r}|), \tag{11.25}$$

where $r = |\boldsymbol{r}|$ and $\boldsymbol{r} = [r_x \; r_y]$.

Angular Spectrum Decomposition of Hankel Functions

The Hankel function approximation (11.25) can be expressed as an integral over directed tonal wave fields in the two dimensional wavenumber domain, i.e., by an angular spectrum decomposition,

$$H_0^{(1)}(\omega, r) = \int_{-\infty}^{\infty} a(\omega, k_{r_y})_{\text{Hankel}} \cdot \exp(-jr_x \cdot \sqrt{k_\omega^2 - k_{r_y}^2}\,) \cdot \exp(-jr_y \cdot k_{r_y}) \cdot dk_{r_y}, \tag{11.26}$$

where

$$a(\omega, k_{r_y})_{\text{Hankel}} = \frac{1}{\pi} \cdot \frac{1}{\sqrt{k_\omega^2 - k_{r_y}^2}}.$$

Equation (11.26) provides a synthesis over directed plane waves in the two dimensional wavenumber domain that forms the circularly symmetric harmonic wave field solution (11.25) of Weber's equation (11.23). The Hankel function decomposition (11.26) gives an expression of a circular phase field that has the same form as the angular spectrum decomposition (4.26) used in slant-plane analysis. The decomposition (11.26) is an alternative to utilizing the spatial Fourier transform (obtained via the principle of stationary phase) of a slant-plane slow-time phase curtain (see Section 15.6).

Weyl Expansion

A Weyl[22] expansion of a circular wave as a superposition of harmonic plane waves [988] provides yet another approach to wave field analysis in two dimensions [262].

[22]Hermann Klaus Hugo Weyl (1885–1955) was a German mathematician and theoretical physicist.

Chapter 12

SAR Algorithms

We overview all major classes of SAR algorithms. We suggest a taxonomy based on primary and secondary algorithm attributes, and we provide detail descriptions of many of these algorithm classes.

Excluded here are detailed descriptions of convolutional back-projection, convolutional back-propagation, direct and filtered Fourier reconstructions, and three dimensional angular spectrum algorithms. These algorithms are addressed in subsequent chapters.

This chapter is outlined as follows: Section 12.1 presents an overview of the major classes of image formation algorithms. Section 12.2 describes algorithms based on coherent summations along range walk traces in the time-space domain of signal measurements. Section 12.3 describes algorithms based on coherent summations along circular back-propagated traces in the image domain. Section 12.4 describes Stolt formatting in spotlight mode. Section 12.5 describes polar formatting in spotlight mode. Section 12.6 describes polar formatting in spotmap mode with non-coherent summing of the spot mosaic. This section also describes Stolt formatting in spotmap mode with coherent summing of the spot mosaic. Section 12.7 describes Stolt formatting in stripmap mode. Section 12.8 explains Huygens-Fresnel inversion without Stolt formatting. Section 12.9 overviews stripmap mode range-Doppler algorithms. Finally, Section 12.10 describes the use of chirp scaling as an enhancement to range-Doppler imaging and as an approximation of Stolt formatting.

12.1 SAR Algorithm Overview

SAR Algorithm Classes

SAR algorithms can be grouped into the following top level classes:

- Hyperbolic Range Walk Signal Domain Coherent Sum Algorithms
- Direct Spherical Back-Propagation Image Domain Coherent Sum Algorithms
- Spotlight Polar Format Algorithms

- Spotmap Polar Format Algorithms
- Spotlight Stolt Format Algorithms
- Spotmap Stolt Format Algorithms
- Stripmap Stolt Format Algorithms
- Rectilinear Spectrum Huygens-Fresnel Inversion Algorithms
- Stripmap Range-Doppler Algorithms
- Stripmap Chirp Scaling Algorithms
- Convolutional Back-Projection Algorithms
- Convolutional Back-Propagation Algorithms
- Direct Fourier Reconstruction Algorithms
- Filtered Fourier Reconstruction Algorithms

All of these algorithms, except for the last four, are covered in this chapter.

SAR Algorithm Classes: Summary Descriptions

The following provides high level descriptions of these algorithm classes.

Hyperbolic Range Walk Time-Space Domain Coherent Sum Algorithm Each pixel in the SAR image is focused by coherent summing along a range walk trace in the (signal measurement) time-space domain. Interpolation of measured time-space data samples to exact range walk trace positions is required. Coherent summations can be performed with uncompressed or pulse compressed data.

Direct Spherical Back-Propagation Image Domain Coherent Sum Algorithms Each range compressed sample is spherically back-propagated to its originating range distance and is replicated in azimuth (because of azimuth ambiguity) along a circle in a coherently summed pixel array. This is repeated for all range samples collected in each pulse and for all pulse positions collected along the synthetic aperture. Interpolation of directly back-propagated echo data to exact pixel locations is required. This class of algorithms is different than a class of algorithms known as convolutional back-propagation algorithms.

Spotlight Polar Format Algorithms Fourier reconstruction of the terrain spectrum guided by a passband form of the Fourier projection theorem is the theoretical basis of polar formatting algorithms [115,679,976]. This class of pivot point algorithms utilizes the Fraunhofer plane wave approximation. Two dimensional interpolation from a polar raster to a rectilinear raster in the wavenumber domain is required before Fourier inversion.

Spotmap Polar Format Algorithms Non-coherent seaming of spotmap subspots, each produced by polar formatting, defines this subclass of spotmap mode algorithms.

Spotlight Stolt Format Algorithms Fourier reconstruction of the terrain spectrum guided by the passband form of the Fourier diffraction theorem is the theoretical basis of Stolt formatting algorithms [40]. This is also the theoretical foundation of holography. Stolt formatting requires Fourier migration and one dimensional interpolation to change the Stolt formatted k-space data raster to a uniformly spaced raster before Fourier inversion.

Spotmap Stolt Format Algorithms Coherent seaming of spotmap subspots, each produced by Stolt formatting, defines this subclass of spotmap mode algorithms.

Stripmap Range-Doppler Algorithms Processing in the time-wavenumber domain guided by a formal fast-time Doppler analysis of the SAR signal is the theoretical basis of range-Doppler algorithms. The signal analysis of range-Doppler algorithms is based on a separation of down-range and along-track concerns. A Taylor series approximation of fast-time signal delay expressed in terms of platform motion parameters, i.e., velocity, acceleration and so on, is for the along-track portion of the problem.

Stripmap Chirp Scaling Algorithms Chirp scaling phase adjustments are applied in the time-wavenumber domain to enhance the performance of range-Doppler algorithms. This class of algorithms requires the use of LFM waveforms. The derivation of the chirp scaling phase adjustments can be shown to be an approximation of the Stolt transform.

Stripmap Stolt Format Algorithms Stolt formatting can be applied to both spotlight and stripmap modes. Application in stripmap mode requires specification of the boundaries of the coherent processing interval extracted from the continuous stream of stripmap mode data.

Huygens-Fresnel Inversion Algorithms without Stolt Formatting This class of algorithms is characterized by Huygens-Fresnel wave field inversions performed in the rectilinear frequency-wavenumber domain without Fourier migrations. Stolt formatting is not required. A set of Huygens-Fresnel transfers functions, with each element in this set designed specifically for one of the cross-range lines spanned by the SAR image pixel array, is required.

Convolutional Back-Projection Algorithms A spatial domain implementation of the inverse Radon transform leads to the class of pivot point convolutional back-projection algorithms. Convolutional back-projection algorithms are based on concepts rooted in a baseband filtered form of the Fourier projection theorem. This class of algorithms is constrained by the assumption of a plane wave Fraunhofer approximation. Convolutional back-projection algorithms are used in medical imaging with ray tomographic sensors. These algorithms have been adapted for SAR with analysis guided by a passband filtered form of the Fourier projection theorem, but with implementation performed in the spatial domain.

Convolutional Back-Propagation Algorithms The Radon transform generalized by concepts from the Fourier diffraction theorem leads to the class of pivot point convolutional back-propagation algorithms. This class of algorithms escapes the plane wave Fraunhofer approximation requirement of convolutional back-projection. Convolution back-propagation algorithms are used in medical imaging with diffraction tomographic sensors.

Direct Fourier Reconstruction Algorithms The class of direct Fourier reconstruction algorithms is a large class of algorithms that includes polar formatting as a subclass [647,679,907]. This class of algorithms seeks to directly interpolate (without a wavenumber domain weighting corresponding to spatial domain convolutional filtering) non-uniformly spaced k-space data to a rectilinear raster amendable to Fourier inversion. Both the Fourier projection and Fourier diffraction theorems are part of the theoretical foundation of this class of algorithms.

Filtered Fourier Reconstruction Algorithms Like the direct Fourier reconstruction algorithms, this class of algorithms seeks to interpolate k-space data to a rectilinear raster, but with the added operation of a wavenumber domain data weighting. Multiplicative operations in k-space correspond to convolutional filtering operations in the spatial domain. These algorithms are k-space equivalents of convolutional back-projection and convolutional back-propagation algorithms. The theoretical basis for the use of a wavenumber domain weighting comes from the Radon transform and principles of Fourier reconstructions under transformation with a change of variables. The filtered Fourier reconstruction class of algorithms includes Stolt formatting as a subclass.

Primary SAR Algorithm Attributes

Rather than classify algorithms based on the origin of their computed imaging disciplines, we seek to categorize the major algorithm classes based on three primary attributes:

- the processing domain [440],
- the wave field model and
- any wave field approximations that might be employed [556].

Time, Space, Frequency and Wavenumber Processing Domains

A SAR imaging problem can often be set up as a wave field inversion problem [309]. Another approach is to set up the problem as a system inversion problem requiring deconvolution [885]. In either case, the signal processing domain options for image formation algorithms are the following:

- time-space domain algorithms [42,221,302],

- time-wavenumber domain algorithms [218, 500, 1001],
- frequency-space domain algorithms [889] and
- frequency-wavenumber domain algorithms [40, 137, 770].

The citations given above are not exhaustive, but provide examples of each processing domain option.

Wave Field Models

In addition to the processing domain, the form of the wave field model used in the SAR signal analysis can be used to assign a class label to a particular SAR algorithm.

- Time-space domain algorithms typically employ signal models based on Huygens wavelets with impulsive or pulse shaped wavefronts [523, 713].
- Frequency-space domain algorithms typically employ signal models based on harmonic Fresnel wave fields.
- Frequency-wavenumber domain algorithms typically employ signal models that leverage tonal wavenumber wave fields resident on Ewald spheres.
- Time-wavenumber domain algorithms typically exploit the formalism of temporal Doppler descriptions of SAR signals gathered by moving sensors.

Wave Structure Approximations

Also related to our SAR algorithm taxonomy are the various wave theory approximations sometimes used to simplify signal models and to aid development of inversion methods. This element of the taxonomy is primarily derived from the discipline of Fourier optics, where the analysis of near field diffraction is made tractable with Fresnel approximations and the analysis of far field diffraction is made tractable with Fraunhofer approximations [395].

Near Fields and Far Fields in Fourier Optics and SAR

In Fourier optics, the Fresnel approximation is employed in the optical near field and the Fraunhofer approximation is employed in the optical far field. The qualifiers of "near field" and "far field" in Fourier optics are not only determined by the distance between the diffracting aperture and the remotely sensed wave field measurements, but also by the relative size of the diffracting (source) aperture and size of the (receiving) sensor array.

The Fresnel and Fraunhofer approximations used in Fourier optics have their analogues in both SAR theory development and SAR practice. But the near field and far field definitions of SAR seem reversed when compared to the near field and far field definitions of Fourier optics. The relative combination of range, synthetic aperture length and ground scene size determines whether a given SAR scenario is classified, in the Fourier optics sense, as far field or near field. Relatively short range airborne spotlight polar formatting systems (which model ground incident wavefronts as plane waves) are usually in an operational realm equivalent to optical far field scenarios. Long range satellite based stripmap range-Doppler systems are usually in the realm of optical near field scenarios (where quadratic approximations of hyperbolic geometries are sometimes used).

Many airborne SARs employ the Fraunhofer approximation and many of the early space based SARs employed the Fresnel approximation. The angular spectrum analysis of Huygens-Fresnel methods are essentially free of the approximations of Fraunhofer and Fresnel. The Huygens-Fresnel signal model is applicable to both near field and far field scenarios in SAR.

Secondary SAR Algorithm Attributes

In addition to the primary attributes of processing domain, wave field models and approximations, it is useful to characterize SAR algorithms in terms of secondary attributes such as:

- bandwidth,
- beamwidth and
- the type of phase stabilization perform by motion compensation.

SAR System Bandwidth

One attribute that is important in characterizing the suitability of a SAR algorithm to a particular radar imaging problem is the *Q-factor*. The Q-factor is defined as the ratio of a passband system's bandwidth divided by the center frequency.

Low Q-factor systems are called *ultra-wide bandwidth* systems [830, 954, 989]. The ultra-wide bandwidth radar system of [873] has a pulse bandwidth of 200 MHz carried by a 300 MHz center frequency. The Q-factor of systems developed and studied in [447] approach unity. Other example low Q-factor systems include [68, 102, 397, 425, 698, 944].

In SAR, the design and implementation of a low Q-factor system is more challenging in many ways than that of a high Q-factor system. Systems that employ polar formatting operate within the realm of high Q-factor.

SAR System Beamwidth and Angular Dwell

Antenna beamwidth and the duration of the slow-time dwell (or extent of angular dwell) can be descriptive attributes of a SAR system [407]. An example of a broad beamwidth and long dwell system is a dual-band VHF-UHF SAR system with an angular dwell of 40 to 50 degrees and a coherent integration time of three to five minutes [477]. This system seeks to form the SAR image through Stolt formatting. Another example large angular dwell system is the work of [351].

SAR System Phase Stabilization

Motion compensation is a critically important SAR process that seeks to accurately assess the difference between the actual flight path and the intended flight path. Estimates of platform position are used to make adjustments to measured phase values of the received fast-time signal. These phase adjustments cause the data to more accurately conform to the anticipated slant-range of the objective flight path.

Motion compensation can make phase adjustments with respect to an objective fixed down-range reference line or make phase adjustments with respect to the changing range to scene center [490]. The first of these is called *motion compensation to a line.* The second is called *motion compensation to a point* [146].

Pivot Point Algorithms

An algorithm that requires motion compensation or phase stabilization with respect to a reference ground point can be called a *pivot point algorithm.* Motion compensation to a pivot point is required in polar format processing [146]. Motion compensation to a line is required in slant-plane Stolt format processing. Examples of slant-plane Stolt formatting combined with phase stabilization to a pivot point can be found [528]. This combination is more appropriate for three dimensional Stolt formatting algorithms (see Section 13.10).

Block versus Single Pixel Processing

“Block processing” versus “single pixel processing” can also be a descriptive attribute for creating a taxonomy of SAR algorithms [407].

Batch versus Progressive Processing

Batch processing requires collective processing to deliver the final image in an all or none delivery mode. Progressive processing is able to incrementally provide SAR imagery, where the progressive nature of SAR product delivery adds either increased fidelity or increased ground coverage.

12.2 Hyperbolic Range Walk Coherent Summing in the Signal Domain

A scatterer's echo signal creates a hyperbolic slow-time trace in the time-space domain of measured data [376]. A pulse compressed, range (only) focused SAR image of sufficient slow-time dwell clearly shows this hyperbolic range walk. One approach to “slow-time focusing” is to coherently sum along the hyperbolic locus of pulse compressed data measurements to form one focused pixel [86, 332, 407, 424, 605, 889, 965, 984, 1010]. After this coherent sum is accumulated, the same processing is repeated with a different hyperbolic trace to produce the next desired image pixel, and so on.

The trace of a given range walk hyperbola may not align perfectly with the array of sampling points in the time-space data collection domain. Interpolations of measured time-space data samples to the objective range walk trace must be performed to be precise [42, 301]. Instead of interpolations, up sampling of the time-space data array is sometimes used [889].

Range walk coherent sum algorithms are "brute force" methods and are computationally too demanding to implement in nearly all applications, except in some off-line calibration tasks [407].

Members of this class of algorithms have been called *exact reconstruction algorithms* [39, 167, 407], *exact transfer functions* [440] or *time domain correlation* algorithms [221, 885]. Algorithms of this class are exact in the sense that approximations such as Fraunhofer and Fresnel are not employed. "Exact" is somewhat of a misnomer since inexact finite order interpolations are often involved to match the continuous range walk hyperbola to the discrete array of point measurements in the time-space data collection domain [407, 440].

Pixel Specific Inverse Huygens-Fresnel Transfer by Convolution

The range walk coherent sum approach to pixel specific SAR image formation processing can be described as an implementation of inverse Huygens-Fresnel transfers by convolution in the time-space domain. For example, if we were to implement the inverse Huygens-Fresnel transfer operations of

$$II\left[k_{\hat{x}}(\omega_p, k_{u_y}), k_{\hat{y}}(\omega_p, k_{u_y})\right]_{\text{SCS}} = H^*_{x_n}(\omega_p, k_{u_y}) \cdot H^*_{y_n}(\omega_p, k_{u_y}) \times \left\{RS(\omega_p, k_{u_y})_{\text{ACS}}\right\} \tag{12.1}$$

in the time-space domain, then the operation would be of the form

$$ii(\hat{x}_n, \hat{y}_n)_{\text{SCS}} = hh^*_{(x_n, y_n)}(-t, -u_y) *_t *_{u_y} \{rs(t, u_y)_{\text{ACS}}\}, \tag{12.2}$$

$$= \int_{u_y}\int_t hh^*_{(x_n, y_n)}[t - t_{d(x_n, y_n)}(u_y)] \cdot rs(t, u_y)_{\text{ACS}} \cdot dt du_y, \tag{12.3}$$

where $*_t$ and $*_{u_y}$ operators are convolutions over time and space, $rs(t, u_y)_{\text{ACS}}$ is the pulse compressed time-space data array and where

$$t_{d(x_n, y_n)}(u_y) = \frac{2}{c}\sqrt{x_n^2 + (u_y - y_n)^2}$$

is the time delay $t_{d(x_n, y_n)}(u_y)$ from aperture position u_y to the particular ground point (x_n, y_n) being focused. This process results in a single focused pixel corresponding to the ground point $(\hat{x}_n, \hat{y}_n)$.

Through an inverse Fourier transform, the dimensionally separated Huygens-Fresnel transfer functions

$$H^*_{x_n}(\omega_p, k_{u_y}) \cdot H^*_{y_n}(\omega_p, u_y)$$

of (12.1) combine into the single two dimensional hyperbolic trace function

$$hh^*_{(x_n, y_n)}(-t, -u_y)$$

in the time-space domain. As contained in (12.3), $hh^*_{(x_n, y_n)}[t - t_{d(x_n, y_n)}(u_y)]$ can be viewed as an impulsive Huygens wavelet centered at (x_n, y_n) with radius in delay time given by $t_{d(x_n, y_n)}$.

The convolutions of (12.2) can also be viewed as a process of time-space domain correlation of measured data with a specific range walk hyperbola [38, 407, 885].

This algorithm has also been characterized as a technique akin to matched filtering specific to each image pixel [324]. An alternative view point is that the operations of (12.2) requires each pixel to be subjected to a temporal-spatial Huygens-Fresnel transfer in a such way that during pixel specific processing, each pixel is at the center of its own scene coordinate system.

Range Walk Coherent Summing with Waveform Matched Filtering

Equation (12.2) can also be modified to implement waveform pulse compression along with single pixel focusing [42]. Range walk coherent summing, combined with fast-time domain pulse compression, can be expressed for a general ground point (x_n, y_n) by [174, 709]

$$ii(\hat{x}_n, \hat{y}_n)_{\text{SCS}} = \int_{u_y}\int_t w^*_{(x_n,y_n)}[t - t_{d(x_n,y_n)}(u_y)] \cdot ss(t, u_y)_{\text{ACS}} \cdot dt du_y,$$

where the impulsive Huygens wavelet

$$hh^*_{(x_n,y_n)}[t - t_{d(x_n,y_n)}(u_y)]$$

of (12.2) is modified to be a generalized pulsed Huygens wavelet

$$w^*_{(x_n,y_n)}[t - t_{d(x_n,y_n)}(u_y)],$$

formed with the waveform $w(t)$ and where $ss(t, u_y)_{\text{ACS}}$ is the raw (not pulse compressed) time-space phase history.

Range Walk Coherent Summing with Frequency-Space Matched Filtering

A form of range walk coherent summing with embedded pulse compression can also be performed in the frequency-space domain [889]. From a generalized form of Parseval's theorem,

$$\begin{aligned}\int_t & w^*_{(x_n,y_n)}[t - t_{d(x_n,y_n)}(u_y)] \cdot ss(t, u_y)_{\text{ACS}} \cdot dt \\ &= \frac{1}{2\pi}\int_{\omega_p} W^*(\omega_p) \cdot Ss(\omega_p, u_y)_{\text{ACS}} \cdot \exp[jt_{d(x_n,y_n)}(u_y) \cdot \omega_p] \cdot d\omega_p,\end{aligned}$$

this (passband) frequency-space expression can be substituted into the time-space expression, giving

$$\begin{aligned}& ii(\hat{x}_n, \hat{y}_n)_{\text{SCS}} \\ &= \frac{1}{2\pi}\int_{u_y}\int_{\omega_p} W^*(\omega_p) \cdot Ss(\omega_p, u_y)_{\text{ACS}} \cdot \exp[jt_{d(x_n,y_n)}(u_y) \cdot \omega_p] \cdot d\omega_p du_y.\end{aligned}$$

This provides a modified range walk coherent summing algorithm where the sum (integral) in time is replaced by a sum (integral) over passband frequencies.

Neighborhood Focusing via Frequency Domain Matched Filtering

Variant algorithms within the range walk coherent sum class have also been developed to achieve localized focusing over a block of pixels. Instead of doing brute force temporal-spatial correlations to achieve exact reconstruction for each pixel, a two dimensional Fourier transform is used to convert measured data corresponding to a small block of pixels to the frequency-wavenumber domain. A matched filter is designed for the hyperbolic range walk of the pixel at the center of the block. Frequency domain matched filtering and Fourier inversion creates a block of SAR pixel data with a two dimensional spatial depth of focus, where the focusing is exact for the center pixel and somewhat focused for the surrounding neighborhood of pixels [950]. This pixel neighborhood focusing method has been called a *fast correlation algorithm* [301, 441, 443].

12.3 Direct Spherical Back-Propagation Coherent Summing in the Image Domain

A pulsed compressed echo, due to azimuth ambiguity, must have originated at one or more scatterers distributed on a semicircle that extends out from the sensor location into the slant-plane [376]. One approach to image formation is coherent summations of pulse compressed data spherically back-propagated into an image domain pixel array [8, 12, 108, 109, 324, 424, 448, 557, 1010]. The back-propagated data forms concentric circles corresponding to each time sample instant within the receive window of a recorded pulse. Coherent integration is performed in the image domain over data from sets of back-propagated circles centered on stop-and-hop sample points along the synthetic aperture.

Circular back-propagation algorithms in the slant-plane can support both stripmap and spotlight modes [493]. Spherical back-propagation can handle nonlinear flight paths, three dimensional scenes, large scene size, wide bandwidth and large angular dwells [99, 703, 833, 959, 960, 1010].

We use the adjective "direct" to distinguish members of this class of algorithms from the "convolutional" back-propagation algorithms presented in Section 14.5.

Range walk coherent summing and direct spherical back-propagation coherent summing are said to be equivalent forms of SAR image formation processing [40, 688, 690]. A risk in application of these two classes of purely temporal-spatial algorithms is violation of Nyquist requirements by coherent summing of overly sparse pulse data sets (see Section 14.7). Frequency-wavenumber domain analysis is required to determine Nyquist requirements of time-space domain algorithms.

Interpolation of Pulse Data in Direct Spherical Back-Propagation

Direct spherical back-propagation requires interpolation of each pulse echo sample to the specific pixel location being coherently summed.

If receive window data are up sampled and indexed, rather than interpolated on the fly, an indexing look up method can be used to select data samples to be accessed and retrieved for coherent summation [889, 1010].

Direct Spherical Back-Propagation of Passband Data

Direct spherical back-propagation must be performed with phase history data at passband [960]. The carrier phase factor must be removed prior and then reinserted after any up sampling or interpolation operations that might be performed.

Pulse by Pulse Processing

Direct spherical back-propagation processing can be performed in a pixel by pixel manner. A pixel is selected, coherent summing is performed with the appropriate back-propagation delay instances for all pulses over the full synthetic aperture, and then the next pixel is selected and processed [1010].

Alternately, the processing can proceed on a pulse by pulse basis, where all delay instant data samples of each focused range window are coherently added to the pixel array before moving to the next pulse of data.

Factorized Direct Back-Propagation Algorithms

Direct back-propagation algorithms are computationally intensive. This has motivated the development of more efficient forms of direct back-propagation algorithms [141]. Note worthy advancements include the factorized forms of direct back-propagation algorithms. Two stage versions [450, 818, 1010] and multiple stage versions [634, 701, 702] have been proposed.

The basis of factorized direct back-propagation methods is the observation that slant-range distances from adjacent slow-time pulse transmission positions are nearly identical for a scatterer location within the ground scene. Hence, there is no need to coherently sum the data from adjacent pulses at a given pixel location if a degree of approximation can be tolerated.

Factorized Direct Back-Propagation and Progressive Beamforming

An association between direct back-propagation and processing for beamforming has oft been noted [407, 470, 475, 493, 494, 513, 634, 688, 836, 837, 867].

Factorized direct back-propagation can be viewed as recursively forming antenna array beam patterns, where a set of adjacent pulses defines a beam of a certain size. Pixels illuminated within the beam need only be coherently summed with a single data value that represents back-propagation that occurs within that beam (and by proxy, the single data value represents the elements of the set of adjacent pulses used to form that beam).

At each stage of factorization, smaller beams are defined with larger subapertures (larger sets of adjacent pulses). The smaller beams are formed recursively with the beam data of the previous stage of factorization. The focused beams at each stage get smaller until they are so narrow they contain only a single pixel. The recursive approach factors out nearly common additions from a brute force approach, thereby reducing complexity.

Weighting the along-track data to compensate the antenna beam pattern is said to be important, but often neglected in direct back-propagation systems [138]. Beam pattern compensation prevents aperture under sampling as discussed in [407].

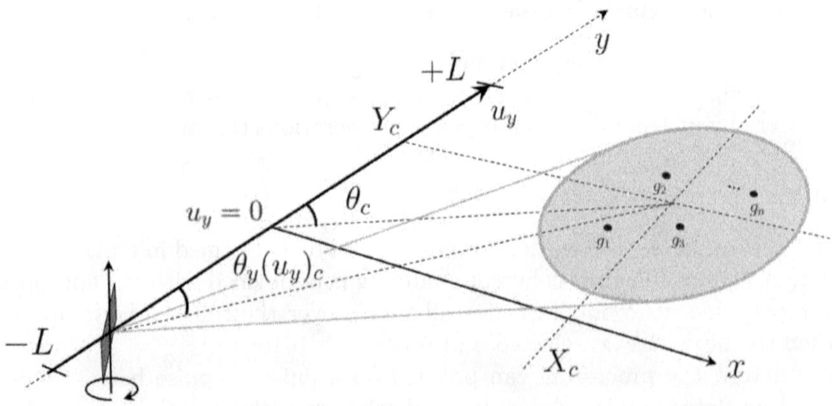

Figure 12.1: Spotlight mode squint imaging geometry in the slant-plane.

12.4 Spotlight Stolt Formatting

In spotlight mode, a mechanical or electronic beam steering capability keeps the boresight of the antenna directed towards scene center (X_c, Y_c) as the sensor moves through a finite synthetic aperture as shown in Fig. 12.1. Pulse echo data recording starts at sensor position $-L$ and ends when the sensor reaches $+L$.

Scene Squint Angle

The fixed *scene squint angle*

$$\begin{aligned} \theta_c &= \theta_y(0)_c, \\ &= \arccos\left(\frac{Y_c}{R_c}\right) \end{aligned}$$

is the angle extended from the center $(u_y = 0)$ of the synthetic aperture to the center (X_c, Y_c) of the spot scene at a slant-range distance of

$$R_c = \sqrt{X_c^2 + Y_c^2}.$$

The scene squint angle is defined in the aperture coordinate system. The scene squint angle does not change with sensor movement. If $Y_c = 0$, then the spot scene is a broadside scene.

Spotlight Instantaneous Boresight Angle

The line of sight vector from each aperture position $(0, u_y)$ to scene center (X_c, Y_c) defines the instantaneous *antenna boresight angle,*

$$\theta_y(u_y)_c = \arctan\left(\frac{X_c}{u_y - Y_c}\right).$$

The antenna boresight angle varies as a function of sensor position u_y within the synthetic aperture. Only the center of the spot is consistently at boresight.

Antenna Footprint Spot Rotation

The antenna beam illumination footprint rotates on the ground patch as the sensor moves along the flight path. Any amplitude modulation induced by this relative scene rotation is usually considered negligible in SAR signal modeling.

Illuminated Spot Size

In our study of spotlight mode Stolt formatting, we assume the use of chirp pulse waveforms and a flat plate surface antenna. The radiated beam pattern determines the illuminated spot size. The antenna's down-range and along-track beamwidths are nominally defined by the null to null angular width of the antenna beam's mainlobe. The size of the illuminated spot varies with the instantaneous frequency of the transmitted chirp signal. Within a three dimensional context and with a flat plate antenna of size $D_z \times D_y$, the center to null (range dependent) spatial widths of the mainlobe of the antenna beam pattern at the ground scene are

$$\begin{aligned} X_o(\omega_p) &= R_c \cdot \arcsin\left(\frac{\lambda_{\omega_p}}{D_z}\right), \\ Y_o(\omega_p) &= R_c \cdot \arcsin\left(\frac{\lambda_{\omega_p}}{D_y}\right). \end{aligned}$$

Constant Beamwidth Approximation

The beam patterns of high frequency radars that transmit relatively small bandwidth, i.e., high Q-factor systems, can sometimes be characterized as frequency independent. This allows the approximation $\omega_p \approx \omega_c$ for all frequencies within the passband. In such cases, antenna beamwidths may be approximated by reference to the center frequency of the transmitted signal,

$$\begin{aligned} X_o &\approx R_c \cdot \frac{\lambda_{\omega_c}}{D_z}, \\ Y_o &\approx R_c \cdot \frac{\lambda_{\omega_c}}{D_y}. \end{aligned} \tag{12.4}$$

A long range, small spot, small angle assumption (i.e., $\theta \approx \sin\theta$, with $\theta = \lambda_{\omega_c}/D$) has been used in the approximations of (12.4).

Spotlight Time-Space Chirp Phase History

We now present the time-space phase history of a spotlight mode SAR in a side by side comparison with the previously developed slant-plane time-space phase history model. The details of signal analysis elements specifically related to spotlight mode are explained

Slant-Plane Time-Space Chirp Phase History

The time-space passband phase history for the slant-plane signal model (7.21) with an infinite synthetic aperture and an isotropic point antenna is

$$
\begin{aligned}
& s_{\Pi[T_s,T_f]}\, s_{\Pi(\infty)}(t, \boldsymbol{u})^{\text{chirp}}_{\text{slant}} \\
&= \sum_n g_n \cdot w_{\Pi(T_p @ t_n)}(t)^{\text{chirp}}_n \; *_t \; hh^2(t, \boldsymbol{u} - \boldsymbol{x}_n),
\end{aligned} \tag{12.5}
$$

where, because of the time impulsive core of

$$
hh^2(t, \boldsymbol{u} - \boldsymbol{x}_n) = \frac{1}{(4\pi)^2} \cdot \frac{1}{|\boldsymbol{u} - \boldsymbol{x}_n|^2} \cdot \delta\left[t - t_d(\boldsymbol{u} - \boldsymbol{x}_n)\right],
$$

the engagement of the generalized chirp wavelet in (12.5) is in terms of a time domain convolution (the prior model given in (7.21) is in terms of a time domain multiplication).

Spotlight Time-Space Chirp Phase History

Spotlight mode analysis is generalized (beyond the analysis of the slant-plane system of Chapter 7) by using a flat plate surface antenna instead of a point antenna and by truncating the synthetic aperture to finite length. The uniformly energized, center referenced, flat plate, time-space beam function (9.36),

$$
\Lambda(t, \boldsymbol{r} - \boldsymbol{a}_c)^{\text{pulse}}_{\text{plate}} \approx \frac{1}{4\pi} \cdot \frac{1}{|\boldsymbol{r} - \boldsymbol{a}_c|} \cdot \int_{-D_y/2}^{+D_y/2} 1 \cdot w\left[t - \frac{1}{c}\sqrt{r_x^2 + (r_y - s_y)^2}\right] \cdot ds_y, \tag{12.6}
$$

is used to develop the spotlight mode model, where the "1" of (12.6) represents the uniform excitation function that energizes the antenna's surface. Modifying (12.5) by using the time-space channel model (10.8), an approximate[1] time-space domain expression for the spotlight mode case is

$$
\begin{aligned}
& s_{\Pi[T_s,T_f]}\, s_{\Pi(2L)}(t, \boldsymbol{u})^{\text{chirp}}_{\text{spot}} \\
&\approx \Pi_{[-L,+L]}(\boldsymbol{u}) \cdot \sum_n g_n \cdot w_{\Pi(T_p @ t_n)}(t)^{\text{chirp}}_n \; *_t \; \Lambda^2(t, \boldsymbol{u} - \boldsymbol{x}_n)^{\text{impulse}}_{\text{plate}}.
\end{aligned} \tag{12.7}
$$

The difference between the slant-plane (12.5) and spotlight (12.7) model for the time-space phase history is inclusion of the finite aperture slow-time window

$$
\Pi_{[-L,+L]}(\boldsymbol{u}) \tag{12.8}
$$

and the chirped surface antenna beam pattern

$$
w_{\Pi(T_p @ t_n)}(t)^{\text{chirp}}_n \; *_t \; \Lambda^2(t, \boldsymbol{u} - \boldsymbol{x}_n)^{\text{impulse}}_{\text{plate}} \tag{12.9}
$$

[1]The approximation comes from specifying the location of the surface antenna by reference to a nominal center point.

that replaces the chirped Huygens wavelet of the point antenna,

$$w_{\Pi(T_p@t_n)}(t)_n^{\text{chirp}} \; *_t \; hh^2(t, \boldsymbol{u} - \boldsymbol{x}_n)_{\text{point}}^{\text{impulse}} .$$

The slow-time dwell afforded by the synthetic aperture slow-time window function (12.8) is different than the fixed field of view afforded by the flat plate antenna's beam pattern (12.9). In spotlight mode SAR, the antenna's beam pattern determines the spot size, not the synthetic aperture length.

Chirped Flat Plate Antenna Induced Weight Function

Equation (9.45) can be used to describe the waveform modulated antenna beam function (12.9) contained within (12.7),

$$\begin{aligned} & w_{\Pi(T_p@t_n)}(t)_n^{\text{chirp}} \; *_t \; \Lambda^2(t, \boldsymbol{u} - \boldsymbol{x}_n)_{\text{plate}}^{\text{impulse}} \\ & = \frac{1}{(4\pi)^2} \cdot \frac{1}{|\boldsymbol{u} - \boldsymbol{x}_n|^2} \cdot \int_{-D_y/2}^{+D_y/2} w_{\Pi(T_p@t_n)} \left[t - t_d(s_y - \boldsymbol{x}_n)\right]_n^{\text{chirp}} \cdot ds_y, \\ & \approx \frac{1}{2\pi} \int_{\beta_p} W(\omega_p) \cdot \underbrace{D_y^2 \cdot \operatorname{sinc}^2 \left[\frac{D_y}{2} \cdot 2k_{\omega_p} \cos\phi_y(\widehat{\boldsymbol{u} - \boldsymbol{x}_n})\right]}_{\mathbb{A}^2(\omega_p, \widehat{\boldsymbol{u} - \boldsymbol{x}_n})_{\text{plate}}} \\ & \qquad \cdot Hh^2(\omega_p, \boldsymbol{u} - \boldsymbol{x}_n) \cdot \exp(j\omega_p t) \cdot d\omega_p, \end{aligned} \tag{12.10}$$

where $t_d(s_y - \boldsymbol{x}_n)$ is the two way signal delay between the scatterer position $\boldsymbol{x}_n$ and points s_y on the flat plate antenna surface. The two way weight function $\mathbb{A}^2(\omega_p, \widehat{\boldsymbol{u} - \boldsymbol{x}_n})_{\text{plate}}$ induced in the frequency-space domain by the flat plate antenna in (12.10) is

$$\mathbb{A}^2(\omega_p, \widehat{\boldsymbol{u} - \boldsymbol{x}_n})_{\text{plate}} = D_y^2 \cdot \operatorname{sinc}^2 \left[\frac{D_y}{2} \cdot 2k_{\omega_p} \cos\phi_y(\widehat{\boldsymbol{u} - \boldsymbol{x}_n})\right].$$

Equation (12.10) gives a time-space domain description of the spatial amplitude modulation of the (frequency dependent) antenna beam pattern. Using this result, the time-space domain model of spotlight mode SAR can be characterized, as an alternative to (12.7), by

$$\begin{aligned} & s_{\Pi[T_s,T_f]} s_{\Pi(2L)}(t, \boldsymbol{u})_{\text{spot}}^{\text{chirp}} \\ & \quad \approx \Pi_{[-L,+L]}(\boldsymbol{u}) \cdot \sum_n \frac{1}{2\pi} \int_{\beta_p} W(\omega_p) \cdot \mathbb{A}^2(\omega_p, \widehat{\boldsymbol{u} - \boldsymbol{x}_n})_{\text{plate}} \cdot \\ & \qquad g_n \cdot Hh^2(\omega_p, \boldsymbol{u} - \boldsymbol{x}_n) \cdot \exp(j\omega_p t) \cdot d\omega_p, \end{aligned} \tag{12.11}$$

which is based on a temporal inverse Fourier transform.

Spotlight Frequency-Space Phase History

The task at hand is to describe the frequency-space phase history of spotlight mode SAR. We do this by comparison with the established slant-plane model.

Slant-Plane Frequency-Space Phase Histories

For the case of a transmitted chirp pulse waveform, the frequency-space passband phase history for the slant-plane signal model with an infinitely long synthetic aperture and a point antenna is

$$S_{\Pi(2\omega_o @ \omega_c)} s_{\Pi(\infty)}(\omega_p, \boldsymbol{u})_{\text{slant}}^{\text{chirp}} = W_{\Pi(2\omega_o @ \omega_c)}(\omega_p)^{\text{chirp}} \cdot \sum_n g_n \cdot Hh^2(\omega_p, \boldsymbol{u} - \boldsymbol{x}_n),$$

where $W_{\Pi(2\omega_o @ \omega_c)}(\omega_p)^{\text{chirp}}$ is the spectrum of the passband chirp pulse waveform.

Spotlight Frequency-Space Phase Histories

Using the frequency-space channel model (10.9), but modified with substitution of a chirp pulse spectrum, the corresponding frequency-space domain model for the spotlight mode system with a flat plate antenna is

$$\begin{aligned} S_{\Pi(2\omega_o @ \omega_c)} s_{\Pi(2L)}(\omega_p, \boldsymbol{u})_{\text{spot}}^{\text{chirp}} = & \\ & \Pi_{[-L,L]}(\boldsymbol{u}) \cdot W_{\Pi(2\omega_o @ \omega_c)}(\omega_p)^{\text{chirp}} \\ & \cdot \sum_n \mathbb{A}^2(\omega_p, \widehat{\boldsymbol{u} - \boldsymbol{x}_n})_{\text{plate}} \cdot g_n \cdot Hh^2(\omega_p, \boldsymbol{u} - \boldsymbol{x}_n). \end{aligned} \tag{12.12}$$

The spotlight frequency-space phase history has a slow-time window function

$$\Pi_{[-L,L]}(\boldsymbol{u})$$

associated with the finite synthetic aperture and a frequency-space weight function

$$\mathbb{A}^2(\omega_p, \widehat{\boldsymbol{u} - \boldsymbol{x}_n})_{\text{plate}} = D_y^2 \cdot \text{sinc}^2\left[\frac{D_y}{2} \cdot 2k_{\omega_p} \cos\phi_y(\widehat{\boldsymbol{u} - \boldsymbol{x}_n})\right]$$

induced by the flat plate antenna. This aperture weight function is dependent on frequency and a direction parameter $\widehat{\boldsymbol{u} - \boldsymbol{x}_n}$ defined within the antenna beam's field of view. This antenna frequency-space weight function comes from the description (9.44) of the (center referenced) antenna beam pattern that illuminates the ground spot.

Spotlight Mode Angular Dwells

If the synthetic aperture length is expressed in terms of angular dwell $\theta_y(u_y)_n$, then slow-time dwells in spotlight mode are different for each scatterer within the illuminated spot[2].

A diagram of how an individual scatterer's angular dwell is determined by its position within the ground spot is shown in Fig. 12.2. Scatterers near the boresight of a broadside scene experience larger angular dwells. The angular dwell decreases as the position of a scatterer moves further off boresight. Angular dwell also decreases as the position of a scatterer moves further down range.

[2]This characteristic does not hold for a SAR system operated in stripmap mode, where scatterer specific angular dwells, as defined by the antenna's angular field of view, are the same for all scatterers illuminated by the passing antenna beam.

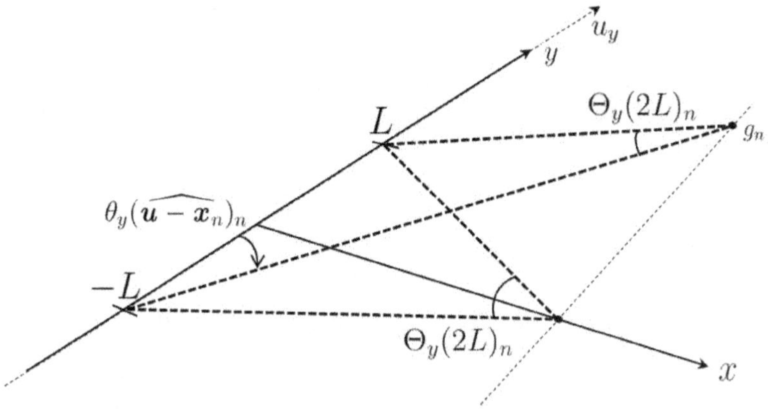

Figure 12.2: Angular dwells of boresight and off boresight scatterers.

Common Synthetic Aperture Length of Spotlight Mode

The along-track spotlight mode window function $\Pi_{[-L,L]}(u_y)$ is the same for all scatterers in the ground scene[3]. To better model the role of a finite synthetic aperture in expressions of the frequency-space signal spectrum, we describe the finite aperture of spotlight mode in terms of scatterer specific angular dwells.

Scatterer Specific Spotlight Angular Dwell and Window

The view angle span

$$\Theta_y(2L)_n = [\theta_y(-L)_n, \theta_y(+L)_n]$$

experienced over a finite aperture of length $2L$ determines the angular dwell of the nth scatterer. We can define scatterer specific slow-time windows in terms of angular dwell,

$$\Pi_{\Theta_y}(\widehat{\boldsymbol{u}-\boldsymbol{x}_n})_n = \begin{cases} 1 & \text{for } \theta_y(\widehat{\boldsymbol{u}-\boldsymbol{x}_n})_n \in \Theta_y(2L)_n, \\ 0 & \text{otherwise.} \end{cases} \tag{12.13}$$

Frequency-Space Phase Histories with Angular Dwell Expressions

Hence, for the nth scatterer, $\Pi_{[-L,+L]}(\boldsymbol{u})$ and $\Pi_{\Theta_y}(\widehat{\boldsymbol{u}-\boldsymbol{x}_n})_n$ are equivalent indicator functions for the $\boldsymbol{u}$ within the synthetic aperture and we have

$$\begin{aligned} & S_{\Pi(2\omega_o @ \omega_c)} s_{\Pi(2L)}(\omega_p, \boldsymbol{u})^{\text{chirp}}_{\text{spot}} \\ = \; & W_{\Pi(2\omega_o @ \omega_c)}(\omega_p)^{\text{chirp}} \cdot \sum_n \Pi_{\Theta_y}(\widehat{\boldsymbol{u}-\boldsymbol{x}_n})_n \cdot \mathbb{A}^2(\omega_p, \widehat{\boldsymbol{u}-\boldsymbol{x}_n})_{\text{plate}} \\ & \cdot g_n \cdot Hh^2(\omega_p, \boldsymbol{u}-\boldsymbol{x}_n). \end{aligned} \tag{12.14}$$

[3]The characteristic of a common slow-time window function for all scatterers does not hold for a SAR system operated in stripmap mode. Scatterers enter and leave the laterally moving stripmap mode antenna beam at different points in slow-time, depending upon scatterer range from the SAR sensor.

The synthetic aperture window function

$$\Pi_{[-L,L]}(\boldsymbol{u})$$

expressed in terms of antenna position $\boldsymbol{u}$ in (12.12) has been replaced by

$$\Pi_{\Theta_y}(\widehat{\boldsymbol{u}-\boldsymbol{x}_n})_n \tag{12.15}$$

in (12.14). The slow-time window function (12.15) is an indicator of scatterer specific angular dwell over the sensor-scatterer directional sweep of $\widehat{\boldsymbol{u}-\boldsymbol{x}_n}$.

Spotlight Frequency-Wavenumber Phase History

To describe the frequency-wavenumber phase history, our approach is to again do a side by side comparison of the slant-plane and spotlight mode signal models. But here, we leverage the slant-plane model with truncation to a finite synthetic aperture as described in terms of scatterer specific angular dwells (see Section 8.5).

Slant-Plane Frequency-Wavenumber Chirp Phase History

The signal spectrum for the slant-plane system with a finite synthetic aperture is

$$\begin{aligned} & S_{\hat{\Pi}(2\omega_o @ \omega_c)} S_{\Pi(\Theta_y)}(\omega_p, k_{u_y}, \boldsymbol{u})^{\text{chirp}}_{\text{slant}} \\ = \; & W_{\Pi(2\omega_o @ \omega_c)}(\omega_p)^{\text{chirp}} \cdot a^2(\omega_p, k_{u_y}) \cdot \sum_n \Pi_{\Theta_y}(\omega_p, k_{u_y})_n \\ & \cdot g_n \cdot H^2_{x_n(\boldsymbol{u})}(\omega_p, k_{u_y}) \cdot H^2_{y_n(\boldsymbol{u})}(k_{u_y}), \end{aligned}$$

where

$$a^2(\omega_p, k_{u_y})$$

is the two way amplitude factor of the (ω_p, k_{u_y}) domain signal spectrum[4] and where

$$\Pi_{\Theta_y}(\omega_p, k_{u_y})_n = \left[k_{\omega_p}\cos\theta_y(+L)_n, k_{\omega_p}\cos\theta_y(-L)_n\right]$$

is the scatterer specific aperture wavenumber window function expressed in terms of angular dwell.

Spotlight Frequency-Wavenumber Chirp Phase History

The corresponding model for spotlight mode is

$$\begin{aligned} & S_{\hat{\Pi}(2\omega_o @ \omega_c)} S_{\Pi(\Theta_y)}(\omega_p, k_{u_y}, \boldsymbol{u})^{\text{chirp}}_{\text{spot}} \\ = \; & W_{\Pi(2\omega_o @ \omega_c)}(\omega_p)^{\text{chirp}} \cdot a^2(\omega_p, k_{u_y}) \\ & \cdot \sum_n \Pi_{\Theta_y}(\omega_p, k_{u_y})_n \cdot \mathbb{A}^2(\omega_p, k_{u_y}, \boldsymbol{u}-\boldsymbol{x}_n)_{\text{plate}} \\ & \cdot g_n \cdot H^2_{x_n(\boldsymbol{u})}(\omega_p, k_{u_y}) \cdot H^2_{y_n(\boldsymbol{u})}(k_{u_y}), \end{aligned} \tag{12.16}$$

[4] Obtained from either the slow-time Fourier transform as evaluated by the method of stationary phase or a weighting induced by a plane wave decomposition within the slant-plane slice of a spherical wave field.

where the antenna in situ field description in the frequency-wavenumber domain is

$$\begin{aligned}
&\mathbb{A}(\omega_p, k_{s_y}, \boldsymbol{x}_n - \boldsymbol{u})_{\text{plate}} \qquad (12.17)\\
&= \int_{\mathbb{S}} a_e[\boldsymbol{s}(\boldsymbol{u})] \cdot \frac{1}{4\pi} \cdot \frac{1}{|\boldsymbol{x}_n - \boldsymbol{s}(\boldsymbol{u})|} \\
&\qquad \cdot \exp[js_x(\boldsymbol{u}) \cdot \breve{k}_x(\omega_p, k_{s_y})] \cdot \exp[js_y(\boldsymbol{u}) \cdot k_{s_y}] \cdot d\boldsymbol{s}, \\
&= \int_{-D_y/2}^{+D_y/2} 1 \cdot \frac{1}{4\pi} \cdot \frac{1}{\sqrt{x_n^2 + [(u_y + s_y) - y_n]^2}} \cdot \exp\left(js_y k_{s_y}\right) \cdot ds_y.
\end{aligned}$$

We replace k_{s_y} of (12.17) with k_{u_y} when substituted into (12.16) because of the flat plate antenna's s_y axis alignment with u_y. We have assumed uniform unity excitation of the flat plate antenna surface,

$$a_e[\boldsymbol{s}(\boldsymbol{u})] = 1.$$

If $\boldsymbol{x}_n$ is sufficiently distant such that

$$\begin{aligned}
\frac{1}{\sqrt{x_n^2 + [(u_y + s_y) - y_n]^2}} &\approx \frac{1}{\sqrt{x_n^2 + (u_y - y_n)^2}}, \\
&= \frac{1}{|\boldsymbol{u} - \boldsymbol{x}_n|},
\end{aligned}$$

then for a flat plate antenna located at $\boldsymbol{u}$,

$$\begin{aligned}
&\mathbb{A}(\omega_p, k_{u_y}, \boldsymbol{u} - \boldsymbol{x}_n)_{\text{plate}} \\
&\approx \frac{1}{4\pi} \cdot \frac{1}{\sqrt{x_n^2 + (u_y - y_n)^2}} \cdot \int_{-\infty}^{+\infty} \Pi_{(D_y)}(s_y) \cdot \exp\left(jk_{s_y} s_y\right) \cdot ds_y, \\
&= \frac{1}{4\pi} \cdot \frac{1}{|\boldsymbol{u} - \boldsymbol{x}_n|} \cdot D_y \cdot \operatorname{sinc}\left(\frac{D_y}{2} \cdot k_{u_y}\right), \qquad (12.18)
\end{aligned}$$

where the unit integrand "1" over s_y of the flat plate antenna with diameter D_y has been interpreted as a rect function $\Pi_{(D_y)}(s_y)$. From (12.18), the induced frequency-wavenumber domain weight function is invariant to transmit frequency and hence we write,

$$\mathbb{A}(\omega_p, k_{u_y}, \boldsymbol{u} - \boldsymbol{x}_n)_{\text{plate}} \mapsto \mathbb{A}(k_{u_y}, \boldsymbol{u} - \boldsymbol{x}_n)_{\text{plate}}.$$

In summary, the frequency-wavenumber phase history model for spotlight mode is

$$\begin{aligned}
&S_{\hat{\Pi}(2\omega_o @ \omega_c)} S_{\Pi(\Theta_y)}(\omega_p, k_{u_y}, \boldsymbol{u})_{\text{spot}}^{\text{chirp}} \qquad (12.19)\\
&\approx W_{\Pi(2\omega_o @ \omega_c)}(\omega_p)^{\text{chirp}} \cdot a^2(\omega_p, k_{u_y}) \cdot \sum_n \Pi_{\Theta_y}(\omega_p, k_{u_y})_n \cdot \mathbb{A}^2(k_{u_y}, \boldsymbol{u} - \boldsymbol{x}_n)_{\text{plate}} \\
&\quad \cdot g_n \cdot H^2_{x_n(\boldsymbol{u})}(\omega_p, k_{u_y}) \cdot H^2_{y_n(\boldsymbol{u})}(k_{u_y}).
\end{aligned}$$

The dependency of (12.19) on $\boldsymbol{u}$ is descriptive of amplitude weights that depend on distances $|\boldsymbol{u} - \boldsymbol{x}_n|$ as contained within $\mathbb{A}(k_{u_y}, \boldsymbol{u} - \boldsymbol{x}_n)_{\text{plate}}$ of (12.18).

Spotlight Angular Spectrum

We next examine the angular spectrum of spotlight mode and contrast it with the model for the modeless slant-plane system.

Slant-Plane Angular Spectrum Chirp Phase History

The slant-plane angular spectrum is

$$\begin{aligned} &\overset{\text{chirp}}{M}_{\Pi(\otimes)} S_{\Pi(\Theta_y)}(\breve{k}_x, k_y, \boldsymbol{u})^{\text{slant}}_{\text{ACS}} \\ &= \sum_n \widehat{\Pi}^{\text{chirp}}_{\otimes}(\breve{k}_x, k_y)_n \cdot \widehat{\Pi}_{\Theta_y}(k_y)_n \cdot g_n \cdot \exp(-jx_n\breve{k}_x) \cdot \exp(-j\hat{y}_n k_y), \end{aligned}$$

where

$$\begin{aligned} &\widehat{\Pi}^{\text{chirp}}_{\otimes}(\breve{k}_x, k_y)_n \cdot \widehat{\Pi}_{\Theta_y}(k_y)_n \\ &= \mathsf{ST}^{k_x}_{\text{1D}}\left\{ W_{\Pi(2\omega_o @ \omega_c)}(\omega_p)^{\text{chirp}} \cdot a^2(\omega_p, k_{u_y}) \cdot \Pi_{\Theta_y}(\omega_p, k_{u_y})_n \right\}. \end{aligned}$$

Spotlight Angular Spectrum Chirp Phase History

The corresponding spotlight mode model for a flat plate antenna is

$$\begin{aligned} &\overset{\text{chirp}}{M}_{\Pi(\otimes)} S_{\Pi(\Theta_y)}(\breve{k}_x, k_y, \boldsymbol{u})^{\text{spot}}_{\text{ACS}} \\ &= \sum_n \mathbb{A}^2(k_y, \boldsymbol{u} - \boldsymbol{x}_n)_{\text{plate}} \cdot \widehat{\Pi}^{\text{chirp}}_{\otimes}(\breve{k}_x, k_y)_n \cdot \widehat{\Pi}_{\Theta_y}(k_y)_n \\ &\quad \cdot g_n \cdot \exp(-jx_n\breve{k}_x) \cdot \exp(-j\hat{y}_n k_y). \end{aligned}$$

Spotlight Huygens-Fresnel Analysis

The spotlight mode angular spectrum in the aperture coordinate system is

$$\overset{\text{chirp}}{M}_{\Pi(\otimes)} S_{\Pi(\Theta_y)}(\breve{k}_x, k_y, \boldsymbol{u})^{\text{spot}}_{\text{ACS}} = \sum_n \mathbb{A}^2(k_y, \boldsymbol{u} - \boldsymbol{x}_n)_{\text{plate}} \cdot \widetilde{G}^{\text{chirp}}_{\widehat{\Pi}(\otimes)} \widetilde{G}_{\widehat{\Pi}(\Theta_y)}(\breve{k}_x, k_y)^n_{\text{ACS}},$$

where a scatterer specific element of the shattered terrain spectrum is

$$\begin{aligned} &\widetilde{G}^{\text{chirp}}_{\widehat{\Pi}(\otimes)} \widetilde{G}_{\widehat{\Pi}(\Theta_y)}(\breve{k}_x, k_y)^n_{\text{ACS}} \\ &= \widehat{\Pi}^{\text{chirp}}_{\otimes}(\breve{k}_x, k_y)_n \cdot \widehat{\Pi}_{\Theta_y}(k_y)_n \cdot g_n \cdot \exp(-jx_n \cdot \breve{k}_x) \cdot \exp(-jy_n \cdot k_y). \end{aligned}$$

The Huygens-Fresnel decomposition of the spotlight mode angular spectrum is

$$\begin{aligned} &\overset{\text{chirp}}{M}_{\widehat{\Pi}(\otimes)} S_{\widehat{\Pi}(\Theta_y)}(\breve{k}_x, k_y, \boldsymbol{u} - \boldsymbol{X}_c)^{\text{spot}}_{\text{ACS}} \\ &\approx H^2_{X_c(u)}(\breve{k}_{\hat{x}}) \cdot H^2_{Y_c(u)}(k_{\hat{y}}) \cdot \mathbb{A}^2(k_{\hat{y}}, \boldsymbol{u} - \boldsymbol{X}_c)_{\text{plate}} \cdot \sum_n \widetilde{G}^{\text{chirp}}_{\widehat{\Pi}(\otimes)} \widetilde{G}_{\widehat{\Pi}(\Theta_y)}(\breve{k}_{\hat{x}}, k_{\hat{y}})^n_{\text{SCS}}, \\ &= H^2_{X_c(u)}(\breve{k}_{\hat{x}}) \cdot H^2_{Y_c(u)}(k_{\hat{y}}) \cdot \mathbb{A}^2(k_{\hat{y}}, \boldsymbol{u} - \boldsymbol{X}_c)_{\text{plate}} \cdot \widetilde{G}^{\text{chirp}}_{\widehat{\Pi}(\otimes)} \widetilde{G}_{\widehat{\Pi}(\Theta_y)}(\breve{k}_{\hat{x}}, k_{\hat{y}})^{\text{spot}}_{\text{SCS}}, \end{aligned} \tag{12.20}$$

where

$$\begin{aligned}&\widetilde{G}^{\text{chirp}}_{\hat{\Pi}(\otimes)}\widetilde{G}_{\hat{\Pi}(\Theta_y)}(\breve{k}_{\hat{x}}, k_{\hat{y}})^{n}_{\text{SCS}}\\&= \widehat{\Pi}^{\text{chirp}}_{\otimes}(\breve{k}_{\hat{x}}, k_{\hat{y}})_n \cdot \widehat{\Pi}_{\Theta_y}(k_{\hat{y}})_n \cdot g_n \cdot \exp(-j\hat{x}_n\breve{k}_{\hat{x}}) \cdot \exp(-j\hat{y}_n k_{\hat{y}})\end{aligned}$$

and

$$\widetilde{G}^{\text{chirp}}_{\hat{\Pi}(\otimes)}\widetilde{G}_{\hat{\Pi}(\Theta_y)}(\breve{k}_{\hat{x}}, k_{\hat{y}})^{\text{spot}}_{\text{SCS}} = \sum_n \widetilde{G}^{\text{chirp}}_{\hat{\Pi}(\otimes)}\widetilde{G}_{\hat{\Pi}(\Theta_y)}(\breve{k}_{\hat{x}}, k_{\hat{y}})^{n}_{\text{SCS}}.$$

Spotlight Huygens-Fresnel Inversion

Application of inverse Huygens-Fresnel transfers to (12.20) gives

$$\begin{aligned}&M^{\text{chirp}}_{\hat{\Pi}(\otimes)} S_{\hat{\Pi}(\Theta_y)}(\breve{k}_{\hat{x}}, k_{\hat{y}}, \boldsymbol{u} - \boldsymbol{X}_c)^{\text{spot}}_{\text{SCS}}\\&= \mathbb{A}^2(k_{\hat{y}}, \boldsymbol{u} - \boldsymbol{X}_c)_{\text{plate}} \cdot \widetilde{G}^{\text{chirp}}_{\hat{\Pi}(\otimes)}\widetilde{G}_{\hat{\Pi}(\Theta_y)}(\breve{k}_{\hat{x}}, k_{\hat{y}})^{\text{spot}}_{\text{SCS}}.\end{aligned}$$

Compensation of the antenna weight function gives

$$M^{\text{chirp}}_{\hat{\Pi}(\otimes)} S_{\hat{\Pi}(\Theta_y)}(\breve{k}_{\hat{x}}, k_{\hat{y}})^{\text{spot}}_{\text{SCS}} = \widetilde{G}^{\text{chirp}}_{\hat{\Pi}(\otimes)}\widetilde{G}_{\hat{\Pi}(\Theta_y)}(\breve{k}_{\hat{x}}, k_{\hat{y}})^{\text{spot}}_{\text{SCS}}.$$

The sensed terrain spectrum has been shattered by the scatterer specific windowing effects of spotlight mode.

Broadside Spotlight Mode Along-Track Presumming

Presumming for a broadside scene in spotlight mode can be performed at various points in the signal processing work flow as explained in Section 7.12.

Squint Spotlight Mode Along-Track Presumming

Convention holds that the PRF must be increased to handle squint geometries in SAR. Compared to the broadside case, a squint imaging geometry creates a slow-time passband signal in the k_{u_y} domain with higher peak aperture wavenumbers, but with smaller aperture wavenumber bandwidth. Hence, the PRF used in a squint spotlight Stolt formatting system can actually be reduced (because of the reduced bandwidth), but at the expense of additional digital processing required in slow-time for aperture wavenumber down conversion from a passband to baseband representation [440, 773, 885, 894].

Spotlight Mode Uniform Resampling

Uniform resampling of the Fourier migrated spectrum for spotlight mode Stolt formatted data proceeds without modification as described for the slant-plane system in Section 7.14.

Spotlight Mode Image Spectrum

After fast-time down conversion, pulse compression, slow-time down conversion (if squinted), beam pattern compensation, presumming, Huygens-Fresnel inversion to the scene coordinate system and uniform resampling of Stolt formatted data arrays, we obtain the spotlight mode image spectrum

$$\begin{aligned} II[n_{\hat{x}}, n_{\hat{y}}]^{\text{spot}}_{\text{SCS}} &= \text{Uniform Resample}\left\{ M^{\text{chirp}}_{\hat{\Pi}(\otimes)} S_{\hat{\Pi}(\Theta_y)}(\breve{k}_{\hat{x}}, k_{\hat{y}})^{\text{spot}}_{\text{SCS}} \right\}, \\ &= I^{\text{chirp}}_{\hat{\Pi}(\otimes)} I_{\hat{\Pi}(\Theta_y)}(\ddot{k}_{\hat{x}}[n_{\hat{x}}], k_{\hat{y}}[n_{\hat{y}}])^{\text{spot}}_{\text{SCS}}. \end{aligned}$$

Offset Fourier Imaging in Stolt Formatting

Typically, the rectilinearly resampled (passband) wavenumber raster is shifted to the origin for Fourier inversion to reduce FFT size. Although this translation within wavenumber space does not change the detected image, it injects a phase carrier into the complex pixel domain, which must be compensated [250,679,682].

Spotlight Mode Imagery

One of the distinguishing properties of the spotlight mode of operation is that the image domain point spread function varies with scatterer location. Spotlight mode imagery is inherently spatially variant. Scatterers more distant from the radar flight line experience smaller angular dwells and hence have broadened point spread functions with less resolution. Scatterers closer to the flight line are resolved with better resolution. This is caused by attributes of the shattered spectrum as induced by the angular dwells of the finite aperture in spotlight mode.

Scatterer Specific Manifold Size and Position Parameters

The size and position of each scatterer specific angular dwell window function

$$\Pi_{\Theta_y}(\omega_p, k_y)_n \tag{12.21}$$

of spotlight mode in the along-track wavenumber phase history domain is dependent on the instantaneous frequency of the chirp signal. The angular dwell wavenumber window function (12.21) for the nth scatterer increases with increasing chirp frequency and is centered at an along-track wavenumber k_{u_y} value determined by

$$\Pi^{\text{center}}_{\Theta_y}(\omega_p, k_y)_n \approx 2k_{\omega_c}\cos\theta_y(0)_n.$$

The slow-time chirp bandwidth of a particular scatterer in a broadside spot is approximately given by [885]

$$\Pi^{\text{width}}_{\Theta_y}(\omega_p, k_y)_n \approx 2 \times 2k_{\omega_p}\cos\theta^{\max}_y(2L)_n \cdot \cos\theta_x(0)_n, \tag{12.22}$$

where $\theta^{\max}_y(2L)_n$ is the largest view angle towards the nth scatterer experienced within the synthetic aperture of length $2L$. As expressed in (12.22), an off boresight scatterer in a broadside scene experiences a bandwidth reduction factor given by $\cos\theta_x(0)_n$. Furthermore, since

$$\Pi^{\text{center}}_{\Theta_y}(\omega_p, k_y)_n \neq 0$$

for an off boresight scatterer in a broadside scene, its phase curtain is a slow-time passband signal.

Spotlight Mode Along-Track Point Spread Function

The one dimensional slow-time inverse discrete Fourier transform of

$$II[n_{\hat{x}}, n_{\hat{y}}]_{\text{SCS}}^{\text{spot}} = II(\ddot{k}_{\hat{x}}[n_{\hat{x}}], k_{\hat{y}}[n_{\hat{y}}])_{\text{SCS}}^{\text{spot}}$$

returns the frequency dependent along-track (only) focused SAR image

$$Ii_{\Pi(\Theta_y)}^{\text{psf}}(\omega_p, \hat{y})_{\text{SCS}}^{\text{spot}} = \sum_n Ii_{\Pi(\Theta_y)}^{\text{psf}}(\omega_p, \hat{y} - \hat{y}_n)_{\text{SCS}}^{n},$$

where the partially focused complex pixels are described by [885]

$$\begin{aligned} & Ii_{\Pi(\Theta_y)}^{\text{psf}}(\omega_p, \hat{y})_{\text{SCS}}^{n} \\ = \; & \underbrace{\sigma_n \cdot \exp\left[j \cdot \underset{\Theta_y}{\Pi}^{\text{center}}(\omega_p, k_{\hat{y}})_n \cdot \hat{y}\right]}_{\text{complex sinusoid}} \\ & \underbrace{\cdot \underset{\Theta_y}{\Pi}^{\text{width}}(\omega_p, k_{\hat{y}})_n \cdot \operatorname{sinc}\left[\frac{1}{2\pi} \underset{\Theta_y}{\Pi}^{\text{width}}(\omega_p, k_{\hat{y}})_n \cdot \hat{y}\right]}_{\text{point spread function}}. \end{aligned} \tag{12.23}$$

The along-track image point spread function is a $\mathsf{sinc}(\cdot)$ function. The resolution and magnitude are inversely proportional to the slow-time bandwidth

$$\underset{\Theta_y}{\Pi}^{\text{width}}(\omega_p, k_{\hat{y}})_n$$

of the off boresight scatterer. There is also an embedded slow-time carrier (complex sinusoid in the $\hat{y}$ dimension of the complex pixel array),

$$\exp\left[j \cdot \underset{\Theta_y}{\Pi}^{\text{center}}(\omega_p, k_{\hat{y}})_n \cdot \hat{y}\right] \tag{12.24}$$

caused by along-track k-space offsets of spectral fragments from terrain spectrum shattering. If the point spread function is sufficiently localized, then (12.24) is more akin to a phase shift at that pixel (point spread function) location, rather than a complex tone that extends over an interval of $\hat{y}$ domain pixels. Since this phase shift or carrier is scatterer specific, compensation may be problematic.

Spotlight Mode Along-Track Resolution

An analytic expression for the along-track resolution can be approximated by the null to null width of the mainlobe of the sinc function:

$$\begin{aligned} \Delta_{\hat{y}_n} &= \frac{2\pi}{\underset{\Theta_y}{\Pi}^{\text{width}}(k_{\hat{y}})_n}, \\ &\approx \frac{\lambda_c}{4} \cdot \frac{1}{\sin\theta_x^{\max}(2L)_n} \cdot \frac{1}{\cos\theta_x(0)_n}, \end{aligned}$$

where

$$\sin\theta_x^{\max}(2L)_n = \cos\theta_y^{\max}(2L)_n.$$

In short range, wide beam, spotlight mode SAR, the $\theta_x^{\max}(2L)_n$ angles vary significantly for different scatterers within the ground scene and the point spread function of spotlight mode SAR imagery is noticeably spatially variant.

In long range, narrow beam, spotlight mode SAR, all scatterer dependent dwells Θ_n are approximated well by the observational dwell Θ_c of the center of the spot. In which case, the spatially variant nature of the SAR imagery may not be perceptually noticeable.

Angular Decomposition of Aperture Wavenumber Bandwidth

We next explore the nature of the spotlight mode window function

$$\Pi_{\Theta_y}(\omega_p, k_{u_y})_{\text{spot}}$$

of the frequency-wavenumber phase history and decompose this manifold into an element related to the antenna's field of view and an element related to the change of position of the antenna as it moves through the synthetic aperture.

Antenna Movement Induced Wavenumber Manifold

An antenna that moves through a linear synthetic aperture induces larger angular dwells than those that are afforded by just the field of view of the stationary antenna's angular beamwidth. These expanded angular dwells in the spatial domain map to larger angular dwell window functions in the frequency-space and frequency-wavenumber domains.

Slow-Time Chirp Window Functions

Window functions on both sides of a slow-time spatial Fourier transform behave like the amplitude modulation windows of temporal chirp signals. The Heisenberg relationship between observation and precision seems suspended. Expanded window functions (achieved through sensor movement) in the spatial domain map to expanded wavenumber manifolds in the Fourier domain.

Broadside Spotlight Aperture Wavenumber Window

For a spotlight mode system, the union of scatterer specific aperture wavenumber window functions is

$$\begin{aligned}\Pi_{\Theta_y}(\omega_p, k_{u_y})_{\text{spot}} &= \bigcup_{n\in\text{spot}} \Pi_{\Theta_y}(\omega_p, k_{u_y})_n, \\ &= [2k_{\omega_p}\cos\theta_y^{\max}, 2k_{\omega_p}\cos\theta_y^{\min}],\end{aligned}$$

where $\theta_y^{\min}$ and $\theta_y^{\max}$ are the extrema of view angles experienced within the slow-time dwell over the entire spot scene. With an abuse of notation, we write this as

$$\Pi_{\Theta_y}(\omega_p, k_{u_y})_{\text{spot}} = 2k_{\omega_p}\cos\Theta_y^{\text{spot}}, \tag{12.25}$$

where

$$\Theta_y^{\text{spot}} = [\theta_y^{\min}, \theta_y^{\max}],$$

with $\theta_y^{\min}$ and $\theta_y^{\max}$ the extrema of view angles between the sensor and all spot scene ground points.

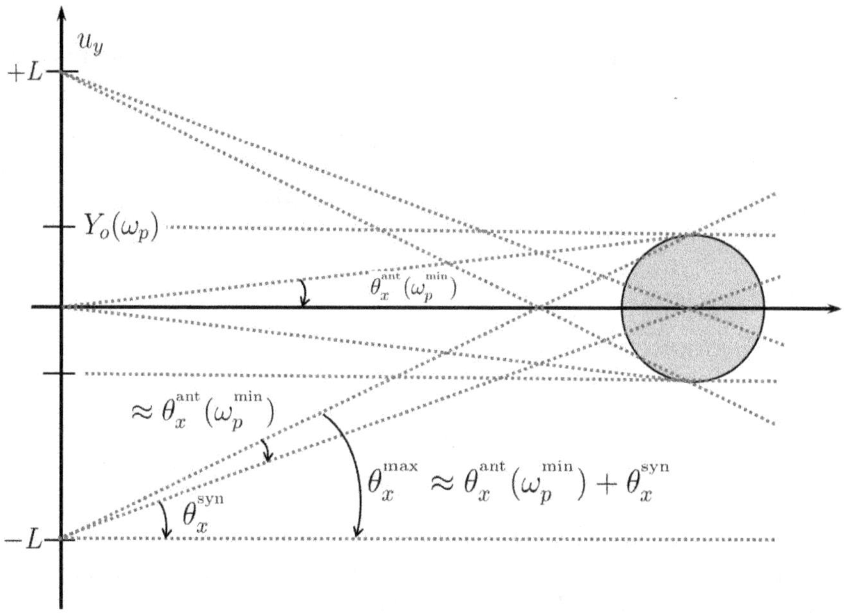

Figure 12.3: Antenna field of view and antenna movement angular dwells.

Long Range Approximations of Broadside Angular Dwells

Note that

$$\cos\theta_y = \sin\theta_x$$

and for the broadside case, also note that

$$\theta_x^{\min} = -\theta_x^{\max}.$$

Furthermore, if we assume that the range is sufficiently long such that

$$\sin\theta_x^{\max}$$

can be simplified with a small angle approximation,

$$\sin\theta_x^{\max} \approx \theta_x^{\max},$$

then, for small angles,

$$\sin\theta_x^{\max} \approx \tan\theta_x^{\max}$$

and as illustrated in Fig. 12.3, we can approximate the extrema of view angles by

$$\begin{aligned}\theta_x^{\max} &\approx \frac{Y_o(\omega_p)+L}{X_c},\\ &= \frac{Y_o(\omega_p)}{X_c}+\frac{L}{X_c}.\end{aligned}$$

Now, since the (stationary) antenna's field of view half-angle is approximately,

$$\theta_x^{\text{ant}}(\omega_p) \approx \frac{Y_o(\omega_p)}{X_c},$$

and since the largest synthetic aperture sensor-scatterer half-angle is approximately,

$$\theta_x^{\text{syn}} \approx \frac{L}{X_c},$$

we have, as illustrated in Fig. 12.3,

$$\theta_x^{\max} \approx \theta_x^{\text{ant}}(\omega_p^{\min}) + \theta_x^{\text{syn}}.$$

Frequency-Wavenumber Window Function Angular Decomposition

The manifold of the frequency-wavenumber phase history can be expressed in terms of the sum of the angular supports due to the field of view of the antenna $\theta_x^{\text{ant}}(\omega_p)$ and due to movement of the antenna θ_x^{syn} through the synthetic aperture. Hence, the broadside version of (12.25) can be approximated by

$$\begin{aligned}&\Pi_{\Theta_y}(\omega_p, k_{u_y})_{\text{spot}}^{\text{broad}}\\ &\quad= 2k_{\omega_p}\cos\Theta_y^{\text{spot}},\\ &\quad\approx \left[-2k_{\omega_p}\sin\left(\theta_x^{\text{ant}}(\omega_p^{\min})+\theta_x^{\text{syn}}\right), +2k_{\omega_p}\sin\left(\theta_x^{\text{ant}}(\omega_p^{\min})+\theta_x^{\text{syn}}\right)\right].\end{aligned} \tag{12.26}$$

Equation (12.26) is useful for characterizing the wavenumber support of phase histories in terms of a stationary antenna's field of view angular beamwidth parameters plus angular dwell parameters associated with antenna movement.

Additional Decomposition Details for Flat Plate Antenna

Now, with

$$\theta_x^{\text{ant}}(\omega_p^{\min}) = \frac{\lambda_{\max}}{D_y}$$

for the beamwidth of a flat plate antenna, we have [885]

$$\Pi_{\Theta_y}(\omega_p, k_{u_y})_{\text{spot}}^{\text{broad}} \approx \left[-2k_{\omega_p}\sin\left(\frac{\lambda_{\max}}{D_y}+\theta_x^{\text{syn}}\right), +2k_{\omega_p}\sin\left(\frac{\lambda_{\max}}{D_y}+\theta_x^{\text{syn}}\right)\right]$$

and with a small angle approximation

$$\Pi_{\Theta_y}(\omega_p, k_{u_y})_{\text{spot}}^{\text{broad}} \approx \left[-2k_{\omega_p} \cdot \left(\frac{\lambda_{\max}}{D_y} + \theta_x^{\text{syn}}\right), +2k_{\omega_p} \cdot \left(\frac{\lambda_{\max}}{D_y} + \theta_x^{\text{syn}}\right)\right]. \tag{12.27}$$

Based on (12.27), the k-space support due to the (stationary) antenna's field of view is estimated by

$$2 \times 2k_{\omega_p} \cdot \left(\frac{\lambda_{\max}}{D_y}\right) = 2 \times \frac{4\pi}{D_y}$$

and likewise, the k-space support due to movement of the antenna is estimated by

$$2 \times 2k_{\omega_p} \cdot \theta_x^{\text{syn}} = 2 \times 2k_{\omega_p} \cdot \frac{L}{X_c}.$$

Squint Spotlight Aperture Wavenumber Window

For a squint spotlight mode of operation, we can write [885]

$$\begin{aligned}\Pi_{\Theta_y}&(\omega_p, k_{u_y})_{\text{spot}}^{\text{squint}} \\ &= \left[2k_{\omega_p}\sin\theta_x^{\text{center}} - 2k_{\omega_p}\sin\theta_x^{\text{spread}}, 2k_{\omega_p}\sin\theta_x^{\text{center}} + 2k_{\omega_p}\sin\theta_x^{\text{spread}}\right],\end{aligned} \tag{12.28}$$

where

$$\theta_x^{\text{center}} = \theta_x(0)_c$$

and where θ_x^{spread} is a measure of angular spread or field of view extent about the squinted boresight. With an increased order of approximation, the expression corresponding to (12.26) for the squint case is

$$\begin{aligned}\Pi_{\Theta_y}(\omega_p, k_{u_y})_{\text{spot}}^{\text{squint}} \approx &\left[2k_{\omega_p}\sin\theta_x^{\text{center}} - 2k_{\omega_p}\sin\left(\theta_x^{\text{ant}}(\omega_p^{\min}) + \theta_x^{\text{syn}}\right),\right. \\ &\left.2k_{\omega_p}\sin\theta_x^{\text{center}} + 2k_{\omega_p}\sin\left(\theta_x^{\text{ant}}(\omega_p^{\min}) + \theta_x^{\text{syn}}\right)\right].\end{aligned}$$

Similar to the broadside case (12.27), we have

$$\begin{aligned}\Pi_{\Theta_y}&(\omega_p, k_{u_y})_{\text{spot}}^{\text{squint}} \\ \approx &\left[2k_{\omega_p}\sin\theta_x^{\text{center}} - 2k_{\omega_p}\sin\left(\theta_x^{\text{ant}}(\omega_p^{\min}) + \theta_x^{\text{syn}}\right),\right. \\ &\left.2k_{\omega_p}\sin\theta_x^{\text{center}} + 2k_{\omega_p}\sin\left(\theta_x^{\text{ant}}(\omega_p^{\min}) + \theta_x^{\text{syn}}\right)\right], \qquad (12.29) \\ \approx &\left[2k_{\omega_p}\sin\theta_x^{\text{center}} - 2k_{\omega_p} \cdot \left(\frac{\lambda_{\max}}{D_y} + \theta_x^{\text{syn}}\right),\right. \\ &\left.2k_{\omega_p}\sin\theta_x^{\text{center}} + 2k_{\omega_p} \cdot \left(\frac{\lambda_{\max}}{D_y} + \theta_x^{\text{syn}}\right)\right].\end{aligned}$$

The along-track bandwidth decompositions (12.27) and (12.29) are useful in the design of system parameters such as the radar's PRF.

Spotlight Image Nyquist Specifications

We give the along-track Nyquist specifications related to the image (pixel) domain for spotlight mode SAR when Stolt formatting is used. This section is an addendum to the general guidelines provided in Section 7.11 for modeless slant-plane models.

Step 1: Specify the Desired Along-Track Image Resolution: There is no change from the slant-plane procedure to specify the desired along-track image resolution

$$\Delta_{\hat{y}}$$

of spotlight mode SAR.

Step 2: Determine Along-Track Image Bandwidth Requirement: There is no change from the slant-plane procedure to specify the along-track image bandwidth requirement

$$\Pi_{k_{\hat{y}}} \geq \frac{2\pi}{\Delta_{\hat{y}}}$$

of spotlight mode SAR.

Step 3: Specify the Along-Track Image Size: In spotlight mode, the antenna beamwidth must be large enough to illuminate the desired ground spot size

$$\Pi_{\hat{y}} = 2Y_o.$$

Hence, a frequency dependent antenna beamwidth for a broadside geometry must satisfy

$$\mathbb{B}(\omega_p^{\max}, X_{\min}) \quad \geq \quad 2Y_o,$$

where $\omega_p^{\max}$ is the highest instantaneous frequency of the transmitted chirp signal and $X_{\min}$ is the minimum range from sensor to points within the ground spot.

Step 4: Determine the Along-Track Image Size Pixel Count: There is no change from the slant-plane procedure to determine the along-track pixel count

$$N_{\hat{y}} = \left\lceil \frac{\Pi_{\hat{y}}}{\Delta_{\hat{y}}} \right\rceil$$

of the spotlight mode image.

Step 5: Determine Along-Track Image Wavenumber Sample Density: There is no change from the slant-plane procedure to determine the required along-track image wavenumber sample density,

$$\Delta_{k_{\hat{y}}} \quad = \quad \frac{2\pi}{N_{\hat{y}}\Delta_{\hat{y}}}.$$

Step 6: Determine Synthetic Aperture Slow-Time Bandwidth Requirement: In general, the aperture wavenumber bandwidth requirement of the finite synthetic aperture in spotlight mode is bounded by

$$2k_{\omega_p}\cos\Theta_y(2L)_{\text{spot}} \le \Pi_{\Theta_y}(\omega_p, k_{u_y})_{\text{spot}} \le 2k_{\omega_p}. \tag{12.30}$$

In broadside mode, we must have

$$2\times 2k_{\omega_p}\cos\theta_y^{\min}(-L)_{\text{spot}} \le \Pi_{\Theta_y}(\omega_p, k_{u_y})_{\text{spot}} \le 2k_{\omega_p}, \tag{12.31}$$

where the first factor of two comes from the symmetry of a broad side spot geometry and where the view angle $\theta_y^{\min}(-L)_{\text{spot}}$ is the minimum view angle θ_y experienced over all scatterer positions within the ground scene from the entry point $-L$ of the synthetic aperture.

Step 7 Determine Synthetic Aperture Sample Density: Given the requirement (12.30) or (12.31) for $\Pi_{\Theta_y}(\omega_p, k_{u_y})_{\text{spot}}$, the derived slow-time sampling interval (spatial pulse repetition interval) must satisfy the Nyquist constraint,

$$\Delta_{u_y} \le \frac{2\pi}{\Pi_{\Theta_y}(\omega_p^{\max}, k_{u_y})_{\text{spot}}}.$$

Hence, the Nyquist sample density along the synthetic aperture flight line requirement for broadside spotlight mode SAR with Stolt formatting is

$$\frac{\lambda_{\min}}{4} \le \Delta_{u_y} \le \frac{\lambda_{\max}}{4}\cdot\frac{1}{\cos\Theta_y(2L)_{\text{spot}}}.$$

Step 8 Determine Required Synthetic Aperture Length Requirement: Considering (12.31), the synthetic aperture half-length L for the broadside case in spotlight mode is

$$\theta_y^{\min}(-L)_{\text{spot}} = \arccos\left[\frac{\frac{1}{2}\Pi_{\Theta_y}(\omega_p^{\max}, k_{u_y})_{\text{spot}}}{2k_{\omega_p^{\max}}}\right]. \tag{12.32}$$

The value of L that satisfies this equation must consider the ground size extent, where we also know from the analysis of the broadside spotlight mode imaging geometry that

$$\theta_y^{\min}(-L)_{\text{spot}} = \arctan\left(\frac{Y_o + L}{X_{\min}}\right). \tag{12.33}$$

Therefore, by (12.32) and (12.33), the required synthetic aperture half-length is

$$L \;=\; X_{\min}\cdot\tan\left[\theta_y^{\min}(-L)_{\text{spot}}\right] - Y_o,$$

with $\theta_y^{\min}(-L)_{\text{spot}}$ given by (12.32).

Step 9 Determine Number of Synthetic Aperture Samples: The number of pulses gathered across a finite synthetic aperture of spotlight mode is

$$M_{u_y} = \left\lceil \frac{2L}{\Delta_{u_y}} \right\rceil.$$

Step 10: Determine Along-Track Aperture Wavenumber Sample Density: The Nyquist sample interval for the aperture wavenumber k_{u_y} is determined by the synthetic aperture length,

$$\Delta_{k_{u_y}} = \frac{2\pi}{M_{u_y}\Delta_{u_y}},$$

where $\Pi_{u_y} = M_{u_y}\Delta_{u_y}$.

Spotlight Stolt Transform Jacobian

To implement the Stolt mapping of a shattered spectrum, the transformation from (k_{ω_p}, k_{u_y}) to $\breve{k}_x$ has a Jacobian determinant with a maximum value of [885]

$$J^{\text{spot}}_{\max} = \frac{1}{\cos \theta^{\text{ax}}_x},$$

where $\theta^{\text{ax}}_x = \max(|\theta^{\min}_x|, |\theta^{\max}_x|)$. For a spotlight geometry with positive squint we have

$$\begin{aligned}\theta^{\text{ax}}_x &= \theta^{\max}_x, \\ &= \arctan\left(\frac{Y_c + Y_o + L}{X_c - X_o}\right).\end{aligned}$$

This angle generates the largest view angle from within the synthetic aperture experienced over a squint spot scene.

Temporal Wavenumber Sample Density Adjustment

The worst case warping of the temporal wavenumber sample interval is

$$\Delta_{\ddot{k}_x} = J^{\text{spot}}_{\max} \cdot \Delta_{k_{\omega_p}}.$$

Hence, we require

$$\Delta_{k_{\omega_p}} \leq \frac{\Delta_{\ddot{k}_x}}{J^{\text{spot}}_{\max}},$$

where we also require,

$$\Delta_{\ddot{k}_x} \leq \frac{2\pi}{2X_o},$$

thus

$$\Delta_{k_{\omega_p}} \leq \frac{1}{J^{\text{spot}}_{\max}} \cdot \frac{2\pi}{2X_o}. \tag{12.34}$$

Fast-Time Frequency Bin Spacing

The fast-time angular frequency sample spacing Δ_{ω_p} derived from (12.34) must satisfy

$$\Delta_{\omega_p} \leq c \cdot \Delta_{k_{\omega_p}}. \tag{12.35}$$

Receive Window Size Expansion to Accommodate Stolt Transformation

In spotlight mode, the echo from the scatterer within the illuminated spot nearest the sensor arrives at a point in time given by

$$T_s = \frac{2}{c} r_{\min}, \tag{12.36}$$

where $r_{\min}$ expresses the minimum sensor-scatterer distance. The last instant of the pulse echo duration from the farthest scatterer arrives at

$$T_f = \frac{2}{c} r_{\max} + T_p, \tag{12.37}$$

where $r_{\max}$ is the maximum sensor-scatterer distance. The receive window duration is

$$\Pi_{(T_{\text{win}})}(t) = [T_s, T_f].$$

Spotlight Fast-Time Domain Nyquist Specifications

The Jacobian determinant of the Stolt transform impacts the Nyquist rate required for fast-time data collections. Fourier migration maps the fast-time angular frequency wavenumber k_{ω_p} from a rectilinear raster in the rectilinear spectrum to a Stolt formatted raster on a semicircle manifold in the angular spectrum. Down-range wavenumbers achieved by Fourier migration may require corresponding adjustments in the fast-time Nyquist sampling rate to achieve Nyquist rates in the migrated domain. The pre-migration fast-time sample density should be increased to accommodate the possible sparsity of samples after the Stolt transform from the k_{u_y} dependent k_{ω_p} domain into the $\breve{k}_x$ dimension [132, 133]. This increase of the Δ_{ω_p} sample density ensures that sufficient $\Delta_{k_{\omega_p}}$ sample density exists in the temporal angular wavenumber domain k_{ω_p} to support interpolation from $\breve{k}_x$ to uniformly spaced down-range wavenumbers $\ddot{k}_x$.

Per the Nyquist requirement (12.35) and with due consideration of (12.34), the fast-time receive window must satisfy

$$\begin{aligned}
\Pi_{(T_{\text{win}})}(t) &\geq \frac{2\pi}{\Delta_{\omega_p}}, \\
&= \frac{2\pi}{c \cdot \Delta_{k_{\omega_p}}}, \\
&= J^{\text{spot}}_{\max} \cdot \frac{2}{c} \cdot 2X_o.
\end{aligned}$$

This can be met by zero padding the fast-time receive window data before the fast-time Fourier transform is applied. The length of the zero pad should achieve [885]

$$\Pi_{(T_{\text{win}})}(t)_{\text{pad}} = \max\left\{T_f - T_s, J^{\text{spot}}_{\max} \cdot \frac{2}{c} \cdot 2X_o\right\}.$$

12.5 Spotlight Polar Formatting

The polar format algorithm is based on the assumption of a far field Fraunhofer plane wave approximation of the true spherical wavefront that propagates across the ground scene. If the scene is sufficiently small and distant, then the approximation is accurate and measured scattering is expressed by integrals of terrain reflectivity over cross-range lines perpendicular to the direction of wave propagation (i.e., parallel to the wavefront) [754]. Spotlight mode polar formatting can be viewed as obtaining line integral measures or *projections* of the terrain function. The polar format image spectrum is reconstructed in the wavenumber domain from Fourier transforms of sets of projections obtained with angular diversity [246, 248, 249].

Fraunhofer Plane Wave Approximation in Polar Formatting

The slant-range distance of the plane wave model of the Fraunhofer approximation can be expressed as an inner product,

$$\begin{aligned} r_n(u_y) &\approx \tilde{r}_n(u_y), \\ &= \left|[\hat{x}_n \ \hat{y}_n] \cdot [\cos\theta_{\hat{x}}(u_y)_c \ \sin\theta_{\hat{x}}(u_y)_c]^T - r_c(u_y)\right|, \\ &= \left|\hat{x}_n \cdot \cos\theta_{\hat{x}}(u_y)_c + \hat{y}_n \cdot \sin\theta_{\hat{x}}(u_y)_c - r_c(u_y)\right|, \end{aligned} \tag{12.38}$$

where $r_n(u_y)$ is the true slant-range of a particular scatterer $g_n(\hat{x}_n, \hat{y}_n)$ and $\tilde{r}_n(u_y)$ is the range of a cross-range line that passes through $g_n(\hat{x}_n, \hat{y}_n)$ and which perpendicularly bisects the sensor to spot center slant-vector $r_c(u_y)$ at $\tilde{r}_n(u_y)$ as illustrated in Fig. 12.4.

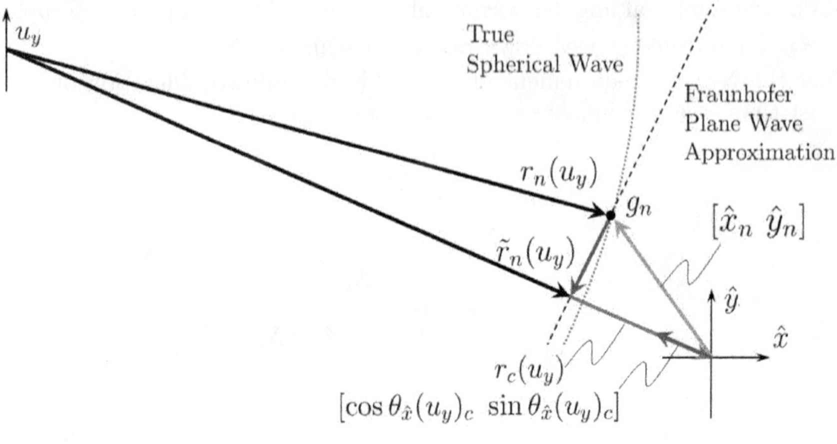

Figure 12.4: Fraunhofer approximation in polar format signal analysis.

Simplification of the Slow-Time Fourier Transform

Without the Fraunhofer approximation, the down converted (monochromatic) frequency-space SAR phase history is

$$\begin{aligned} Ss(\omega_p, u_y)_{\text{ACS}} &= \sum_n g_n \cdot \exp\left[-jk_{\omega_p} \cdot 2r_n(u_y)\right], \\ &= \sum_n g_n \cdot \exp\left\{-jk_{\omega_p} \cdot 2\sqrt{(X_c + \hat{x}_n)^2 + \left[u_y - (Y_c + \hat{y}_n)\right]^2}\right\}. \end{aligned} \tag{12.39}$$

The non-linearity of $\sqrt{\cdot}$ in (12.39) seemingly prevents us (if we did not have the method of stationary phase in hand) from easily finding the slow-time Fourier transform of $Ss(\omega_p, u_y)$. As an alternative approach, use of the Fraunhofer approximation (12.38) yields a simplified expression,

$$\begin{aligned} & Ss(\omega_p, u_y)_{\text{ACS}} \\ &\approx \sum_n g_n \cdot \exp\left[-jk_{\omega_p} \cdot 2\tilde{r}_n(u_y)\right], \qquad (12.40) \\ &\approx \sum_n g_n \cdot \exp[-j\hat{x}_n \cdot 2k_{\omega_p} \cos\theta_{\hat{x}}(u_y)_c] \cdot \exp[-j\hat{y}_n \cdot 2k_{\omega_p} \cos\theta_{\hat{y}}(u_y)_c] \\ &\qquad \cdot \exp[+jk_{\omega_p} \cdot 2r_c(u_y)]. \end{aligned}$$

The Polar Format

If the scatterer non-informative phase factor $\exp\left[jk_{\omega_p} \cdot 2r_c(u_y)\right]$ is compensated by

$$\exp[-jk_{\omega_p} \cdot 2r_c(u_y)] \times \{Ss(\omega_p, u_y)_{\text{ACS}}\}$$

and if we define a pair of *polar formatted wavenumbers*,

$$\begin{aligned} \check{k}_x(\omega_p, u_y) &= 2k_{\omega_p} \cos\theta_{\hat{x}}(u_y)_c, \\ \check{k}_y(\omega_p, u_y) &= 2k_{\omega_p} \cos\theta_{\hat{y}}(u_y)_c, \end{aligned} \tag{12.41}$$

then a phase history expressed in a polar formatted domain is obtained,

$$Ss(\omega_p, u_y)_{\text{SCS}} = \sum_n g_n \cdot \exp[-j\hat{x}_n \check{k}_{\hat{x}}(\omega_p, u_y)] \cdot \exp[-j\hat{y}_n \check{k}_{\hat{y}}(\omega_p, u_y)]. \tag{12.42}$$

Since $\exp[-jk_{\omega_p} \cdot 2r_c(u_y)]$ of (12.40) can be viewed as a transfer with respect to the center of the spot, the resulting image spectrum (12.42) is resident in the scene coordinate system.

Coordinate Transfers by Pivot Point Motion Compensation

Compensation of

$$\exp[+jk_{\omega_p} \cdot 2r_c(u_y)]$$

by a spot center referenced transfer function

$$\exp[-jk_{\omega_p} \cdot 2r_c(u_y)]$$

can be achieved by phase stabilization through motion compensation to a point. Hence, the polar format algorithm is a pivot point algorithm.

Polar Format Frequency-Wavenumber Phase History

The frequency-wavenumber phase history of polar formatted data is indicated by

$$Ss(\omega_p, u_y)_{\text{SCS}} \mapsto SS[\check{k}_{\hat{x}}(\omega_p, u_y), \check{k}_{\hat{y}}(\omega_p, u_y)]_{\text{SCS}}.$$

The phase history (12.42) is approximately equivalent to the desired terrain spectrum,

$$\begin{aligned} & SS[\check{k}_{\hat{x}}(\omega_p, u_y), \check{k}_{\hat{y}}(\omega_p, u_y)]_{\text{SCS}} \\ &= \sum_n g_n \cdot \exp[-j\hat{x}_n \check{k}_{\hat{x}}(\omega_p, u_y)] \cdot \exp[-j\hat{y}_n \check{k}_{\hat{y}}(\omega_p, u_y)], \\ &= GG(\check{k}_{\hat{x}}, \check{k}_{\hat{y}})_{\text{SCS}}, \\ &\approx GG(k_{\hat{x}}, k_{\hat{y}})_{\text{SCS}} \end{aligned}$$

and comprises the polar formatted image spectrum,

$$SS[\check{k}_{\hat{x}}(\omega_p, u_y), \check{k}_{\hat{y}}(\omega_p, u_y)]_{\text{SCS}} \mapsto II(\check{k}_{\hat{x}}, \check{k}_{\hat{y}})_{\text{SCS}}.$$

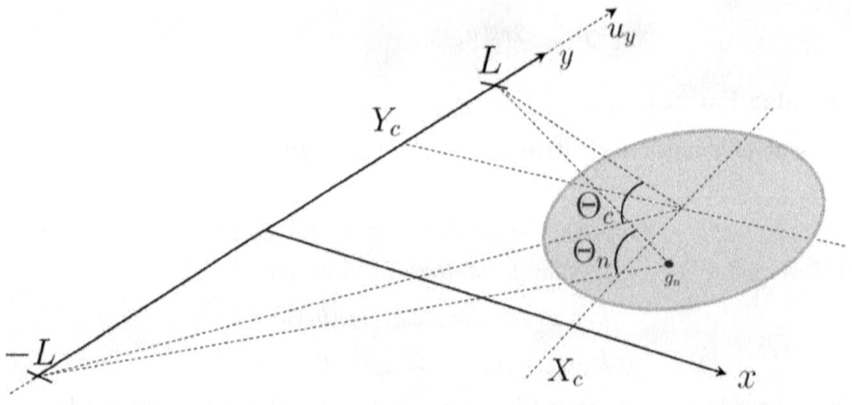

Figure 12.5: Scatterer specific angular dwell versus spot center angular dwell.

Polar Format versus Stolt Format Comparison

The scatterer dependent angular dwell model of Stolt formatting has subtle but important differences compared to the angular dwell model of polar formatting. In polar formatting, the angular dwells Θ_n of all scatterers are approximated as equal to the particular angular dwell Θ_c of the spot center [490]. The angular manifolds of all scatterers within the spot scene, as shown in Fig. 12.5, are assumed to satisfy

$$\Theta_n \approx \Theta_c.$$

Image formation processing with Stolt formatting makes no such approximation [147].

Stolt Formatted Wavenumbers The mapping between scatterer specific view angles and the dimensional wavenumbers achieved through Stolt formatting are

$$\begin{aligned}\breve{k}_x(\omega_p, u_y) &= 2k_{\omega_p}\cos\theta_x(u_y)_n,\\ \breve{k}_y(\omega_p, u_y) &= 2k_{\omega_p}\cos\theta_y(u_y)_n.\end{aligned}$$

Polar Formatted Wavenumbers The mappings to dimensional wavenumbers produced by polar formatting seem almost trivially different (at least from a notational perspective),

$$\begin{aligned}\breve{k}_{\hat{x}}(\omega_p, u_y) &= 2k_{\omega_p}\cos\theta_{\hat{x}}(u_y)_c,\\ \breve{k}_{\hat{y}}(\omega_p, u_y) &= 2k_{\omega_p}\cos\theta_{\hat{y}}(u_y)_c.\end{aligned} \tag{12.43}$$

Angular dwell is referenced to scene center in polar formatting, e.g., $\theta_{\hat{x}}(u_y)_c$ instead of $\theta_{\hat{x}}(u_y)_n$. Polar formatting is precise for the very center of the ground scene. Polar formatting is approximate for all other scatterers. The fidelity of the approximation degrades with scatterer distance from scene center.

Polar Format and Shattered Terrain Spectrum Models

Since there is no scatterer specific angular dwell model in polar formatting, the notion of a shattered terrain spectrum does not apply to polar formatting [147].

Transmitted and Received LFM Pulse Signals

The use of an LFM chirp waveform is core to the conventional signal analysis of polar formatting. Our use of LFM chirp couplets is central to our approach to LFM signal analysis. We begin our analysis by describing the transmitted and received LFM signals.

Baseband LFM Waveforms

A baseband LFM pulse waveform can be expressed in the various forms previously presented in Chapter 6,

$$\begin{aligned}\overline{p_{\Pi(T_p)}(t)}^{\text{LFM}} &= \overline{\Pi_{(T_p)}(t)}\cdot\exp[j\cdot\overline{\omega_b(t)}^{\text{LFM}}\cdot t],\\ &= \overline{\Pi_{(T_p)}(t)}\cdot\exp(j\cdot\overline{\alpha t}\cdot t),\\ &= \overline{\Pi_{(T_p)}(t)}\cdot\exp(j\alpha t^2).\end{aligned}$$

Transmitted Passband LFM Signals

The transmitted passband LFM pulse signal is

$$\begin{aligned}\overline{s_{\Pi(T_p)}(t)}^{\text{LFM}} &= \exp(j\omega_c t)\times\{\overline{p_{\Pi(T_p)}(t)}^{\text{LFM}}\},\\ &= \overline{\Pi_{(T_p)}(t)}\cdot\exp[j\cdot\overline{\omega_p(t)}^{\text{LFM}}\cdot t],\\ &= \overline{\Pi_{(T_p)}(t)}\cdot\exp(j\omega_c t + j\alpha t^2),\end{aligned}$$

where the carrier frequency ω_c within the passband can be arbitrarily assigned as convenience requires. For example, the carrier frequency can be set to be either the minimum passband frequency

$$\omega_c \equiv \omega_p^{\min},$$

the middle passband frequency

$$\omega_c \equiv \omega_p^{\text{mid}}$$

or the maximum passband frequency

$$\omega_c \equiv \omega_p^{\max}.$$

We find occasions to use all of these options.

Received Passband LFM Signals

The received LFM echo signal from the nth scatterer is

$$\begin{aligned}
& s_{\Pi(T_p)} s_{\Pi(\infty)}(t, u_y)_n^{\text{LFM}} \\
&= \overline{\Pi_{(T_p)}(t - t_n)} \cdot \exp[j \cdot \overline{\omega_p(t - t_n)}^{\text{LFM}} \cdot (t - t_n)], \\
&= \overline{\Pi_{(T_p)}(t - t_n)} \cdot \exp[j \cdot \overline{\omega_p(t - t_n)}^{\text{LFM}} \cdot t] \cdot \exp[-j \cdot \overline{\omega_p(t - t_n)}^{\text{LFM}} \cdot t_n], \\
&= \overline{\Pi_{(T_p)}(t - t_n)} \cdot \exp[j \cdot \overline{\omega_p(t - t_n)}^{\text{LFM}} \cdot t] \\
&\qquad \cdot \exp(-j\omega_c t_n) \cdot \exp[-j \cdot \overline{\alpha \cdot (t - t_n)} \cdot t_n], \\
&= \overline{\Pi_{(T_p)}(t - t_n)} \cdot \exp[j \cdot \overline{\omega_p(t - t_n)}^{\text{LFM}} \cdot t] \\
&\qquad \cdot \exp[-j\omega_c t_n(u_y)] \cdot \exp[-j \cdot \overline{\alpha t_n(u_y)} \cdot t] \cdot \exp[j\alpha t_n^2(u_y)]. \qquad (12.44)
\end{aligned}$$

We seek to understand each of the phase history elements contained in the LFM pulse echo decomposition (12.44).

LFM Time-Space Phase History

The time-space domain decomposition of (12.44) contains the following:

- an LFM pulse waveform with a delayed chirp couplet,

$$w_{\Pi(T_p @ t_n)}(t)^{\text{LFM}} = \overline{\Pi_{(T_p)}(t - t_n)} \cdot \exp[j \cdot \overline{\omega_p(t - t_n)}^{\text{LFM}} \cdot t], \qquad (12.45)$$

- the carrier phase curtain

$$\exp\left[-j\omega_c t_n(u_y)\right],$$

- the complex sinusoid

$$\exp[-j \cdot \overline{\alpha t_n(u_y)} \cdot t]$$

  with a chirp couplet frequency determined by the time instant $t_n(u_y)$,

$$\overline{\alpha t_n(u_y)} = \overline{\omega_b[t_n(u_y)]}^{\text{LFM}} \qquad (12.46)$$

- and a quadratic phase factor

$$\exp\left[j\alpha t_n^2(u_y)\right]. \qquad (12.47)$$

Chirp Couplet Scatterer Locator Frequency

As explained in Chapter 6, we choose to call

$$\overline{\omega_b[t_n(u_y)]}^{\text{LFM}} = \overline{\alpha t_n(u_y)}$$

of (12.46) a *chirp couplet scatterer locator frequency.*

Chirp Couplet Scatterer Locator Sinusoid

Likewise, we call the complex sinusoid,

$$\exp[-j \cdot \overline{\omega_b[t_n(u_y)]}^{\text{LFM}} \cdot t] = \exp[-j \cdot \overline{\alpha t_n(u_y)} \cdot t]$$

the *chirp couplet scatterer locator sinusoid.*

Residual Phase

The phase factor (12.47) is time independent and is determined by the square of the signal delay $t_n^2(u_y)$ between sensor and scatterer. The time independent phase factor $\exp\left(j\alpha t_n^2\right)$ of (12.47) is conventionally called the *residual video phase.* We shall more accurately called this phase factor the *residual quadratic phase* or sometimes more concisely, the *residual phase.*

Pulse Compression of LFM Signals

There are two options for achieving pulse compression of LFM signals. One option is matched filtering as explained in Section 5.3. The other is a process known as *dechirping, deramping* or *stretch processing* [46, 534].

Chirp Receivers and Stretch Receivers

There are two common types of radar hardware subsystems available to transmit, receive and digitize chirp signals: *chirp receivers* and *stretch receivers* [23, 26, 534]. Stretch receivers are also called *dechirp* or *deramp* receivers. The coherent receivers described in Section 2.8 can process echo signals which may or may not be chirped, and for reasons that will become increasingly clearer, are sometimes called "chirp receivers" (the chirp structure of the received echo signal is preserved by a chirp receiver). Stretch receivers can only handle chirp waveforms (stretch receivers do not preserve the chirp structure of the received echoes).

Stretch Processing History

The development of stretch processing began in 1951 as classified work in digital implementations of wide bandwidth radar systems with frequency modulated waveforms [144, 534]. Before the development of stretch processing, performance gains in range resolution where sought after by increasing peak transmitted power and decreasing pulse length. The development of frequency modulated waveforms and pulse compression techniques achieved the goal of increased range resolution. The resulting methods also provided increased transmitted signal power gains (without increasing the peak transmitted power level, but by increasing the pulse duration) and thereby improved the received signal SNR [104].

Two-Pass Processing

Conventional matched filtering performed in the frequency domain employs two discrete Fourier transforms. This approach to range imaging is said to require *two-pass processing.* The drawback of two-pass processing is the high computational complexity when the time-bandwidth product of the LFM signal is large (the time-bandwidth product determines the required FFT size).

One-Pass Processing

Stretch processing only requires a single discrete Fourier transform. A Fourier transform is not required to take the measured fast-time data to the wavenumber domain. Stretch processing directly yields spectral data. Hence, deramping is said to require *one-pass processing* for stretch waveform pulse compression.

Stretch processing provides sampling rate efficiencies [221, 1002]. One-pass stretch processing can achieve fine resolution over a range swath with a restricted down-range extent. The efficiencies of stretch processing are lost if the range swath size becomes too large.

All Range Processing

In two-pass processing, the range resolution of the system will be limited by the rate of the analog-to-digital converter, but there is no limit on range swath size. The two-pass approach to chirp pulse compression is sometimes called "all range processing."

Local Oscillator Reference Signals in Stretch and Chirp Receivers

The primary difference between the hardware of a chirp receiver and that of a stretch receiver is the nature of the signal utilized in the local oscillator:

- a chirp receiver uses a sinusoid waveform for the local oscillator reference signal,
- a stretch receiver uses a chirp waveform for the local oscillator reference signal.

Stretch processing of an LFM signal is accomplished by mixing (i.e., multiplying) the received signal with the local oscillator's reference LFM signal. This chirp on chirp mixing changes the functional form of the received chirp echoes signals into echoes with a simple sinusoidal structure. Stretch processing "dechirps" a chirp echo into a tone signal [142, 144, 160].

A dechirped tone signal has a frequency proportional to the illuminated scatterer's position relative to a range position associated with the local oscillator chirp signal's time reference. Hence, the terminology "locator tones" and "locator sinusoids" to describe the output signals of a stretch receiver.

Stretch Up-Chirp and Down-Chirp Differences

With up-chirp LFM waveforms and digital processing, the received chirp echo signal is conjugated and mixed (multiplied) to dechirp the signal. This is followed by a low pass filter and a forward discrete Fourier transform. Down range scatterer offsets relative to spot center produce complex locator tones with positive locator frequencies. Up range scatterer offsets produce negative locator frequencies. With down-chirp LFM waveforms, conjugation is not needed, but to have positive frequencies correspond to down range offsets, an inverse discrete Fourier transform must be used for one-pass processing.

Point Spread Function of Stretch Receivers

Since the frequency of the locator tone signal at the output of a stretch receiver is determined by the range position of the scatterer, the Fourier transform of the square pulse tone signal produces a sinc like impulse response at the scatterer's range position within the range profile. The LFM pulse envelop or temporal weight function

$$\widehat{\Pi}_{[0,T_p]}(t)^{\text{LFM}}$$

is a time indicator function for the LFM waveform pulse interval $[0, T_p]$. With stretch processing, the Fourier transform of $\widehat{\Pi}_{[0,T_p]}(t)^{\text{LFM}}$ determines the *stretched* point spread function,

$$\text{psf}_{\omega_s}(\omega_s)^{\text{one}}_{\text{pass}} = \mathsf{FT}^{t}_{\text{1D}}\left\{\widehat{\Pi}^{*}_{[0,T_p]}(t)^{\text{LFM}}\right\}. \tag{12.48}$$

Since the one-pass Fourier transform is applied in the time domain to the locator tone signal, this stretch pulsed compressed point spread function resides in the ω_s domain. We sometimes add an "s" subscript to emphasize that this version of the ω domain is entered into through a one-pass "stretch" process.

The fast-time stretch point spread function can be scaled to convert one-pass range compressed data from the ω_s to x_s domain via

$$x_s = \frac{c}{2}\cdot\frac{1}{2\alpha}\cdot\omega_s.$$

Point Spread Functions of Chirp Receivers

Compare the fast-time stretch point spread function (12.48) to the point spread function obtained by fast-time matched filtering,

$$\text{psf}_t(t)^{\text{two}}_{\text{pass}} = \mathsf{FT}_{\text{1D}}\left\{|P(\omega_b)|^2\right\}.$$

In two-pass processing, the spectral shape governs the nature of the point spread function obtained with matched filtering. In one-pass processing, the time envelop of the LFM chirp pulse signal determines the nature of the point spread function.

Resolution of Stretch Processing

Compare the down-range resolution achieved with two-pass pulse compression with a pulse of duration of T_p,

$$\Delta_x = \frac{c}{2}\cdot\frac{1}{T_p} \tag{12.49}$$

to the resolution of stretch processing,

$$\Delta_{x_s} = \frac{c}{2} \cdot \frac{1}{\alpha T_p}, \tag{12.50}$$

where α is the quadratic phase parameter of the LFM waveform. The α parameter is viewed as providing a "stretch" of the pulse duration T_p of (12.50) when compared to (12.49). A deramp receiver is said to do *"stretch"* processing because of an effective expansion of the transmitted pulse duration [144],

$$T_p \mapsto \alpha T_p$$

in an an expression that describes the range resolution achieved with one-pass processing.

Time-Bandwidth Products of Stretch Waveforms

For a given time-bandwidth product, a stretch receiver trades time expansion for bandwidth reduction in a way that maintains time-bandwidth product. The bandwidth of a chirp waveform may be much larger than that which can be directly processed by the current state of the art in analog-to-digital converter (ADC) technology. Stretch processing can be used to reduce the bandwidth of a waveform of a given time-bandwidth product to be within reach of the supportable bandwidth of the current state of the art analog-to-digital converter technologies.

Stretch Processing Distortions

Deramp pulse compression is a good choice when time-bandwidth product is high because of reduced sampling rate requirements. However, stretch processing systems must tolerate a degree of SNR loss (due to skew effects) and systemic phase distortions. Residual quadratic phase is natively resident in the LFM pulse echo as shown by (12.44). Two-pass processing removes this residual phase. One-pass processing leaves it. Furthermore, the reduced bandwidth advantage of stretch processing comes with the loss of the ability to do "all range" processing.

Receive Window Offset Compensation and Huygens Fresnel Transfers

The two-pass receive window center time reference is sometimes incorporated into the design of the pulse compression matched filter by setting $p(t-T_c)$ as the reference pulse. The receive window time offset $T_c = \frac{2}{c}X_c$ is equivalent to the Huygens-Fresnel transfer of X_c that describes the terrain function in the aperture coordinate system. Receive window time offset compensation by designing the two-pass pulse compression matched filter with $p(t - T_c)$ implicitly converts the received echo data representation to the scene coordinate system. A similar result is achieved in stretch processing by setting the time reference of the local oscillator's reference chirp signal to a nominal center position within the receive window.

Fast-Time Stretch Processing

We now provide a detailed description of the deramp processing performed by a stretch receiver.

Echo and Reference Chirp Signal Decomposition

Consider the echo signal of a single scatterer based on a time-space phase history factorization (12.44) given by

$$\begin{aligned}
& s_{\Pi(T_p @ t_n)} s_{\Pi(\infty)} (t, u_y)_n^{\text{LFM}} \\
&= \overline{\Pi_{(T_p)} (t - t_n)} \cdot \exp[j \cdot \overline{\omega_p (t - t_n)^{\text{LFM}}} \cdot (t - t_n)], \\
&= w_{\Pi(T_p @ t_n)} (t)^{\text{LFM}} \cdot \exp(-j\omega_c t_n) \cdot \exp\left(-j \cdot \overline{\alpha t_n} \cdot t\right) \cdot \exp\left(j\alpha t_n^2\right).
\end{aligned} \tag{12.51}$$

The deramp receiver's local oscillator uses a replica of the LFM chirp signal with a time reference of T_c,

$$\begin{aligned}
& s_{\Pi(T_{\text{win}} @ T_c)} (t - T_c)_{\text{LO}}^{\text{LFM}} \\
&= \overline{\Pi_{(T_{\text{win}})} (t - T_c)^{\text{LFM}}} \cdot \exp[j \cdot \overline{\omega_p (t - T_c)^{\text{LFM}}} \cdot (t - T_c)], \\
&= w_{\Pi(T_{\text{win}} @ T_c)} (t)^{\text{LFM}} \cdot \exp(-j\omega_c T_c) \cdot \exp\left(-j \cdot \overline{\alpha T_c} \cdot t\right) \cdot \exp\left(j\alpha T_c^2\right),
\end{aligned} \tag{12.52}$$

where T_{win} indicates the radar's receive window duration. The time duration T_{win} of the reference signal is longer than the transmitted pulse duration T_p by at least T_{swath} seconds. This guarantees that the full length of all pulse reflections will be mixed with the reference chirp signal.

Deramped Signal Notation

We indicate that a scatterer specific echo signal has been deramped with the notation

$$s_{\Pi(T_p @ t_n)} s_{\Pi(\infty)} (t, u_y)_n^{\text{LFM}} \mapsto D_{\Pi(T_p @ t_n)} s_{\Pi(\infty)} (t, u_y)_n^{\text{LFM}},$$

where an upper case "D" is used in $D_{\Pi(T_p @ t_n)} s_{\Pi(\infty)} (\cdot, \cdot)_n^{\text{LFM}}$ because the deramp receiver outputs data that has the form of spectral data (i.e, locator sinusoids), even though the stretch receiver's output data remains within the time domain.

Deramped Phase History

To model the deramp operation of the stretch receiver's local oscillator, we conjugate (12.52) and multiply the result with (12.51) to obtain

$$\begin{aligned}
& D_{\Pi(T_p @ t_n)} s_{\Pi(\infty)} (t, u_y)_n^{\text{LFM}} \\
&= \left[s_{\Pi(T_p @ t_n)} s_{\Pi(\infty)} (t, u_y)_n^{\text{LFM}} \right] \cdot \left[s_{\Pi(T_{\text{win}} @ T_c)} (t - T_c)_{\text{LO}}^{\text{LFM}} \right]^*, \\
&= \left[w_{\Pi(T_p @ t_n)} (t)^{\text{LFM}} \right] \cdot \left[w_{\Pi(T_{\text{win}} @ T_c)} (t)^{\text{LFM}} \right]^* \\
&\quad \cdot \exp\left[-j\omega_c (t_n - T_c)\right] \cdot \exp[-j \cdot \overline{\alpha \cdot (t_n - T_c)} \cdot t] \cdot \exp\left[j\alpha \cdot (t_n^2 - T_c^2)\right].
\end{aligned} \tag{12.53}$$

The carrier phase curtain

$$\exp\left[-j\omega_c (t_n - T_c)\right]$$

and the couplet locator sinusoid

$$\exp[-j \cdot \alpha \cdot \overline{(t_n - T_c)} \cdot t] \tag{12.54}$$

of (12.53) are now dependent on the time difference $t_n - T_c$.

Multiplication of Delayed Couplet Phase Factors

Let us examine more closely the product

$$\left[w_{\Pi(T_p @ t_n)}(t)^{\text{LFM}}\right] \cdot \left[w_{\Pi(T_{\text{win}} @ T_c)}(t)^{\text{LFM}}\right]^*$$

within (12.53) of LFM waveforms with delayed chirp couplets, one chirp couplet delayed by t_n, the other delay set in the stretch receiver's hardware to the local oscillator time reference T_c. The phase factor that results from this product is

$$\begin{aligned}
&\measuredangle \left[w_{\Pi(T_p @ t_n)}(t)^{\text{LFM}}\right] \cdot \left[w_{\Pi(T_{\text{win}} @ T_c)}(t)^{\text{LFM}}\right]^* \\
&= \exp\left[j\alpha \cdot (t - t_n) \cdot t\right] \cdot \exp\left[-j\alpha \cdot (t - T_c) \cdot t\right], \\
&= \exp\left[-j\alpha \cdot (t_n - T_c) \cdot t\right].
\end{aligned}$$

This product yields an *additional* couplet locator sinusoid

$$\exp[-j \cdot \alpha \cdot \overline{(t_n - T_c)} \cdot t]$$

that is distinct from that of (12.54) as contained in (12.53). Hence, the deramped signal of (12.53) can be written as

$$\begin{aligned}
&\measuredangle D_{\Pi(T_p @ t_n)} s_{\Pi(\infty)}(t, u_y)_n^{\text{LFM}} \\
&= \underbrace{\exp\left[-j\omega_c(t_n - T_c)\right]}_{\text{carrier phase curtain}} \cdot \underbrace{\exp[-j \cdot \overline{2\alpha \cdot (t_n - T_c) \cdot t}]}_{\text{locator tone}} \cdot \underbrace{\exp\left[j\alpha \cdot (t_n^2 - T_c^2)\right]}_{\text{residual phase}}.
\end{aligned} \tag{12.55}$$

The *stretch locator sinusoid* of (12.55),

$$\exp[-j \cdot 2\alpha \cdot \overline{(t_n - T_c)} \cdot t]$$

has a locator frequency $\omega_s(t_n - T_c)$ equal to

$$\omega_s(t_n - T_c) = 2\alpha \cdot (t_n - T_c),$$

which is defined by the time instant difference $t_n - T_c$. By using a stretch receiver with a delayed reference signal set to T_c, the dechirped returns have locator frequencies reduced from $2\alpha t_n$ to $2\alpha \cdot (t_n - T_c)$.

Stretch Locator Sinusoid Frequency Reduction by Range Swath Referencing

The value of this reduction can be illustrated by example. Consider a stretch receiver where α is set to

$$\alpha = \pi \frac{B_{\text{LFM}}}{T_p}$$

with $B_{\text{LFM}} = 200$ MHz and $T_p = 10$ microseconds. If the center of the range swath is 20 km away, then

$$\begin{aligned} T_c &= \frac{2}{c} X_c, \\ &= 1.33 \text{ microseconds.} \end{aligned}$$

With this choice of a timing reference in the stretch receiver's local oscillator, a scatterer at a range of 21 km will have a dechirped frequency of 133 MHz, instead of a deramped frequency of 2.8 GHz obtained if T_c had been set to $T_c = 0$. This represents a 21-fold reduction in the frequency of the stretch locator tone.

Intermediate Frequency Bandwidth

The sampling rate of the stretch receiver's analog-to-digital converter is selected based on the expected frequency range (bandwidth) of possible locator tones. The bandwidth of the output signal of the deramp local oscillator is governed by the range of time delays experienced between the input chirp signal echoes and the time reference of the local oscillator chirp waveform. For example, the locator frequency of the dechirped return from the nth scatterer is

$$\begin{aligned} \omega_s(t_n - T_c) &= 2\alpha \cdot (t_n - T_c), \\ &= 2\alpha \cdot \frac{2}{c} \cdot (x_n - X_c). \end{aligned}$$

Using the spatial limits of the range swath, the minimum and maximum values for the locator frequencies of the dechirped returns are

$$\omega_s^{\min} = 2\alpha \cdot \frac{2}{c} \cdot (X_o - X_c) \tag{12.56}$$

and

$$\omega_s^{\max} = 2\alpha \cdot \frac{2}{c} \cdot (X_o + X_c). \tag{12.57}$$

The bandwidth

$$\beta_{\text{stretch}} = [\omega_s^{\min}, \omega_s^{\max}]$$

of all dechirped returns, which determines the intermediate frequency bandwidth of the stretch receiver, is

$$\begin{aligned} \beta_{\text{stretch}} &= 2\alpha \cdot \frac{2}{c} \cdot 2X_o, \\ &= 2\pi\gamma \cdot T_{swath}, \\ &= 2\pi \frac{B_{\text{chirp}}}{T_p} \cdot T_{\text{swath}}, \\ &= \beta_{\text{chirp}} \frac{T_{\text{swath}}}{T_p}. \end{aligned}$$

For cases when the swath width echo duration T_{swath} is less than the pulse duration T_p, deramp on receive will reduce the fast-time sampling rate requirement.

Down-Range Huygens-Fresnel Transfer by Deramp Receiver Time Reference

Another result of using a delayed referenced signal in the stretch receiver is that the swath center reference of the echo signal is shifted to range zero in the scene coordinate system. This shift to a zero range offset is an implicit implementation of a Huygens-Fresnel transfer in the down-range dimension. In this case, the inverse Huygens-Fresnel transfer from the aperture to scene coordinate system is performed by the polar formatting stretch receiver hardware subsystem.

Slow-Time Stretch Processing

Stretch processing can be used in both fast-time and slow-time domains [530,740].

Slow-time stretch, also called SPECAN (SPECtral ANalysis) SAR [1017], can be achieved either by digital processing in software, or more often, by pivot point phase stabilization in hardware. Motion compensation to a point effectively performs what can be called "along-track stretch" or "slow-time dechirp" processing [146].

A key advantage of slow-time stretch processing is a reduction in the required PRF for slow-time data gathering [407, 885]. The disadvantages of slow-time stretch processing are a restriction on the along-track ground scene size, spatially dependent degradation of along-track resolution and phase distortions.

"Stretch-Stretch" and "Chirp-Chirp" Approaches to SAR

It is fair to say that polar formatting is a form of "stretch-stretch" processing with one-pass image formation in both fast-time and slow-time domains. It is also fair to say that Stolt formatting is a form of "chirp-chirp" processing with two-pass image formation processing (but not necessarily matched filtering) in both fast-time and slow-time domains.

Fast-Time Two-Pass and Slow-Time One-Pass Processing

It is also possible to perform two-pass processing in fast-time and one-pass processing slow-time. We overview this approach.

Slow-Time Stretch Reference Phase Curtain

Slow-time stretch requires a slow-time reference chirp signal to use for the along-track deramp multiply. Slow-time stretch is a (hyperbolic) chirp on chirp multiply in the u_y domain. As a candidate reference signal for slow-time stretch processing, consider the hyperbolic phase curtain of scene center (X_c, Y_c),

$$Ss(\omega_p, u_y)_c = \exp\left[-jk_{\omega_p} \cdot 2\sqrt{X_c^2 + (u_y - Y_c)^2}\right]. \tag{12.58}$$

Also consider the polychromatic frequency-space phase history,

$$Ss(\omega_p, u_y) = W(\omega_p) \cdot \sum_n g_n \cdot \exp\left[-jk_{\omega_p} \cdot 2\sqrt{x_n^2 + (u_y - y_n)^2}\right] \tag{12.59}$$

returned by all scatterers that comprise the ground scene. Begin by expressing (12.59) in terms of both the aperture and scene coordinate system parameters,

$$
\begin{aligned}
&Ss(\omega_p, u_y) \\
&= W(\omega_p) \cdot \sum_n g_n \cdot \exp\left\{-jk_{\omega_p} \cdot 2\sqrt{(X_c + \hat{x}_n)^2 + [u_y - (Y_c + \hat{y}_n)]^2}\right\}, \\
&= W(\omega_p) \cdot \sum_n g_n \cdot \exp\left[-jk_{\omega_p} \cdot 2\sqrt{(X_c + \hat{x}_n)^2 + (Y_c - u_y + \hat{y}_n)^2}\right].
\end{aligned}
$$

Slow-Time Stretch or Motion Compensation to a Point

Multiplication in the u_y domain of the measured slow-time signal (12.59) with the conjugate of the reference signal (12.58), combined with a matched filtering multiply with the conjugated pulse spectrum $W^*(\omega_p)$, produces a range focused, dechirped, frequency-space phase history,

$$
\begin{aligned}
&RD(\omega_p, u_y) \qquad (12.60)\\
&= \{W^*(\omega_p) \cdot Ss(\omega_p, u_y)_c^*\} \times_{u_y} Ss(\omega_p, u_y), \\
&= \exp\left[+jk_{\omega_p} \cdot 2\sqrt{X_c^2 + (u_y - Y_c)^2}\right] \\
&\quad \times_{u_y} |W(\omega_p)|^2 \cdot \sum_n g_n \cdot \exp\left[-jk_{\omega_p} \cdot 2\sqrt{(X_c + \hat{x}_n)^2 + (Y_c - u_y + \hat{y}_n)^2}\right].
\end{aligned}
$$

By using the Fraunhofer approximation of (12.38) for the slow-time phase curtain produced by a scatterer at $(\hat{x}_n, \hat{y}_n)$ [885],

$$
\begin{aligned}
&\sqrt{(X_c + \hat{x}_n)^2 + (Y_c - u_y + \hat{y}_n)^2} \\
&\quad = \sqrt{X_c^2 + (Y_c - u_y)^2 + 2X_c\hat{x}_n + 2Y_c\hat{y}_n + \cdots} \\
&\quad = \sqrt{X_c^2 + (Y_c - u_y)^2} + \hat{x}_n \cdot \cos\theta_{\hat{x}}(u_y)_c + \hat{y}_n \cdot \cos\theta_{\hat{y}}(u_y)_c + \cdots \\
&\quad \approx \sqrt{X_c^2 + (Y_c - u_y)^2} + \hat{x}_n \cdot \cos\theta_{\hat{x}}(u_y)_c + \hat{y}_n \cdot \cos\theta_{\hat{y}}(u_y)_c, \qquad (12.61)
\end{aligned}
$$

where

$$
\begin{aligned}
\theta_{\hat{x}}(u_y)_c &= \arccos\left(\frac{X_c}{\sqrt{X_c^2 + (Y_c - u_y)^2}}\right), \\
\theta_{\hat{y}}(u_y)_c &= \arccos\left(\frac{u_y - Y_c}{\sqrt{X_c^2 + (Y_c - u_y)^2}}\right),
\end{aligned}
$$

are sensor position dependent direction cosine angles with respect to scene center, the slow-time stretch signal (12.60) can be approximated by

$$\begin{aligned} &RD(\omega_p, u_y) \\ &\approx \quad |W(\omega_p)|^2 \cdot \sum_n g_n \cdot \exp\left[-j\hat{x}_n \cdot 2k_{\omega_p} \cos\theta_{\hat{x}}(u_y)_c\right] \\ &\qquad \cdot \exp\left[-j\hat{y}_n \cdot 2k_{\omega_p} \cos\theta_{\hat{y}}(u_y)_c\right], \\ &= \quad |W(\omega_p)|^2 \cdot \sum_n g_n \cdot \exp\left(-j\hat{x}_n \cdot \check{k}_{\hat{x}}\right) \cdot \exp\left(-j\hat{y}_n \cdot \check{k}_{\hat{y}}\right). \end{aligned} \tag{12.62}$$

The polar mapping of wavenumbers in (12.62) is defined by

$$\begin{aligned} \check{k}_{\hat{x}} &= 2k_{\omega_p} \cos\theta_{\hat{x}}(u_y)_c, \\ \check{k}_{\hat{y}} &= 2k_{\omega_p} \cos\theta_{\hat{y}}(u_y)_c. \end{aligned}$$

This transformation from the (ω_p, u_y) domain to the $(\check{k}_y, \check{k}_y)$ is the polar formatting process of the polar format algorithm. Also, the conjugate reference phase curtain $\exp[+jk_{\omega_p} \cdot 2\sqrt{X_c^2 + (u_y - Y_c)^2}]$ of (12.60) cancels the spot center phase factor $\exp[-jk_{\omega_p} \cdot 2\sqrt{X_c^2 + (Y_c - u_y)^2}]$ corresponding to part of the phase function of (12.61).

Stretch-Stretch Processed Phase History

If a stretch receiver (instead of a chirp receiver) is used, then the processing in fast-time is one-pass and we would have the result of stretch-stretch processing,

$$SS(t, u_y) \mapsto DD(t, u_y).$$

The stretch-stretch phase history $DD(t, u_y)$ is deramped in both the fast-time and slow-time dimensions.

Slow-Time LFM Signals

The slow-time phase center of the hyperbolic phase curtain of the nth scatterer occurs at $u_y = y_n$ within the synthetic aperture, where the slant-range distance $r_n(u_y)$ is a minimum.

Quadratic Approximations of Hyperbolic Phase Curtains

A signal delay hyperbola can be well approximated by a parabola near its phase center. The Fresnel or quadratic approximation can be obtained through truncation of a Taylor series expansion of

$$r_n(u_y) = \sqrt{x_n^2 + (u_y - y_n)^2}$$

about the point (x_n, y_n),

$$r_n(u_y) = x_n + \frac{(u_y - y_n)^2}{2x_n} + \cdots. \tag{12.63}$$

If we truncate to second order, then we obtain the Fresnel approximation,

$$\begin{aligned} r_n(u_y) &\approx x_n + \frac{(u_y - y_n)^2}{2x_n}, \\ &= \breve{r}_n(u_y), \end{aligned} \tag{12.64}$$

where the notation $\breve{r}_n(u_y)$ is suggestive of a spatial domain parabolic approximation (rather than a wavenumber domain Stolt mapping). The second order approximation (12.64) of the hyperbolic phase function allows the range walk phase curtain to be decomposed with a Fresnel approximation,

$$\begin{aligned} &\exp\left[-jk_{\omega_p} \cdot 2r_n(u_y)\right] \\ &\approx \exp\left[-jk_{\omega_p} \cdot 2\breve{r}_n(u_y)\right], \\ &= \exp\left(-j2k_{\omega_p} \cdot x_n\right) \cdot \exp\left[-j2k_{\omega_p} \cdot \frac{(u_y - y_n)^2}{2x_n}\right]. \end{aligned} \tag{12.65}$$

The phase factor

$$\exp(-j2k_{\omega_p} \cdot x_n)$$

is descriptive of scatterer position down-range. The Fresnel phase factor

$$\exp\left[-j2k_{\omega_p} \cdot \frac{(u_y - y_n)^2}{2x_n}\right]$$

has a phase function

$$\phi(\omega_p, u_y)_n^{\text{Fresnel}} = -2k_{\omega_p} \cdot \frac{(u_y - y_n)^2}{2x_n} \tag{12.66}$$

that forms the slow-time chirp that will be slow-time deramped.

Slow-Time Linear Phase Functions

As a quadratic phase signal, the approximated slow-time phase curtain can be described as a *slow-time linear frequency modulated signal* (near its phase center). Since the slow-time phase function is quadratic, the slow-time derivative of the phase function (12.66) yields a slow-time LFM signal defined as a function of u_y with slow-time signal characteristics determined by the scatterer's down-range position x_n and the fast-time signal frequencies ω_p. The linear phase function of the nth scatterer at y_n is

$$\partial_{u_y}\left\{\phi(\omega_p, u_y)_n^{\text{Fresnel}}\right\} = -2k_{\omega_p} \cdot \frac{(u_y - y_n)}{x_n}. \tag{12.67}$$

The linear phase function (instantaneous frequency) of the broadside boresight point $(X_c, 0)$ is

$$\partial_{u_y}\left\{\phi(\omega_p, u_y)_c^{\text{Fresnel}}\right\} = -2k_{\omega_p} \cdot \frac{u_y}{X_c}. \tag{12.68}$$

Slow-Time Stretch

Slow-time stretch of all scatterer echoes is performed by

$$SD(\omega_p, u_y) = Ss^*(\omega_p, u_y)_c \times_{u_y} \sum_n Ss(\omega_p, u_y)_n.$$

If we consider a narrow range swath where $x_n \approx X_c$ holds, then the output of slow-time stretch processing yields the difference of (12.67) and (12.68),

$$\partial_{u_y} \left\{ \phi(\omega_p, u_y)_n^{\text{Fresnel}} - \phi(\omega_p, u_y)_c^{\text{Fresnel}} \right\} \approx -2k_{\omega_p} \cdot \frac{y_n}{X_c}. \tag{12.69}$$

Slow-Time One-Pass Imagery

The slow-time one-pass Fourier transform of this slow-time locator signal is

$$Sd(\omega_p, k_{u_y}) = \sum_n \cdot g_n \cdot \exp\left[-j2k_{\omega_p} \cdot (x_n - X_c)\right] \cdot 2L \cdot \text{sinc}\left[L \cdot \left(k_{u_y} - 2k_{\omega_p} \cdot \frac{y_n}{X_c}\right)\right],$$

which carries information on the along-track location of scatterers with respect to the spot center and thereby yields an image in the along-track dimension in the k_{u_y} domain. Note that the one-pass slow-time Fourier transform gives,

$$SD(\omega_p, u_y) \mapsto Sd(\omega_p, k_{u_y}).$$

Slow-Time Locator Frequency

The along-track point spread function

$$2L \cdot \text{sinc}\left[L \cdot \left(k_{u_y} - 2k_{\omega_p} \cdot \frac{y_n}{X_c}\right)\right] \tag{12.70}$$

of y_n produced by a slow-time stretch process appears at

$$k_{u_y} = 2k_{\omega_p} \cdot \frac{y_n}{X_c}.$$

We can invert this mapping to create an along-track image that resides in the y domain,

$$y = \frac{X_c}{2k_{\omega_p}} k_{u_y}.$$

Note that the resolution of the slow-time stretched along-track point spread function of (12.70) is governed by the synthetic aperture length.

Slow-Time Stretch Locator Tones Bandwidth

The bandwidth of the dechirped signal is determined by the band generated by the union of possible slow-time locator sinusoids (12.69),

$$\exp\left(-j2k_{\omega_p} \cdot \frac{y_n}{X_c} \cdot u_y\right),$$

which is defined by the (broadside) spot size in the along-track direction,

$$\Pi_{\Theta_{\text{spot}}}\left(\omega_p, k_{u_y}\right)_{\text{pass}}^{\text{one}} \approx \left[-2k_{\omega_p} \cdot \frac{Y_0}{X_c}, +2k_{\omega_p} \cdot \frac{Y_0}{X_c}\right].$$

The bandwidth of the set of along-track locator tones is proportional to the along-track dimension of the ground patch. This bandwidth is independent of the synthetic aperture length.

Two-Pass Slow-Time Aperture Wavenumber Bandwidth

At long ranges, recall that the broadside wavenumber support region (with Stolt formatting) is

$$\Pi_{\Theta_{\text{spot}}}\left(\omega_p, k_{u_y}\right)_{\text{pass}}^{\text{two}} = \left[-2k_{\omega_p} \sin\Theta_{(Y_o+L,X_c)}, +2k_{\omega_p} \sin\Theta_{(Y_o+L,X_c)}\right],$$

where the angular dwell $\Theta_{(Y_o+L,X_c)}$ depends on the sum $Y_o + L$ of the along-track dimension of the ground patch and the synthetic aperture length.

One-Pass Slow-Time Aperture Wavenumber Bandwidth

In the same way that fast-time stretch processing reduces the intermediate frequency bandwidth for chirp echoes from small range swaths, slow-time stretching reduces the PRF for small along-track ground spots. The slow-time stretched signal bandwidth specification is not governed by the view angle extremes of the synthetic aperture, but is determined only by spot size. Hence, we can write for the broadside case for polar formatting,

$$\Pi_{\Theta_{\text{spot}}}\left(\omega_p, k_{u_y}\right)_{\text{pass}}^{\text{one}} = \left[-2k_{\omega_p} \sin\Theta_{(Y_o,X_c)}, +2k_{\omega_p} \sin\Theta_{(Y_o,X_c)}\right],$$

where the angular dwell $\Theta_{(Y_o,X_c)}$ depends on only the Y_o along-track dimension of the ground patch.

PRF Differences Between Polar and Stolt Formatting

Spotlight Mode Stolt Formatting The Nyquist sample density requirement along the synthetic aperture flight line for broadside spotlight mode SAR with Stolt formatting is

$$\Delta_{u_y}^{\text{Solt}} \le \frac{\lambda}{4} \cdot \frac{1}{\sin\Theta_{(Y_0+L,X_c)}}.$$

Spotlight Mode Polar Formatting The Nyquist sample density requirement along the synthetic aperture flight line for broadside spotlight mode SAR with polar formatting is

$$\Delta_{u_y}^{\text{polar}} \le \frac{\lambda}{4} \cdot \frac{1}{\sin\Theta_{(Y_0,X_c)}}.$$

The polar format PRI is only governed by the size of the spot size Y_o through

$$\Theta_{(Y_0,X_c)},$$

not by the sum of spot size and the synthetic aperture length $Y_0 + L$ through

$$\Theta_{(Y_0+L,X_c)}.$$

The PRF can be reduced when slow-time stretch processing is used for small spot scenes.

Slow-Time Aliasing in Stolt and Polar Formatting

Stolt format signal analysis predicts a spotlight mode slow-time signal bandwidth that is proportional to the antenna's angular field of view added to the slow-time dwell angle. As a result, SAR hardware built specifically for polar formatting systems, from the perspective of Stolt formatting signal analysis, aliases the gathered slow-time data (at least until slow-time stretch is performed). The PRF of a polar format hardware system is typically too low to directly support measurement of the slow-time phase curtain required as input with Stolt formatting.

Space-Wavenumber Chirp Couplets

Chirp couplets of non-linear phase functions are revealed when chirp phase signals are expressed in a pseudo-linear form represented by

$$\exp\left[j\phi(t)^{\text{chirp}}\right] = \exp\left[j \cdot \overline{\omega_b(t)} \cdot t\right],$$

where $\overline{\omega_b(t)}$ describes the time-frequency coupling of the chirped slow-time signal. Analogous to fast-time chirp signals, a slow-time chirp couplet can be defined in spatial dimensions. We anticipate finding a formulation of a spatial chirp couplet such that

$$\exp\left[j\phi(u_y)^{\text{chirp}}\right] = \exp\left[j \cdot \overline{k_{u_y}(u_y)} \cdot u_y\right] \tag{12.71}$$

holds, where $\overline{k_{u_y}(u_y)}$ is a general description of the space-wavenumber coupling. We call $\overline{k_{u_y}(u_y)}$ a *spatial chirp couplet* or a *space-wavenumber chirp couplet*.

Spatial Chirp Couplet Formulation Challenge

The spatial chirp signal of the slow-time hyperbolic phase curtain of linear flight SAR is

$$\begin{aligned} \exp\left[j\phi(u_y)^{\text{chirp}}\right] &= \exp\left[-jk_{\omega_p} \cdot 2r_n(u_y)\right], \\ &= \exp\left[-jk_{\omega_p} \cdot 2\sqrt{x_n^2 + (u_y - y_n)^2}\right]. \end{aligned} \tag{12.72}$$

It will take some work to express (12.72) in the chirp couplet form of (12.71).

Approximation of Hyperbolic Phase Curtain in Chirp Couplet Form

One approach is to use the Taylor series expansion of (12.72) and truncate to second order, which reveals a spatial linear frequency modulated chirp couplet structure. In this case, the along-track LFM would suggest, for the center of the spot at $(X_c, 0)$,

$$\exp\left[-j \cdot \overline{k_{u_y}(u_y)_c} \cdot u_y\right] = \exp\left(-j \cdot \overline{2k_{\omega_p} \cdot \frac{u_y}{2X_c}} \cdot u_y\right), \qquad (12.73)$$

where we define the *slow-time space-wavenumber chirp couplet* as

$$\overline{k_{u_y}(u_y)_c} = \overline{2k_{\omega_p} \cdot \frac{u_y}{2X_c}}.$$

The spatial chirp couplet formulation (12.73), obtained with a truncated Taylor series, is less than satisfactory in several ways, including its range dependency. We shall eventually develop an alternative definition for spatial chirp couplets in a way that does not involve the Fresnel approximation. This formulation of a spatial chirp couplet shall prove useful for three dimensional Stolt formatting (see Section 13.6).

Polar Format Uniform Resampling

With phase stabilization to a point reference, collected frequency-wavenumber phase history data reside on a polar raster [115, 585]. Fourier domain reconstruction algorithms utilized in scene center pivot point data acquisition systems must perform polar to rectilinear interpolations to obtain uniformly sampled data in the $(\check{k}_x, \check{k}_y)$ domain [250, 682].

Two Dimensional Interpolation in Polar Formatting

Two dimensional interpolation is required in polar formatting. This is usually carried out by a pair of one dimensional interpolations [26, 488]. An interpolation is performed first in the radial wavenumber direction, thereby producing data on a keystone grid. Interpolation is next performed in the azimuth wavenumber direction, utilizing non-uniformly spaced input data [250]. Two dimensional, but separable, windowed sinc interpolation kernels or spline interpolators, for example, can be used [628, 700].

Stretch-Stretch and Chirp-Chirp Data Volume Comparison

The increased complexity of two dimensional interpolation (or two separate one dimensional interpolations) of polar formatting, compared to the one dimensional interpolation required in Stolt formatting, is typically offset by the comparatively reduced volume of data in polar formatting because of the stretch processing performed in both fast-time and slow-time domains.

Polar Formatting Image Spectrum

As is typical in offset Fourier imaging systems [680, 682], a complex polar format image is natively a passband image that is down converted (by multiplication with $\exp(-j\omega_c t)$ in the signal domain) from measurements made by a passband sensor. Phase artifacts from this down conversion can be compensated, as required.

If the passband Fourier offset rectilinearly resampled data set is shifted (by the L.shift(k_{ω_c}) operator applied in k-space) to the origin of the $(\check{k}_x, \check{k}_y)$ domain preparatory to Fourier inversion, then the resulting image will have the same down conversion phase artifact in the complex pixel domain, but induced by the shift in the frequency domain. This injected sinusoid should be viewed as a phase distortion and should be compensated (along with the residual phase distortion of deramp-deramp processing) if complex pixel data are to be exploited.

Polar Format Image

Both polar formatting and Stolt formatting produce spatially variant spotlight mode imagery [883]. In polar formatting, resolution is higher for scatterers nearer spot center. In Stolt formatting, resolution is higher for a scatterer closer to boresight and nearer in range due to increased angular dwell.

The causes of spatially variant imaging in polar formatting are multiple, including Fraunhofer approximation plane wave mismatch and reference hyperbola range walk mismatch in slow-time stretch processing. The degree of mismatch is usually acceptable for long range, high frequency radars with narrow beam antennas, but may be unacceptable in near range, low frequency radars with wide beam antennas. These signal model approximation errors can flow into a design limit on the SAR frequency band (it must be sufficiently high) and on the antenna beamwidth (must be sufficiently small to illuminate just the size restricted ground spot size) for the polar format algorithm to be effective [490, 495].

Image Distortions in Polar Format Processing

Similar to the inaccuracies that result with fast-time stretch processing [101], analogous inaccuracies exist in along-track data when slow-time stretch processing is used. Fast-time and slow-time distortions in polar formatting are caused by the following:

- the fast-time Fraunhofer plane wave approximation [277, 280, 566],
- differential range walk distortions from slow-time stretch reference signal mismatch,
- fast-time skew and
- residual quadratic phase noise from both fast-time and slow-time stretch processing.

Some of these can be fully compensated [493]. Others can only be approximately compensated.

Fraunhofer Plane Wave Approximation - Spatially Variant Blurring

As shown in Fig. 12.6 for a broadside geometry, wavefront curvature approximations cause the along-track resolution to increasingly degrade towards the azimuthal edges of the spot scene [346]. These visual degradations can be limited by restricting scene diameter in the cross-range dimension [26, 146, 492]. For the

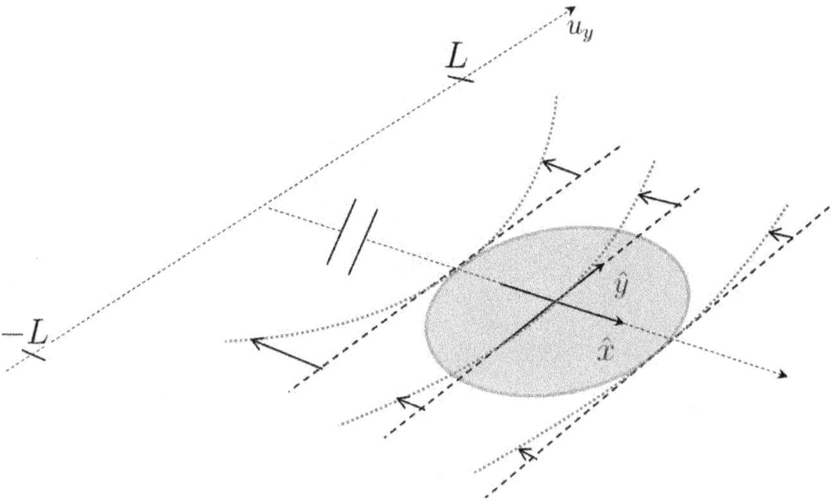

Figure 12.6: Fraunhofer plane wave mismatch.

Fraunhofer approximation to be acceptable, it is said that the along-track wavefront curvature error near the edge of the spot should be less than about a range resolution cell.

Hyperbolic Reference Phase Curtain Mismatch - Keystone Distortion

The polar format algorithm's required pivot point motion compensation induces an approximation that causes a depth of field focus characteristic in the down-range dimension. Use of motion compensation to a point (slow-time stretch) imposes a limit on the allowed ground spot size in the down-range dimension [270]. This depth of field blurring limits spot size in a way that is different from the impact of the Fraunhofer approximation.

The slow-time stretch reference hyperbola is perfectly matched to scene center. A slow-time reference hyperbola mismatch exits for all scatterers with different up-range and down-range positions, as illustrated in Fig. 12.7. Scatterers nearer in range are under compensated. Scatterers further in range are over compensated [397].

Geometric Key Stone Image Warping

Differential hyperbola mismatch in pivot point phase stabilization causes a distortion over the entire image called keystone distortion [13,493,495], which geometrically warps the entire spotlight mode image as illustrated in Fig. 12.8. The source of the keystone distortion in polar formatted imagery is sometimes erroneously attributed to wavefront curvature.

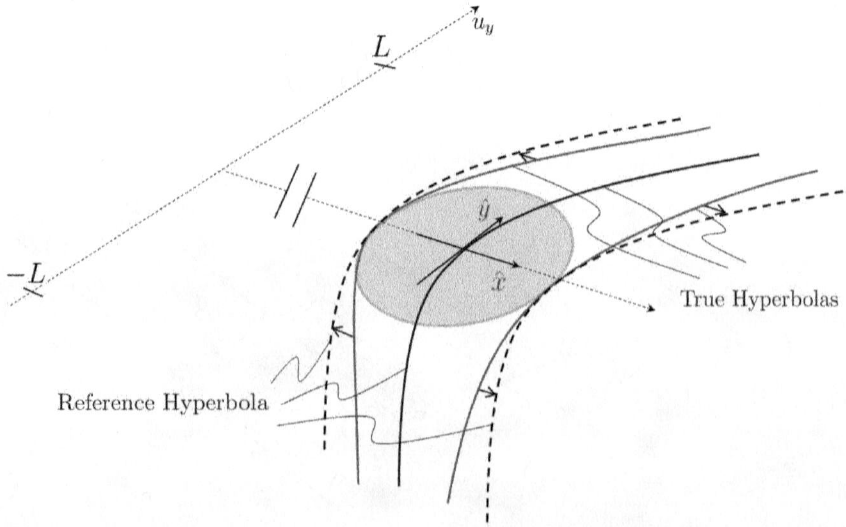

Figure 12.7: Slow-time stretch reference hyperbola mismatch.

Polar Format Distortion Compensations

Post processing steps have been proposed for polar format processing to compensate distortions caused by approximation errors. Typically, some form of spatially variant processing is used [753].

Key Stone Warp Compensation

Keystone distortion can be corrected by various means [221, 975], including a spatial warping of the entire SAR image after polar format image formation processing is complete [492]. Of course, such warping is problematic in itself, because of induced pixel domain point spread function distortions.

Fraunhofer Plane Wave Approximation Error Compensation

A degree of Fraunhofer approximation error can be corrected with an aperture function that is quadratic [492]. Another approach used in [617] is to divide the scene into multiple subpatches and correct each subpatch separately. The size of the pixel domain subpatches are based on the local severity of the distortion [283, 495].

Post Processing for "Image Enhancement"

Other forms of image post processing can compensate for SAR image formation processing deficiencies in ways that make visually pleasing SAR products. It has been noted that shadows can be made darker by pixel histogram remapping or by using spectral shaping to boost mid-range and attenuate high frequencies. Although the visual quality of detected imagery can sometimes be improved by

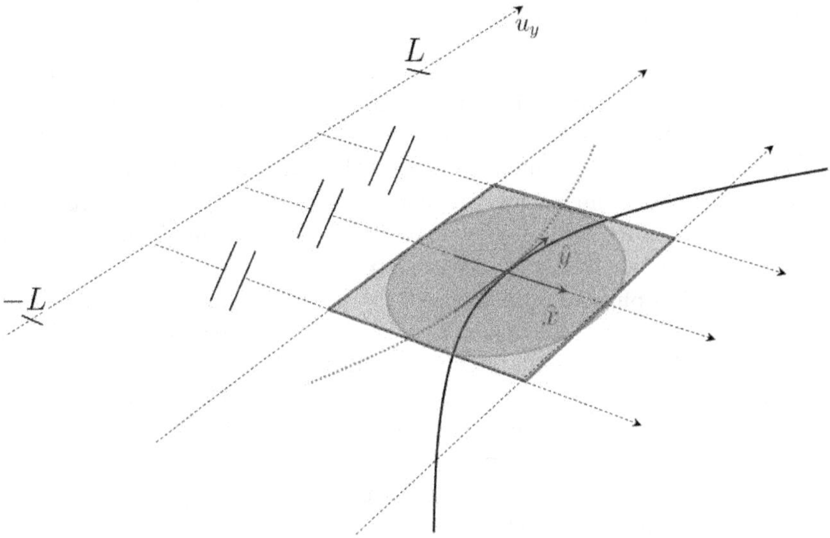

a. Spot image without keystone warping.

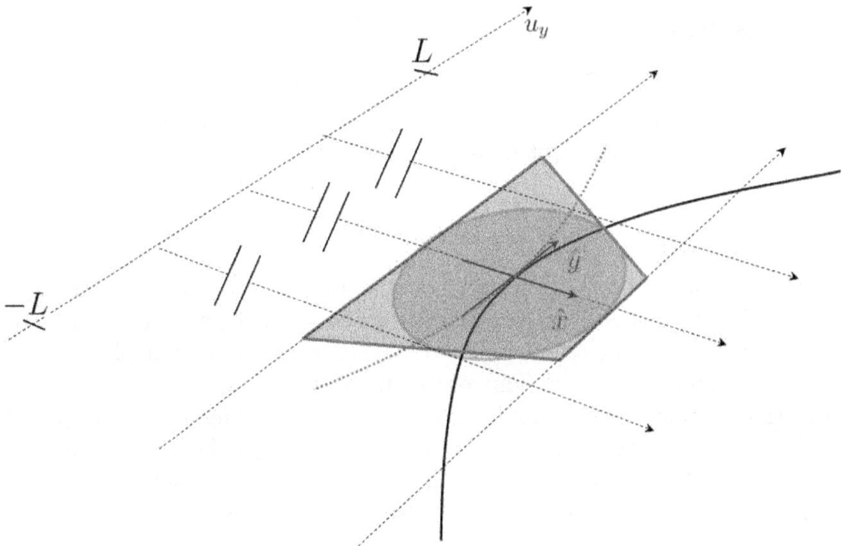

b. Spot image with keystone warping.

Figure 12.8: Image keystone warping in polar formatted SAR image.

such post processing methods, the quality of the coherent data (associated phase information) may actually degrade.

Fast-Time Skew Distortions and Deskew

Fast-time pulse duration mismatch in reference to receive window duration is called *skew.* That is, the duration of stretch locator tones do not span the entire received window duration and the stretched echoes from scatterers with different down-range positions do not align in time within the receive window.

Deskew processing can be used to compensate the skew of fast-time stretch processing. Deskew is a frequency dependent time alignment of locator tone signals. Deskew can be achieved in various ways.

One way is by a phase modification of the Fourier transform of the deramped data. The appendix of [146] describes a method that achieves time alignment of deramp pulsed echoes in the receive window. If deskew is performed in the analog domain, then an all-pass filter with linear frequency dependent phase can be used. A linear phase delay requires a large time-bandwidth product for good filter characteristics.

As an alternative to deskew, a time gate can be used to clip the center part of the receive window data to avoid skew effects, but with the expense of a reduction in resolution and SNR [490].

Slow-Time Skew

Slow-time skew does not exist in polar formatting because all scatterer produced slow-time stretched locator echo signals extend throughout the full extent of the synthetic aperture in spotlight mode.

Residual Quadratic Phase Error

Residual quadratic phase and associated distortions occur in both fast-time and slow-time dimensions with stretch-stretch based polar format processing. Residual quadratic phase "noise" impacts the accuracy of measured phase values of complex scatterer signatures. This phase distortion can be troublesome in applications that exploit coherent SAR imagery, such as moving object imaging [617]. This distortion is potentially a show-stopper in using polar formatting to form the spot images of spotmap imagery if high quality coherent spotmap imagery is desired (see Section 12.6). Non-coherent (detected) imagery is not affected by residual quadratic phase noise.

Residual Quadratic Phase Compensation

Ideally, residual quadratic phase can be compensated. However, residual quadratic phase is only present at the time instant t_n if a scatterer exists at x_n. If an echo is absent in any given range bin, compensation applied to such a range bin actually injects, instead of eliminates, phase distortion. This is particularly problematic in ground to air imaging scenarios (i.e., in ISAR).

Confusion of Skew and Residual Quadratic Phase

Terminology for the artifacts of skew and residual quadratic phase seem to be somewhat interchanged or confused in the literature [146, 490, 655].

12.6 Spotmap Algorithms

Spotlight mode achieves resolution beyond that obtainable in stripmap mode. Spotlight mode, in an intentional design trade off, looses the ability to directly provide wide area coverage. Spotmap processing uses spotlight mode to create a high resolution wide area image with a mosaic of multiple subspots [26, 611, 648].

Multiple approaches for producing wide area spotmap imagery exist, including non-coherent and coherent approaches to subspot seaming. Coherent spotmap imagery can be formed with Stolt formatted subimages. Non-coherent spotmap imagery can be formed with Stolt or polar formatted subimages.

Non-Coherent Fusion of Polar Formatted Spot Images

Non-coherent spotmap imaging with polar formatting divides the ground scene into subspots, forms separately focused polar format subimages and then non-coherently patches them together with pixel domain post processing [26, 148, 649]. The resulting wide area image unavoidably contains seaming artifacts [147].

Coherent Fusion of Stolt Formatted Spot Images

Coherent spotmap imaging with Stolt formatting divides the ground scene into subspots, forms separately focused Stolt formatted subimages and then coherently patches them together.

With Stolt formatting, coherent fusion can be performed in either the complex pixel domain with post processing or in the Fourier migrated wavenumber domain during spotmap image formation processing [870, 887]. With judicious application of inverse and forward Huygens-Fresnel transfers, multiple subspots can be coherently fused in an integrated (angular spectrum) k-space referenced to an overall (mosaic) scene coordinate system. The merged scene can be coherently correct without any seaming artifacts after the global inverse discrete Fourier transformation [885]. Padding with zeros is required to accommodate the mosaic image size before the wide area inverse (two dimensional) Fourier transform is performed, but no post processing is required for mosaic seaming.

Reformatting of Polar Format Spot Images for Coherent Fusion

Stretch processing incurs various approximation losses and phase distortions, including residual quadratic phase, skew and slow-time stretch hyperbolic reference signal mismatch. This section addresses the possibility of simultaneously correcting all of these artifacts by a single software process that essentially "unstretches" the fast-time stretch done by deramp and "unstretches" the slow-time stretch done by motion compensation to a point [378, 885]. Unstretch processing can be used to recondition stretch-stretch polar formatted data for subsequent chirp-chirp processing with Stolt formatting.

These software processes recover chirp echoes from fast-time stretch hardware and recovers scatterer specific slow-time hyperbolic phase curtains from point referenced motion compensation hardware. Chirp recovery in fast-time is called "time domain decompression" in [885]. More appropriate names may be "reramping," "rechirping" or "unstretch" processing.

Fast-Time Chirp Recovery from Stretched LFM Signals

The first step of fast-time unstretch processing is up sampling of stretched digital data. The second step rechirps the up sampled stretched data with a chirp-on-tone multiply. The data are then sufficiently reconditioned to proceed with two-pass matched filtering for pulse compression. This yields the same range profile data that would have been produced by a chirp receiver [899]. These processing steps are now illustrated with an example.

Direct Sampling of Chirp Echoes at a Stretch Sample Rate Shown in Fig. 12.9a is an illustration of what would be measured if a baseband chirp echo were directly sampled with a data acquisition rate designed for a stretch receiver. Note that the sampling rate is sufficient only for the part of the LFM chirp signal near the phase center, where at least two (real) samples are gathered during each oscillation of the chirp waveform. Samples further away from the phase center are measured at a sub-Nyquist sampling rate and are aliased.

Chirp-on-Chirp Stretch Multiply After a stretch process, which is a chirp-on-chirp multiply operation, the resulting stretched tone and sample points are shown in Fig. 12.9b. The deramped data are unaliased across the entire receive window.

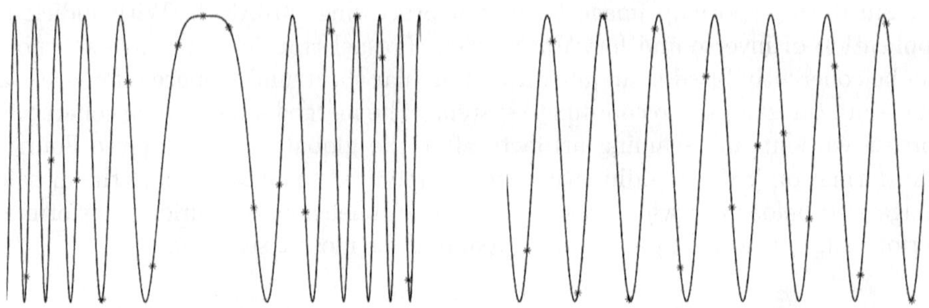

a. Chirp sampling at stretch rate. b. Stretched tone and sample points.

Figure 12.9: Steps one and two of unstretch processing.

Up Sampling of the Stretched Locator Tones Shown in Fig. 12.10a is the sampled tone waveform obtained after up sampling of the stretch tone of

Fig. 12.9b.

Chirp-on-Tone Unstretch (Rechirp) Multiply Shown in Fig. 12.10b is the sampled chirp waveform obtained after rechirping of the up sampled stretch tone of Fig. 12.10a, where the unstretch operation is implemented with a chirp-on-tone multiply operation. As can be seen in Fig. 12.10b, the chirp phase function has been fully recaptured with sufficient sampling density, even at the edges of the chirp signal far away from the phase center.

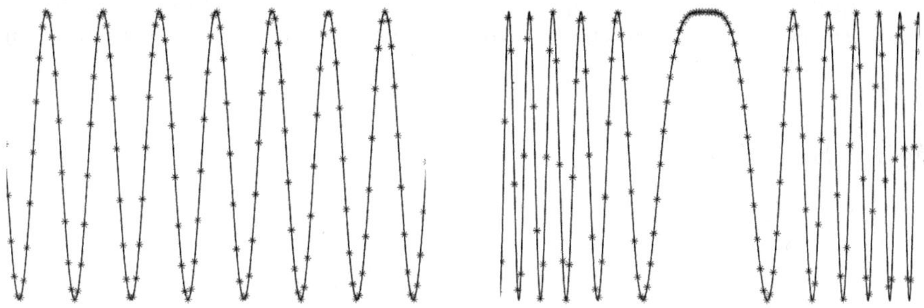

a. Up-sampled stretched tone. b. Rechirped signal.

Figure 12.10: Steps three and four of unstretch processing.

Computational Trade Offs in Unstretch Processing

The implementation costs of fast-time rechirp processing are increased data rates and increased computations. However, computations are no longer required for a deskew operation and residual quadratic phase compensation.

Slow-Time Chirp Recovery from Stretched Phase Curtains

If motion compensation to a point is being used, similar to inverse processing to reramp the signal in fast-time, phase adjustments performed by motion compensation to a point can be reversed or backed out of the measured slow-time data by using *slow-time rechirp* or *slow-time unstretch* processing [885]. This recovers the hyperbolic chirp of previously stretched slow-time data [397]. The slow-time data are up sampled and re-compensated to a down-range reference line (instead of a pivot point motion compensation reference) [495]. Once this is done, Stolt formatting can proceed as if motion compensation to a line was originally used to acquire the SAR data. Rechirping in slow-time counters challenges in wide beamwidth, wide bandwidth imaging scenarios, but where legacy stretch-stretch hardware must be used to acquire SAR data.

PRF Reduction with Stolt Formatting

Stretch processing in both fast-time and slow-time enables sampling rates to be reduced. Fast-time stretch allows a lower ADC clock rate to be used in the receiver. Slow-time stretch allows a lower PRF to be used for pulse data collections. The lower ADC and PRF data collection rates of polar formatting are not directly usable in Stolt formatting.

In some cases, the lower PRF of slow-time stretch processing may be desired, but Stolt formatting may be more desirable to enhance the quality of along-track image formation. In these cases, slow-time stretch in hardware can be used with slow-time unstretch processing in software. With rechirp reconditioning of polar formatted data, Stolt formatting can proceed as if a chirp receiver and motion compensation to a line system had originally collected the data. The drawbacks of this approach are the inescapable ground scene size constraints imposed by stretch-stretch processing in the front-end hardware system. This size constraint is determined by the intermediate frequency passband of the receiver hardware.

Digital Spotlighting

Processing with a pivot point stretch receiver can be used to filter echo signals from terrain segments which are illuminated, but not desired. Stretch locator frequencies from unwanted range gates are filtered out at the output of the stretch receiver [26]. A similar process in slow-time stretch processing can be used to suppress unwanted along-track returns [378, 887]. Fast-time stretch processing and range gating, combined with slow-time stretch processing and cross-range gating, provides the basis of a process called *digital spotlighting* [407, 774, 885] or *scene spotlighting* [27]. Digital spotlighting isolates echoes returned from a specific ground patch of interest and suppresses echoes returned from unwanted nearby areas. The geometric complexity of the two dimensional gating function may increase due to squint geometries and other considerations [885].

One reason for performing digital spotlighting is to suppress unwanted echo returns that leak in through antenna sidelobes and suppress along-track ambiguities. Uncertainty or impreciseness in the physical antenna's beamwidth may cause required slow-time Nyquist rates (PRFs) to be underestimated. This type of aliasing can be suppressed by digital spotlighting.

12.7 Stripmap Stolt Formatting

In stripmap mode SAR, there is no beam steering to extend the angular dwell of the SAR data collection. The radar is held fixed as the platform moves along a linear flight path as illustrated in Fig. 12.11.

Antenna Induced Slow-Time Amplitude Modulations

We assume in our discussion of stripmap mode SAR that the antenna is configured in a fixed broad side geometry. The fly-by movement of a stripmap mode SAR induces a slow-time amplitude modulation on the gathered phase history data. The induced antenna weight function is most naturally described in the time-space domain, but can also be expressed in corresponding partial and full Fourier transform domains. This type of antenna induced amplitude modulation does not occur when the antenna is steered towards a spot of ground.

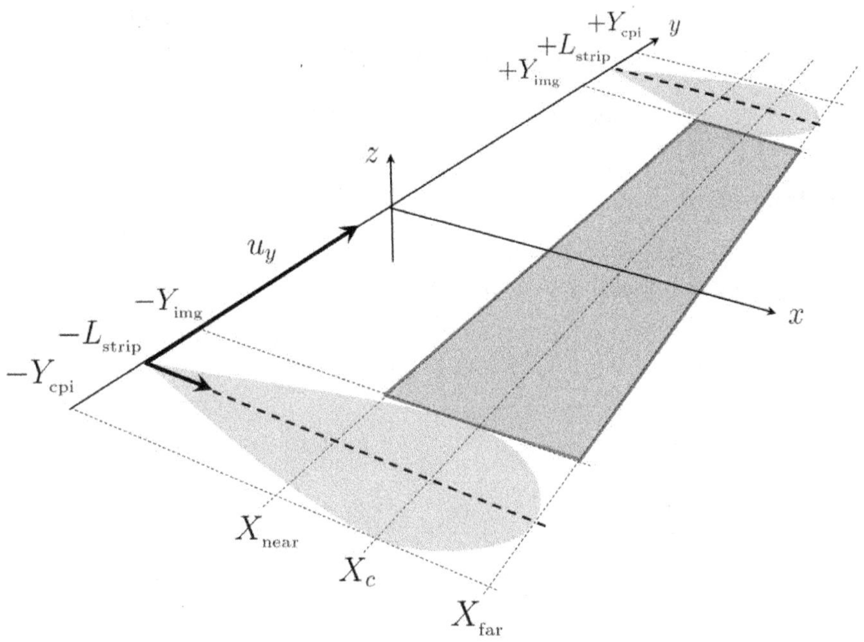

Figure 12.11: Imaging geometry of stripmap mode.

Stripmap Mode Pulse Count Variance

The along-track portion of the down-range swath that is illuminated by the antenna beam varies from pulse to pulse as the radar moves down the flight line. The number of pulses, or equivalently, the synthetic aperture length experienced by any given scatterer in stripmap mode is determined by the width of the antenna beam at that scatterer's down-range position. Because the antenna beamwidth increases with range, scatterers at the far edge of the swath are illuminated by more pulses than scatterers at the near edge. In stripmap mode, synthetic aperture length is range dependent.

Stripmap Mode Range Dependent Beamwidth

The range dependent antenna beamwidth in the far field at a range of x_n is

$$\mathbb{B}(x_n) = x_n \cdot 2 \sin \phi_x^{\text{ant}} .$$

Frequency Dependent Flat Plate Stripmap Beam Patterns

The antenna's boresight to first null half-beam angle ϕ_x^{ant} is frequency dependent for a flat plate antenna,

$$\begin{aligned} \phi_x^{\text{ant}} &= \arcsin\left(\frac{\lambda_{\omega_p}}{D_y}\right), \\ &= \phi_x(\omega_p)_{\text{plate}}. \end{aligned} \qquad (12.74)$$

For pulsed chirp signals, the frequency dependence (12.74) spatially modulates the beamwidth as a function of instantaneous chirp frequency,

$$\mathbb{B}(\omega_p, x_n) = x_n \cdot 2\sin\phi_x(\omega_p)_{\text{plate}}.$$

Down-Range Dependent Stripmap Mode Synthetic Aperture Length

A scatterer at (x_n, y_n) is observable over a slow-time interval of u_y determined by the frequency and range dependent beamwidth,

$$u_y \in \left[y_n - \frac{1}{2}\mathbb{B}(\omega_p, x_n), y_n + \frac{1}{2}\mathbb{B}(\omega_p, x_n)\right].$$

The weight function

$$\widehat{\Pi}_{\mathbb{B}(\omega_p, x_n)}(u_y - y_n) = \left[y_n - \frac{1}{2}\mathbb{B}(\omega_p, x_n), y_n + \frac{1}{2}\mathbb{B}(\omega_p, x_n)\right] \tag{12.75}$$

is descriptive of the antenna beam induced amplitude modulation. The weight function (12.75) is nominally truncated to just the mainlobe of the beam pattern to obtain a weighted on-off window function. Only when the along-track sensor position is within $\widehat{\Pi}_{\mathbb{B}(\omega_p, x_n)}(u_y - y_n)$ is the scatterer considered illuminated. Clearly, this is an approximation in that this model ignores the sidelobe structure of the passing antenna beam pattern that falls outside of the spatial window (12.75).

Stripmap Mode Slow-Time Coherent Processing Interval Definition

From a signal processing perspective, especially for satellite based SAR, dealing with the continuous stream of radar data produced by stripmap mode is a challenge. A finite along-track segment must be extracted for image formation processing to proceed. The beginning and end boundaries of the extracted segment are free parameters of the system's signal processing work flow design. These boundaries define the *coherent processing interval* (CPI) of stripmap mode image formation processing.

There are a couple of questions that must be addressed in dealing with the signal processing of stripmap mode:

- How does one guarantee that each scatterer within a CPI ground segment is illuminated with a full suite of pulses within the nominal field of view of the passing antenna beam pattern, especially for scatterers near the azimuthal edges of the CPI segment?

- How does one prevent echo data from scatterers outside of the extracted CPI ground segment from cluttering the echo data originating from within the CPI?

We address these challenges in our stripmap mode signal model development.

Ground Swath in Aperture Coordinate System

The Fourier transform pair of the fully illuminated stripmap ground segment in the aperture coordinate system is

$$g_{\Pi(2X_o@X_c)} g_{\Pi(2Y_{\text{img}}@Y_c)}(x,y)_{\text{ACS}} \quad \Leftrightarrow \quad G_{\Pi(\infty)} G_{\Pi(\infty)}(k_x,k_y)_{\text{ACS}}, \tag{12.76}$$

where the usual along-track distance $2Y_o$ that would have been part of the along-track subscript of (12.76) is replaced with the extracted along-track stripmap image segment $2Y_{\text{img}}$,

$$\Pi(2Y_{\text{img}}@Y_c),$$

which is a subset of a longer along-track coherent processing interval indicated by $2Y_{\text{cpi}}$ in

$$\Pi(2Y_{\text{cpi}}@Y_c)$$

of

$$g_{\Pi(2X_o@X_c)} g_{\Pi(2Y_{\text{cpi}}@Y_c)}(x,y)_{\text{ACS}}.$$

The segments $2Y_{\text{img}}$ and $2Y_{\text{cpi}}$ are centered at Y_c within the terrain strip. The nominal center point Y_c can be used to locate a scene centered coordinate system (at least for that segment). A sequence of extracted segments exists with different segment center points defined as the sensor moves along the endless (space based) flight line.

Ground Swath in Scene Coordinate System

The fully illuminated ground swath Fourier transform pair in the scene (CPI centered) coordinate system is

$$g_{\Pi[-X_o,+X_o]} g_{\Pi[-Y_{\text{img}},+Y_{\text{img}}]}(\hat{x},\hat{y})_{\text{SCS}} \quad \Leftrightarrow \quad G_{\Pi(\infty)} G_{\Pi(\infty)}(k_{\hat{x}},k_{\hat{y}})_{\text{SCS}}.$$

Full Beam Illuminated Along-Track Scatterers

The size Y_{img} of an along-track strip is a free parameter of the signal processing design task. As shown in Fig. 12.11, scatterers at the far (down-range) edge within the *along-track image ground swath segment*

$$y_n \in [-Y_{\text{img}}, Y_{\text{img}}]$$

are fully illuminated over a frequency and range dependent slow-time interval given by

$$\begin{aligned} u_y &\in [-L_{\text{strip}}(\omega_p, X_{\text{far}}), +L_{\text{strip}}(\omega_p, X_{\text{far}})], \\ &= \left[-\frac{1}{2}\mathbb{B}(\omega_p, X_{\text{far}}) - Y_{\text{img}}, +Y_{\text{img}} + \frac{1}{2}\mathbb{B}(\omega_p, X_{\text{far}})\right]. \end{aligned}$$

The frequency and range dependent half-length of the synthetic aperture is defined by

$$L_{\text{strip}}(\omega_p, X_{\text{far}}) = Y_{\text{img}} + \frac{1}{2}\mathbb{B}(\omega_p, X_{\text{far}}).$$

The slow-time coherent processing interval where ground scatterers are at least partially illuminated is specified by

$$\begin{aligned} u_y &\in [-Y_{\text{cpi}}(\omega_p, X_{\text{far}}), +Y_{\text{cpi}}(\omega_p, X_{\text{far}})], \\ &= \left[-\mathbb{B}(\omega_p, X_{\text{far}}) - Y_{\text{img}}, +Y_{\text{img}} + \mathbb{B}(\omega_p, X_{\text{far}})\right]. \end{aligned}$$

The frequency and range dependent half-length of the coherent processing interval is

$$Y_{\text{cpi}}(\omega_p, X_{\text{far}}) = Y_{\text{img}} + \mathbb{B}(\omega_p, X_{\text{far}}).$$

Partial Beam Illuminated Along-Track Scatterers

Pulses gathered from scatterers near the azimuthal edges of the fully illuminated stripmap image segment

$$[-Y_{\text{img}}, +Y_{\text{img}}]$$

also contain echo contributions from scatterers in adjacent out-of-image areas. We describe these two problematic and neighboring areas with a union,

$$[-Y_{\text{cpi}}, -Y_{\text{img}}] \cup [+Y_{\text{img}}, +Y_{\text{cpi}}]. \tag{12.77}$$

Ground points within the CPI defined by (12.77) are only illuminated by a portion of the moving beam pattern. Digital spotlighting can be used to suppressed echoes from these partially illuminated segments of the terrain strip [869, 870, 885].

Image and CPI Synthetic Aperture Lengths

In summary, scatterers within

$$y_n \in [-Y_{\text{img}}, +Y_{\text{img}}] \tag{12.78}$$

are guaranteed to be fully illuminated over the stripmap synthetic aperture segment

$$u_y \in [-L_{\text{img}}(\omega_p, X_{\text{far}}), +L_{\text{img}}(\omega_p, X_{\text{far}})].$$

Scatterers within

$$y_n \in [-Y_{\text{cpi}}, +Y_{\text{cpi}}]$$

are only partially illuminated.

Angular Dwell in Stripmap Mode

The angular field of view of a stationary antenna with a frequency dependent beamwidth is

$$\Phi_x(\omega_p)_{\text{ant}} = [-\phi_x(\omega_p)_{\text{ant}}, +\phi_x(\omega_p)_{\text{ant}}].$$

In stripmap mode, the sweep of slow-time view angle, measured with respect to the x axis, is

$$\begin{aligned} \theta_x(u_y - y_n) &\in \Theta_x(\omega_p)_{\text{strip}}, \\ &\in [-\phi_x(\omega_p)_{\text{ant}}, +\phi_x(\omega_p)_{\text{ant}}], \\ &= \Phi_x(\omega_p)_{\text{ant}}. \end{aligned}$$

The slow-time observational angular dwell can be expressed in terms of the antenna's field of view,

$$\Theta_x(\omega_p)_{\text{strip}} = \Phi_x(\omega_p)_{\text{ant}},$$

or if expressed in terms of a direction angle defined with respect to the y axis, as

$$\Theta_y(\omega_p)_{\text{strip}} = \Phi_y(\omega_p)_{\text{ant}}. \tag{12.79}$$

Stripmap Mode Angular Dwell Spatial Invariance

Although the beamwidth increases with scatterer range, the angular dwell experienced by each scatterer in stripmap mode is independent of range.

Stripmap Time-Space Chirp Phase History

Our approach for presenting the time-space phase history of a stripmap mode SAR is to present a side by side comparison with the previously developed spotlight mode time-space phase history. The details of the specific elements uniquely related to stripmap mode are contrasted and explained.

Spotlight Time-Space Chirp Phase History

The time-space phase history of a spotlight mode system that employs chirp waveforms and uses a flat plate antenna is

$$\begin{aligned}
& s_{\Pi[T_s,T_f]}\, s_{\Pi(2L)}(t,\boldsymbol{u})_{\text{spot}}^{\text{chirp}} \\
&\quad \approx \Pi_{[-L,+L]}(\boldsymbol{u}) \cdot \sum_n g_n \cdot w_{\Pi(T_p @ t_n)}(t)_n^{\text{chirp}} *_t \Lambda^2(t, \boldsymbol{u} - \boldsymbol{x}_n)_{\text{plate}}^{\text{impulse}}, \\
&\quad = \Pi_{[-L,+L]}(\boldsymbol{u}) \cdot \sum_n \frac{1}{2\pi} \int_{\beta_p} W_{\Pi(2\omega_o @ \omega_c)}(\omega_p)^{\text{chirp}} \cdot \mathbb{A}^2(\omega_p, \widehat{\boldsymbol{u} - \boldsymbol{x}_n})_{\text{plate}} \\
&\qquad\qquad \cdot g_n \cdot Hh^2(\omega_p, \boldsymbol{u} - \boldsymbol{x}_n) \cdot \exp(j\omega_p t) \cdot d\omega_p.
\end{aligned}$$

Stripmap Time-Space Chirp Phase History

The corresponding model for stripmap mode is

$$\begin{aligned}
& s_{\Pi[T_s,T_f]}\, s_{\Pi(2L_{\text{cpi}})}(t,\boldsymbol{u})_{\text{strip}}^{\text{chirp}} \\
&\quad \approx \sum_n \widehat{\Pi}_{\mathbb{B}(\omega_p,x_n)}(\boldsymbol{u} - \boldsymbol{y}_n) \cdot g_n \cdot w_{\Pi(T_p @ t_n)}(t)_n^{\text{chirp}} *_t \Lambda^2(t, \boldsymbol{u} - \boldsymbol{x}_n)_{\text{plate}}^{\text{impulse}}, \\
&\quad = \sum_n \frac{1}{2\pi} \int_{\beta_p} \widehat{\Pi}_{\mathbb{B}(\omega_p,x_n)}(\boldsymbol{u} - \boldsymbol{y}_n) \cdot W_{\Pi(2\omega_o @ \omega_c)}(\omega_p)^{\text{chirp}} \cdot \mathbb{A}^2(\omega_p, \widehat{\boldsymbol{u} - \boldsymbol{x}_n})_{\text{plate}} \\
&\qquad\qquad \cdot g_n \cdot Hh^2(\omega_p, \boldsymbol{u} - \boldsymbol{x}_n) \cdot \exp(j\omega_p t) \cdot d\omega_p.
\end{aligned}$$

In the time-space phase history of stripmap mode SAR, the beam pattern weight function

$$\widehat{\Pi}_{\mathbb{B}(\omega_p,x_n)}(\boldsymbol{u} - \boldsymbol{y}_n)$$

defines the scatterer dependent synthetic aperture length and also describes the amplitude modulation induced by a passing non-steered SAR antenna.

Stripmap Frequency-Space Chirp Phase History

We present a side by side comparison with the previously developed spotlight mode frequency-space phase history model. This highlights the modal differences (and similarities) between stripmap and spotlight modes.

Spotlight Frequency-Space Phase History

The frequency-space phase history of a spotlight mode system that employs a chirp waveform is

$$\begin{aligned} & S_{\Pi(2\omega_o @\omega_c)} s_{\Pi(2L)} (\omega_p, \boldsymbol{u})^{\text{chirp}}_{\text{spot}} \\ &= \Pi_{[-L,L]}(\boldsymbol{u}) \cdot W_{\Pi(2\omega_o @\omega_c)}(\omega_p)^{\text{chirp}} \\ &\quad \cdot \sum_n \mathbb{A}^2(\omega_p, \widehat{\boldsymbol{u}-\boldsymbol{x}_n})_{\text{plate}} \cdot g_n \cdot Hh^2(\omega_p, \boldsymbol{u}-\boldsymbol{x}_n). \end{aligned} \tag{12.80}$$

Stripmap Frequency-Space Phase History

The corresponding model for stripmap mode is

$$\begin{aligned} & S_{\Pi(2\omega_o @\omega_c)} s_{\Pi(2L_{\text{strip}})} (\omega_p, \boldsymbol{u})^{\text{chirp}}_{\text{strip}} \\ &= W_{\Pi(2\omega_o @\omega_c)}(\omega_p)^{\text{chirp}} \cdot \sum_n \widehat{\Pi}_{\mathbb{B}(\omega_p, x_n)}(\boldsymbol{u}-\boldsymbol{y}_n) \cdot \mathbb{A}^2(\omega_p, \widehat{\boldsymbol{u}-\boldsymbol{x}_n})_{\text{plate}} \\ &\quad \cdot g_n \cdot Hh^2(\omega_p, \boldsymbol{u}-\boldsymbol{x}_n). \end{aligned}$$

The antenna induced beam pattern window function $\widehat{\Pi}_{\mathbb{B}(\omega_p, x_n)}(\boldsymbol{u}-\boldsymbol{y}_n)$ defines the scatterer dependent synthetic aperture length in the frequency-space phase history of stripmap mode SAR.

Stripmap Frequency-Space Phase History with Angular Dwell

We now express the frequency-space phase history in terms of angular dwell. We compare the stripmap mode case with the previously established spotlight mode model.

Spotlight Frequency-Space History with Angular Dwell

The frequency-space phase history of a spotlight mode system expressed in terms of angular dwell is

$$\begin{aligned} & S_{\Pi(2\omega_o @\omega_c)} s_{\Pi(2L)} (\omega_p, \boldsymbol{u})^{\text{chirp}}_{\text{spot}} \\ &= W_{\Pi(2\omega_o @\omega_c)}(\omega_p)^{\text{chirp}} \cdot \sum_n \Pi_{\Theta_y}(\widehat{\boldsymbol{u}-\boldsymbol{x}_n})_n \cdot \mathbb{A}^2(\omega_p, \widehat{\boldsymbol{u}-\boldsymbol{x}_n})_{\text{plate}} \\ &\quad \cdot g_n \cdot Hh^2(\omega_p, \boldsymbol{u}-\boldsymbol{x}_n). \end{aligned}$$

Stripmap Frequency-Space History with Angular Dwell

From (12.79),

$$\begin{aligned}\widehat{\Pi}_{\mathbb{B}(\omega_p, x_n)}(\boldsymbol{u}-\boldsymbol{y}_n) &= \widehat{\Pi}_{\Phi_y(\omega_p)_{\text{plate}}}(\boldsymbol{u}-\boldsymbol{y}_n), \\ &= \widehat{\Pi}_{\Phi_y}(\omega_p, \boldsymbol{u}-\boldsymbol{y}_n)_{\text{plate}}\end{aligned}$$

and hence

$$\begin{aligned}&S_{\Pi(2\omega_o @ \omega_c)} s_{\Pi(2L_{\text{strip}})}(\omega_p, \boldsymbol{u})^{\text{chirp}}_{\text{strip}} \\ &= W_{\Pi(2\omega_o @ \omega_c)}(\omega_p)^{\text{chirp}} \\ &\quad \cdot \sum_n \widehat{\Pi}_{\Phi_y}(\omega_p, \boldsymbol{u}-\boldsymbol{y}_n)_{\text{plate}} \cdot \mathbb{A}^2(\omega_p, \widehat{\boldsymbol{u}-\boldsymbol{x}_n})_{\text{plate}} \cdot g_n \cdot Hh^2(\omega_p, \boldsymbol{u}-\boldsymbol{x}_n).\end{aligned}$$

The slow-time window function is now expressed in terms of the frequency dependent angular dwell

$$\Pi_{\Theta_y}(\widehat{\boldsymbol{u}-\boldsymbol{x}_n})_n \quad \mapsto \quad \widehat{\Pi}_{\Phi_y}(\omega_p, \boldsymbol{u}-\boldsymbol{y}_n)_{\text{plate}}$$

afforded by the real aperture of the flat plate antenna.

Stripmap Frequency-Wavenumber Phase History

To analyze the stripmap mode frequency-wavenumber phase history, we present a side by side comparison with the previously developed spotlight mode frequency-wavenumber phase history.

Spotlight Frequency-Wavenumber Phase History

The frequency-wavenumber phase history of a spotlight mode system that employs a chirp waveform is

$$\begin{aligned}&S_{\widehat{\Pi}(2\omega_o @ \omega_c)} S_{\Pi(\Theta_y)}(\omega_p, k_{u_y}, \boldsymbol{u})^{\text{chirp}}_{\text{spot}} \\ &= W_{\Pi(2\omega_o @ \omega_c)}(\omega_p)^{\text{chirp}} \cdot a^2(\omega_p, k_{u_y}) \\ &\quad \cdot \sum_n \Pi_{\Theta_y}(\omega_p, k_{u_y})_n \cdot \mathbb{A}^2(k_{u_y}, \boldsymbol{u}-\boldsymbol{x}_n)_{\text{plate}} \\ &\quad \cdot g_n \cdot H^2_{x_n(\boldsymbol{u})}(\omega_p, k_{u_y}) \cdot H^2_{y_n(\boldsymbol{u})}(k_{u_y}),\end{aligned}$$

where the sensor-scatterer range dependent antenna weight function is

$$\mathbb{A}(k_{u_y}, \boldsymbol{u}-\boldsymbol{x}_n)_{\text{plate}} = \frac{1}{4\pi} \cdot \frac{1}{|\boldsymbol{u}-\boldsymbol{x}_n|} \cdot D_y \cdot \operatorname{sinc}\left(\frac{D_y}{2} k_{u_y}\right)$$

and where the coefficients of the round trip angular spectrum decomposition are represented by $a^2(\omega_p, k_{u_y})$.

Stripmap Frequency-Wavenumber Phase History

The corresponding model for stripmap mode is

$$
\begin{aligned}
& S_{\widehat{\Pi}(2\omega_o @ \omega_c)} S_{\Pi(2L_{\text{strip}})}(\omega_p, k_{u_y}, \boldsymbol{u})^{\text{chirp}}_{\text{strip}} \\
&= W_{\Pi(2\omega_o @ \omega_c)}(\omega_p)^{\text{chirp}} \cdot a^2(\omega_p, k_{u_y}) \cdot \widehat{\Pi}_{\Phi_y}(k_{u_y})_{\text{plate}} \\
&\quad \cdot \sum_n \mathbb{A}^2(k_{u_y}, \boldsymbol{u} - \boldsymbol{x}_n)_{\text{plate}} \cdot g_n \cdot H^2_{x_n(\boldsymbol{u})}(\omega_p, k_{u_y}) \cdot H^2_{y_n(\boldsymbol{u})}(k_{u_y}),
\end{aligned}
$$

where we equate k_{u_y} with k_{s_y} in

$$
\begin{aligned}
\mathbb{A}^2(k_{s_y}, \boldsymbol{u} - \boldsymbol{x}_n)_{\text{plate}} &\approx \frac{1}{4\pi} \cdot \frac{1}{\sqrt{x_n^2 + y_n^2}} \cdot \int_{-\infty}^{+\infty} \Pi_{(D_y)}(s_y) \cdot \exp\left(j s_y k_{s_y}\right) \cdot ds_y, \\
&= \frac{1}{4\pi} \cdot \frac{1}{|\boldsymbol{x}_n|} \cdot D_y \cdot \operatorname{sinc}\left(\frac{D_y}{2} k_{s_y}\right), \\
&= \mathbb{A}(k_{s_y}, \boldsymbol{x}_n)_{\text{plate}}
\end{aligned}
$$

and since s_y is aligned with u_y,

$$
\mathbb{A}(k_{s_y}, \boldsymbol{x}_n)_{\text{plate}} = \mathbb{A}(k_{u_y}, \boldsymbol{x}_n)_{\text{plate}}.
$$

In stripmap mode, the wavenumber support for all scatterers share the same frequency independent manifold (for a flat plate antenna) [885]. Hence, there is no need for a scatterer specific wavenumber support function. The spotlight mode window

$$
\Pi_{\Theta_y}(\omega_p, k_{u_y})_n
$$

has been replaced by

$$
\widehat{\Pi}_{\Phi_y}(k_{u_y})_{\text{plate}} = \left[-\frac{2\pi}{D_y}, +\frac{2\pi}{D_y}\right].
$$

Stripmap Angular Spectrum

We next examine the angular spectrum of stripmap mode and contrast it with the model for the spotlight mode system.

Spotlight Angular Spectrum Chirp Phase History

The spotlight mode model is

$$
\begin{aligned}
& M^{\text{chirp}}_{\Pi(\otimes)} S_{\Pi(\Theta_y)}(\breve{k}_x, k_y, \boldsymbol{u})_{\text{spot}} \\
&= \sum_n \widehat{\Pi}^{\text{chirp}}_{\otimes}(\breve{k}_x, k_y)_n \cdot \widehat{\Pi}_{\Theta_y}(k_y)_n \cdot \mathbb{A}^2(k_y, \boldsymbol{u} - \boldsymbol{x}_n)_{\text{plate}} \\
&\qquad \cdot g_n \cdot H^2_{x_n(\boldsymbol{u})}(\breve{k}_x) \cdot H^2_{y_n(\boldsymbol{u})}(k_y),
\end{aligned}
$$

where

$$
\begin{aligned}
& \widehat{\Pi}^{\text{chirp}}_{\otimes}(\breve{k}_x, k_y)_n \cdot \widehat{\Pi}_{\Theta_y}(k_y)_n \\
&= \mathsf{ST}^{\breve{k}_x}_{1\text{D}} \left\{ W_{\Pi(2\omega_o @ \omega_c)}(\omega_p)^{\text{chirp}} \cdot a^2(\omega_p, k_{u_y}) \cdot \Pi_{\Theta_y}(\omega_p, k_{u_y})_n \right\}.
\end{aligned}
$$

Stripmap Angular Spectrum Chirp Phase History

The corresponding stripmap mode model is

$$\begin{aligned} &M^{\text{chirp}}_{\widehat{\Pi}(\otimes)} S_{\widehat{\Pi}(\Phi_y)}(\breve{k}_x, k_y, \boldsymbol{u})_{\text{strip}} \\ &= \widehat{\Pi}^{\text{chirp}}_{\otimes}(\breve{k}_x, k_y)_{\text{strip}} \cdot \widehat{\Pi}_{\Phi_y}(k_y)_{\text{plate}} \\ &\qquad \cdot \sum_n \mathbb{A}^2(k_y, \boldsymbol{u} - \boldsymbol{x}_n)_{\text{plate}} \cdot g_n \cdot H^2_{x_n(\boldsymbol{u})}(\breve{k}_x) \cdot H^2_{y_n(\boldsymbol{u})}(k_y), \end{aligned}$$

where

$$\begin{aligned} &\widehat{\Pi}_{\otimes}(\breve{k}_x, k_y)^{\text{chirp}}_{\text{strip}} \cdot \widehat{\Pi}_{\Phi_y}(k_y)_{\text{plate}} \\ &= \mathsf{ST}^{\breve{k}_x}_{1\text{D}} \left\{ W_{\Pi(2\omega_o @ \omega_c)}(\omega_p)^{\text{chirp}} \cdot a^2(\omega_p, k_{u_y}) \cdot \Pi_{\Phi_y}(k_{u_y})_{\text{plate}} \right\}. \end{aligned}$$

The sensor-scatterer range dependent spectral weight is approximated by the weight function referenced to spot center,

$$\begin{aligned} \mathbb{A}(k_y, \boldsymbol{u} - \boldsymbol{x}_n)_{\text{plate}} &\approx \mathbb{A}(k_y, \boldsymbol{u} - \boldsymbol{X}_c)_{\text{plate}}, \\ &= \frac{1}{4\pi} \cdot \frac{1}{|\boldsymbol{u} - \boldsymbol{X}_c|} \cdot D_y \cdot \operatorname{sinc}\left(\frac{D_y}{2} k_y\right) \end{aligned}$$

and using this as an approximate expression for stripmap mode, we change the notation,

$$\mathbb{A}(k_y, \boldsymbol{u} - \boldsymbol{X}_c)_{\text{plate}} \mapsto \mathbb{A}_{X_c(\boldsymbol{u})}(k_y)_{\text{plate}}.$$

and we can reduce functional dependence down to just k-space,

$$M^{\text{chirp}}_{\widehat{\Pi}(\otimes)} S_{\widehat{\Pi}(\Phi_y)}(\breve{k}_x, k_y, \boldsymbol{u})^{\text{strip}}_{\text{ACS}} \mapsto M^{\text{chirp}}_{\widehat{\Pi}(\otimes)} S_{\widehat{\Pi}(\Phi_y)}(\breve{k}_x, k_y)^{\text{strip}}_{\text{ACS}},$$

where

$$\begin{aligned} &M^{\text{chirp}}_{\widehat{\Pi}(\otimes)} S_{\widehat{\Pi}(\Phi_y)}(\breve{k}_x, k_y)^{\text{strip}}_{\text{ACS}} \\ &\approx \widehat{\Pi}_{\otimes}(\breve{k}_x, k_y)^{\text{chirp}}_{\text{strip}} \cdot \widehat{\Pi}_{\Phi_y}(k_y)_{\text{plate}} \cdot \mathbb{A}^2_{X_c(\boldsymbol{u})}(k_y)_{\text{plate}} \\ &\qquad \cdot \sum_n g_n \cdot H^2_{x_n(\boldsymbol{u})}(\breve{k}_x) \cdot H^2_{y_n(\boldsymbol{u})}(k_y). \end{aligned}$$

Stripmap Huygens-Fresnel Analysis

The Huygens-Fresnel decomposition of the stripmap mode angular spectrum is

$$\begin{aligned} &M^{\text{chirp}}_{\widehat{\Pi}(\otimes)} S_{\widehat{\Pi}(\Phi_y)}(\breve{k}_x, k_y)^{\text{strip}}_{\text{ACS}} \\ &\approx \widehat{\Pi}^{\text{chirp}}_{\otimes}(\breve{k}_x, k_y)_{\text{strip}} \cdot \widehat{\Pi}_{\Phi_y}(k_y)_{\text{plate}} \cdot \mathbb{A}^2_{X_c(\boldsymbol{u})}(k_y)_{\text{plate}} \\ &\qquad \cdot \sum_n g_n \cdot \exp(-jx_n\breve{k}_x) \cdot \exp(-jy_n k_y), \\ &= H^2_{X_c(\boldsymbol{u})}(\breve{k}_{\hat{x}}) \cdot H^2_{Y_c(\boldsymbol{u})}(k_{\hat{y}}) \cdot \mathbb{A}^2_{X_c(\boldsymbol{u})}(k_{\hat{y}})_{\text{plate}} \cdot \widetilde{G}^{\text{chirp}}_{\widehat{\Pi}(\otimes)} \widetilde{G}_{\widehat{\Pi}(\Phi_y)}(\breve{k}_{\hat{x}}, k_{\hat{y}})^{\text{strip}}_{\text{SCS}}, \end{aligned}$$

where

$$\widetilde{G}^{\text{chirp}}_{\widehat{\Pi}(\otimes)} \widetilde{G}_{\widehat{\Pi}(\Phi_y)} (\breve{k}_{\hat{x}}, k_{\hat{y}})^{\text{strip}}_{\text{SCS}} = \widehat{\Pi}^{\text{chirp}}_{\otimes} (\breve{k}_{\hat{x}}, k_{\hat{y}})_{\text{strip}} \cdot \widehat{\Pi}_{\Phi_y} (k_{\hat{y}})_{\text{plate}} \cdot \underbrace{\sum_n g_n \cdot \exp(-j\hat{x}_n \breve{k}_{\hat{x}}) \cdot \exp(-j\hat{y}_n k_{\hat{y}})}_{G_{\Pi(\infty)} G_{\Pi(\infty)} (\breve{k}_{\hat{x}}, k_{\hat{y}})^{\text{strip}}_{\text{SCS}}}$$

and

$$\sum_n g_n \cdot \exp(-j\hat{x}_n \breve{k}_{\hat{x}}) \cdot \exp(-j\hat{y}_n k_{\hat{y}}) = \sum_n g_n \cdot H^2_{\hat{x}_n(u)}(\breve{k}_{\hat{x}}) \cdot H^2_{\hat{y}_n(u)}(k_{\hat{y}}).$$

Stripmap Huygens-Fresnel Inversion

Application of inverse Huygens-Fresnel transfers in stripmap mode gives

$$M^{\text{chirp}}_{\widehat{\Pi}(\otimes)} S_{\widehat{\Pi}(\Phi_y)} (\breve{k}_{\hat{x}}, k_{\hat{y}})^{\text{strip}}_{\text{SCS}} = \mathbb{A}^2_{X_c(u)} (k_{\hat{y}})_{\text{plate}} \cdot \widetilde{G}^{\text{chirp}}_{\widehat{\Pi}(\otimes)} \widetilde{G}_{\widehat{\Pi}(\Phi_y)} (\breve{k}_{\hat{x}}, k_{\hat{y}})^{\text{strip}}_{\text{SCS}}.$$

If known, the antenna beam pattern can be compensated [447], leaving

$$M^{\text{chirp}}_{\widehat{\Pi}(\otimes)} S_{\widehat{\Pi}(\Phi_y)} (\breve{k}_{\hat{x}}, k_{\hat{y}})^{\text{strip}}_{\text{SCS}} = \widetilde{G}^{\text{chirp}}_{\widehat{\Pi}(\otimes)} \widetilde{G}_{\widehat{\Pi}(\Phi_y)} (\breve{k}_{\hat{x}}, k_{\hat{y}})^{\text{strip}}_{\text{SCS}}.$$

Stripmap Mode Along-Track Presumming

Since the beamwidth restricted synthetic aperture length is shorter than the CPI ground swath segment when $Y_{\text{img}} > \mathbb{B}(\omega_p, X_{\text{far}})$, slow-time presumming cannot be employed to reduce data volumes and computational load in stripmap mode.

Stripmap Mode Uniform Resampling

Uniform resampling of the Fourier migrated spectrum for stripmap mode proceeds without modification as described in Section 7.14 for the slant-plane case. The sample density of $\ddot{k}_{\hat{x}}$ is determined by the range extent of the CPI ground segment,

$$\Delta_{\ddot{k}_{\hat{x}}} = \frac{2\pi}{2X_o}.$$

Stripmap Image Spectrum

In stripmap mode, the image spectrum is

$$I^{\text{chirp}}_{\widehat{\Pi}(\otimes)} I_{\widehat{\Pi}(\Phi_y)} (\ddot{k}_{\hat{x}}[n_{\hat{x}}], k_{\hat{y}}[n_{\hat{y}}])^{\text{strip}}_{\text{SCS}} = \mathbb{A}^2_{X_c(u)} (k_{\hat{y}}[n_{\hat{y}}])_{\text{plate}} \cdot \widetilde{G}^{\text{chirp}}_{\widehat{\Pi}(\otimes)} \widetilde{G}_{\widehat{\Pi}(\Phi_y)} (\ddot{k}_{\hat{x}}[n_{\hat{x}}], k_{\hat{y}}[n_{\hat{y}}])^{\text{strip}}_{\text{SCS}}.$$

Stripmap Image Spectrum Bandwidth

The frequency-wavenumber domain description of the two way stripmap mode flat plate antenna weight function can be described in terms of the aperture wavenumber support afforded by the beam pattern

$$\begin{aligned}\widehat{\Pi}_{\Phi_y}(k_y)_{\text{plate}} &= 2 \cdot \widehat{\Pi}_{\Phi_y}(k_{u_y})_{\text{plate}}, \\ &= \left[-\frac{4\pi}{D_y}, +\frac{4\pi}{D_y}\right]. \end{aligned} \tag{12.81}$$

The stripmap frequency-wavenumber window function (12.81) is independent of scatterer position and independent of signal frequency. This slow-time wavenumber window function has a center wavenumber given by

$$\widehat{\Pi}^{\text{center}}_{\Phi_y}(k_y)_{\text{plate}} = 0$$

and an along-track image wavenumber bandwidth

$$\widehat{\Pi}^{\text{width}}_{\Phi_y}(k_y)_{\text{plate}} = \frac{8\pi}{D_y}.$$

Stripmap Image

Since the operative length of the synthetic aperture in stripmap mode SAR is fully defined by the fixed angular dwell afforded by the field of view of the antenna's beam pattern, the along-track point spread function is spatially invariant in stripmap Stolt formatting.

Stripmap Invariant Point Spread Functions

As before, to determine the two dimensional resolution of a stripmap mode SAR, we examine the two dimensional wavenumber support of the stripmap mode SAR image spectrum

$$I^{\text{chirp}}_{\widehat{\Pi}(\otimes)} I_{\widehat{\Pi}(\Phi_y)} (\ddot{k}_{\hat{x}}[n_{\hat{x}}], k_{\hat{y}}[n_{\hat{y}}])^{\text{strip}}_{\text{SCS}}.$$

This manifold is expressed in terms of a bounding box of angular dwell and passband bandwidth as shown in Fig. 12.12. The range wavenumber dimension of this k-space bounding box is

$$\Pi_{k_{\hat{x}}} = 2(k^{\max}_{\omega_p} - k^{\min}_{\omega_p})$$

and the along-track dimension of the bounding box is

$$\Pi_{k_{\hat{y}}} = \frac{8\pi}{D_{\hat{y}}}.$$

When an inverse two dimensional Fourier transform is applied to the table top function

$$\Pi_{k_{\hat{x}}} \cdot \Pi_{k_{\hat{y}}}$$

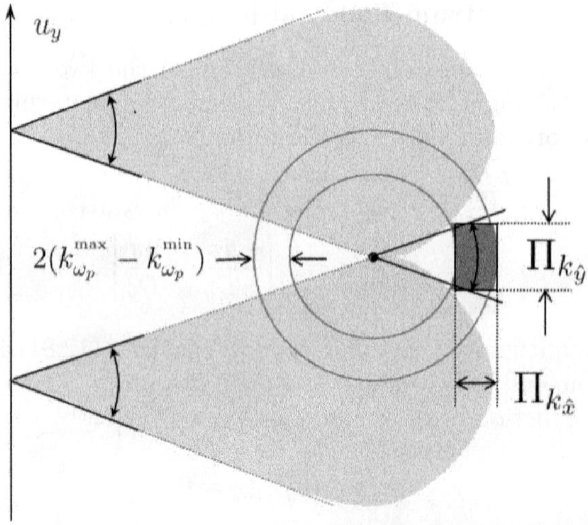

Figure 12.12: Stripmap mode k-space bounding box resolution analysis.

of the bounding box, the resulting point spread function is modeled as a two dimensional sinc function,

$$\underbrace{\Pi_{k_{\hat{x}}} \cdot \text{sinc}\left(\frac{\Pi_{k_{\hat{x}}}}{2} \cdot \hat{x}\right)}_{\mathsf{psf}_{\hat{x}}(\hat{x})} \cdot \underbrace{\Pi_{k_{\hat{y}}} \cdot \text{sinc}\left(\frac{\Pi_{k_{\hat{y}}}}{2} \cdot \hat{y}\right)}_{\mathsf{psf}_{\hat{y}}(\hat{y})}.$$

Using null to null distance of the sinc function as a measure of resolution, the estimate of down-range resolution is

$$\begin{aligned} \Delta_{\hat{x}} &= \frac{2\pi}{\Pi_{k_{\hat{x}}}}, \\ &= \frac{2\pi}{2(k_{\omega_p}^{\max} - k_{\omega_p}^{\min})}, \\ &= \frac{c}{2B} \end{aligned}$$

and the along-track resolution is

$$\begin{aligned} \Delta_{\hat{y}} &= \frac{2\pi}{\Pi_{k_{\hat{y}}}}, \\ &= \frac{D_y}{4}. \end{aligned}$$

Stripmap Image Nyquist Specifications

We specify Nyquist requirements related to the image domain for stripmap mode SAR when Stolt formatting is used. This section is an addendum to general guidelines provided in Section 7.11 for modeless slant-plane models of SAR. Only the specifications that differ from the slant-plane case are elaborated.

Step 1: Specify the Desired Along-Track Image Resolution: There is no change from the slant-plane procedure to specify the desired along-track image resolution for stripmap mode,

$$\Delta_{\hat{y}}.$$

Since there is an induced requirement between the desired resolution and the acceptable stripmap mode antenna size,

$$D_y \leq 4\Delta_{\hat{y}}.$$

The flat plate antenna diameter D_y can be no larger than the upper bound $4\Delta_{\hat{y}}$ when used in stripmap mode. This is required to ensure the antenna beam pattern is large enough to collect sufficient sample counts to achieve the objective resolution.

Step 2: Determine Along-Track Image Bandwidth Requirement: There is no change from the slant-plane procedure to specify the along-track image bandwidth requirement for stripmap mode,

$$\Pi_{k_{\hat{y}}} \geq \frac{2\pi}{\Delta_{\hat{y}}}.$$

Step 3: Specify the Along-Track Image Size: Typically, the stripmap mode customer has no interest in the along-track "size" of stripmap mode imagery, but is more concerned about the revisit scan time required for wide area surveillance. In stripmap mode, the along-track image size

$$\Pi_{\hat{y}} = \left[-Y_{\text{img}}, +Y_{\text{img}}\right]$$

used in signal processing work flow design is a free parameter. The along-track image size parameter determines the coherent processing interval (CPI) of the batch processing.

Step 4: Determine the Along-Track Image Size Pixel Count: There is no change from the slant-plane procedure to determine the along-track pixel count for stripmap mode,

$$N_{\hat{y}} = \left\lceil \frac{\Pi_{\hat{y}}}{\Delta_{\hat{y}}} \right\rceil.$$

Step 5: Determine Along-Track Image Wavenumber Sample Density: There is no change from the slant-plane procedure to determine the along-track image wavenumber sample density for stripmap mode,

$$\begin{aligned}
\Delta_{k_{\hat{y}}} &= \frac{2\pi}{N_{\hat{y}}\Delta_{\hat{y}}}, \\
&\approx \frac{2\pi}{\Pi_{\hat{y}}}.
\end{aligned}$$

Step 6: Determine Synthetic Aperture Slow-Time Bandwidth Requirement: We have

$$\begin{aligned}\Pi_{k_{u_y}} &\geq \frac{1}{2}\cdot\Pi_{k_{\hat{y}}},\\ &= \frac{1}{2}\cdot\frac{2\pi}{\Delta_{\hat{y}}},\\ &= \frac{4\pi}{D_{\hat{y}}}.\end{aligned}$$

Step 7 Determine Synthetic Aperture Sample Density: The slow-time sampling interval (pulse repetition spatial interval) must satisfy

$$\begin{aligned}\Delta_{u_y} &\leq \frac{2\pi}{\Pi_{k_{u_y}}}\\ &= \frac{D_y}{2}.\end{aligned} \tag{12.82}$$

As noted in [407], the $D_y/2$ requirement sometimes stated for along-track sampling seems to originate in [942], where analogies were made with the grating lobe suppression of real aperture radiation patterns. This sampling rate has been noted as marginal in several works [217, 793] and it is recommended in [407] that the along-track sampling interval be set to $D_y/4$.

Step 8 Determine Required Synthetic Aperture Length Requirement: The synthetic aperture half-length L_{strip} in stripmap mode is

$$L_{\text{strip}} = Y_{\text{img}} + \frac{1}{2}\mathbb{B}(\omega_p^{\text{min}}, X_{\text{far}}),$$

with Y_{img} a free parameter of the image formation process work flow design.

Step 9 Determine Number of Synthetic Aperture Samples: The number of pulses consumed in a stripmap mode CPI is

$$M_{u_y} = \left\lceil \frac{2L_{\text{strip}}}{\Delta_{u_y}} \right\rceil.$$

Step 10: Determine Along-Track Aperture Wavenumber Sample Density: The Nyquist sample interval for the aperture wavenumber k_{u_y} is determined by the angular dwell of the beam pattern,

$$\Delta_{k_{u_y}} \leq \frac{2\pi}{\mathbb{B}(\omega_p^{\text{min}}, X_{\text{far}})}.$$

Stripmap Fast-Time Domain Nyquist Specifications

Complex Nyquist sampling in fast-time requires a sample interval upper bounded by

$$\Delta_t \leq \frac{2\pi}{2\omega_0}.$$

To specify when fast-time sampling begins in stripmap mode, the echo signal from the closest scatterer arrives at a time instant of

$$T_s = \frac{2}{c} r_{\min}$$

and the last of the pulse echo from the farthest scatterer arrives at

$$T_f = \frac{2}{c} r_{\max} + T_p.$$

In stripmap mode, the minimum and maximum ranges are dependent on beam-width, which in turn, is dependent upon the signal frequency. Thus,

$$r_{\min}(\omega_p^{\min}) \approx X_c - X_0(\omega_p^{\min})$$

and

$$r_{\max}(\omega_p^{\min}) \approx \sqrt{[X_c + X_0(\omega_p^{\min})]^2 + \mathbb{B}^2(\omega_p^{\min}, X_{\text{far}})}.$$

The number of receive window samples collected from each pulse echo is

$$N_t = \frac{T_f - T_s}{\Delta_t}.$$

After a fast-time discrete Fourier transform, the frequency bin size is

$$\Delta_{\omega_p} = \frac{2\pi}{N_t \Delta_t}.$$

Range Swath Constraints

We next determine a constraint on range swath width that prevents range ambiguities. The along-track sampling interval is $\Delta_{u_y} \leq D_y/4$ meters, so the temporal interval in true slow-time is

$$\tau_u \leq \frac{1}{v_{\text{radar}}} \cdot \frac{D_y}{4} \text{ seconds.}$$

Data collection for one pulse should be finished before the next pulse is ready to be collected (to prevent range ambiguity), this limits the maximum range that is covered before the sensor travels $D_y/4$ meters. Combining these requirements gives a stripmap down-range swath size constraint,

$$R_{\text{swath}}^{\max} \leq \frac{c}{2} \left(\frac{1}{v_{\text{radar}}} \cdot \frac{D_y}{4} \right).$$

Sample Reconciliation in Fast-Time and Down-Range

Care must also be taken in stripmap mode to determine if up sampling is required in fast-time in order to support the Fourier migrated wavenumber manifold with sufficient sample density. The angle

$$\phi_x^{\max} = \arcsin\left(\frac{\lambda_{\max}}{D_y}\right)$$

generates the largest instantaneous view angle and the associated largest value of the determinant of the Stolt transform is

$$J_{\max}^{\text{strip}} = \frac{1}{\cos\phi_x^{\max}}.$$

Hence, the length of the receive window in fast-time must be padded to be at least

$$T_{\min} \geq \max\left[T_f - Ts, J_{\max}^{\text{strip}} \cdot \frac{2}{c} \cdot 2X_0\right].$$

12.8 Huygens-Fresnel Inversion without Stolt Formatting

We now explore a form of SAR image formation processing called *range stacking* [881, 885]. Range stacking is Huygens-Fresnel inversion without Stolt formatting. Range stacking algorithms are performed within the rectilinear frequency-wavenumber domain.

Huygens-Fresnel inversion of the rectilinear spectrum offers several advantages. For one, range stacking does not require Fourier migration into an angular spectrum and does not require uniform resampling of Fourier migrated wavenumber data. Range stacking does not suffer from truncation errors in data interpolations. Range stacking algorithms are easily parallelized. Range stacking can be used in both stripmap and spotlight mode systems.

Range Stacking Overview

Consider the image spectrum $II(\breve{k}_{\hat{x}}, k_{\hat{y}})_{\text{SCS}}$ expressed in the scene coordinate system and the two dimensional inverse Fourier transform that yields the SAR image,

$$ii(\hat{x},\hat{y})_{\text{SCS}} = \frac{1}{4\pi^2}\int_{k_{\hat{x}}}\int_{k_{\hat{y}}} II(\breve{k}_{\hat{x}}, k_{\hat{y}})_{\text{SCS}} \cdot \exp(j\hat{x}\breve{k}_{\hat{x}})\exp(j\hat{y}k_{\hat{y}}) \cdot d\breve{k}_{\hat{x}}dk_{\hat{y}}. \quad (12.83)$$

Also consider the Stolt mapping from the rectilinear spectrum to the Fourier migrated angular spectrum,

$$\begin{aligned} 2\sqrt{k_{\omega_p}^2 - k_{u_y}^2} &\mapsto \breve{k}_{\hat{x}}(\omega_p, k_{u_y}), \\ 2k_{u_y} &\mapsto k_{\hat{y}}(k_{u_y}). \end{aligned}$$

We seek to express the inverse Fourier transform of (12.83), not by Fourier inversion from the $(\breve{k}_{\hat{x}}, k_{\hat{y}})$ domain, but by Fourier inversion from the (ω_p, k_{u_y}) domain. This can be done by a change of variables using

$$II(\breve{k}_{\hat{x}}, k_{\hat{y}})_{\text{SCS}} = MS[\breve{k}_{\hat{x}}(\omega_p, k_{u_y}), k_{\hat{y}}(k_{u_y})]_{\text{SCS}} \cdot |J(\omega_p, k_{u_y})|$$

in a substitution into (12.83),

$$\begin{aligned} &ii(\hat{x},\hat{y})_{\text{SCS}} \\ &= \frac{1}{4\pi^2}\int_{k_{u_y}}\int_{\beta_p} MS[\breve{k}_{\hat{x}}(\omega_p, k_{u_y}), k_{\hat{y}}(k_{u_y})]_{\text{SCS}} \quad (12.84) \\ &\quad \cdot \exp[j\hat{x} \cdot \breve{k}_{\hat{x}}(\omega_p, k_{u_y})] \cdot \exp[j\hat{y} \cdot k_{\hat{y}}(k_{u_y})] \cdot |J(\omega_p, k_{u_y})| \cdot d\omega_p dk_{u_y}, \end{aligned}$$

where $J(\omega_p, k_{u_y})$ is the Jacobian of the Stolt transformation [881],

$$|J(\omega_p, k_{u_y})| = \frac{k_{\omega_p}}{\breve{k}_{\hat{x}}(\omega_p, k_{u_y})}$$

and where we equate the following phase factors in (12.83) and (12.84),

$$\begin{aligned} \exp(j\hat{x}\breve{k}_{\hat{x}}) &= \exp[j\hat{x} \cdot \breve{k}_{\hat{x}}(\omega_p, k_{u_y})], \\ \exp(j\hat{y}k_{\hat{y}}) &= \exp[j\hat{y} \cdot k_{\hat{y}}(k_{u_y})]. \end{aligned}$$

The range compressed and Fourier migrated spectrum

$$MS[\breve{k}_{\hat{x}}(\omega_p, k_{u_y}), k_{\hat{y}}(k_{u_y})]_{\text{SCS}}$$

can be decomposed further in terms of the spectrum of the measured signal as expressed within the aperture coordinate system and subject to inverse Huygens-Fresnel transfers in rectilinear form,

$$\begin{aligned} &MS[\breve{k}_{\hat{x}}(\omega_p, k_{u_y}), k_{\hat{y}}(k_{u_y})]_{\text{SCS}} \\ &\quad = H^*_{X_c}[\breve{k}_x(\omega_p, k_{u_y})] \cdot H^*_{Y_c}[k_y(k_{u_y})] \times \left\{P^*(\omega_p) \cdot SS(\omega_p, k_{u_y})_{\text{ACS}}\right\}. \end{aligned} \tag{12.85}$$

Substitute (12.85) into (12.84) to obtain,

$$\begin{aligned} &ii(\hat{x}, \hat{y})_{\text{SCS}} \\ &\quad = \frac{1}{4\pi} \int_{k_{u_y}} \int_{\beta_p} \underbrace{H^*_{X_c}[\breve{k}_x(\omega_p, k_{u_y})] \cdot H^*_{Y_c}[k_y(k_{u_y})] \cdot P^*(\omega_p) \cdot SS(\omega_p, k_{u_y})_{\text{ACS}}}_{RS[\breve{k}_{\hat{x}}(\omega_p, k_{u_y}), k_{\hat{y}}(k_{u_y})]_{\text{SCS}}} \\ &\qquad \cdot \exp[j\hat{x} \cdot \breve{k}_{\hat{x}}(\omega_p, k_{u_y})] \cdot \exp[j\hat{y} \cdot k_{\hat{y}}(k_{u_y})] \cdot |J(\omega_p, k_{u_y})| \cdot d\omega_p dk_{u_y}, \end{aligned}$$

where the un-migrated, pulse compressed, range focused spectrum is

$$\begin{aligned} &RS[\breve{k}_{\hat{x}}(\omega_p, k_{u_y}), k_{\hat{y}}(k_{u_y})]_{\text{SCS}} \\ &\quad = H^*_{X_c}[\breve{k}_x(\omega_p, k_{u_y})] \cdot H^*_{Y_c}[k_y(k_{u_y})] \cdot P^*(\omega_p) \cdot SS(\omega_p, k_{u_y})_{\text{ACS}}. \end{aligned}$$

Range Line Specific Huygens-Fresnel Transfers

Consider the problem of forming a part of the SAR image comprised of just one cross-range line at the down-range position of

$$x = X_c - \hat{x}_i, \tag{12.86}$$

where x is a variable of the aperture coordinate system and where the cross-range line indicated by $-\hat{x}_i$ (for $\hat{y} \in [-Y_o, +Y_o]$) is a variable of the scene coordinate system. By substituting (12.86) into $\exp(jx\breve{k}_x)$,

$$\exp(jx\breve{k}_x) = \exp(jX_c\breve{k}_x) \cdot \exp(-j\hat{x}_i\breve{k}_x), \tag{12.87}$$

a decomposition of the Huygens-Fresnel transfer function associated with (12.87) is obtained,

$$H_x^*(\breve{k}_x) = H_{X_c}^*(\breve{k}_x) \cdot H_{\hat{x}_i}(\breve{k}_{\hat{x}}). \tag{12.88}$$

If we are only interested in a subset of the processing that will focus just one cross-range line, we express the down-range Huygens-Fresnel transfer of (12.85) in terms of the decomposed version (12.88),

$$H_x^*[\breve{k}_x(\omega_p, k_{u_y})] = H_{X_c}^*[\breve{k}_x(\omega_p, k_{u_y})] \cdot H_{\hat{x}_i}[\breve{k}_x(\omega_p, k_{u_y})]. \tag{12.89}$$

Image Formation of a Single Cross-Range Line

There are two algorithms for focusing the SAR frequency-wavenumber phase history data for this one specific range line [885].

Range Stacking Algorithm I

The focused image for the single cross-range line at $-\hat{x}_i$ in the scene coordinate system can be expressed as

$$\begin{aligned} &ii(-\hat{x}_i, \hat{y})_{\text{SCS}} \\ &= \frac{1}{2\pi}\int_{k_{u_y}} \frac{1}{2\pi}\int_{\beta_p} RS[\breve{k}_{\hat{x}}(\omega_p, k_{u_y}), k_{\hat{y}}(k_{u_y})]_{\text{SCS}} \\ &\quad \cdot \underbrace{\exp[-j\hat{x}_i \cdot \breve{k}_{\hat{x}}(\omega_p, k_{u_y})]}_{\text{range line specific}} \cdot \exp[j\hat{y} \cdot k_{\hat{y}}(k_{u_y})] \cdot |J(\omega_p, k_{u_y})| \cdot d\omega_p dk_{u_y}. \end{aligned} \tag{12.90}$$

Note that $\exp[-j\hat{x}_i \cdot \breve{k}_{\hat{x}}(\omega_p, k_{u_y})]$ of (12.90) is a forward Huygens-Fresnel transfer function

$$H_{\hat{x}_i}[\breve{k}_{\hat{x}}(\omega_p, k_{u_y})] \quad = \quad \exp[-j\hat{x}_i \cdot \breve{k}_{\hat{x}}(\omega_p, k_{u_y})] \tag{12.91}$$

that propagates the in situ field referenced to scene center to the particular cross-range line at $-\hat{x}_i$. With this view point, we have

$$\begin{aligned} &ii(-\hat{x}_i, \hat{y})_{\text{SCS}} \\ &= \frac{1}{2\pi}\int_{k_{u_y}} \\ &\quad \underbrace{\cdot \frac{1}{2\pi}\int_{\beta_p} H_{\hat{x}_i}[\breve{k}_{\hat{x}}(\omega_p, k_{u_y})] \cdot RS[\breve{k}_{\hat{x}}(\omega_p, k_{u_y}), k_{\hat{y}}(k_{u_y})]_{\text{SCS}} \cdot |J(\omega_p, k_{u_y})| \cdot d\omega_p}_{iI[-\hat{x}_i, k_{\hat{y}}(k_{u_y})]_{\text{line}}} \\ &\quad \cdot \exp[j\hat{y} \cdot k_{\hat{y}}(k_{u_y})] \cdot dk_{u_y}, \end{aligned} \tag{12.92}$$

where the pulse compressed phase history $RS[\breve{k}_{\hat{x}}(\omega_p, k_{u_y}), k_{\hat{y}}(k_{u_y})]_{\text{SCS}}$ has been subjected to a forward Huygens-Fresnel transfer $H_{\hat{x}_i}[\breve{k}_{\hat{x}}(\omega_p, k_{u_y})]$ in the frequency-wavenumber domain to change from the scene coordinate system to a system referenced to a specific cross-range line of interest. Coherent integration of

$$\frac{1}{2\pi}\int_{\beta_p} H_{\hat{x}_i}[\breve{k}_{\hat{x}}(\omega_p, k_{u_y})] \cdot RS[\breve{k}_{\hat{x}}(\omega_p, k_{u_y}), k_{\hat{y}}(k_{u_y})]_{\text{SCS}} \cdot |J(\omega_p, k_{u_y})| \cdot d\omega_p$$

over all passband frequencies yields the along-track spectrum

$$iI[-\hat{x}_i, k_{\hat{y}}(k_{u_y})]_{\text{line}}$$

of one focused cross-range line at $-\hat{x}_i$.

Fourier Inversion of Focused Range Line Spectrum

Performing the slow-time inverse Fourier transform

$$ii(-\hat{x}_i, \hat{y})_{\text{SCS}} = \frac{1}{2\pi} \int_{k_{u_y}} iI[-\hat{x}_i, k_{\hat{y}}(k_{u_y})]_{\text{line}} \cdot \exp[j\hat{y} \cdot k_{\hat{y}}(k_{u_y})] \cdot dk_{u_y}$$

results in a fully focused image (in both the down-range and cross-range dimensions) containing just the one cross-range line at $-\hat{x}_i$. This process can be repeated or performed simultaneously for all cross-range lines.

Range Stacking Algorithm II

The other option for a range stacking algorithm comes from considering the other permutation of integration order,

$$\begin{aligned} &ii(-\hat{x}_i, \hat{y})_{\text{SCS}} \\ &= \frac{1}{2\pi} \int_{\beta_p} P^*(\omega_p) \cdot \frac{1}{2\pi} \int_{k_{u_y}} H^*_{X_c}[\breve{k}_x(\omega_p, k_{u_y})] \cdot H^*_{Y_c}[k_y(k_{u_y})] \cdot SS(\omega_p, k_{u_y})_{\text{ACS}} \\ &\quad \cdot \exp[-j\hat{x}_i \cdot \breve{k}_{\hat{x}}(\omega_p, k_{u_y})] \cdot \exp[j\hat{y} \cdot k_{\hat{y}}(k_{u_y})] \cdot \left|J(\omega_p, k_{u_y})\right| \cdot dk_{u_y} d\omega_p, \end{aligned}$$

If we define

$$\begin{aligned} &Ii(\omega_p, -\hat{x}_i, \hat{y})_{\text{SCS}} \\ &= \frac{1}{2\pi} \int_{k_{u_y}} \underbrace{H^*_{X_c}[\breve{k}_x(\omega_p, k_{u_y})] \cdot H^*_{Y_c}[k_y(k_{u_y})] \cdot SS(\omega_p, k_{u_y})_{\text{ACS}}}_{SS(\omega_p, k_{u_y})_{\text{SCS}}} \cdot H_{\hat{x}_i}[\breve{k}_{\hat{x}}(\omega_p, k_{u_y})] \\ &\quad \cdot \exp[j\hat{y} \cdot k_{\hat{y}}(k_{u_y})] \cdot \left|J(\omega_p, k_{u_y})\right| \cdot dk_{u_y}, \end{aligned}$$

then

$$\begin{aligned} &Ii(\omega_p, -\hat{x}_i, \hat{y})_{\text{SCS}} \\ &= \frac{1}{2\pi} \int_{k_{u_y}} H_{\hat{x}_i}[\breve{k}_{\hat{x}}(\omega_p, k_{u_y})] \cdot SS(\omega_p, k_{u_y})_{\text{SCS}} \cdot \exp(jyk_{u_y}) \cdot \left|J(\omega_p, k_{u_y})\right| \cdot dk_{u_y}. \end{aligned}$$

A focused cross-range line can be obtained through coherent integration over the sensor's passband,

$$ii(-\hat{x}_i, \hat{y}) = \frac{1}{2\pi} \int_{\beta_p} P^*(\omega_p) \cdot Ii(\omega_p, -\hat{x}_i, \hat{y})_{\text{SCS}} \cdot d\omega_p.$$

12.9 Stripmap Range-Doppler Algorithms

Image formation processing in the time-wavenumber domain [172, 592, 637] is occupied by the class of range-Doppler algorithms [407]. Detailed descriptions of range-Doppler processing can be found in [40, 167, 500]. Range-Doppler algorithms were the first widely used SAR image formation methods, with application predominantly in long-range, narrow beam, stripmap mode, space based SAR systems [632]. Range-Doppler processing is suitable in SAR applications that accept modest to course levels of image resolution typical of stripmap modality.

Range-Doppler Algorithm Overview

With range-Doppler processing, the SAR along-track image formation challenge is divided into two separate problems: range cell migration correction and along-track matched filtering [42, 214].

Range Cell Migration Correction

A range-Doppler algorithm starts with the Fourier transform in the slow-time dimension to take $ss(t, u_y)$ to $sS(t, k_{u_y})$. Then range-Doppler formatting is performed to remove the fast-time delay or "range migration" associated with hyperbolic range walk [407]. This formatting is done as a function of the aperture or "Doppler wavenumber" k_{u_y} and can be viewed as "flattening" the hyperbolic range walk [770]. This process is sometimes called *"range walk correction."*

Range Walk Correction Methods

Range walk correction can be performed by interpolation from the hyperbolic trace to a straight line contour [770]. Range walk correction can also be achieved by phase shifting and summing operations with parameters shared across a down-range depth of focus region.

Along-Track Matched Filtering

After range walk correction processing, the fast-time and slow-time processing steps are separable one dimensional matched filtering steps. Along-track matched filtering is performed with a flatten reference hyperbola matched to a particular cross-range line within a localized subset of cross-range lines. A bank of along-track matched filters are used to cover the entire range swath [407].

Range-Doppler processing has a depth of focus characteristic in along-track image formation processing [221]. The subswath range gated structure of the range-Doppler approach has been interpreted as a bank of matched filters tuned to the diffraction patterns of reference point scatterers within each of the range subswaths [219].

Down-Range Regions of Azimuth Depth of Focus

The sizes of the depth of focus regions control complexity trade offs. Such design compromises may impact the quality of coherent imagery and hinder subsequent autofocus processing [407]. More generally, there have been stated concerns about the quality of phase information produced by range-Doppler processing [771]. There has been work to reduce the spatial variability of SAR imagery produced by range-Doppler processing [353].

Range-Doppler Dwell Restrictions

Range-Doppler algorithms are constrained to relatively short slow-time dwells to minimize hyperbolic range migration in the down-range direction. The imposition of a dwell constraint in stripmap mode is achieved by limiting the width of the antenna beam, which limits the number of pulse echoes collected from each ground point as the stripmap sensor flies past. This in turn limits the along-track resolution.

Range-Doppler Algorithms and Fresnel Approximations

Range-Doppler algorithms are sometimes formulated and implemented in ways that employ the Fresnel (truncated series) approximation [786]. With the Fresnel approximation, a quadratic phase function is used to approximate a hyperbolic slow-time phase signal. Early versions of the range-Doppler algorithm were typically based on the Fresnel approximation. The Fresnel approximation requires a constrained operational concept to restrict the number of pulses on each scatterer to fit within the simplifying signal model assumptions [26].

Range-Doppler Processing without Range Walk Correction

Interpolation for range migration correction is not required when range cell migration is less than about one third of a range resolution cell [407]. In this case, the interpolation of range migration correction can be dispensed with and the range-Doppler algorithm reduces to a pair of one dimensional imaging problems [407]. When range walk is sufficiently small to ignore, range-Doppler processing has been called "Fresnel-approximation based inversion" [871].

Range-Doppler Processing and Stolt Formatting

By employing a series expansion of the Stolt mapping with truncation to low order, the relationship between the along-track matched filter of range-Doppler processing and an approximation of the Stolt transform can be obtained [137].

The Huygens-Fresnel transfers of omega-k processing have been viewed as providing a "bulk focusing" that is exact only for the range of the corresponding reference hyperbola. The Stolt formatting subprocess has been viewed as providing a focusing operation for nearby cross-range lines [218, 407].

But not all Stolt formatting methods employ Huygens-Fresnel transfers. In either case, with or without Huygens-Fresnel transfers, Stolt formatting provides exact "range migration correction" (to use the language of range-Doppler analysis) for all scatterers in the ground scene [397].

The range dependent depth of focus effect known to exist in range-Doppler algorithms and sometimes sought after in omega-k processing [9, 218, 241, 963], will not be found in Stolt formatting. There is no range dependent, along-track processing used in Stolt format image formation [734]. The "bulk focusing" concept and the concerns of range dependent depth of focus [218] do not apply to angular spectrum inversion methods.

Range-Doppler Processing Concepts

Formulation of range-Doppler signal processing relies on exploiting fast-time (temporal) dependencies in describing slow-time (spatial) parameters [215, 497, 573, 585]. In such analysis, the aperture wavenumber k_{u_y} is often referred to as the "Doppler wavenumber" [40, 440]. Examples of SAR analysis based on a Doppler formulation include [167, 218].

It is possible to *formally* view the aperture wavenumber k_{u_y} as related to a scatterer specific, fast-time dependent, Doppler frequency via the definition,

$$\begin{aligned}\omega_{\text{Doppler}}(t)_n &= 2k_{\omega_c} \cdot v_{\text{radar}} \sin\theta_x(t)_n, \\ &= 2k_{\omega_c} \sin\theta_x(t)_n \cdot v_{\text{radar}}, \\ &= k_{u_y}(t)_n \cdot v_{\text{radar}},\end{aligned}$$

where v_{radar} is the sensor platform's velocity, $\theta_x(t)_n$ is the line of sight angle (defined with respect to the x axis) between sensor and scatterer and where we have defined

$$k_{u_y}(t)_n = 2k_{\omega_c} \sin\theta_x(t)_n. \tag{12.93}$$

The scatterer specific line of sight angle $\theta_x(t)_n$ and corresponding "Doppler shift" frequency $\omega_{\text{Doppler}}(t)_n$ evolve with fast-time.

Note that with (12.93), instead of defining a scatterer specific range closure rate between sensor and scatterer,

$$v_{\text{cl}}(t)_n = v_{\text{radar}} \sin\theta_x(t)_n,$$

we have define a scatterer specific, fast-time dependent "Doppler" wavenumber,

$$\begin{aligned}k_{u_y}(t)_n &= 2k_{\omega_c} \sin\theta_x(t)_n, \\ &= k_{u_y}[\theta_x(t)_n].\end{aligned}$$

Working with the slant-range closure velocity $v_{\text{cl}}(t)_n$, instead of the fast-time dependent along-track wavenumber $k_{u_y}[\theta_x(t)_n]$, is more common in conventional range-Doppler analysis.

Range-Doppler Analysis of Squint Configurations

When squinted, along-track points do not share the same range dependent slow-time hyperbola shapes and complications arise in the range-Doppler formulation [166,233,1012]. The mathematics of range-Doppler formulations can become quite tedious, especially for squint geometries. Some have viewed the formal temporal Doppler analysis of this class of algorithms as "being overloaded by unnecessarily lengthy calculations" [137].

The Formalism of a Fast-Time Analysis of Slow-Time Signals

We now give a brief overview of a formalized temporal Doppler analysis for the simpler broadside case. Assume the transmitted signal $w(t)$ is a passband pulse waveform,

$$w(t) = p(t) \cdot \exp(j\omega_c t).$$

The received time-space echo signal $s(t)$ of a single scatterer is

$$s\left[t - \frac{2}{c}r_n(u_y)\right] = p\left[t - \frac{2}{c}r_n(u_y)\right] \cdot \exp\left\{j\omega_c\left[t - \frac{2}{c}r_n(u_y)\right]\right\},$$

where the distance between the surface antenna's phase center located at (x_s, y_s) and scatterer position (x_n, y_n) is

$$r_n(u_y) = \sqrt{[x_s(u_y) - x_n]^2 + [y_s(u_y) - y_n]^2}.$$

Since we want to formulate the problem in terms of a temporal Doppler signal, we use one more layer of functional composition and express the distance between sensor and scatterer as a function of fast-time,

$$r_n[u_y(t)] \quad = \quad \sqrt{\{x_s[u_y(t)] - x_n\}^2 + \{y_s[u_y(t)] - y_n\}^2},$$

which we more compactly express as

$$r_n(t) \quad = \quad \sqrt{[x_s(t) - x_n]^2 + [y_s(t) - y_n]^2}.$$

First Order Range-Doppler Approximation of Slant-Range

To proceed with a range-Doppler formulation, expand the sensor-scatterer slant-range equation, expressed as a function of fast-time, into a Taylor series about the fast-time instant $t = 0$,

$$r_n(t) = r_n(0) + \dot{r}_n(0) \cdot t + \frac{1}{2}\ddot{r}_n(0) \cdot t^2 + \cdots .$$

A first order fast-time based approximation of a scatterer's echo signal received by a moving radar is

$$r_n(t)^{\text{1st}} = r_n(0) + \dot{r}_n(0) \cdot t. \tag{12.94}$$

If we assume the sensor-scatterer down-range distance $r_n(0)$ at the time instant $t = 0$ is approximated sufficiently well for all scatterers within some range segment of the scene illuminated by the antenna beam, then we can use a down-range reference point X_{ref} within the range focus segment in making the following approximation for $r_n(0)$ in (12.94) [74],

$$\begin{aligned} r_n(0) \quad &\approx \quad r_{(X_{\text{ref}},0)}(0), \\ &= \quad X_{\text{ref}}. \end{aligned} \tag{12.95}$$

The reference range X_{ref} does not change with fast-time because of stripmap mode operations.

The rate of change of sensor-scatterer slant-range $\dot{r}_n(0)$ at $t = 0$ is of interest. We can use the sensor-scatterer view angle at $t = 0$ to characterize the range rate by

$$\dot{r}_n(0) \approx v_{\text{radar}} \sin\theta_x(0)_n. \tag{12.96}$$

The approximation for range (12.95) and the expression for range rate (12.96) at the time instant $t = 0$ gives a 1st order approximation for range closure with respect to fast-time,

$$r_n(t)^{\text{1st}} = X_{\text{ref}} + v_{\text{radar}} \sin\theta_x(0)_n \cdot t.$$

Using this first order approximation in the received echo signal model,

$$\begin{aligned} s\left[t - \frac{2}{c} r_n(t)\right] &\approx s\left[t - \frac{2}{c} r_n(t)^{\text{1st}}\right], \\ &= p\left[t - \frac{2}{c} r_n(t)^{\text{1st}}\right] \cdot \exp\left\{j\omega_c\left[t - \frac{2}{c} r_n(t)^{\text{1st}}\right]\right\}, \end{aligned}$$

gives

$$\begin{aligned} s\left[t - \frac{2}{c} r_n(t)^{\text{1st}}\right] &= p\left\{t - \frac{2}{c}\left[X_{\text{ref}} + v_{\text{radar}} \sin\theta_x(0)_n \cdot t\right]\right\} \\ &\quad \cdot \exp\left\{j\omega_c\left[t - \frac{2}{c}[X_{\text{ref}} + v_{\text{radar}} \sin\theta_n(0) \cdot t]\right]\right\}. \end{aligned}$$

Next, regroup the elements in a way that seeks to physically motivate (from the Doppler perspective) an understanding of the different time-space phase history elements,

$$\begin{aligned} s\left[t - \frac{2}{c} r_n(t)^{\text{1st}}\right] &= p\left[\underbrace{\left(1 - \frac{2}{c} v_{\text{radar}} \sin\theta_x(t)_n\right) t}_{\text{time dilation}} - \underbrace{\frac{2}{c} X_{\text{ref}}}_{\text{two way delay}}\right] \cdot \underbrace{\exp(j\omega_c t)}_{\text{carrier}} \\ &\quad \cdot \underbrace{\exp\left[-j\frac{2}{c}\omega_c X_{\text{ref}}\right]}_{\text{down-range phase factor}} \cdot \underbrace{\exp\left(-j\frac{2}{c}\omega_c v_{\text{radar}} \sin\theta_n(0) \cdot t\right)}_{\text{cross-range phase factor}}. \end{aligned} \tag{12.97}$$

The cross-range phase factor can be associated with a Doppler shift and can be rewritten as

$$\begin{aligned} \exp\left(-j\frac{2}{c}\omega_c v_{\text{radar}} \sin\theta_x(0)_n \cdot t\right) &= \exp\left(-j2k_{\omega_c} \sin\theta_x(0)_n v_{\text{radar}} \cdot t\right), \\ &= \exp\left(-j\omega_{\text{Doppler}}(0)_n \cdot t\right), \end{aligned}$$

where we have defined a scatterer dependent "Doppler" shift frequency in reference to the time instant $t = 0$,

$$\omega_{\text{Doppler}}(0)_n = 2k_{\omega_c} \sin\theta_x(0)_n \cdot v_{\text{radar}}.$$

Another, maybe more familiar form of the cross-range phase factor is

$$\begin{aligned}\exp\left(-j\omega_{\text{Doppler}}(0)_n \cdot t\right) &= \exp\left[-jk_{u_y}[\theta_x(0)_n] \cdot v_{\text{radar}} t\right], \\ &= \exp\left[-jk_{u_y}[\theta_x(0)_n] \cdot \tau\right],\end{aligned}$$

with

$$k_{u_y}[\theta_x(t)_n] = 2k_{\omega_c} \sin\theta_x(t)_n$$

and with the temporal slow-time variable

$$\tau = v_{\text{radar}} t$$

being associated with the spatial version u_y of "slow-time."

Designing a cross-range matched filter referenced to the fast-time instant $t = 0$ and based on the down-range reference point $(X_{\text{ref}}, 0)$ within a range subswath provides means to focus that range segment in azimuth via range-Doppler processing.

Range-Doppler Quadratic Focusing

The first order approximation models the azimuth chirp as a linear frequency chirp. The true tangent function that governs the azimuth chirp can be better approximated by adding higher order terms. We next explore the addition of a "quadratic focusing factor."

Second Order Range-Doppler Approximation of Slant-Range

The Fresnel approximation in range-Doppler signal modeling can be made more precise by including the next higher order term of the Taylor series expansion of slant-range closure, which involves not just a slant-range velocity term, but also a slant-range acceleration term. Start by considering the second order approximation of scatterer specific slant-range as a function of fast-time about the time instant $t = 0$,

$$r_n(t)^{\text{2nd}} = r_n(0) + \dot{r}_n(0) \cdot t + \frac{1}{2}\ddot{r}_n(0) \cdot t^2.$$

If we assume the scatterer specific slant-range acceleration $a_{\text{radar}} \sin\theta_x(t)_n$ at the time instant $t = 0$ is approximately the same for all scatterers within a certain range subswath within the illuminating beam, then we can use

$$\begin{aligned}r_n(0) &\approx X_{\text{ref}}, \\ \dot{r}_n(0) &\approx v_{\text{radar}} \sin\theta_x(0)_n, \\ \ddot{r}_n(0) &\approx a_{\text{radar}} \sin\theta_x(0)_n\end{aligned}$$

and we obtain

$$r_n(t)^{\text{2nd}} = X_{\text{ref}} + v_{\text{radar}} \sin\theta_x(0)_n \cdot t + \frac{1}{2} a_{\text{radar}} \sin\theta_x(0)_n \cdot t^2.$$

The last term of this expansion adds the following phase factor to the received signal model of (12.97),

$$\exp\left(-j\frac{2}{c}\omega_c\frac{1}{2}a_{\text{radar}} \sin\theta_x(0)_n t^2\right) = \exp\left(-j2k_{\omega_c} \cdot \frac{1}{2}a_{\text{radar}} \sin\theta_x(0)_n t^2\right). \tag{12.98}$$

This phase factor has been given various names, including the name of "focusing compensation." These approximate range-Doppler formulations of scatterer informative phase factors provide a k-space description of the fast-time evolution of the position of scatterers. Range-Doppler imaging employs matched filters (based on these first order, second order, etc. approximations) to focus the cross-range element of the SAR image by matched filtering in the so called "Doppler" domain of k-space.

Range Walk Correction by Chirp Scaling

Another approach to range walk correction is to warp the signal space by changing the scale of fast-time chirps in the down-range domain as a function of aperture position u_y. This cannot be done in the fast-time/slow-time signal domain, but is possible in the time-wavenumber domain. This "chirp scaling" approach to the "range walk correction" part of range-Doppler processing is described next.

12.10 Stripmap Chirp Scaling Algorithms

One of the historical threads in the evolution of SAR algorithm development has been the effort to find approaches that avoid the computational load of interpolation. This challenge was especially important when digital hardware solutions provided, compared to today's standards, only modest computational resources.

Polar formatting and Stolt formatting require interpolation in the frequency-wavenumber domain. The range walk correction of range-Doppler algorithms also requires interpolation in the time-wavenumber domain. These operations can be computationally expensive.

Chirp scaling is a type of range-Doppler algorithm that does not require interpolations.

Chirp scaling algorithms were developed independently around 1992 by two groups, the Canada Center for Remote Sensing (CCRS) and the German Aerospace (Deutsche Forschungsanstalt fuer Luft und Raumfahrt (DLR)) [217, 498, 660, 764–768, 770, 806]. An early joint paper by these two groups suggested the name of *chirp scaling* for this class of algorithms. Chirp scaling algorithms have also been called *differential range deramp algorithms* [217].

Chirp Scaling and Range-Doppler Algorithms

Chirp scaling is a form of range migration correction applicable to range-Doppler processing. Conventional range-Doppler processing performs range walk correction to straighten out slow-time hyperbolas by doing a time domain interpolation with pulse compressed data. Chirp scaling performs this range walk correction with equivalent processing of uncompressed LFM waveform data in the time-wavenumber domain [146].

Chirp Scaling and LFM Waveforms

Chirp scaling requires linear frequency modulated waveforms and compensates the differential range variability of the slow-time hyperbolic range walk for scatterers with different range values. Chirp scaling algorithms exploit the linear frequency modulation structure of a transmitted LFM pulse waveform to achieve this objective [407].

The chirp scaling algorithm seeks to make the hyperbolic loci of scatterers at different down-range values trace the same shape. Once this is accomplished, the range curvature migration correction becomes range independent and can be performed in bulk for the entire measured data set [40]. For LFM fast-time signals, the phase adjustment is an aperture wavenumber dependent quadratic phase modulation, causing a change in scale in the fast-time dimension [144]. This multiplication by a chirp scaling phase factor also adjusts the phase centers of the echo chirps such that all scatterers trace the same hyperbolic profile, regardless of range.

Chirp Scaling Limitations

Range-Doppler algorithms, even with chirp scaling enhancements, have approximations that begin to fail as along-track bandwidths increase (e.g., if the antenna beamwidth increases) [218, 764, 765, 806]. Complications also arise when chirp scaling is applied to squint geometries due to couplings between down-range and along-track signal characteristics [589]. When designed for operations in a squint geometry, range-Doppler approximations must be refined with higher order Taylor's series terms.

SAR Data Simulation via Inverse Chirp Scaling

It has been suggested that since the chirp scaling algorithm does not require interpolators, chirp scaling can be "run backwards" over a "perfectly focused" scene to produce raw phase history data for SAR simulations and studies [440, 764].

Secondary Range Compression

When range-Doppler and chirp scaling algorithms are stressed with high spatial bandwidths or high squint imaging geometries, the broken approximations of these algorithms require additional phase compensations. A common approach is *secondary range compression* [39, 40, 146, 221, 234, 500, 661, 770, 994, 999, 1001]. The corrective phase factor of secondary range compression is determined for a reference cross-range line, hence this form of compensation has a down-range depth of focus limitation [146]. The effectiveness of secondary range compression decreases as the squint angle increases [1012]. Secondary range compression has also been viewed as a non-linear FM chirp scaling operation [234, 770].

Secondary range compression is usually performed before range walk correction and can be applied during pulse compression [500, 1001]. An approach for combined range cell migration correction and secondary range compression has been proposed [551]. This method is applied in the (ω, k_{u_y}) domain and uses the chirp-z transform. Motion compensation operations and Doppler centroid tracking operations can also be integrated into the chirp scaling function [661].

Secondary range compression improves the fidelity of range-Doppler coherent data products. One study showed that the primary advantage is with the phase element of range-Doppler SAR imagery, with little perceptual difference in the detected SAR imagery [407, 442].

Secondary range compression models are viewed by some as sufficiently accurate that approximation residuals are deemed as minimal in some stripmap SAR operational and imaging geometry scenarios [440].

Chirp Scaling and Stolt Formatting Algorithms

The chirp scaling algorithm can be developed as either a refinement of range-Doppler processing or as an approximation of the Stolt transform [146, 593, 1010]. The approximation of Stolt formatting provided by chirp scaling breaks down for high-resolution images at low frequencies [746]. Strangely (since Stolt formatting is inherently approximation free), some work has explored ways to modify Stolt formatting to include a notion of secondary range compression [963].

Overview of Chirp Scaling Analysis

The mathematical basis of chirp scaling is related to time scaling of quadratic phase functions of large time-bandwidth product [734, 770].

Consider the time-wavenumber phase history of a SAR system with LFM pulse echo signals where the pulses are not yet range compressed. A quadratic phase function is multiplied in fast-time to cause fast-time LFM chirp echoes to be slightly altered to achieve slow-time hyperbolic range walk consistency in the down-range direction. More specifically, consider the t_o time referenced phase signal of an LFM pulse echo,

$$\exp\left[-j\alpha(t-t_o)^2\right] = \exp\left[-j\pi\gamma(t-t_o)^2\right].$$

The phase structure can be slightly reshaped by multiplication with another quadratic phase function with a chirp rate γ_s that is a small fraction c_s of γ,

$$\gamma_s = c_s\gamma.$$

The chirp scaling phase signal used for (quadratic) range walk correction is

$$\exp[-j\pi\gamma_s t^2] = \exp[-j\pi c_s\gamma t^2].$$

Multiplication of the two LFM chirp signals gives

$$\begin{aligned}
\exp[-j\pi c_s\gamma t^2]\cdot\exp\left[-j\pi\gamma(t-t_o)^2\right] & \\
&= \exp\{-j\pi\gamma[(1+c_s)t^2 - 2t_o t + t_o^2]\}, \\
&= \exp[-j\pi\gamma(1+c_s)t^2]\cdot\exp(-j\pi\gamma\cdot 2t_o t)\cdot\exp(-j\pi\gamma t_o^2),
\end{aligned}$$

which contains a scaled chirp phase factor

$$\exp[-j\pi\gamma(1+c_s)t^2]$$

that can be viewed as having a slightly shifted time reference

$$t_o' = \frac{1}{1+c_s}t_o$$

that relocates the phase center of the quadratic phase function and produces a new chirp rate

$$\begin{aligned}\gamma' &= \gamma(1 + c_s), \\ &= \gamma + \gamma_s.\end{aligned}$$

The position of the new phase center is proportional to t_o and thus scales linearly as a function of range delay. The adjusted chirp rate that results from chirp scaling is approximately constant as a function of range delay and hence the compensated range-Doppler phase history can be universally focused or "compressed" in the along-track dimension.

The compensated data

$$sS(t, k_{u_y})$$

are then fast-time Fourier transformed to yield

$$SS(\omega, k_{u_y}),$$

where a single bulk range walk correction is performed. Along-track compression is performed after the down-range inverse Fourier transform, but before the final along-track inverse Fourier transform. Two dimensional inverse Fourier transforms with coordinate scaling yields the SAR image.

Chirp Scaling Approximation Error Compensations

Although the slow-time phase curtains of the time-wavenumber phase history become range invariant, chirp scaling is not approximation free and induces an artifact of signal variability in the aperture wavenumber domain [442]. Residual errors related to this range invariant assumption can be approximately compensated [770]. For example, the extraneous phase at the phase center after chirp scaling is

$$-\pi\gamma_s \frac{c_s}{1 + c_s} t_o^2.$$

Chirp Scaling and Approximate Stolt Formatting

Chirp scaling can be shown to warp the frequency-wavenumber phase history into a Stolt-like manifold. We present here part of a formulation of chirp scaling as derived in careful detail in [440]. The primary value of this exercise is to show that chirp scaling is as much an approximation seeking to approach the form of Stolt formatting as it is an enhancement of range-Doppler processing. To begin, consider

$$\begin{aligned}\sqrt{k_{\omega_p}^2 - k_{u_y}^2} &= \sqrt{(k_{\omega_c}^2 + k_{\omega_c}k_{\omega_b} + k_{\omega_b}^2) - k_{u_y}^2}, \\ &= k_{\omega_c} \cdot \left[\left(1 - \frac{k_{u_y}^2}{k_{\omega_c}^2}\right) + \left(\frac{k_{\omega_b}}{k_{\omega_c}} + \frac{k_{\omega_b}^2}{k_{\omega_c}^2}\right)\right]^{1/2}, \\ &= \underbrace{k_{\omega_c} \cdot \sqrt{k_{\omega_c}^2 - k_{u_y}^2}}_{\beta_c} \cdot \left[1 + \frac{\left(\frac{k_{\omega_b}}{k_{\omega_c}} + \frac{k_{\omega_b}^2}{k_{\omega_c}^2}\right)}{\left(1 - \frac{k_{u_y}^2}{k_{\omega_c}^2}\right)}\right]^{1/2}, \qquad (12.99)\end{aligned}$$

where we have used an explicit decomposition of the temporal passband wavenumber into carrier and baseband elements,

$$k_{\omega_p} = k_{\omega_c} + k_{\omega_b}.$$

Representing the Stolt mapping of just the carrier element within (12.99) by

$$\beta_c \quad = \quad \sqrt{k_{\omega_c}^2 - k_{u_y}^2}$$

and using the expansion

$$\sqrt{1+\epsilon} = 1 + \epsilon/2 - \epsilon^2/8 + \cdots$$

to expand the second term of (12.99) and keeping only terms up to $k_{\omega_b}^2$ gives an approximation of the (passband) Stolt mapping,

$$\begin{aligned} 2\sqrt{k_{\omega_p}^2 - k_{u_y}^2} &\approx k_{\omega_c}\beta_c \left[1 + \frac{1}{2\beta_c^2}\left(\frac{k_{\omega_b}}{k_{\omega_c}} + \frac{k_{\omega_b}^2}{k_{\omega_c}^2}\right) - \frac{1}{8\beta_c^4}\left(\frac{k_{\omega_b}}{k_{\omega_c}}\right)^2 \right], \\ &= k_{\omega_c}\beta_c + \frac{k_{\omega_b}}{\beta_c} + \frac{k_{\omega_b}^2}{k_{\omega_c}\beta_c} - \frac{k_{\omega_b}^2}{k_{\omega_c}\beta_c^3}, \\ &= \frac{2\omega_c\beta_c}{c} + \frac{2\omega_b}{c\beta_c} + \frac{\omega_b^2}{\omega_c c\beta_c} - \frac{\omega_b^2}{\omega_c c\beta_c^3}. \end{aligned}$$

The expression above provides an approximation of the Stolt mapping. It is the multiplication of this expression by a down-range value x_o that yields the phase factor of secondary range compression used in range-Doppler chirp scaling.

Chapter 13

Volumetric Stolt Formatting

Three dimensional Stolt formatting, or put more generally, *three dimensional angular spectrum processing* for *volumetric synthetic aperture radar* (VolSAR) is emerging as a comparatively new concept in SAR. Three dimensional angular spectrum processing produces SAR imagery potentially focused in three dimensions. VolSAR is doable with SAR systems that synthesize or coherently fuse linear, curvilinear, surface (flat and curved) and volume synthetic apertures.

VolSAR processing yields volumetric data, but the degree of focusing in each dimension depends on flight parameters and induced angular dwell excursions in each dimension. It is possible to simultaneously achieve aperture synthesis in the down-range, along-track and above-ground dimensions with non-linear flight operations [865, 968].

Volumetric imaging with SAR can be indicative of intrinsic side-looking geometries, multiple aspect imaging or due to the penetration of electromagnetic radiation (at some frequencies) into the imaged scene volume [627].

This chapter is outlined as follows: Section 13.1 gives an overview of VolSAR based on Fourier migrations in three dimensions. Section 13.2 sets up the imaging geometry of VolSAR flight operations. Section 13.3 develops a model for the VolSAR time-space phase history. Section 13.4 explores the VolSAR frequency-space phase history. Section 13.5 examines the nature of three dimensional phase curtains. Section 13.6 describes the rectilinear VolSAR signal spectrum of the frequency-wavenumber domain. Section 13.7 generalizes the analysis of angular dwell to three dimensions. Section 13.8 introduces multi-dimensional aperture synthesis with Fourier migrations. Section 13.9 establishes general principles involved with Fourier reconstructions under a change of variables transformation. Finally, Section 13.10 presents simulation studies that demonstrate various VolSAR imaging concepts.

13.1 Angular Spectrum Processing Overview

Stolt formatting was an advancement in SAR that escaped the Fresnel (truncated Taylor series) and Fraunhofer (plane wave) constraints of range-Doppler and polar formatting algorithms. Three dimensional angular spectrum processing is a further advancement that escapes the slant-plane constraint of two dimensional Stolt formatting. Furthermore, unlike coherent summing approaches, VolSAR processing brings the computational efficiency of the fast Fourier transform algorithm to the challenges and opportunities of non-linear flight SAR image formation processing.

VolSAR from Surface Apertures

Slant-plane SAR forms a two dimensional image by synthesizing a one dimensional linear aperture with a wide bandwidth radar. Volumetric SAR imaging forms a three dimensional image by synthesizing at least a surface or volume aperture, or by combining multiple linear, curvilinear or surface subapertures through coherent fusion [557,960]. If a curvilinear or surface k-space trace on an Ewald sphere spans three dimensions, then three dimensional imaging, to some extent, is possible. The achievable resolution at a given scatterer position in each dimension is determined by the angular dwell in each dimension. Resolution attributes in VolSAR are inherently scatterer specific.

Stolt formatting for linear flight SAR migrates wavenumber data out of a straight line trace in k-space into a two dimensional semicircle manifold. Stolt formatting for VolSAR migrates wavenumber data out of a plane in k-space into a three dimensional semisphere manifold [131, 827]. If a synthetic aperture is three dimensional, Fourier migrations are not required.

Range Migration, Range Walk, Range Curvature: Hurdles or Opportunities?

The polar format algorithm uses motion compensation to a reference ground point to compensate range migrations perfectly for spot center and approximately for nearby scatterers. The signal processing steps within the family of range-Doppler algorithms likewise seek to straighten out any range migrations that might occur across the synthetic aperture [26, 526].

However, is it possible that the range walk migration of along-track aperture synthesis is really an opportunity in itself to synthesize an aperture in the down-range dimension?

Likewise, it is claimed that Stolt formatting does not tolerate out-of-plane motions very well [336]. Could this indicate a motion sensitivity that is exploitable to simultaneously synthesize an aperture in the out-of-slant-plane direction, i.e., in the above-ground dimension?

The answer to each of these questions is yes.

After decades of seeking to mitigate down-range or out of slant-plane motion deviations, a major change in perspective will be required by some practicing SAR communities to now consider the opportunities afforded by range migrations and its multi-dimensional generalizations in non-linear flight concepts.

Long Apertures, Large Ground Spots and Short Ranges

Huygens–Fresnel wave theory formulations of SAR systems and signals provide a foundation for aperture synthesis in three dimensions. Down-range and above-ground aperture synthesis can be achieved in scenarios that have been traditionally considered as rather stressing, e.g., SAR scenarios that have combinations of long synthetic apertures, large ground scene sizes and short down-range distances. These stressing imaging geometries are opportunities to be capitalized on in advanced and emerging image formation processing algorithms [969].

Are Surface Apertures Required for Three Dimensional Imaging?

Wave field inversion with multi-dimensional synthetic apertures has been considered for several inverse imaging scenarios [799].

One model states that remote sensing of a diffracted field is a linear transformation that maps n-dimensional object profiles into $n-1$ dimensional observations. The inverse problem is to map the $n-1$ dimensional observations back to the n dimensional object [257].

It is reasonable to ask if observations can have dimension lower than $n-1$ or can multiple, lower dimensional data collections be fused to form an aperture of $n-1$ dimensions preparatory to forming an image in n dimensions?

Affirmative examples can be found. For instance, knowledge of ground topology can be used to proceed with the three dimensional imaging problem based on only a one dimensional aperture [960]. More generally, is it possible to form a three dimensional image from one dimensional data when such ancillary information is not available?

Coherent Fusion of Subapertures

One concept of a surface aperture is where pulse echo data are collected by multiple passes of a sensor and the pulses are viewed as a collective sample over a two dimensional (planar) or three dimensional (curved) surface [265, 705, 705, 1031]. Three dimensional synthetic apertures can be coherently fused from two dimensional apertures; which likewise, can be fused from one dimensional apertures.

It has been shown that coherent integration of multiple subapertures (flight contours) will not only improve image resolution, but will also increase signal to noise ratios [883].

Fourier Migration in VolSAR

VolSAR sensor movement is not required in all three dimensions to form images with some degree of resolution in all three dimensions. Three dimensional Pythagorean relationships among direction cosines of sensor-scatterer slant-vectors enable generalized Stolt transformations of one or two dimensional subapertures. Dimensions without (independent) sensor movement can sometimes have sufficient Fourier domain support achieved by Fourier migrations.

The three dimensional nature of the wavenumber manifold of spotlight mode SAR has long been recognized [26, 490], but flight excursions in the third dimension were typically not sufficient to provide meaningful resolution in the third dimension [26]. These deficiencies evaporate when highly maneuvering flight paths are considered and tailored specifically for efficient data gathering operations for volumetric SAR [536, 537, 587, 687].

Coherent Fusion in Angular Spectrum Wavenumber Domain

Coherent fusion can be performed in the k-space of an angular spectrum domain. Wavenumber domain fusion is performed after Fourier migration of each subaperture and after subaperture specific Huygens-Fresnel transfers to a common ground referenced coordinate system. An inverse three dimensional Fourier transform of the fused k-space data yields voxel data.

Unique Characteristics of Elevation Aperture Synthesis

It is important to remember that it is the density of samples in each dimension of k-space that determines the supportable size in each dimension of the imaged ground scene.

Since the elevation extent of a typically ground scene is typically much less than the down-range and along-track extent, the required sample density in the above-ground wavenumber dimension is typically less. This observation should guide flight design concepts for specific VolSAR imaging objectives.

Volumetric SAR Resolution

It is the size of the volume that encloses the synthetic aperture that determines achievable resolution. It is the sample density within the volume that determines the supportable size of the volumetric scene. Although the grid compensated data of a single flight path may be sparse, fusion scenarios can be envisioned that yield a measure of coherent processing gain in all three dimensions of volumetric image formation processing.

Motion Compensation to a Grid

Fourier transforms require uniform rectilinear data sampling. A non-linear flight path fails to meet this requirement. To compensate this deficiency, a type of motion compensation that we call *motion compensation to a grid* may be employed.

Phase adjustments are based on sensor position estimates with phase adjustments made to one *or more* near grid points. This signal domain grid provides uniform spatial samples for the multi-dimensional slow-time Fourier transforms that take the spatial phase curtain to the wavenumber domain.

The objective slow-time sampling grid is set up to make phase adjustments to grid points within a two or three dimensional volume that encloses the synthetic aperture flight path, regardless of the flight path shape. There need not be an objective flight path with the VolSAR motion compensation to a grid concept.

As an example, if VolSAR processing is used with a circular flight aperture, received signal sampling points in space can be phase adjusted to nearby (one or more) rectilinear grid points within an enclosing square that contains the circular flight path.

13.2 Three Dimensional Imaging Geometry

Box Bounded Ground Scene

Assume the scene of interest is centered at

$$(X_c, Y_c, Z_c)$$

within the aperture coordinate system and is bounded within a box of down-range extent $[-X_o, +X_o]$, along-track extent $[-Y_o, +Y_o]$ and vertical extent $[-Z_o, +Z_o]$.

Box Bounded Synthetic Aperture

Assume a VolSAR synthetic aperture of some form is centered at

$$(L_{u_{xc}}, L_{u_{yc}}, L_{u_{zc}})$$

within the aperture coordinate system and is bounded within a box of down-range extent $[-L_{u_x}, +L_{u_x}]$, along-track extent $[-L_{u_y}, +L_{u_y}]$ and vertical extent $[-L_{u_z}, +L_{u_z}]$.

Scene Centered Volumetric Synthetic Aperture

There is nothing that precludes

$$(X_c, Y_c, Z_c) = (0, 0, 0)$$

in considering various simulated imaging scenarios of potential interest, in which case, the aperture coordinate system coincides with the scene coordinate system.

13.3 VolSAR Time-Space Phase History

A broadband signal is not required in the formulation of a VolSAR imaging system. We deal with both the monochromatic continuous wave and general pulse waveform cases, with the latter viewed as a polychromatic extension of the former. We shall ignore the influence of the beam pattern in our VolSAR analysis.

VolSAR with Continuous Wave Signals

Starting with the monochromatic continuous wave case, an isotropic point antenna is assumed to transmit a continuous wave tone signal with infinite time duration,

$$w_{\Pi(\infty)}(t)^{\text{cw}} = \exp(j\omega_c t).$$

Single Scatterer Continuous Wave Echo Signal

The radar echo signal received at the point

$$\boldsymbol{u} = [u_x \; u_y \; u_z]$$

within the box bounded VolSAR synthetic aperture and which was scattered by a discrete scatterer located at

$$\boldsymbol{x}_n = [x_n \; y_n \; z_n]$$

is

$$\begin{aligned} s_{\Pi(\infty)}\,\boldsymbol{s}_{\Pi(\boldsymbol{u})}\,(t,\boldsymbol{u})_n^{\text{cw}} &= g_n \cdot \exp\left\{j\omega_c\left[t - t_n(\boldsymbol{u})\right]\right\}, \\ &= g_n \cdot \exp\left(j\omega_c t\right) \cdot \exp\left[-j\omega_c t_n(\boldsymbol{u})\right], \\ &= g_n \cdot w_{\Pi(\infty)}(t)^{\text{cw}} \cdot \measuredangle H h^2\left[\omega_c, r_n(\boldsymbol{u})\right], \end{aligned}$$

where the echo delay is determined by the sensor-scatterer slant-range,

$$t_n(\boldsymbol{u}) = \frac{2}{c} \cdot r_n(\boldsymbol{u}).$$

We have assumed no constraints on temporal support of the continuous wave signal (hence the $\Pi(\infty)$ subscript). $\Pi(\boldsymbol{u})$ is an indicator function of the three dimensional box function that encloses the VolSAR synthetic aperture. The slow-time phase curtain is expressed as a squared (round trip) harmonic Fresnel phase field,

$$\measuredangle H h^2\left[\omega_c, r_n(\boldsymbol{u})\right] = \exp\left[-jk_{\omega_c} \cdot 2r_n(\boldsymbol{u})\right]$$

produced by the nth scatterer.

Ground Scene Continuous Wave Echo Signal

The echo signal collected from all scatterers within the volumetric scene is

$$s_{\Pi(\infty)}\,\boldsymbol{s}_{\Pi(\boldsymbol{u})}\,(t,\boldsymbol{u})^{\text{cw}} = \sum_n s_{\Pi(\infty)}\,\boldsymbol{s}_{\Pi(\boldsymbol{u})}\,(t,\boldsymbol{u})_n^{\text{cw}}.$$

VolSAR with Pulse Signals

The polychromatic version of the received echo signal gathered with a transmitted pulse waveform

$$w_{\Pi(T_p)}(t)^{\text{pulse}} = \exp(j\omega_c t) \cdot p_{\Pi(T_p)}(t)$$

is

$$\begin{aligned} s_{\Pi(T_s,T_f)}\,\boldsymbol{s}_{\Pi(\boldsymbol{u})}\,(t,\boldsymbol{u})^{\text{pulse}} &= \sum_n g_n \cdot w_{\Pi(T_p)}\left[t - t_n(\boldsymbol{u})\right]^{\text{pulse}}, \\ &= \sum_n g_n \cdot p_{\Pi(T_p)}\left[t - t_n(\boldsymbol{u})\right] \cdot \exp\left\{j\omega_c\left[t - t_n(\boldsymbol{u})\right]\right\}, \\ &= \sum_n g_n \cdot p_{\Pi(T_p)}\left[t - t_n(\boldsymbol{u})\right] \cdot hh^2\left[t, r_n(\boldsymbol{u})\right], \end{aligned}$$

where the radar's receive window is $t \in \Pi\left(T_s, T_f\right)$ and where

$$hh^2\left[t, r_n(\boldsymbol{u})\right] = \exp\left\{j\omega_c\left[t - t_n(\boldsymbol{u})\right]\right\}.$$

Down Converted VolSAR Time-Space Phase History

Down conversion gives

$$S_{\Pi(\omega_c)}\, s_{\Pi(u)}(t, \boldsymbol{u})^{\text{cw}} = \sum_n g_n \cdot \measuredangle Hh^2\left[\omega_c, r_n(\boldsymbol{u})\right]$$

for the monochromatic case and gives

$$\begin{aligned} s^{\text{carrier}(-\omega_c)}_{\Pi(T_s,T_f)}\, s_{\Pi(u)}(t, \boldsymbol{u})^{\text{pulse}} &= \sum_n g_n \cdot p_{\Pi(T_p)}\left[t - t_n(\boldsymbol{u})\right] \cdot \exp\left[-j\omega_c t_n(\boldsymbol{u})\right], \\ &= \sum_n g_n \cdot p_{\Pi(T_p)}\left[t - t_n(\boldsymbol{u})\right] \cdot \measuredangle Hh^2\left[\omega_c, r_n(\boldsymbol{u})\right] \end{aligned}$$

for the polychromatic case.

Pulse Compression

With VolSAR continuous wave signals, pulse compression is not required to form the image in the down-range dimension for some VolSAR imaging scenarios. Use of phase or frequency modulated pulse waveforms with pulse compression, however, will provide increased signal-to-noise ratios through coherent processing gains.

13.4 VolSAR Frequency-Space Phase History

Monochromatic Frequency-Space Phase History

After down conversion for the continuous waveform case, the measured slow-time signal is fast-time independent and is comprised of three dimensional, center frequency dependent, slow-time phase curtains scaled by scatterer strength,

$$\begin{aligned} S_{\Pi(\omega_c)}\, s_{\Pi(u)}(\omega_c, \boldsymbol{u}) &= \sum_n g_n \cdot \exp\left[-jk_{\omega_c} \cdot 2r_n(\boldsymbol{u})\right], \\ &= \sum_n g_n \cdot \measuredangle Hh^2[\omega_c, r_n(\boldsymbol{u})], \\ &= Ssss(\omega_c, u_x, u_y, u_z). \end{aligned}$$

As indicated by the notation $Ssss(\omega_c, u_x, u_y, u_z)$, volumetric SAR image formation processing of frequency-space phase history data is essentially four dimensional: three spatial dimensions coupled with the temporal frequency parameter.

Polychromatic Frequency-Space Phase History

For the polychromatic case, the frequency-space phase history of the down converted VolSAR pulse signal for the entire ground scene is

$$\begin{aligned} S^{\text{L.shift}(\omega_c)}_{\Pi(2\omega_o @ \omega_c)}\, s_{\Pi(u)}(\omega_p, \boldsymbol{u})^{\text{pulse}} &= P_{\Pi(2\omega_o @ 0)}(\omega_p - \omega_c) \cdot \sum_n g_n \cdot \exp\left[-jk_{\omega_p} \cdot 2r_n(\boldsymbol{u})\right], \\ &= P_{\Pi(2\omega_o @ 0)}(\omega_b) \cdot \sum_n g_n \cdot Hh^2[\omega_p, r_n(\boldsymbol{u})], \end{aligned}$$

where $\omega_p = \omega_c + \omega_b$ with $\omega_b \in [-\omega_o, \omega_o]$. This frequency-space VolSAR signal spectrum is essentially the same as the phase history from the continuous wave case, except for the weighting imposed by the pulse spectrum and the change of the slow-time phase curtain from dependency on the single center frequency to multiple passband frequencies.

Assuming that compensation of down conversion phase artifacts will take place, the frequency-space phase history can also be dealt with in purely passband form,

$$S_{\Pi(2\omega_o @\omega_c)} \boldsymbol{s}_{\Pi(\boldsymbol{u})} (\omega_p, \boldsymbol{u})^{\text{pulse}} = W_{\Pi(2\omega_o @\omega_c)} (\omega_p) \cdot \sum_n g_n \cdot Hh^2[\omega_p, r_n(\boldsymbol{u})].$$

Pulse Compression Matched Filtering

Pulse compression can be performed before or after slow-time Fourier transforms. To proceed with pulse compression matched filtering at this point in the frequency-space domain, multiplication of

$$S^{\text{shift}(-\omega_c)}_{\Pi(2\omega_o @\omega_c)} \boldsymbol{s}_{\Pi(\boldsymbol{u})} (\omega_p, \boldsymbol{u})^{\text{pulse}}$$

by

$$P^*_{\Pi(2\omega_o @0)} (\omega_b)$$

gives

$$P^*_{\Pi(2\omega_o @0)} (\omega_b) \times \left\{ S^{\text{L.shift}(\omega_c)}_{\Pi(2\omega_o @\omega_c)} \boldsymbol{s}_{\Pi(\boldsymbol{u})} (\omega_p, \boldsymbol{u}) \right\} = |P_{\Pi(2\omega_o @0)} (\omega_b)|^2 \cdot \sum_n g_n \cdot Hh^2[\omega_p, r_n(\boldsymbol{u})].$$

Pulse compression cancels the phase of $P_{\Pi(2\omega_o @0)} (\omega_b)$ and leaves a residual amplitude modulation weight function comprised of the modulus squared pulse spectrum,

$$|P_{\Pi(2\omega_o @0)} (\omega_b)|^2.$$

13.5 Three Dimensional Phase Curtains

We introduce a general description of spatial chirp signals and their Fourier transforms. Similar to the development of time-frequency chirp couplets presented in Section 6.1, *space-wavenumber chirp couplets* prove useful in our approach to volumetric SAR signal analysis.

The truncated Taylor series chirp couplet formulation introduced in Section 12.5 is deficient in several aspects. For one, it was specific to linear aperture SAR and does not easily generalize to curvilinear, surface or volume synthetic apertures.

We provide a formulation of slow-time chirp signals that is of utility in describing the spatial chirp phase curtain of general three dimensional flight operations. The generalized formulation has a simple geometric interpretation.

Generalized Slow-Time Chirp Couplets

A useful expression of a slow-time phase curtain in terms of a space-wavenumber chirp couplet is of the form

$$\exp\left[-j \cdot \overline{k_{u_y}(u_y)_n} \cdot (u_y - y_n)\right] = \exp\left[-j \cdot \overline{k_{\omega_c}\cos\theta_y(u_y)_n} \cdot (u_y - y_n)\right], \quad (13.1)$$

where $\theta_y(u_y)_n$ is the direction cosine angle between the nth scatterer and the y axis as a function of sensor position u_y within the multi-dimensional synthetic aperture. We define the scatterer specific *slow-time chirp couplet* in (13.1) as

$$\overline{k_{u_y}(u_y)_n} \equiv \overline{k_{\omega_c}\cos\theta_y(u_y)_n}. \quad (13.2)$$

This approach is motivated by our prior study of the slant-plane angular spectrum and the role of the Pythagorean quadruple among the wavenumbers of a directed harmonic plane wave. This definition can be made scatterer independent by relying on the isomorphism between view angle and angular spectrum angle,

$$\overline{k_{\omega_c}\cos\theta_y(u_y)_n} = \overline{k_{\omega_c}\cos\phi_y(u_y)}.$$

Chirp Couplet Phase Curtains in Three Dimensions

The geometrically descriptive form (13.2) of a slow-time chirp couplet for the y dimension can be easily generalized to curvilinear, surface or volume synthetic apertures when augmented with additional slow-time chirp couplets defined for the x and z dimensions,

$$\begin{aligned}\overline{k_{u_x}(u_x)_n} &= \overline{k_{\omega_c}\cos\theta_x(u_x)_n}, \\ &= \overline{k_{\omega_c}\cos\phi_x(u_x)},\end{aligned}$$

$$\begin{aligned}\overline{k_{u_z}(u_z)_n} &= \overline{k_{\omega_c}\cos\theta_z(u_z)_n}, \\ &= \overline{k_{\omega_c}\cos\phi_z(u_z)},\end{aligned}$$

where

$$[\theta_x(u_x)_n \; \theta_y(u_y)_n \; \theta_z(u_z)_n] \equiv [\phi_x(u_x) \; \phi_y(u_y) \; \phi_z(u_z)]$$

is a full suite of direction cosine angles and corresponding isomorphically mapped angular spectrum angles.

Three Dimensional Phase Curtain Spatial Fourier Transform

Based on experience gained in two dimensional angular spectrum analysis, we postulate that the core of the spatial Fourier transform of a generalized, non-linear, two way, phase curtain of a single scatterer in three dimensions is

$$\begin{aligned}\exp\left[-jk_{\omega_c} \cdot 2r_n(\boldsymbol{u})\right] \;\rightleftharpoons\; & \exp\left[-jx_n(u_x) \cdot 2\overline{k_{\omega_c}\cos\theta_x(u_x)_n}\right] \\ & \cdot \exp\left[-jy_n(u_y) \cdot 2\overline{k_{\omega_c}\cos\theta_y(u_y)_n}\right] \\ & \cdot \exp\left[-jz_n(u_z) \cdot 2\overline{k_{\omega_c}\cos\theta_z(u_z)_n}\right],\end{aligned}$$

where we have defined,

$$\begin{aligned} x_n(u_x) &= x_n - u_x, \\ y_n(u_y) &= y_n - u_y, \\ z_n(u_z) &= z_n - u_z. \end{aligned}$$

What we call the "core" of a Fourier transform pair is the mapping between phase factors on either side of a more fully expressed Fourier transform pair.

Slow-Time Chirp Phase Curtains with Fast-Time Chirp Couplets

The slow-time chirp phase curtain,

$$\exp\left[-jk_{\omega_c} \cdot 2r_n(\boldsymbol{u})\right]$$

can also contain a fast-time chirp signal element if

$$k_{\omega_c} = \frac{1}{c} \cdot \omega_c$$

is replaced with

$$k_{\overline{\omega_p[t-t_n(\boldsymbol{u})]}} = \frac{1}{c} \cdot \overline{\omega_p[t-t_n(\boldsymbol{u})]}.$$

Hence, we can add a fast-time chirp couplet to our slow-time chirp phase curtain expressions to handle the case of a polychromatic SAR system that uses chirp waveforms,

$$\exp\left[-jk_{\omega_c} \cdot 2r_n(\boldsymbol{u})\right] \mapsto \exp\left[-jk_{\overline{\omega_p[t-t_n(\boldsymbol{u})]}} \cdot 2r_n(\boldsymbol{u})\right].$$

Combining notation for both fast-time and slow-time chirp couplets, we expect the core of the spatial Fourier transform of a spatial chirp phase curtain gathered with a polychromatic temporal chirp waveform to be

$$\begin{aligned} \exp\left[-jk_{\overline{\omega_p[t-t_n(\boldsymbol{u})]}} \cdot 2r_n(\boldsymbol{u})\right] \rightleftharpoons \; & \exp\left[-jx_n(u_x) \cdot \overline{2k_{\omega_p} \cos\theta_x(u_x)_n}\right] \\ & \cdot \exp\left[-jy_n(u_y) \cdot \overline{2k_{\omega_p} \cos\theta_y(u_y)_n}\right] \\ & \cdot \exp\left[-jz_n(u_z) \cdot \overline{2k_{\omega_p} \cos\theta_z(u_z)_n}\right]. \end{aligned} \tag{13.3}$$

In the frequency-wavenumber domain, the monochromatic case deals only with ω_c, while the polychromatic case must deal with all ω_p within the passband. Other than that, the functional expressions are essentially the same.

It was noted in [889] that angular spectrum analysis and the isomorphism between view angle and wavenumber angle should enable the application of Stolt formatting methods in non-linear SAR systems. Consistent with this concept, the postulated Fourier transform pair (13.3) may be key.

13.6 VolSAR Frequency-Wavenumber Phase History

The continuous waveform version of the VolSAR frequency-space phase history is

$$\begin{aligned}
&S_{\Pi(\omega_c)}\boldsymbol{s}_{\Pi(\boldsymbol{u})}(\omega_c,\boldsymbol{u}) \\
&\quad = \sum_n g_n \cdot \exp\left[-jk_{\omega_c}\cdot 2r_n(\boldsymbol{u})\right] \\
&\quad = \sum_n g_n \cdot \exp\left[-jk_{\omega_c}\cdot 2\sqrt{(u_x-x_n)^2+(u_y-y_n)^2+(u_z-z_n)^2}\,\right], \\
&\quad = \sum_n g_n \cdot \exp\left[-jk_{\omega_c}\cdot 2\sqrt{x_n^2(u_x)+y_n^2(u_y)+z_n^2(u_z)}\,\right].
\end{aligned}$$

We enter the frequency-wavenumber domain with a three dimensional slow-time Fourier transform performed over a box bounded volume (containing a uniform rectilinear sample grid) that encloses the synthetic aperture flight path. The core of the three dimensional Fourier transform in the slow-time dimensions of the radar echo signal (the phase curtain) from one scatterer is

$$\begin{aligned}
&SSSS\left(\omega_c,k_{u_x},k_{u_y},k_{u_z}\right)_n \\
&\quad = \mathsf{FT}^{u}_{3D}\left\{\exp\left[-jk_{\omega_c}\cdot 2\sqrt{x_n^2(u_x)+y_n^2(u_y)+z_n^2(u_z)}\,\right]\right\}, \\
&\quad \rightleftharpoons \exp\left[-jx_n(u_x)\cdot 2\overline{k_{\omega_c}\cos\theta_x(u_x)_n}\,\right] \\
&\qquad \cdot\exp\left[-jy_n(u_y)\cdot 2\overline{k_{\omega_c}\cos\theta_y(u_y)_n}\,\right] \\
&\qquad \cdot\exp\left[-jz_n(u_z)\cdot 2\overline{k_{\omega_c}\cos\theta_z(u_z)_n}\,\right], \\
&\quad = SS(\omega_c,k_{u_x})_{x_n}\cdot SS(\omega_c,k_{u_y})_{y_n}\cdot SS(\omega_c,k_{u_z})_{z_n}.
\end{aligned}$$

The VolSAR signal's frequency-wavenumber spectrum of the entire ground scene is tonal and dimensionally separable,

$$SSSS\left(\omega_c,k_{u_x},k_{u_y},k_{u_z}\right) \equiv \sum_n SS(\omega_c,k_{u_x})_{x_n}\cdot SS(\omega_c,k_{u_y})_{y_n}\cdot SS(\omega_c,k_{u_z})_{z_n},$$

where $SS(\omega_c,k_{u_x})_{x_n}$, $SS(\omega_c,k_{u_y})_{y_n}$ and $SS(\omega_c,k_{u_z})_{z_n}$ are the dimension specific phase factors of the frequency-wavenumber phase history,

$$\begin{aligned}
SS(\omega_c,k_{u_x})_{x_n} &= \exp\left[-jx_n(u_x)\cdot 2\overline{k_{\omega_c}\cos\theta_x(u_x)_n}\right], \\
SS(\omega_c,k_{u_y})_{y_n} &= \exp\left[-jy_n(u_y)\cdot 2\overline{k_{\omega_c}\cos\theta_y(u_y)_n}\right], \\
SS(\omega_c,k_{u_z})_{z_n} &= \exp\left[-jz_n(u_z)\cdot 2\overline{k_{\omega_c}\cos\theta_z(u_z)_n}\right].
\end{aligned}$$

If a polychromatic chirp pulse waveform system is involved, the analysis requires that ω_c be replaced by ω_p.

13.7 Three Dimensional Angular Dwell

Angular manifolds in VolSAR are traced out as sensor position is varied. Spatial angular manifolds are geometrically determined by changes in relative position between sensor and scatterers. The angular spectrum manifold is obtained by an isomorphic mapping from view angle sweeps in the spatial domain to the wavenumber domain. Here, we generalize measures of angular dwell to three dimensions as expressed in both spatial and wavenumber domains.

View Angle to Wavenumber Angle Mapping

The generalized Stolt transformation is defined by the direction cosines of the sensor-scatterer slant-vectors that reach to all points within the ground scene. The complexity of the view angle analysis is simplified with the definition of isomorphic mappings to slow-time aperture wavenumber vectors that sweep through isomorphically induced submanifolds resident on an Ewald sphere [40, 314].

Isomorphism Between Spatial and Wavenumber Angular Dwell Manifolds

We define an isomorphic mapping of a sensor-scatterer slow-time view angle vector

$$\boldsymbol{\theta}(\boldsymbol{u})_n = [\theta_x(u_x)_n \; \theta_y(u_y)_n \; \theta_z(u_z)_n]$$

within the spatial domain angular manifold $\boldsymbol{\Theta}$,

$$\boldsymbol{\theta}(\boldsymbol{u})_n \in \boldsymbol{\Theta},$$

to equally valued k-space angles

$$\boldsymbol{\phi}(\boldsymbol{k}_u) = [\phi_x(u_x) \; \phi_y(u_y) \; \phi_z(u_z)]$$

within an induced angular manifold $\boldsymbol{\Phi}$ in the aperture wavenumber $\boldsymbol{k}_u$ domain,

$$\boldsymbol{\phi}(\boldsymbol{k}_u) \in \boldsymbol{\Phi}.$$

The view angle dwell $\boldsymbol{\Theta}$ defines a spatial domain manifold subtended by view angles that evolve with the changing position of the SAR sensor throughout the volumetric synthetic aperture. The spatial domain manifold $\boldsymbol{\Theta}$ is mapped to a wavenumber domain angular manifold $\boldsymbol{\Phi}$ by the three dimensional slow-time Fourier transform. This isomorphism yields the bridging relationship of equal angular manifolds,

$$\boldsymbol{\Theta} \equiv \boldsymbol{\Phi}, \tag{13.4}$$

between the range of sensor-scatterer view angles $\boldsymbol{\theta}(\boldsymbol{u})_n$ in the spatial $\boldsymbol{u}$ domain and wavenumber angles $\boldsymbol{\phi}(\boldsymbol{k}_u)$ in the three dimensional $\boldsymbol{k}_u$ aperture wavenumber domain.

The dependence of SAR spatial domain angular dwells on particular scatterers is no longer explicit after slow-time (spatial) Fourier transforms.

This functional isomorphism provides an approach to SAR signal analysis that models the SAR phase history in a way that does not destroy sensor-scatterer specific geometries (such as occurs in polar formatting), but where these details are implicitly retained in the wavenumber domain phase history analysis.

Volumetric Inverse Huygens-Fresnel Transfers

Fourier transform computational costs are reduced if the VolSAR voxel volume

$$vvv(\hat{x}, \hat{y}, \hat{z})_{\text{SCS}}$$

is centered in the scene coordinate system. If there is an offset between the aperture and scene coordinate systems, then wavenumber domain Nyquist sample densities required for

$$VVV(k_x, k_y, k_z)_{\text{ACS}}$$

are higher than those required for $VVV(k_{\hat{x}}, k_{\hat{y}}, k_{\hat{z}})_{\text{SCS}}$. However, it is not always the case that there is an offset set up in the signal analysis model between the aperture coordinate system and scene coordinate system.

Volumetric Presumming

A Huygens-Fresnel shift from aperture to scene coordinate systems, combined with *presumming* (if the scene volume is smaller than the aperture volume), either by coherent integration or by subsampling, can be applied to

$$VVV(k_x, k_y, k_z)_{\text{ACS}}$$

to yield

$$VVV(k_{\hat{x}}, k_{\hat{y}}, k_{\hat{z}})_{\text{SCS}}$$

and thereby reduce sample counts.

Inverse Huygens-Fresnel Transfers of VolSAR

Consider multiplication of $VVV(k_x, k_y, k_z)_{\text{ACS}}$ by the inverse Huygens-Fresnel transfer functions,

$$\begin{aligned} H^*_{X_c}(k_x) &= \exp\left[j\left(L_{x_c} - X_c\right)k_x\right], \\ H^*_{Y_c}(k_y) &= \exp\left[j\left(L_{y_c} - Y_c\right)k_y\right], \\ H^*_{Z_c}(k_z) &= \exp\left[j\left(L_{z_c} - Z_c\right)k_z\right], \end{aligned}$$

for a slow-time coordinate shift to scene center,

$$VVV(k_{\hat{x}}, k_{\hat{y}}, k_{\hat{z}})_{\text{SCS}} = H^*_{X_c}(k_x) \cdot H^*_{Y_c}(k_y) \cdot H^*_{Z_c}(k_z) \cdot VVV(k_x, k_y, k_z)_{\text{ACS}}.$$

The Huygens-Fresnel transfers shift scatterer specific phase referencing from the center of the box volume synthetic aperture located at $(L_{x_c}, L_{y_c}, L_{z_c})_{\text{ACS}}$ to

$$(X_c, Y_c, Z_c)_{\text{ACS}} = (0, 0, 0)_{\text{SCS}}.$$

"Huygens-Fresnel Transfers" of Co-Centered Aperture and Scene

If the aperture and scene are co-centered, then

$$\begin{aligned} (L_{x_c}, L_{y_c}, L_{z_c})_{\text{ACS}} &= (X_c, Y_c, Z_c)_{\text{ACS}}, \\ &= (0, 0, 0)_{\text{SCS}}. \end{aligned}$$

In this case, where phase-only Huygens-Fresnel transfers are not required, an angular spectrum, dare we say, "matched filter" or *pivot point reference filter* is required to phase compensate the data on the Ewald sphere referenced to $(0,0,0)_{\mathrm{SCS}}$. Application of this matched filter ensures that Fourier inversion of terrain spectral data resident on the Ewald sphere yields a volumetric point spread function. Without this matched filter, Fourier inversion would return the frequency-wavenumber domain data to an in situ frequency-space wave field without formation of VolSAR voxel point spread functions.

Angular Spectrum Pivot Point Reference Filter

The required angular spectrum pivot point reference filter can be obtained by a motion compensation to a point process and subsystem. The motion compensation subsystem generates position induced phase data throughout the flight profile referenced to a single ground point. The three dimensional spatial Fourier transform of this flight profile phase reference yields the pivot point reference filter. The conjugate of this volumetric flight profile reference filter is used as a "matched filter" to focus the volumetric imagery when no Huygens-Fresnel transfer functions are required for a coordinate shift.

Huygens-Fresnel Transfers After Volumetric Presumming

Huygens-Fresnel transfers can occur post presumming [885], as long as the transfer functions themselves are also formulated in the presummed data domain. This processing order, as explained in Section 7.13, intentionally aliases high frequency portions of chirp phase curtain that are canceled in the Huygens-Fresnel transfers, but preserves unaliased representations of scatterer positions expressed by phase factors of the scene coordinate system.

13.8 Fourier Migrations in Three Dimensions

The volumetric imaging version of Stolt formatting was developed for a two dimensional planar synthetic aperture in [600].

Stolt Formatting in Three Dimensions

In three dimensional Stolt formatting, we must express a Stolt mapping from the aperture wavenumbers

$$\boldsymbol{k}_u = \left[k_{u_x}\ k_{u_y}\ k_{u_z}\right]$$

to Fourier migrated terrain wavenumbers,

$$\breve{\boldsymbol{k}}_x = \left[\breve{k}_x\ \breve{k}_y\ \breve{k}_z\right].$$

Fourier Migrated Wavenumbers

The angular spectrum for a slant-plane system satisfies a two dimensional Pythagorean relationship in a wavenumber domain as described in Section 4.6. Hence, there exists a ϕ_x in the two dimensional wavenumber domain such that

$$\breve{k}_x = 2k_{\omega_c} \cos\phi_x,$$

where ϕ_x is related by isomorphism to $\theta_x(u_x)$. A straight forward generalization suggests we expand our set of angular spectrum wavenumbers to form a three dimensional wavenumber space that includes along-track and above-ground aperture wavenumbers,

$$\begin{aligned}\breve{k}_y &= 2k_{\omega_c} \cos\phi_y, \\ \breve{k}_z &= 2k_{\omega_c} \cos\phi_z.\end{aligned}$$

Three Dimensional Fourier Migrated and Aperture Wavenumbers

With this full suite of angular spectrum wavenumbers, we define a three dimensional *angular spectrum wavenumber vector,*

$$\breve{\boldsymbol{k}} = [\breve{k}_x \ \breve{k}_y \ \breve{k}_z]$$

to complement the *aperture wavenumber vector,*

$$\boldsymbol{k}_u = [k_{u_x} \ k_{u_y} \ k_{u_z}].$$

In angular spectrum processing, one or more of the aperture wavenumbers $\boldsymbol{k}_u$ are obtained by slow-time Fourier transforms in the dimensions of

$$\boldsymbol{u} = [u_x \ u_y \ u_z]$$

that support sensor movement. One of the remaining angular wavenumbers (if any) of $\breve{\boldsymbol{k}}$ are obtained by Fourier migration with angular spectrum geometries controlled by data collection geometries.

Pythagorean Foundation of Three Dimensional Fourier Migrations

We have a Pythagorean constraint in the three dimensional angular spectrum wavenumber domain,

$$(2k_{\omega_c})^2 = (2k_{\omega_c} \cos\phi_x)^2 + (2k_{\omega_c} \cos\phi_y)^2 + (2k_{\omega_c} \cos\phi_z)^2, \tag{13.5}$$

which, in turn, can be expressed as

$$|\breve{\boldsymbol{k}}|^2 = \breve{k}_x^2 + \breve{k}_y^2 + \breve{k}_z^2. \tag{13.6}$$

On the other side of the angular spectrum isomorphism, the Pythagorean constraint expressed in the aperture wavenumber domain is

$$(k_{\omega_c})^2 = [k_{\omega_c} \cos\theta_x(u_x)_n]^2 + [k_{\omega_c} \cos\theta_y(u_y)_n]^2 + [k_{\omega_c} \cos\theta_z(u_z)_n]^2, \tag{13.7}$$

which, in turn, can be expressed as,

$$|\boldsymbol{k}_u|^2 = k_{u_x}^2 + k_{u_y}^2 + k_{u_z}^2.$$

These relationships provide a foundation for angular spectrum descriptions of Fourier migration in three dimensional k-space.

Pythagorean Constraints Among Aperture and Migrated Wavenumbers

Pythagorean relationships of dimensional aperture wavenumbers in terms of the view angles of individual scatterers gives

$$\begin{aligned} k_{u_x} &= k_{\omega_c}\sqrt{1-\cos^2\theta_y(u_y)_n-\cos^2\theta_z(u_z)_n}, \\ k_{u_y} &= k_{\omega_c}\sqrt{1-\cos^2\theta_x(u_x)_n-\cos^2\theta_z(u_z)_n}, \\ k_{u_z} &= k_{\omega_c}\sqrt{1-\cos^2\theta_x(u_x)_n-\cos^2\theta_y(u_y)_n}. \end{aligned} \tag{13.8}$$

The isomorphism between $\boldsymbol{\Theta}$ and $\boldsymbol{\Phi}$ provides similar equations in terms of angular spectrum wavenumbers,

$$\begin{aligned} \breve{k}_x &= 2k_{\omega_c}\sqrt{1-\cos^2\phi_y-\cos^2\phi_z}, \\ \breve{k}_y &= 2k_{\omega_c}\sqrt{1-\cos^2\phi_x-\cos^2\phi_z}, \\ \breve{k}_z &= 2k_{\omega_c}\sqrt{1-\cos^2\phi_x-\cos^2\phi_y}. \end{aligned} \tag{13.9}$$

These rules identify the functional relationships between dimensional Stolt formatted wavenumbers and dimensional aperture wavenumbers,

$$\begin{aligned} \breve{k}_x\left(k_{\omega_c},k_{u_y},k_{u_z}\right) &= 2\sqrt{k_{\omega_c}^2-k_{u_y}^2-k_{u_z}^2}, \\ \breve{k}_y\left(k_{\omega_c},k_{u_x},k_{u_z}\right) &= 2\sqrt{k_{\omega_c}^2-k_{u_x}^2-k_{u_z}^2}, \\ \breve{k}_z\left(k_{\omega_c},k_{u_x},k_{u_y}\right) &= 2\sqrt{k_{\omega_c}^2-k_{u_x}^2-k_{u_y}^2}. \end{aligned} \tag{13.10}$$

These relationships define a mapping from aperture wavenumbers obtained via slow-time Fourier transforms to angular wavenumbers obtainable via Fourier migrations.

The foundation of VolSAR image formation processing is the migration from available aperture wavenumbers $\boldsymbol{k}_u$ to absent angular wavenumbers $\breve{\boldsymbol{k}}_u$ in the manner defined by (13.10).

Given sensor movement in some or all dimensions (u_x, u_y, u_z) in the synthetic aperture, slow-time Fourier transforms produce the mappings $u_i \mapsto k_{u_i}$ for each dimension $i \in \{x, y, z\}$ within the span of sensor movement. We have also seen that we can define isomorphic mappings that yield $\breve{k}_i$ with $i \in \{x, y, z\}$ where (13.5) holds, and importantly, it follows that (13.10) also holds true. Hence, we have the following equivalent expression that describes the VolSAR aperture wavenumber phase history,

$$\begin{aligned} SSSS(\omega_p,k_{u_x},k_{u_y},k_{u_z}) \equiv \; & \textstyle\sum_n \; \exp\left[-jx_n(u_x)\cdot 2\sqrt{k_{\omega_c}^2-k_{u_y}^2-k_{u_z}^2}\,\right] \\ & \times \; \exp\left[-jy_n(u_y)\cdot 2\sqrt{k_{\omega_c}^2-k_{u_x}^2-k_{u_z}^2}\,\right] \\ & \times \; \exp\left[-jz_n(u_z)\cdot 2\sqrt{k_{\omega_c}^2-k_{u_x}^2-k_{u_y}^2}\,\right]. \end{aligned} \tag{13.11}$$

Each of the radical expressions within (13.11) describes the required Stolt formatting when one of the k_{u_i} cannot be obtained directly by Fourier transformation (e.g., when no sensor movement occurs in that dimension).

Pythagorean Constraints and Three Dimensional Fourier Migrations

The constraint of the Pythagorean triplet finds a solution to a two dimensional wave equation given knowledge of the wavelength of a tonal wave field along a line. The three dimensional version of a Pythagorean quadruple finds a solution to a three dimensional problem given knowledge of the tonal wave field over a plane [513].

Fourier migration in two dimensions migrates the known k-space field distributed over a line to the semicircle manifold of the passband version of the Fourier diffraction theorem (see Section 14.4). Fourier migration in three dimensions migrates the known k-space field distributed over a portion of a plane to some subset of a semisphere of the Ewald sphere.

Ewald Sphere Angular Extent and Sample Density

The main challenge of Ewald sphere based imaging is the task of populating manifolds on the Ewald sphere with sufficient data in extent and density to support volumetric imaging [837]. Good resolution in all dimensions requires angular extent in all dimensions. The ability to image a scene of a given size depends on the dimensional sample densities that populate traces on the Ewald sphere. The required data can be obtained with non-linear flight paths or fusion of multiple linear and non-linear flight paths [28]. Various sensor-scene configurations can be envisioned for expanding the k-space manifold of data support and achieving the required wavenumber domain sample density [674, 976].

One Dimensional Measurements Migrated into Two Dimensions

For the one dimensional case with data gathered along u_y, we have the two dimensional Stolt transform for Fourier migration into $\breve{k}_x$ to enable two dimensional Fourier reconstructions,

$$\begin{aligned} \breve{k}_x &= 2\sqrt{k_{\omega_p}^2 - k_{u_y}^2}, \\ \breve{k}_y &= 2k_{u_y}. \end{aligned}$$

Two Dimensional Measurements Migrated into Three Dimensions

For the two dimensional case with data gathered in both the u_x and u_y signal domain dimensions, we have the three dimensional Stolt transform for Fourier migration from (k_{u_x}, k_{u_y}) into $\breve{k}_z$ to enable three dimensional Fourier reconstructions,

$$\begin{aligned} \breve{k}_x &= 2k_{u_x}, \\ \breve{k}_y &= 2k_{u_y}, \\ \breve{k}_z &= 2\sqrt{k_{\omega_p}^2 - k_{u_x}^2 - k_{u_y}^2}. \end{aligned}$$

Three Dimensional Measurements Do Not Require Fourier Migration For the three dimensional case with data gathered along u_x, u_y and u_z, there is no need for Fourier migration to enable three dimensional Fourier reconstructions. Hence, we use the direct assignments

$$\begin{aligned}\breve{k}_x &= 2k_{u_x},\\ \breve{k}_y &= 2k_{u_y},\\ \breve{k}_z &= 2k_{u_z},\end{aligned}$$

obtained without Fourier migration.

Fourier Migration onto the Ewald Sphere from One Dimension

Fourier migrations can also be performed from one dimensional (linear) synthetic apertures onto the three dimensional Ewald sphere. A Fourier migration from a linear subaperture differs from a migration from a plate subaperture in the nature of the manifold trace. A plate subaperture creates a two dimensional surface projection on the Ewald sphere. A linear subaperture creates a line trace on the Ewald sphere angled in a way that corresponds to the depression angle of the slant-plane of data collection.

13.9 Fourier Reconstructions Under Change of Variables Transformation

The following is a general description of Fourier reconstructions when a change of variables transformation (such as with Fourier migration) is involved. This framework is key to correct angular spectrum algorithm development based in Fourier reconstruction methods [896, 898].

Inverse Fourier Transform of Fourier Migrated Spectrum

Consider the inverse Fourier transform that gives VolSAR voxel data from angular spectrum data,

$$\begin{aligned}&vvv(x,y,z)_{\text{ACS}}\\ &= \frac{1}{8\pi^3}\int_{\breve{k}_x}\int_{\breve{k}_y}\int_{\breve{k}_z} MMM(\breve{k}_x,\breve{k}_y,\breve{k}_z)_{\text{ACS}}\\ &\qquad\cdot\exp(jx\breve{k}_x)\exp(jy\breve{k}_y)\exp(jy\breve{k}_y)\cdot d\breve{k}_x d\breve{k}_y d\breve{k}_z.\end{aligned} \tag{13.12}$$

A correct formulation of the inverse Fourier transform performed when under a general transformation that requires a change of variables (e.g., Fourier migration from the measured signal spectrum) is important.

Consider the general case where the three wavenumbers $\breve{k}_x$, $\breve{k}_y$ and $\breve{k}_z$ are expressed in terms of transformations T_x, T_y and T_z from the four signal domain wavenumbers: k_{ω_p}, k_{u_x}, k_{u_y} and k_{u_z},

$$\begin{aligned}\breve{k}_x &= T_x(k_{\omega_p}, k_{u_x}, k_{u_y}, k_{u_z}),\\ \breve{k}_y &= T_y(k_{\omega_p}, k_{u_x}, k_{u_y}, k_{u_z}),\\ \breve{k}_z &= T_z(k_{\omega_p}, k_{u_x}, k_{u_y}, k_{u_z}).\end{aligned}$$

When expressed in terms of a change of variables from the signal collection domain, the migrated spectrum

$$MMM(\breve{k}_x, \breve{k}_y, \breve{k}_z)_{\text{ACS}}$$

of (13.12) is described by the measured signal spectrum

$$SSSS(k_{\omega_p}, k_{u_x}, k_{u_y}, k_{u_z})_{\text{ACS}}$$

with wavenumbers subject to the transformations

$$[T_x(\cdot), T_y(\cdot), T_z(\cdot)]. \tag{13.13}$$

Inverse Fourier Transform of Jacobian Determinant Weighted Signal Spectrum

When the inversion (13.12) is expressed in terms of the transformation from $SSSS(\cdot)$ to $MMM(\cdot)$, then the determinant of the Jacobian matrix,

$$|JJJJ(k_{\omega_p}, k_{u_z}, k_{u_y}, k_{u_z})|$$

must be included. Hence,

$$\begin{aligned} &vvv(x,y,z)_{\text{ACS}}\\ &= \int_{k_{\omega_p}} \frac{1}{8\pi^3} \int_{k_{u_x}} \int_{k_{u_y}} \int_{k_{u_z}} \\ &\qquad SSSS(k_{\omega_p}, k_{u_x}, k_{u_y}, k_{u_z})_{\text{ACS}} \cdot |JJJJ(k_{\omega_p}, k_{u_z}, k_{u_y}, k_{u_z})| \\ &\qquad \cdot \exp\left[jx \cdot T_x(k_{\omega_p}, k_{u_x}, k_{u_y}, k_{u_z})\right] \\ &\qquad \cdot \exp\left[jy \cdot T_y(k_{\omega_p}, k_{u_x}, k_{u_y}, k_{u_z})\right] \\ &\qquad \cdot \exp\left[jz \cdot T_z(k_{\omega_p}, k_{u_x}, k_{u_y}, k_{u_z})\right] \\ &\qquad \cdot dk_{u_x} dk_{u_y} dk_{u_z} \\ &\quad \cdot dk_{\omega_p}, \end{aligned} \tag{13.14}$$

where the integral over k_{ω_p} represents a polychromatic coherent sum (instead of an inverse Fourier transform, as represented by the other integrals).

Box Bounded Scene

Consider the terrain spectrum obtained through a three dimensional Fourier transform when a bounding box manifold

$$\Pi_{(2X_o @ 0)} \Pi_{(2Y_o @ 0)} \Pi_{(2Z_o @ 0)}(x, y, z) \tag{13.15}$$

encloses the ground scene. This bounding box manifold serves as an indicator function for the neighborhood of the ground scene [76]. We have set

$$(X_c, Y_c, Z_c) = (0, 0, 0).$$

Forward Fourier Transformation of Box Bounded Scene

The *forward* Fourier transform of the box bounded scene is

$$\begin{aligned} & VVV(\ddot{k}_x, \ddot{k}_y, \ddot{k}_z)_{\text{ACS}} \\ &= \int_{-\infty}^{+\infty}\int_{-\infty}^{+\infty}\int_{-\infty}^{+\infty} \Pi_{(2X_o @ 0)}\Pi_{(2Y_o @ 0)}\Pi_{(2Z_o @ 0)}(x, y, z) \\ &\qquad \cdot vvv(x, y, z)_{\text{ACS}} \\ &\qquad \cdot \exp(-j\ddot{k}_x x)\exp(-j\ddot{k}_y y)\exp(-j\ddot{k}_z z) \\ &\quad \cdot dxdydz, \end{aligned} \tag{13.16}$$

where we have introduced un-migrated (and uniformly sampled) wavenumbers $\ddot{k}_x$, $\ddot{k}_y$ and $\ddot{k}_z$.

Fourier Inversion Under Change of Variables Transformation

Consider substitution of (13.14) into (13.16) to describe the volumetric image spectrum characterized by the measured signal spectrum data subject to a change of variables transformation,

$$\begin{aligned} & VVV(\ddot{k}_x, \ddot{k}_y, \ddot{k}_z)_{\text{ACS}} \\ &= \int_{-\infty}^{+\infty}\int_{-\infty}^{+\infty}\int_{-\infty}^{+\infty} \Pi_{(2X_o @ 0)}\Pi_{(2Y_o @ 0)}\Pi_{(2Z_o @ 0)}(x, y, z) \\ &\quad \int_{k_{\omega_p}} \frac{1}{8\pi^3}\int_{k_{u_x}}\int_{k_{u_y}}\int_{k_{u_z}} \\ &\qquad SSSS(k_{\omega_p}, k_{u_x}, k_{u_y}, k_{u_z})_{\text{ACS}} \cdot |JJJJ(k_{\omega_p}, k_{u_z}, k_{u_y}, k_{u_z})| \\ &\qquad \cdot \exp\left[jx \cdot T_x(k_{\omega_p}, k_{u_x}, k_{u_y}, k_{u_z})\right] \\ &\qquad \cdot \exp\left[jy \cdot T_y(k_{\omega_p}, k_{u_x}, k_{u_y}, k_{u_z})\right] \\ &\qquad \cdot \exp\left[jz \cdot T_z(k_{\omega_p}, k_{u_x}, k_{u_y}, k_{u_z})\right] \\ &\qquad \cdot dk_{u_x} dk_{u_y} dk_{u_z} \\ &\quad \cdot dk_{\omega_p} \\ &\quad \cdot \exp(-j\ddot{k}_x x)\exp(-j\ddot{k}_y y)\exp(-j\ddot{k}_z z) \\ & \cdot dxdydz. \end{aligned} \tag{13.17}$$

Switching the order of integration and combining exponential transform kernels,

$$
\begin{aligned}
&VVV(\ddot{k}_x, \ddot{k}_y, \ddot{k}_z)_{\mathrm{ACS}} \\
&= \int_{k_{\omega_p}} \frac{1}{8\pi^3} \int_{k_{u_x}} \int_{k_{u_y}} \int_{k_{u_z}} \\
&\qquad SSSS(k_{\omega_p}, k_{u_x}, k_{u_y}, k_{u_z})_{\mathrm{ACS}} \cdot |JJJJ(k_{\omega_p}, k_{u_z}, k_{u_y}, k_{u_z})| \\
&\qquad \cdot \int_{-\infty}^{+\infty} \int_{-\infty}^{+\infty} \int_{-\infty}^{+\infty} \Pi_{(2X_o@0)} \Pi_{(2Y_o@0)} \Pi_{(2Z_o@0)}(x, y, z) \\
&\qquad\qquad \cdot \exp\left\{ j[T_x(k_{\omega_p}, k_{u_x}, k_{u_y}, k_{u_z}) - \ddot{k}_x] \cdot x \right\} \\
&\qquad\qquad \cdot \exp\left\{ j[T_y(k_{\omega_p}, k_{u_x}, k_{u_y}, k_{u_z}) - \ddot{k}_y] \cdot y \right\} \\
&\qquad\qquad \cdot \exp\left\{ j[T_z(k_{\omega_p}, k_{u_x}, k_{u_y}, k_{u_z}) - \ddot{k}_z] \cdot z \right\} \\
&\qquad \cdot dxdydz \\
&\qquad \cdot dk_{u_x} dk_{u_y} dk_{u_z} \\
&\quad \cdot dk_{\omega_p}.
\end{aligned}
\tag{13.18}
$$

From the Fourier transform pair of rect and sinc functions, and from

$$
\Pi_{(2X_o)}(x) = \Pi_{[-1,+1]}\left(\frac{1}{2X_o} \cdot x\right),
$$

we have for the inner integral involving the three dimensional box indicator function,

$$
\begin{aligned}
&\int_{-\infty}^{+\infty} \int_{-\infty}^{+\infty} \int_{-\infty}^{+\infty} \Pi_{(2X_o@0)} \Pi_{(2Y_o@0)} \Pi_{(2Z_o@0)}(x, y, z) \\
&\qquad \cdot \exp\left[j[T_x(k_{\omega_p}, k_{u_x}, k_{u_y}, k_{u_z}) - \ddot{k}_x] \cdot x \right] \\
&\qquad \cdot \exp\left[j[T_y(k_{\omega_p}, k_{u_x}, k_{u_y}, k_{u_z}) - \ddot{k}_y] \cdot y \right] \\
&\qquad \cdot \exp\left[j[T_z(k_{\omega_p}, k_{u_x}, k_{u_y}, k_{u_z}) - \ddot{k}_z] \cdot z \right] \cdot dxdydz \\
&= \int_{-\infty}^{+\infty} \left[\Pi_{[-1,+1]}\left(\frac{1}{2X_o} \cdot x\right) \cdot \exp(-j\ddot{k}_x x) \right] \\
&\qquad \cdot \exp\left[jT_x(k_{\omega_p}, k_{u_x}, k_{u_y}, k_{u_z}) \cdot x \right] \cdot dx \\
&\quad \cdot \int_{-\infty}^{+\infty} \left[\Pi_{[-1,+1]}\left(\frac{1}{2Y_o} \cdot y\right) \cdot \exp(-j\ddot{k}_y y) \right] \\
&\qquad \cdot \exp\left[jT_y(k_{\omega_p}, k_{u_x}, k_{u_y}, k_{u_z}) \cdot y \right] \cdot dy \\
&\quad \cdot \int_{-\infty}^{+\infty} \left[\Pi_{[-1,+1]}\left(\frac{1}{2Z_o} \cdot z\right) \cdot \exp(-j\ddot{k}_z z) \right] \\
&\qquad \cdot \exp\left[jT_z(k_{\omega_p}, k_{u_x}, k_{u_y}, k_{u_z}) \cdot z \right] \cdot dz.
\end{aligned}
\tag{13.19}
$$

From

$$\Pi_{[-1,+1]}\left(\frac{1}{2X_o}\cdot x\right) \Leftrightarrow 2X_o \cdot \mathrm{sinc}(X_o \cdot k_x),$$

the Fourier transform (13.19) of the box indicator function (for the volumetric ground scene) evaluates to

$$\begin{aligned}
&2X_o \cdot \mathrm{sinc}\{X_o \cdot [T_x(k_{\omega_p}, k_{u_x}, k_{u_y}, k_{u_z}) - \ddot{k}_x]\}\\
&\cdot 2Y_o \cdot \mathrm{sinc}\{Y_o \cdot [T_y(k_{\omega_p}, k_{u_x}, k_{u_y}, k_{u_z}) - \ddot{k}_y]\}\\
&\cdot 2Z_o \cdot \mathrm{sinc}\{Z_o \cdot [T_z(k_{\omega_p}, k_{u_x}, k_{u_y}, k_{u_z}) - \ddot{k}_z]\}.
\end{aligned} \tag{13.20}$$

Inserting (13.20) into (13.18), we finally obtain

$$\begin{aligned}
&VVV(\ddot{k}_x, \ddot{k}_y, \ddot{k}_z)_{\mathrm{ACS}}\\
&= \int_{k_{\omega_p}} \frac{1}{8\pi^3} \int_{k_{u_x}} \int_{k_{u_y}} \int_{k_{u_z}}\\
&\qquad SSSS(k_{\omega_p}, k_{u_x}, k_{u_y}, k_{u_z})_{\mathrm{ACS}} \cdot |JJJJ(k_{\omega_p}, k_{u_z}, k_{u_y}, k_{u_z})|\\
&\qquad\quad \cdot 2X_o \cdot \mathrm{sinc}\{X_o \cdot [T_x(k_{\omega_p}, k_{u_z}, k_{u_y}, k_{u_z}) - \ddot{k}_x]\}\\
&\qquad\quad \cdot 2Y_o \cdot \mathrm{sinc}\{Y_o \cdot [T_y(k_{\omega_p}, k_{u_z}, k_{u_y}, k_{u_z}) - \ddot{k}_y]\}\\
&\qquad\quad \cdot 2Z_o \cdot \mathrm{sinc}\{Z_o \cdot [T_z(k_{\omega_p}, k_{u_z}, k_{u_y}, k_{u_z}) - \ddot{k}_z]\}\\
&\qquad \cdot dk_{u_x} dk_{u_y} dk_{u_z}\\
&\quad \cdot dk_{\omega_p}.
\end{aligned} \tag{13.21}$$

Not only is the Jacobian determinant

$$|JJJJ(k_{\omega_p}, k_{u_z}, k_{u_y}, k_{u_z})|$$

required for proper scaling for *Fourier reconstruction* of the terrain spectrum

$$VVV(\ddot{k}_x, \ddot{k}_y, \ddot{k}_z)_{\mathrm{ACS}}$$

from the measured signal spectrum

$$SSSS(k_{\omega_p}, k_{u_x}, k_{u_y}, k_{u_z})_{\mathrm{ACS}},$$

but correctly scaled and normalized sinc interpolation kernels are required for Fourier reconstructions done preparatory to Fourier inversion of the box bounded scene.

Fourier Reconstruction of Sphere Bounded Terrain Functions

Note that if a different bounding manifold for the volumetric terrain scene were used (e.g., a sphere to enclose and bound the ground scene), then the sinc functions would change to some other functional (e.g., interpolations based on Bessel functions).

Huygens-Fresnel Transfer to Change the Remote Field to the In Situ Field

The next step is to apply the inverse Huygens-Fresnel transfer functions to obtain the wavenumber domain description of the in situ field of the volumetric scene,

$$VVV(\ddot{k}_{\hat{x}}, \ddot{k}_{\hat{y}}, \ddot{k}_{\hat{z}})_{\text{SCS}} = H^*H^*H^*(\ddot{k}_x, \ddot{k}_y, \ddot{k}_z)_{\text{ACS}} \times VVV(\ddot{k}_x, \ddot{k}_y, \ddot{k}_z)_{\text{ACS}}. \quad (13.22)$$

This inverse Huygens Fresnel transfer is required even if $(X_c, Y_c, Z_c) = (0, 0, 0)$. This ensures that all complex amplitude coefficients related to the spatial Fourier transform of the phase curtain are compensated to leave pure locator sinusoids (on the Ewald sphere) with wavenumber domain frequencies descriptive of scatterer positions.

Fourier Inversion of Reconstructed Terrain Spectrum Under Transformation

Fourier inversion of the terrain spectrum (13.22) gives

$$\begin{aligned} vvv(\hat{x}, \hat{y}, \hat{z})_{\text{SCS}} &= \frac{1}{8\pi^3} \int_{-\infty}^{+\infty} \int_{-\infty}^{+\infty} \int_{-\infty}^{+\infty} VVV(\ddot{k}_{\hat{x}}, \ddot{k}_{\hat{y}}, \ddot{k}_{\hat{z}})_{\text{SCS}} \\ &\quad \cdot \exp(j\hat{x}\ddot{k}_{\hat{x}}) \exp(j\hat{y}\ddot{k}_{\hat{y}}) \exp(j\hat{z}\ddot{k}_{\hat{z}}) \cdot d\ddot{k}_{\hat{x}} d\ddot{k}_{\hat{y}} d\ddot{k}_{\hat{z}}. \end{aligned}$$

Jacobian Determinant of Three Dimensional Stolt Transform

The Jacobian matrix of the four dimensional to three dimensional Stolt transformation is

$$JJJJ(k_{\omega_p}, k_{u_x}, k_{u_y}, k_{u_z}) = \frac{\partial T(\breve{k}_x, \breve{k}_y, \breve{k}_z)}{\partial(k_{\omega_p}, k_{u_x}, k_{u_y}, k_{u_z})}.$$

Jacobian Determinant of Elevation Stolt Transform

As an example, the Jacobian determinant of a Stolt transform in three dimensions, where k_x and k_y are obtained by Fourier transformation and $\breve{k}_z$ is obtained by Fourier migration, is

$$\begin{aligned} &\left|JJJ(k_{\omega_p}, k_{u_x}, k_{u_y})\right| \\ &= \frac{k_{\omega_p}}{\sqrt{k_{\omega_p}^2 - k_{u_x}^2 - k_{u_y}^2}}, \\ &= \frac{k_{\omega_p}}{\breve{k}_{u_z}}, \\ &= \frac{1}{\cos\phi_z(k_{u_x}, k_{u_y})}, \end{aligned}$$

where ϕ_z is the angular spectrum angle defined with respect to the $\breve{k}_z$ axis and is expressed as a function of k_{u_x} and k_{u_y}. The trigonometric form of this expression was recognized in [334]. The Jacobian determinant can also be interpreted in terms of cardinal axis projections of uniformly distributed masses on semicircle line and hemisphere surface integrals [722].

Uniform Resampling Under Three Dimensional Stolt Transform

Planar Subaperture Mapping to Uniform Samples on Ewald Sphere

An above-ground or elevation Fourier migration of

$$SSS\left(k_{\omega_p}[\cdot], k_{\hat{x}}[\cdot], k_{\hat{y}}[\cdot]\right)_{\text{SCS}}$$

is performed to form

$$SSM\left(k_{\hat{x}}[\cdot], k_{\hat{y}}[\cdot], \breve{k}_{\hat{z}}[n_{\hat{z}}]\right)_{\text{SCS}}.$$

The resampled version of the array is

$$SSV\left(k_{\hat{x}}[\cdot], k_{\hat{y}}[\cdot], \ddot{k}_{\hat{z}}[n_{\hat{z}}]\right)_{\text{SCS}}.$$

Data samples of the rectilinear spectrum $SSS(\cdot)$ are uniformly spaced in the k_{ω_p}, k_x and k_y domains. Stolt formatted data migrated to polychromatic concentric spheres of radii $2k_{\omega_p}$ for $\omega_p \in [\overset{\min}{\omega_p}, \overset{\max}{\omega_p}]$ need to be interpolated and resampled to uniformly spaced positions $\ddot{k}_{\hat{z}}$ along the $\breve{k}_z$ axis [827]. The desired rectilinear output points of the resampling operation are

$$\ddot{k}_{\hat{z}}[n_{\hat{z}}] \quad = \quad \left(n_{\hat{z}} - \frac{N_{\hat{z}}}{2} - 1\right)\Delta_{\ddot{k}_{\hat{z}}} \quad \text{for} \quad n_{\hat{z}} = 1, 2, \ldots, N_{\hat{z}},$$

where

$$\Delta_{\ddot{k}_{\hat{z}}} = \frac{2\pi}{\Pi_{(2Z_o)}(z)}.$$

The weighted coherent summation of the interpolation for each $n_{\hat{x}}$ and $n_{\hat{y}}$, and for a given $\ddot{k}_{\hat{z}}[n_{\hat{z}}]$ is over all frequency bins n_{ω_p}, where the ingested input sample positions

$$\breve{k}_{\hat{z}}[n_{\omega_p}, n_{\hat{x}}, n_{\hat{y}}] = \sqrt{k^2_{\omega_p}[n_{\omega_p}] - k^2_{\hat{x}}[n_{\hat{x}}] - k^2_{\hat{y}}[n_{\hat{y}}]} \tag{13.23}$$

are within a neighborhood defined by

$$\left|\breve{k}_{\hat{z}}[n_{\omega_p}, n_{\hat{x}}, n_{\hat{y}}] - \ddot{k}_{\hat{z}}[n_{\hat{z}}]\right| \leq N_s \Delta \ddot{k}_{\hat{z}}. \tag{13.24}$$

In (13.24), n_{ω_p} is the free variable and N_s determines the size of the finite interpolation kernel with a span that resides in the $\ddot{k}_{\hat{z}}$ dimension. Interpolation and coherent integration (over n_{ω_p}) to achieve the required above-ground spherical Fourier migration for a fixed $n_{\hat{x}}$ and $n_{\hat{y}}$ is given by

$$\begin{aligned} &SSV(k_{\hat{x}}[n_{\hat{x}}], k_{\hat{y}}[n_{\hat{y}}], \ddot{k}_{\hat{z}}[n_{\hat{z}}])_{\text{SCS}} \\ &= \sum\nolimits_{n_{\omega_p}: \left|\breve{k}_{\hat{z}}[n_{\omega_p}, n_{\hat{x}}, n_{\hat{y}}] - \ddot{k}_{\hat{z}}[n_{\hat{z}}]\right| \leq N_s \Delta \ddot{k}_{\hat{z}}} \\ &\quad SSS\left(k_{\omega_p}[n_{\omega_p}], k_{\hat{x}}[n_{\hat{x}}], k_{\hat{y}}[n_{\hat{y}}]\right)_{\text{SCS}} \\ &\quad \cdot 2Z_o \cdot \text{sinc}\{Z_o \cdot [T_z(k_{\omega_p}[n_{\omega_p}] k_{\hat{x}}[n_{\hat{x}}], k_{\hat{y}}[n_{\hat{y}}]) - \ddot{k}_{\hat{z}}]\} \\ &\quad \cdot \left|JJJ[k_{\omega_p}[n_{\omega_p}], k_{\hat{x}}[n_{\hat{x}}], k_{\hat{y}}[n_{\hat{y}}]]\right|. \end{aligned}$$

The Jacobian determinant is

$$\left|JJJ[k_{\omega_p}[n_{\omega_p}], k_{\hat{x}}[n_{\hat{x}}], k_{\hat{y}}[n_{\hat{y}}]]\right| = \frac{k_{\omega_p}[n_{\omega_p}]}{\breve{k}_{\hat{z}}[n_{\omega_p}, n_{\hat{x}}, n_{\hat{y}}]}.$$

The sinc interpolation kernel can be tapered by a window function such as a Hamming window [885]. If the number of polychromatic frequency bins is rather limited or sparse, this windowing over the frequency bins of the coherent sum is not recommended.

Huygens-Fresnel Transfer from Remote to In Situ

A valid processing order includes the option of applying Fourier migrated inverse Huygens-Fresnel transfer functions to Fourier migrated phase history data to obtain the in situ field of the volumetric scene, as described in the wavenumber domain,

$$SSV(\breve{k}_{\hat{x}}, \breve{k}_{\hat{y}}, \ddot{k}_{\hat{z}})_{\mathrm{SCS}} = H^*H^*H^*(\breve{k}_x, \breve{k}_y, \ddot{k}_z)_{\mathrm{ACS}} \times SSV(\breve{k}_x, \breve{k}_y, \ddot{k}_z)_{\mathrm{ACS}}.$$

Volumetric SAR Imagery

The volumetric image spectrum $SSV(k_{\hat{x}}, k_{\hat{y}}, \ddot{k}_{\hat{z}})_{\mathrm{SCS}}$ obtained by Fourier reconstruction is inverse Fourier transformed to obtain the VolSAR voxel block,

$$vvv[n_{\hat{x}}, n_{\hat{y}}, n_{\hat{z}}]_{\mathrm{SCS}} \approx \mathsf{iFT}_{3\mathrm{D}}\{SSV(k_{\hat{x}}[n_{\hat{x}}], k_{\hat{y}}[n_{\hat{y}}], \ddot{k}_{\hat{z}}[n_{\hat{z}}])_{\mathrm{SCS}}\}.$$

No Slant-Plane Effects in VolSAR Imagery

VolSAR image formation processing should yield SAR image products that do not suffer from unwanted distortions of layover, foreshortening and out of slant-plane blurring.

VolSAR Voxel Characteristics

The point spread function of a VolSAR image may not be symmetric due to the different sizes of the synthetic aperture in different dimensions. The resolution of VolSAR in each slow-time dimension is dependent on the dimensional extent of data that populates a submanifold of the Ewald sphere.

Presentation of VolSAR Image Products

Image formats can be extracted from a voxel volume that are quite diverse [667, 668]. Two dimensional SAR imagery can be obtained along any slice of the VolSAR voxel volume. SAR image products can be extracted and presented in various formats, including movie like sequences of two dimensional cross sectional SAR imagery extracted from desired slice directions. Other novel VolSAR imagery presentation formats include computer generated fly-through or walk-through [631], or fixed perspective visualizations of three dimensional SAR scenes rendered from the VolSAR voxel block. Although not very desirable, two dimensional SAR image products can also be formed from flattened voxel volumes [885]. In such cases, pixel intensities represent integrations of stacked voxel intensities.

13.10 VolSAR Concept Simulation Studies

Various simulations of VolSAR are performed to illustrate concepts of three dimensional angular spectrum processing and volumetric imaging. Some of the simulation scenarios require Fourier migrations. Others do not. Some are of practical interest. Others are theoretical curiosities.

Simulated VolSAR Scenarios

The following VolSAR scenarios are simulated and discussed:

- Solid Cube Synthetic Aperture without Fourier Migration
- Overhead Square Plate Aperture with Fourier Migrations
- Fused Overhead Square Plate and Standoff Square Wall Aperture
- Fused Overhead Square Plate and Two Orthogonal Square Wall Apertures
- Cubic Shell Synthetic Aperture with Fourier Migrations
- Spherical Shell Synthetic Aperture without Fourier Migration
- Dome Synthetic Aperture without Fourier Migration
- Dome Super Resolution Extension through Ewald Sphere Symmetries
- Polychromatic Overhead Square Plate Aperture with Fourier Migrations
- Fused L, C and X Band Overhead Square Plate Aperture with Fourier Migrations

In each case, the imaging geometry will be described. Processed VolSAR imagery from simulated VolSAR signals will be presented to allow a qualitative assessment of the nature of the VolSAR imagery. In lieu of describing in a general way the system engineering process for VolSAR system design, we step through the design of our VolSAR simulation experiments for each scenario. Not all simulated scenarios are practical, but they do provide insights into the possibilities and limitations of VolSAR.

Solid and Shell Synthetic Apertures

We explore the "solid" and the "shell" volumetric synthetic apertures. A solid synthetic aperture is a bounded region of space that contains a full three dimensional rectilinear raster of data samples collected within the volume. Never mind at this stage how this synthetic aperture would be implemented in practice. A solid synthetic aperture formed within a three dimensional volume does not require Fourier migrations. A study of this theoretical aperture provides a baseline against which the image quality of other more practical (e.g., shell or fused) synthetic apertures can be compared.

A shell synthetic aperture is a flat or curved surface that is populated with surface data samples. Some shell synthetic aperture types may have practical possibilities.

Shell synthetic apertures can be formed, with varying degrees of data sparseness, from fused linear or curvilinear subapertures. Volume synthetic apertures can be formed from fused surface subapertures

Achieving Diffraction Limit with Finite Volumetric Synthetic Apertures

If a synthetic aperture is constrained to be a straight line, then an infinitely long straight flight path is required to obtain the angular dwell of π radians that achieves the two way diffraction limit of $\lambda_c/4$. If the radar carrying platform is allowed to maneuver, in a semicircle for example, then a diffraction limit angular dwell can be gather with a flight path of finite length.

The diffraction limit can also be achieved with fused piecewise linear synthetic apertures. The requirement is that the composite piecewise linear synthetic aperture provides a collective angular dwell of at least π radians.

The manifold of a combined two segment (finite and fused) piecewise linear synthetic aperture that achieves the diffraction limit is illustrated in Fig. 13.1a. If the finite lengths of each linear subaperture are sufficiently long, then both linear subapertures achieve an angular dwell of at least $\pi/2$ radians for every scatterer in the ground scene. Combined, they provide the π radians of dwell required to achieve the diffraction limit.

Monochromatic and Polychromatic VolSAR Simulations

Most of our simulated VolSAR scenarios are based on the use of continuous wave signals that demonstrate the performance of monochromatic systems. Simulated VolSAR systems with pulsed waveforms show the improved performance that comes with the coherent processing gain of polychromatic sensing. This is demonstrated in a couple of cases.

Solid Cube Synthetic Aperture Simulation

We consider first the simulation of a solid cube (volume) synthetic aperture that encloses the simulated terrain function. The scenario of this simulation cannot be achieved in practice because it requires data collections above, below and within the scene. However, this simulation provides a baseline for performance comparisons with scenarios that have practical possibilities.

SAR image formation processing with a solid cube aperture has been called "three dimensional Fourier transform synthetic aperture focusing" [631].

Set Fast-Time Signal Domain Parameters We start with the design of fast-time signal parameters.

- Set the frequency f_c of the simulated continuous wave signal and determine λ_c. Here, we use $f_c = 10.0$ GHz, giving a wavelength of $\lambda_c = 3.0$ cm.

- The diffraction limit $\lambda_c/4$ is 0.75 cm in this simulation.

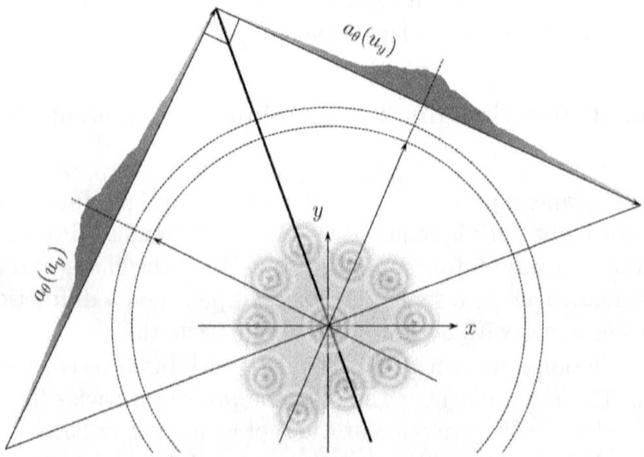

a. Piecewise linear manifold.

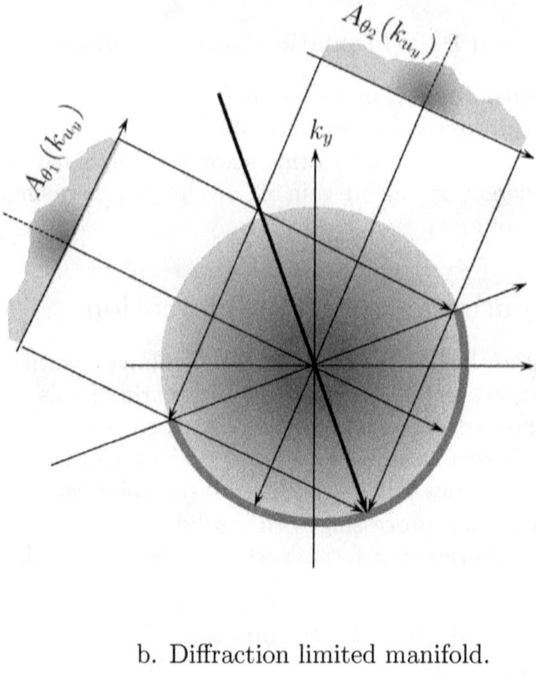

b. Diffraction limited manifold.

Figure 13.1: Passband sensor diffraction limit achieved with piecewise linear subapertures.

Set Desired Image and Scene Parameters In the following, we only explicitly state parameters and associated values for the x dimension. The associated parameters in the y and z dimensions are set to be the same, unless otherwise indicated.

- The desired image resolution $\Delta_{\hat{x}}$ is set to the diffraction limit,

$$\Delta_{\hat{x}} = \frac{\lambda_c}{4}.$$

- Set the scene size by specifying the number of diffraction limited resolution cells that exist along each side (dimension) of a bounding box that encloses the simulated ground scene,

$$2X_o = N_{\hat{x}} \times \Delta_{\hat{x}}.$$

We set $N_{\hat{x}}$ to be 63 resolution cells in this simulation.

- Set the center of the scene coordinate system (X_c, Y_c, Z_c) in the aperture coordinate system. We set $(X_c, Y_c, Z_c) = (0, 0, 0)$ for this solid cube synthetic aperture (i.e., the aperture and scene coordinate systems coincide).

Set Slow-Time Synthetic Aperture Parameters

- Since we desire to achieve the diffraction limit, set the spatial PRI in each dimension to be equal to the round trip diffraction limit,

$$\Delta_{u_x} = \frac{\lambda_c}{4}.$$

- Set the center of the volumetric synthetic aperture $(L_{c_x}, L_{c_y}, L_{c_z})$ in the aperture coordinate system. We set $(L_{c_x}, L_{c_y}, L_{c_z}) = (0, 0, 0)$ for the solid cube synthetic aperture simulation. This volumetric synthetic aperture surrounds the set of discrete scatterers.

- In each dimension, set the dimensional synthetic aperture half-lengths equal to the dimensional ground scene size,

$$L_x = 2X_o - \Delta_x. \qquad (13.25)$$

The adjustment of L_x by subtracting Δ_x in (13.25) places each sample point at the center of its resolution cell. The size of the box synthetic aperture in each dimension is

$$\begin{aligned} 2L_x &= 2X_o - 2\Delta_x, \\ &\approx 4X_o. \end{aligned}$$

A little work shows that the geometry of this volumetric synthetic aperture gives an angular dwell for each point in the ground scene of at least $\pi/2$ radians in each dimension. The box volume with sides of length $2L_i \approx 4X_i$ is the smallest box aperture sufficient to obtain $\pi/2$ radians of angular dwell in each dimension for each scatterer in the box bounded ground scene. Data collections over any larger aperture box volume wastes effort (more data may improve coherent processing gain, but will not improve resolution).

Since the volumetric synthetic aperture encompasses the ground scene, each scatterer is observed with a full 2π radians of angular dwell in each dimension. The wavenumber domain data support manifold is the full Ewald sphere.

- Determine the number of samples gathered in each dimension,

$$M_{u_x} = \left\lceil \frac{2L_x}{\Delta_{u_x}} \right\rceil .$$

If M_{u_x} is even, then add 1 to M_{u_x} to make it odd. This adjustment makes display of results easier (each scatterer will be at the center of its resolution cell).

- Set the size

$$M_{\text{radix2}} \geq M_{u_x}$$

of the radix-two FFT that will be used for Fourier transformations. The aperture data will be padded to the specified radix-two FFT size.

- The aperture wavenumbers in each dimension must have a support that spans from $-k_{\omega_c}$ to $+k_{\omega_c}$ in a system that seeks to achieve the diffraction limit. Hence, determine the padded sample density in the aperture wavenumber domain,

$$\Delta_{k_{u_x}} = \frac{2k_{\omega_c}}{M_{\text{radix2}}} .$$

Set Simulated Ground Scene Parameters

- Define the positions (x_n, y_n, z_n) of a set of scatterers $\boldsymbol{g}_n$. We choose to simulate nineteen individual point scatterers, all with equal radar cross section magnitude. Give the radar cross section constant magnitude of each scatterer $\boldsymbol{g}_n$ a random phase assignment.

- The wavenumber grid density in the scene coordinate system is specified by Nyquist sampling rates determined by the size of each dimension of a bounding box designed to enclose the ground scene. For the down-range dimension, for example,

$$\begin{aligned} \Delta_{k_{\hat{x}}} &\leq \frac{2\pi}{\Pi_{(2X_o)}(\hat{x})}, \\ &= \frac{2\pi}{2X_o}. \end{aligned} \tag{13.26}$$

- No presumming is performed in these simulations. Conversely, to better see attributes of the point spread function, the data are padded up to an over sampling factor of two,

$$M^{\text{padded}}_{\text{radix2}} = 2 \times M_{\text{radix2}} .$$

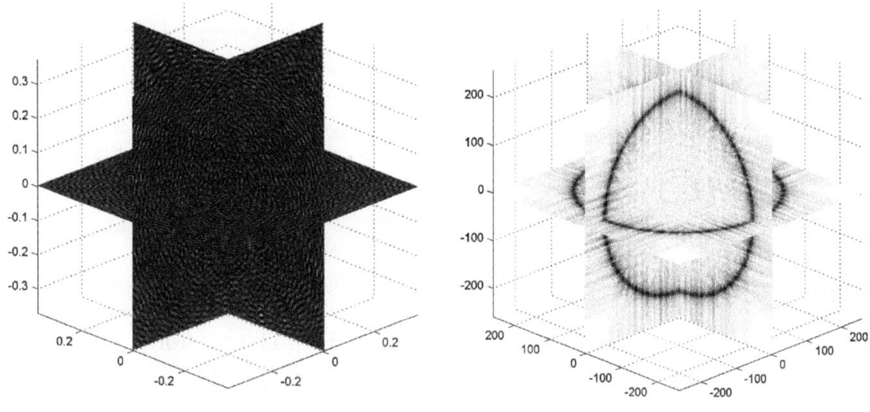

a. Time-space phase history. b. Frequency-wavenumber history.

Figure 13.2: Solid cube synthetic aperture simulation.

Simulate Down Converted Time-Space Phase History.

- Simulate the monochromatic down converted continuous wave time-space phase history by computing

$$sss[m_{u_x}, m_{u_y}, m_{u_z}]_{\mathrm{ACS}} = g_n[n_s] \cdot \exp(-jk_{\omega_c} \cdot 2r_n[m_{u_x}, m_{u_y}, m_{u_z}]), \quad (13.27)$$

where

$$r_n[\boldsymbol{m}] = \sqrt{(u_x[m_{u_x}] - x_n[n_s])^2 + (u_y[m_{u_y}] - y_n[n_s])^2 + (u_z[m_{u_z}] - z_n[n_s])^2}$$

with $\boldsymbol{m} = [m_{u_x}, m_{u_y}, m_{u_z}]$ and n_s indexing individual scatterers. The gathered phase curtain samples (13.27) are coherently summed over all aperture positions

$$\boldsymbol{u}[\boldsymbol{m}] = (u_x[m_{u_x}], u_y[m_{u_y}], u_z[m_{u_z}])$$

and for all simulated scatterers $\boldsymbol{g}_n[n_s]$. Fig. 13.2a shows[1] the simulated time-space phase history gathered within the solid cube synthetic aperture. These images are plotted with an inverted gray scale. Here, dark corresponds to more energetic data samples.

Setup Huygens-Fresnel Transfer Functions

- Preparatory to determining the Huygens-Fresnel transfer function for this aperture-scene interaction, determine

$$hhh[m_{u_x}, m_{u_y}, m_{u_z}]_{\mathrm{ACS}} = \exp(-jk_{\omega_c} \cdot 2r_c[m_{u_x}, m_{u_y}, m_{u_z}])$$

[1]These images have been processed with a gamma correction factor of between 0.2 and 0.4 to allow more detail to be seen. This gamma correction darkens the image in printed format.

where

$$r_c[\boldsymbol{m}] = \sqrt{(u_x[m_{u_x}] - X_c)^2 + (u_y[m_{u_y}] - Y_c)^2 + (u_z[m_{u_z}] - Z_c)^2}.$$

In practice, this reference phase history can be obtained in a way analogous to methods used to produce a phase reference in conventional motion compensation to a point processing employed by slant-plane pivot point SAR systems.

Compute the Frequency-Wavenumber Phase History.

- Compute the (in place) three dimensional Fourier transforms of both the scattered time-space phase history shown in Fig. 13.2a,

$$SSS[\boldsymbol{M}]_{\text{ACS}} = \mathsf{FT}_{\text{3D}}\left\{sss[\boldsymbol{m}]_{\text{ACS}}\right\}$$

and the pivot point referenced time-space phase history,

$$HHH[\boldsymbol{M}]_{\text{ACS}} = \mathsf{FT}_{\text{3D}}\left\{hhh[\boldsymbol{m}]_{\text{ACS}}\right\},$$

where $\boldsymbol{M} = [M_{u_x}, M_{u_y}, M_{u_z}]$ with M_{u_x}, M_{u_y} and M_{u_z} as wavenumber domain indexes, e.g.,

$$M_{u_x} = 1, \ldots, M_{\text{radix2}}^{\text{padded}}.$$

Fig. 13.2b shows the resulting frequency-wavenumber phase history. The monochromatic Ewald sphere is clearly evident. If the transmitted waveform is monochromatic, then the Ewald sphere reduces to a shell of radius k_{ω_c} [26, 168].

Perform Inverse Huygens-Fresnel Transfers

- Conjugate $HHH[\boldsymbol{M}]_{\text{ACS}}$ and perform the inverse Huygens-Fresnel transfer with a sample by sample multiplication,

$$VVV[\boldsymbol{M}]_{\text{SCS}} = H^*H^*H^*[\boldsymbol{M}]_{\text{ACS}} \times SSS[\boldsymbol{M}]_{\text{ACS}}.$$

No Fourier migrations of $SSS[\boldsymbol{M}]_{\text{ACS}}$ are required to obtain $VVV[\boldsymbol{M}]_{\text{SCS}}$ for the solid volume synthetic aperture case.

Perform Inverse Fourier Transfers to Form VolSAR Image

- Perform an inverse three dimensional Fourier transform of $VVV[\boldsymbol{M}]_{\text{SCS}}$,

$$vvv[\boldsymbol{m}]_{\text{SCS}} = \mathsf{iFT}_{\text{3D}}\left\{VVV[\boldsymbol{M}]_{\text{SCS}}\right\}.$$

Fig. 13.3 gives various views of the resulting VolSAR voxel block $vvv[\boldsymbol{m}]_{\text{SCS}}$. Cross sectional slices through the center planes of the VolSAR voxel volume of Fig. 13.3a are shown in Figs. 13.3b-d. The solid cube synthetic aperture produces diffraction limited imagery focused in all three dimensions.

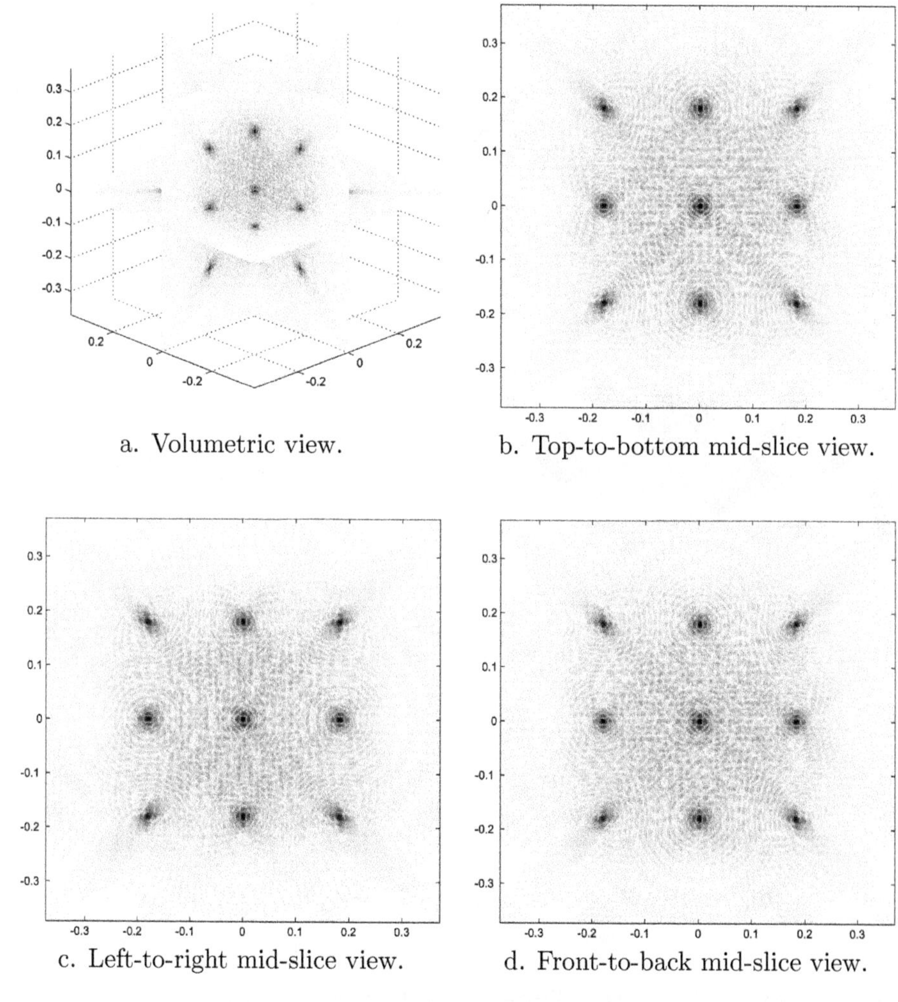

a. Volumetric view.

b. Top-to-bottom mid-slice view.

c. Left-to-right mid-slice view.

d. Front-to-back mid-slice view.

Figure 13.3: Solid cube synthetic aperture VolSAR imagery.

Cubic Shell Synthetic Aperture Simulation

The next case examined is the formation of a "box surface" or "cubic shell" synthetic aperture. This shell or surface aperture can be created incrementally by coherent fusion of each side of the box with data collected as a flat plate subaperture. We demonstrate the incremental coherent fusion of the planar subapertures that collectively form the fused box aperture. We begin with the overhead square plate subaperture.

Overhead Square Plate Subaperture

The overhead square plate subaperture can be formed, with varying degrees of data spareness, by a swarm of sensors flying at the same height over the scene or by a single sensor making multiple passes over the scene [86].

Since this synthetic aperture is only two dimensional, conversion from the time-space phase history to the frequency-wavenumber phase history involves Fourier transformation for the two dimensions spanned by sensor movement and requires Fourier migration into the one dimension not spanned by sensor movement.

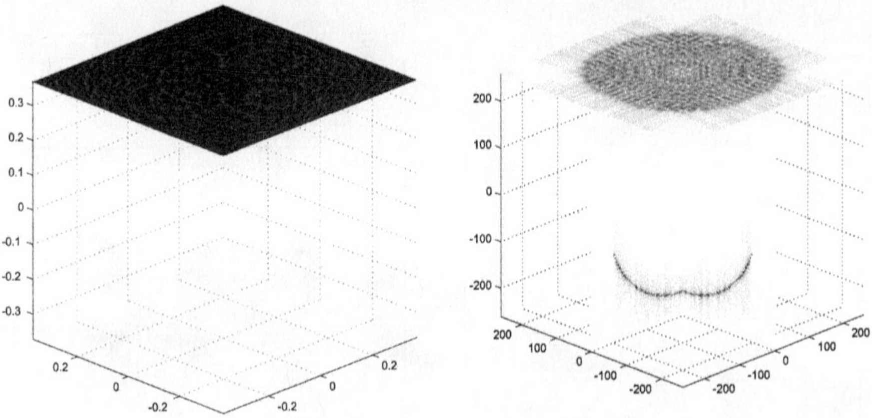

a. Time-space phase history. b. Frequency-wavenumber history.

Figure 13.4: Overhead planar subaperture simulation.

Simulate Down Converted Time-Space Phase History

- Simulate the down converted continuous wave time-space phase history by computing

$$ss[m_{u_x}, m_{u_y}]_{\text{ACS}} = g_n[n_s] \cdot \exp(-jk_{\omega_c} \cdot 2r_n[m_{u_x}, m_{u_y}]),$$

where

$$\begin{aligned} r_n[m_{u_x}, m_{u_y}] & \\ = & \sqrt{(u_x[m_{u_x}] - x_n[n_s])^2 + (u_y[m_{u_y}] - y_n[n_s])^2 + (L_z - z_n[n_s])^2}. \end{aligned}$$

Shown in Fig. 13.4a is the time-space phase history of the overhead square plate subaperture. The two dimensional Fourier transform of the data of this flat synthetic subaperture is shown near the top of Fig. 13.4b.

Setup Huygens-Fresnel Transfer Functions

- Preparatory to determining the Huygens-Fresnel transfer function for this aperture-scene interaction, determine

$$hh[m_{u_x}, m_{u_y}]_{\text{ACS}} = \exp(-jk_{\omega_c} \cdot 2r_c[m_{u_x}, m_{u_y}])$$

where

$$r_c[m_{u_x}, m_{u_y}] \;=\; \sqrt{(u_x[m_{u_x}] - X_c)^2 + (u_y[m_{u_y}] - Y_c)^2 + (L_z - Z_c)^2}.$$

Compute the Frequency-Wavenumber Phase History

- Compute the two dimensional Fourier transform of both the scattered time-space phase history shown in Fig. 13.4a,

$$SS[M_{u_x}, M_{u_y}]_{\text{ACS}} = \mathsf{FT}_{\text{2D}}\left\{ss[m_{u_x}, m_{u_y}]_{\text{ACS}}\right\}$$

and the pivot point referenced time-space phase history,

$$HH[M_{u_x}, M_{u_y}]_{\text{ACS}} = \mathsf{FT}_{\text{2D}}\left\{hh[m_{u_x}, m_{u_y}]_{\text{ACS}}\right\}.$$

Perform Fourier Migration into Above-Ground Wavenumber

- The three dimensional wavenumber phase history

$$SSM(k_{u_x}[M_{u_x}], k_{u_y}[M_{u_y}], \breve{k}_z[M_{u_x}, M_{u_y}])_{\text{ACS}}$$

is obtained via Fourier migration of

$$SS[M_{u_x}, M_{u_y}]_{\text{ACS}}$$

into above-ground wavenumber values located at

$$\breve{k}_{u_z}[M_{u_x}, M_{u_y}] = \sqrt{k_{\omega_c}^2 - k_{u_x}^2[M_{u_x}] - k_{u_y}^2[M_{u_y}]},$$

where k_{u_x} and k_{u_y} are the "free parameters" obtained from down-range and along-track Fourier transform aperture wavenumbers of the planar subaperture. The operations of Stolt formatting that produce the required Fourier migration (for all $[M_{u_x}, M_{u_y}]$ and the $\ddot{k}_{u_z}[M_{u_z}]$ that span the Ewald semi-sphere) are

$$\begin{aligned} &SSM[M_{u_x}, M_{u_y}, M_{u_z}]_{\text{ACS}} \\ &\quad = SS[M_{u_x}, M_{u_y}]_{\text{ACS}} \\ &\qquad \times 2Z_o \cdot \mathsf{sinc}[2Z_o \cdot (\ddot{k}_{u_z}[M_{u_z}] - \breve{k}_z[M_{u_x}, M_{u_y}]) \\ &\qquad \times \frac{k_{\omega_c}}{\breve{k}_{u_z}[M_{u_x}, M_{u_y}]}, \end{aligned} \tag{13.28}$$

with the requirement that the assignment is performed only for view angles within the field of view of the plate aperture. This can be achieved by requiring that the assignment (13.28) only occurs for specific $[M_{u_x}, M_{u_y}]$ values where the Jacobian determinant satisfies,

$$\frac{k_{\omega_c}}{\breve{k}_{u_z}[M_{u_x}, M_{u_y}]} \leq \frac{1}{\cos\theta_z^{\max}}$$

with

$$\theta_z^{\max} = \arctan\left(\frac{L_z + Z_o}{Z_o}\right).$$

Otherwise, sample values of $SSM[M_{u_x}, M_{u_y}, M_{u_z}]_{\text{ACS}}$ outside the isomorphically mapped field of view angular dwell is set to zero. Shown near the bottom of Fig. 13.4b is the result of Fourier migration that reveals an angular submanifold of an Ewald semisphere.

Perform Inverse Huygens-Fresnel Transfers

- The Fourier migrated version $HHM[M_{u_x}, M_{u_y}, M_{u_z}]_{\text{ACS}}$ of the Huygens-Fresnel transfer function is obtained in the same manner that yielded

$$SSM[M_{u_x}, M_{u_y}, M_{u_z}]_{\text{ACS}}.$$

Conjugate $HHM[\boldsymbol{M}]_{\text{ACS}}$ and perform the inverse Huygens-Fresnel transfer with the sample by sample multiplication,

$$VVV[\boldsymbol{M}]_{\text{SCS}} = H^*H^*M^*[\boldsymbol{M}]_{\text{ACS}} \times SSM[\boldsymbol{M}]_{\text{ACS}}.$$

Perform Inverse Fourier Transfers to Form VolSAR Image

- Perform an inverse three dimensional Fourier transform,

$$vvv[\boldsymbol{m}]_{\text{SCS}} = \mathsf{iFT}_{\text{3D}}\left\{VVV[\boldsymbol{M}]_{\text{SCS}}\right\}.$$

The images of Figs. 13.5a-d show various views of the VolSAR voxel block obtained after Fourier inversion. The voxel slice of the top view of Fig. 13.5b is unique, and both side views of Fig. 13.5c-d are nearly identical in point spread function structure. From the top view perspective, the point spread function is nearly ideal. The side view point spread functions have high resolution in along-track and down-range dimensions, but have less resolution in the height dimension. This is due to lack of observational extent in the elevation dimension obtained from the allowed view angle perspectives obtained within the overhead flat plate synthetic aperture. Resolution in elevation can be improved by expanding the size of the overhead plate aperture.

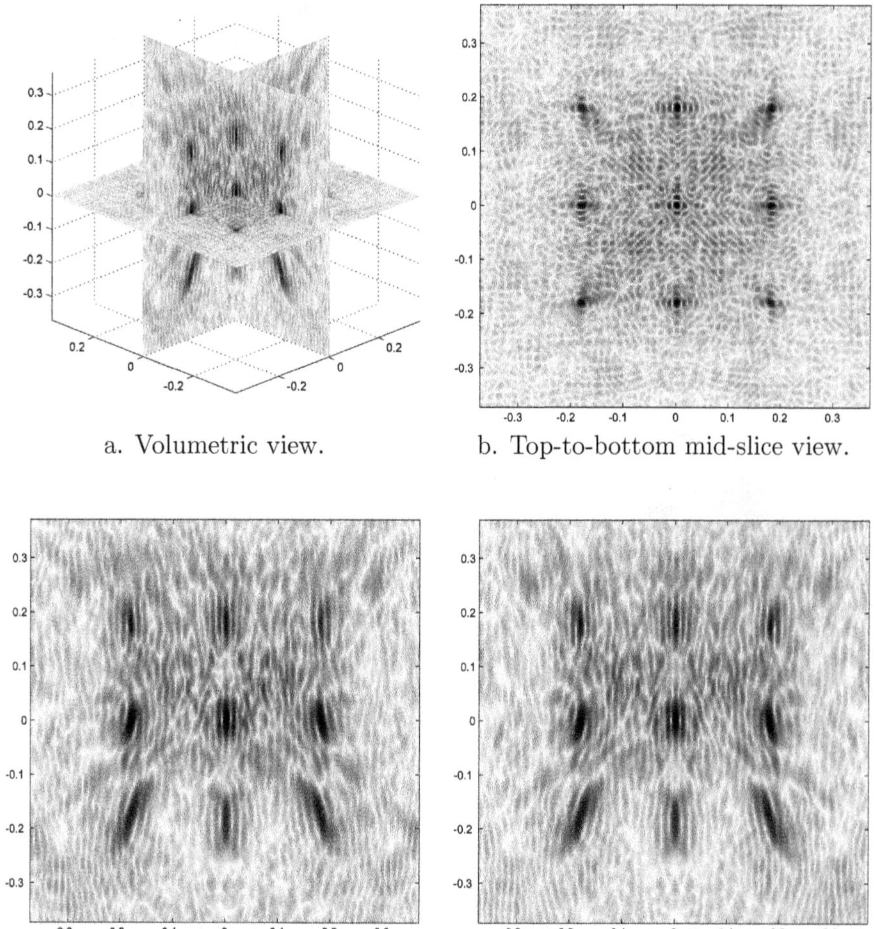

a. Volumetric view.

b. Top-to-bottom mid-slice view.

c. Left-to-right mid-slice view.

d. Front-to-back mid-slice view.

Figure 13.5: Overhead planar subaperture VolSAR imagery.

Standoff Wall Subaperture

The next case we examine is the augmentation of the overhead square plate subaperture with the addition of a fused standoff square wall subaperture. This additional planar subaperture is one side of the eventual complete cubic shell synthetic aperture.

A standoff wall subaperture can be formed, with varying degrees of data spareness, by a swarm of sensors flying in a stack at different heights or by a single sensor making multiple passes with changing altitude on each pass.

Since this synthetic aperture is only two dimensional, conversion from the time-space phase history to the frequency-wavenumber phase history involves Fourier transformation for the two dimensions spanned by sensor movement and requires Fourier migration for the dimension not spanned by sensor movement.

Down-Range Fourier Migration

If the planar subaperture is formed in the y and z dimensions, then the Fourier migration dimension is into a semisphere oriented in the down-range $\breve{k}_x$ dimension. The migration proceeding the same way as described for the overhead plate subaperture.

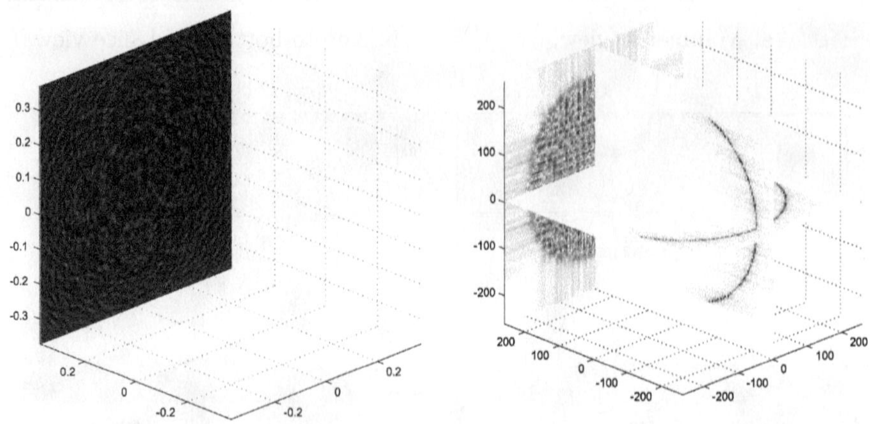

a. Time-space phase history. b. Frequency-wavenumber history.

Figure 13.6: Planar wall subaperture simulation.

Shown in Fig. 13.6a is the time-space phase history of the wall subaperture. The two dimensional Fourier transform of the data of this flat plate synthetic aperture is shown in Fig. 13.4b. Also shown in Fig. 13.4b is the result of monochromatic Fourier migration onto a solid angle segment of an Ewald semisphere.

Overhead and Wall Subaperture Fusion

The wall subaperture Fourier migrated Ewald sphere data of Fig. 13.6b is coherently added (in the Fourier migrated wavenumber domain before Fourier inversion) to the overhead subaperture Fourier migrated Ewald sphere data of

Fig. 13.4b. This gives a fused synthetic aperture that comprises two sides of the eventual complete cubic shell synthetic aperture. The slice images of Figs. 13.7a-d

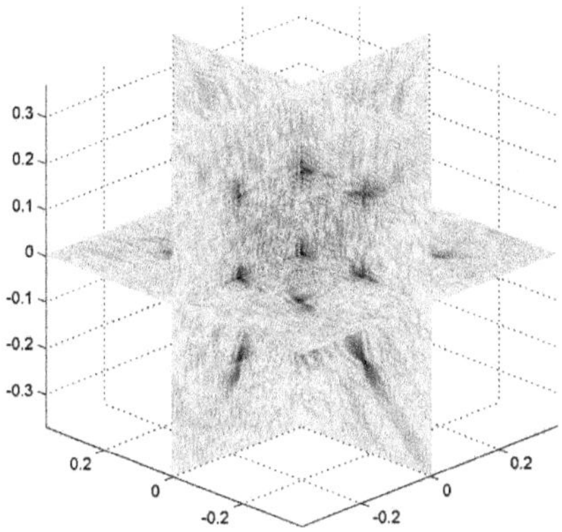

a. Volumetric view.

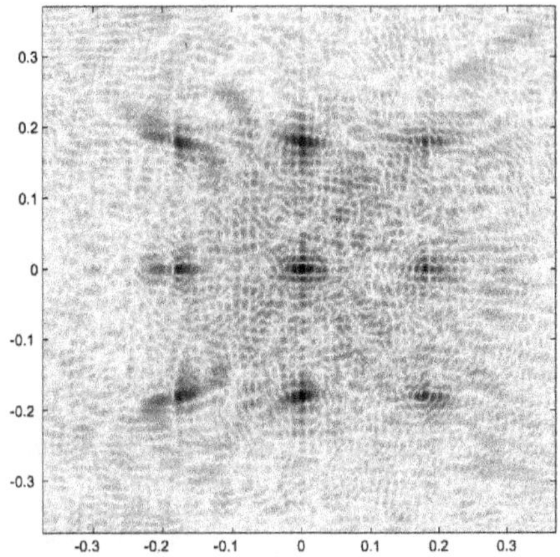

b. Top-to-bottom mid-slice view.

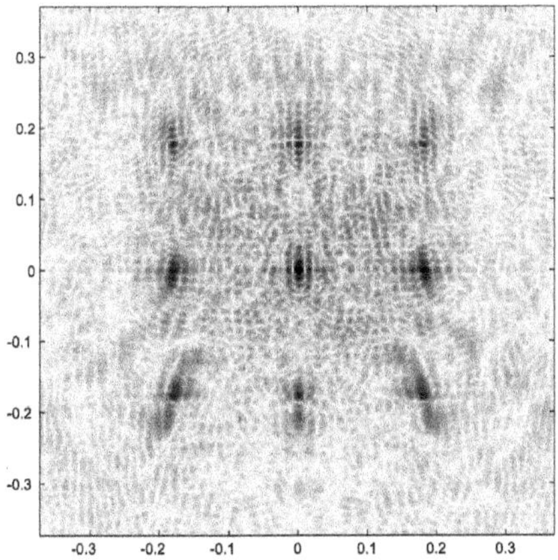

c. Left-to-right mid-slice view.

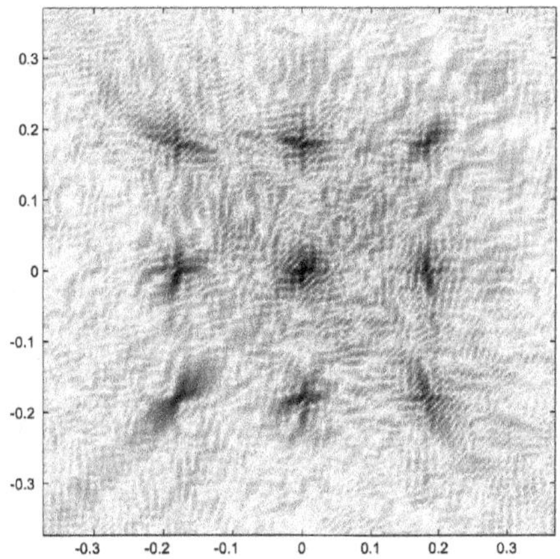

d. Front-to-back mid-slice view.

Figure 13.7: Planar wall fused subaperture VolSAR imagery.

show that the down-range and along-track resolution has been improved by fusion of the wall subaperture with the overhead subaperture.

Overhead and Two Orthogonal Wall Subapertures

The next case we examine is the augmentation of the fused overhead and single wall subaperture with yet another wall subaperture.

Along-Track Fourier Migration

If the planar subaperture is formed in the x and z dimensions, then the Fourier migration dimension is into a semisphere oriented in the along-track k_y dimension. The migration proceeds in the same way as described for the overhead plate and the other wall subaperture.

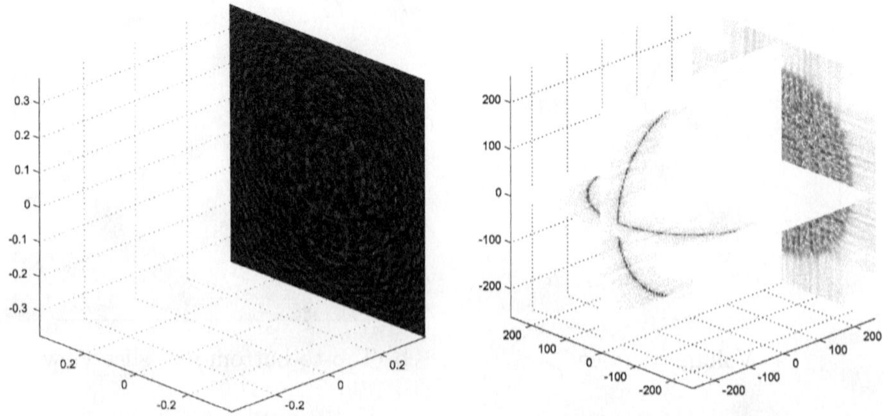

a. Time-space phase history. b. Frequency-wavenumber history.

Figure 13.8: Orthogonal wall fused subaperture simulation.

Shown in Fig. 13.8a is the time-space phase history of the added orthogonal standoff wall aperture. The two dimensional Fourier transform of the data of this flat synthetic aperture is shown on the left of Fig. 13.8b. Also shown in this figure is Fourier migration onto a solid angle of an Ewald semisphere.

Overhead and Two Orthogonal Wall Subaperture Fusion

The wall Fourier migrated Ewald sphere data of Fig. 13.8b is coherently added (in Fourier migrated k-space) to the Fourier migrated Ewald sphere data of the overhead and other wall subaperture. This gives a synthetic aperture that comprises one half of the eventual complete box VolSAR synthetic aperture. Two dimensional Fourier transformation and Fourier migration is required of all three plate subapertures. The slice images of Figs. 13.9a-d show that the down-range and along-track resolution has been improved by fusion of the additional orthogonal wall aperture.

Coherent Fusion of the Full Cubic Shell Synthetic Aperture

We next examine the volumetric imagery produced by the complete cubic shell. All six faces of the cube require two dimensional Fourier transformation and one dimensional Fourier migration.

Figs. 13.10a-d show that resolution has been improved to about the diffraction limit by inclusion of all sides of the box shell synthetic aperture. The slice images of Figs. 13.10b-d should be compared with the slice images of the cubic solid

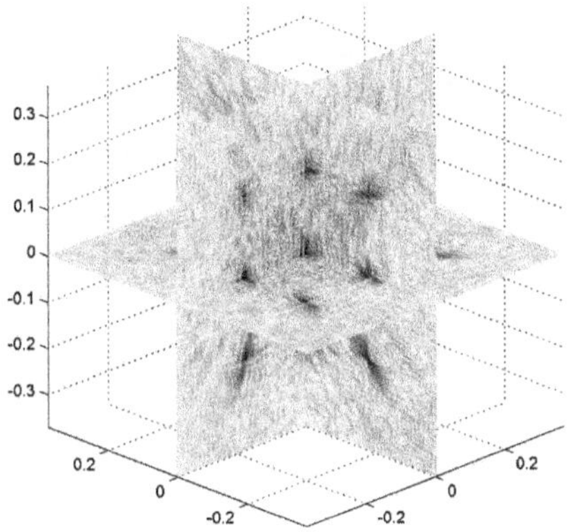

a. Volumetric view.

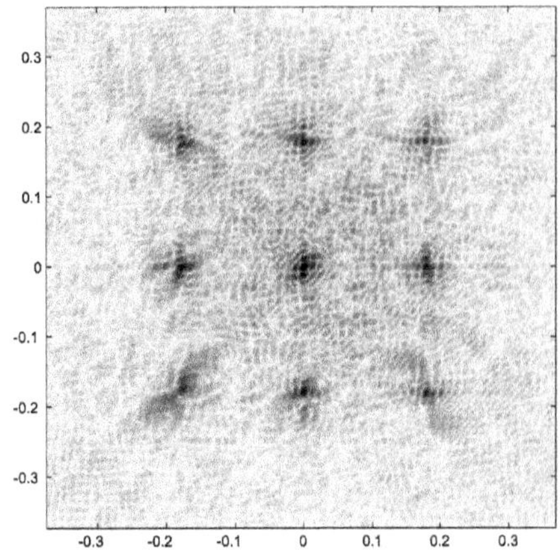

b. Top-to-bottom mid-slice view.

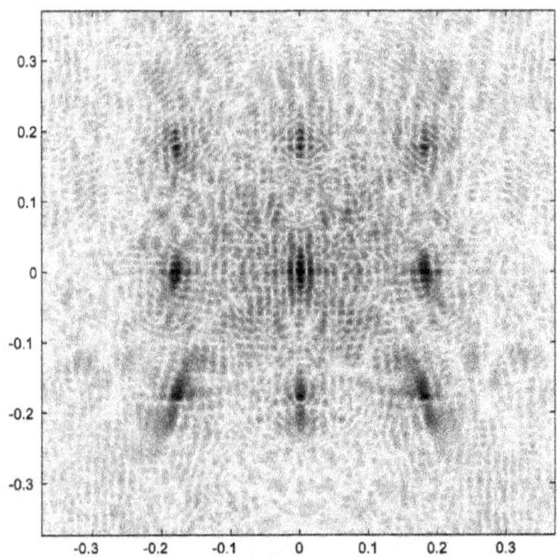

c. Left-to-right mid-slice view.

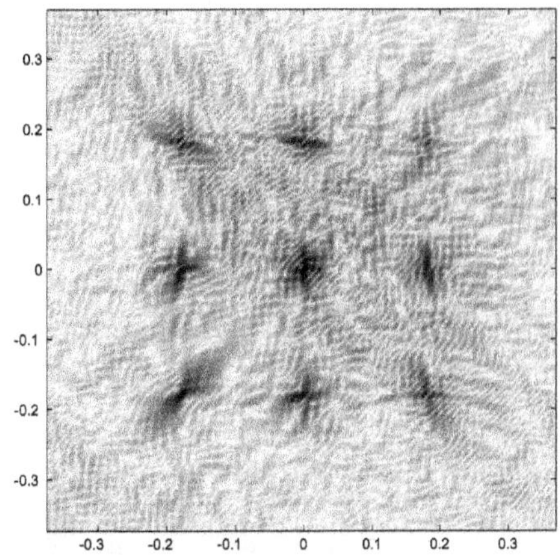

d. Front-to-back mid-slice view.

Figure 13.9: Orthogonal wall fused subaperture VolSAR imagery.

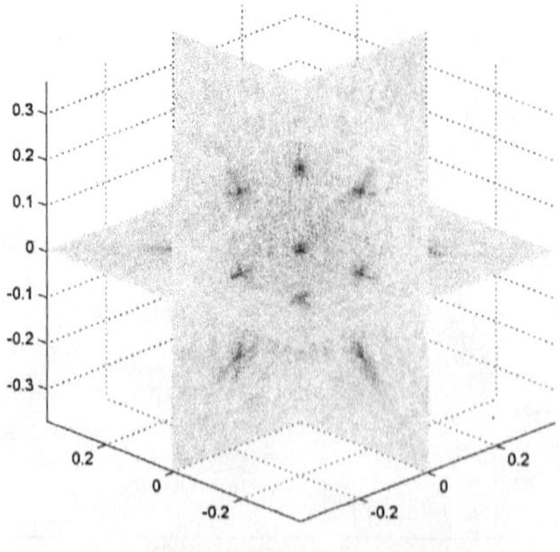

a. Volumetric view.

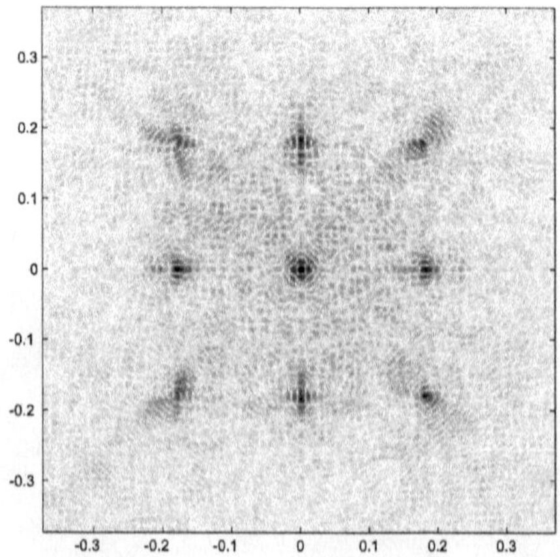

b. Top-to-bottom mid-slice view.

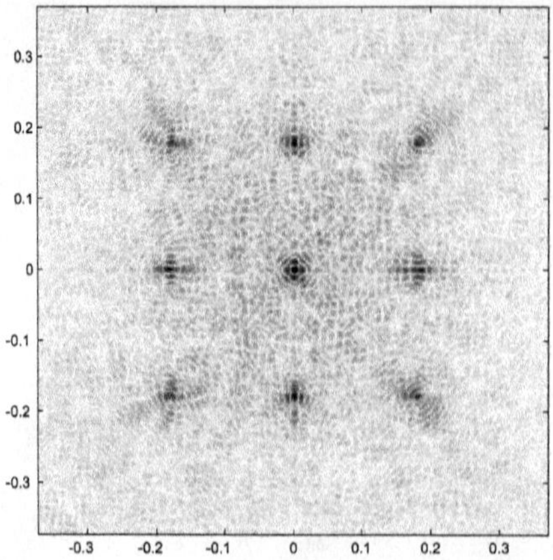

c. Left-to-right mid-slice view.

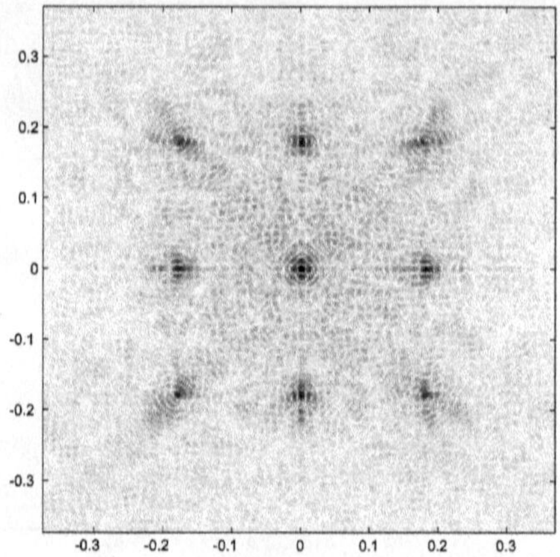

d. Front-to-back mid-slice view.

Figure 13.10: Cubic shell synthetic aperture VolSAR imagery.

volume synthetic imagery shown in Figs. 13.3b-d. Resolution is about the same. Reduction from a box volume to a box shell aperture seems to cause an increase in the sidelobe structure and an increase of speckle within the imagery. An alternative view point is that inclusion of interior data samples increases the coherent processing gain of the volumetric imagery in a way that suppresses volumetric sidelobes, or equivalently, increases the coherent processing gain that builds the mainlobe.

Spherical Shell Synthetic Aperture Simulation

Our next case is the formation of a full three dimensional spherical shell synthetic aperture.

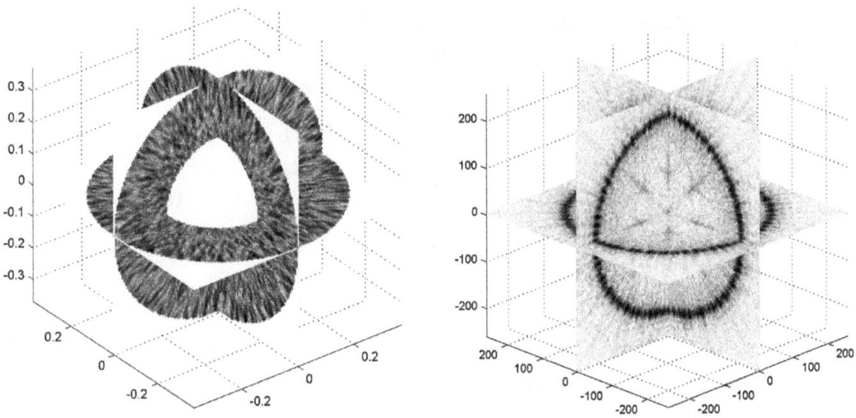

a. Time-space phase history. b. Frequency-wavenumber history.

Figure 13.11: Spherical shell aperture simulation.

Simulate Down Converted Spherical Shell Time-Space Phase History.

- Simulate the down converted continuous wave spherical shell time-space phase history by computing

$$sss[m_{u_x}, m_{u_y}, m_{u_z}]_{\text{ACS}} = g_n[n_s] \cdot \exp(-jk_{\omega_c} \cdot 2r_n[m_{u_x}, m_{u_y}, m_{u_z}]),$$

  where

$$r_n[\boldsymbol{m}] = \sqrt{(u_x[m_{u_x}] - x_n[n_s])^2 + (u_y[m_{u_y}] - y_n[n_s])^2 + (u_z[m_{u_z}] - z_n[n_s])^2},$$

  is computed and stored with coherent summation if the $\boldsymbol{m}$ of $r_n[\boldsymbol{m}]$ is such that $r_c[\boldsymbol{m}]$ satisfies

$$\eta \cdot L \leq r_c[\boldsymbol{m}] \leq L$$

where

$$r_c[m] = \sqrt{(u_x[m_{u_x}] - X_c)^2 + (u_y[m_{u_y}] - Y_c)^2 + (u_z[m_{u_z}] - Z_c)^2}$$

is the radial distance from aperture positions to scene center. The Huygens-Fresnel transfer function referenced to the volumetric scene center pivot point is set up in like manner.

Shown in Fig. 13.11a is the time-space phase history of the full spherical shell aperture. The three dimensional Fourier transform of the data is shown in Fig. 13.11b. This surface aperture is three dimensional and no Fourier migrations are required to achieve formatting to the Ewald sphere.

We observed in this empirical study that the thickness of the monochromatic shell in the time-space domain must be increased to about $\eta \approx 0.5$ to be able to obtain a thin Ewald sphere. A thin Ewald sphere after Fourier transformation is expected for monochromatic sensing. This behavioral dependency of the spherical shell aperture thickness should be explored further in future work for both monochromatic and polychromatic cases.

Fig. 13.12 shows slices of volumetric imagery of the spherical shell synthetic aperture. Compared to the imagery of the solid cube aperture (Fig. 13.3), the imagery is quite similar, except for the sidelobe structure of the spatially variant point spread function.

Dome Shell Synthetic Aperture Simulation

Clearly, a scene centered spherical shell synthetic aperture is not practical, but half of this aperture, the overhead dome shell aperture formed from a full upper 2π steradian hemisphere is practical [687]. Single or multiple sensors flying on an ascending or descending spherical spiral can conceivably collect data to form a spherical dome synthetic aperture.

Shown in Fig. 13.13a is the time-space phase history of the overhead spherical dome shell aperture. The three dimensional Fourier transform of the spherical dome data is shown in Fig. 13.13b. The hemisphere of the Ewald sphere is clearly evident.

Figs. 13.14a-d show slices of volumetric imagery of the overhead spherical dome shell synthetic aperture. Compared to the imagery of the full spherical shell aperture (Fig. 13.12), resolution seems only to be lost on scatterers below the base of the dome.

Super Resolution through Ewald Sphere Symmetries

If a scatterer is isotropic, then one would expect the gathered phase history collected from one side of the scatterer to match the scatter collected from the other side. We propose simulating this effect by reflection of data from one side of the Ewald sphere to the other side and merging the data through coherent fusion. This process of bandwidth extrapolation or extension is a form of *super resolution* [478, 783, 863].

Fig. 13.15a shows the data extended from the dome synthetic aperture to the bottom half of the spherical shell aperture. Shown in Figs. 13.16a-d is the super

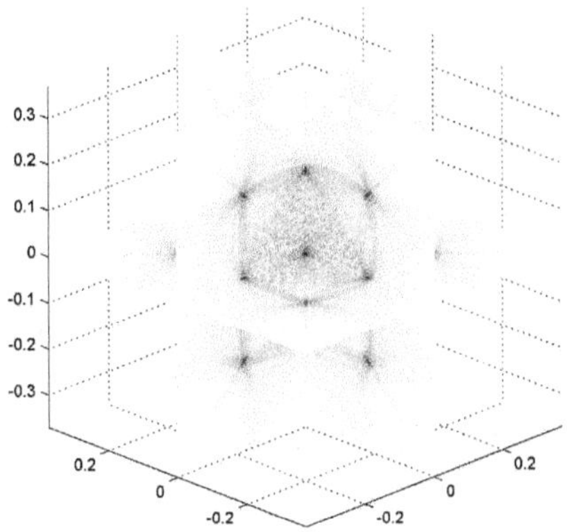

a. Volumetric view.

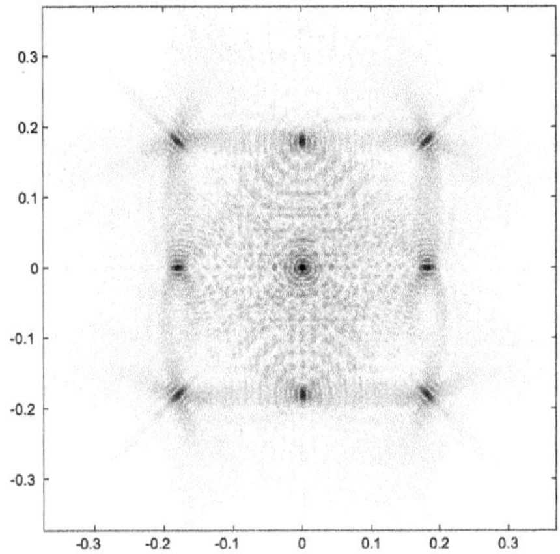

b. Top-to-bottom mid-slice view.

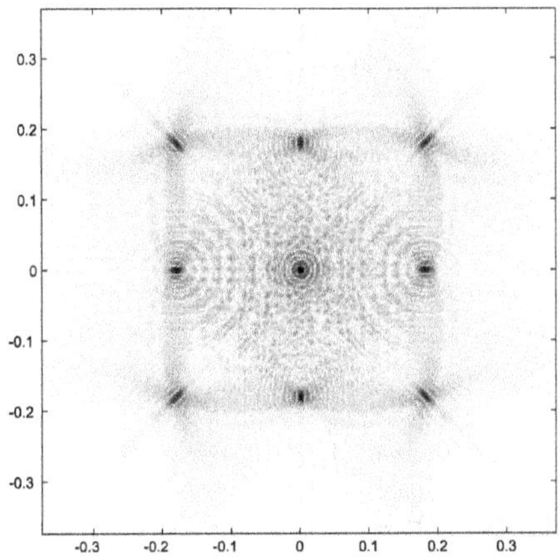

c. Left-to-right mid-slice view.

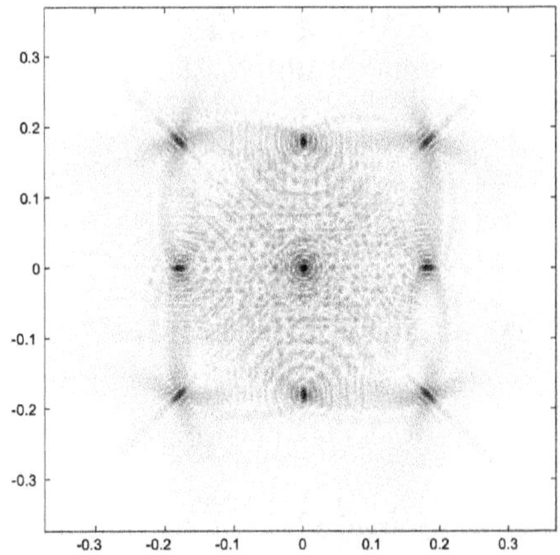

d. Front-to-back mid-slice view.

Figure 13.12: Spherical shell aperture VolSAR imagery.

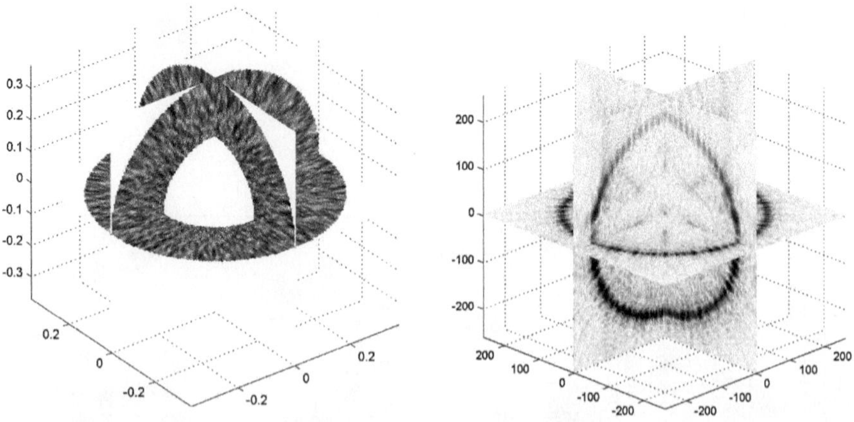

a. Time-space phase history. b. Frequency-wavenumber history.

Figure 13.13: Dome shell synthetic aperture simulation.

resolution imagery. Subterranean scatterers show improved resolution compared to the half-dome imagery of Figs. 13.14a-d.

Although subterranean data cannot be collected (assuming no signal penetration) from an airborne sensor, with the isotropic assumption, forward scattered data can be estimated from backward scattered data. The price paid for this assumption is the destruction of phase information in the image domain. Fourier transforms that are symmetrically even come from real data. Transforming odd data (one half of the spherical shell) to even data (via super resolution spectrum reflection), creates an even spectrum (in the k_z dimension). Hence, the super resolution voxel data will be real and provide no phase information.

Polychromatic Simulations

It is reasonable to expect that polychromatic sensing increases the information content and quality of VolSAR imagery [911].

Polychromatic Overhead Square Plate Subaperture

The number of frequency bins used in our polychromatic scenario is 64. As explained in Sections 12.4, the density of the fast-time frequency bins must satisfy

$$\Delta_{\omega_p} \leq \frac{1}{\cos\theta_z^{\max}} \cdot \frac{c}{2} \cdot \Delta_{\ddot{k}_z},$$

where

$$\Delta_{\ddot{k}_z} = \frac{2\pi}{2Z_o}.$$

The passband of this simulation has frequency bins located at

$$10.0 \text{ GHz } - \frac{1}{2\pi}\left(n_{\omega_p} \cdot \Delta_{\omega_p}\right)$$

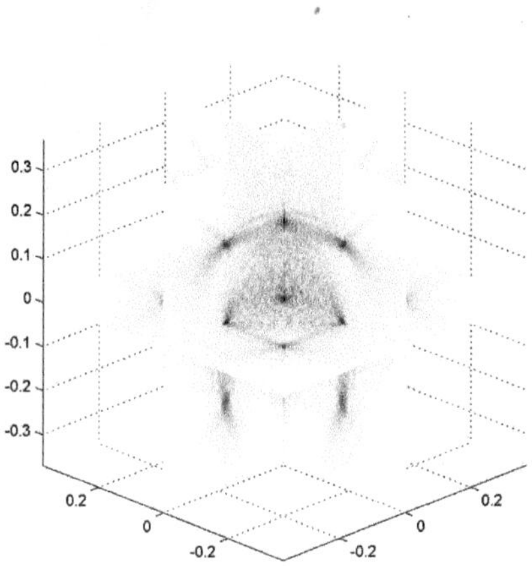

a. Volumetric view.

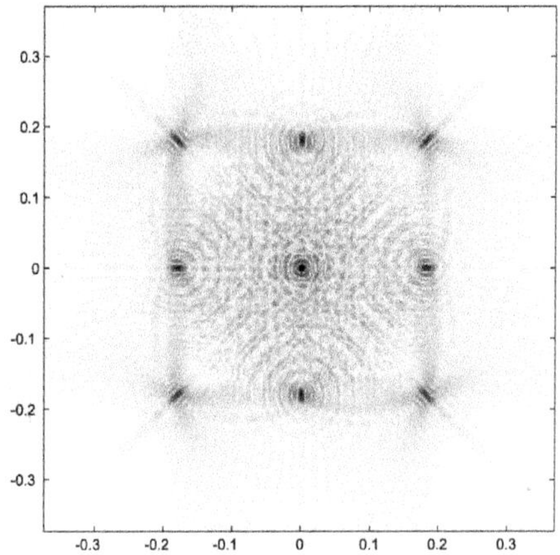

b. Top-to-bottom mid-slice view.

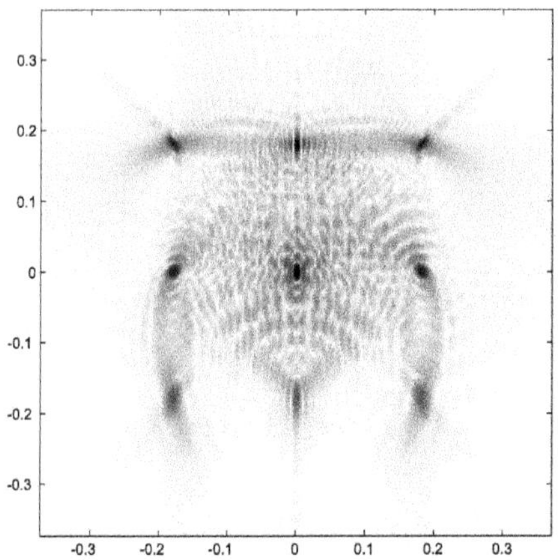

c. Left-to-right mid-slice view.

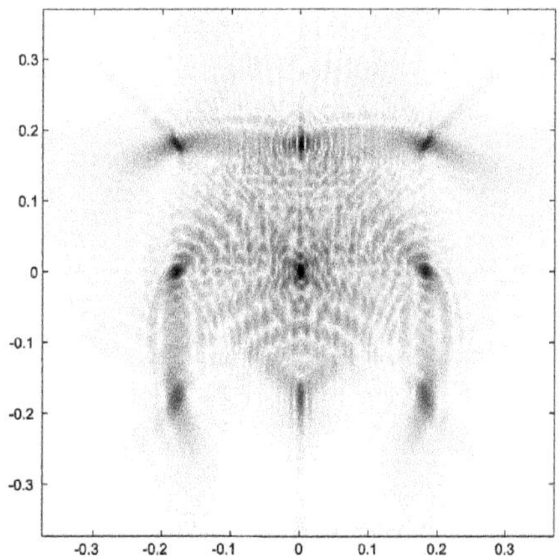

d. Front-to-back mid-slice view.

Figure 13.14: Dome shell synthetic aperture VolSAR imagery.

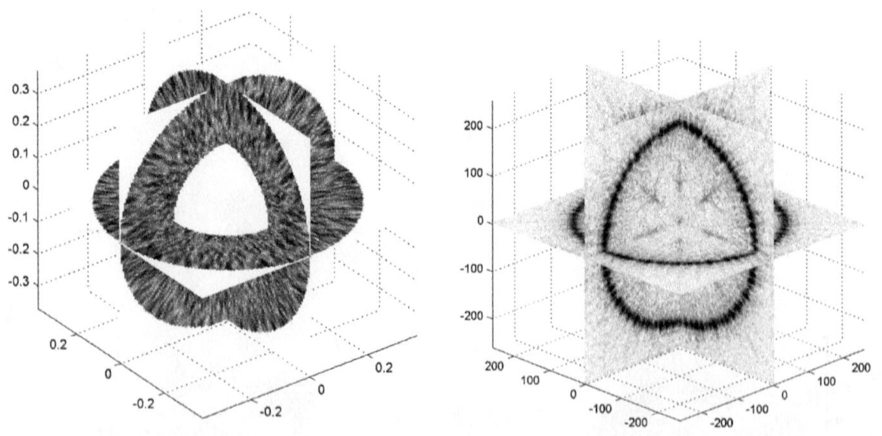

a. Time-space phase history. b. Frequency-wavenumber history.

Figure 13.15: Dome shell wavenumber extrapolation.

for

$$n_{\omega_p} = 0, \ldots, 63.$$

These parameters and requirements yield a frequency bin spacing of about 126.7 MHz with 64 discrete frequency bins, giving a system bandwidth of about 8.1 GHz. The polychromatic simulation spans the spectrum starting at about 2 GHz and ending at about 10 GHz.

Simulate Down Converted Polychromatic Time-Space Phase History

- Simulate the down converted polychromatic time-space phase history by computing

$$Ss[n_{\omega_p}, m_{u_x}, m_{u_y}]_{\mathrm{ACS}} = g_n[n_s] \cdot \exp(-jk_{\omega_p}[n_{\omega_p}] \cdot 2r_n[m_{u_x}, m_{u_y}]),$$

where

$$\begin{aligned} & r_n[m_{u_x}, m_{u_y}] \\ = & \sqrt{(u_x[m_{u_x}] - x_n[n_s])^2 + (u_y[m_{u_y}] - y_n[n_s])^2 + (L_z - z_n[n_s])^2}. \end{aligned}$$

Although the subaperture is only two dimensional, the gathered time-space phase history is three dimensional because of the polychromatic extension (one temporal wavenumber dimension and two spatial wavenumber dimensions). The polychromatic Huygens-Fresnel transfer functions are set up in like manner. Shown in Fig. 13.17a is the time-space phase history of one frequency of the overhead square plate aperture gathered by a polychromatic sensor system.

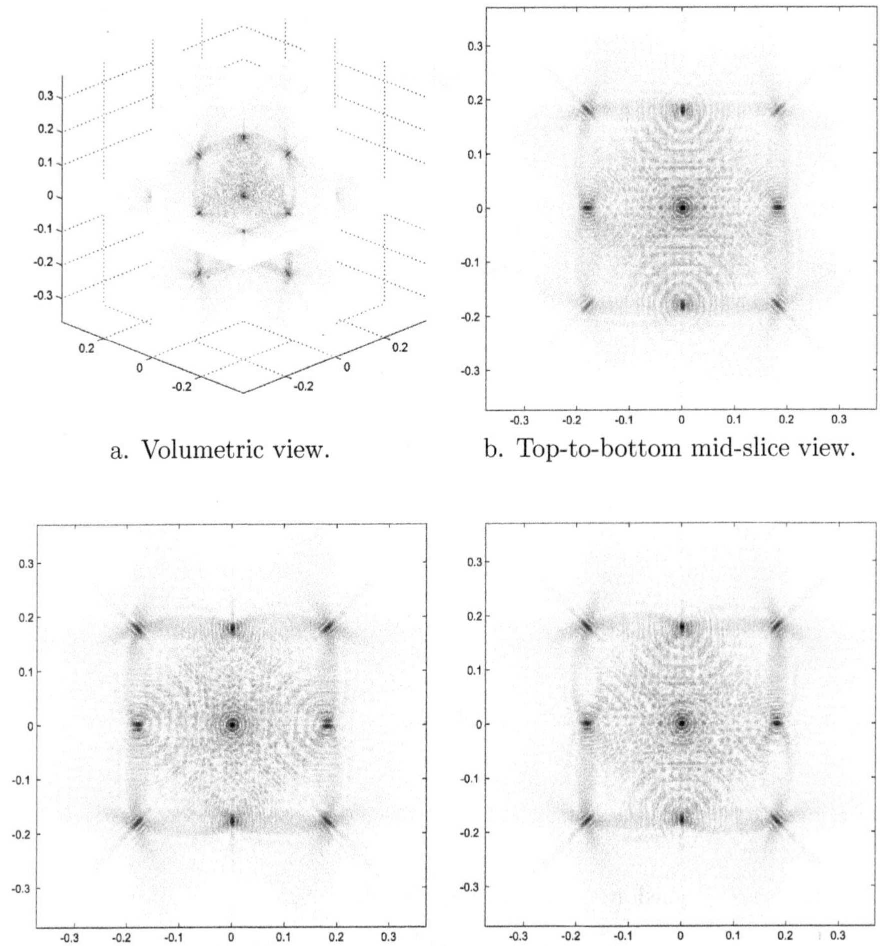

a. Volumetric view.

b. Top-to-bottom mid-slice view.

c. Left-to-right mid-slice view.

d. Front-to-back mid-slice view.

Figure 13.16: Extrapolated dome shell synthetic aperture VolSAR imagery.

Compute the Polychromatic Frequency-Wavenumber Phase History

- Compute the two dimensional Fourier transform of both the scattered time-space phase history shown in Fig. 13.4a,

$$SSS[n_{\omega_p}, M_{u_x}, M_{u_y}]_{\text{ACS}} = \mathsf{FT}_{\text{2D}} \left\{ Sss[n_{\omega_p}, m_{u_x}, m_{u_y}]_{\text{ACS}} \right\}$$

and the pivot point referenced time-space phase history,

$$HHH[n_{\omega_p}, M_{u_x}, M_{u_y}]_{\text{ACS}} = \mathsf{FT}_{\text{2D}} \left\{ Hhh[n_{\omega_p}, m_{u_x}, m_{u_y}]_{\text{ACS}} \right\}.$$

The two dimensional Fourier transform of the data of this square plate synthetic aperture for one frequency bin is shown near the top of Fig. 13.17b.

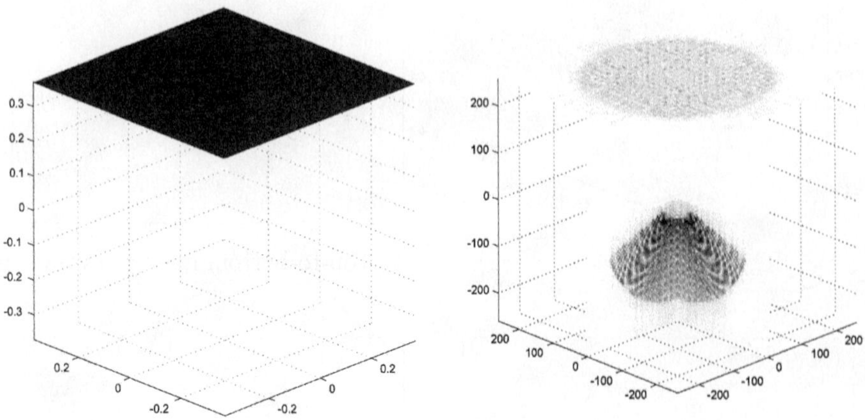

a. Time-space phase history. b. Frequency-wavenumber history.

Figure 13.17: Polychromatic overhead planar subaperture simulation.

Perform Polychromatic Fourier Migration into Above-Ground Wavenumber

- The three dimensional Fourier migrated phase history

$$SSM(k_{u_x}[M_{u_x}], k_{u_y}[M_{u_y}], \breve{k}_z[M_{u_x}, M_{u_y}])_{\text{ACS}}$$

is obtained via Stolt formatting of

$$SSS[n_{\omega_p}, M_{u_x}, M_{u_y}]_{\text{ACS}}$$

into above-ground wavenumber values located at

$$\breve{k}_{u_z}[n_{\omega_p}, M_{u_x}, M_{u_y}] = \sqrt{k^2_{\omega_p}[n_{\omega_p}] - k^2_{u_x}[M_{u_x}] - k^2_{u_y}[M_{u_y}]},$$

where k_{u_x} and k_{u_y} are parameters obtained from down-range and along-track Fourier transform aperture wavenumbers of the planar subaperture.

The operations of Stolt formatting produce the required polychromatic Fourier migration (for all n_{ω_p}, for all $[M_{u_x}, M_{u_y}]$ and for the $\ddot{k}_{u_z}[M_{u_z}]$ that span the Ewald semisphere),

$$\begin{aligned} &SSM[M_{u_x}, M_{u_y}, M_{u_z}]_{\text{ACS}} \\ &= \sum_{n_{\omega_p}} SSS[n_{\omega_p}, M_{u_x}, M_{u_y}]_{\text{ACS}} \\ &\qquad \times 2Z_o \cdot \mathsf{sinc}[2Z_o \cdot (\ddot{k}_{u_z}[M_{u_z}] - \breve{k}_z[n_{\omega_p}, M_{u_x}, M_{u_y}]) \\ &\qquad \times \frac{k_{\omega_c}}{\breve{k}_{u_z}[n_{\omega_p}, M_{u_x}, M_{u_y}]}, \end{aligned} \tag{13.29}$$

with the requirement that the assignment is only for view angle or angular spectrum angles within the field of view of the plate aperture. This can be achieved by updating the coherent sum (13.29) only if

$$\frac{k_{\omega_c}}{\breve{k}_{u_z}[n_{\omega_p}, M_{u_x}, M_{u_y}]} \le \frac{1}{\cos\theta_z^{\max}}$$

is satisfied, with

$$\theta_z^{\max} = \arctan\left(\frac{L_z + Z_o}{Z_o}\right).$$

Shown in the lower half of Fig. 13.17b is the result of polychromatic Fourier migration. Fourier migration of this polychromatic data forms a thick conic section (i.e., a solid angle) of the Ewald sphere. When a passband signal is transmitted, this conic section of the Ewald sphere has a radial thickness equal to the temporal wavenumber bandwidth $k_{\omega_p} \in \beta_p/c$ of the transmitted signal.

Perform Inverse Huygens-Fresnel Transfers

- The Fourier migrated version

$$HHM[M_{u_x}, M_{u_y}, M_{u_z}]_{\text{ACS}}$$

of the Huygens-Fresnel transfer function is obtained in the same manner that yielded

$$SSM[M_{u_x}, M_{u_y}, M_{u_z}]_{\text{ACS}}.$$

Conjugate $HHM[\boldsymbol{M}]_{\text{ACS}}$ and perform the inverse Huygens-Fresnel transfer with the sample by sample multiplication,

$$VVV[\boldsymbol{M}]_{\text{SCS}} = H^*H^*M^*[\boldsymbol{M}]_{\text{ACS}} \times SSM[\boldsymbol{M}]_{\text{ACS}}.$$

Perform Inverse Fourier Transfers to Form VolSAR Image

- Perform an inverse three dimensional Fourier transform of $VVV[\boldsymbol{M}]_{\text{SCS}}$,

$$vvv[\boldsymbol{m}]_{\text{SCS}} = \mathsf{iFT}_{\text{3D}}\left\{VVV[\boldsymbol{M}]_{\text{SCS}}\right\}.$$

The images of Figs. 13.18a-d show various views of the resulting VolSAR voxel block. Compared to the monochromatic version shown in Figs. 13.5a-d, resolution is improved and the sidelobe structure is suppressed.

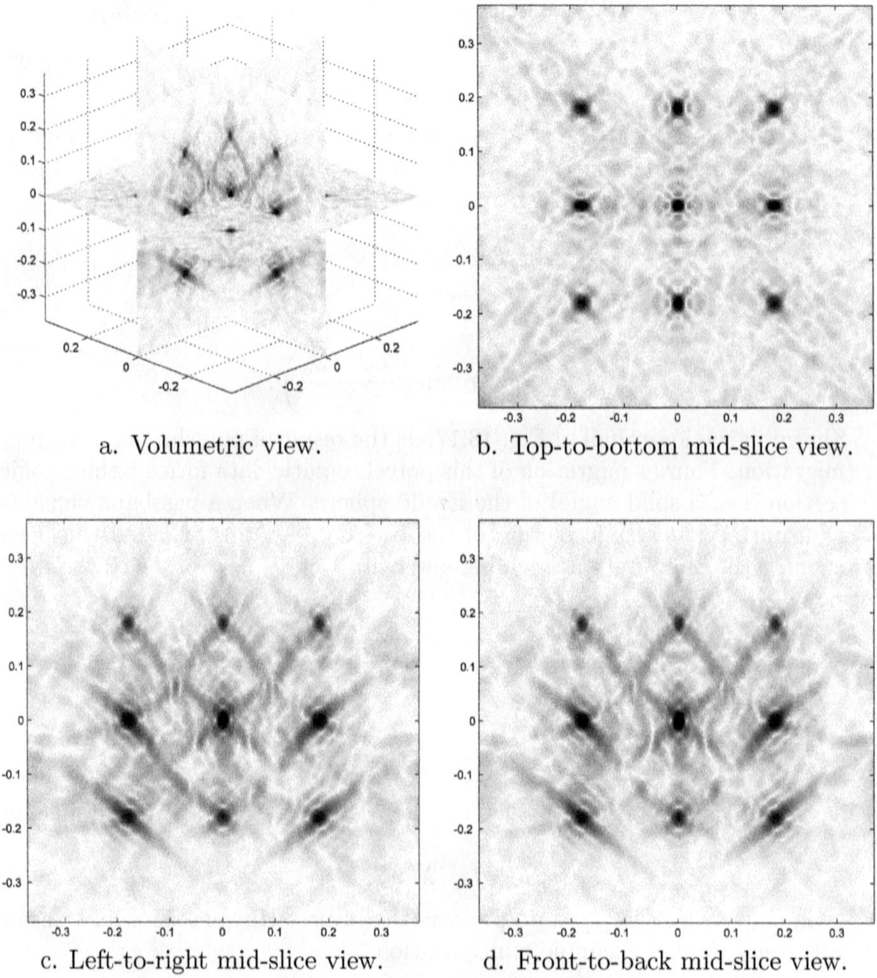

a. Volumetric view.

b. Top-to-bottom mid-slice view.

c. Left-to-right mid-slice view.

d. Front-to-back mid-slice view.

Figure 13.18: Polychromatic overhead planar subaperture VolSAR imagery.

Coherent Fusion of Monochromatic L, C, and X Band Simulations

Different data collections of the same scene with different passband frequencies can be coherently fused by complex addition performed in a Fourier migrated wavenumber domain. These passband continuous wave frequencies can be widely separated.

Ewald Spheres and Signal Passband Bandwidths

As given in Table 2.1, radius values of Ewald spheres for P, L, S, C, X, Ka and Ku band radar systems are, respectively, about 10, 21, 63, 105, 210, 315 and 733 radians/meter.

Overhead Square Plate Subaperture

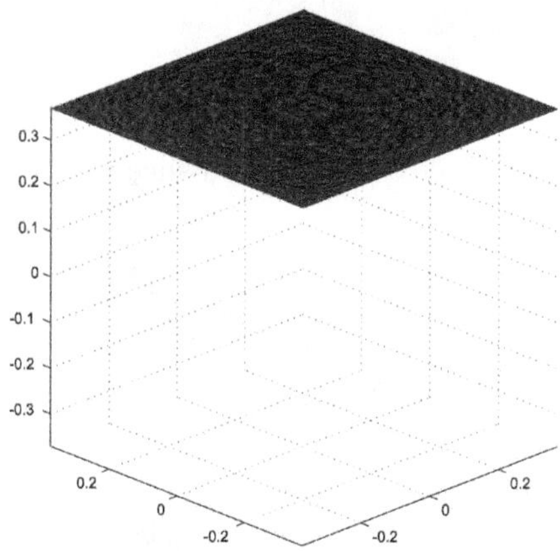

a. Time-space phase history.

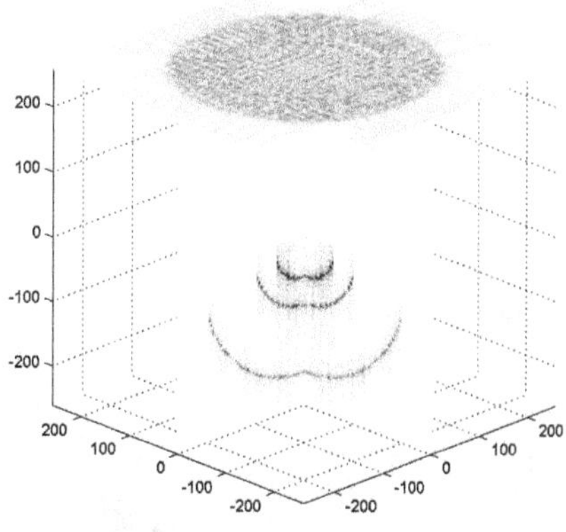

b. Frequency-wavenumber history.

Figure 13.19: Fused L, C and X band overhead planar subaperture simulation.

Shown in Fig. 13.19a is the time-space phase history for the L band signal of the overhead square plate aperture. The two dimensional Fourier transform of the data of this flat plate synthetic aperture is shown near the top of Fig. 13.19b. Also shown in this figure is the result of Fourier migration of L, C and X band sensing, yielding three Ewald shells corresponding to the L, C and X band continuous wave signals with temporal wavenumbers of 21, 105 and 210. The Fourier migrated data forms thin spherical shells in the three dimensional wavenumber domain.

The images of Figs. 13.20a-d show various views of the resulting fused L, C and X band VolSAR voxel block. The diffraction limit of each continuous wave signal is obtained and the corresponding point spread functions of the three continuous wave frequencies appear superimposed.

A more interesting scenario to study would be to view the imagery of a scene comprised of frequency dependent scatterers.

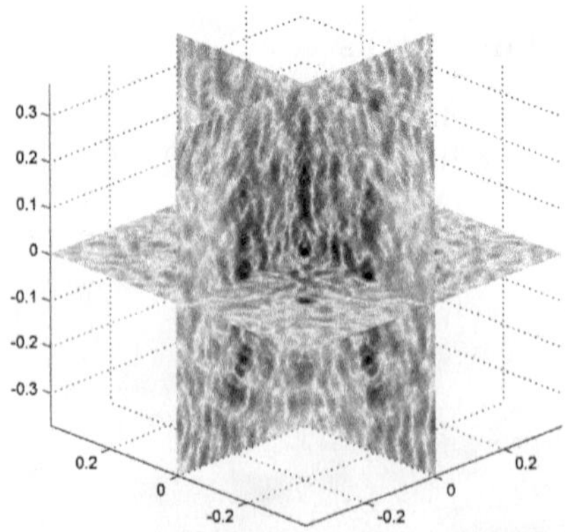

a. Volumetric view.

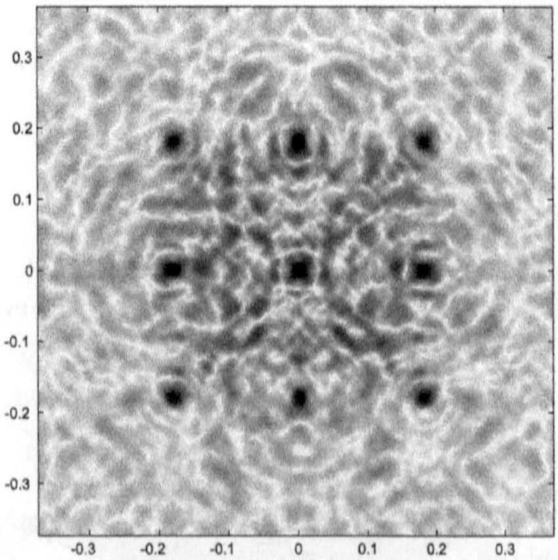

b. Top-to-bottom mid-slice view.

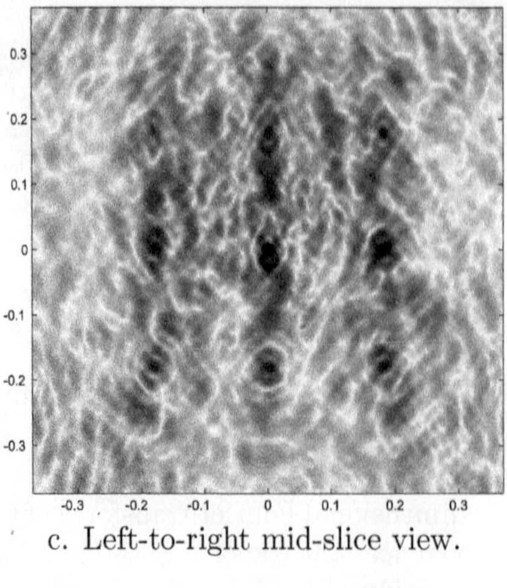

c. Left-to-right mid-slice view.

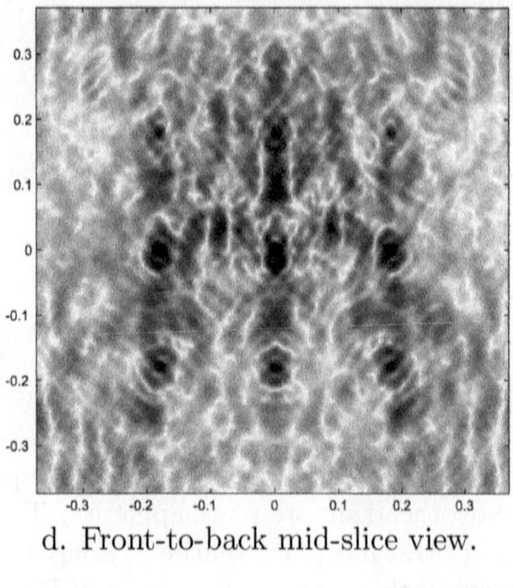

d. Front-to-back mid-slice view.

Figure 13.20: Fused L, C and X band overhead planar subaperture VolSAR imagery.

Additional VolSAR Concepts

One can imagine many concepts for generalized volumetric synthetic aperture imaging. These include the notion of an overhead, multiple sensor, arch-tunnel synthetic aperture, a single sensor surrounding cylinder (barrel) synthetic aperture, etc.

Of particular interest is the notion of a meandering sensor swarm aperture formed with multiple sensor platforms. This is a practical scenario, where a swarm of sensors simultaneously image a spot area of interest with flight paths that move either in a coordinated or in a random way. VolSAR processing can be used to coherently fuse and process the collective data. Such a sensor swarm should be able to achieve sufficient data sample density and angular extent to deliver useful three dimensional imagery.

Aperture Sparseness

Three dimensional imaging is sometimes investigated in the context of relatively small ground scene, large stand off ranges and planar wavefronts [27]. There is an inherent mismatch in this concept of operation and three dimensional imaging concepts. These assumptions make volumetric imaging rather challenging (because of lack of extended angular dwell) and cause the problem to become more akin to sparse inversion challenges [689].

The elevation imaging challenge is also typically different than the along-track and down-range imaging tasks because the number of required passes is much less than that needed to support the spatial manifold in the vertical dimension, for example, urban areas in a region with flat topology [345]. This observation can be used to advantage when formulating VolSAR flight plans for imaging.

The issue of aperture sparseness caused by limited sensor resources within a VolSAR synthetic aperture volume or subspace is not extensively explored here. Much work remains to be done in regards to the various trades of subaperture geometries and sample densities for a multitude of conceivable volumetric imaging concepts [328].

Chapter 14

SAR Related Computed Imaging Algorithms

In seeking to trace the origins and relationships of SAR algorithms to other types of computed imaging algorithms, it is often the case that the same algorithm class emerges out of different application domains, e.g., medical imaging, geophysical exploration, remote sensing, etc. [407]. We explore in this chapter the broad scope of computed imaging algorithms that are in some way related to algorithms used in SAR imaging. These include imaging methods of holography, ray tomography, diffraction tomography and seismology [135, 502].

Two methods dominant in computed tomography for medical imaging, convolutional back-projection (CBj) and convolutional back-propagation (CBp), are given special attention. We take care to identify differences between implementations of CBj and CBp algorithms in baseband and passband sensor systems. The class of convolutional back-propagation algorithms is different than the class of direct back-propagation (DBp) algorithms described previously in Section 12.3.

SAR's polar formatting is a *direct* Fourier reconstruction (dFR) method related to CBj. SAR's Stolt formatting is a *filtered* Fourier reconstruction (fFR) method related to CBp. Polar formatting and CBj are based on unfiltered and filtered versions of the Fourier projection theorem, respectively. Stolt formatting and CBp are based on a filtered version of the Fourier diffraction theorem.

We also touch on some remaining wave theory principles that are sometimes used in non-SAR computed imaging applications, especially in acoustic medical imaging. For example, we look at the homomorphic wave motion equations of Riccati and wave field approximations of Rytov. We shall see that not all principles commonly applied in acoustic imaging are applicable to the electromagnetic spectrum of SAR systems.

This chapter is outlined as follows: Section 14.1 overviews the relationship between seismic imaging methods and SAR. Section 14.2 overviews the relationship between holographic imaging methods and SAR, and gives a survey of holographic imaging techniques. Section 14.3 overviews the relationships of ray and diffraction tomographic imaging methods to SAR imaging, and gives a survey of tomographic imaging techniques. Section 14.4 overviews direct (unfiltered) Fourier reconstruction methods and presents the Fourier projection and Fourier diffraction theorems. Section 14.5 gives an overview of the convolutional back-projection and convolutional back-propagation algorithms that form the core of computed tomography algorithms used in many medical imaging systems. This section presents the Hilbert transform and discusses analytic signals. This section also presents the Radon transform and cross-sectional image reconstructions based on the inverse Radon transform from sets of baseband projections. Section 14.6 overviews filtered Fourier reconstruction methods that are wavenumber domain equivalents of convolutional spatial reconstruction methods. Section 14.7 surveys some of the many variations of SAR algorithms that can be found in the literature. Finally, Section 14.8 gives a list of various applications of SAR.

14.1 Seismic Imaging

The reemergence of wave equation formulations of SAR in the late 1980s and early 1990s seems to have come from the discipline of seismology [137,474,708,752,790]. The "exploding reflectors" that are sometimes used to model seismic reflections of acoustic source explosions [193,591,949] are equivalent to the Huygens-Fresnel point singularities we have used to model the reaction of SAR scatterers to an incident electromagnetic wave field.

Kirchhoff Migration in Seismic Imaging

Seismic migration was performed with graphical methods until the 1960s. The introduction of wave equation migration in the 1970s led a revolution in applied methods in seismic imaging. Seismic inversion algorithms were developed based on finite difference approximations of wave equations. During this time frame, the diffraction summation migration method was improved with a Kirchhoff integral representation of the solution of the seismic wave equation [580]. The improved algorithm is known as *Kirchhoff migration.* Recorded acoustic wave fields, as measured along an acoustic sensor array, are used as either initial conditions or boundary conditions for an acoustic wave field model described with Kirchhoff integral formulations.

Fourier transform methods were also introduced into seismic imaging at this time. The frequency domain methods proved more accurate than the finite difference time-space domain methods [375,376].

Seismic Imaging and Direct Back-Propagation

With a wave field theory foundation, inversion based on partial differential equations can be solved in either the signal domain or image domain. Image space (direct back-propagation) and signal space (range walk) migration schemes differ from each other only in the form of the wave equations used for wave extrapolation [193]. In seismic inversion, migrations in image space are more popular for historical and practical reasons [376]. These approaches set up a computational grid in image space and do wave equation migration (direct spherical back-propagation). A spherical *averaging* approach (since seismic sensing is non-coherent) [324] is used in seismic imaging.

Signal Domain Hyperbolic Migration in Seismic Imaging

The signature of an expanding seismic spherical wavefront, as captured along a line of seismic acoustic sensors laid along the ground, is a hyperbola in the recorded time-space data matrix [65, 813]. One form of seismic migration seeks to use the observed data in the time-space data matrix to estimate subterranean point scatterers that exist in a vertical slice that forms a subterranean cut (equivalent to the SAR slant-plane) [376]. The hyperbola of signal delays is associated with "normal move-out" trajectories (sensor array element offsets referenced from the ground surface nadir point over the scatterer being focused) [656, 728, 789]. For each point on a subterranean planar cut of interest, a fast-time signal delay hyperbola is computed. The points of intersection of the "traces" (the equivalent notion in seismology of radar pulse data in SAR) and a reference hyperbola are determined. Values near intersection points are interpolated and non-coherently summed to form the value of the point reflectivity at the nadir centered apex of the signal delay hyperbola. Summation along hyperbolic signal delay profiles is also known as *diffraction summation migration* in seismology [424, 789].

Seismic Imaging and Fourier Migration

The time-space domain process of "downward continuation" in seismic imaging [82] corresponds to direct back-propagation of pulse compressed time-space domain phase history data in SAR imaging [137]. When implemented in the frequency-wavenumber domain, this downward continuation is equivalent to the operations achieved by Huygens-Fresnel transfers.

Huygens-Fresnel transfer [71] methods in seismology [742] are sometimes called *field continuation* methods [137, 790, 791, 812]. Field continuation methods are based on the theory [522] that knowledge of a part of a wave field as sensed on a manifold can be used to estimate the field by propagation either upwards of downwards onto any other surface.

The Stolt formatting approach is sometimes referred to as the *F-K method,* as first reported by Stolt [309, 376]. The "F-K" terminology of seismology corresponds to the "omega-k" terminology used by some in the SAR community.

Convolutional Back-Propagation in Seismic Imaging

Generalized diffraction tomography applied to seismic reflections has also been formulated [656], where convolutional back-propagation [256] is called *geophysical diffraction tomography.* This tomographic seismic inversion problem depends on determining the "plane wave response" of a geophysical formation.

14.2 Holographic Imaging

Gabor invented holography in 1947 [369]. The holograms he was able to produce were of low quality because coherent light sources (lasers) did not yet exist.

Interest in holography waned within a few years of its invention [372].

Einstein theorized around 1916 concepts of stimulated emission that later proved to make laser possible [297, 299]. The first microwave laser was built in 1954. The first optical laser was built in 1960, after which, holography experienced a resurgence. High quality two dimensional holograms were produced by Leith and Upatnieks in 1960 [575]. They demonstrated three dimensional holography in 1964 [576–578].

Holography and SAR

Some SAR image formation algorithms have long been recognized as holographic [502, 539, 562, 572, 573, 921]. One view point interprets Stolt formatting as a wide band generalization of single frequency "backward-wave" (as opposed to forward scattered wave) holographic reconstruction [827].

Work in the area of optical holography [213, 227, 229, 575] at the University of Michigan, Ann Arbor during the time of the Wolverine Project may have been a catalyst for early explorations of the synergism between SAR and holography [372]. It is significant in the history of SAR that Leith and Upatnieks were active in both early SAR research and in the resurgence of interest in optical holography. This early interaction in the 1960s created a foundation for the return to holographic concepts in SAR in the 1990s, especially in the extension of SAR to three dimensional imaging problems.

Given the reemergence of Stolt formatting methods in SAR during the early 1990s, SAR and holography have a *There and Back Again*[1] relationship [539, 572–574]. So, in a very real sense, the Stolt formatting methods "introduced" in the late 1980s and early 1990s were a return to earlier methods in SAR, rather than something new in SAR. The primary difference was use of optical computing in the 1960s and use of high performance digital computing in the 1990s.

The flow of innovative concepts has begun to move in both directions. The circular SAR concepts of [866, 868, 871] have been extended to three dimensional imaging in millimeter wave holography [827–829].

Holographic Sensor Systems

Applications in holography span sensor systems that are optical, microwave and acoustic. We give a brief overview of some of these systems and application domains.

Optical Holography

Optical holography [155, 570] requires illumination with a coherent (usually) light source. Holographic phase history recording is typically achieved with a two dimensional (planar) sensor such as that provided by optical films. Interferometry between the transmitted reference signal and scattered signal is recorded. An optical holographic image can be recovered through optical or computed inversion approaches.

[1] Part of the full title of the abbreviated title of *The Hobbit,* by J. R. R. Tolkein.

Holographic Imaging Geometries

The relative geometry between an illumination source and the plane of the recording sensor surface defines various types of holographic imaging. The major classes are on-axis (or in-line) and off-axis optical holography. Both are considered extensions of original holographic techniques developed by both Gabor [369] and Upatnieks [576]. These techniques have been described within the context of Fourier optics [392, 393, 395].

Off-Axis Optical Holography

The interference pattern of an off-axis, single frequency, reference wave field with the scattered wave field can be recorded on a film. This interference pattern is then used as a diffraction grating by back-illuminating the planar recording with a wave source of the same frequency to produce the hologram [369, 576].

Multiple Hologram Recording with Sensor View Angle Diversity

Off-axis holography has the ability to record multiple holographic images into the same hologram film. Different images are acquired with illumination produced at different angles. This angular diversity in illumination can lead to tens or even hundreds of images to be recorded on a single film [372].

Simultaneous Source-Receiver Holographic Scanning

Instead of direct capture of the interferogram, holographic data collections can be configure to operate in a mode where the transmitted signal is recorded along with the received signal. This mode moves both transmitter and receiver (to extend angular dwell) during data acquisition and is called *simultaneous source-receiver scanning* [828].

Temporal Diversity in Holographic Imaging

Temporal diversity can be exploited in holography. A single holographic interferogram has been produced with time lapsed data recordings [802]. A double exposure is used to obtain the reference and scattered phase fields. For the first exposure there is no object in the system. The object is placed into the field for the second exposure [674].

Diffusers

A laser beam creates a plane wave. A laser diffuser can be used to create angular diversity of the incident wave field in holographic imaging. The diffuser creates variability in the angle of incidence of rays that "originate" from each point on the diffuser's surface and which converge on each surface point of the object being imaged.

Hence, the hologram can be reconstructed from the range of view angles afforded by the effect of the diffuser. The role of the diffuser is a substitute for relative object rotation as required in plane wave tomography [674] and as employed in holography with angular diversity in illumination. Some view the innate speckle of coherent SAR sensing as a substitute for the diffuser used in holography [681].

Non-Coherent Optical Holography

Holographic imaging with non-coherent light sources has been investigated [1,242]. These holograms function with non-coherent white light because the hologram itself operates as a narrow band filter to create an effective monochromatic (but non-coherent) illumination.

Microwave Holography

Holography can also be applied to millimeter and microwave frequencies [319–321, 460,461,558,635,636,948]. Three dimensional wavefront reconstruction techniques for breast microwave imaging have been investigated [333].

Acoustic Holography

Acoustic holography can be achieved with systems that employ monochromatic plane wave illumination, planar sensors in the near field with angle diversity, and which use pitch-catch mode sensing [514, 673, 674, 993].

Early acoustic holography performed data capture by scanning the scattered wave field's phase and amplitude over a receiving aperture. In ultrasonic applications, "two dimensional" recording apertures can be planar, cylindrical or spherical recording surfaces surrounding the scattering region [713].

Acoustic holographic inversion with echo mode sensing has been called *backward-wave reconstruction* [93]. Both spherical and planar sensors with echo mode sensing have been considered [36, 710, 713]. The planar sensor version requires a full circular sweep in holography [482, 483].

Synthetic aperture acoustic imaging from airborne systems has been studied [877]. Range walk migration summation has been called "multifrequency holography" when applied to acoustic imaging [308].

14.3 Tomographic Imaging

Tomography [106,258,260,459,513,555,674,690] is a cross sectional imaging technique that relies on view angle diversity. The first mathematical development of tomographic imaging was done by Radon[2] in 1917 [756]. The first practical application of Radon's work was done by Bracewell [95] in radio astronomy in the mid 1950s [95]. A subsequent development merged Doppler concepts to obtain radar images of the Moon via "radar aperture synthesis" [430, 744, 940]. This resulted in a form of Doppler beam sharpening that exploited the relative rotation between the Earth and Moon to resolve lunar surface scatterers.

There has been much follow on work that explored tomographic imaging in many different applications, especially in medical imaging [51, 57, 67, 206, 207, 245, 266, 411, 430, 457, 487, 511, 550, 718, 732, 744, 757, 758, 802, 856, 940, 946, 947, 962].

[2]Johann Karl August Radon (1887-1956) was an Austrian mathematician.

Clinical tomographic medical imaging prototypes first appeared in the early 1960s [206, 207], with the systems becoming mainstream in the early 1970s.

X-ray tomography is achieved by passing parallel sets of narrow x-ray beams through an object and detecting the attenuated x-ray beams with a plate of x-ray detectors [257].

Optical tomography uses light pulses to probe an object in biomedical imaging applications [20]. Optical tomographic sensors can be built to measure time of flight (on a femto-second time scale) or measure frequency, and can be built to be polarization and phase sensitive.

Microwave medical tomographic imaging has also been investigated [819].

Ray and Diffraction Tomography

There are two major classes of tomographic sensor systems. One implements *ray tomography,* the other implements *diffraction tomography.* "Ray tomography" and "diffraction tomography" terminology is used to distinguished between computed imaging sensors with short wavelengths (x-ray) and sensors with comparatively long wavelengths (e.g., ultrasonic acoustic waves) [517].

Ray Tomography

Non-diffracting x-rays travel in straight lines. Ray tomography gathers signal attenuation measurements characterized by line integrals over straight paths [730]. The integrative attenuation factors are described in terms of "projections" [928]. Ray tomography is a problem of projection or straight line integral inversion [399, 401, 405, 519]. The theoretical foundation of ray tomography is the Radon transform, or equivalently, a filtered and baseband version of the Fourier projection theorem.

Short versus Long Wavelength Tomography

When wavelengths become large, wave travel through matter is not along straight lines. Waves are refracted or diffracted as the wave propagates through the object being scanned. With microwave or ultrasound imaging problems, wave-matter interactions such as diffraction, refraction and even multiple bounce scattering can become dominant [837]. In these cases, the straight line model of ray tomography will fail, along with ray tomographic reconstruction techniques [255, 413, 414, 510, 560]. With long wavelengths, wave-matter interactions can be modeled with the more general framework of diffraction theory.

Diffraction Tomography

Tomographic imaging with ultrasound is a problem of inverse diffraction [149, 509, 514, 515, 891]. The extension of ray tomography to diffraction tomography moves beyond principles of the Fourier projection theorem and is based on solutions of wave motion equations [177, 183, 184, 257, 417, 481, 482, 641, 674, 679, 692, 745, 919, 919, 934, 992, 993]. The Fourier diffraction theorem is part of the theoretical basis of diffraction tomography.

Tomographic Algorithm Overview

The family of tomographic algorithms is very broad, including both spatial [469] and wavenumber domain approaches [727].

Convolution Back-Projection and Convolutional Back-Propagation

The major spatial domain algorithm for short wavelength ray tomography is the *convolutional back-projection* (CBj) algorithm [37,255,257,581,690,730,898].

The major spatial domain algorithm for long wavelength diffraction tomography is the *convolutional back-propagation* (CBp) algorithm [252,253,255,257,912].

CBp is a generalization of CBj, where the straight line contours of back-projected data is generalized to circular contours [4,4,252,252,257,509,557,560, 757,843].

Direct and Convolutional Back-Propagation

There are two forms of back-propagation. We label one as "direct" and the other as "convolutional." CBp algorithms are different than the DBp (direct back-propagation) algorithms of Section 12.3. Careful use of these adjectives helps minimize confusion. CBj, CBp and DBp are all tomographic inversion algorithms performed in the spatial domain.

Fourier Reconstructive Tomography

There is a large class of tomographic algorithms performed in k-space that we choose to call *Fourier reconstruction* (FR) algorithms. Both polar formatting and Stolt formatting are Fourier reconstruction methods. Fourier reconstruction methods have been extensively explored in a variety of computed imaging problems [95,169,187,245,700,730,738,831,896,898,927,947].

The two major classes of spatial domain computed tomography algorithms, the convolutional back-projection and convolutional back-propagation algorithms, with their baseband and passband forms, each have corresponding *filtered Fourier reconstruction* (fFR) equivalents.

There are Fourier reconstruction methods that are implemented without filtering, we call these *direct Fourier reconstruction* (dFR) methods. Direct Fourier reconstruction methods include the k-space equivalent of spatial domain direct back-propagation.

The theoretical foundations used in the development of various tomographic Fourier reconstruction methods (both direct and filtered) are

- k-space descriptions of Hilbert and Radon transforms,
- the Fourier projection theorem and
- the Fourier diffraction theorem.

Fourier Reconstructive Holography

When the k-space manifold of the passband form of the Fourier diffraction theorem is generalized to the Ewald sphere, then the Fourier reconstruction problem steps out of the domain of tomography and into the realm of holography. Tomographic FR methods are the two dimensional versions of more general three dimensional holographic FR methods.

Spatial and Wavenumber Domain "Traces" and Contours"

CBj and CBp involve reconstructive straight line and circular traces in the spatial domain. The Fourier reconstruction methods also involve straight line and circular traces, but in the wavenumber domain. Arbitrarily shaped traces or contours used in Fourier reconstruction methods are addressed in [519, 858, 896].

"Projections" and "Propagations"

The term "projection" is often interchanged with the term "propagation" in the literature of computed imaging [258, 493, 713]. The term "projection" originates in the application domain of ray tomography with its straight line integral. But more generally, the term "projection" is often used in a causal way in the literature to describe any general line integral trace or k-space curvilinear manifold contour, whatever the shape of the "projection."

For spatial domain algorithms, it may be best to use the term "projections" for straight line integrals related to planar wavefronts and use the term "propagations" for circular line integrals related to two dimensional slices of back-propagated spherical wavefronts (i.e., circular wavefronts).

We seek to consistently use the term "projection" for linear traces and the term "propagation" for circular traces to distinguish among various spatial domain convolutional algorithms.

We suggest that "projection" and "propagation" not be used to describe elements of Fourier reconstruction algorithms. We recommend that, generally, the terms of "trace," "contour" or "manifold" be used instead for k-space manifolds of data support. It may be sometimes appropriate to use the term "projections" for straight line contours associated with the Fourier projection theorem as applied in some wavenumber domain algorithms.

Computed Tomography Algorithm Taxonomy

Taxonomies of tomographic algorithms can be found in [400, 556]. The major subclasses of computed tomographic (CT) algorithms are the following:

- direct Fourier reconstruction with k-space linear contours ($\mathrm{dFR_l}$),
- direct Fourier reconstruction with k-space circular contours ($\mathrm{dFR_c}$),
- convolutional back-projection (CBj) with spatial linear contours,
- convolutional back-propagation (CBp) with spatial circular contours,
- filtered Fourier reconstruction with k-space linear contours ($\mathrm{fFR_l}$) and

- filtered Fourier reconstruction with k-space circular contours ($\mathrm{fFR_c}$).

Spatial Tomography and SAR

Relationships between various approaches to cross-sectional tomographic imaging and two dimensional SAR have been studied by many [158, 163, 323, 490, 538, 623, 676, 677, 686, 837]. Both CBj and CBp algorithms have been adapted for SAR. The adaption and application of parallel beam (linear projection) tomographic processing to SAR seems to originate in [679].

Direct and Filtered Fourier Reconstruction Algorithm Taxonomy

The sometimes overlooked challenge in adapting CBj/CBp and dFR/fFR algorithms to SAR is the difference between algorithms designed for non-coherent (baseband) sensors and algorithms designed for coherent (passband) sensors [250, 679]. The major subclasses of Fourier reconstruction algorithms used in both baseband and passband sensor systems, including the k-space equivalents of CBj and CBp, can be categorized as follows.

- Direct Fourier reconstruction on straight contours at baseband ($\mathrm{dFR_{l@b}}$)
  - corresponds to a baseband form of the Fourier projection theorem.
- Direct Fourier reconstruction on straight contours at passband ($\mathrm{dFR_{l@p}}$)
  - corresponds to polar formatting in SAR.
- Direct Fourier reconstruction on circular contours at baseband ($\mathrm{dFR_{c@b}}$)
  - corresponds to a baseband form of the Fourier diffraction theorem.
- Direct Fourier reconstruction on circular contours at passband ($\mathrm{dFR_{c@p}}$)
  - corresponds to Stolt formatting in SAR.
- Filtered Fourier reconstruction on straight contours at baseband ($\mathrm{fFR_{l@b}}$)
  - corresponds to CBj for ray tomography, but implemented in k-space.
- Filtered Fourier reconstruction on straight contours at passband ($\mathrm{fFR_{l@p}}$)
  - corresponds to CBj adapted for SAR, but implemented in k-space.
- Filtered Fourier reconstruction on circular contours at baseband ($\mathrm{fFR_{c@b}}$)
  - corresponds to CBp for diffraction tomography, but implemented in k-space.
- Filtered Fourier reconstruction on circular contours at passband ($\mathrm{fFR_{c@p}}$)
  - corresponds to CBp adapted for SAR and can be viewed as a form of Stolt formatting.

The filtered versions of these algorithms seek to improve performance by using optimized wavenumber domain weight functions [328,484]. These weight functions are the primary difference between "direct" and "filtered" versions of Fourier reconstruction methods.

Designs of weights or filters for filtered Fourier reconstructions are motivated by various theoretical considerations, including:

- wavenumber domain multiplicative "weights" corresponding to spatial domain convolutional filters derived from the Radon transform and
- wavenumber domain interpolation kernels for Fourier migrations prescribed by analysis of Fourier reconstructions under a change of variables transformation.

Fourier reconstructions with irregularly or non-uniformly spaced wavenumber domain samples is a major theme within the dFR and fFR classes of algorithms [329].

14.4 Direct Fourier Reconstruction

Fourier Projection Theorem

The *Fourier projection theorem* [95,454,647,817], also known as the Fourier slice theorem or projection slice theorem [462,588,645,674], establishes that the one dimensional Fourier transform of a (straight line) *projection* of a two dimensional function is equal to a (straight line) "slice" of the two dimensional Fourier transform of the function.

To establish this theorem, we begin by revisiting the definition of the Fourier transform, but expressed in terms of unrotated and rotated coordinate systems.

Rotated Cartesian Coordinate Systems

Let the system (x', y') be equivalent to the scene centered[3] Cartesian coordinate system (x, y) rotated by an angle $\theta \in [0, 2\pi)$. Given a point (x, y) in the unrotated system, the corresponding coordinate (x', y') in the rotated coordinate system is [260]

$$\begin{bmatrix} x' \\ y' \end{bmatrix} = \begin{bmatrix} +\cos\theta & +\sin\theta \\ -\sin\theta & +\cos\theta \end{bmatrix} \cdot \begin{bmatrix} x \\ y \end{bmatrix}.$$

The inverse transformation that takes the rotated coordinates back to the unrotated system is

$$\begin{bmatrix} x \\ y \end{bmatrix} = \begin{bmatrix} +\cos\theta & -\sin\theta \\ +\sin\theta & +\cos\theta \end{bmatrix} \cdot \begin{bmatrix} x' \\ y' \end{bmatrix}.$$

Two Dimensional Fourier Transform in Cartesian Coordinates

The Fourier transform of a two dimensional terrain function $gg(x, y)$ in terms of rectilinear parameters in an (unrotated) Cartesian coordinate system is

$$GG(k_x, k_y) = \int_{-\infty}^{+\infty} \int_{-\infty}^{+\infty} gg(x, y) \cdot \exp\left[-j\left(k_x x + k_y y\right)\right] \cdot dxdy.$$

[3] In this chapter, we drop the caret "$\hat{x}$" to simplify notation for scene centered parameters.

The inverse Fourier transform in the (unrotated) coordinate system is

$$gg(x, y) = \frac{1}{2\pi} \int_{-\infty}^{+\infty} \int_{-\infty}^{+\infty} GG(k_x, k_y) \cdot \exp\left[+j\left(xk_x + yk_y\right)\right] \cdot dk_x dk_y.$$

Two Dimensional Fourier Transforms in Rotated Coordinate System

The Fourier transform of a two dimensional function in the rotated coordinate system is

$$GG(k_{x'}, k_{y'}) = \int_{-\infty}^{\infty} \int_{-\infty}^{\infty} gg(x', y') \cdot \exp\left[-j\left(k_{x'}x' + k_{y'}y'\right)\right] \cdot dx'dy'.$$

The inverse Fourier transform in the rotated coordinate system is

$$gg(x', y') = \frac{1}{2\pi} \int_{-\infty}^{+\infty} \int_{-\infty}^{+\infty} GG(k'_x, k'_y) \cdot \exp\left[+j\left(xk'_x + yk'_x\right)\right] \cdot dk'_x dk'_y.$$

Projections

A linear projection is defined as a straight line integral at an angle θ through a two dimensional function $gg(x, y)$,

$$p_\theta(l) = \int_{\text{line}} gg(x, y) \cdot dl.$$

This projection can be expressed in various forms, if the angled line corresponds to the angle of the rotated coordinate system, then we can write

$$p_\theta(x') = \int_{-\infty}^{+\infty} gg(x', y') \cdot dy'.$$

The one dimensional function $p_\theta(x')$ expressed in terms of points along the rotated axis x' for a fixed value of θ defines the integrative projection along y' [446].

Range Profiles as Projections in SAR

A primary difference between the application of the Fourier projection theorem in ray tomography compared to its applications in SAR is that the integrative straight line projection is perpendicular to the direction of wave travel in SAR (reflected energy is integrated along the wavefront) [679], whereas in ray tomography, the integrative straight line projection is along the line of x-ray travel (integration represents accumulated attenuation of the penetrative x-ray) [509].

Projections and Range Profiles

When applied to radar, the line integral of a projection is defined in a direction orthogonal to the line of sight of the radar. The data sample at x' of a projection in SAR is equivalent to the range bin data sample at x' of a range profile collected by a single pulse. At far ranges, the range profile projection corresponds to the aggregate return of a cross-range line of scatterers under a nearly planar wavefront that propagates across the scene. Shown in Fig. 14.1a, for example, is a range profile created by a planar wavefront that propagates along the direction of the rotated x' axis. The Fraunhofer range profile

$$r_\theta(x') \quad = \quad \int_{-Y_o'}^{+Y_o'} gg(x',y') \cdot dy'$$

is an integrative measure of terrain scattering as sensed along lines perpendicular to x' and parallel to y'. The range profile or projection $r_\theta(x')$ is a set of line integrals along y' at each of the down-range points x'.

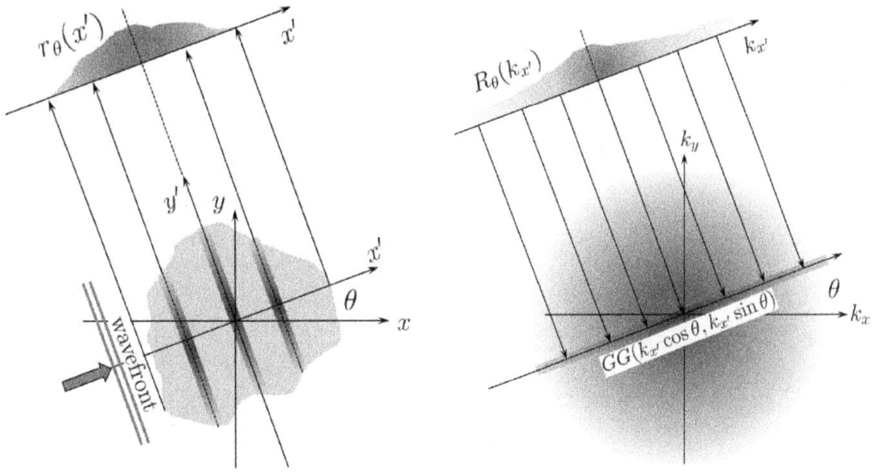

a. Range profile at angle θ. b. Fourier transform of range profile.

Figure 14.1: Fourier projection theorem in SAR.

When employed in time-space domain phase history models, the integrative measure $r_\theta(x')$ over a continuous terrain function can be used to replace the discrete scatterer reflectivity coefficient g_n, e.g.,

$$g_n \cdot w\left(t - \frac{2}{c}x'\right) \mapsto r_\theta(x') \cdot w\left(t - \frac{2}{c}x'\right).$$

Fourier Transforms of Projections

In ray tomography, Fourier data are obtained by computing one dimensional Fourier transforms of projection data. In SAR, Fourier data can often be directly obtained through stretch processing in both fast-time and in slow-time [679]. Otherwise, in SAR, the Fourier transform of a Fraunhofer range profile projection in the rotated coordinate system is required,

$$R_\theta(k_{x'}) \quad = \quad \int_{-X_o'}^{+X_o'} r_\theta(x') \cdot \exp(-jk_{x'}x') \cdot dx'.$$

If $r_\theta(x')$ has finite support in the x' dimension (which is always the case in physical systems), then the Fourier transform $R_\theta(k_{x'})$ has unbounded support in the $k_{x'}$ dimension. A projection with finite support is illustrated in Fig. 14.1a. The Fourier transform of this projection and its unbounded support is conceptually illustrated in Fig. 14.1b.

Fourier Projection Theorem Proof

The Fourier projection theorem establishes that

$$R_\theta(k_{x'}) = GG(k_{x'}\cos\theta, k_{x'}\sin\theta).$$

This is proved by considering,

$$\begin{aligned} R_\theta(k_{x'}) &= \int_{-X_o'}^{+X_o'} r_\theta(x')\cdot\exp(-jk_{x'}x')\cdot dx', \\ &= \int_{-X_o'}^{+X_o'}\int_{-Y_o'}^{+Y_o'} gg(x',y')\cdot dy'\cdot\exp(-jk_{x'}x')\cdot dx', \\ &= \int_{-X_o'}^{+X_o'}\int_{-Y_o'}^{+Y_o'} gg(x',y')\cdot\exp[-jk_{x'}\cdot(x\cos\theta+y\sin\theta)]\cdot dy'dx', \end{aligned}$$

which is the same as

$$\begin{aligned} &R_\theta(k_{x'}) \\ &= \int_{-X_o}^{+X_o}\int_{-Y_o}^{+Y_o} gg(x,y)\cdot\exp(-jx\cdot k_{x'}\cos\theta)\exp(-jy\cdot k_{x'}\sin\theta)\cdot dxdy, \\ &= GG(k_{x'}\cos\theta, k_{x'}\sin\theta). \end{aligned}$$

The Fourier transform $R_\theta(k_{x'})$ of the one dimensional range profile $r_\theta(x')$ obtained from $gg(x',y')$ at an angle θ is equal to the one dimensional slice

$$GG(k_{x'}\cos\theta, k_{x'}\sin\theta)$$

at the same angle of the two dimensional Fourier transform $GG(k_x,k_y)$ of $gg(x,y)$.

Fourier Projection Theorem and Baseband Sensors

Assume the range profile $r_\theta(x')$ and range profile spectrum $R_\theta(k_{x'})$ are comprised of baseband data. If range profile projections from all view angles are available, the baseband terrain spectrum $GG(k_x,k_y)$ can be reconstructed from an angular sweep of the Fourier transforms $R_\theta(k_{x'}) = GG(k_{x'}\cos\theta, k_{x'}\sin\theta)$ of the projections $r_\theta(x')$. The process of reconstructing the two dimensional terrain spectrum $GG(k_x,k_y)$ with an angular sweep of baseband projection spectra $GG(k_{x'}\cos\theta, k_{x'}\sin\theta)$ is illustrated for two projections in Fig. 14.2b.

Fourier Projection Theorem and Passband Sensors

SAR sensors are passband systems. Hence, application of the Fourier projection theorem must be adapted to a passband format in developing algorithms for

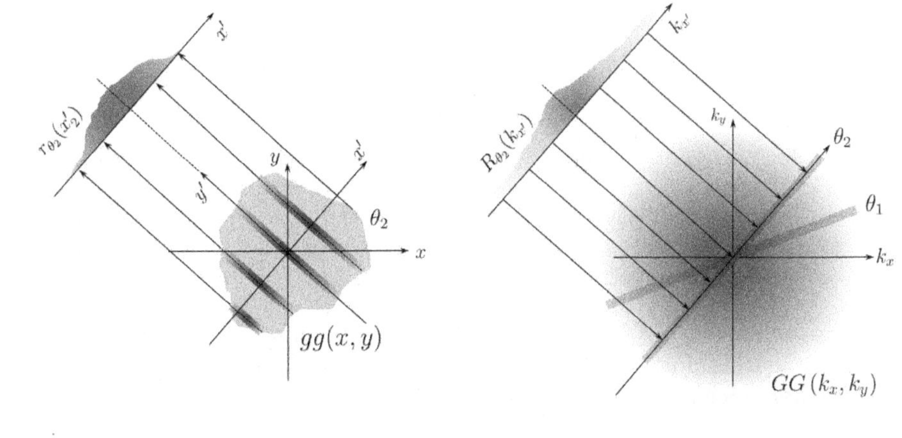

a. Projection at angle θ_2 b. Two projection reconstruction.

Figure 14.2: Construction of terrain spectrum from multiple projections.

SAR. As shown in Fig. 14.3, the portion of the terrain spectrum that can be reconstructed with data gathered by passband polar format SAR projections is limited to just the passband of the sensor and the pivot point angular dwell Θ_c afforded by movement of the sensor across the synthetic aperture. Because of the coherent nature of SAR, it is possible to have the wavenumber support annulus reduce to a single monochromatic circle if a continuous wave signal is used [168]. Furthermore, the coherent nature of SAR allows reconstructions with polar manifolds of relatively small angular sweep.

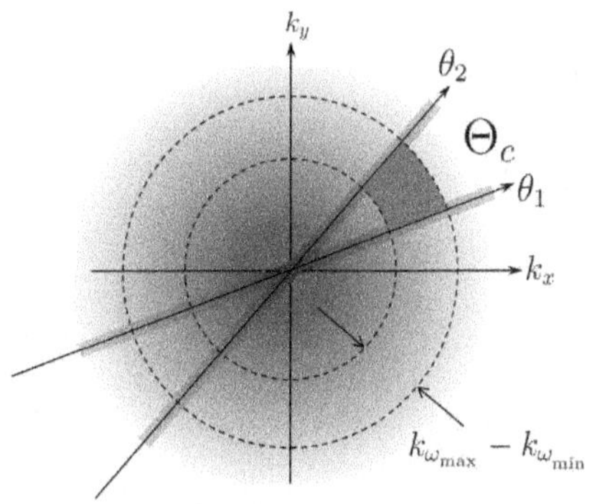

Figure 14.3: Fourier projection theorem manifold of passband sensors.

Fourier Diffraction Theorem

The Fourier diffraction theorem describes Fourier manifolds for

- the case of arbitrary (non-planar) wave incidence and wave scattering that occurs in all directions, and

- data collected on an infinitely long linear aperture with an angular orientation of θ.

In contrasting the Fourier projection and Fourier diffraction theorems, we see that

- the (baseband) Fourier projection theorem requires a finite linear array to gather data measurements and hence the linear trace in the wavenumber domain is unbounded as shown in Figs. 14.1a-b, and
- the (baseband) Fourier diffraction theorem requires an infinite linear array to gather data measurements and produces a bounded semicircle wavenumber trace as shown in Figs. 14.4a-b.

Fourier Diffraction Theorem Baseband Manifold

The baseband version of the Fourier diffraction theorem establishes that when an object $gg(x, y)$ is illuminated with a general (not necessarily planar) incident waveform, the Fourier transform of the (isotropically) forward scattered field, as measured on an infinite linear sensor array oriented at an angle of θ as shown in Fig. 14.4a, gives values of the two dimensional Fourier transform $GG(k_x, k_y)$ of the object along a semicircular trace as shown in Fig. 14.4b.

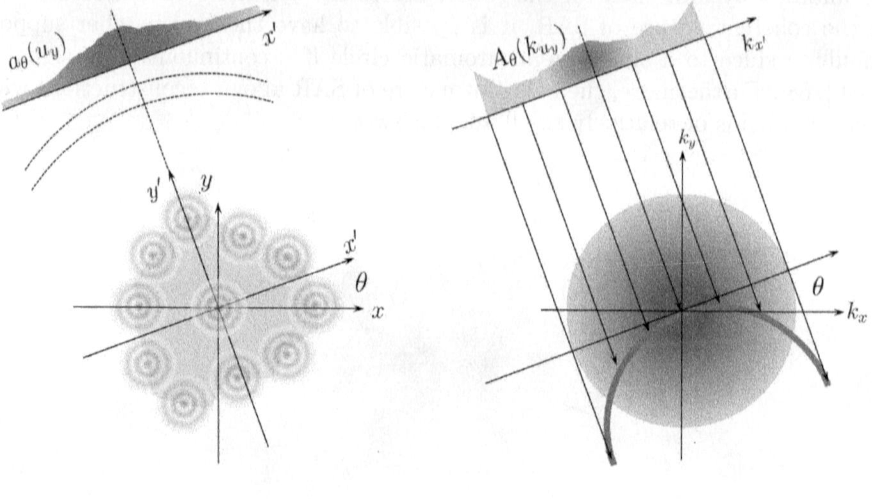

a. Infinite linear measurement. b. Finite baseband manifold.

Figure 14.4: Baseband Fourier diffraction theorem of forward scattered field.

Fourier Diffraction Theorem and Baseband Sensors

In medical imaging, the non-coherent nature of some sensors locates the semicircle wavenumber trace such that it intersects the origin in (k'_x, k'_y)-space [256,837,928].

The position of the semicircle trace shown in Fig. 14.4b is for a baseband sensor capturing the forward scattered field. Since the support of the data measurement domain along the infinitely long linear array of Fig. 14.4a is unbounded, the Fourier transform of the measured data is finite and limited to within the radius of the semicircle trace. The radius of this semicircle manifold is determined by the frequency of the monochromatic signal that is received and detected by non-coherent means.

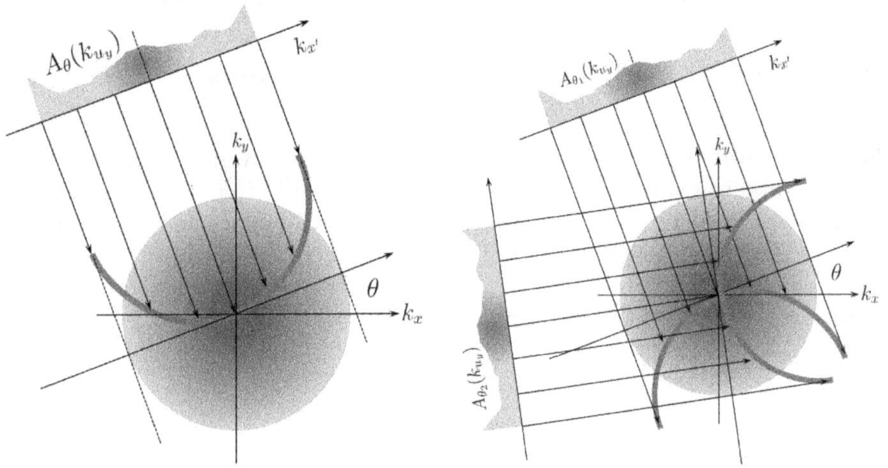

a. Single look baseband manifold. b. Multiple look baseband manifolds.

Figure 14.5: Baseband Fourier diffraction theorem of backward scattered field with angular diversity.

The Fourier diffraction theorem also has a pulse echo version that describes the data captured from the backward scattered field. The position of the semicircle manifold for the backward scattered field collected with a non-coherent sensor is shown in Fig. 14.5a.

Forward Scattered and Backward Scattered Wave Fields

In non-coherent sensor systems operating in pitch-catch mode, the semicircular trace (achieved by Fourier migration) of the Fourier transform of data collected with an infinite linear array is defined as

$$\breve{k}_x(\omega_c, k_y) = k_{\omega_c} - \sqrt{k_{\omega_c}^2 - k_y^2}.$$

The temporal angular frequency wavenumber k_{ω_c} determines the nominal speed of (e.g., acoustic wave) propagation within the medium surrounding the object. The Stolt mapping $-\sqrt{k_{\omega_c}^2 - k_{y'}^2}$ is negative because the imaging modality is transmissive and the forward scattered field is captured (i.e., the left semicircle manifold of a slice of the Ewald sphere is traced and down converted by a shift to the origin of the k-space domain.).

In echo mode baseband sensor systems the circular trace is

$$\breve{k}_x(\omega_c, k_y) = +\sqrt{k_{\omega_c}^2 - k_y^2} - k_{\omega_c}. \tag{14.1}$$

The one way Stolt mapping $+\sqrt{k_{\omega_c}^2 - k_y^2}$ is positive because the backward scattered field is captured (i.e., the right semicircle manifold of the Ewald sphere is traced and down converted by a shift to the origin of the k-space domain.).

Fourier Diffraction Theorem and Baseband Sensors with Angular Diversity

With transmissive non-coherent sensors [482], angular diversity in the imaging geometry gathers Fourier domain traces as illustrated in Fig. 14.5b. Tomographic dFR inversion on this manifold is achieved by employing an interpolation procedure to resample the non-uniformly spaced k-space data with a rectilinear sampling grid and then performing a two dimensional inverse Fourier transform [907]. A baseband version of the Fourier diffraction theorem can be viewed as the k-space basis of CBp algorithms [252].

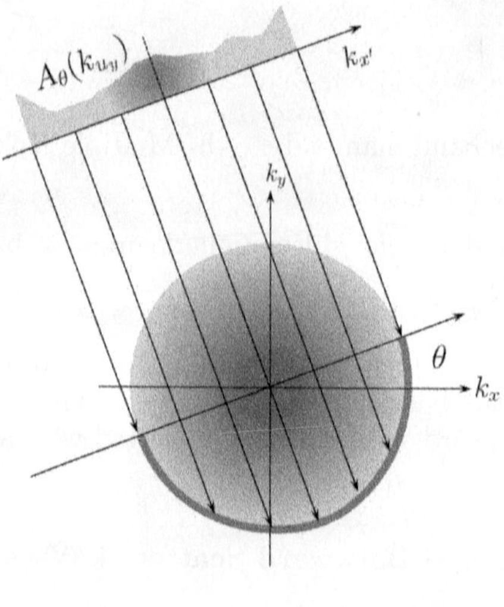

Figure 14.6: Fourier diffraction theorem manifold of passband sensor.

Fourier Diffraction Theorem and Passband Sensors

It has been noted [406] that because of similar methods in ray tomography, there was a common practice in the application of Stolt formatting methods in early SAR research to down convert the SAR passband signal to baseband by defining the Stolt mapping as

$$\breve{k}_x(\omega_p, k_{u_y}) = 2\sqrt{k_{\omega_p}^2 - k_{u_y}^2} - 2k_{\omega_c}.$$

This SAR Stolt mapping corresponds to the baseband Stolt mapping of (14.1) used with echo mode, non-coherent sensor systems. Typically, the shift to baseband of the SAR Stolt mapping was viewed as not substantively changing the implementation of Stolt formatting, but only changing the axis against which the $\breve{k}_x$ wavenumber data are plotted. But this is really not the case, as can be seen if the semicircle manifolds of baseband and passband systems are compared when angular diversity exists. Angular diversity with a baseband system "fills in" the Ewald sphere with wavenumber data measurements as illustrated in Fig. 14.5b [482, 837]. Angular diversity with a passband system does not expand the wavenumber data support beyond the passband shell of the Ewald sphere as illustrated in Figs. 14.6 and 14.7d. In the passband system, there is no extension or spread of the radial wavenumber with diversity of view angle.

Fourier Projection and Fourier Diffraction Theorems with Baseband Sensors

In the limit of vanishing wavelength (i.e., in the limit of increasing frequency), the Fourier diffraction theorem's semicircular manifolds of baseband sensor systems relax into the straight line manifolds of the Fourier projection theorem [6,255,256]. That is, in the limit of small wave length, the semicircle arcs of Fig. 14.7b near the k-space origin warp to straight lines passing through the origin as shown in Fig. 14.7a [256]. Hence, diffraction tomography is said to "reduce" to ray tomography in the short wavelength limit.

Based on this observation, the Fourier diffraction theorem is sometimes viewed as related to the CBp algorithm [252]. $\mathrm{FR_c}$ methods (direct and filtered) of diffraction tomography are also sometimes viewed as generalizations of the $\mathrm{FR_l}$ (direct and filtered) methods of ray tomography [647].

Direct Fourier Reconstructions on Linear Contours at Baseband

The Fourier projection theorem establishes that the Fourier transform of a projection gathered at an angle θ is a radial slice through the two dimensional Fourier transform of the scene along the same direction θ [690]. Direct Fourier reconstruction algorithms with baseband data rasters along linear contours are based on a baseband form of the Fourier projection theorem. Direct Fourier reconstruction on linear contours at baseband takes the Fourier transform of each projection and formats the data samples onto a polar raster as conceptually illustrated in Fig. 14.7a. An angular sweep of polar raster of data is interpolated to a rectilinear grid and then a two dimensional inverse Fourier transform yields the image. The algorithms of this class are sometimes called Fourier-Hankel methods [12,324,448].

Direct Fourier Reconstructions on Linear Contours at Passband

Spotlight mode SAR with polar format processing can be modeled as acquiring passband Fourier data as shown in Fig. 14.7c [187].

The first to recognize the relationship of spotlight mode SAR to the Fourier projection theorem seems to be Brown [115], in his extension of the work of Walker [976]. Polar formatting has also been considered for non-SAR tomographic applications based on dFR [906, 907].

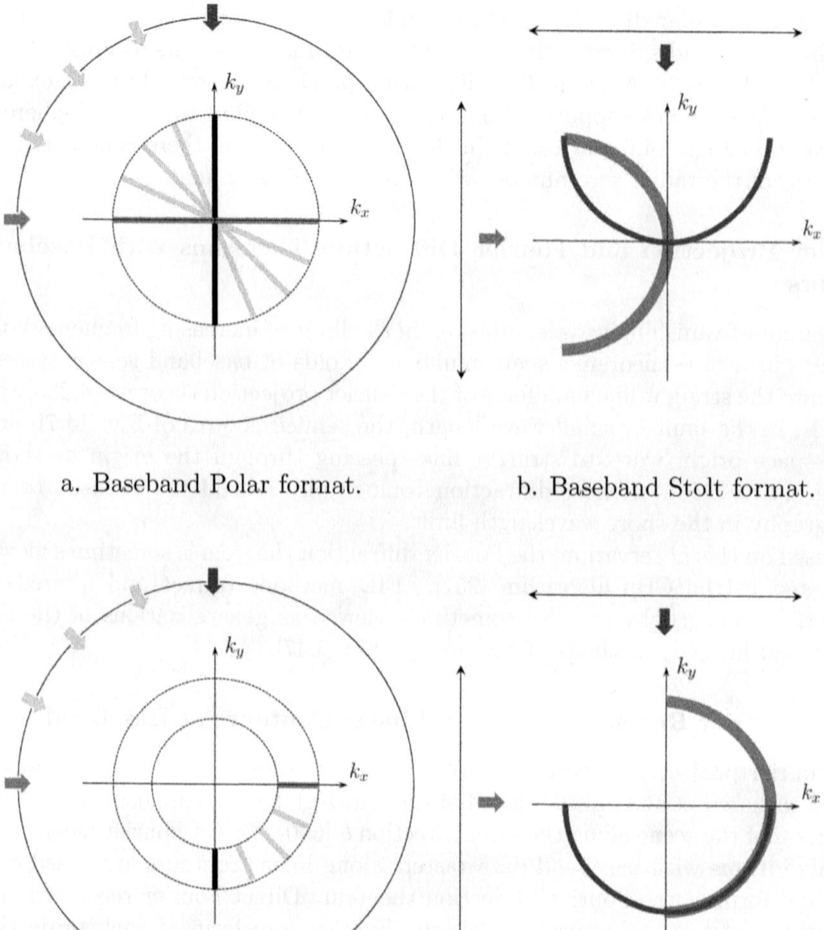

a. Baseband Polar format.

b. Baseband Stolt format.

c. Passband Polar format.

d. Passband Stolt format.

Figure 14.7: Stolt and polar formats in passband and baseband forms.

Direct Fourier Reconstructions on Circular Contours at Baseband

Diffraction tomography is described by semicircle traces in k-space [468, 858]. Through Fourier migration, linear array sensor data recording produces a semicircle trace of data in the Fourier domain. The position of the semicircular trace in the wavenumber domain depends on whether the sensor is a non-coherent baseband or a coherent passband sensor, and whether the sensor system is capturing backward scatter (as in echo mode imaging) [38, 555, 928] or forward scatter (as in the pitch-catch mode of some forms of medical sensing) [509, 891]. Basebanded angular spectrum semicircles are obtained from baseband sensors as shown in Fig. 14.7b. The corresponding FR imaging problem is reconstructing the image spectrum from a sweep of basebanded semicircular arcs [256].

Direct Fourier Reconstructions on Circular Contours at Passband

Direct Fourier reconstruction at passband guided by the Fourier diffraction theorem can be described as reconstructions on circular contours at passband as shown in Fig. 14.6d. The Fourier projection theorem may seem a specialization of the Fourier diffraction theorem [557, 560, 730], but this association is problematic with passband sensor systems, where the passband shell of the Ewald sphere is not shifted to bisect the origin of k-space.

14.5 Convolutional Spatial Reconstruction

Convolutional back-projection is based on the Radon transform. To be able to formulate the inverse Radon transform, we need the Hilbert transform.

Hilbert Transform

The Hilbert transform is a tool used to generate complex *analytic* (single sideband) representations of passband signals. An analytic signal is a complex signal with its real part equal to a given real signal and its imaginary part equal to the Hilbert transform of the real signal [309].

Real and Causal Signals are Analytic Real and causal signals (that "start" at time 0) can be formulated as analytic signals.

Single Sided Up Converted Signals are Analytic Up converted (passband) forms of analytic baseband signals are analytic and have single sided spectra.

Coherent Receivers and Analytic Signals A coherent receiver generates a complex analytic representation of a real passband signal (see Section 2.8).

Forward Hilbert Transform

Given the real signal $s_{\rm rl}(t)$, the forward Hilbert[4] transform outputs a quadrature phase signal $s_{\rm qd}(t)$,

$$s_{\rm qd}(t) = \frac{1}{\pi} \int_{-\infty}^{+\infty} s_{\rm rl}(\tau) \cdot \frac{1}{t-\tau} \cdot d\tau. \tag{14.2}$$

The Hilbert transform $s_{\rm qd}(t)$ is the convolution of the real signal $s_{\rm rl}(t)$ with $1/\pi t$,

$$s_{\rm qd}(t) = s_{\rm rl}(t) * \frac{1}{\pi t}.$$

The Hilbert transform shifts the phase of the real signal $s_{\rm rl}(t)$ by $\pi/2$ to produce $j \cdot s_{\rm qd}(t)$ in phase quadrature. Hence, $j \cdot s_{\rm qd}(t)$ is purely imaginary.

There is a singularity in the integrand of (14.2), so the forward Hilbert transform is more properly expressed as

$$s_{\rm qd}(t) = \lim_{\epsilon \mapsto 0} \left[\frac{1}{\pi} \int_{-\infty}^{t-\epsilon} s_{\rm rl}(\tau) \cdot \frac{1}{t-\tau} \cdot d\tau + \frac{1}{\pi} \int_{t+\epsilon}^{+\infty} s_{\rm rl}(\tau) \cdot \frac{1}{t-\tau} \cdot d\tau \right].$$

Single Sided Analytic Signal Spectrum

Let a signal in analytic form be written as

$$s_{\rm an}(t) = s_{\rm rl}(t) + j s_{\rm qd}(t).$$

Let the Fourier transform of $s_{\rm an}(t)$ be

$$S_{\rm an}(\omega_p) = S_{\rm rl}(\omega_p) + j S_{\rm qd}(\omega_p).$$

Fourier Transform of Hilbert Transformed Signal

The Fourier transform of the quadrature element obtained from the Hilbert transform is

$$S_{\rm qd}(\omega_p) = S_{\rm rl}(\omega_p) \cdot HT(\omega_p),$$

where

$$\begin{aligned} HT(\omega_p) &= \mathsf{FT}_{\rm 1D} \left\{ \frac{1}{\pi t} \right\}, \\ &= -j \cdot \mathsf{sgn}(\omega_p) \end{aligned}$$

and where the signum function $\mathsf{sgn}(\omega_p)$ is

$$\mathsf{sgn}(\omega_p) = \begin{cases} 1 & \omega_p > 0, \\ 0 & \omega_p = 0, \\ -1 & \omega_p < 0. \end{cases}$$

Since

$$S_{\rm qd}(\omega_p) = -j \cdot \mathsf{sgn}(\omega_p) \cdot S_{\rm rl}(\omega_p),$$

[4]David Hilbert (1862-1943) was a German mathematician.

we have [309]

$$S_{\rm an}(\omega_p) = \begin{cases} 2S_{\rm rl}(\omega_p) & \omega_p > 0, \\ S_{\rm rl}(\omega_p) & \omega_p = 0, \\ 0 & \omega_p < 0. \end{cases}$$

Analytic signals $s_{\rm an}(t)$ have single sided spectra equal to essentially twice the positive frequency part of the spectrum of the real signal $s_{\rm rl}(t)$.

Inverse Hilbert Transform

The inverse Hilbert transform is

$$\begin{aligned} s_{\rm rl}(t) &= \frac{1}{\pi}\int_{-\infty}^{+\infty} s_{\rm qd}(\tau)\cdot\frac{-1}{t-\tau}\cdot d\tau, \\ &= s_{\rm qd}(t) * \frac{-1}{\pi t}. \end{aligned}$$

Analytic Expressions in Magnitude-Phase Form

Analytic signals can be expressed in a magnitude-phase form [309],

$$s_{\rm an}(t) = |s_{\rm an}(t)|\cdot\exp[j\angle s_{\rm an}(t)],$$

where

$$\begin{aligned} |s_{\rm an}(t)| &= \sqrt{s_{\rm rl}^2(t) + s_{\rm qd}^2(t)}, \\ \angle s_{\rm an}(t) &= \arctan\left[\frac{s_{\rm qd}(t)}{s_{\rm rl}(t)}\right]. \end{aligned}$$

Polar Coordinates in Spatial and Wavenumber Domains

Within the problem domain of CBj algorithms, it is sometimes useful to formulate a spatial two dimensional Fourier transform in terms of polar coordinates. We begin by defining the relationships between polar and rectilinear parameters in spatial and wavenumber domains.

The radial range r and angle θ of a polar description can be specified in terms of the Cartesian coordinate system parameters (x, y),

$$\begin{aligned} r &= \sqrt{x^2 + y^2}, \\ \theta &= \arctan\left(\frac{y}{x}\right). \end{aligned}$$

The wavenumbers corresponding to the Cartesian system are k_x and k_y. The associated *radial wavenumber* k_r and *wavenumber angle* ϕ of the polar Fourier domain are

$$\begin{aligned} k_r &= \sqrt{k_x^2 + k_y^2}, \\ \phi &= \arctan\left(\frac{k_y}{k_x}\right). \end{aligned}$$

The inverse relationships are

$$\begin{aligned} x &= r\cos\theta, \\ y &= r\sin\theta, \end{aligned}$$

and

$$\begin{aligned} k_x &= k_r\cos\phi, \\ k_y &= k_r\sin\phi. \end{aligned}$$

Note that [734]

$$\begin{aligned} k_x \cdot x + k_y \cdot y &= k_r \cdot r\cos\theta\cos\phi + k_r \cdot r\sin\theta\sin\phi, \\ &= k_r \cdot r\cos(\theta - \phi). \end{aligned} \tag{14.3}$$

Forward Fourier Transform in Polar Coordinates

From (14.3), the forward Fourier transform expressed in polar coordinates is

$$GG(k_r, \phi) = \int_0^{2\pi}\int_0^{\infty} gg(r,\theta) \cdot \exp[-jk_r \cdot r\cos(\theta - \phi)] \cdot r \cdot dr d\theta.$$

Forward Fourier Transform from Polar to Cartesian Coordinates

The polar form of the forward Fourier transform that maps to a Cartesian wave-number domain is

$$\begin{aligned} &GG(k_x, k_y) \\ &= GG(k_r\cos\phi, k_r\sin\phi), \\ &= \int_0^{2\pi}\int_{-\infty}^{+\infty} gg(r,\theta) \cdot \exp[-j(r\cos\theta \cdot k_r\cos\phi + r\sin\theta \cdot k_r\sin\phi)] \cdot r \cdot dr d\theta, \\ &= \int_0^{2\pi}\int_{-\infty}^{+\infty} gg(r,\theta) \cdot \exp[-j(r\cos\theta \cdot k_x + r\sin\theta \cdot k_y)] \cdot r \cdot dr d\theta, \end{aligned} \tag{14.4}$$

where r can be viewed as the Jacobian of the rectilinear to polar transformation.

Inverse Fourier Transform in Polar Coordinates

The inverse Fourier transform expressed in polar coordinates is

$$gg(r,\theta) = \frac{1}{4\pi^2}\int_0^{2\pi}\int_{-\infty}^{+\infty} GG(k_r, \phi) \cdot \exp[jr \cdot k_r\cos(\theta - \phi)] \cdot k_r \cdot dk_r d\phi.$$

Inverse Fourier Transform From Polar to Cartesian Coordinates

From (14.3), the inverse Fourier transform from polar to Cartesian coordinates is [186, 187]

$$\begin{aligned} gg(x,y) & \qquad\qquad (14.5) \\ &= gg(r\cos\theta, r\sin\theta), \\ &= \frac{1}{4\pi^2}\int_0^{2\pi}\int_{-\infty}^{+\infty} GG(k_r,\phi) \\ &\qquad \cdot \exp[j(r\cos\theta\cdot k_r\cos\phi + r\sin\theta\cdot k_r\sin\phi)]\cdot k_r\cdot dk_r d\phi, \\ &= \frac{1}{4\pi^2}\int_0^{2\pi}\int_{-\infty}^{+\infty} GG(k_r,\phi) \\ &\qquad \cdot \exp[j(x\cdot k_r\cos\phi + y\cdot k_r\sin\phi)]\cdot k_r\cdot dk_r d\phi, \end{aligned}$$

where k_r can be viewed as the Jacobian of the polar to rectilinear transformation in k-space.

Radon Transform

An early mathematical basis for a solution to reconstructive tomography in the spatial domain was solved by Radon. Radon proved that the cross section of any two dimensional object can be reconstructed from a set of straight ray directed projections captured by a finite linear sensor array. The non-coherent projections of the Radon transform must have a continuous full sweep of 2π radians of angular diversity to enable exact reconstruction.

Forward Radon Transform

The two dimensional function $pp(x',\theta)$ defined by a full sweep of linear projections $p_\theta(x')$ collected over all possible view angles $\theta \in [0,\pi)$ is the Radon transform of $gg(x,y)$,

$$pp(x',\theta) = \{p_\theta(x')\} \quad \text{for all} \quad \theta \in [0,\pi).$$

Inverse Radon Transform

The Radon inversion problem is the reconstruction of $gg(x,y)$ from the $\{p_\theta(x')\}$. To develop the inverse Radon transform, we start with the inverse Fourier transform from polar to Cartesian coordinates.

Inverse Fourier Transform with a Semicircle Angular Sweep

Integration over the wavenumber angle ϕ in (14.5) can be reduced to a semicircle sweep by breaking the integral of the inverse Fourier transform from polar to

Cartesian coordinates into two parts,

$$\begin{aligned} gg(x,y) & \\ = & \frac{1}{4\pi^2}\int_0^{\pi}\int_0^{+\infty} GG(k_r,\phi)\cdot\exp(jx\cdot k_r\cos\phi + jy\cdot k_r\sin\phi)\cdot k_r\cdot dk_r d\phi \\ & +\frac{1}{4\pi^2}\int_0^{\pi}\int_0^{+\infty} GG(k_r,\phi+\pi) \\ & \quad \cdot\exp[jx\cdot k_r\cos(\phi+\pi) + jy\cdot k_r\sin(\phi+\pi)]\cdot k_r\cdot dk_r d\phi. \end{aligned}$$

Using

$$GG(k_r,\phi+\pi) = GG(-k_r,\phi),$$

we obtain

$$gg(r,\theta) = \frac{1}{4\pi^2}\int_0^{\pi}\int_{-\infty}^{+\infty} GG(k_r,\phi)\cdot\exp(jx\cdot k_r\cos\phi + jy\cdot k_r\sin\phi)\cdot|k_r|\cdot dk_r d\phi, \tag{14.6}$$

where k_r can be viewed as the Jacobian of the polar to rectilinear transform and where the absolute value $|k_r|$ comes from the semicircle angular sweep. In order to find a connection with the Hilbert transform, this absolute value can be expressed in terms of the signum function,

$$|k_r| = k_r\cdot\mathsf{sgn}(k_r).$$

Semicircle Angular Sweep in Polar-Polar Form

We next formulate the inverse Fourier transform (14.6) in terms of the mapping from polar to Cartesian coordinates,

$$\begin{aligned} gg(r\cos\theta, r\sin\theta) & \\ = & \frac{1}{4\pi^2}\int_0^{\pi}\int_{-\infty}^{+\infty} GG(k_r\cos\phi, k_r\sin\phi)\exp[-jr\cdot k_r\cos(\phi-\theta)]\cdot|k_r|\cdot dk_r d\phi, \end{aligned}$$

where we have used (14.3). From the Fourier projection theorem, the Fourier transform of a projection is a one dimensional slice of the two dimensional Fourier transform,

$$P_\phi(k_r) = GG(k_r\cos\phi, k_r\sin\phi).$$

Hence,

$$\begin{aligned} gg(r\cos\theta, r\sin\theta) & \\ = & \frac{1}{4\pi^2}\int_0^{\pi}\int_{-\infty}^{+\infty} k_r\cdot P_\phi(k_r)\cdot\mathsf{sgn}(k_r) \\ & \quad \cdot\exp[j(r\cos\theta\cdot k_r\cos\phi + r\sin\theta\cdot k_r\sin\phi)]\cdot dk_r d\phi, \\ = & \frac{1}{4\pi^2}\int_0^{\pi}\int_{-\infty}^{+\infty} k_r\cdot P_\phi(k_r)\cdot\mathsf{sgn}(k_r)\cdot\exp[j(x\cos\phi + y\sin\phi)k_r]\cdot dk_r d\phi \end{aligned}$$

and since $x' = x\cos\phi + y\sin\phi$,

$$\begin{aligned} & gg(r\cos\theta, r\sin\theta) \\ = & \frac{1}{2\pi}\int_0^{\pi}\frac{1}{2\pi}\int_{-\infty}^{+\infty}[k_r \cdot P_\phi(k_r)] \cdot [\mathsf{sgn}(k_r)] \cdot \exp(jx'k_r) \cdot dk_r d\phi, \\ = & gg(x,y), \end{aligned}$$

where we have obtained a Fourier inversion from a polar to Cartesian system. The inner integral over k_r, involving a multiplication in the frequency domain by k_r and by $\mathsf{sgn}(k_r)$, can be expressed as a convolution in the spatial domain [309],

$$\begin{aligned} & \frac{1}{2\pi}\int_{-\infty}^{+\infty}[k_r \cdot P_\phi(k_r)] \cdot [\mathsf{sgn}(k_r)] \cdot \exp(jx'k_r) \cdot dk_r \\ = & \frac{1}{j} \cdot \left[\frac{\partial p_\phi(x')}{\partial_{x'}}\right] * \left[\frac{-1}{j\pi x'}\right], \\ = & \int_{-\infty}^{+\infty}\left[\frac{\partial p_\phi(\chi)}{\partial_\chi}\right] \cdot \left[\frac{1}{\pi(x'-\chi)}\right] d\chi, \end{aligned} \tag{14.7}$$

which is equivalent to the Hilbert transform of

$$\frac{\partial p_\phi(x')}{\partial_{x'}}. \tag{14.8}$$

Hence, an image of the terrain function can be reconstructed with spatial domain processing in terms of the derivative of projections subjected to the Hilbert transform,

$$ii(x,y) = \frac{1}{2\pi}\int_0^{\pi}\left[\frac{\partial p_\phi(x')}{\partial_{x'}} * \frac{1}{\pi x'}\right] \cdot d\phi.$$

Filtered Projections: Spatial Domain

The partial derivative of the projection $p_\phi(x')$ with respect to x', followed by the Hilbert transform, leads to what is called a *filtered projection* [960, 1010],

$$q_\phi(x') = \frac{\partial p_\phi(x')}{\partial_{x'}} * \frac{1}{\pi x'}. \tag{14.9}$$

Inverse Radon Transform: Spatial Domain Filtering

The inverse Radon transform [237, 462, 756], when implemented in the spatial domain, is comprised of a spatial domain reconstruction obtained from a semicircle sweep of correctly positioned (within the image array) of filtered projections,

$$ii(x,y) = \frac{1}{2\pi}\int_0^{\pi} q_\phi(x') \cdot d\phi. \tag{14.10}$$

Convolutional Back-Projection

One approach to pivot point SAR signal inversion through spatial domain reconstruction is by positioning and interpolation of $q_\phi(x')$ into an image pixel array as suggested by (14.10). Each spatial domain filtered projection is rotated appropriately and sample values are assigned to all pixels that fall on the angled back-projected line. The filtered projections progressively reconstruct the image over the full sweep of reconstruction angles [250]. This processing is the basis of convolutional back-projection.

Convolutional Back-Projection Reconstruction

Some CBj methods require interpolation to convert data reconstructed in a spatial back-projection polar format to a rectilinear pixel array [860,906]. Even within the polar array, a nearby pixel location of x' may not correspond to one of the discrete values for which the filtered projection is known. In this case, interpolation is necessary to fully populate the pixel array, often, with simple linear interpolation methods being adequate [509]. One approach for eliminating the computations required for interpolation, is to up sample the filtered projections by a factor in the range of 10 to 100 [509].

Filtered Projections from Differentiation and Hilbert Transforms

The conventional CBj process applies the Hilbert transform followed by differentiation [619]. In [509], the back-projection algorithm is implemented with differentiation first, followed by the Hilbert transform.

Filtered Projections from Wavenumber Weighting

In addition to the spatial domain expression (14.9), filtered projections can also be obtained by Fourier inversion of weighted (filtered) Fourier projection slices,

$$q_\phi(x') = \frac{1}{2\pi} \int_{-\infty}^{+\infty} |\breve{k}'_x| \cdot P_\phi(\breve{k}'_x) \cdot \exp(jx'\breve{k}'_x) \cdot d\breve{k}'_x.$$

The $|\breve{k}'_x|$ factor in convolutional back-projection behaves as a high pass filter [968].

Filtered Projections from Spatial Convolution

For each angle ϕ, the $q_\phi(x')$ is equivalent to convolving $p_\phi(x')$ with the impulse response of a filter with frequency response $|\breve{k}'_x|$ [250]. The filtered projections are related to measured projections $p_\phi(x')$ by

$$\begin{aligned} q_\phi(x') &= p_\phi(x') * h(x'), \\ &= \int_{-\infty}^{\infty} p_\phi(\chi) \cdot h(x' - \chi) \cdot d\chi, \end{aligned}$$

where the filter impulse response $h(x')$ is the inverse Fourier transform of $|k_{x'}|$,

$$h(x') = \frac{1}{2\pi} \int |k_{x'}| \cdot \exp(jx'k_{x'}) \cdot dk_{x'}.$$

Convolutional Back-Propagation

The straight line integrals of short wavelength tomography are not appropriate for inversion of diffracted wave fields [468, 556, 865], hence the development of CBp with circular contours in spatial reconstructions.

The diffractive time-space domain inversion problem, from the perspective of CBp processing, is a problem of reconstructing scattering functions from generalized (curvilinear) line integrals [206, 207, 702, 794]. This family of algorithms is equivalent to "inversion of integral geometry" in applied mathematics [175, 519, 702, 703, 794].

In some forms of CBp image formation processing, the phase measurements for each incident plane wave (view angle) is filtered (with the same filter of CBj) and summed into the pixel array as a function of view angle [257].

One version of a CBp algorithm [252, 261] uses different filters for different regions of the image. Work has been done to extend convolutional back-projection to convolutional back-propagation in SAR [53].

14.6 Filtered Fourier Reconstruction

Convolutional spatial reconstructions can be formulated with wavenumber domain "filtering," where the filter is implemented as a wavenumber domain multiplication of $R_\theta(k_{x'})$ along a slice at angle θ with $|k_{x'}|$. This formulation of the inverse Radon transform is closely aligned with the Fourier projection theorem, the difference being the "filtering" factor of $|k_{x'}|$.

Filtered Fourier Reconstructions on Linear Traces in SAR

In SAR filtered Fourier reconstruction algorithms $\text{fFR}_{\text{l@p}}$ based on linear traces at passband, each deramped radial line is a passband projection centered at ω_c. Each projection $P_\theta(\omega_p)$ is multiplied by the filter function $|\omega_p|$ to provide a form of filtered polar formatting. Filtered polar formatting is considered equivalent to CBj when Jacobian weighted two dimensional sinc-kernel interpolators [860] are used to obtain uniformly resampled data from polar format data raster [187].

Filtered Fourier Reconstructions on Circular Traces in SAR

Filtered Fourier domain reconstruction methods $\text{fFR}_{\text{c@p}}$ based on circular traces have been used in SAR passband image formation processing [679].

Failures in Filtered Fourier Reconstruction Approaches

Are Spatial and Wavenumber Domain Methods Equivalent? Filtered Fourier reconstruction methods have typically yielded poorer results than their "equivalent" spatial domain convolutional methods.

For example, Munson states, but without demonstrated examples, that his initial experiments on spatial domain passband $\text{CB}_{\text{j@p}}$ algorithms produced images of somewhat better quality than wavenumber domain passband $\text{fFR}_{\text{l@p}}$ algorithms [679].

There is a general consensus that filtered Fourier reconstruction algorithms are inferior to their spatial domain convolutional counterparts.

Differences Caused by Inexact Interpolation? Inexact interpolation has often been blamed for the apparent deficiencies of filtered Fourier reconstruction methods compared to the "equivalent" spatial convolutional methods [95,97,187, 212,255,582,645,853,907,936]. For example, it has been claimed in computed tomography that there exists "extreme sensitivity of the quality of the reconstructed image to interpolation errors, (and) the interpolation must be exact" [738].

Inexact interpolation as the source of this fault is a curious suspect, given that interpolations are involved in both $\text{fFR}_\text{l}/\text{fFR}_\text{c}$ and CBj/CBp algorithms, and problematic symptoms related to inexact interpolation should be evident in either case.

Differences Caused by Sparsity of Polar Raster Samples? It has also been suggested that the increasing sparsity of polar raster samples with increasing distance from the wavenumber domain origin led to CBj algorithms becoming dominant over fFR methods in computed tomography [288, 730].

Again, this is curious, given the fact that the sparsity of polar raster radial data occurs in both spatial and wavenumber domain versions of these "equivalent" algorithms.

Relative Success of Passband Fourier Reconstruction in SAR In comparisons of baseband and passband forms of polar raster Fourier reconstructions, it is thought that Fourier reconstruction methods perform better in SAR imaging because the polar raster of SAR is inherently much closer to a Cartesian raster than the polar raster of computed tomography. Thus, some have argued that low order FR interpolators are adequate in coherent passband SAR sensing, but insufficient in non-coherent baseband computed tomography [187].

The True Culprit! Soumekh and Kaveh assert that premises used in the development of many filtered Fourier reconstructions methods have not been correctly formulated, and that this has been the cause of the performance gap of wavenumber domain $\text{fFR}_\text{l}/\text{fFR}_\text{c}$ algorithms relative to the "equivalent" spatial domain CBj/CBp methods [898].

Fourier Reconstructions under Change of Variable Transformations Soumekh and Kaveh assert that most errors in FR algorithm development are based on a failure to account for the correct nature of Fourier reconstructions when a change of variables transformation is involved. They assert [898] that this failure is due to the common mistake of ignoring the role of the change of variables that defines coordinate transformations or Fourier migrations. A correct handling of this change of variables is crucial in determining which non-unform samples to ingest into the kernel of finite interpolators when creating the uniform Cartesian raster preparatory to Fourier inversion.

Soumekh and Kaveh have formulated a correct framework of Fourier migration followed by uniform resampling, where the form and size of the interpolation kernel is determined by the bounding volume of the imaged scene (see Section 13.9).

Single Sample "Interpolation" in CW VolSAR We note from the demonstrated VolSAR results of Chapter 13 that single valued interpolations to a thin shell Ewald sphere can yield effective Fourier reconstructions. In these VolSAR examples, the Fourier reconstruction interpolation is based on single sample inputs!

Interpolation in Aperture and Scene Coordinate Systems Another possible factor that might contribute to failures to achieve good results in Fourier reconstructions may be related to interpolation applied to wavenumber data in an aperture coordinate system that has not been preprocessed with Huygens-Fresnel transfers. This forces the interpolator to work on high frequency phase oscillations resident in data expressed in the aperture coordinate system. Interpolation systems should be formulated and implemented within the context of a scene coordinate system [891].

Riccati Wave Motion Equations and Rytov Approximations

Inverse diffractive imaging has been investigated within the framework of small perturbations (the Born approximation) and the Helmholtz wave equation [993]; and similarly, within the framework of smooth perturbations (the Rytov[5] approximation) and the Riccati[6] phase function equations of wave motion [481]. The Riccati-Rytov approach has been utilized mostly in acoustic imaging. We shall give an overview of the Riccati-Rytov formulation of inverse diffractive imaging and shall discuss its suitability for inversion in electromagnetic imaging.

Wave Motion Equations in Terms of Phase Functions (not Phase Factors)

If we ignore the amplitude elements of complex wave fields, then in the context of phase field inversion, a frequency-space phase field can be described in terms of a spatial map of the phase factor [251, 807, 934],

$$\measuredangle Ss(\omega, \boldsymbol{r}) = \exp[j\theta(\omega, \boldsymbol{r})].$$

This is the approach of the Helmholtz wave motion equation. If the wave field is described in terms of a spatial map of the phase factor's phase function,

$$\angle Ss(\omega, \boldsymbol{r}) = \theta(\omega, r),$$

where

$$\measuredangle Ss(\omega, \boldsymbol{r}) = \exp[j\angle Ss(\omega, \boldsymbol{r})].$$

then the resulting "wave field" models are attributed to Riccati.

[5] S. M. Rytov (1908–1996) was a Russian mathematician.

[6] Jacopo Francesco Riccati (1676–1754) was an Italian mathematician.

Fresnel's Harmonic Phase Field and Riccati's Wave Motion Equation

Fresnel's harmonic phase field

$$\measuredangle Hh(\omega, \boldsymbol{r}) \quad = \quad \exp(jk_\omega \cdot \boldsymbol{r}) \tag{14.11}$$

is a solution of the Helmholtz wave motion equation,

$$\nabla^2\{Hh(\omega, \boldsymbol{r})\} + k_\omega^2 \cdot \nabla Hh(\omega, \boldsymbol{r}) = F(\omega, \boldsymbol{r}). \tag{14.12}$$

It can be shown that the phase function of Fresnel's phase factor (14.11),

$$\theta(\omega, \boldsymbol{r}) = k_\omega \cdot \boldsymbol{r}$$

is a solution of the Riccati equation, which for harmonic wave motion is

$$\nabla^2\{\theta(\omega, \boldsymbol{r})\} + k_\omega^2 \cdot \nabla\theta(\omega, \boldsymbol{r}) = F'(\omega, \boldsymbol{r}). \tag{14.13}$$

The Helmholtz equation is in terms of a phase factor. The Riccati equation is in terms of a phase function (the phase function of the phase factor of the Helmholtz equation).

To derive the Riccati equation, a logarithm is used on the Helmholtz equation to reduce phase factors to phase functions. The later can be viewed as a homomorphic formulation of the former.

The Helmholtz (14.13) and Riccati (14.12) equations are identical in form. However, the wave field models associated with these equations are not identical.

Rytov's Approximation

The weak scattering assumption often utilized within the context of the Helmholtz equation is an approximation due to Born. A smooth scattering assumption within the context of Riccati's equation is an approximation due to Rytov [920]. Rytov's approximation is based on perturbation models of weak scattering. Born or Rytov approximations [692] are typically used with diffraction tomography, convolutional back-projection [252] and filtered Fourier reconstruction algorithms [513].

The higher order terms ignored in the small scattering models of Born and Rytov differ [674]. Hence, what is "small scattering" for one may not be "small scattering" for the other. The Rytov approximation is called the *smooth scattering approximation* [891].

The Born approximation linearizes Helmholtz equations of incident and scattered fields. The Rytov approximation linearizes Riccati equations of incident and scattered fields [256]. If the incident field is a plane wave, the linearized Riccati-Rytov form is analogous to the Rayleigh-Sommerfeld diffraction formula [64, 262, 557]. The Riccati-Rytov approach to wave field modeling requires unwrapped phase measurements [730].

Early work on the use of Riccati-Rytov formulations in inverse diffractive imaging in ultrasonics includes [251, 481, 482, 514, 515, 521, 993]. In some non-SAR applications, use of the Rytov approximation and the Riccati equation appears to be more accurate than use of Born approximations and the Helmholtz equation [251, 257, 518, 521, 726]. Moreover, in various non-SAR computed imaging applications, a common restriction on object volume is relaxed by replacing the Helmholtz-Born approach with the Riccati-Rytov approach [672, 919].

Rytov-Riccati Failure in Echo Mode Imaging

The Rytov approximation proves superior to the Born approximation in transmissive ultrasonic imaging, but fails rapidly in echo mode imaging systems [883]. An example reflection mode diffractive acoustic imaging system includes the multi-frequency synthetic aperture system [859] as suggested in [685]. The Rytov approximation becomes progressively worse as the scattering angle approaches 180 degrees and seems to fail entirely for backscatter [713].

The Riccati formulation and the Rytov approximation, although commonly used for acoustic imaging, have not been recommended for algorithm development in SAR image formation processing. Formulations based on the Riccati equation have been used for estimating and compensating some forms of phase errors in SAR [892].

14.7 SAR Algorithms: Variations on Themes

We survey the SAR literature and briefly summarize a wide variety of algorithms found in the literature. Although thorough, this literature survey is not exhaustive.

Doppler Processing in SAR

Doppler SAR Advanced forms of high resolution Doppler processing in SAR are an area of active research [86]. Such methods are called *Doppler SAR* and these algorithms span the space from static spatial SAR analysis to Doppler induced temporal analysis.

Velocity Induced SAR Velocity induced *intra-pulse* Doppler shifts were ignored in the original formulation of Doppler beam sharpening. More recent versions of DBS consider these modulations. SARs that use frequency modulated continuous wave (FM-CW) inversion methods can extend Doppler beam sharpening processing to a formulation that is called *velocity induced SAR* [407, 440, 871].

SAR Imaging of Moving Objects Spectral methods that seek to integrate SAR methods for imaging moving objects have been investigated [179]. Four dimensional space-velocity imaging is performed to separate interfering scatterers in elevation and to detect slow movers in [343].

Factorized Direct Back-Propagation Algorithms

Aperture Factorized Direct Back-Propagation In one form of factorized direct back-propagation, the synthetic aperture is divided into subapertures. A single pulse representation of each subaperture is coherently summed across a sparsely populated pixel array to create a full scale low resolution image. Since the low resolution images contain fewer pixels, there is an overall reduction in complexity. All full scale low resolution images formed from different subapertures are then coherently summed to produce a full scale, higher resolution image [449,

957, 958, 960, 960, 1010]. The low resolution images must be up sampled before they are coherently summed.

Aperture Factorized Direct Back-Propagation in Polar Coordinates A polar pixel array can be used to accumulate the coherent data of low resolution images before interpolation into a rectilinear pixel array [1010]. A lower Nyquist sample density requirement is exploited by working on a polar grid to locate subaperture image pixels in uniform increments of radius and angle (relative to the center of each subaperture). The required Nyquist sample density in the polar pixel array increases with subaperture length.

Aperture and Scene Factorized Direct Back-Propagation Other variants within the class of factorized direct back-propagation algorithms divide the image, or divide both the aperture and image [47,48,267,368,839,959,1004]. For example, a dyadic quad-tree approach can be used for the spatial decomposition of the image domain [634]. The number of data array samples required to build a given subimage increases with the length of the subaperture. The subimage size may be reduced as the length of the subaperture increases to keep the number of required data samples constant [960]. The image factorization of [701,702] divides subimages based on regions defined by intersections of back-propagation circular arcs.

Spotmap Step Transform Algorithms

Non-Coherent Spotmap Imaging with Step Transforms Stretch processing applied to a multiplicity of terrain subspots can be achieved via the *step transform* [612, 926]. The step transform is a hardware centric form of polar format spotmap imaging. Step transform range processing performs multiple deramps in range. Step transform azimuth processing performs multiple deramps in azimuth. Both processes are followed with subspot specific one-pass image formation Fourier transforms. The step transform seeks to provide spotmap data products more directly with the sensor system hardware, rather than relying on digital post processing to mosaic multiple spot images. The step transform must be modified to accommodate non-linearities if applied in a squint imaging geometry [1002].

Down-Range Profiling with Step Transforms and Coherent Summing Some forms of the step transform are implemented by having the deramp receiver use a saw tooth reference signal instead of a single LFM waveform. The output of the saw tooth deramp receiver for a single pulse echo is a stepped frequency response. The beat frequency of each subwindow within the receive window represents a low resolution range profile. The high resolution image is formed by coherent summing of the low resolution profiles.

Cross-Range Profiling with Step Transforms and Coherent Summing A similar approach in the along-track dimension divides the beam into subapertures [1012]. Subaperture images are slow-time stretched processed to form low resolution images, then the subaperture images are combined coherently to produce a high resolution cross-range profile [622, 808].

Subaperture Polar Format Algorithms

Subaperture processing within many classes of algorithms has been explored [26, 649, 655, 739, 958]. The following subaperture approaches are specific to polar formatting.

Progressive Subaperture Fraunhofer Corrections An early approach to using subapertures and progressive image formation processing with polar formatting was introduced in [130]. A three step image formation process was employed with coarse resolution first obtained in initial along-track image formation. Fine resolution is then obtained with down-range processing, and lastly, the along-track processing is revisited to produce fine resolution in the along-track dimension. Local phase deviations due to Fraunhofer approximations are partially compensated after the first coarse along-track image formation processing step.

Tiered Subaperture Approaches to Fraunhofer Compensations One approach to mitigating the effects of spatial blurring in polar formatting is partitioning the slow-time aperture into subapertures. Then iterating the decomposition by dividing subapertures into sub-subapertures. This tiered subaperture approach increases the achievable spot size in polar formatting [271–273]. Each added tier of subaperture processing expands the supported patch diameter, but with diminishing returns.

Overlapped Subaperture Algorithms Another approach to plane wave error compensation in polar formatting is *overlapped subaperture processing* (OSA) [492]. The slow-time phase history is divided into subapertures that overlap. A slow-time phase ramp correction is applied between coarse and fine slow-time compression steps that corrects Fraunhofer approximation errors.

In more detail, the initial stage of OSA processing does a coarse one-pass process in the along-track dimension with a reduced along-track sampling rate. Multiple along-track stretch reference functions specific to particular down-range and along-track reference points are employed.

The overlapped slow-time deramp processes yield coarse along-track wavenumber frequency bands that collectively form a set of along-track wavenumber subapertures. Each subaperture is small enough to limit wave curvature error to one range cell. The pulses of each subaperture are processed by their own range walk correction vector that compensates the linear element of the range variation over the subaperture as a function of course azimuth position.

Since the range data are deramped, but not yet one-passed in range, a range walk correction corresponds to a simple frequency shift. Each along-track subaperture is shifted in frequency for Fraunhofer correction, windowed and Fourier transform to perform coarse one-pass image formation in the along-track dimension.

The course one-pass image data are corrected for along-track beam pattern variation and digital spotlighting can be performed. The correction factors can also include quadratic focus, higher order focus and autofocus for motion induced error correction. At this point, the one-pass of range profiling is performed to produce the fine range image. To form fine resolution in the along-track dimension, processing is performed across subapertures.

One reason for the development of the OSA was to accommodate certain computing architectures circa 1990 [495].

Subaperture Stolt Formatting Algorithms

Coherent Fusion of Subapertures with Stolt Formatting Subaperture processing can be used to reduce FFT size in Stolt formatting. Stolt formatting preserves coherence information among the low resolution subaperture images. The set of low resolution images are coherently summed to produce a high resolution image [885].

Subaperture Range-Doppler Algorithms

Range-Doppler Subaperture Processing Subaperture methods have been developed for range-Doppler processing. Subaperture processing divides the received SAR signal into subapertures, possibly overlapped. Each subaperture is correlated with a matched reference function, yielding a low resolution image. Other approaches use piecewise linear approximations to the slow-time range walk. The subapertures images are coherently summed [659,967].

Extended Correlation Subaperture Processing Range-Doppler processing can be used with subaperture and subpatch processing in an algorithm called *extended correlation processing* [26]. The subapertures are chosen such that each has insignificant range cell migration. The subimages are phase compensated and spatially rotated, up sampled and coherently summed to form the final image. Such approaches must deal with depth of focus issues, often done by segmentation of the range swath into subswaths [984].

Alternate Approaches to Rechirping

Polar Format Data Reconditioning There are methods that provide alternatives to slow-time rechirping for spanning the gap between polar and Stolt formatting methods [147]. Some methods seek to perform along-track data alignments and reformatting by introducing spatially variant modifications of the SAR phase history. Variants of this preprocessing subsystem and subsequent approaches of image formation processing have been called *differential Doppler algorithms, wide field polar format algorithms* and the *Stolt polar format algorithms.* The wide field polar format algorithm works to remove the range curvature phase errors in the polar format algorithm to produce images of large ground scenes without geometric keystone distortion [147].

Advanced Antenna Types

Phased Array Antennas Phase array antennas vary the phase of signals from individual antenna transmit and receive elements to generate and steer multiple, distinctly directed, radar beams. If beam control is achieved via electronic beam steering with phased array antennas [287], then an additional amplitude modulation associated with the radiating elements of the phased array should be considered in SAR phase history modeling [885].

Waveform Design for Phased Array Antennas Narrow bandwidth signals can be steered from phased array antennas with simple phase shifters. Electronically steered phased array antennas struggle with steering wide bandwidth signals without the use of expensive and complex time delay hardware elements. One approach to solving this challenge is the use of modified LFM signals, where a wide bandwidth LFM chirp waveform is assembled from subchirp waveforms staggered in time and frequency [281]. This allows each relatively narrow band subpulse to be steered with its own steering phase operation. A cost of using this subpulse approach is an increase in the PRF by a factor determined by the number of subpulses used to synthesis the wide bandwidth waveform.

Modified Fast Fourier Transform Algorithms in SAR

Fast Fourier Transforms for Sparse or Non-Uniform Data Expressing the fast Fourier transform in a multilevel fashion can lead to fast implementations for the Fourier transform of sparse or non-uniformly spaced data [185].

Chirp-Z Transform in SAR Algorithms

Chirp-Z Transforms to Avoid Fourier Reconstruction Interpolation A modified chirp-z transform has been proposed to replace polar format interpolation processing [3, 657]. In some instances, the chirp-z transform has been used in one dimension and then the discrete Fourier transform is applied in the other. The chirp-z transform can be used to interpolate uniformly spaced data from one sample grid to another [17, 291, 292].

Chirp-Z Transform in OSA Processing OSA can be applied to the trapezoidal data raster collected by a center frequency and chirp rate agile transceiver [495]. The coarse along-track deramp FFT can be replaced by a chirp-z transform [725]. The purpose of the chirp-z transform is to keep the along-track pixel spacing constant as a function of range. Each chirp-z is computed with three FFTs.

Chirp-Z Transform in Stolt Formatting The chirp-z transform has been examined as a basis for SAR image formation processing for Stolt formatting used in spotlight mode [551, 1016].

Chirp-Z Transform as a Form of Chirp Scaling The chirp-z transform has been shown to be equivalent to a particular implementation of the chirp scaling algorithm [552].

Advanced Waveforms

Non-Linear Frequency Modulation The power spectral density of an LFM chirp waveform is nearly rectangular. Hence, its pulse compressed inverse Fourier transform is nearly a sinc function. Reducing the sidelobes can be done with conventional application of a window function for data tapering [435]. But such approaches have challenges with stretch processing. The receive window duration mismatch between the received echo signal pulse length and the reference local oscillator chirp signal duration reduces the achieved SNR of matched filtering by a decibel or two [277, 504]. One advantage of non-linear frequency modulated signals is the ability to avoid SNR loss associated with pulse shaping and sidelobe suppression [276,278]. Another purpose for non-linear frequency modulated chirp waveforms is that signal power spectral density can be advantageously shaped such that the matched filter response has reduced sidelobes compared to the LFM waveform (with the expense of some mainlobe widening) [279, 520].

Waveform Design for Convolutional Back-Projection The option of modifying the transmitted waveform in a way that eliminates the need of filter processing in CBj has been investigated [15, 16]. The transmitted waveform can be designed as the impulse response of the filter of CBj. Thereby, the CBj algorithm is simplified by eliminating the filtering operation since filtered projections are obtained directly from the receiver.

Hilbert Transform Waveforms for Convolutional Back-Projection Another approach incorporates a derivative of the Hilbert transform into the waveform design. Hence, the need to apply the Hilbert transform followed by a differentiation is eliminated in CBj [619].

Enhanced Waveform Design for LFM One-Pass Compression The LFM pulse weight function can be tailored to enhance point spread function characteristics [157,435,534,715]. Radar system impulse response characteristics can be captured in a reference waveform, sometimes called a *replica*. A replica waveform can be used in weight function designs that compensate hardware imperfections [651].

Stretched LFM Waveforms and Electronic Protection It has been noted that if advanced chirp waveforms are used, for example, for electronic protection [388], then a straight forward approach to residual quadratic phase compensation might not be effective [899].

FM-CW Waveforms and Continuous Phase Curtains The continuous version of a slow-time phase curtain can be obtained with the use of frequency modulated continuous wave signals in non-pulsed SAR systems [46, 86, 871, 1026, 1027]. Use of FM-CW signals requires large one dimensional Fourier transforms of data from an extended receive window. The one dimensional output of the large discrete Fourier transform is then parsed and formatted into a two dimensional array of frequency-wavenumber domain data. Angular spectrum image formation processing then proceeds as if pulsed data were originally collected [407, 870, 985, 986].

PRF Design Concepts

Along-Track Ambiguity-to-Signal Ratio The required along-track sampling density of the synthetic aperture in squinted space based SARs are sometimes intentionally relaxed to obtain increased coverage in the down-range dimension [221]. In other cases, with ultra-wide beamwidth imaging scenarios, the "working" angular dwell has been intentionally expanded for various reasons [397]. If expanded too much without increasing the PRF, strongly scattering objects can cause along-track ghosting in the SAR image [221,407,586,793]. The aliased signal echoes in the along-track dimension can be analyzed as noise and an *along-track ambiguity to signal ratio* (AASR) can be defined as a technical performance metric. Such approaches have been met with criticism [886,899].

Scan SAR Concepts

Multiple Beam SAR Multiple beam SAR concepts can be used to overcome radar ambiguity constraints to increase the range swath width in satellite SAR systems [192].

Stripmap SPECAN (SPECtrum ANalysis) By using a two step approach, the along-track deramp processing of slow-time stretch spotlight mode SPECAN SAR has been applied to stripmap mode SAR [553,554].

Kirchhoff Integral Equation and SAR

Range-Doppler and Kirchhoff Integral Equation of Wave Motion The range-Doppler algorithm has been viewed as a "matched filter" solution expressed in the time-wavenumber domain of the Kirchhoff integral equation of wave motion [433,812].

Kirchhoff Integral Methods and Basebanded SAR Data It has been noted that basebanded SAR data cannot be used in Kirchhoff integral formulations of SAR imaging [424].

Chirp Scaling Algorithms

Chirp Scaling with Non-LFM Waveform Systems Chirp scaling was designed specifically for range-Doppler systems that use LFM waveforms. Chirp scaling can be indirectly applied to other waveform types [442]. If LFM waveforms are not used, then the gathered data must be pulse compressed based on the native transmitted waveform characteristic and rechirped by convolution with a LFM pulse waveform [407,440,442]. Once this reconditioning has been performed, chirp scaling can proceed as if LFM pulses were originally employed in the SAR sensor system.

However, some of the computational savings of chirp scaling (in removing the need for range migration correction interpolation) is lost in dealing with rechirped pulse echo data.

Accelerated Chirp Scaling In many narrow range swath SAR scenarios, the pulse echo duration is larger than the range swath echo duration [442]. Hence, any processing required in the chirped data domain, in some sense, is wasted effort in these cases.

The standard chirp scaling approach performs chirp scaling before pulse compression. There is a form of chirp scaling applied to pulsed compressed data. Since this approach has computational efficiencies, it is called *accelerated chirp scaling* [440].

In accelerated chirp scaling, the time span of the chirp scaling operation need only be large enough to support a shift of the phase center over a distance given by the differential range across the comparatively narrow range swath.

Accelerated chirp scaling exploits the extraneous fast-time phase chirp that is typically removed by secondary range compression and adjusts secondary range compression's LFM signal in the range-Doppler domain of pulse compressed data [440, 442]. Since the temporal support of the secondary range compression chirp is typically much smaller than the support of the transmitted pulsed chirp, processing is substantially reduced.

Extended Chirp Scaling Chirp scaling algorithms designed for squint geometries use higher order expansions of slant-range equations [234, 654, 661–663] and are called *extended chirp scaling algorithms.*

Extended chirp scaling is effective with squint geometries of up to about 30 degrees in some C-band SAR systems [1012]. Chirp scaling can also be used in bistatic configurations [995].

Since chirp scaling and accelerated chirp scaling are approximate, care must be taken to remove a residual phase error that will exist in the image spectrum. This residual phase error is more severe in squinted systems [442].

Frequency Scaling An alternative to chirp scaling when non-chirped waveforms are used is frequency scaling [655].

SAR Based on Spectral Estimation

Long Dwell SAR Spectral Estimation SAR methods based on spectral estimation over large temporal spans have been developed [341].

Imaging of Scatterers Separated in Elevation by Spectral Estimation Scatterers at different elevations within a shared down-range and along-track resolution cell can be resolved with multiple component signal reconstructions via spectral estimation [982].

Another work proposes a three dimensional imaging technique based on a linear minimum mean square error approach [342].

Linear Algebra Approaches

Matrix Inversion One approach that seeks to eliminate interpolation in Stolt formatting is to map points on a circular arc to points on a rectilinear grid through matrix inversion [212]. This approach tends to be computationally expensive and numerically unstable [730]. This approach is different than the use of a coordinate transformation as done in [194].

Fast approaches for computing the Householder transformation has been investigated for a form of C-SAR inversion [124].

Gridding Another approach in Fourier reconstructions is called "gridding" [727]. (The term "gridding" is also used in seismology [205].) A grid (non-uniform wavenumber sample set) is sought such that a weighting based on sample density and convolution with a finite interpolation kernel provides an accurate "grid" of reconstructed data in the image domain. The gridded values in the transform domain do not necessarily equal the values that would have been obtained had the Fourier transform been performed. Several different gridding functions have been evaluated [484]. There are claims that there are no Nyquist requirements on the sample density in the Fourier domain or on the support of the imaged scene [727]. A comparison of gridding and various polar format interpolation methods has been performed [404].

Gridding in Radio Astronomy Early forms of gridding were used by radio astronomers, where k-space was divided into a grid and the point at the center of each cell was set equal to the sum of all data points within the cell [629]. Variations of this include weighting the sample points based on distance from the center point [936] and inverse filtering after a convolution with a window function [814]. Along the lines of Section 13.9, it was shown that a sinc function of infinite extent, followed by uniform resampling is optimal in some sense [727].

Parallelized Omega-K Processing

Range Stacking Algorithms in Computed Tomography An algorithm essentially equivalent to a range-stacking algorithm for computed tomography, but where each partial reconstruction is for a different plane wave illumination angle, was given in [256]. A similar algorithm called "fast-correlation" is presented in [440]. Some view range-stacking as processing that applies a "matched filter" to focus cross-range lines one at a time in a SAR image [440].

Parallelized Omega-K Generalization of Stolt migration for parallel implementation on an array processor was investigated in [83]. Other approaches to parallelizing omega-k algorithms have been developed [1013].

Alternatives to Presumming

Spectrum Replication and Data Down Sampling The PRF can be selected such that the sampled version of squint frequency-wavenumber phase history suffers no overlap aliasing. The baseband element of the replicated spectrum of the sampled data can be extracted to achieve a form of slow-time down conversion processing [217, 218]. A low-pass filter is used to extract the recovered baseband spectrum. This can be done for both Stolt formatting [885] and polar format image formation processing [492]. Frequency domain replication with down sampling [211] has been applied to Stolt formatting for squint mode imaging [389, 748, 751]. With polar formatting, the sample rate is set high enough to sample the field of view afforded by the real beam, but not sufficiently high to sample the slow-time phase curtain required for Stolt format processing.

Inverse SAR

ISAR Angular Spectrum Processing The application of angular spectrum processing to ISAR has been explored [900].

ISAR Range-Doppler Processing Range-Doppler algorithms can also be applied in inverse SAR imaging of rotating or tracked objects [26,638,801]. The first step is to remove the effects of gross range changes to object center by performing motion compensation to a (moving) point reference [26, 142]. Range-Doppler imaging of rotating objects [976] represents measured data on a three dimensional Fourier domain manifold. But unlike the Ewald sphere of Stolt formatting, the k-space sphere is specific to the center of the object (instead of being descriptive of the angular support manifolds of each scatterer). This form of image formation processing for inverse SAR also requires compensation of motion through resolution cells. One approach uses a quadratic approximation to the circular motion of off center object points and a polar format for recorded data [117].

Multistatic SAR Methods

Quasistatic Analysis Methods Sometimes, the signal analysis of multistatic SAR systems is simplified with a "quasi-static" model, where transmitting and receiving sensors are separate, but nearly collocated. One approach for the signal analysis of such systems is to assume an imagined "equivalent sensor" to be resident at the midpoint between the two antennas [827].

Geosynchronous SAR

If the altitude of a satellite SAR increases until a point of geosynchronous orbit is reached, the angular velocity of the Earth's rotation is nearly equal to that of the satellite, but if the rotation direction is different, then aperture synthesis can be achieved. Geosynchronous SAR algorithms must be tailored to the particular challenges and opportunities of relative sensor-scatterer movements afforded by geosynchronous orbits [473, 941].

Interferometric SAR

The terrain slope is used in [747] to achieve improved altimetric resolution in satellite based SAR interferometry.

Motion Compensation

Compared to spaceborne SAR systems, motion compensation is more critical as a system level design element in airborne SAR systems [142, 339].

Low Q-Factor Algorithm Motion Compensation Designs Analytic work in motion compensation tailored to specific algorithms used in low frequency and wide bandwidth SAR systems and applications includes [9, 366, 367, 527, 532].

Agile Waveform Generation and Motion Compensation Motion compensation can seek to phase compensate to evenly spaced intervals in distance, time or angle [142]. Motion compensation can be achieved with agile waveform generation capable of pulse to pulse adjustments of the radar frequency, phase, slow-time PRF and fast-time ADC rates [130].

Agile Waveform Generation and Polar Formatting If an agile radar waveform generator changes the center frequency and chirp rate on a pulse-to-pulse basis, then the polar raster of polar formatting can be changed to a trapezoidal raster to eliminate the required resampling in the down-range wavenumber dimension [187, 495]. Agile waveform generation can also be designed to compensate third order range curvature terms in the expansion of the distance between reference points in a large ground scene and the center of spot reference point.

Stolt Formatting with Integrated Motion Compensation An algorithm that integrates multiple step, range-adaptive, motion compensation into Stolt formatting has been developed [779, 781].

Dual Format Algorithm An algorithm called the *dual format algorithm* has been proposed [403] that allows motion compensation phase error corrections with image formation to an arbitrary grid, typically spherical or polar, instead of a rectilinear grid.

Turbulent Atmospheric Induced Phase Errors and Compensation An iterative methods for correction of phase affected by turbulent atmosphere or motion instability has been explored [479]. The approach minimizes a functional using partial phase information and an initial guess.

ISAR and Motion Compensation Motion compensation has been applied to the ISAR problem where the quality of object motion estimation is judged with a focal quality indicator function [854].

Autofocus Algorithms

Major approaches for autofocus processing include [113, 275, 293, 295, 338, 381, 382, 421, 422, 800, 971, 972, 974, 996].

Post Processing for Image Enhancement

One of the most important measures for SAR image quality is the contrast ratio. Well formed SAR imagery has dark shadows [407, 437]. Histogram remapping is not a fair approach to obtaining dark shadows in true "high quality" SAR image formation processing.

Digital Spotlighting

Uses of Digital Spotlighting Digital spotlighting can be used to suppress slow-time grating or sidelobe images [407, 870]. Digital spotlighting can also be used to perform polar formatting with data obtained from a stripmap system [683]. Digital spotlighting has proved useful in direct back-propagation processing [697, 698]. Digital spotlighting has also been used to extract coherent signatures of moving objects [873]. Another potential use of digital spotlight processing is in spatially variant processing to mitigate local approximation errors [569]. Digital spotlighting has been used in the elevation dimension of Stolt formatting methods for three dimensional SAR [933].

Digital Spotlighting and Spatially Variant Motion Compensation Digital spotlighting can be the basis of an approach to performing motion compensation where motion error characteristics vary over different subregions of the ground scene [698]. Different phase error corrections can be applied to the multiplicity of subregions. Subaperture digital spotlighting does not require Fourier processing of the entire synthetic aperture or imaging of the entire ground scene. Subapertures and subpatches allow the particular implementation of the algorithm to be tailored both to the imaging geometry and to the architecture of the computing hardware [887].

Computed Tomography Algorithms

Fan Beam Tomography Fan beam projections are modeled by line integrals along rays that emanate from a point source [31, 98, 456, 509, 581]. Fan beam projections can be more efficient since all measurements in one fan are acquired simultaneously. Fan beam tomography can employ preprocessing to artificially create plane wave data [256]. The process of associating fan beam projections with parallel beam projections is called *rebinning.*

Time of Flight Tomography Geometric optics requires that the signal wavelength be much smaller than scatterer size [418]. This assumption does not always hold in ultrasonics.

Acoustic reflective and refraction effects can be modeled as ultrasound velocity variations in the inhomogeneous body being imaged, leading to variations in time of flight of the sound wave signals [386, 418].

The small scatterer assumption is not required for ray based time of flight tomography methods [255, 313, 331, 416, 674, 729]. This is also common with modeling and reconstruction based on variations of refraction index in optics [482, 927, 993].

Reconstructions of propagation delay in ultrasonics may employ a convolutional model [758, 759]. One form of a convolutional approach in projection or propagation delay measurements has been suggested in [757–759].

Tomography with Sparse Angular Sampling A temptation with tomographic methods formulated in the temporal-spatial domain is to seek ways to form the image with sparse angular sampling [107, 525, 557, 594, 857, 861, 931, 932].

For example, an early study in acoustic medical imaging examined performance as a function of wavelength and number of view angles [255]. The concept of angular bandlimiting in a polar to rectilinear interpolation was explored in [906,907]. Such methods tend to be iterative and incorporate a priori information to compensate for data deficiencies [240, 418].

The risk in these approaches is that Nyquist sampling requirements are often ignored [535]. For example, see [757], where tomographic reconstructions with as few as three projections is attempted.

As noted in [885], the Nyquist issues germane to a particular data set may sometimes not be sufficiently considered by some practitioners of CBj methods. Spatial domain algorithms such as CBj may not require Fourier domain processing for image formation, but Fourier analysis should be performed in all such cases to support analysis for proper signal sampling rate and sample density considerations.

Severely under sampled data in the view angle domain, although occasionally investigated, is not suitable for SAR imaging [885].

Ray Tomography with ART Tomographic inversion methods which do not employ small or smooth scattering approximations include [90, 200, 426]. These algorithms are iterative in nature and are based on wave field modeling in forward and inverse directions. These iterative approaches are known as ART (Algebraic Reconstruction Techniques) in ray tomography [56, 455, 457, 458, 690]. ART is an iterative approach to describing the measured wave field, given some initial guess of the in situ field. The estimate of the in situ field is back-projected to the sensor plane and a difference field is computed from the measurement and the back-projected estimate. The difference field is used to update the estimate of the in situ field with a new fore-projected estimate. This iterative inverse scattering solution is formulated as an optimization problem requiring the minimization of the error between a scattering object model and the measured data [556]. ART [398] is equivalent [731] to methods based on projections onto convex sets [1018].

Diffraction Tomography with ART An ART approach to plane wave ultrasonic tomography has been formulated by replacing the forward-projection and back-projection processes of ART in x-ray tomography with forward-propagation and back-propagation processes with a finite difference marching scheme [692].

Strong Scatterer Tomography The Born approximation deals with the case where scatterer effects are weak, but not negligible. In other inverse imaging problems, strong scattering effects must be considered. Approaches for dealing with strong scatterer wave fields have been addressed in various applications [34, 90, 162, 263, 286, 420, 427, 928]. One approach is to consider different orders of multiple scattering. For example, a model of order $n + 1$ considers the incident field and fields scattered by n surrounding scatterers [34]. Other approaches exist that modify the formulation of the diffraction based tomographic imaging problem [743]. Diffraction tomography for strong scattering has been formulated in [258] with an approach called an iterative "Fredholm integral transform" that reduces to the Radon transform in the limit of small wavelength. This formulation leads to a generalized algebraic reconstruction technique, which reduces to CBj in the limit of small wavelength.

Variations of Convolutional Back-Projection A wavenumber domain CBj method was introduced in [51]. This method uses the radial wavenumber variable and involves back-projection followed by two dimensional spatial filtering. The operations are similar to CBj, but performed in reverse order [581]. A fast CBj algorithm that surrenders some image quality for increased computational efficiency is presented in [402].

Multiple Stage Convolutional Back-Propagation In an effort to speed up CBp algorithms, two stage algorithms [450,818,1010] and multiple stage algorithms [476,634,701,702] have been proposed. Recursive algorithms [449,957,958, 960] have also been developed. Algorithms related to factorized back-propagation applied to medical tomography have been proposed [47,48,100] and are called fast hierarchical back-propagation algorithms. At each stage of the fast hierarchical back-propagation algorithm, each of the subimages is at the full resolution level of the final image [47,48]. This is not true of the multilevel inversion method [100], the link based method [701], quad-tree back-propagation algorithm [634,716] and factorized back-propagation [818,958], where resolution of each stage is also refined along with the increasing scene size during reconstruction.

This class of algorithms can be consider the dual of the class of multilevel domain decomposition algorithms [79], which is based on a hierarchical decomposition of the Radon transform.

Factorized back-propagation reduces the computational complexity of standard convolutional back-propagation by a divide and conquer strategy of partitioning both the aperture and image domain, and using redundancy among elements of the decomposition. The level of decomposition can be traded for processing speed or processing artifacts. A fast polar back-projection algorithm for synthetic aperture sonar has been compared with fast factorized back-projection and FR algorithms [833].

Processing Speedup versus Accuracy Fast factorized convolutional back-propagation provides a controlled compromise between computation reduction and tunable levels of approximation [138]. The approximation error comes from scatterers being off boresight relative to the center line of each subpatch. Off boresight scatterers incur a phase error proportional to distance from the boresight line [960]. This phase error causes grating lobes of ambiguous imagery and incorrect height estimates in down-range interferometry [138].

Bistatic Convolutional Back Propagation In bistatic sensing, the time-space model becomes more complex since the integration of reflectivity takes place over more complicated surfaces like prolate spheroid functions, rather than simple spheres [713].

Convolutional Back-Projection for Bistatic SAR Adaptation of CBj for bistatic SAR has been investigated [54,55].

Fourier Reconstruction Algorithms

Interpolation in Fourier Reconstructions One approach to interpolation in Fourier reconstruction is the use of a weighted average. Zeroth order or linear interpolation uses the surrounding four polar samples (up sampled two or three times) in a weighted average. The weighting can be set to be inversely proportional to the Euclidian distance. Some frequency domain zeroth order and bilinear interpolation methods [259] are different than Fourier mode imaging systems that interpolate the spatial Fourier transform of the object function from sets of unevenly spaced sample data [516, 860, 891].

Interpolation in Three Dimensional Fourier Reconstructions Three dimensional versions of Fourier reconstructions are explored in [645].

Interpolations from a Trapezoidal Raster Fourier reconstruction methods in SAR can be based on polar rasters or trapezoidal rasters. The Fourier samples of the polar raster (ω, θ) are interpolated onto a rectilinear raster with coordinates $(\check{k}_{\hat{x}}, \check{k}_{\hat{y}})$ and then the two dimensional inverse FFT is applied [682]. Reconstructions from a polar raster using linear interpolation, from a polar expansion using an approximation of a Fourier inversion formulation and using a trapezoidal rule were all investigated in [643, 644, 646, 647].

An analysis of artifacts due to Fourier domain interpolations between Cartesian to Cartesian sample sets has been conducted [678].

Interpolations from a Baseband Concentric Squares Raster Fourier reconstruction methods can be based on rasters built from concentric squares of samples [563].

Interpolations from a Baseband Circular Raster Bilinear interpolation in Fourier reconstructions from a baseband circular raster for diffraction tomography, with results comparable to CBp, was obtained in [730], but proved computationally more intensive than the CBp methods.

Alternatives to Wavenumber Domain Interpolation in Fourier Reconstructions Another approach is to acquire data in a way that one circular arc (wavenumber domain) passes through each desired rectilinear k-space sample point [326]. This approach tends to not be practically possible because of the large number of projections required [730].

Higher Dimensional Spatially Filtered and Direct FR Methods In three dimensional formulations of back-propagation and coherent summing over spheres, there is often a k^2 factor, where the k^2 arises out of the spherical coordinate representation of the integral formulation [208, 711–713]. It has been noted that with spherical back-propagation this k^2 factor can be canceled as the equations are worked to develop an algorithm. This results in a back-propagation algorithm that does not seem to required "filtering" as demanded by the $|k|$ factor as required in two dimensional CBj with straight path direct back-projection [713].

Advanced Radon Transform Concepts

Higher Dimensional Radon Transforms Higher dimensional and progressive Radon transforms exist and have been proposed for electromagnetic inverse scattering [80], electron microscopy [212] and SAR image formation processing [488,588]. These methods build on cascaded lower dimensional partial Radon transforms and allow multistage CBj to be accomplished.

Discrete Radon Transform A discrete Radon transform has been developed [29–31,47,519,724,775,898,905]. Issues related to data sampling requirements in discrete Radon transforms have been explored.

Three Dimensional Imaging with Stolt Formatting

Three Dimensional Fourier Diffraction Theorem with Planar Sensor Array The Fourier diffraction theorem has two and three dimensional versions. The two dimensional form requires a linear sensor array. The three dimensional form requires a planar sensor array [258,631]. Example systems include [81,993]. Three dimensional Stolt formatting is said to be the Fourier domain equivalent of three dimensional direct and convolutional back-propagation algorithms [373,556].

Three Dimensional Stolt Formatting with Elevation Digital Spotlighting A three dimensional range migration approach with digital spotlighting in elevation is presented in [933]. It is noted that the range swath is usually larger than the height variation within the illuminated scene in three dimensional SAR. In spite of this difference, digital spotlighting can be performed in elevation to seek alias free recovery of wavenumber data because the elevation spatial support is typically larger than that supported by the elevation wavenumber sample density. The length of the elevation synthetic aperture is compared with the height of the scene and zero padded, if needed, to avoid circular convolution [885].

Three Dimensional Image Formation in Wavenumber Space An algorithm for three dimensional imaging has been proposed [828, 871] that is a blending of direct back-propagation techniques [93, 395] with a single frequency wavefront reconstruction technique [866]. The spatial Fourier transforms are viewed as providing a plane wave decomposition of measured spherical wavefronts. The Huygens-Fresnel transfers of the wavenumber domain are viewed as back-propagation operators in k-space of measured wave field data from the aperture to the scattering scene. The algorithm concludes with a three dimensional inverse Fourier transform to convert the wavenumber data to spatial imagery.

Three Dimensional Image Formation with Range Stacking Methods Range stacking has been used to form three dimensional images with two dimensional azimuth-elevation synthetic apertures obtained with a SAR on a rail track system [887].

Cylindrical Stolt Formatting One approach to image formation processing with cylindrical scanning is to seek to formulate wave motion equation solutions in cylindrical coordinates, where Stolt formatting is performed in polar coordinates [334] or where the Green's function specific to the cylindrical scan geometry is utilized [335]. Cylindrical scanning that synthesizes a two dimensional aperture is studied in [348]. This algorithm first back-propagates scattered data onto a planar aperture in the vicinity of the non-planar measurement aperture and then proceeds with three dimensional Stolt formatting. The "three dimensional range migration" algorithm of [600] required the elevation synthetic aperture to be symmetric about the scene in the elevation dimension.

Back-Propagation to a Planar Aperture Combined with Stolt Formatting A study of various three dimensional SAR image formation algorithms in the near field with planar, cylindrical and spherical scanning geometries was performed in [347]. Back-propagation or motion compensation to a planar reference aperture for subsequent three dimensional Stolt formatting was explored.

Three Dimensional Direct Back-Propagation

Direct back-propagation with coherent summation over three dimensional surfaces in an image voxel space has also been explored [713]. Volumetric direct back-propagation requires knowledge of the three dimensional coordinates of every pixel being coherently summed in the image array [1010].

Non-Linear Flight Direct Back-Propagation Direct back-propagation is applicable to image formation processing with data collected from curvilinear synthetic aperture flight paths. Coherent summing in the pixel domain on a common slant-plane or on the ground plane for linear flight SAR and for non-linear flight SAR has been investigated [617]. Three dimensional imagery can be formed with strongly non-linear apertures such as a circular flight path or multiple flight tracks [361–363, 365, 960, 968].

Three Dimensional Polar Formatting

Three Dimensional Fourier Projection Theorem The Fourier projection theorem deals with one direction of plane wave incidence and a finite linear array measurement of straight line integrative projections. The two dimensional Fourier projection theorem can be generalized to three dimensions [150, 387, 993]. With single frequency plane wave illumination in three dimensions, the two dimensional spatial Fourier transform of the measured forward scattered field, as captured over a planar measurement surface, provides planar (instead of linear) "slices" or "cuts" of the three dimensional Fourier transform of the object [252, 256, 257, 674, 686, 993].

Three Dimensional Stereoscopic SAR

Slant-plane SAR imagery has the characteristic that out-of-plane scatterers appear at pixel locations closer in range than their true positions. This is called *foreshortening* or *layover*. Different SAR images, taken with different squint fly-by geometries, will have differences in the layover characteristic. Processing that estimates scatterer height from the different layover of pairs of detected SAR images is called *stereo SAR* [247, 274, 284, 452, 491, 750, 1014]. Hard copy of stereo SAR imagery can be viewed stereoscopically because of the layover dependent parallax.

Three Dimensional Tomographic Methods

Three Dimensional Ray Tomographic Methods Early work in three dimensional image formation from two dimensional projections includes [231, 232, 409]. A fast CBj technique for the three dimensional Radon transform has been developed [49]. A study to extend CBj to a three dimensional pivot point algorithm with circular flight SAR data was conducted in [310]. Back-projection has been employed for three dimensional tomographic image formation [183]. Work to improve elevation focusing by reconstructing a set of uniform baselines in the interpolation step was investigated [599]. Ancillary information about the extension of the height sector being imaged was employed.

Three Dimensional Diffraction Tomographic Methods A three dimensional approach based in diffraction tomography was introduced in [686]. Fourier reconstructions on circular contours [674] and spherical contours for three dimensional imaging have also been investigated [993].

Three Dimensional Tomographic Methods with Multiple Pass Sensing Tomographic imaging has been extended to three dimensional SAR with multiple baseline aperture synthesis [780]. In the two pass case, a second synthetic aperture was formed in the normal direction perpendicular to the slant-plane of the first aperture [600].

Convolutional Back-Propagation with Two Dimensional Apertures A wave theory based formulation of CBp for two dimensional synthetic apertures has been developed [705]. This approach improves the image by adding a spatially variable filter that compensates the antenna beam pattern and includes a geometric scaling factor based on the Jacobian determinant.

Three Dimensional Linear Algebra Approaches Singular value decomposition methods that employ a priori information about radiation penetration depth has been applied to three dimensional SAR with multiple pass data collections [61]. Three dimensional multipass tomographic reconstruction with singular value decomposition and volumetric scattering has been investigated [345].

Three Dimensional Spectral Analysis

A multipass volumetric spaceborne SAR system that uses stacks of several subaperture acquisitions to reconstruct the elevation data through spectral analysis has been proposed [1032]. This system requires estimates of the number of scatterers present in each cell.

Three Dimensional Beamforming Methods

Beamforming in Three Dimensional SAR A three dimensional beamforming approach to synthetic aperture diffraction tomography has also been proposed [837]. Beamforming in the elevation dimension with single pass SAR imagery in down-range and along-track dimensions is investigated in [824]. Three dimensional SAR in a deployed system has been achieved with beamforming methods applied in the elevation dimension [825].

Three Dimensional Multistatic Imaging

Three Dimensional Bistatic SAR Flight path strategies for three dimensional bistatic SARs have been developed [785].

Volumetric SAR Algorithms and Flight Concepts

Example VolSAR System Parameters

A three dimensional data set of an urban environment has been collected for volumetric imaging studies [151], where $\Delta_{el} = 0.18^o$ in $[43.7^o, 45.0^o]$ with $f_c = 9.6$ GHz and a bandwidth of 640 MHz [310]. Another example of parameters used in a VolSAR investigation includes a flight path with a 100 m radius at 10 m above a ground reference point [968]. Another set of simulation parameters for VolSAR is given in [933].

Circular SAR VolSAR processing is applicable to SAR systems with two dimensional aperture synthesis based on circular flight SAR [126, 128, 885]. Circular SAR processing has been explored in the context of tomographic [317] and holographic [885] imaging. The notion of an annular coded aperture has been suggested [838].

"Slant-Plane" Circular SAR A key difference between linear flight SAR and circular flight C-SAR is the "slant-plane" geometry. A slant-plane for linear flight SAR is well defined. In C-SAR, no such *plane* exists. Hence, the notion of "slant-plane C-SAR" [885] may be somewhat of a misnomer. Rather, the "slant-plane" of C-SAR is more accurately viewed as an inverted conic surface, where the inverted cone's tip is located at the scene's center point, and where the circular end of the inverted cone coincides with the flight path of the radar. A spotlight mode signal is sometimes modeled as being produced from a circular flight line, although in reality, the platform flies in a straight line [146]. In these cases, the circular flight model seeks to characterize the effect of motion compensation to a point on SAR phase history data characteristics.

Cylindrical Scanning Carl Wiley postulated volumetric SAR with a synthetic cylindrical array in 1985 [990]. Spotlight mode SAR extended to include the concept of a cylindrical scan has been called *elevation circular SAR* [875, 885].

Planar Spiral Scanning Operational modes which fly in two dimensional spiral flight patterns have been considered [828].

Spherical Scanning Walker [976] suggested recording of data in a spherical format for pivot point algorithms. Spherical scanning that synthesizes a two dimensional aperture is studied in [348]. The algorithm back-propagates backscattered data onto a planar subaperture in the vicinity of the non-planar measurement aperture and then proceeds with three dimensional Stolt formatting.

Multiple Pass Flight Concepts Multiple passes of the sensor can be used to gather data for three dimensional imaging [61].

Planar Arrays from Transverse Linear Arrays One approach to obtaining a planar array from a single pass data collection is to use a linear (physical) array with transverse platform motion [613, 614].

Forward Looking SAR Forward looking SAR sensor configurations have been investigated for a variety of applications [132,133,230,613,615,872]. Forward looking flight empowers elements of volumetric imaging. One study used slant-plane image formation processing to detect ground aircraft that may be obstructing the landing strip on approach [132, 133]. Another study of forward-looking SAR [615] used a transverse linear array with one element transmitting and all receiving. A volumetric model was developed with a third order three dimensional Taylor series expansion.

Down Looking SAR Three dimensional imaging with data collected from a down looking SAR with a single transmitting and multiple receiving linear array antenna has been explored [285].

Sensor Swarm Interferometry Concepts for three dimensional SAR imaging via a swarm of sensors are also emerging [396, 547, 549]. One concept for a sensor swarm in space based SAR is the "cartwheel" configuration [627].

Meandering Flight Paths Another possibility is the use of curved or even meandering flight paths [33, 707], such as flight paths obtainable from a helicopter platform [43]. Basis pursuit algorithms [171, 330] have been applied to linear flight SAR for two dimensional imaging [161, 545, 670] and to three dimensional imaging with circular and pseudo-random flight paths [28, 311, 312, 327, 1003].

SAR Related Topics

Other important SAR related topics and example references are given in Table 14.1.

Table 14.1: SAR Related Topics

SAR Related Topics	
Apodization:	[153, 154, 264, 736, 904, 1006]

Continued on next page...

Table 14.1 – Continued

SAR Related Topics	
Calibration:	[357, 359]
Computation Reduction:	[19, 776, 777]
Canadian RADARSAT:	[761, 762]
Image Processing in Speckled Images:	[92, 574]
Low Cost SAR Systems:	[938, 939]
Polarimetric SAR:	[188, 778]
Radio Frequency Interference Suppression:	[698, 699]
SAR Data Compression:	[18, 19, 436, 437, 506]
SAR Image Quality:	[652]
Space Shuttle Imaging Radar:	[191, 219, 302, 508, 820]
Space Time Adaptive Processing and SAR:	[884]
Speckle and Speckle Reduction:	[5, 294, 303, 394, 568, 952, 961]
Spotlight Mode Processing of Stripmap Mode Data:	[602]
Synthetic Aperture Sonar:	[152, 224, 601, 774]
Turbulence/Propagation Induced Distortion	[318, 380, 951]

14.8 SAR Applications: Beginnings

The applications of SAR are diverse and continue to be discovered [72, 451]. Beyond the obvious defense applications, both strategic and tactical, there exist many non-defense applications of SAR in remote sensing. Table 14.2 lists some of these and provide a few supporting references. SAR remains an active frontier of research, both in the facets of image formation algorithms and in applications that seek to exploit SAR imagery [175].

Table 14.2: List of SAR Applications

SAR Applications	
Abandon Barge Detection:	[633]
Agriculture:	[44]
Cartography:	[156]
Change Detection:	[282, 772]
Diaster Management:	[608]
Earthquake Assessment:	[626]
Flood Monitoring:	[379]
Foliage Penetration:	[268, 489, 826]
Forestry Studies:	[21, 429]
Geology:	[735]

Continued on next page...

Table 14.2 – Continued

SAR Applications	
Glacier Motion:	[787]
Wind Speed Estimation:	[540]
Ground Subsurface Imaging:	[304]
Landing Site Selection:	[567]
Land Cover Classification:	[44]
Land Surface Classification:	[22]
Oil Slick Detection:	[855]
Moving Object Imaging:	[210, 617, 883, 884, 888, 895, 901]
Moving Object Detection:	[760, 878–880, 1008, 1019]
Object Identification:	[122, 616, 714, 893, 924, 945, 998]
Ocean Currents:	[390]
Ocean Stratified Internal Wave Motion:	[633]
Oceanography:	[391, 633]
Reconnaissance:	[505]
Remote Sensing of Oceans:	[11, 116, 942]
Runway Imaging:	[874]
Sea Elevation Mapping:	[815]
Sea Ice Mapping:	[220, 769]
Sea Surface Velocity Vector Mapping:	[598]
Soil Moisture Monitoring:	[608]
Tectonic Mapping:	[608]
Terrain Classification:	[608]
Topography:	[428, 564, 571, 707, 1029]
Topography–Lunar:	[822, 1033, 1034]
Topography–Planetary:	[430, 432]
Unmanned Aerial Vehicle Sensor:	[1025]
Urban Development Monitoring:	[103, 151, 209, 310]
Vegetation Measurement:	[608]

Chapter 15

Fourier Analysis Details

We now provide a degree of mathematical detail by deriving the Fourier transforms of windowed and unwindowed LFM chirp signals [733]. We determine the Fourier transform of an LFM pulse signal by two different methods. The first relies on Fresnel integrals to exactly describe the weight function of the frequency domain. The second relies on a truncated Taylor series approximation and the principle of stationary phase. We also use the principle of stationary phase to approximate the spatial Fourier transform of an unwindowed slow-time phase curtain of a linear flight SAR.

This chapter is outlined as follows: Section 15.1 derives the Fourier transform of the unwindowed (infinite time duration) LFM signal. Section 15.2 derives the exact Fourier transform of the windowed LFM pulse signal. This derivation is accomplished without an approximation provided by a truncated Taylor series expansion. Section 15.3 seeks a simpler (and approximate) expression of the window function of the spectrum of the LFM pulse. Section 15.4 presents the principle of stationary phase and Taylor series based methods for approximately deriving the Fourier transform of chirp signals. Section 15.5 uses the principle of stationary phase to derive an approximate Fourier transform of the windowed LFM pulse signal. Finally, Section 15.6 uses the principle of stationary phase to derive an approximate spatial Fourier transform of the slow-time phase curtain of SAR.

15.1 Unwindowed LFM Waveform Fourier Transform

Key to an exact derivation of the Fourier transform of an unwindowed LFM phase signal is the use of *normalized double sided indefinite Fresnel integrals* to exactly express the infinite duration weight function of the frequency domain.

Fourier Transform Integral of the Unwindowed LFM Waveform

We derive the Fourier transform pair (6.14) previously presented in Section 6.3,

$$\Pi_{[-\infty,+\infty]}(t) \cdot \exp\left(j \cdot 2\alpha \cdot \frac{t^2}{2}\right)$$
$$t \Leftrightarrow \omega \quad \Pi_{[-\infty,+\infty]}(\omega) \cdot \sqrt{\frac{\pi}{\alpha}} \cdot e^{j\frac{\pi}{4}} \cdot \exp\left(-j \cdot \frac{1}{2\alpha} \cdot \frac{\omega^2}{2}\right),$$

which is the exact Fourier transform of an LFM chirp phase signal with infinite time duration.

Fourier Transform Integral of the Unwindowed LFM Waveform

The indefinite Fourier transform integral of an unwindowed LFM waveform is formulated as

$$\begin{aligned} P_{\Pi(?)}(\omega)^{\text{LFM}} &= \mathsf{FT}^t_{1\text{D}}\left\{p_{\Pi(\infty)}(t)^{\text{LFM}}\right\}, \\ &= \int_{-\infty}^{+\infty} \exp(j\alpha t^2) \cdot \exp(-j\omega t) \cdot dt, \\ &= \int_{-\infty}^{+\infty} \exp\left[j\alpha\left(t^2 - t \cdot \frac{\omega}{\alpha}\right)\right] dt. \qquad (15.1) \end{aligned}$$

We have indicated lack of foresight in the expected bandwidth of the result by placing a "?" in the $\Pi(\cdot)$ indicator function of $P_{\Pi(?)}(\omega)^{\text{LFM}}$. As explained in Section 15.1, it is reasonable to expect that the infinite support in time might lead to infinitesimal localization (an impulse) in the frequency domain. But we shall see, contrary to this rule, an infinite time duration LFM chirp signal has infinite bandwidth.

Unwindowed LFM Phase Factor

We proceed with evaluation of the Fourier transform. By completing the square of

$$t^2 - t \cdot \frac{\omega}{\alpha}$$

in (15.1), we obtain

$$t^2 - t \cdot \frac{\omega}{\alpha} + \frac{\omega^2}{(2\alpha)^2} = \left(t - \frac{\omega}{2\alpha}\right)^2.$$

Using the completed square, the phase function of the integrand of (15.1) can be written as

$$t^2 - t \cdot \frac{\omega}{\alpha} = \left(t - \frac{\omega}{2\alpha}\right)^2 - \frac{\omega^2}{(2\alpha)^2}$$

and the Fourier transform becomes

$$P_{\Pi(?)}(\omega)^{\text{LFM}} = \underbrace{\exp\left(-j \cdot \frac{1}{2\alpha} \cdot \frac{\omega^2}{2}\right)}_{\text{quadratic phase factor}} \cdot \underbrace{\int_{-\infty}^{+\infty} \exp\left[j\alpha\left(t - \frac{\omega}{2\alpha}\right)^2\right] dt.}_{?} \qquad (15.2)$$

The quadratic phase factor,

$$\exp\left(-j\cdot\frac{1}{2\alpha}\cdot\frac{\omega^2}{2}\right),$$

which is outside the time integral of (15.2), is the phase factor of the Fourier transform of the LFM time signal expressed in (6.14). We shall now show that the remaining integral element of (15.2),

$$\int_{-\infty}^{+\infty}\exp\left[j\alpha\left(t-\frac{\omega}{2\alpha}\right)^2\right]dt \tag{15.3}$$

reduces to a simple complex constant *valid over all* ω – implying the existence of an infinite duration window function in the ω domain.

Unwindowed LFM Complex Amplitude

To determine the complex amplitude element that results from (15.3), proceed with a change of variables by defining

$$\begin{aligned}\tau &= \sqrt{\alpha}\left(t-\frac{\omega}{2\alpha}\right),\\ d\tau &= \sqrt{\alpha}\cdot dt\end{aligned}$$

and the time integral embedded in (15.3) changes to

$$\int_{-\infty}^{+\infty}\exp\left[j\alpha\left(t-\frac{\omega}{2\alpha}\right)^2\right]dt \quad\Rightarrow\quad \frac{1}{\sqrt{\alpha}}\int_{-\infty}^{+\infty}\exp(j\tau^2)d\tau.$$

Using

$$\exp(j\tau^2)=\cos(\tau^2)+j\sin(\tau^2),$$

together with integral tables that specify the constant values that come from double sided indefinite Fresnel integrals,

$$\begin{aligned}\int_{-\infty}^{+\infty}\cos(\tau^2)d\tau &= \sqrt{\frac{\pi}{2}},\\ \int_{-\infty}^{+\infty}\sin(\tau^2)d\tau &= \sqrt{\frac{\pi}{2}},\end{aligned}$$

we have

$$\frac{1}{\sqrt{\alpha}}\int_{-\infty}^{+\infty}\exp(j\tau^2)d\tau=\frac{1}{\sqrt{\alpha}}\left(\sqrt{\frac{\pi}{2}}+j\sqrt{\frac{\pi}{2}}\right).$$

We can now write the Fourier transform of the unwindowed LFM signal as

$$P_{\Pi(?)}(\omega)^{\text{LFM}} = \frac{1}{\sqrt{\alpha}}\left(\sqrt{\frac{\pi}{2}}+j\sqrt{\frac{\pi}{2}}\right)\cdot\exp\left(-j\cdot\frac{1}{2\alpha}\cdot\frac{\omega^2}{2}\right).$$

The amplitude element reduces to a complex constant with amplitude and phase given by

$$\frac{1}{\sqrt{\alpha}}\left(\sqrt{\frac{\pi}{2}}+j\sqrt{\frac{\pi}{2}}\right)=\sqrt{\frac{\pi}{\alpha}}\cdot e^{j\frac{\pi}{4}}.$$

Hence, the complete Fourier transform is

$$P_{\Pi(?)}(\omega)^{\text{LFM}} = \sqrt{\frac{\pi}{\alpha}} \cdot e^{j\frac{\pi}{4}} \cdot \exp\left(-j\frac{\omega^2}{4\alpha}\right). \tag{15.4}$$

Equation (15.4) is valid for all ω and we conclude that the bandwidth of an infinite time duration LFM chirp signal is infinite and we write

$$\Pi(?) = \Pi(\infty)$$

in the subscript of $P_{\Pi(?)}(\omega)^{\text{LFM}}$ of (15.4).

15.2 LFM Pulse Exact Fourier Transform

An exact derivation of the Fourier transform of an LFM pulse signal is obtained with the use of *normalized single sided definite Fresnel integrals* to express the window function of the LFM chirp pulse spectrum.

Fourier Transform Integral of the Windowed LFM Waveform

Proceeding, our task is to evaluate the definite integral of the Fourier transform of a windowed LFM chirp pulse signal [170, 309, 445],

$$\begin{aligned} P_{\Pi(?)}(\omega)^{\text{LFM}} &= \mathsf{FT}^t_{1\text{D}}\left\{p_{\Pi(T_p)}(t)^{\text{LFM}}\right\}, \\ &= \int_{-\infty}^{+\infty} \Pi_{(T_p)}(t) \cdot \exp(j\alpha t^2) \cdot \exp(-j\omega t) \cdot dt, \\ &= \int_{-T_p/2}^{+T_p/2} \exp\left[j\alpha\left(t^2 - t \cdot \frac{\omega}{\alpha}\right)\right] \cdot dt, \end{aligned}$$

where we have indicated no prior knowledge about the bandwidth of $P_{\Pi(?)}(\omega)^{\text{LFM}}$ with the question mark, "?" (although we have already stated in Section 6.4 that the "?" should be replaced by "$2\alpha T_p$"). By completing the square (see Section 15.1), we have

$$P_{\Pi(?)}(\omega)^{\text{LFM}} = \underbrace{\exp\left(-j\frac{\omega^2}{4\alpha}\right)}_{\text{phase modulation}} \cdot \underbrace{\int_{-T_p/2}^{+T_p/2} \exp\left[j\alpha\left(t - \frac{\omega}{2\alpha}\right)^2\right] dt.}_{?} \tag{15.5}$$

Making the change of variables

$$\begin{aligned} \tau &= \sqrt{\alpha}\left(t - \frac{\omega}{2\alpha}\right), \\ d\tau &= \sqrt{\alpha} \cdot dt \end{aligned}$$

in (15.5) and using the symmetry of the zero centered time pulse to divide the integral into two parts, we have

$$\begin{aligned} P_{\Pi(?)}(\omega)^{\text{LFM}} & \\ = & \frac{1}{\sqrt{\alpha}} \cdot \exp\left(-j\frac{\omega^2}{4\alpha}\right) \\ & \times \left[\int_{\sqrt{\alpha}\left(-\frac{T_p}{2}-\frac{\omega}{2\alpha}\right)}^{0} \exp(j\tau^2)\cdot d\tau + \int_{0}^{\sqrt{\alpha}\left(+\frac{T_p}{2}-\frac{\omega}{2\alpha}\right)} \exp(j\tau^2)\cdot d\tau\right]. \end{aligned}$$

We change the order of the limits of the first integral,

$$\begin{aligned} P_{\Pi(?)}(\omega)^{\text{LFM}} & \\ = & \frac{1}{\sqrt{\alpha}} \cdot \exp\left(-j\frac{\omega^2}{4\alpha}\right) \qquad (15.6) \\ & \times \left[-\int_{0}^{\sqrt{\alpha}\left(-\frac{T_p}{2}-\frac{\omega}{2\alpha}\right)} \exp(j\tau^2)\cdot d\tau + \int_{0}^{\sqrt{\alpha}\left(+\frac{T_p}{2}-\frac{\omega}{2\alpha}\right)} \exp(j\tau^2)\cdot d\tau\right]. \end{aligned}$$

We need tools provided by definite Fresnel integrals to extend our analysis. The complex Fresnel integral is defined as

$$F(\omega) = [C(\omega) + jS(\omega)],$$

where the normalized forms of the real and imaginary parts are

$$\begin{aligned} C(\omega) &= \sqrt{\frac{2}{\pi}} \cdot \int_0^{\omega} \cos\left(\sqrt{\frac{\pi}{2}}x^2\right)\cdot dx, \\ S(\omega) &= \sqrt{\frac{2}{\pi}} \cdot \int_0^{\omega} \sin\left(\sqrt{\frac{\pi}{2}}x^2\right)\cdot dx. \qquad (15.7) \end{aligned}$$

The definition of the normalized Fresnel integrals (15.7) suggests that we add $\sqrt{2/\pi}$ scale factors to (15.6) and adjust the limits of the integral to make explicit the required $\sqrt{\pi/2}$ scale factors contained in the integrands of (15.7),

$$\begin{aligned} P_{\Pi(?)}(\omega)^{\text{LFM}} = & \frac{1}{\sqrt{\alpha}} \cdot \exp\left(-j\frac{\omega^2}{4\alpha}\right) \\ & \times\sqrt{\frac{\pi}{2}} \cdot \left[-\sqrt{\frac{2}{\pi}} \int_0^{\sqrt{\frac{\pi}{2}}\cdot\frac{1}{\sqrt{2\pi\alpha}}(-\alpha T_p-\omega)} \exp(j\tau^2)\cdot d\tau \right. \\ & \left. + \sqrt{\frac{2}{\pi}} \int_0^{\sqrt{\frac{\pi}{2}}\cdot\frac{1}{\sqrt{2\pi\alpha}}(+\alpha T_p-\omega)} \exp(j\tau^2)\cdot d\tau\right]. \end{aligned}$$

Since the Fresnel integral is an odd function, $F(-\omega) = -F(\omega)$, we exploit this odd symmetry in the first integral,

$$\begin{aligned} P_{\Pi(?)}(\omega)^{\text{LFM}} &= \sqrt{\frac{\pi}{\alpha}} \cdot \exp\left(-j\frac{\omega^2}{4\alpha}\right) \\ &\times \frac{1}{\sqrt{2}} \cdot \left[+ \sqrt{\frac{2}{\pi}} \int_0^{\sqrt{\frac{\pi}{2}} \cdot \frac{1}{\sqrt{2\pi\alpha}}(+\alpha T_p + \omega)} \exp(j\tau^2) d\tau \right. \\ &\left. + \sqrt{\frac{2}{\pi}} \int_0^{\sqrt{\frac{\pi}{2}} \cdot \frac{1}{\sqrt{2\pi\alpha}}(+\alpha T_p - \omega)} \exp(j\tau^2) d\tau \right]. \end{aligned}$$

We can now express the spectrum of the LFM pulse in terms of definite Fresnel integral functions,

$$\begin{aligned} &P_{\Pi(?)}(\omega)^{\text{LFM}} \\ &= \sqrt{\frac{\pi}{\alpha}} \cdot \exp\left(-j\frac{\omega^2}{4\alpha}\right) \\ &\quad \underbrace{\cdot \left\{ \frac{1}{\sqrt{2}} \cdot \left\{ F\left[\frac{1}{2\sqrt{\alpha}} \cdot (\omega + \alpha T_p)\right] + F\left[\frac{1}{2\sqrt{\alpha}} \cdot (-\omega + \alpha T_p)\right] \right\} \right\}}_{\omega \text{ domain window function}}, \end{aligned} \tag{15.8}$$

Equation (15.8) is an exact expression of the Fourier transform of the windowed LFM chirp pulse signal. The spectral window function of (15.8) has a nominal width of $2\alpha T_p$. If we express the frequency domain LFM pulse spectrum's exact weight function as

$$\widehat{\Pi}^{\text{exact}}_{(2\alpha T_p)}(\omega)^{\text{LFM}} = \frac{1}{\sqrt{2}} \cdot \left\{ F\left[\frac{1}{2\sqrt{\alpha}} \cdot (\omega + \alpha T_p)\right] + F\left[\frac{1}{2\sqrt{\alpha}} \cdot (-\omega + \alpha T_p)\right] \right\}, \tag{15.9}$$

then the exactly expressed Fourier transform pair of the LFM chirp pulse signal is

$$\Pi_{(T_p)}(t) \cdot \exp(j\alpha t^2) \;\; {}_t\!\Leftrightarrow_\omega \;\; \widehat{\Pi}^{\text{exact}}_{(2\alpha T_p)}(\omega)^{\text{LFM}} \cdot \sqrt{\frac{\pi}{\alpha}} \cdot \exp\left(-j\frac{\omega^2}{4\alpha}\right).$$

An example plot of $\widehat{\Pi}^{\text{exact}}_{(2\alpha T_p)}(\omega)^{\text{LFM}}$ is shown in Fig. 15.1 for an LFM signal with $\alpha T_p = 50$ MHz. The spectrum of an LFM waveform approximates a rectangular shaped spectrum. This holds true if the time-bandwidth product $B \times T_p$ is sufficiently large.

15.3 LFM Pulse Approximate Fourier Transform: Form I

We next seek an approximation of (15.9) that provides a simpler expression of the window function that defines the support of the spectrum of the LFM pulse.

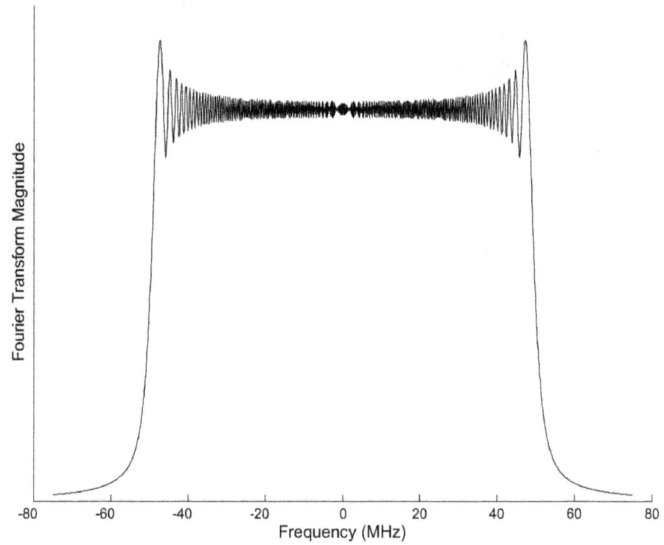

Figure 15.1: The magnitude of the Fourier transform of an LFM signal.

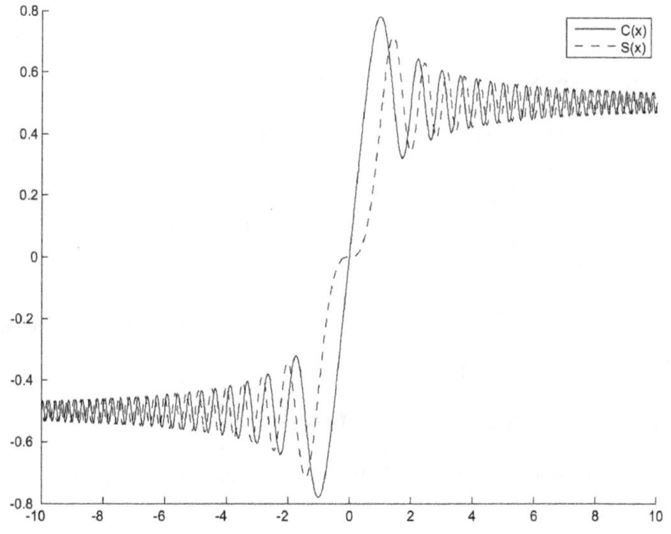

Figure 15.2: Fresnel integral real and imaginary parts.

LFM Pulse Spectrum Window Function Amplitude

Plots of the real and imagery parts of the double sided Fresnel integral are shown in Fig. 15.2. When the range of integration of the definite Fresnel integrals is large, we have the asymptotic limits,

$$\begin{aligned} C(\omega) &\mapsto \frac{1}{2}, \\ S(\omega) &\mapsto \frac{1}{2} \end{aligned}$$

for large ω. Hence, with sufficient integration support, the value of the definite Fresnel integral approaches the complex value

$$\begin{aligned} F(\omega) &\approx \left(\frac{1}{2} + j \cdot \frac{1}{2}\right), \\ &= \frac{1}{\sqrt{2}} \cdot e^{j\frac{\pi}{4}}. \end{aligned}$$

Thus, the exact weight function can be approximated by a window function scaled by a complex weight of $e^{j\frac{\pi}{4}}$,

$$\begin{aligned} \widehat{\Pi}^{\text{exact}}_{(2\alpha T_p)}(\omega)^{\text{LFM}} &\approx \Pi^{\text{approx}}_{(2\alpha T_p)}(\omega)^{\text{LFM}} \cdot \frac{1}{\sqrt{2}} \cdot \left(\frac{1}{\sqrt{2}} \cdot e^{j\frac{\pi}{4}} + \frac{1}{\sqrt{2}} \cdot e^{j\frac{\pi}{4}}\right), \\ &= \Pi^{\text{approx}}_{(2\alpha T_p)}(\omega)^{\text{LFM}} \cdot e^{j\frac{\pi}{4}} \end{aligned}$$

and our first expression of an approximation of the Fourier transform pair of the LFM chirp pulse signal is

$$\Pi_{(T_p)}(t) \cdot \exp(j\alpha t^2) \;\; {}_t\!\Leftrightarrow_\omega \;\; \Pi^{\text{approx}}_{(2\alpha T_p)}(\omega)^{\text{LFM}} \cdot \sqrt{\frac{\pi}{\alpha}} \cdot e^{j\frac{\pi}{4}} \cdot \exp\left(-j\frac{\omega^2}{4\alpha}\right).$$

LFM Pulse Spectrum Window Function Width

Our next task is to show that the width of the LFM pulse spectrum approximate window function is given by

$$\begin{aligned} \Pi^{\text{approx}}_{(?)}(\omega)^{\text{LFM}} &\approx [-\alpha T_p, +\alpha T_p], \\ &= \Pi^{\text{approx}}_{(2\alpha T_p)}(\omega)^{\text{LFM}}. \end{aligned}$$

This window function provides an estimate of the bandwidth of the spectrum of an LFM pulse signal.

Double Sided Normalized Fresnel Integrals

Using approximations of the Fresnel integral based on asymptotic limits of integration,

$$CC\left(\frac{1}{2\sqrt{2\alpha}} \cdot \omega\right) \approx \begin{cases} +\frac{1}{2} & \text{for} \quad \omega \gg 0, \\ -\frac{1}{2} & \text{for} \quad \omega \ll 0. \end{cases}$$

$$SS\left(\frac{1}{2\sqrt{2\alpha}} \cdot \omega\right) \approx \begin{cases} +\frac{1}{2} & \text{for} \quad \omega \gg 0, \\ -\frac{1}{2} & \text{for} \quad \omega \ll 0, \end{cases}$$

we define

$$FF(\cdot) = CC(\cdot) + jSS(\cdot).$$

Shifted Double Sided Normalized Fresnel Integrals

When viewed in terms of double sided expressions, the magnitude of the complex function

$$FF\left[\frac{1}{2\sqrt{2\alpha}} \cdot (\omega + \alpha T_p)\right]$$

is a left shifted complex "up step" function that steps up at $\omega = -\alpha T_p$ from values of about $-\sqrt{2}/2$ to a rise of about $+\sqrt{2}/2$. Likewise, the magnitude of

$$FF\left[\frac{1}{2\sqrt{2\alpha}} \cdot (-\omega + \alpha T_p)\right]$$

is a right shifted "down step" function that steps down at $\omega = +\alpha T_p$ from about $+\sqrt{2}/2$ to about $-\sqrt{2}/2$. That is, the left shifted double sided Fresnel integral function of (15.8) has a magnitude function approximated by

$$\left|FF\left[\frac{1}{2\sqrt{2\alpha}} \cdot (\omega + \alpha T_p)\right]\right| \approx \begin{cases} +\frac{\sqrt{2}}{2} & \text{for} \quad \omega > -\alpha T_p, \\ -\frac{\sqrt{2}}{2} & \text{for} \quad \omega < -\alpha T_p. \end{cases}$$

The magnitude of the right shifted Fresnel integral function is approximated by

$$\left|FF\left[\frac{1}{2\sqrt{2\alpha}} \cdot (-\omega + \alpha T_p)\right]\right| \approx \begin{cases} -\frac{\sqrt{2}}{2} & \text{for} \quad \omega > +\alpha T_p, \\ +\frac{\sqrt{2}}{2} & \text{for} \quad \omega < +\alpha T_p. \end{cases}$$

The left shifted Fresnel integral can be interpreted as turning "on" the window function of the chirp signal spectrum. The other subtracted, right shifted Fresnel integral can be viewed as turning "off" the window function. The resulting "on" duration of the composite window function

$$\Pi_{(?)}^{\text{LFM}}(\omega)^{\text{approx}}$$

is

$$-\alpha T_p \leq \omega \leq \alpha T_p.$$

We express the approximate window of the LFM pulse spectrum as

$$\Pi_{[-\alpha T_p, \alpha T_p]}^{\text{approx}}(\omega)^{\text{LFM}}$$

or as

$$\Pi_{(2\alpha T_p)}^{\text{approx}}(\omega)^{\text{LFM}}.$$

Hence, we have obtained an approximate form of the Fourier transform of the LFM pulse waveform,

$$\Pi_{(T_p)}(t) \cdot \exp(j\alpha t^2) \quad {}_t\Leftrightarrow_\omega \quad \Pi_{(2\alpha T_p)}^{\text{approx}}(\omega)^{\text{LFM}} \cdot \sqrt{\frac{\pi}{\alpha}} \cdot e^{j\frac{\pi}{4}} \cdot \exp\left(-j\frac{\omega^2}{4\alpha}\right).$$

15.4 Method of Stationary Phase

A principle of stationary phase [202, 307, 407, 903] can be used to approximately evaluate Fourier expressions involving a subclass of chirp signals. The principle of stationary phase is stated as follows [734]:

Given an amplitude function $a(t)$ that varies slowly (by comparison) and given a phase function $\phi(t)$ that is highly oscillating within the limits of the integral, and which is twice differentiable, the following can be used to approximately determine the Fourier transform:

$$\int_a^b a(t) \cdot \exp\left[-j\phi(t)\right] \cdot dt \approx \sqrt{\frac{2\pi}{\ddot{\phi}(t^*)}} \cdot e^{j\frac{\pi}{4}} \cdot a(t^*) \cdot \exp\left[-j\phi(t^*)\right]. \tag{15.10}$$

The method of stationary phase as expressed by (15.10) holds sufficiently accurate provided that t^* is the only critical point within the interval $[a, b]$ where

$$\partial_t \left\{\phi(t^*)\right\} = 0.$$

The point t^* is the only point of stationary phase within the limits of the integral. Rapid oscillations away from t^* contribute little to the integral. The principle of stationary phase is derived from a truncated Taylor series expansion of $\phi(t)$ about the point t^*.

15.5 LFM Pulse Approximate Fourier Transform: Form II

We next evaluate the Fourier transform of the LFM pulse with a Taylor series expansion and application of the principle of stationary phase. This development presented here follows closely the development presented in [445]. We express results in terms of both angular and cyclic frequency parameters. For LFM waveforms, the cyclic and angular frequency bandwidths are indicated by

$$\begin{aligned} B_{\text{LFM}} &= \gamma T_p, \\ \beta_{\text{LFM}} &= 2\alpha T_p. \end{aligned}$$

LFM Pulse Fourier Transform

Our task is to evaluate

$$\begin{aligned} P_{\Pi(B_{\text{LFM}})}(f)^{\text{LFM}} &= \int_{-\infty}^{+\infty} \Pi_{(T_p)}(t) \cdot \exp\left(j\pi\gamma t^2\right) \cdot \exp\left(-j2\pi f t\right) \cdot dt, \\ &= \int_{-\infty}^{+\infty} \Pi_{(T_p)}(t) \cdot \exp\left[j\left(\pi\gamma t^2 - 2\pi f t\right)\right] \cdot dt, \\ &= \int_{-\infty}^{+\infty} \Pi_{(T_p)}(t) \cdot \exp\left[j\phi\left(t\right)\right] \cdot dt, \end{aligned} \tag{15.11}$$

where we have used foresight to indicate that we know the bandwidth of the LFM pulse signal is B_{LFM}, since by definition, the LFM pulse is dependent upon the chirp rate parameter γ that is defined by the pulse bandwidth and time duration parameters,

$$\gamma = \frac{B_{\text{LFM}}}{T_p}.$$

Phase Function of Fourier Transform of Baseband LFM Pulse

The phase function of the Fourier transform (15.11) is

$$\phi(t) = \pi\gamma t^2 - 2\pi f t. \tag{15.12}$$

First Derivative

As required in the application of the principle of stationary phase, we must find the first derivative of $\phi(t)$,

$$\partial_t \{\phi(t)\} = 2\pi\gamma t - 2\pi f,$$

set it equal to zero to identify the solo stationary point,

$$2\pi\gamma t^* - 2\pi f = 0$$

and solve for t^*. We require $t^* \in [-T_p/2, +T_p/2]$. There is a coupling between time and frequency. Hence, the stationary point is characterized in various useful forms by

$$\begin{aligned} t^* &= \frac{1}{\gamma} f, & (15.13) \\ &= \frac{T_p}{B_{\text{LFM}}} f, \\ &= \frac{2\pi f}{2\pi\gamma}, \\ &= \frac{\omega}{2\alpha}. & (15.14) \end{aligned}$$

Hence, t^* can be characterized as a function of angular frequency ω,

$$t^*(\omega) = \frac{\omega}{2\alpha} \text{ for } \omega \in \beta_{\text{LFM}}, \tag{15.15}$$

or as expressed in (15.13), as a function of cyclic frequency,

$$t^*(f) = \frac{f}{\gamma} \text{ for } f \in \beta_{\text{LFM}}.$$

Considering this coupling between time and frequency, and considering the time limits of the LFM pulse, there must exist a window function over the set of candidate time instances t^* that we can write as

$$\Pi_{(T_p @ 0)}(t^*),$$

where we consider t^* as a free parameter as $\omega \in \beta_{\text{LFM}}$ or $f \in \beta_{\text{LFM}}$ is allowed to vary. If we consider the dual couplet that specifies frequency as a function of time, then implicitly from (15.15), windows also exist in the f and ω domains,

$$\Pi_{(\gamma T_p)}(f),$$
$$\Pi_{(2\alpha T_p)}(\omega).$$

Second Derivative

The second partial derivative of the phase function with respect to time gives

$$\begin{aligned} \partial_t^2 \{\phi(t)\} &= \pm 2\pi\gamma, \\ &= \pm 2\alpha. \end{aligned} \tag{15.16}$$

This result is positive if the stationary point is at a phase minimum (which occurs with an up chirp LFM pulse) and negative if at a phase maximum (which occurs with a down chirp LFM pulse). All higher order derivatives of the LFM chirp signal's quadratic phase function equate to zero (by definition of the LFM waveform).

Taylor Series Expansion

Expanding the phase function (15.12) about t^* with a Taylor series gives

$$\phi(t) = \phi(t^*) + \dot{\phi}(t^*) \cdot (t - t^*) + \frac{1}{2}\ddot{\phi}(t^*) \cdot (t - t^*)^2 .$$

Since the method of stationary phase requires that t^* be such that

$$\dot{\phi}(t^*) = 0,$$

we loose the first order term of the Taylor series expansion,

$$\phi(t) = \phi(t^*) + \frac{1}{2}\ddot{\phi}(t^*) \cdot (t - t^*)^2 .$$

Hence, the Fourier transform of (15.11) can be written as

$$\begin{aligned} & P_{\Pi(B_{\mathrm{LFM}}@0)}(f)^{\mathrm{LFM}} \\ &= \int_{-\infty}^{+\infty} \Pi_{(T_p@0)}(t) \cdot \exp[j\phi(t)] \cdot dt, \\ &= \int_{-\infty}^{+\infty} \Pi_{(T_p@0)}(t^*) \cdot \exp\left[j\phi(t^*) + j\frac{1}{2}\ddot{\phi}(t^*) \cdot (t - t^*)^2\right] dt, \\ &= \exp[j\phi(t^*)] \int_{-\infty}^{+\infty} \Pi_{(T_p@0)}(t^*) \cdot \exp\left[j\frac{1}{2}\ddot{\phi}(t^*) \cdot (t - t^*)^2\right] dt. \end{aligned}$$

By interchanging the indefinite integral (over t) and the window function in terms of t^*, we obtain, with an acceptable degree of approximation [445], the following,

$$P_{\Pi(B_{\mathrm{LFM}})}(f)^{\mathrm{LFM}} \approx \exp[j\phi(t^*)] \cdot \Pi_{(T_p@0)}(t^*) \cdot \int_{-\infty}^{+\infty} \exp\left[j\frac{1}{2}\ddot{\phi}(t^*) \cdot (t - t^*)^2\right] dt.$$

Substituting

$$t - t^* = \tau$$

into the integrand,

$$P_{\Pi(B_{\mathrm{LFM}})}(f)^{\mathrm{LFM}} \approx \exp[j\phi(t^*)] \cdot \Pi_{(T_p@0)}(t^*) \cdot \int_{-\infty}^{+\infty} \exp\left[j\frac{1}{2}\ddot{\phi}(t^*) \cdot \tau^2\right] d\tau.$$

Setting

$$v^2 = \frac{1}{2}\ddot{\phi}(t^*) \cdot \tau^2,$$

and using

$$\begin{aligned} v &= \frac{1}{\sqrt{2}} \cdot \sqrt{\ddot{\phi}(t^*)} \cdot \tau, \\ du &= \frac{\sqrt{2}}{\sqrt{\ddot{\phi}(t^*)}} \cdot dv, \end{aligned}$$

we obtain an integral expression in terms of the new variable v,

$$P_{\Pi(B_{\text{LFM}})}(f)^{\text{LFM}} \approx \frac{\sqrt{2}}{\sqrt{\ddot{\phi}(t^*)}} \cdot \exp\left[j\phi(t^*)\right] \cdot \Pi_{(T_p@0)}(t^*) \cdot \int_{-\infty}^{+\infty} \exp(jv^2)dv.$$

This expression contains a double sided indefinite Fresnel integral,

$$\int_{-\infty}^{+\infty} \exp(jv^2)dv = \sqrt{\frac{\pi}{2}} + j\sqrt{\frac{\pi}{2}}.$$

Hence,

$$P_{\Pi(B_{\text{LFM}})}(f)^{\text{LFM}} \approx \frac{1}{\sqrt{\ddot{\phi}(t^*)}} \exp\left[j\phi(t^*)\right] \cdot \Pi_{(T_p@0)}(t^*) \cdot \sqrt{\pi} \cdot (1 + j1).$$

From (15.13),

$$t^* = \frac{1}{\gamma}f$$

and from (15.16),

$$\ddot{\phi}(t^*) = 2\pi\gamma,$$

we obtain

$$\begin{aligned} \Pi_{(T_p@0)}(t^*) &= \Pi_{(\gamma T_p@0)}(f), \\ \phi(t^*) &= \pi\gamma\left(\frac{f}{\gamma}\right)^2 - 2\pi f\frac{f}{\gamma}, \\ &= -\pi\frac{f^2}{\gamma}, \\ &= -\frac{1}{2} \cdot \frac{\omega^2}{2\alpha}. \end{aligned}$$

These expressions yield the following approximation for the Fourier transform of a baseband LFM pulse:

$$\begin{aligned} P_{\Pi(B_{\text{LFM}})}(f)^{\text{LFM}} &\approx \Pi_{(\gamma T_p@0)}(f) \cdot (1 + j) \cdot \frac{1}{\sqrt{\pi} \cdot \sqrt{2}} \cdot \frac{\sqrt{\pi}}{\sqrt{\gamma}} \cdot \exp\left(-j\frac{\pi}{\gamma}f^2\right), \\ &\approx \Pi_{(\gamma T_p@0)}(f) \cdot e^{j\frac{\pi}{4}} \cdot \frac{1}{\sqrt{\gamma}} \cdot \exp\left(-j\frac{\pi}{\gamma}f^2\right). \end{aligned}$$

Summarizing, we have obtained the Fourier transform pair in terms of cyclic frequency f,

$$\Pi_{(T_p@0)}(t) \cdot \exp\left(j\pi \cdot \frac{\gamma}{1} \cdot t^2\right) \ {}_t\Leftrightarrow_f \ \sqrt{\frac{1}{\gamma}} \cdot e^{j\frac{\pi}{4}} \cdot \Pi_{(\gamma T_p@0)}(f) \cdot \exp\left(-j\pi \cdot \frac{1}{\gamma} \cdot f^2\right),$$

and in terms of angular frequency ω,

$$\Pi_{(T_p@0)}(t) \cdot \exp\left(j\frac{1}{2} \cdot \frac{2\alpha}{1} \cdot t^2\right) \ {}_t\Leftrightarrow_\omega \ \sqrt{\frac{\pi}{\alpha}} \cdot e^{j\frac{\pi}{4}} \cdot \Pi_{(2\alpha T_p@0)}(\omega) \cdot \exp\left(-j\frac{1}{2} \cdot \frac{1}{2\alpha} \cdot \omega^2\right).$$

15.6 Phase Curtain Fourier Transform: Stationary Phase Approximate Form

A stationary phase approach [78] can be used to determine the Fourier transform of the two way slow-time phase curtain or along-track chirp signal [407, 771],

$$\begin{aligned} &\mathsf{FT}_{1\mathrm{D}}^{u_y}\left\{\exp\left[-jk_{\omega_c} \cdot 2\sqrt{x_n^2 + (u_y - y_n)^2}\right]\right\}, \\ &\quad = \int_{-\infty}^{+\infty} \exp\left[-jk_{\omega_c} \cdot 2\sqrt{x_n^2 + (u_y - y_n)^2}\right] \cdot \exp(-j2k_{u_y} u_y) \cdot du_y, \end{aligned}$$

where we have used foresight in modifying the kernel of the Fourier transform

$$\exp(-jk_{u_y} u_y) \mapsto \exp(-j2k_{u_y} u_y)$$

to account for two way signal travel. To determine the Fourier transform (15.17), the method of stationary phase depends on the following derivatives,

$$\begin{aligned} \phi(u_y) &= -k_{\omega_c} \cdot 2\sqrt{x_n^2 + (u_y - y_n)^2} - 2k_{u_y} u_y, \\ \dot{\phi}(u_y) &= -k_{\omega_c} \cdot 2\frac{(u_y - y_n)}{\sqrt{x_n^2 + (u_y - y_n)^2}} - 2k_{u_y}, \\ \ddot{\phi}(u_y) &= -k_{\omega_c} \cdot 2\frac{x_n^2}{[x_n^2 + (u_y - y_n)^2]^{3/2}}. \end{aligned}$$

Solving $\dot{\phi}(u_y^*) = 0$ for the stationary point u_y^* gives

$$u_y^* = y_n + x_n \cdot \frac{k_{u_y}}{\sqrt{k_{\omega_c}^2 - k_{u_y}^2}}.$$

The phase function at the stationary point is

$$\phi(u_y^*) = -x_n \cdot 2\sqrt{k_{\omega_c}^2 - k_{u_y}^2} - y_n \cdot 2k_{u_y}.$$

The amplitude function of the result is obtained by inserting

$$\ddot{\phi}(u_y^*) = -2\frac{(k_{\omega_c}^2 - k_{u_y}^2)^{3/2}}{x_n^2 k_{\omega_c}^2}$$

into

$$\begin{aligned}\sqrt{\frac{j2\pi}{\ddot{\phi}(u_y^*)}} &= \sqrt{\frac{-j\pi k_{\omega_c}^2 x_n^2}{(k_{\omega_c}^2 - k_{u_y}^2)^{3/2}}} \\ &= \sqrt{\pi}\cdot\sqrt{-j}\cdot x_n \cdot \frac{k_{\omega_c}}{\sqrt{(k_{\omega_c}^2 - k_{u_y}^2)^{3/2}}}.\end{aligned}$$

Hence, evaluation of the Fourier transform gives,

$$\begin{aligned}&SS(\omega_c, k_{u_y})_n \\ &= \sqrt{\pi}\cdot e^{-j\frac{\pi}{4}}\cdot x_n \cdot \frac{k_{\omega_c}}{\sqrt{(k_{\omega_c}^2 - k_{u_y}^2)^{3/2}}} \qquad (15.17)\\ &\quad \cdot \exp(-jx_n\cdot 2\sqrt{k_{\omega_c}^2 - k_{u_y}^2}\,)\cdot\exp(-jy_n\cdot 2k_{u_y}), \\ &= g_n\cdot a(\omega_c, k_{u_y}, x_n)_{\text{psp}}\cdot\exp(-jx_n\cdot 2\sqrt{k_{\omega_c}^2 - k_{u_y}^2}\,)\cdot\exp(-jy_n\cdot 2k_{u_y}). \qquad (15.18)\end{aligned}$$

The *frequency-wavenumber weight function* provided by a principle of stationary phase is

$$a(\omega_c, k_{u_y}, x_n)_{\text{psp}} = \sqrt{\pi}\cdot e^{-j\frac{\pi}{4}}\cdot x_n\cdot\frac{k_{\omega_c}}{\sqrt{(k_{\omega_c}^2 - k_{u_y}^2)^{3/2}}}$$

and is specific to scatterer down-range position x_n.

Bibliography

[1] N. Abramson. Incoherent holography. In *Proceedings of SPIE, Holography 2000*, volume 4149, pages 153–164, 3 October 2000.

[2] J. W. Adams and R. W. Bayma. Digital filters for synthetic aperture radar. In *IEEE National Radar Conference*, pages 123–128, Ann Arbor, Michigan, 20–21 April 1988.

[3] J. W. Adams, R. W. Bayma, M. E. Lawrence, and L. Petrosian. On the modified chirp-z transform for synthetic aperture radar systems. *IEEE Transactions on Signal Processing*, 39(4):953–955, April 1991.

[4] M. F. Adams and A. P. Anderson. Synthetic aperture tomographic (SAT) imaging for microwave diagnostics. *Proceedings of the IEE, Part H*, 129:83–88, 1982.

[5] M. Aggarwal and U. B. Desai. Modeling and restoration of multilook corrected speckled images. In *Proceedings of the IEEE International Conference on Image Processing*, pages 498–501, Washington, DC, October 1995.

[6] M. Agranovsky and E. T. Quinto. Injectivity sets for the Radon transform over circles and complete systems of radial functions. *Journal of Functional Analysis*, 139:383–414, 1996.

[7] S. Aks and D. J. Vezzetti. Ultrasonic scattering theory I: Scattering by single objects. *Ultrasonic Imaging*, 2(85–101), 1980.

[8] M. Albuquerque, P. Prats, and R. Scheiber. Applications of time-domain back-projection SAR processing in the airborne case. In *Proceedings of the European Conference on Synthetic Aperture Radar EuSAR*, 2008.

[9] E. Alivizatos, A. Potsis, A. Reigber, A. Moreira, and N. Uzunoglu. SAR processing with motion compensation using the extended wavenumber algorithm. In *Proceedings of the 5th European Conference on Synthetic Aperture Radar EuSAR*, volume 1, pages 157–160, 2004.

[10] M. R. Allen and L. E. Hoff. Wide-angle wideband SAR matched filter image formation for enhance detection performance. In *Proceedings of SPIE, Algorithms for Synthetic Aperture Radar Imagery*, volume 2230, pages 302–314, Orlando, Florida, 4–8 April 1994.

[11] W. R. Alpers and C. L. Rufenach. The effect of orbital motions on synthetic aperture radar imagery of ocean waves. *IEEE Transactions on Antennas and Propagation*, 27(5):685–690, September 1979.

[12] L.-E. Andersson. On the determination of a function from spherical averages. *SIAM Journal of Mathematical Analysis*, 19(1):214–232, 1988.

[13] O. Arikan and D. C. Munson, Jr. An analysis of keystone format spotlight mode synthetic aperture radar. In *Proceedings of the 21st Asilomar Conference on Signals, Systems, and Computers*, pages 690–694, Pacific Grove, California, 2–4 November 1987.

[14] O. Arikan and D. C. Munson, Jr. A tomographic formulation of bistatic synthetic aperture radar. In *Proceedings ComCon'88*, pages 418–431, Baton Rouge, Louisiana, October 1988.

[15] O. Arikan and D. C. Munson, Jr. Analysis and simulation of a new algorithm for spotlight-mode synthetic aperture radar. In *IEEE International Conference on Acoustics, Speech, and Signal Processing*, volume 3, pages 1453–1456, Glasgow, Scotland, 23–26 May 1989.

[16] O. Arikan and D. C. Munson, Jr. A new backprojection algorithm for spotlight-mode SAR and ISAR. In *Proceedings of SPIE, Symposium on Innovative Science and Technology for Government and Civilian Applications*, pages 1453–1456, Los Angeles, California, May 1989.

[17] O. Arikan and D. C. Munson, Jr. Interpolation between two uniform grids. In *Proceedings of the Bilkent International Conference on New Trends in Communication, Control and Signal Processing*, pages 1044–1058, Ankara, Turkey, 2-5 July 1990.

[18] D. V. Arnold. Vector quantization of synthetic array radar data. Master's thesis, Brigham Young University, Provo, Utah, 1987.

[19] D. V. Arnold, D. M. Chabries, and P. L. Jackson. Reduction of SAR digital processing time using vector quantization. In *IEEE 1988 National Radar Conference*, pages 134–139, Ann Arbor, Michigan, April 1988.

[20] S. R. Arridge and J. S. Schotland. Optical tomography: Forward and inverse problems. *Inverse Problems*, 25(123010):1–59, 2009.

[21] J. Askne, P. B. G. Dammert, L. M. H. Ulander, and G. Smith. C-band repeat-pass interferometric SAR observations of the forest. *IEEE Transactions on Geoscience and Remote Sensing*, 35(1):25–35, January 1997.

[22] J. Askne and J. O. Hagberg. Potential of interferometric SAR for classification of land surfaces. In *Proceedings of the International Geoscience and Remote Sensing Symposium*, volume 3, pages 985–987, 1993.

[23] D. A. Ausherman. Digital vs. optical techniques in synthetic aperture radar data processing. In *Proceedings of SPIE, Application of Digital Image Processing*, volume 119, pages 238–257, Bellingham, Washington, 25–26 August 1977.

[24] D. A. Ausherman. Digital versus optical techniques in synthetic aperture radar (SAR) data processing. *Optical Engineering*, 19:34–44, March/April 1980.

[25] D. A. Ausherman. SAR digital image formation processing. In *Proceedings of SPIE, Digital Image Processing*, volume 528, pages 118–133, Los Angeles, California, 22–23 January 1985.

[26] D. A. Ausherman, A. Kozma, J. L. Walker, H. M. Jones, and E. C. Poggio. Developments in radar imaging. *IEEE Transactions on Aerospace and Electronic Systems*, 20(4):363–400, July 1984.

[27] C. D. Austin, E. Ertin, and R. L. Moses. Sparse multipass 3D imaging: Applications to the GOTCHA data. In *Proceedings of SPIE, Algorithms for Synthetic Aperture Radar Imagery XVI*, Orlando, Florida, 13–17 April 2009.

[28] C. D. Austin and R. L. Moses. Wide-angle sparse 3D synthetic aperture radar imaging for nonlinear flight paths. In *IEEE National Aerospace and Electronics Conference (NAECON)*, pages 330–336, 16–18 July 2008.

[29] A. Averbuch, R. R. Coifman, D. L. Donoho, M. Israeli, and Y. Shkolnisky. A framework for discrete integral transformations I–the pseudo-polar Fourier transform. *SIAM Journal of Scientific Computing*, 30(2):764–784, February 2008.

[30] A. Averbuch, R. R. Coifman, D. L. Donoho, M. Israeli, Y. Shkolnisky, and I. Sedelnikov. A framework for discrete integral transformations I–the discrete Radon transform. *SIAM Journal of Scientific Computing*, 30(2):785–803, February 2008.

[31] A. Averbuch, I. Sedelnikov, and Y. Shkolnisky. CT reconstruction from parallel and fan-beam projections by a 2-D discrete Radon transform. *IEEE Transactions on Image Processing*, 21(2):733–741, February 2012.

[32] D. Avitzour. SNR/bandwidth tradeoff in coherent radar sampling. *IEEE Transactions on Aerospace and Electronic Systems*, 26(2):403–405, March 1990.

[33] S. R. J. Axelsson. Beam characteristics on three-dimensional SAR in curved or random paths. *IEEE Transactions on Geoscience and Remote Sensing*, 42(10):2324–2334, October 2004.

[34] M. Azimi and K. C. Kak. Distortion in diffraction tomography caused by multiple scattering. *IEEE Transactions on Medical Imaging*, 2:176–195, 1983.

[35] B. B. Baker and E. T. Copson. *The Mathematical Theory of Huygens' Principle*. Oxford: Claredon Press, 1950. Third Edition, 1987.

[36] J. Ball, S. A. Johnson, and F. Stenger. Explicity inversion of the Helmholtz equation for ultrasound insonifcation and spherical detection. *Acoustical Holography*, 8, 1980.

[37] R. Bamler. Diffraction and diffraction tomography: A signal theoretical description. In *Proceedings of Image Science*, volume Ph 150, pages 281–284, Helsinki, Finland, 1985. Acta Polytechnia Scandinavica.

[38] R. Bamler. Can one see the Ewald sphere? *Applied Optics*, 27(10):2012–2016, May 1988.

[39] R. Bamler. A systematic comparison of SAR focusing algorithms. In *Proceedings of the IEEE International Geoscience Remote Sensing Symposium*, volume 2, pages 1005–1009, Helsinki, Finland, June 1991.

[40] R. Bamler. A comparison of range-Doppler and wavenumber domain SAR focusing algorithms. *IEEE Transactions on Geoscience and Remote Sensing*, 30(4):706–713, July 1992.

[41] R. Baraniuk and P. Steeghs. Compressive radar imaging. In *Proceedings of the IEEE International Radar Conference*, Waltham, Massachusetts, April 2007.

[42] B. C. Barber. Theory of digital imaging from orbital synthetic aperture data. *International Journal of Remote Sensing*, 6(7):1009–1057, July 1985.

[43] M. W. Barclay and N. K. Williams. High resolution SAR/ISAR imaging from a helicopter platform. In *IEE Radar Conference*, volume 449, 14-16 October 1997.

[44] C. F. Barnes and J. Burki. Late-season rural land-cover estimation with polarmetric SAR intensity pixel blocks and sigma-tree structured near neighbor classifiers. *IEEE Transactions on Geoscience and Remote Sensing*, 44(9):2384–2392, September 2006.

[45] C. W. Barnes. Object restoration in a diffraction-limited imaging system. *Journal of the Optical Society of America*, 56(5):575–578, May 1966.

[46] D. Barrick. FM/CW radar signal and digital processing. *NOAA Technical Report*, ERL 283-WPL, July 1973.

[47] S. Basu and Y. Bresler. $O(N^2 \log N)$ filtered backprojection reconstruction algorithm for tomography. *IEEE Transactions on Image Processing*, 9(10):1760–1773, October 2000.

[48] S. Basu and Y. Bresler. Error analysis and performance optimization of fast hierarchical backprojection algorithms. *IEEE Transactions on Image Processing*, 10(7):1103–1117, July 2001.

[49] S. Basu and Y. Bresler. $O(N^2 \log N)$ filtered backprojection algorithm for 3-D Radon transform. *IEEE Transactions on Medical Imaging*, 21(2):76–88, February 2002.

[50] Henry Bateman. *The Mathematical Analysis of Electrical and Optical Wave-Motion of the Basis of Maxwell's Equations*. Cambridge, University Press, 1915.

[51] R. H. T. Bates and T. M. Peters. Towards improvements in tomography. *New Zealand Journal of Science*, 14:883–896, 1971.

[52] R. H. T. Bates, V. A. Smith, and R. D. Murch. Manageable multidimensional inverse scattering theory. *Physics Reports*, 201(4):187–277, April 1991.

[53] J. L. Bauck and W. K. Jenkins. Tomographic processing of spotlight-mode synthetic aperture radar signals with compensation for wavefront curvature. In *IEEE International Conference on Acoustics, Speech, and Signal Processing*, volume 2, pages 1192–1195, April 1988.

[54] J. L. Bauk and W. K. Jenkins. Convolution-backprojection image reconstruction for bistatic synthetic aperture radar. In *Proceedings of the IEEE International Symposium on Circuits and Systems*, pages 1512–1515, Portland, Oregon, May 1989.

[55] J. L. Bauk and W. K. Jenkins. Convolution-backprojection image reconstruction for bistatic synthetic aperture radar with correction for wavefront curvature and propagation attenuation. In *Proceedings of SPIE, Technical Symposium on Aerospace Sensing*, volume 1101, pages 11–18, Orlando, Florida, March 1989.

[56] S. H. Bellman, R. Bender, R. Gordon, and J. E. Rowe. ART is science, being a defense of algebraic reconstruction techniques for three-dimensional electron microscopy. *Journal of Theoretical Biology*, 32:205–216, 1971.

[57] R. Bender, S. H. Bellman, and R. Gordon. ART and the robosome: A preliminary report on the three dimensional structure of individual ribosomes determined by an algebraic reconstruction technique. *Journal of Theoretical Biology*, 29:483–487, 1970.

[58] J. R. Bennett and I. G. Cumming. A digital processor for the production of SEASAT synthetic aperture imagery. In *Proceedings of the SURGE Workshop*, volume ESA-SP-154, Frascati, Italy, 1979.

[59] J. R. Bennett, I. G. Cumming, and R. Deane. The digital processing of SEASAT synthetic aperture radar data. In *Proceedings of the IEEE International Radar Conference*, pages 168–175, April 1980.

[60] J. R. Bennett, I. G. Cumming, P. R. McConnell, and L. Gutteridge. Features of a generalized digital synthetic aperture radar processor. In *Proceedings of the Fifteenth International Symposium on Remote Sensing of the Environment*, Ann Arbor, Michigan, May 1981.

[61] P. Berardino, G. Fornaro, R. Lanari, E. Sansosti, F. Serafino, and F. Soldovieri. Multi-pass synthetic aperture radar for 3-D focusing. In *Proceedings of the IEEE International Geoscience and Remote Sensing Symposium*, volume 1, pages 176–178, 2002.

[62] P. Berens. Extended range migration algorithm for squinted spotlight SAR. In *Proceedings of the IEEE International Geoscience and Remote Sensing Symposium*, volume 6, pages 4053–4055, Toulouse, France, July 2003.

[63] P. Berens. Efficient wave number domain processing for squinted spotlight SAR. In *Proceedings of the 5th European Conference on Synthetic Aperture Radar*, volume 1, pages 171–174, Ulm, Germany, May 2004.

[64] M. Berger, D. Bruck, M. Fischer, K. J. Langenberg, J. Oberst, and V. Schmitz. Potential and limits of holographic reconstruction algorithms. *Journal of Nondestructive Evaluation*, 2:85–93, 1981.

[65] A. J. Berkhout. Wavefield extrapolation techniques in seismic migration, a tutorial. *Geophysics*, 46(12):1638–1656, 1981.

[66] A. J. Berkhout. *Seismic Migration–Imaging of Acoustic Energy by Wavefield Extrapolation.* Elsevier Science, 1982.

[67] M. V. Berry and D. F. Gibbs. The interpretation of optical projections. *Proceedings of the Royal Society A.*, 314:143–152, 1970.

[68] L. A. Bessette, S. M. Crooks, and S. Ayasli. P-3 ultra-wideband SAR, target and clutter phenomenology. In *Proceedings of Radar Conference*, pages 125–129, Grayling, Michigan, 1999.

[69] G. Beylkin. Imaging of discontinuities in the inverse scattering problem by inversion of a causal generalized Radon transform. *Journal of Mathematical Physics*, 26:99–108, 1985.

[70] G. Beylkin and R. Burridge. Linearized inverse scattering problems in acoustics and elasticity. *Wave Motion*, 12:15–52, 1990.

[71] B. K. Bhattacharyya. Some general properties of potential fields in space and frequency domain: A review. *Geoexploration*, 5:127–143, 1967.

[72] R. Birk, W. Camus, E. Valenti, and W. J. McCandles. Synthetic aperture radar imaging systems. *IEEE Aerospace and Electronic Systems Magazine*, 10(11):15–23, November 1995.

[73] D. Blacknell, A. Freeman, S. Quegan, I. A. Ward, I. P. Finley, C. J. Oliver, R. G. White, and J. W. Wood. Geometric accuracy in airborne SAR images. *IEEE Transactions on Aerospace and Electronic Systems*, 25(2):241–258, March 1989.

[74] R. E. Blahut. *Theory of Remote Image Formation.* Cambridge University Press, 2004.

[75] R. E. Blahut, W. Miller Jr, and C. Wilcox (eds). *Radar and Sonar, Part I.* Springer Verlag, New York, 1991.

[76] N. Bleistein. Physical optics far-field inverse scattering in the time domain. *The Journal of the Acoustical Society of America*, 60:1249–1255, 1976.

[77] N. Bleistein, J. K. Cohen, and J. W. Stockwell. *The Mathematics of Multidimensional Seismic Inversion.* Springer, New York, 2000.

[78] N. Bleistein and R. A. Handelsman. *Asymptotic Expansions of Integrals.* Holt, Rinehart and Winston, New York, 1975.

[79] S. Boag, Y. Bresler, and E. Michielssen. A multilevel domain decomposition algorithm for fast $O(N^2 \log N)$ reprojection of tomographic images. *IEEE Transactions on Image Processing*, 9(9):1573–1582, September 2000.

[80] W.-M. Boerner, C.-M. Ho, and B.-Y. Foo. Use of Radon's projection theory in electromagnetic inverse scattering. *IEEE Transactions on Antennas Propagation*, 29(2):336–341, March 1981.

[81] N. N. Bojarski. 3-D electromagnetic short pulse inverse scattering. Technical report, Syracuse University Research Corporation, 1967.

[82] G. Bolondi, F. Rocca, and S. Savelli. A frequency domain approach to two-dimensional migration. *Geophysical Prospecting*, 26(4):750–772, December 1978.

[83] V. Bonanomi and M. Guzzi. *Stolt Migration Generalized for Electromagnetics in Array Processor: E.E. Thesis.* PhD thesis, Milan Politechnic, 1988.

[84] H. G. Booker and P. C. Clemmow. The concept of an angular spectrum of plane waves, and its relation to that of a polar diagram and aperture distribution. *Proceedings of the IEE, Radio Section*, 97, Part III(922):11–17, January 1950.

[85] B. Borden. Mathematical problems in radar inverse scattering. *Inverse Problems*, 18:1–28, 2002.

[86] B. Borden and M. Cheney. Synthetic-aperture imaging from high-Doppler-resolution measurements. *Inverse Problems*, 21:1–11, 2005.

[87] F. Bordoni, F. Lombardini, F. Gini, and A. Jakobsson. Multibaseline cross-track SAR interferometry using interpolated arrays. *IEEE Transactions on Aerospace and Electronic Systems*, 41(4):1473–1482, 2005.

[88] M. Born. *Optik.* Springer, Berlin, 1933.

[89] M. Born and E. Wolf. *Principles of Optics.* Cambridge University Press, Cambridge, United Kingdom, seventh edition, 1999.

[90] D. T. Borup, S. A. Johnson, W. W. Kim, and M. J. Berggren. Nonperturbative diffraction tomography via Gauss-Newton iteration applied to the scattering integral equation. *Ultrasonic Imaging*, 14(69–85), 1992.

[91] C. J. Bouwkamp. Diffraction theory. *Reports on Progress in Physics IV*, 17:35–100, 1954.

[92] A. C. Bovik and D. C. Munson, Jr. Boundary detection in speckle images. In *Proceedings of the IEEE International Conference on Acoustics, Speech, and Signal Processing*, pages 893–896, 1985.

[93] A. L. Boyer. Reconstruction of ultrasonic images by backward propagation. *Acoustic Holography*, 3:333–384, 1970.

[94] Robert Boyle. *The Philosophical Works of Robert Boyle*, volume II. Abridge by P. Shaw, London, second edition, 1738.

[95] R. N. Bracewell. Strip integration in radio astronomy. *Australian Journal of Physics*, 9:198–217, November 1956.

[96] R. N. Bracewell. *The Fourier Transform and its Applications.* McGraw-Hill, New York, 1965.

[97] R. N. Bracewell. Image reconstruction in radio astronomy. In G. T. Herman, editor, *Image Reconstruction from Projections: Implementation and Applications*, pages 81–104, Berlin, 1979. Springer.

[98] R. N. Bracewell and A. C. Riddle. Inversion of fan-beam scans in radio astronomy. *Astrophysical Journal*, 150(2):427–434, 1967.

[99] M. Brandfass and L. F. Lobianco. Modified fast factorized backprojection as applied to X-band data for curved flight paths. In *Proceedings of the European Conference on Synthetic Aperture Radar EuSAR*, page 4, June 2008.

[100] A. Brandt, J. Mann, M. Brodski, and M. Galun. A fast and accurate multilevel inversion of the Radon transform. *SIAM Journal on Applied Mathematics*, 60(2):437–462, 1999.

[101] T. Brehm, A. Wahlen, and H. Essen. High resolution millimeterwave SAR. In *Proceedings of the 1st European Radar Conference (EURAD '04)*, pages 217–219, Amsterdam, The Netherlands, 14–15 October 2004.

[102] A. R. Brenner and J. H. G. Ender. Very wideband radar imaging with the airborne SAR sensor PAMIR. In *Proceedings of the IEEE International Geoscience and Remote Sensing Symposium*, volume 1, pages 544–535, 2003.

[103] A. R. Brenner and L. Roessing. Radar imaging of urban areas by means of very high-resolution SAR and interferometric SAR. *IEEE Transactions on Geoscience and Remote Sensing*, 46(10):2971–2982, October 2008.

[104] D. R. Bromaghim and J. P. Perry. A wideband linear FM ramp generator for the long-range imaging radar. *IEEE Transactions on Microwave Theory and Techniques*, 26(5):322, May 1980.

[105] E. Brookner. *Radar Technology.* Artech House, 1977.

[106] R. A. Brooks and G. DiChiro. Principles of computer assisted tomography (CAT) in radiographic and radioisotropic imaging. *Physics in Medicine and Biology*, 21:689–732, 1976.

[107] R. A. Brooks, G. H. Weiss, and A. J. Talbert. A new approach to interpolation in computed tomography. *Journal of Computer Assisted Tomography*, 2:577–585, 1978.

[108] A. Broquetas, L. Jofre, and A. Cardama. A near field spherical wave inverse synthetic aperture radar technique. In *Proceedings of the IEEE Antennas Propagation Society International Symposium*, volume 2, pages 114–117, Chicago, Illinois, July 1992.

[109] A. Broquetas, J. Palau, L. Jofre, and A. Cardama. Spherical wave near field imaging and radar cross-section measurement. *IEEE Transactions on Antennas and Propagation*, 46(5):730–735, May 1998.

[110] W. N. Brouw. *Aperture Synthesis*, volume 14, pages 131–175. Academic, New York, 1975.

[111] A. Brown and B. Wolt. Digital L-band receiver architecture with direct RF sampling. In *Position, Location and Navigation Symposium*, 1994.

[112] R. Brown, W. Camus, and E. Valenti. Synthetic aperture radar imaging systems. *IEEE Aerospace and Electronic Systems Magazine*, pages 15–23, November 1995.

[113] W. D. Brown and D. C. Ghiglia. Some methods for reducing propagation-induced phase errors in coherent imaging systems. *Journal of the Optical Society of America*, 5(6):924–938, June 1988.

[114] W. M. Brown, Jr. Synthetic aperture radar. *IEEE Transactions on Aerospace and Electronic Systems*, 3(2):217–229, March 1967.

[115] W. M. Brown, Jr. Walker model for radar sensing of rigid target fields. *IEEE Transactions on Aerospace and Electronic Systems*, 16:104–107, January 1980.

[116] W. M. Brown, Jr, C. Elachi, and T. W. Thompson. Radar imaging of ocean surface patterns. *Journal of Geophysical Research*, 81:2656–2667, 1976.

[117] W. M. Brown, Jr and R. J. Fredericks. Range-Doppler imaging with motion through resolution cells. *IEEE Transactions on Aerospace and Electronic Systems*, 5(1):98–102, January 1969.

[118] W. M. Brown, Jr, G. G. Houser, and R. G. Jenkins. Synthetic aperture processing with limited storage and presumming. *IEEE Transactions on Aerospace and Electronic Systems*, 9:166–176, 1973.

[119] W. M. Brown, Jr and L. J. Porcello. Theory of coherent systems. *IRE Transactions on Military Electronics*, 6:187–196, April 1962.

[120] W. M. Brown, Jr and L. J. Porcello. An introduction to synthetic aperture radar. *IEEE Spectrum*, 6(9):52–62, September 1969.

[121] D. Bruno and S. E. Hobbs. Radar imaging from geosynchronous orbit: Temporal decorrelation aspects. *IEEE Transactions on Geoscience and Remote Sensing*, 38(7):2924–2929, 2010.

[122] M. L. Bryant, L. L. Gostin, and M. Soumekh. 3D E-CSAR imaging of a T-72 tank and synthesis of its SAR reconstructions. *IEEE Transactions on Aerospace and Electronic Systems*, 39(1):211–227, January 2003.

[123] S. Buckreuss. Motion compensation for the experimental SAR system of the DLR. In *IEEE International Geoscience and Remote Sensing Symposium*, pages 1971–1974, Pasadena, California, 1994.

[124] J. Burki. *Fast Circular Aperture Sysnthesis in SAR All-Aspect Target Imaging.* PhD thesis, Georgia Institute of Technology, Atlanta, Georgia, 2008.

[125] J. Burki and C. F. Barnes. Beam raster scanning 2D flyover SAR. In *The Fourth IEEE Workshop on Sensor Array and Multi-channel Processing*, pages 339–343, Waltham, Massachusetts, 12–14 July 2006.

[126] J. Burki and C. F. Barnes. Interpolation-free algorithm for SAR 2D aperture synthesis. In *Proceedings of SPIE, Algorithms for Synthetic Aperture Radar Imagery XIII*, volume 6237, Orlando, Florida, 17–19 May 2006.

[127] J. Burki and C. F. Barnes. Fast C-SAR algorithm. In *Proceedings of SPIE, Algorithms for Synthetic Aperture Radar Imagery XV*, Orlando, Florida, 16–20 March 2007.

[128] J. Burki and C. F. Barnes. SAR with two-dimensional aperture synthesis. In *IEEE International Radar Conference*, pages 1418–1422, Waltham, Massachusetts, 17–21 April 2007.

[129] J. Burki and C. F. Barnes. Slant plane C-SAR processing using Householder transform. *IEEE Transactions on Image Processing*, 17(10):1900–1907, 2008.

[130] B. L. Burns and J. T. Cardaro. A SAR image formation algorithm that compensates for the spatially variant effects of antenna motion. In *Proceedings of SPIE, Algorithms for Synthetic Aperture Radar Imagery*, volume 2230, Orlando, Florida, 4–8 April 1994.

[131] L. J. Busse. Three-dimensional imaging using a frequency-domain synthetic aperture focusing technique. *IEEE Transactions on Ultrasonics, Ferroelectromagnetics, Frequency and Control*, 39:174–179, March 2008.

[132] N. Cadalli and D. C. Munson, Jr. A comparison of ω-k and generalized SAR inversion for runway imaging. In *IEEE International Conference on Image Processing*, volume 1, pages 693–696, September 2000.

[133] N. Cadalli and D. C. Munson, Jr. A simulation study of ω-k SAR algorithms for the highly squinted case with application to runway imaging. In *IEEE International Conference on Acoustics, Speech and Signal Processing*, volume 5, pages 3025–3028, Istanbul, Turkey, June 5–9, 2000.

[134] C. Cafforio, C. Prati, and E. Rocca. Full resolution focusing of SEASAT SAR images in the frequency-wavenumber domain. In *Proceedings of the 9th EARSel Sympoisum*, pages 336–355, 1988.

[135] C. Cafforio, C. Prati, and E. Rocca. Synthetic aperture radar: A new application for wave equation techniques. *Geophysical Prospecting*, 7:809–830, 1989.

[136] C. Cafforio, C. Prati, and E. Rocca. Full resolution focusing of SEASAT SAR images in the frequency-wavenumber domain. *International Journal of Remote Sensing*, 12(3):491–510, 1991.

[137] C. Cafforio, C. Prati, and E. Rocca. SAR data focusing using seismic migration techniques. *IEEE Transactions on Aerospace and Electronic Systems*, 27(2):194–207, March 1991.

[138] H. J. Callow, R. E. Hansen, and T. O. Saebo. Effect of approximations in fast factorized backprojection in synthetic aperture imaging of spot regions. In *OCEANS 2006*, pages 1–6, Boston, Massachusetts, 2006.

[139] G. Calvagno and D. C. Munson, Jr. A frequency-domain approach to interpolation from a nonuniform grid. *Signal Processing*, 52:1–21, 1996.

[140] H. M. J. Cantalloube, E. Colin-Koeniguer, and H. Oriot. High resolution SAR imaging along circular trajectories. In *Proceedings of the IEEE International Geoscience and Remote Sensing Symposium*, pages 850–853, July 2007.

[141] H. M. J. Cantalloube and P. Dubois-Fernandez. Airborne X-band SAR imaging with 10 cm resolution — technical challenges and preliminary results. In *Proceedings of the IEEE International Geoscience and Remote Sensing Symposium*, volume 1, pages 185–187, July 2003.

[142] H. M. J. Cantalloube and C. Nahum. Motion compensation and autofocus of range/Doppler or bidimensional processing for airborne synthetic aperture radar. *Aerospace Science and Technology*, 4:251–263, 1998.

[143] J. Capon. High-resolution frequency-wavenumber spectrum analysis. *Proceedings of the IEEE*, 57(8):1408–1418, August 1969.

[144] W. J. Caputi, Jr. Stretch: A time-transformation technique. *IEEE Transactions on Aerospace and Electronic Systems*, 7(2):269–278, March 1971.

[145] R. E. Carande. Dual baseline and frequency along-track interferometry. In *Proceedings of the IEEE International Geoscience Remote Sensing Symposium*, pages 1585–1588, Houston, Texas, 1992.

[146] W. G. Carrara, R. S. Goodman, and R. M. Majewski. *Spotlight Synthetic Aperture Radar: Signal Processing Algorithms.* Artech House, Boston, Massachusetts, 1995.

[147] W. G. Carrara, R. S. Goodman, and M. A. Ricoy. New algorithms for widefield SAR image formation. *IEEE International Radar Conference*, 60:38–43, 2004.

[148] W. G. Carrara, S. Tummala, and R. S. Goodman. Motion compensation algorithm for widebeam stripmap SAR. In *Proceedings of SPIE, Algorithms for Synthetic Aperture Radar Imagery II*, volume 2487, Orlando, Florida, 1995.

[149] P. L. Carson, T. V. Oughton, and W. R. Hendee. Ultrasonic transaxial tomography by reconstruction. *Ultrasound In Medicine*, 2:341–350, 1976.

[150] W. H. Carter. Computational reconstruction of scattering objects from holograms. *Journal of the Optical Society of America*, 60(3):306–314, 1970.

[151] C. H. Casteel, L. A. Gorham, M. J. Minardi, S. Scarborough, and K. D. Naidu. A challenge problem for 2D/3D imaging of targets from a volumetric data set in an urban environment. In *Proceedings of SPIE, Algorithms for Synthetic Aperture Radar Imagery XIV*, Orlando, Florida, 9–13 May 2007.

[152] F. R. Castella. Application of one-dimensional holographic techniques to a mapping sonar system. *Acoustic Holography*, 3, 1971.

[153] C. Castillo-Rubio, S. Llorente-Romano, and M. Burgos-Garcia. Robust SVA method for every sampling rate condition. *IEEE Transactions on Aerospace and Electronic Systems*, 43(2):571–580, April 2007.

[154] C. Castillo-Rubio, S. Llorente-Romano, and M. Burgos-Garcia. Spatially variant apodization for squinted synthetic aperture radar images. *IEEE Transactions on Image Processing*, 16(8):2023–2027, August 2007.

[155] W. T. Cathey. *Optical Information Processing and Holography.* Wiley, New York, 1974.

[156] R. G. Caves, P. J. Harley, and S. Quegan. Matching map features to synthetic aperture radar SAR images using template matching. *IEEE Transactions on Geoscience and Remote Sensing*, 30(4):680–685, July 1992.

[157] A. D. Cenzo. The effect of weighting on time sidelobe suppression. In *Proceedings of Synthetic Aperture Radar Technology Conference*, Las Cruces, 8–10 December 1978.

[158] A. D. Cenzo. A comparison of resolution for spotlight synthetic-aperture radar and computer-aided tomography. *Proceedings of the IEEE*, 74:1165–1166, August 1986.

[159] A. D. Cenzo. A new look at nonseparable synthetic aperture radar processing. *IEEE Transactions on Aerospace and Electronic Systems*, 24(3):218–223, May 1988.

[160] A. D. Cenzo. Strip mode processing of spotlight synthetic aperture radar data. *IEEE Transactions on Aerospace and Electronic Systems*, 24(3):225–230, May 1988.

[161] M. Cetin and W. Karl. Feature-enhanced synthetic aperture radar image formation based on nonquadratic regularization. *IEEE Transactions on Image Processing*, 10:623–631, April 2001.

[162] S. Cha and C. M. Vest. Computer tomography of strongly refracting objects. In *Proceedings of SPIE, Applications of Mathematics in Modern Optics*, volume 358, pages 149–155, San Diego, California, 7 December 1982.

[163] C. K. Chan and N. H. Farhat. Frequency swept tomographic imaging of three-dimensional perfectly conducting objects. *IEEE Transactions on Antennas Propagation*, 29(2):312–319, March 1981.

[164] Y. K. Chan and V. C. Koo. An introduction to synthetic aperture radar. *Progress In Electromagnetics Research B*, 2:27–60, 2008.

[165] R. D. Chaney, A. S. Willsky, and L. M. Novak. Coherent aspect-dependent SAR image formation. In *Proceedings of SPIE, Algorithms for Synthetic Aperture Radar Imagery*, volume 2230, pages 256–274, Orlando, Florida, 4–8 April 1994.

[166] C. Y. Chang, M. Jin, and J. C. Curlander. Squint mode SAR processing algorithms. In *Proceedings of the IEEE International Geoscience Remote Sensing Symposium*, volume 1, pages 2584–2587, Vancouver, Canada, July 1989.

[167] C. Y. Chang, M. Y. Jin, and J. C. Curlander. SAR processing based on the exact two-dimensional transfer function. In *Proceedings of the IEEE International Geoscience Remote Sensing Symposium*, volume 1, pages 355–359, May 1992.

[168] C. Chen and H. C. Andrews. Multifrequency imaging of radar turnable data. *IEEE Transactions on Aerospace and Electronic Systems*, 16(1):15–22, January 1980.

[169] D. S. Chen and J. P. Allebach. Analysis of error in reconstruction of two-dimensional signals from irregularly spaced samples. *IEEE Transactions on Acoustics, Speech and Signal Processing*, 35:173–180, 1987.

[170] H.-C. Chen and C. D. McGillem. Target motion compensation in synthetic aperture radar. In *IEEE Digital Avionics Systems Conference*, pages 14–19, Virginia Beach, Virginia, October 1991.

[171] S. Chen, D. Donoho, and M. Suanders. Atomic decomposition by basis pursuit. *SIAM Journal on Scientific Computing*, 20(1):33–61, 1998.

[172] V. C. Chen and H. Ling. *Time-Frequency Transforms for Radar Imaging and Signal Anlaysis.* Artech House, Boston, Massachusetts, 2002.

[173] M. Cheney. Introduction to synthetic aperture radar (SAR) and SAR interferometry. In C. K. Chui, L.L. Schumaker, and J. Stoeckler, editors, *Approximation Theory X: Wavelets, Splines, and Applications*, pages 167–177, Nashville, Tennessee, 2001. Vanderbilt University Press.

[174] M. Cheney. A mathematical tutorial on synthetic aperture radar. *SIAM Review*, 43(2):301–312, 2001.

[175] M. Cheney. Tomography problems arising in synthetic aperture radar. *Radon Transforms and Tomography: 2000 AMS-IMS SIAM Joint Summer Research Contemporary Mathematics*, 278:15–27, 2001.

[176] M. Cheney. Imaging frequency-dependent synthetic-aperture radar. *Inverse Problems*, 29(5), 2013.

[177] M. Cheney, G. Beylkin, E. Somersalo, and R. Burridge. Three-dimensional inverse scattering for the wave equation with variable speed: Near-field formulae using point sources. *Inverse Problems*, 5:1–6, 1989.

[178] M. Cheney and B. Borden. Microlocal structure of inverse synthetic aperture data. *Inverse Problems*, 19:173–194, 2003.

[179] M. Cheney and B. Borden. Imaging moving targets from scattered waves. *Inverse Problems*, 24, 2008.

[180] M. Cheney and B. Borden. *Fundamentals of Radar Imaging.* SIAM, Philadelphia, 2009.

[181] M. Cheney and B. Borden. Problems in synthetic-aperture imaging. *Inverse Problems*, 25:1–18, 2009.

[182] M. Cheney and C. J. Nolan. Synthetic-aperture imaging through a dispersive layer. *Inverse Problems*, 20:507–532, 2004.

[183] M. Cheney and J. H. Rose. Three-dimensional inverse scattering for the wave equation: Weak scattering approximations with error estimates. *Inverse Problems*, 4:435–447, 1988.

[184] L. A. Chernov. *Wave Propagation in a Random Medium*. McGraw-Hill, New York, 1969.

[185] W. C. Chew and J. M. Song. Fast Fourier Transform of sparse spatial data to sparse Fourier data. In *IEEE Antennas and Propagation International Symposium*, volume 4, pages 2324–2327, July 2000.

[186] H. Choi and D. C. Munson, Jr. On the optimality and exactness of wavenumber-domain SAR data processing. In *Proceedings of the IEEE International Conference on Image Processing*, volume 1, pages 456–460, 1994.

[187] H. Choi and D. C. Munson, Jr. Direct-Fourier reconstruction in tomography and synthetic aperture radar. *International Journal of Imaging Systems and Technology*, 9(1):1–13, 1998.

[188] A. Chu, Y. Kim, J. V. Zyl, W. Tung, E. O'Leary, and Y. Lou. The NASA/JPL AIRSAR integrated processor. In *Proceedings of the IEEE International Geoscience Remote Sensing Symposium Conference*, pages 1908–1910, June 1998.

[189] J. H. Chun and C. A. Jacowitz. Fundamentals of frequency domain migration. *Geophysics*, 46:717–733, 1981.

[190] C. Chung-Chien and H. C. Andrews. Multi-frequency imaging of radar turntable data. *IEEE Transactions on Aerospace and Electronic Systems*, 16(1):15–22, January 1980.

[191] J. Cimino, C. Elachi, and M. Settle. SIR-B the second shuttle imaging radar experiment. *IEEE Transactions on Geoscience and Remote Sensing*, 24(4):445–452, 1986.

[192] J. P. Claassen and J. Eckerman. Proceedings of the synthetic aperture radar technology conference. In *A System for Wide Swath Constant Incident Angle Coverage*, Las Cruces, New Mexico, 1978.

[193] J. F. Claerbout. Toward a unified theory of reflector mapping. *Geophysics*, 36:467–481, 1971.

[194] J. Clark, M. Palmer, and P. Lawrence. A transformation method for the reconstruction of functions from nonuniformly spaced samples. *IEEE Transactions on Acoustics, Speech and Signal Processing*, 33(4):1151–1165, October 1985.

[195] P. C. Clemmov. A method for the exact solution of a class of two-dimensional diffraction problems. *Proceedings of the Royal Society A.*, 205:286, 1951.

[196] P. C. Clemmov. *The Plane Wave Spectrum Representation of Electromagnetic Fields*. Pergamon, Oxford, 1966.

[197] M. C. Cobb and J. H. McClellan. Developments in repeat pass interferometric radar for Earth and planetary sciences. In *Proceedings of the IEEE International Radar Conference*, volume 1, pages 207–211, Apr 26–29 2004.

[198] D. Colton, H. Haddar, and M. Piana. The linear sampling method in inverse electromagnetic scattering theory. *Inverse Problems*, 19:105–107, 2003.

[199] D. Colton and R. Kress. *Integral Equation Methods in Scattering Theory*. John Wiley, New York, 1983.

[200] D. Colton and P. Monk. A modified dual space method for solving the electromagnetic inverse scattering problem for an infinite cylinder. *Inverse Problems*, 10:87–108, 1994.

[201] C. E. Cook. Pulse compression–key to more efficient radar transmission. *Proceedings of the IRE*, 48:310–316, March 1960.

[202] C. E. Cook and M. Bernfeld. *Radar Signals, An Introduction to Theory and Application*. Academic, New York, 1967.

[203] C. E. Cook, M. Bernfeld, and C. A. Palmieri. Matched filtering, pulse compression and waveform design–IV. *Microwave Journal*, 8(1):73–88, January 1965.

[204] J. W. Cooley and J. W. Tukey. An algorithm for the machine calculation of complex Fourier series. *Mathematics of Computation*, 19:297–301, 1965.

[205] L. Cordell. A scattered equivalent-source method for interpolation and gridding of potential-field data in three dimensions. *Geophysics*, 57:629–636, 1992.

[206] A. M. Cormack. Representation of a function by its line integrals, with some radiological applications, I. *Journal of Applied Physics*, 34(9):2722–2727, September 1963.

[207] A. M. Cormack. Representation of a function by its line integrals, with some radiological applications, II. *Journal of Applied Physics*, 35:2908–2913, October 1964.

[208] A. M. Cormack. Reconstruction of densities from their projections, with applications in radiological physics. *Physics in Medicine and Biology*, 18:195–207, 1973.

[209] D. G. Corr. Coherence change detection for urban development monitoring. In *Proceedings of the IEE Colloquium on Radar Interferometry*, volume 6, pages 1–6, 1997.

[210] D. G. Corr and A. Rodriques. Coherent change detection of vehicle movements. In *Proceedings of the IEEE International Geoscience and Remote Sensing Symposium*, volume 5, pages 2451–2453, 1998.

[211] R. Crochiere and L. R. Rabiner. Interpolation and decimation of digital signals–A tutorial review. *Proceedings of the IEEE*, 69(3):300–331, 1981.

[212] R. A. Crowther, D. J. DeRosier, and A. Klug. The reconstruction of a three-dimensional structure from projections and its application to electron microscopy. *Proceedings of the Royal Society of London A.*, 317:319–340, 1970.

[213] L. J. Cultrona, E. N. Leith, L. J. Porcello, and W. E. Vivian. On the application of coherent optical processing techniques to synthetic-aperture radar. *Proceedings of the IEEE*, 54:1026–1032, August 1966.

[214] I. Cumming and J. R. Bennett. Digital processing of SEASAT SAR data. In *Proceedings of the IEEE International Conference on Acoustics, Speech and Signal Processing*, volume 4, pages 710–718, Washington, DC, 2–4 April 1979.

[215] I. Cumming, P. F. Kavanagh, and M. R. Ito. Resolving the Doppler ambiguity for spaceborne synthetic aperture radar. In *Proceedings of the IEEE International Geoscience and Remote Sensing Symposium*, volume 3, pages 1639–1643, 1986.

[216] I. Cumming, Y. L. Neo, and F. H. Wong. Interpretations of the omega-k algorithm and comparisons with other algorithms. In *Proceedings of the IEEE International Geoscience and Remote Sensing Symposium*, volume 2, pages 1455–1458, July 2003.

[217] I. Cumming, F. Wong, and K. Raney. A SAR processing algorithm with no interpolation. In *Proceedings of the IEEE International Geoscience and Remote Sensing Symposium*, volume 1, pages 376–379, Houston, Texas, May 1992.

[218] I. G. Cumming and F. H. Wong. *Digital Processing of Synthetic Aperture Radar Data: Algorithm and Implementation.* Artech House, Boston, Massachusetts, 2005.

[219] J. C. Curlander. Performance of the SIR-B digital imaging processing subsystem. *IEEE Transactions on Geoscience and Remote Sensing*, 24(4):649–652, 1986.

[220] J. C. Curlander, B. Holt, and K. J. Hussey. Determination of sea ice motion using digital SAR imagery. *IEEE Journal of Oceanic Engineering*, 10(4):358–367, 1985.

[221] J. C. Curlander and R. N. McDonough. *Synthetic Aperture Radar: Systems and Signal Processing.* John Wiley & Sons, New York, 1991.

[222] A. Currie and M. A. Brown. Wide-swath SAR. *IEE Proceedings on Radar and Signal Processing*, 139(2):122–135, April 1992.

[223] L. J. Cutrona. Comparison of sonar system performance achievable using synthetic-aperture techniques with the performance achievable by more conventional means. *Journal of the Acoustical Society of America*, 58:336–348, 1975.

[224] L. J. Cutrona. Additional characteristics of synthetic-aperture sonar systems and a further comparison with nonsynthetic-aperture sonar systems. *Journal of the Acoustical Society of America*, 61:1213–1217, 1977.

[225] L. J. Cutrona. Means to achieve wide swath widths in synthetic aperture satellite borne radars. In *Proceedings of the Synthetic Aperture Radar Technology Conference*, pages 251–256, Las Cruces, New Mexico, 2-10 March 1978.

[226] L. J. Cutrona. Synthetic aperture radar. In M. Skolnik, editor, *Radar Handbook*, New York, 1990. McGraw-Hill.

[227] L. J. Cutrona, E. N. Leith, and L. J. Porcello. Optical data processing and filtering systems. *IRE Transactions on Information Theory*, 6:386–400, June 1960.

[228] L. J. Cutrona, E. N. Leith, L. J. Porcello, and W. E. Vivian. On the application of coherent optical processing techniques to synthetic aperture radar. *Proceedings of the IEEE*, 54:1026–1032, July 1966.

[229] L. J. Cutrona, W. E. Vivian, E. N. Leith, and G. O. Hall. A high-resolution radar combat-surveillance system. *IRE Transactions on Military Electronics*, 5(2):127–131, April 1961.

[230] S. Dai and W. Wiesbeck. The imaging mode of forward looking SAR with two receiving antennas. In *Proceedings of the IEEE International Geoscience Remote Sensing Symposium*, volume 3, pages 1776–1778, Hamburg, Germany, October 1999.

[231] Y. Das and W. M. Boerner. Application of algorithms for 3-D image reconstruction from 2-D projections to E.M. inverse scattering. In *USNC/URSI, Annual Meeting*, Boulder, Colorado, October 1975.

[232] Y. Das and W. M. Boerner. On radar target shape estimation using algorithms for reconstruction from projections. *IEEE Transactions on Antennas Propagation*, 26:274–279, February 1978.

[233] G. W. Davidson and I. G. Cumming. Signal properties of spaceborne squint mode SAR. *IEEE Transactions on Geoscience and Remote Sensing*, 35:611–617, 1997.

[234] G. W. Davidson, I. G. Cumming, and M. R. Ito. A chirp scaling approach for processing squint mode SAR data. *IEEE Transactions on Aerospace and Electronic Systems*, 32(1):121–133, January 1996.

[235] D. F. Davis. Tomographic image processing of inverse synthetic aperture radar data. Technical report, Naval Weapons Center, China Lake, California, September 1991.

[236] Pierre de Fermat. Published personal letter to Cureau de la Chambre. *Oeuvres de Fermat*, 2:354, 1891.

[237] S. R. Deans. *The Radon Transform and Some of Its Applications.* Wiley, New York, 1983.

[238] D. W. Dearholt and W. R. McSpadden. *Electromagnetic Wave Propagation.* McGraw-Hill, New York, 1973.

[239] R. Decartes. Dioptrique, météores. *Principia Philosophiae*, 1644. Published (anonymously) in Leyden in 1637 with prefactory essay, Discours de la méthodé.

[240] A. H. Delaney and Y. Bresler. A fast iterative tomographic reconstruction algorithm. In *Proceedings of the Acoustics, Speech and Signal Processing Conference*, volume 4, pages 2295–2298, 1995.

[241] D. Dendal and J. L. Marchand. Omega-k techniques advantages and weaker aspects. In *Proceedings of the IEEE International Geoscience and Remote Sensing Symposium*, pages 366–368, May 1992.

[242] Y. N. Denisyuk. Photographic reconstruction of the optical properties of an object in its own scattered radiation field. *Soviet Physics-Doklady*, 7(6):543–545, 1962.

[243] A. G. Derneryd and A. Lagerstedt. Novel slotted waveguide antenna with polarimetric capabilities. In *Proceedings of the IEEE International Geoscience Remote Sensing Symposium Conference*, pages 2054–2056, Firenze, Italy, 1995.

[244] A. G. Derneryd, R. N. O. Petersson, and P. Ingvarson. Slotted waveguide antennas for remote sensing satellites. In *Proceedings of the PIERS Conference*, Noordwikj, Netherlands, July 1994.

[245] D. J. DeRosier and A. Klug. Reconstruction of three dimensional structures from electron micrographs. *Nature*, 217:130–134, January 1968.

[246] M. D. Desai. *A New Method of Synthetic Aperture Radar Image Reconstruction Using Modified Convolution Backprojection Algorithm.* PhD thesis, University of Illinois, 1985.

[247] M. D. Desai. Spotlight mode SAR stereo technique for height computation. *IEEE Transactions on Image Processing*, 6(10):1400–1411, October 1997.

[248] M. D. Desai and W. K. Jenkin. Performance evaluation and architectural implications of convolution backprojection image reconstruction for synthetic aperture radar. In *IEEE International Conference on Circuits and Systems*, Montreal, Canada, May 1984.

[249] M. D. Desai and W. K. Jenkin. Convolution backprojection image reconstruction for synthetic aperture radar. In *IEEE International Conference on Circuits and Systems*, Tokyo, Japan, May 1985.

[250] M. D. Desai and W. K. Jenkins. Convolution backprojection image reconstruction for spotlight mode synthetic aperture radar. *IEEE Transactions on Image Processing*, 1(4):505–517, October 1992.

[251] A. J. Devaney. Inverse scattering within the Rytov approximation. *Optics Letters*, 6:374–376, 1981.

[252] A. J. Devaney. A filtered backpropagation algorithm for diffraction tomography. *Ultrasonic Imaging*, 4:336–350, 1982.

[253] A. J. Devaney. Inverse scattering as a form of computed tomography. *Proceedings of SPIE, Applications of Mathematics in Modern Optics*, 358:10–16, 1982.

[254] A. J. Devaney. Inversion formula for inverse scattering within the Born approximation. *Optics Letters*, 7:111–112, 1982.

[255] A. J. Devaney. A computer simulation study of diffraction tomography. *IEEE Transactions on Biomedical Engineering*, 30(7):377–386, July 1983.

[256] A. J. Devaney. Geophysical diffraction tomography. *IEEE Transactions on Geoscience and Remote Sensing*, 22(1):3–13, January 1984.

[257] A. J. Devaney. Propagation emission tomography. *Proceedings of SPIE, Inverse Optics II*, 558:43–46, 1985.

[258] A. J. Devaney. Reconstructive tomography with diffracting wavefields. *Inverse Problems*, 2:161–183, 1986.

[259] A. J. Devaney. A fast filtered backpropagation algorithm for ultrasonic tomography. *IEEE Transactions on Ultrasonics, Ferroelectrics and Frequency Control*, 34(3):330–340, May 1987.

[260] A. J. Devaney. *Mathematical Foundations of Imaging, Tomography and Wavefield Inversion.* Cambridge University Press, 2012.

[261] A. J. Devaney and G. Beylkin. Diffraction tomography using arbitrary transmitter and receiver surfaces. *Ultrasonic Imaging*, 6:181–193, 1984.

[262] A. J. Devaney, H. J. Liff, and S. Apsell. Spectral representations for free space propagation of complex phase perturbations of optical fields. *Optics Communications*, 15(1):1–5, September 1975.

[263] A. J. Devaney and E. Wolf. A new perturbation expansion for inverse scattering from three-dimensional finite range potentials. *Physics Letters*, 89A:269–272, 1982.

[264] F. M. Dickey, L. A. Romero, and A. W. Doerry. Window functions for imaging radar: A maximum energy approach to continuous and notched spectrums. *Optical Engineering*, 42(7):2113–2128, 1972.

[265] R. Dilsavor, A. Mitra, M. Hensel, and M. Soumekh. GPS-based spatial and spectral registration of delta-heading multipass SAR imagery for coherent change detection. In *Proceedings of U.S. Army Workshop on Synthetic Aperture Radar Technology*, Redstone Arsenal, Alabama, October 2002.

[266] K. A. Dines and R. J. Lytle. Computerized geophysical tomography. *Proceedings of the IEEE*, 67:471–480, 1979.

[267] Y. Ding and D. C. Munson, Jr. A fast back-projection algorithm for bistatic SAR imaging. In *Proceedings of the IEEE International Conference on Image Processing*, volume 2, pages 449–452, Rochester, New York, 22-25 September 2002.

[268] R. DiPietro, R. Fante, M. Soumekh, R. Perry, and L. Trump. FOPEN planar sub-array processing SAR image formation. In *Proceedings of European Conference on Synthetic Aperture Radar*, Munich, May 2000.

[269] P. Djuric and S. Kay. Parameter estimation of chirp signals. *IEEE Transactions on Acoustics, Speech, and Signal Processing*, 38(12):2118–2126, December 1990.

[270] A. Doerry. Patch diameter limitations due to high chirp rates in focused synthetic aperture radar images. *IEEE Transactions on Aerospace and Electronic Systems*, 30(4):1125–1129, October 1994.

[271] A. Doerry. Patch diameter limits for tiered subaperture SAR image formation algorithms. In *Proceedings of SPIE, Algorithms for Synthetic Aperture Radar Imagery*, volume 2487, pages 132–143, 1994.

[272] A. Doerry. Synthetic aperture radar processing with polar formatted subapertures. In *Proceedings of the 28th Asilomar Conference on Circuits, Systems and Computing*, Pacific Grove, California, 31 October – 2 November 1994.

[273] A. Doerry. *Synthetic Aperture Radar Processing with Tiered Subapertures.* PhD thesis, University of New Mexico, May 1995.

[274] A. Doerry. High-precision stereoscopic 3-D mapping accuracy. In E. G. Zelnio, editor, *Proceedings of SPIE, Algorithms for Synthetic Aperture Radar Imagery VIII*, volume 4382, Orlando, Florida, 16 April 2001.

[275] A. Doerry. Autofocus correction of SAR images exhibiting excessive residual migration. In *Proceedings of SPIE, Radar Sensor Technology IX*, volume 5788, pages 34–45, Bellingham, Washington, 2005.

[276] A. Doerry. Generating non-linear FM chirp waveforms for radar. Technical report, Sandia National Laboratories, September 2006.

[277] A. Doerry. Performance limits for synthetic aperture radar–second edition. Technical report, Sandia National Laboratories, February 2006.

[278] A. Doerry. SAR processing with non-linear FM chirp waveforms. Technical report, Sandia National Laboratories, December 2006.

[279] A. Doerry. Generating precision nonlinear FM chirp waveforms. In *Proceedings of SPIE, Radar Sensor Technology XI*, volume 6574, 2007.

[280] A. Doerry. Wavefront curvature limitations and compensation to polar format processing for synthetic aperture radar images. Technical report, Sandia National Laboratories, January 2007.

[281] A. Doerry. Wideband SAR processing with segmented chirps for phased-array radars. In *Proceedings of SPIE, Radar Sensor Technology XI*, volume 6574, 2007.

[282] A. W. Doerry. SAR data collection and processing requirements for high quality coherent change detection. In *Proceedings of SPIE, Radar Sensor Technology XII*, volume 6947, 2008.

[283] N. E. Doren, C. V. Jakowatz, Jr, D. E. Wahl, and P. A. Thompson. General formulation for wavefront curvature correction in polar-formatted spotlight-mode SAR images using space-variant post-filtering. In *IEEE Conference on Image Processing*, volume 1, Santa Barbara, California, 1997.

[284] I. Dowman. The geometry of SAR images for geocoding and stereo applications. *International Journal on Remote Sensing*, 13(9), 1992.

[285] L. Du, Y. Wang, W. Hong, W. Tan, and Y. Wu. A three-dimensional range migration algorithm for downward-looking 3D-SAR with single-transmitting and multiple-receiving linear array antennas. *EURASIP Journal on Advances in Signal Processing*, 957916:15pp, 2010.

[286] B. Duchene, D. Lesselier, and W. Tabbara. Diffraction tomography approach to acoustical imaging and media characterization. *Journal of the Optical Society of America*, 2(11):1943–1953, November 1985.

[287] G. Duchossors and C. Honvault. The first ESA remote sensing satellite system ERS-1. In *15th International Symposium on Remote Sensing of the Environment*, Ann Arbor, Michigan, May 1981.

[288] D. E. Dudgeon and R. M. Mersereau. *Multidimensional Digital Signal Processing.* Prentice-Hall, Englewood Cliffs, New Jersey, 1995.

[289] E. Durand. Academie des sciences. *Comptes rendus de V*, 226:1812, 1948.

[290] B. Edde. *Radar: Principles, Technology and Applications.* Prentice-Hall, Upper Saddle River, New Jersey, 1995.

[291] P. R. Edholm and G. T. Herman. Linograms in image reconstruction from projections. *IEEE Transactions on Medical Imaging*, 6(4):301–307, December 1987.

[292] P. R. Edholm, G. T. Herman, and D. A. Roberts. Image reconstruction from linograms: Implementation and evaluation. *IEEE Transactions on Medical Imaging*, 7:239–246, 1988.

[293] P. H. Eichel, D. C. Ghiglia, and C. V. Jakowatz, Jr. A new phase correction method for synthetic aperture radar. In *Digital Signal Processing Workshop*, pages 2.12.1–2.12.2, Stanford Sierra Camp, South Lake Tahoe, California, 15–17 September 1988.

[294] P. H. Eichel, D. C. Ghiglia, and C. V. Jakowatz, Jr. Speckle processing method for synthetic aperture radar phase correction. *Optics Letters*, 14(1):1–3, January 1989.

[295] P. H. Eichel and C. V. Jakowatz, Jr. Phase-gradient algorithm as an optimal estimator of the phase derivative. *Optics Letters*, 14(20):1101, 1989.

[296] A. Einstein. On the electrodynamics of moving bodies. *Annalen der Physik*, (ser. 4), 17(10):891–921, 1905.

[297] A. Einstein. Strahlungs-emission und -absorption nach der quantentheorie (Emission and absorption of radiation in quantum theory). *Verhandl. D. Deutsche Physikalische Gesellschaft*, 18:318–323, 1916.

[298] A. Einstein. Cosmological considerations in the general theory of relativity. *Preussische Akademie der Wissenschaften, Sitzungsberichte*, part 1:142–152, 1917.

[299] A. Einstein. Zur quantentheorie der strahlung (On the quantum theory of radiation). *Physikalische Zeitschrift*, 18:121–128, 1917.

[300] C. Elachi, , W. E. Brown, J. B. Cimino, T. Dixon, D. L. Evans, J. P. Ford, R. S. Saunders, C. Breed, H. Masursky, J. F. McCauley, G. Shaber, L. Delliwig, A. England, H. MacDonald, P. Martin-Kaye, and F. Sabins. Shuttle imaging radar experiment. *Science*, 218:996, 1982.

[301] C. Elachi. *Spaceborne Radar Remote Sensing: Applications and Techniques.* IEEE Press, New York, 1987.

[302] C. Elachi, T. Bichnell, R. L. Jordan, and C. Wu. Spaceborne synthetic-aperture imaging radars: Applications, techniques, and technology. *Proceedings of the IEEE*, 70(10):1174–1209, October 1982.

[303] C. Elachi and D. D. Evans. Effects of random phase changes on the formation of synthetic aperture radar imagery. *IEEE Transactions on Antennas and Propagation*, 25:149–153, January 1977.

[304] C. Elachi, L. E. Roth, and G. C. Scaber. Spaceborne radar subsurface imaging in hyperarid regions. *IEEE Transactions on Geoscience and Remote Sensing*, 22:382–387, 1984.

[305] T. M. Elfouhaily and C.-A. Guerin. A critical survey of approximate scattering wave theories from random rough surfaces. *Institute of Physics Publishing*, Waves in Random Media 14:1–40, 2004.

[306] J. H. G. Ender. The meaning of k-space for classical and advanced SAR techniques. In *International Symposium Physics in Signal and Image Processing*, pages 23–38, Marseille, January 2001.

[307] A. Erdelyi. *Asymptotic Expansions.* Dover Publications, New York, 1956.

[308] H. Ermert and R. Karg. Multifrequency acoustic holography. *IEEE Transactions on Sonics and Ultrasonics*, 26(4):279–286, July 1979.

[309] O. Ersoy. *Difffraction, Fourier Optics and Imaging.* Wiley-Interscience, 2007.

[310] E. Ertin, C. D. Austin, S. Sharma, R. L. Moses, and L. C. Potter. GOTCHA experience report: Three-dimensional SAR imaging with complete circular apertures. In *Proceedings of SPIE, Algorithms for Synthetic Aperture Radar Imagery XIV*, Orlando, Florida, 9-13 April 2007.

[311] E. Ertin, R. L. Moses, and L. C. Potter. Interferometric methods for 3-D target reconstruction with multi-pass circular SAR. In *7th European Conference on Synthetic Aperture Radar*, Friedrichshafen, Germany, 2–5 June 2008.

[312] E. Ertin, L. C. Potter, and R. L. Moses. Enhanced imaging over complete circular apertures. In *Fortieth Asilomar Conference on Signals, Systems and Computers*, pages 1580–1584, 29 October – 1 November 2006.

[313] C. Esmersoy, M. L. Oristaglio, and B. C. Levy. Multidimensional Born velocity inversion: Single wideband point source. *The Journal of the Acoustical Society of America*, 78:1052–1057, 1985.

[314] G. Evans, J. Blackedge, and P.Yardley. *Analytic Methods for Partial Differential Equations*. Springer-Verlag, 2001.

[315] P. P. Ewald. Zur begrundung der kristalloptik; teil i: Dispersionstheorie. *Annalen der Physik*, 49(1):1, 1916.

[316] P. P. Ewald. Zur begrundung der kristalloptik; teil ii: Theorie der reflexion und brechung. *Annalen der Physik*, 49(2):117, 1916.

[317] D. Falconner and G. Moussally. Tomographic imaging of radar data gathered on a circular flight path about a three-dimensional target zone. In *Proceedings of SPIE, Algorithms for Synthetic Aperture Radar Imagery II*, volume 2487, Orlando, Florida, 5 June 1995.

[318] R. L. Fante. Turbulence-induced distortion of synthetic aperture radar images. *IEEE Transactions on Geoscience and Remote Sensing*, 32(4):958–961, July 1994.

[319] N. H. Farhat. Microwave holography and its applications in modern aviation. In *Proceedings of SPIE, Engineering Applications of Holography*, pages 295–314, 1972.

[320] N. H. Farhat. High resolution microwave holography and the imaging of remote objects. *Optical Engineering*, 14(5):499–505, September 1975.

[321] N. H. Farhat. Principles of broad-band coherent imaging. *Journal on Optics Society of America*, 67:1015–1021, 1977.

[322] N. H. Farhat and C. K. Chan. Three-dimensional imaging by wave-vector diversity. In A. Metherell, editor, *Acoustical Imaging*, volume 8, 1979.

[323] N. H. Farhat, C. L. Werner, and T. H. Chu. Prospects of three-dimensional and tomographic imaging radar networks. *Radio Science*, 19(5):1347–1355, 1984.

[324] J. A. Fawcett. Inversion of N-dimensional spherical means. *SIAM Journal of Applied Mathematics*, 45:336–341, 1985.

[325] B. L. Felsen and N. Marcuvitz. *Radiation and Scattering of Waves*. McGraw-Hill, New York, 1973.

[326] A. F. Fercher, H. Bartelt, H. Becker, and E. Wiltschko. Image formation by inversion of scattered data: Experiments and computational simulation. *Applied Optics*, 18:2427–2439, 1979.

[327] M. Ferrara, J. A. Jackson, and C. Austin. Enhancement of multi-pass 3D circular SAR images using sparse reconstruction techniques. In *Proceedings of SPIE, Algorithms for Synthetic Aperture Radar Imagery XVI*, Orlando, Florida, 13–17 April 2009.

[328] M. Ferrara, J. T. Parker, and M. Cheney. Resolution optimization with irregularly sampled Fourier data. *Inverse Problems*, 29(054008), 2013.

[329] J. Fessler and B. Sutton. Nonuniform Fast Fourier Transform using min-max interpolation. *IEEE Transactions on Signal Processing*, 51(2):560–574, February 2003.

[330] M. Figueiredo, R. Nowak, and S. Wright. Gradient projection for sparse reconstruction: Application to compressed sensing and other inverse problems. *IEEE Journal of Selected Topics in Signal Processing*, 1:589–597, December 2007.

[331] M. Fink. Time reversal of ultrasonic fields–Part I: Basic principles. *IEEE Transactions on Ultrasonics, Ferroelectrics and Frequency Control*, 30(5):555–566, 1992.

[332] J. P. Fitch. *Synthetic Aperture Radar (Signal Processing and Digital Filtering)*. Springer-Verlag, New York, 1988.

[333] D. Flores-Tapia, G. Thomas, A. Ashtari, and S. Pistorius. 3D breast microwave imaging based on wavefront reconstruction. In *28th Annual International Conference on IEEE Engineering in Medicine and Biology Society*, 2006.

[334] D. Flores-Tapia, G. Thomas, and S. Pistorius. A wavefront reconstruction for 3-D cylindrical subsurface radar imaging. *IEEE Transactions on Image Processing*, 17(10):1908–1925, October 2008.

[335] D. Flores-Tapia, G. Thomas, and S. Pistorius. Wavefront reconstruction of elevation circular synthetic aperture radar imagery using a cylindrical Green's function. *EURASIP Journal on Advances in Signal Processing*, 657323:12pp, 2010.

[336] T. J. Flynn. Wavenumber-domain SAR focusing from a nonuniform synthetic aperture. In *IEEE International Conference on Acoustics, Speech, and Signal Processing*, volume 3, pages 1–4, 23-26 March 1992.

[337] V. A. Fock. *Electromagnetic Diffraction and Propagation of Waves*. Pergamon Press, New York, 1965.

[338] G. Fornaro. Trajectory deviations in airborne SAR: Analysis and compensation. *IEEE Transactions on Aerospace and Electronic Systems*, 35(3):997–1009, July 1995.

[339] G. Fornaro, G. Franceschetti, and S. Perna. Motion compensation of squinted airborne SAR raw data: Role of processing geometry. *Proceedings of IEEE International Geoscience and Remote Sensing Symposium*, 2:1518–1521, September 2004.

[340] G. Fornaro, G. Franceschetti, and G. Shirinzi. Motion compensation errors: Effects on the accuracy of airborne SAR images. *IEEE Transactions on Aerospace and Electronic Systems*, 41(4):1338–1352, 2005.

[341] G. Fornaro, F. Lombardini, and F. Serafino. Three-dimensional multipass SAR focusing: Experiments with long-term spaceborne data. *IEEE Transactions on Geoscience and Remote Sensing*, 43(4):702–714, April 2005.

[342] G. Fornaro and A. Pauciullo. LMMSE 3-D SAR focusing. *IEEE Transactions on Geoscience and Remote Sensing*, 47(1):214–223, January 2009.

[343] G. Fornaro, D. Reale, and F. Serafino. Four-dimensional SAR imaging for height estimation and monitoring of single and double scatterers. *IEEE Transactions on Geoscience and Remote Sensing*, 47(1):224–237, January 2009.

[344] G. Fornaro, E. Sansosti, R. Lanari, and M. Tesauro. Role of processing geometry in SAR raw data focusing. *IEEE Transactions on Aerospace and Electronic Systems*, 38(2):441–454, April 2002.

[345] G. Fornaro, F. Serafino, and F. Soldovieri. Three-dimensional focusing with multipass SAR data. *IEEE Transactions on Geoscience and Remote Sensing*, 41(3):507–517, March 2003.

[346] J. Fortuny. An efficient 3-D near field ISAR algorithm. *IEEE Transactions on Aerospace and Electronic Systems*, 34(4):1261–1270, October 1998.

[347] J. Fortuny-Guasch. *Efficient Algorithms for Three-Dimensional Near-Field Synthetic Aperture Radar Imaging*. PhD thesis, University of Karlsruhe, Karlsruhe Germany, 2001.

[348] J. Fortuny-Guash and J. M. Lopez-Sanchez. Extension of the 3-D range migration algorithm to cylindrical and spherical scanning geometries. *IEEE Transactions on Antennas and Propagation*, 49(10):1434–1443, October 2001.

[349] G. Franceschetti and R. Lanari. *Synthetic Aperture Radar Processing*. CRC Press, New York, 1999.

[350] G. Franceschetti, R. Lanari, and E. S. Marzouk. Efficient and high precision space-variant processing of SAR data. *IEEE Transactions on Aerospace and Electronic Systems*, 31(1):227–237, January 1995.

[351] G. Franceschetti, R. Lanari, V. Pascazio, and G. Schirinizi. WASAR: A wide-angle SAR processor. *Proceedings of the IEE, Part F*, 139(2):107–114, April 1992.

[352] G. Franceschetti and G. Shirinzi. A SAR processor based on two-dimensional FFT codes. *IEEE Transactions on Aerospace and Electronic Systems*, 26(2):356–365, March 1990.

[353] G. Franceschetti and G. Shirinzi. Aberration free SAR raw data processing via transformed grid predeformation. In *IEEE International Geoscience and Remote Sensing Symposium*, pages 1593–1595, 1993.

[354] G. Toraldo Di Francia. *Introduction to the Theory of Fourier Integrals*. Clarendon Press, Oxford, 1948.

[355] W. Franz. Zur formulierung des huygensschen prinzips. *Zeitschrift für Naturforschung A*, 3a:500–506, 1948.

[356] D. Fraser. Interpolation by the FFT revisited – an experimental investigation. *IEEE Transactions on Acoustics, Speech and Signal Processing*, 37(5):665–675, May 1989.

[357] A. Freeman. SAR calibration: An overview. *IEEE Transactions on Geoscience and Remote Sensing*, 30(6):1107–1121, November 1992.

[358] A. Freeman, W. T. K. Johnson, B. Honeycutt, R. Jordan, S. Hensley, P. Siqueira, and J. Curlander. The myth of the minimum SAR antenna area constraint. *IEEE Transactions on Geoscience and Remote Sensing*, 38(1):320–324, January 2000.

[359] A. Freeman and M. Zink. An application of the monopulse principle to determining elevation angles in SAR images. *IEEE Transactions on Geoscience and Remote Sensing*, 32(3):616–625, May 1994.

[360] A.-J. Fresnel. Memoire sur la diffraction de la lumiere. *Memories de l'Academie des Sciences*, 5:339–487, 1826. Submitted 20 April 1818.

[361] O. Frey, C. Magnard, M. Ruegg, and E. Meier. Focusing SAR data acquired from non-linear sensor trajectories. In *IEEE International Geoscience and Remote Sensing Symposium*, pages 415–418, 2008.

[362] O. Frey, C. Magnard, M. Ruegg, and E. Meier. Non-linear SAR data processing by time-domain back-projection. In *Proceedings of the European Conference on Synthetic Aperture Radar*, pages 165–168, 2008.

[363] O. Frey, C. Magnard, M. Ruegg, and E. Meier. Focusing of airborne synthetic aperture radar data from highly nonlinear flight tracks. *IEEE Transactions on Geoscience and Remote Sensing*, 47(6):1844–1858, June 2009.

[364] O. Frey, F. Morsdorf, and E. Meier. Tomographic processing of multiple baseline P-band SAR data for imaging of a forested area. In *IEEE International Geoscience and Remote Sensing Symposium*, pages 156–159, 2007.

[365] O. Frey, F. Morsdorf, and E. Meier. Tomographic imaging of a forested area by airborne multi-baseline P-band SAR. *Sensors, Special Issue on Synthetic Aperture Radar*, 8(9):5884–5896, September 2008.

[366] P.-O. Frolind and L. M. H. Ulander. Motion compensation effects in wavelength-resolution SAR. In *Geoscience and Remote Sensing Symposium Proceedings*, pages 436–438, Singapore, 3–8 August 1997.

[367] P.-O. Frolind and L. M. H. Ulander. Motion compensation effects for repeat-pass processing in wavelength-resolution SAR. In *Geoscience and Remote Sensing Symposium Proceedings*, pages 2637–2639, Seattle, Washington, 6–10 July 1998.

[368] P.-O. Frolind and L. M. H. Ulander. Evaluation of angular interpolation kernels in fast back-projection SAR processing. In *IEE Proceedings on Radar, Sonar and Navigation*, volume 153, pages 243–249, June 2006.

[369] D. Gabor. A new microscope principle. *Nature*, 161(4098):777–778, 1948.

[370] D. Gabor. Microscopy by reconstructed wavefronts. *Proceedings of the Royal Society (London)*, A 197:454–487, 1949.

[371] D. Gabor. Microscopy by reconstructed wavefronts: II. *Proceedings of the Royal Society (London)*, 64 (Pt. 6)(378 B):449–469, 1951.

[372] D. Gabor. Holography. *Nobel Lecture, Imperial Colleges of Science and Technology*, 1971.

[373] S. Ganapathy, W. S. Wu, and B. Schumult. Analysis and design for a real-time system for nondestructive evaluation in the nuclear industry. *Ultrasonics*, 20:249, 1982.

[374] J. D. Gaskill. *Linear Systems, Fourier Transforms, and Optics.* Wiley, New York, 1978.

[375] J. Gazdag. Wave equation migration with the phase shift method. *Geophysics*, 43:1342–1351, 1978.

[376] J. Gazdag and P. Sguazzero. Migration of seismic data. *Proceedings of the IEEE*, 72(10):1302–1315, October 1984.

[377] I. M. Gel'fund and G. E. Shilov. *Generalized Functions, Volume I: Properties and Operations.* Academic, New York, 1968.

[378] G. J. Genello, Jr, M. C. Wicks, and M. Soumekh. Alias-free processing of P-3 data. In *Proceedings of SPIE, Algorithms for Synthetic Aperture Radar Imagery VI*, volume 3721, pages 189–200, Orlando, Florida, April 1999.

[379] D. Geudtner, R. Winter, and P. W. Vachon. Flood monitoring using ERS-1 SAR interferometry coherence maps. In *Proceedings of the International Geoscience and Remote Sensing Symposium*, volume 2, pages 966–968, 1996.

[380] D. C. Ghiglia and W. D. Brown. Some methods for reducing propagation-induced phase errors in coherent imaging systems. II. Numerical results. *Journal of the Optical Society of America*, 5(6):942–957, 1988.

[381] D. C. Ghiglia and G. A. Mastin. Two-dimensional phase correction of synthetic-aperture-radar imagery. *Optics Letters*, 14(20):1104–1106, October 1989.

[382] D. C. Ghiglia and L. A. Romero. Direct phase estimation from phase differences using fast elliptic partial differential equation solvers. *Optics Letters*, 14(20):1107–1109, October 1989.

[383] C. H. Gierull. On a concept for an airborne downward-looking imaging radar. *International Journal of Electronics and Communications*, 53(6):295–304, 1999.

[384] F. Gini and F. Lombardinio. Multibaseline cross-track SAR interferometry: A signal processing perspective. *IEEE Aerospace and Electronic Systems Magazine*, 20(8):71–93, August 2005.

[385] R. Giret, H. Jeuland, and P. Enert. A study of a 3D-SAR concept for a millimeter-wave imaging radar onboard an UAV. In *Proceedings of the European Radar Conference*, pages 201–204, Amsterdam, The Netherlands, October 2004.

[386] G. H. Glover and J. L. Sharp. Reconstruction of ultrasound propagation speed distribution in soft tissue: Time-of-flight tomography. *IEEE Transactions on Sonics and Ultrasonics*, 24(4):229–234, July 1977.

[387] M. Goiten. Three dimensional density reconstruction from a series of two dimensional projections. *Nuclear Instruments and Methods*, 101:509–518, 1972.

[388] W. Goj. *Synthetic Aperture Radar and Electronic Warfare.* Artech House, Dedham, Massachusetts, 1989.

[389] A. Golden, Jr, S. C. Wei, K. K. Ellis, and S. Tummala. Migration processing of spotlight SAR data. In *Proceedings of SPIE, Algorithms for Synthetic Aperture Radar Imagery*, volume 2230, pages 25–35, Orlando, Florida, 4–8 April 1994.

[390] R. M. Goldstein, T. P. Barnett, and H. A. Zebker. Remote sensing of ocean currents. *Science*, 246:1282–1285, 1989.

[391] R. M. Goldstein and H. A. Zebker. Interferometric radar measurements of ocean surface currents. *Nature*, 328(20):707–709, 1987.

[392] J. W. Goodman. Digital image formation from detected holographic data. *Acoustical Holography*, 1969.

[393] J. W. Goodman. An introduction to the principles and application of holography. *Proceedings of the IEEE*, 59:1292–1304, September 1971.

[394] J. W. Goodman. Some fundamental properties of speckle. *Journal of the Optical Society of America*, 66:1145–1150, 1976.

[395] J. W. Goodman. *Introduction to Fourier Optics.* Roberts & Company, Colorado, 5th edition, 2005, 1st (1968).

[396] N. A. Goodman, S. C. Lin, D. Rajakrishna, and J. M. Stiles. Processing of multiple-receiver spaceborne arrays for wide-area SAR. *IEEE Transactions on Geoscience and Remote Sensing*, 40(4):841–852, April 2002.

[397] R. Goodman, S. Tummala, and W. Carrara. Issues in ultra-wideband, widebeam SAR image formation. In *Proceedings of the IEEE International Radar Conference*, pages 479–485, Alexandria, Virginia, 8–11 May 1995.

[398] R. Gordon, R. Bender, and G. T. Herman. Algebraic reconstruction techniques (ART) for three-dimensional electron microscopy and X-ray photography. *Journal of Theoretical Biology*, 29:471–481, December 1970.

[399] R. Gordon and G. T. Herman. Reconstruction of pictures from their projections. *Communications of the ACM*, 14(12):759–768, December 1971.

[400] R. Gordon and G. T. Herman. Three-dimensional reconstruction from projections: A review of algorithms. *International Review of Cytology*, 38:111–151, 1974.

[401] R. Gordon, G. T. Herman, and S. A. Johnson. Image reconstruction from projections. *Scientific American*, 233(4):56–68, October 1975.

[402] L. A. Gorham, U. K. Majumder, P. Buxa, M. J. Backues, and A. C. Lindgren. Implementation and analysis of a fast backprojection algorithms. In *Proceedings of SPIE, Algorithms for Synthetic Aperture Radar Imagery XIII*, volume 6237, Orlando, Florida, 17 May 2006.

[403] L. A. Gorham and B. D. Rigling. Dual format algorithm for monostatic SAR. In *Proceedings of SPIE, Algorithms for Synthetic Aperture Radar Imagery XVII*, volume 7699, 2010.

[404] L. A. Gorham, B. D. Rigling, and E. G. Zelnio. A comparison between imaging radar and medical imaging polar format algorithm implementations. In *Proceedings of SPIE, Algorithms for Synthetic Aperture Radar Imagery XIV*, volume 6568, 2007.

[405] P. T. Gough and R. H. T. Bates. Computer generated holograms for processing radiographic data. *Computers and Biomedical Research, an International Journal*, 5:700–708, 1972.

[406] P. T. Gough and D. W. Hawkins. A unified framework for modern synthetic aperture imaging algorithms. *International Journal of Imaging Systems and Technology*, 8(4):343–358, 1996.

[407] P. T. Gough and D. W. Hawkins. Imaging algorithms for strip-map synthetic aperture sonar: Minimizing the effects of aperture errors and aperture undersampling. *IEEE Journal of Oceanic Engineering*, 22(1):27–39, January 1997.

[408] L. C. Graham. Synthetic interferometer radar for topographic mapping. *Proceedings of the IEEE*, 62(6):763–768, June 1974.

[409] D. G. Grant. TOMOSYNTHESIS: A three-dimensional radiographic imaging technique. *IEEE Transactions on Biomedical Engineering*, 19(1):20–28, January 1972.

[410] A. L. Gray and P. J. Farris-Manning. Repeat-pass interferometry with airborne synthetic aperture radar. *IEEE Transactions on Geoscience and Remote Sensing*, 31(1):180–191, January 1993.

[411] P. E. Green, Jr. Radar astronomy measurement techniques. Technical report, Lincoln Laboratory, MIT, Cambridge, Massachusetts, 1962.

[412] L. Greengard and J.-Y. Lee. Accelerating the nonuniform Fast Fourier Transform. *SIAM Review*, 43(3):443–454, 2004.

[413] J. F. Greenleaf. Computerized transmission tomography. *Methods of Experimental Physics–Ultrasound*, 19:563–589, 1981.

[414] J. F. Greenleaf. Computerized tomography with ultrasound. *Proceedings of the IEEE*, 71(3):330–337, March 1983.

[415] J. F. Greenleaf and R. C. Bahn. Clinical imaging with transmissive ultrasonic computerized tomography. *IEEE Transactions on Biomedical Engineering*, 28:177–185, 1981.

[416] J. F. Greenleaf, S. A. Johnson, S. L. Lee, G. T. Herman, and E. H. Wood. Algebraic reconstruction of spatial distributions of acoustic absorption within tissue from their two-dimensional acoustic projections. *Acoustical Holography*, 5:591–603, 1974.

[417] J. F. Greenleaf, S. A. Johnson, and A. H. Lent. Measurement of spatial distribution of refractive index in tissues by ultrasonic computer assisted tomography. *Ultrasound in Medicine and Biology*, 3:327–339, 1978.

[418] J. F. Greenleaf, S. A. Johnson, W. F. Samayoa, and F. A. Duck. Algebraic reconstruction of spatial distributions of acoustic velocities in tissue from their time-of-flight profiles. *Acoustical Holography*, 6:71–90, 1975.

[419] F. M. Grimaldi. Physico-mathesis de lumine, coloribus, et iride. *Bologna*, 1665.

[420] J. Groot and M. Otten. SAR imaging of corner reflectors larger than the spatial resolution. *IEEE Transactions on Geoscience and Remote Sensing*, 32(3):721–724, May 1994.

[421] A. M. Guarnieri. Residual SAR focusing: An application to coherence improvement. *IEEE Transactions on Geoscience and Remote Sensing*, 34(1):201–211, January 1996.

[422] A. M. Guarnieri and C. Prati. Autofocusing and residual migration of SAR images. In *Proceedings of the IEEE International Conference on Acoustics, Speech, and Signal Processing*, pages 1953–1956, 1990.

[423] A. Gunawardena and D. Longstaff. Ultra-wideband widebeam SAR processing in the time-space domain. In *Proceedings of the IEEE International Radar Conference*, volume 36, Edinburgh, United Kingdom, 1997.

[424] A. Gunawardena and D. Longstaff. Wave equation formulation of synthetic aperture radar (SAR) algorithms in the time-space domain. *IEEE Transactions on Geoscience and Remote Sensing*, 36(6):1995–1999, November 1998.

[425] A. Gustavsson, L. M. H. Ulander, B. H. Flood, P.-O. Frolind, H. Hellsten, T. Jonsson, B. Larsson, and G. Stenstrom. Development and operation of an airborne VHF SAR system–lessons learned. In *Proceedings of the IEEE International Conference on Geoscience and Remote Sensing Symposium*, volume 1, pages 458–462, Seattle, Washington, July 1998.

[426] S. Gutman and M. Klibanov. Regularized quasi-Newton method for inverse scattering problems. *Mathematical and Computer Modeling*, 18:5–31, 1993.

[427] T. M. Habashy, R. W. Groom, and B. R. Spies. Beyond the Born and Rytov approximations: A nonlinear approach to electromagnetic scattering. *Journal of Geophysical Research*, 98:1759–1775, 1993.

[428] J. O. Hagberg and L. M. H. Ulander. On the optimization of SAR for topographic mapping. *IEEE Transactions on Geoscience and Remote Sensing*, 31(1):303–306, January 1993.

[429] J. O. Hagberg, L. M. H. Ulander, and J. Askne. Repeat-pass SAR interferometry over forested terrain. *IEEE Transactions on Geoscience and Remote Sensing*, 33(3):331–340, March 1995.

[430] T. Hagfors and D. B. Campbell. Mapping of planetary surfaces by radar. *Proceedings of the IEEE*, 61(9):1219–1225, September 1973.

[431] J. E. Hansen. *Spherical Near-Field Antenna Measurements*. Peter Peregrinus Ltd., Stevenage, United Kingdom, 1988.

[432] R. O. Harger. Mapping of planetary surfaces by radar. *IRE Transactions on Military Electronics*, 9:264–278, July–October 1965.

[433] R. O. Harger. *Synthetic Aperture Radar Systems: Theory and Design*. Academic Press, New York, 1970.

[434] R. F. Harrington. *Time-Harmonic Electromagnetic Fields*. McGraw-Hill, New York, 1961.

[435] F. J. Harris. On the use of windows for harmonic analysis with the discrete Fourier transform. *Proceedings of the IEEE*, 66(1):51–83, January 1978.

[436] J. E. Harris. Image quality measurements for comparison of synthetic aperture radar processors. Master's thesis, Brigham Young University, Provo, Utah, 1987.

[437] J. E. Harris, R. S. Ostler, D. M. Chabries, and R. W. Christiansen. Quality measures for SAR images. In *Proceedings of the IEEE International Conference on Acoustics, Speech, and Signal Processing*, pages 1064–1067, April 1988.

[438] J. E. Harvey. Fourier treatment of near-field scaar diffraction theory. *American Journal of Physics*, 47:974, 1979.

[439] G. Haslam and B. Reid. Motion sensing requirements for synthetic aperture radar. *Proceedings of the IEEE*, 1:126–131, S. 1983.

[440] D. W. Hawkins. *Synthetic Aperture Imaging Algorithms: With Application to Wide Bandwidth Sonar*. PhD thesis, Electrical and Electronic Engineering, University of Canterbury, Christchurch, New Zealand, October, 1996.

[441] D. W. Hawkins and P. T. Gough. Recent sea trials of a synthetic aperture sonar. *Proceedings of the Institute of Acoustics*, 17(8):1–10, December 1995.

[442] D. W. Hawkins and P. T. Gough. An accelerated chirp scaling algorithm for synthetic aperture imaging. In *IEEE International Geoscience and Remote Sensing*, volume 1, pages 471–473, Singapore, 3-8 August 1997.

[443] M. P. Hayes and P. T. Gough. Broad-band synthetic aperture sonar. *IEEE Journal of Oceanic Engineering*, 17(1):80–94, January 1992.

[444] S. Haykin. *Communication Systems*. John Wiley, third edition, 1994.

[445] A. Hein. *Processing of SAR Data: Fundamentals, Signal Processing, Interferometry*. Artech House, 2004.

[446] S. Helgason. *The Radon Transform*. Birkhauser, Boston, Massachusetts, 1980.

[447] H. Hellsten. Inverse scattering analysis of diffraction limited SAR. *IEEE Transactions on Antennas and Propagation*, 38(10):1517–1522, 1990.

[448] H. Hellsten and L. E. Andersson. An inverse method for the processing of synthetic aperture radar data. *Inverse Problems*, 3(1):111–124, 1987.

[449] H. Hellsten and L. M. H. Ulander. Airborne array aperture UWB UHF radar motivation and system considerations. In *Proceedings of the IEEE International Radar Conference*, pages 47–53, Waltham, Massachusetts, 20–21 April 1999.

[450] H. Hellsten, L. M. H. Ulander, A. Gustavsson, and B. Larsson. Development of VHF CRABAS II SAR. In G. S. Ustach, editor, *Proceedings of SPIE, Radar Sensor Technology*, volume 2747, pages 48–60, Orlando, Florida, 8–9 April 1996.

[451] F. M. Henderson and editors A. J. Lewis. *Manual of Remote Sensing, Volume 2: Principles and Applications of Imaging Radar*. John Wiley, New York, 3rd edition, 1998.

[452] S. Hensley and S. Shaffer. Automatic DEM generation using Magellan stereo data. In *Proceedings of the IEEE Geoscience and Remote Sensing Symposium*, volume 3, 1994.

[453] G. T. Herman. Two methods for reconstructing pictures from their projections: A comparative study. *Computer Graphics and Image Processing*, 1:123–144, 1973.

[454] G. T. Herman. *Image Reconstruction from Projections: The Fundamentals of Computerized Tomography*. Academic Press, New York, 1980.

[455] G. T. Herman, A. Lent, and S. Rowland. ART: Mathematics and applications (a report on the mathematical foundations and on the applicability to real data of the algebraic reconstruction technique). *Journal of Theoretical Biology*, 32:1–32, 1971.

[456] G. T. Herman and A. Naparstek. Fast image reconstruction based on Radon inversion formula appropriate for rapidly collected data. *SIAM Journal on Applied Mathematics*, 33:511, November 1977.

[457] G. T. Herman and S. Rowland. Resolution in ART: An experimental investigation of the resolving power of an algebraic picture reconstruction technique. *Journal of Theoretical Biology*, 33:213–233, 1971.

[458] G. T. Herman and S. Rowland. Three methods for reconstructing objects from X-rays: A comparative study. *Computer Graphics and Image Processing*, 2:151–178, 1973.

[459] G. T. Herman, H. K. Tuy, K. J. Langenberg, and P. Sabatier. *Basic Methods of Tomography and Inverse Problems.* Adam Hilger, Bristol, 1987.

[460] B. P. Hildebrand and K. A. Haines. Holography by scanning. *Journal on Optics Society of America*, 59:1–6, 1969.

[461] B. P. Hildebrand and K. A. Haines. *An Introduction to Acoustical Holography.* Plenum, New York, 1972.

[462] H. P. Hiriyannaiah. X-ray computed tomography for medical imaging. *IEEE Signal Processing Magazine*, pages 42–59, March 1997.

[463] J. Homer, I. D. Longstaff, and C. Callagan. High resolution 3D SAR via multi-baseline interferometry. In *Proceedings of the IEEE International Geoscience Remote Sensing Symposium*, pages 796–798, 1996.

[464] J. Homer, I. D. Longstaff, and Z. She. Improved digital elevation models via multi-baseline interferometry. In *Proceedings of the IEEE International Geoscience Remote Sensing Symposium*, pages 1579–1581, 1997.

[465] J. Homer, I. D. Longstaff, Z. She, and D. Gray. High resolution 3-D imaging via multi-pass SAR. *IEE Proceedings on Radar, Sonar and Navigation*, 149(1):45–50, February 2002.

[466] H. Honl. Eine strenge formulierung des klassichen beugungsproblems. *Zeitschrift fur Physik*, 131(3):290–304, 1952.

[467] R. Hooke. Micrographia, 47, 1665.

[468] B. Horn. Density reconstruction using arbitrary ray sampling schemes. *Proceedings of the IEEE*, 66(5):551–562, May 1978.

[469] G. N. Hounsfield. *A Method of and Apparatus for Examination of a Body by Radiation such as X-ray or Gamma Radiation.* British Patent, Number 1,283,915, 1972.

[470] K. M. Houston. A fast beamforming algorithm. In *IEEE Oceans Conference*, volume 1, pages 211–216, 1994.

[471] S. A. Hovanessian. *Introduction to Synthetic Array and Imaging Radars.* Artech House, Dedham, Massachusetts, 1980.

[472] S. A. Hovanessian. *Synthetic Array and Imaging Radars.* Springer-Verlag, 1980.

[473] C. Hu, T. Long, T. Zeng, F. Liu, and Z. Liu. The accurate focusing and resolution analysis method in geosynchronous SAR. *IEEE Transactions on Geoscience and Remote Sensing*, 2(5):1–15, May 2011.

[474] J. Hui and H. C. Huan. Spaceborne SAR imaging in wavenumber domain. In *Proceedings of the International Geoscience and Remote Sensing Symposium*, pages 2141–2144, 1995.

[475] A. J. Hunter, M. Hayes, and P. T. Gough. A comparison of fast factored back-projection and wavenumber algorithms for SAS reconstruction. In *Fifth World Congress on Ultrasonics*, pages 527–530, Paris, 2003.

[476] W. B. Hyslop, R. K. Woods, and P. C. Lauterbur. Four-dimensional spectral-spatial imaging using projection reconstruction. *IEEE Transactions on Medical Imaging*, 14:374–383, 1995.

[477] J. Isenman, E. Jones, and D. Caliga. Signal/data processor implementation and algorithms for realtime wide-angle ultra-wideband SAR image formation. In *HPEC*, 2009.

[478] T. Isernia, V. Pascazio, R. Pierri, and G. Schirinzi. Image reconstruction from Fourier transform magnitude with applications to synthetic aperture radar imaging. *Journal of the Optical Society of America*, 13(5):922–934, 1996.

[479] T. Isernia, V. Pascazio, R. Pierri, and G. Schirinzi. Synthetic aperture radar imaging from phase corrupted data. *IEE Proceedings on Radar, Sonar and Navigation*, 143(4):268–274, August 1996.

[480] A. Ishimaru. *Wave Propagation and Scattering in Random Media.* Academic Press, New York, 1978.

[481] K. Iwata and R. Nagata. Calculation of three-dimensional refractive index distribution from interferograms. *Journal of the Optical Society of America*, 133(60), 1970.

[482] K. Iwata and R. Nagata. Calculation of refractive index distribution from interferograms using the Born and Rytov's approximation. *Proceedings of the ICO Conference on Optical Methods in Scientific and Industrial Measurements, Tokyo, 1974, and Japan Journal of Applied Physics*, 14(1):379–383, 1975.

[483] K. Iwata and R. Nagata. Straight path approximation and Rytov's approximation in the reconstruction of refractive index distribution. In *Digest of Papers, Topical Meeting Image Processing for 2-D and 3-D Reconstructions from Projections*, volume Paper TUB1-1, Stanford, California, 1975.

[484] J. I. Jackson, C. H. Meyer, D. G. Nishimura, and A. Macovski. Selection of a convolution function for Fourier inversion using gridding. *IEEE Transactions on Medical Imaging*, 10(3):473–478, September 1991.

[485] A. Jain. Multibeam synthetic aperture radar for global oceanography. *IEEE Transactions on Antennas Propagation*, 27:535–538, July 1979.

[486] C. V. Jakowatz, Jr and N. Doren. Comparison of polar formatting and back-projection algorithms for spotlight-mode SAR image formation. In *Proceedings of SPIE, Algorithms for Synthetic Aperture Radar Imagery XIII*, volume 6237, Orlando, Florida, 17 May 2006.

[487] C. V. Jakowatz, Jr and A. C. Kak. Computerized tomography using X-rays and ultrasound. Technical report, Perdue University, West Lafayette, Indiana, 1976.

[488] C. V. Jakowatz, Jr and P. A. Thompson. A new look at spotlight mode synthetic aperture radar as tomography: Imaging 3-D targets. *IEEE Transactions on Image Processing*, 4(5):699–703, May 1995.

[489] C. V. Jakowatz, Jr and D. E. Wahl. Three-dimensional tomographic imaging for foliage penetration using multiple-pass spotlight mode SAR. In *Proceedings of the 35st Asilomar Conference on Circuits, Systems and Computing*, volume 1, pages 121–125, Pacific Grove, California, 2001.

[490] C. V. Jakowatz, Jr, D. E. Wahl, P. H. Eichel, D. C. Ghiglia, and P. A. Thompson. *Spotlight-Mode Synthetic Aperture Radar: A Signal Processing Approach.* Kluwer, Boston, Massachusetts, 1996.

[491] C. V. Jakowatz, Jr, D. E. Wahl, and P. A. Thompson. Three-dimensional SAR imaging using cross-track coherent stereo collections. In *Proceedings of the 31st Asilomar Conference on Circuits, Systems and Computing*, volume 2, pages 1199–1203, Pacific Grove, California, 1997.

[492] C. V. Jakowatz, Jr, D. E. Wahl, P. A. Thompson, and N. E. Doren. Space-variant filtering for correction of wavefront curvature effects in spotlight-mode SAR imagery formed via polar formatting. In *Proceedings of SPIE, Algorithms for Synthetic Aperture Radar Imagery IV*, volume 3070, pages 33–42, April 1997.

[493] C. V. Jakowatz, Jr, D. E. Wahl, and D. A. Yocky. Beamforming as a foundation for spotlight-mode SAR image formation by backprojection. In *Proceedings of SPIE, Algorithms for Synthetic Aperture Radar Imagery XV*, volume 6970, Orlando, Florida, 16 March 2008.

[494] C. V. Jakowatz, Jr, D. E. Wahl, and D. A. Yocky. A beamforming algorithm for bistatic SAR image formation. In *Proceedings of SPIE, Algorithms for Synthetic Aperture Radar Imagery XVII*, volume 7699, 2010.

[495] C. V. Jakowatz, Jr, D. E. Wahl, D. A. Yocky, B. K. Bray, W. J. Bow, and J. A. Richards. Comparison of algorithms for use in real-time spotlight-mode SAR image formation. In *Proceedings of SPIE, Algorithms for Synthetic Aperture Radar Imagery XI*, volume 5427, pages 108–116, Bellingham, Washington, 12–16 April 2004.

[496] H. Jensen, L. C. Graham, L. J. Porcello, and E. N. Leith. Side-looking airborne radar. *Scientific American*, pages 84–95, October 1977.

[497] M. Y. Jin. Optimal Doppler centroid estimation for SAR data from quasi-homogeneous source. *IEEE Transactions on Geoscience and Remote Sensing*, 24(6):1022–1025, 1986.

[498] M. Y. Jin, F. Cheng, and M. Cheng. Chirp scaling algorithms for SAR processing. In *Proceedings of the IEEE International Geoscience Remote Sensing Symposium*, pages 1169–1172, August 1993.

[499] M. Y. Jin and M. Cheng. Analysis and simulation for a spotlight-mode aircraft SAR in circular flight path. In *Proceedings of the IEEE International Geoscience Remote Sensing Symposium*, volume 1, August 1993.

[500] M. Y. Jin and C. Wu. A SAR correlation algorithm which accommodates large range migration. *IEEE Transactions on Geoscience and Remote Sensing*, 22:592–597, November 1984.

[501] N. Joachimowicz, C. Pichot, and J. P. Hugonin. Inverse scattering: An iterative numerical method for electromagnetic imaging. *IEEE Transactions on Antennas and Propagation*, 39:1742–1751, 1991.

[502] R. B. Johnson. Radar analogies for optics. In *Proceedings of SPIE, Effective Utilization of Optics in Radar Systems*, volume 128, pages 75–83, 28 December 1977.

[503] S. A. Johnson, J. F. Greenleaf, M. Tanaka, B. Rajagopalan, and R. C. Bahn. Quantitative synthetic aperture reflection imaging with correction for refraction and attenuation: Application of seismic techniques in medicine. In *Proceedings of the Biomedical Symposium*, volume 17, pages 337–349, San Diego, California, 1978.

[504] J. A. Johnston and A. C. Fairhead. Waveform design and Doppler senstivity analysis for nonlinear FM chirp pulses. *IEE Proceedings F: Communications, Radar and Signal Processing*, 133(2):163–175, April 1986.

[505] J. H. Jolley and C. Dotson. Synthetic aperture radar improves reconnaissance. *Defense Electronics*, 13(9):111, September 1981.

[506] R. V. Jones, V. D. Vaughn, R. L. Frost, and D. M. Chabries. The effect of codebook size on the vector quantization of SAR data. In *IEEE 1988 National Radar Conference, 3rd*, pages 129–133, Ann Arbor, Michigan, 20–21 April 1988.

[507] R. L. Jordan. The Seasat—a synthetic aperture radar aperture. *IEEE Journal of Oceanic Engineering*, 5(5):154, April 1980.

[508] R. L. Jordan, B. L. Honeycutt, and M. Werner. The SIR-C/X SAR synthetic aperture radar system. *IEEE Transactions on Geoscience and Remote Sensing*, 33(4):829–839, 1995.

[509] A. C. Kak. Computerized tomography with X-ray, emission and ultrasound sources. *Proceedings of the IEEE*, 67(9):1245–1272, September 1979.

[510] A. C. Kak. Tomographic imaging with diffracting and nondiffracting sources. In S. Haykin, editor, *Array Signal Processing*, 1984.

[511] A. C. Kak, C. V. Jakowatz, Jr, N. A. Baily, and R. A. Keller. Computerized tomography using video recorded fluoroscopic images. *IEEE Transactions on Biomedical Engineering*, 24:157–169, March 1977.

[512] A. C. Kak and M. Slaney. *Principles of Computerized Tomographic Imaging*. IEEE Press, 1988.

[513] A. C. Kak and M. Slaney. *Principles of Computerized Tomographic Imaging*. Society of Industrial and Applied Mathematics, New York, 2001.

[514] M. Kaveh, R. K. Mueller, and R. D. Inverson. Ultrasonic tomography based on perturbation solutions of the wave equation. *Computer Graphics and Image Processing*, 9:105–116, 1979.

[515] M. Kaveh, R. K. Mueller, R. Rylander, T. R. Coulter, and M. Soumekh. Experimental results in ultrasonic diffraction tomography. *Acoustical Imaging*, 9:433, 1980.

[516] M. Kaveh and M. Soumekh. Computer-assisted diffraction tomography. In H. Stark, editor, *Image Recovery: Theory and Application*. Academic Press, New York, 1986.

[517] M. Kaveh, M. Soumekh, and J. E. Greenleaf. Signal processing for diffraction tomography. *IEEE Transactions on Sonics and Ultrasonics*, 31:230–238, July 1984.

[518] M. Kaveh, M. Soumekh, and R. K. Mueller. A comparison of Born and Rytov approximations in acoustic tomography. *Acoustical Imaging*, 10:325–335, 1981.

[519] M. Kaveh, M. Soumekh, and R. K. Mueller. Tomographic imaging via wave equation inversion. In *Proceedings of the IEEE International Conference on Acoustics, Speech and Signal Processing*, volume 3, pages 1553–1556, Paris, France, May 1982.

[520] B. M. Keel, J. A. Saffold, M. R. Walbridge, and J. Chadwick. Non-linear stepped chirp waveforms with sub-pulse processing for range sidelobe suppression. In *Proceedings of SPIE, Algorithms for Radar Sensor Technology III*, volume 3395, pages 87–98, Orlando, Florida, 16 April 1998.

[521] J. B. Keller. Accuracy and validity of the Born and Rytov approximations. *Journal of the Optical Society of America*, 59:1003–1004, 1969.

[522] O. D. Kellog. *Foundations of Potential Theory*. Dover Publications, 1953.

[523] E. M. Kennaugh and R. L. Cosgriff. The use of impulse response in electromagnetic scattering problem. In *IRE National Convention Record, Part I*, pages 72–77, 1958.

[524] T. A. Kennedy. Strapdown inertial measurement units for motion compensation for synthetic aperture radars. *IEEE AES Magazine*, pages 32–35, October 1988.

[525] S. K. Kenue and J. F. Greenleaf. Limited angle multifrequency diffraction tomography. *IEEE Transactions on Sonics and Ultrasonics*, 29(8):213–217, July 1982.

[526] J.-H. Kim and L.-S. Kim. An ASIC implementation of range cell migration compensation algorithm for synthetic aperture radar signal processing. In *IEEE Workshop on VLSI Signal Processing*, volume VIII, pages 135–144, 1995.

[527] D. R. Kirk and R. P. Maloney. A new autofocus technique for wide-band wide-angle synthetic aperture radar. In *Proceedings Aerosense Conference*, Orlando, Florida, April 1998.

[528] D. R. Kirk and R. P. Maloney. Impact of platform motion on wide-angle synthetic aperture radar (SAR). In *IEEE International Radar Conference*, pages 41–46, 1999.

[529] J. C. Kirk, Jr. Digital synthetic aperture radar technology. In *IEEE International Radar Conference Record*, pages 482–487, 1975.

[530] J. C. Kirk, Jr. A discussion of digital processing in synthetic aperture radar. *IEEE Transactions on Aerospace and Electronic Systems*, 11(3):326–337, May 1975.

[531] J. C. Kirk, Jr. Motion compensation for synthetic aperture radar. *IEEE Transactions on Aerospace and Electronic Systems*, 11(3):338–348, May 1975.

[532] J. C. Kirk, Jr. Signal based motion compensation. In *IEEE International Radar Conference*, pages 463–468, 2000.

[533] J. C. Kirk, Jr, D. Woods, and J. Salzman. Efficient motion tolerant FOPEN SAR processing. In *Proceedings of EUSAR - 5th European Conference on Synthetic Aperture Radar*, volume 1, pages 179–182, Ulm, Germany, 25–27 May 2004.

[534] J. R. Klauder, A. C. Price, S. Darlington, and W. J. Albersheim. The theory and design of chirp radars. *The Bell System Technical Journal*, 39(4):745–808, July 1960.

[535] A. Klug and R. A. Crowther. Three-dimensional image reconstruction from the viewpoint of information theory. *Nature*, 238:435–440, August 1972.

[536] K. Knaell. Three-dimensional SAR from curvilinear apertures. In *Technical Report, Carerock Division, Naval Surface Warfare Center*, 1994.

[537] K. Knaell. Progress in three-dimensional SAR from curvilinear apertures. In *Proceedings of SPIE, Radar Processing, Technology and Applications III*, volume 3462, pages 110–121, San Diego, California, 19 July 1998.

[538] K. K. Knaell and G. P Cardillo. Radar tomography for the generation of three-dimensional images. *IEE Proceedings–Radar, Sonar and Navigation*, 142(2):54–60, April 1995.

[539] W. E. Kock. *Radar, Sonar, and Holography*. Academic Press, New York, 1973.

[540] E. Korsbakken, J. A. Johannessen, and O. M. Johannessen. Coastal wind field retrievals from ERS synthetic aperture radar. *Journal Geophysical Research*, 103(C4):7857–7874, 15 April 1998.

[541] J. J. Kovaly. *Synthetic Aperture Radar*. Academic, New York, 1970.

[542] J. J. Kovaly. *Synthetic Aperture Radars*. Artech House, Dedham, Massachusetts, 1976.

[543] J. J. Kovaly. High resolution radar fundamentals. In Eli Broakner, editor, *Radar Technology*. Artech House, 1977.

[544] A. Kozma. Tilted-plane optical processor. *Applied Optics*, 11:1766–1777, August 1972.

[545] T. Kragh and A. Kharbouch. Monotonic iterative algorithms for SAR image restoration. In *Proceedings of the IEEE International Conference on Image Processing*, pages 645–648, October 2006.

[546] G. Krieger, N. Gebert, and A. Moreira. Multidimensional waveform encoding: A new digital beamforming technique for synthetic aperture radar remote sensing. *IEEE Transactions on Geoscience and Remote Sensing*, 46(1):31–46, January 2008.

[547] G. Krieger, I. Hajnsek, K. P. Papathanassiou, M. Younis, and A. Moreira. Interferometric synthetic aperture radar (SAR) missions employing formation flying. *Proceedings of the IEEE*, 98(5):816–843, May 2010.

[548] G. Krieger and A. Moreira. Multistatic SAR satellite formations: Multistatic SAR. In *Proceedings of the IEEE International Geoscience Remote Sensing Symposium*, pages 2680–2684, Seoul, Korea, 2005.

[549] G. Krieger, A. Moreira, H. Fiedler, I. Hajnsek, M. Werner, M. Younis, and M. Zink. TanDEM-X: A satellite formation for high-resolution SAR interferometry. *IEEE Transactions on Geoscience and Remote Sensing*, 45(11):3317–3341, November 2007.

[550] D. E. Kuhl and R. Q. Edwards. Image separation radio-isotope scanning. *Radiology*, 80:653–661, 1963.

[551] R. Lanari. A new method for the compensation of the SAR range cell migration based on the chirp-z transform. *IEEE Transactions on Geoscience and Remote Sensing*, 33(5):1296–1299, September 1995.

[552] R. Lanari and G. Fornaro. A short discussion on the exact compensation of the SAR range-dependent range cell migration effect. *IEEE Transactions on Geoscience and Remote Sensing*, 35(6):1446–1452, November 1997.

[553] R. Lanari, M. Tesauro, E. Sansosti, and G. Fornaro. Spotlight SAR data focusing based on a two-step processing approach. *IEEE Transactions on Geoscience and Remote Sensing*, 39(9):1993–2004, September 2001.

[554] R. Lanari, S. Zoffoli, E. Sansosti, G. Fornaro, and F. Serafino. New approach for hybrid strip-map/spotlight SAR data focusing. *IEE Proceedings on Radar, Sonar and Navigation*, 148(6):363–372, December 2001.

[555] K. J. Langenberg. Applied inverse problems. In P.C. Sabatier and A. Hilger, editors, *Basic Methods of Tomography and Inverse Problems.* Bristol, 1987.

[556] K. J. Langenberg. Introduction to the special issue on inverse problems. *Wave Motion*, 11:99–112, 1989.

[557] K. J. Langenberg, M. Berger, T. Kreutter, K. Mayer, and V. Schmitz. Synthetic aperture focusing technique signal processing. *NDT International*, 19(3):177–189, June 1986.

[558] K. J. Langenberg, M. Brandfass, K. Mayer, T. Kreutter, A. Brull, P. Felinger, and D. Huo. Principles of microwave imaging and inverse scattering. *EARSeL Advances in Remote Sensing Symposium*, 2:163–186, 1993.

[559] K. J. Langenberg and T. Gurke. Toward a unified theory of vector diffraction tomography. In W. M. Boerner and et al., editors, *Direct and Inverse Methods in Radar Polarimetry.* Reidel, Dordrecht, 1989.

[560] K. J. Langenberg and V. Schmitz. Generalized tomography as a unified approach to linear inverse scattering. *Acoustical Imaging*, 14, 1985.

[561] J. Larmor. On the electrodynamic and thermal relations of energy of magnetisation. *Proceedings of the London Mathematical Society*, 1(2):1–13, 1903.

[562] R. W. Larson, J. Zelenka, and E. L. Johansen. Microwave hologram radar system. *IEEE Transactions on Aerospace and Electronic Systems*, 8:208–216, March 1972.

[563] W. Lawton. A new polar Fourier transform for computer-aided tomography and spotlight synthetic aperture radar. *IEEE Transactions on Acoustics, Speech and Signal Processing*, 36(6):931–933, June 1988.

[564] F. W. Leberl, G. Domik, and M. Kobrick. Radar stereomapping techniques and application to SIR-B images of Mt. Shasta. *IEEE Transactions on Geoscience and Remote Sensing*, 24(9), September 1986.

[565] J. A. C. Lee, O. Arikan, and D. C. Munson, Jr. Formulation of a general imaging algorithm for high-resolution synthetic aperture radar. In *International Conference on Acoustics, Speech, and Signal Processing*, volume 4, pages 2092–2095, May 7–10, 1996.

[566] J. A. C. Lee and D. C. Munson, Jr. Effect of a nonplanar wavefront in spotlight-mode synthetic aperture radar. In *Proceedings of the IEEE International Conference on Image Processing*, pages 2092–2095, Austin, Texas, November 13–16, 1994.

[567] J. A. C. Lee and D. C. Munson, Jr. Runway imaging from an approaching aircraft using synthetic aperture radar. In *Proceedings of the IEEE International Conference on Image Processing*, September 16–19, 1996.

[568] J. S. Lee. Speckle analysis and smoothing of synthetic aperture radar images. *Computer Graphics Image Processing*, 17:24–32, 1981.

[569] R. Lee, J. Verdi, and M. Soumekh. Enhancements of NP-3 UHF image quality using digital spotlighting technique. In *Proceedings of IEEE International Conference on Radar Systems*, pages 1–6, Atlanta, Georgia, May 2001.

[570] W. H. Lee. Sampled Fourier transform hologram generated by computer. *Applied Optics*, 9:639–643, March 1970.

[571] J. M. N. Leitao and M. A. T. Figueiredo. Interferometric image reconstruction as a nonlinear Bayesian estimation problem. In *Proceedings of the IEEE International Conference on Image Processing*, pages 453–456, Washington, DC, October 1995.

[572] E. N. Leith. Quasi-holographic techniques in the microwave region. *Proceedings of the IEEE*, 59(9):1305–1318, September 1971.

[573] E. N. Leith. Range-azimuth-coupling abberations in pulse-scanned imaging systems. *Journal of the Optical Society of America*, 63(2):119–126, February 1973.

[574] E. N. Leith. Complex spatial filters for image deconvolution. *Proceedings of the IEEE*, 65(1):18–28, January 1977.

[575] E. N. Leith and J. Upatnieks. New techniques in wavefront reconstruction. *Journal of the Optical Society of America*, 51:1469, 1961.

[576] E. N. Leith and J. Upatnieks. Reconstructed wavefronts and communication theory. *Journal of the Optical Society of America*, 52:1123–1130, 1962.

[577] E. N. Leith and J. Upatnieks. Wavefront reconstruction with continuous tone objects. *Journal of the Optical Society of America*, 53:1377–1381, 1963.

[578] E. N. Leith and J. Upatnieks. Wavefront reconstruction with diffused illumination and three-dimensional objects. *Journal of the Optical Society of America*, 54(11):1295–1301, November 1964.

[579] H. Levine and J. Schwinger. On the theory of diffraction by an aperture in an infinite plane screen. *Physical Review*, 74:958–974, 1948.

[580] R. M. Lewis. Physical optics inverse diffraction. *IEEE Transactions on Antennas Propagation*, 17(3):308–314, March 1969.

[581] R. M. Lewitt. Reconstruction algorithms: Transform methods. *Proceedings of the IEEE*, 71(3):390–408, March 1983.

[582] R. M. Lewitt and R. H. T. Bates. Image reconstruction from projections: I: General theoretical considerations. *Optik*, 50:19–33, 1978.

[583] A. Li. Algorithms for the implementation of Stolt interpolation in SAR processing. In *Proceedings of the IEEE International Geoscience Remote Sensing Symposium*, volume 1, pages 360–362, 1992.

[584] A. Li and O. Loeffeld. Two-dimensional SAR processing in the frequency domain. In *Proceedings of the IEEE International Geoscience Remote Sensing Symposium*, pages 1005–1009, Helsinki, 1991.

[585] F. K. Li, D. N. Held, J. C. Curlander, and C. Wu. Doppler parameter estimation for spaceborne synthetic-aperture radars. *IEEE Transactions on Geoscience and Remote Sensing*, 23(1):47–56, January 1985.

[586] F. K. Li and W. T. K. Johnson. Ambiguities in spaceborne synthetic aperture radar systems. *IEEE Transactions on Aerospace and Electronic Systems*, 19(3):389–397, May 1983.

[587] J. Li, Z. Bi, Z.-S. Liu, and K. Knaell. Use of curvilinear SAR for three-dimensional target feature extraction. *IEE Proceedings on Radar, Sonar and Navigation*, 144(5):275–283, October 1997.

[588] Z.-P. Liang and D. C. Munson, Jr. Partial Radon transform. *IEEE Transactions on Image Processing*, 6(10):1467–1469, October 1997.

[589] M. Y. Lin and C. Wu. A SAR correlation algorithm which accommodates large range migration. *IEEE Transactions on Geoscience and Remote Sensing*, 22:592–597, November 1984.

[590] W. J. Van De Lindt. Digital technique for generating synthetic aperture radar images. *IBM Journal of Research and Development*, 21:415–432, September 1977.

[591] D. Loewenthal, L. Lou, R. Robertson, and J. W. Sherwood. The wave equation applied to migration. *Geophysical Prospecting*, 24:380–399, 1976.

[592] D. Loewenthal and I. R. Mufti. Reversed time migration in spatial frequency domain. *Geophysics*, 48:627–635, May 1983.

[593] O. Loffeld, A. Hein, and F. Schneider. SAR focusing: Scaled inverse Fourier transformation and chirp scaling. In *Proceedings of the IEEE International Geoscience and Remote Sensing Symposium*, pages 630–632, June 1998.

[594] B. F. Logan. The uncertainty principle in reconstructing functions from projections. *Duke Math Journal*, 42:666–706, 1975.

[595] A. W. Lohmann. Three-dimensional properties of wave-fields. *Optik*, 51:105, 1978.

[596] F. Lombardini. Differential tomography: A new framework for SAR interferometry. In *IEEE Geoscience and Remote Sensing Symposium*, volume 2, pages 1206–1208, July 2003.

[597] F. Lombardini. Differential tomography: A new framework for SAR interferometry. *IEEE Transactions on Geoscience and Remote Sensing*, 43(1):37–44, January 2005.

[598] F. Lombardini, F. Gini, and P. Matteucci. Multibaseline ATI-SAR for robust ocean surface velocity estimation in presence of bimodal Doppler spectrum. In *IEEE Geoscience and Remote Sensing Symposium*, volume 1, pages 578–580, July 2001.

[599] F. Lombardini and M. Pardini. 3-D SAR tomography: The multibaseline sector interpolation approach. *IEEE Transactions on Geoscience and Remote Sensing Letters*, 5(4):630–634, October 2008.

[600] J. M. Lopez-Sanchez and J. Fortuny-Guash. 3-D radar imaging using range migration techniques. *IEEE Transactions on Antennas and Propagation*, 48(5):728–737, May 2000.

[601] A. Louis and E. T. Quinto. Local tomographic methods in SONAR. In D. Colton, H. W. Engl, A. K. Louis, J. R. McLaughlin, and W. Rundell, editors, *Surveys on Solution Methods for Inverse Problems*, New York, 2000. Springer.

[602] Y. H. Lu, T. S. Yeo, N. L. Tan, and C. B. Zhang. A stripmap to spotlight data converting algorithm. In *Proceedings of the International Geoscience and Remote Sensing Symposium*, pages 1168–1170, Seattle, July 1998.

[603] R. L. Lucke. Rayleigh-Sommerfeld diffraction and Poisson's spot. *European Journal of Physics*, 27:193–204, 2006.

[604] J. E. Luminati. *Wide-Angle Multistatic Synthetic Aperture Radar: Focused Image Formation and Aliasing Artifact Mitigation*. PhD thesis, Air Force Institute of Technology, Wright-Patterson, Ohio, May 2004.

[605] P. Maass. Wideband radar: the hyperbolic transform. *Inverse Problems*, 5:849–857, 1989.

[606] M. D. MacLeod. Fast interpolation by FFT with greatly increased accuracy. *Electronics Letters*, 29(13):1200–1201, June 1993.

[607] A. Macovski. Ultrasonic imaging using arrays. *Proceedings of the IEEE*, 67(4):484–495, April 1979.

[608] S. N. Madsen, W. Edelstein, L. D. Didomenico, and J. LaBrecque. A geosynchronous synthetic aperture radar: for tectonic mapping, disaster management and measurements of vegetation and soil moisture. In *Proceedings of the IEEE International Geoscience Remote Sensing Symposium*, pages 447–449, 2001.

[609] S. N. Madsen, H. Z. Zebker, and J. Martin. Topographic mapping using radar interferometry: Processing techniques. *IEEE Transactions on Geoscience and Remote Sensing*, 31(1):246–256, January 1993.

[610] R. D. Mager and N. Bleistein. An examination of the limited aperture problem of physical optics inverse scattering. *IEEE Transactions on Antennas and Propagation*, 26(5):695–699, 1978.

[611] C. Magnard, O. Frey, M. Ruegg, and E. Meier. Improved airborne SAR data processing by blockwise focusing, mosaicing and geocoding. In *Proceedings of the 7th European Conference on Synthetic Aperture Radar (EUSAR '08)*, volume 1, pages 375–378, June 2008.

[612] N. Magotra and M. M. Puri. Synthetic aperture radar image formation. In *Proceedings of the IEEE International Conference on Acoustics, Speech, and Signal Processing*, pages 1869–1872, 1990.

[613] B. R. Mahafza. *Two- and Three-Dimensional Synthetic Aperture Radar Imaging Using a Linear Array with Transverse Motion.* PhD thesis, University of Alabama in Huntsville, Department of Electrical and Computer Engineering, March 1989.

[614] B. R. Mahafza. Two-dimensional SAR imaging using linear arrays with transverse motion. *Journal of the Franklin Institute*, 330(1):95–102, 1993.

[615] B. R. Mahafza, D. L. Knight, and N. F. Audeh. Forward-looking SAR imaging using a linear array with transverse motion. In *IEEE SOUTHEASTCON*, 1993.

[616] A. D. Maio, G. Fornaro, and A. Pauciullo. Detection of single scatterers in multidimensional SAR imaging. *IEEE Transactions on Geoscience and Remote Sensing*, 47(7):2284–2297, July 2009.

[617] U. Majumder, M. Soumekh, M. Minardi, and S. Scarborough. Synthetic aperture radar moving target indication processing of along-track monopulse nonlinear GOTCHA data. In *IEEE International Radar Conference*, pages 1–6, Pasadena, California, 2009.

[618] E. L. Malus. Nouveau bulletin des sciences. *par la Socieété philomathique de Paris*, 1:266, 1809. Mémoires de la Society d'Arcueil, volume 2, 1809.

[619] J. Mann and R. Hummel. Spotlight-mode synthetic aperture radar without Fourier transforms. In *Proceedings of SPIE, Algorithms for Synthetic Aperture Radar Imagery*, volume 2230, pages 2–13, Orlando, Florida, 4–8 April 1994.

[620] N. Marechal. Tomographic formulation of interferometric SAR for terrain elevation mapping. *IEEE Transactions on Geoscience and Remote Sensing*, 33(3):726–739, May 1995.

[621] S. L. Marple, Jr. Computing the discrete-time analytic signal via FFT. In *Proceedings of the 15th Asilomar Conference on Circuits, Systems and Computing*, volume 2, pages 1322–1325, Pacific Grove, California, 1997.

[622] P. Martinson. Radar matched filtering. In *Radar Technology*, pages 163–169, New York, 1977. Artech House.

[623] W. Masri, M. Mina, S. Upda, L. Upda, T. Xue, and W. Lord. Synthetic aperture focusing techniques applied in the near field of a focused transducer. In *Proceedings of the Ultrasonics Symposium*, pages 783–786, Seattle, Washington, November 1995.

[624] D. Massonnet, F. Adragna, and M. Rossi. CNES general-purpose SAR correlator. *IEEE Transactions on Geoscience and Remote Sensing*, 32(3):636–643, May 1994.

[625] D. Massonnet, K. Feigl, M. Rossi, and F. Adragna. Radar interferometric mapping of deformation in the year after the Landers Earthquake. *Nature*, 369(6477):227–230, 19 May 1994.

[626] D. Massonnet, M. Rossi, C. Carmona, F. Adragna, G. Peltzer, K. Feigl, and T. Rabaute. The displacement field of the Landers Earthquake mapped by radar interferometry. *Nature*, 364:138–142, 1993.

[627] D. Massonnet and J.-C. Souyris. *Imaging with Synthetic Aperture Radar.* CRC Press, 2008.

[628] S. Matej and I. Bajla. A high-speed reconstruction from projections using direct Fourier method with optimized parameters – an experimental analysis. *IEEE Transactions on Medical Imaging*, 9:421–429, 1990.

[629] N. C. Mathur. A pseudodynamic programming technique for the design of correlator supersynthesis arrays. *Radio Science*, 4(3):235–244, 1969.

[630] A. W. Maue. Toward formulator of a general diffraction problem via an integral equation. *Zeitschrift fur Physik*, 126:601–618, 1949.

[631] K. Mayer, R. Marklein, K. J. Langenberg, and T. Kreutter. Three-dimensional imaging system based on Fourier synthetic aperture focusing technique. *Ultrasonics*, 28:241–255, July 1990.

[632] S. W. McCandless. *SAR in Space–The Theory, Design, Engineering and Application of a Space-Based SAR Systems*, chapter 4. Artech House, Norwood, Massachusetts, 1989.

[633] S. W. McCandless and C. R. Jackson. *SAR Marine Users Manual.* NOAA, 2004.

[634] J. McCorkle and M. Rofheart. An order $n^2 \log(n)$ backprojector algorithm for focusing wide-angle wide-bandwidth arbitrary-motion synthetic aperture radar. In G. S. Ustach, editor, *Proceedings of SPIE, Radar Sensor Technology*, volume 2747, pages 25–36, Orlando, Florida, 8–9 April 1996.

[635] D. L. McMakin, D. M. Sheen, H. D. Collins, T. E. Hall, and R. H. Seversten. Wideband, millimeter-wave, holographic weapons surveillance system. In *Proceedings of SPIE, European Symposium on Optics for Environmental Public Safety*, volume 2511, pages 131–141, 1995.

[636] D. L. McMakin, D. M. Sheen, H. D. Collins, T. E. Hall, and R. R. Smith. Millimeter-wave high resolution holographic surveillance system. In *Proceedings of SPIE, International Symposium on Substance Identification Technology*, volume 2092, pages 525–535, 1993.

[637] J. G. Mehlis. Synthetic aperture radar range-azimuth ambiguity design constraints. In *IEEE International Radar Conference*, pages 143–152, 1980.

[638] D. L. Mensa. *High Resolution Radar Imaging.* Artech House, Dedham, Massachusetts, 1981.

[639] D. L. Mensa. *High Resolution Radar Cross-Section Imaging.* Artech House, Boston, Massachusetts, 1991.

[640] D. L. Mensa, S. Halevy, and G. Wade. Coherent Doppler tomography for microwave imaging. *Proceedings of the IEEE*, 71(2):254–261, February 1983.

[641] D. L. Mensa, G. Heidbreder, and G. Wade. Aperture synthesis by object rotation in coherent imaging. *IEEE Transactions on Nuclear Science*, 27:989–998, April 1980.

[642] J. R. Mentzer. *Scattering and Diffraction of Radio Waves.* Pergamon, New York, 1955.

[643] R. M. Mersereau. The digital reconstruction of multidimensional signals from their projections. In *Proceedings of the 10th Annual Allerton Conference on Circuit and System Theory*, pages 326–334, 4–6 October 1972.

[644] R. M. Mersereau. Recovering multidimensional signals from their projections. *Computer Graphics and Image Processing*, 1(6):179–195, October 1973.

[645] R. M. Mersereau. Direct Fourier transform techniques in 3-D image reconstruction. *Computers in Biology and Medicine*, 6:247–258, 1976.

[646] R. M. Mersereau and D. E. Dudgeon. The representation of two-dimensional sequences as one-dimensional sequences. *IEEE Transactions on Acoustics, Speech and Signal Processing*, 22:320–325, October 1974.

[647] R. M. Mersereau and A. V. Opppenheim. Digital reconstruction of multidimensional signals from their projections. *Proceedings of the IEEE*, 62(10):1319–1338, October 1974.

[648] A. S. Milman. SAR imaging by omega-k migration. *International Journal of Remote Sensing*, 14(10):1965–1979, 1993.

[649] A. S. Milman. The hyperbolic geometry of SAR imaging. *Radio Science*, 2004.

[650] J. H. Mims and J. F. Farrell. Synthetic aperture imaging with maneuvers. *IEEE Transactions on Aerospace and Electronic Systems*, 8(4):410–418, July 1972.

[651] T. Misra and A. Moreira. A new method for generation of optimum matched filter from replica data for pulse compression. In *Proceedings of the Geoscience and Remote Sensing Symposium*, volume 2, pages 1011–1014, 1991.

[652] R. H. Mitchel and S. Marder. Synthetic aperture radar SAR image quality considerations. *Optical Engineering*, 21:48–55, January 1982.

[653] R. L. Mitchell. Creating complex signal samples from a band-limited real signal. *IEEE Transactions on Aerospace and Electronic Systems*, 25(3):425–427, 1989.

[654] J. Mittermayer, A. Moreira, and O. Loffeld. High precision processing of spotlight SAR data using the extended chirp scaling algorithm. In *Proceedings of the of EUSAR – European Conference on Synthetic Aperture Radar*, pages 561–564, Friedrichshafen, Germany, May 1998.

[655] J. Mittermayer, A. Moreira, and O. Loffeld. Spotlight SAR data processing using the frequency scaling algorithm. *IEEE Transactions on Geoscience and Remote Sensing*, 37(5):2198–2214, September 1999.

[656] P. Mora. Inversion = migration + tomography. *Geophysics*, 54(12):1575–1586, December 1982.

[657] K. J. Moraski and D. C. Munson, Jr. Fast tomographic reconstruction using chirp-z interpolation. In *Proceedings of the 25st Asilomar Conference on Signals, Systems, and Computers*, pages 1052–1056, Pacific Grove, California, November 1991.

[658] A. Moreira. A new method of aircraft motion error extraction from radar raw data for real time motion compensation. *IEEE Transactions on Geoscience and Remote Sensing*, 28(4):620–626, July 1990.

[659] A. Moreira. Real-time synthetic aperture radar SAR processing with a new subaperture approach. *IEEE Transactions on Geoscience and Remote Sensing*, 30(4):714–722, July 1992.

[660] A. Moreira. *Method of Image Generation by Means of Two-Dimensional Data processing in Connection with Radar with Synthetic Aperture.* Canadian Patent Application 2,155,502, 1995.

[661] A. Moreira and Y. Huang. Airborne SAR processing of highly squinted data using a chirp scaling approach with integrated motion compensation. *IEEE Transactions on Geoscience and Remote Sensing*, 32(5):1029–1040, September 1994.

[662] A. Moreira, J. Mittermayer, and R. Scheiber. Processing of ScanSAR imaging modes using the extended chirp scaling algorithm. In *Proceedings of the of EUSAR – European Conference on Synthetic Aperture Radar*, pages 557–560, Friedrichshafen, Germany, May 1998.

[663] A. Moreira, J. Mittermayer, and R. Schiber. Extended chirp scaling algorithm for air- and spaceborne SAR data processing in stripmap and ScanSAR imaging modes. *IEEE Transactions on Geoscience and Remote Sensing*, 42(5):1123–1136, September 1996.

[664] G. Morris and L. Harkness, editors. *Airborne Pulsed Doppler Radar.* Artech House, 1996.

[665] P. M. Morse and H. Feshbach. *Methods of Theoretical Physics.* McGraw-Hill, New York, 1953.

[666] P. M. Morse and K. U. Ingard. *Theoretical Acoustics.* McGraw-Hill, New York, 1968.

[667] R. L. Moses, P. Adams, and T. Biddlecome. Three-dimensional target visualization from wide-angle IFSAR data. In *Proceedings of SPIE, Algorithms for Synthetic Aperture Radar Imagery XII*, volume 5808, pages 76–85, Bellingham, Washington, 2005.

[668] R. L. Moses, E. Ertin, and C. Austin. Synthetic aperture radar visualization. In *Proceedings of the 38th Asilomar Conference on Signals, Systems and Computers*, Pacific Grove, California, 7–10 November 2004.

[669] R. L. Moses and L. C. Potter. Noncoherent 2D and 3D SAR reconstruction from wide-angle measurements. In *Thirteenth Annual Adaptive Sensor Array Processing Workshop*, MIT Lincoln Laboratory, Lexington, Massachusetts, 7–8 June 2005.

[670] R. L. Moses, L. C. Potter, and M. Cetin. Wide angle SAR imaging. In *Proceedings of SPIE, Algorithms for Synthetic Aperture Radar Imagery XI*, volume 5427, Orlando, Florida, 12–16 April 2004.

[671] A. N. Movshovich. Digital demodulation of an FM signal. *Radioteknika*, 1:54–57, 1987.

[672] R. K. Mueller. Diffraction tomography I: The wave equation. *Ultrasonic Imaging*, 2:213–222, 1980.

[673] R. K. Mueller, M. Kaveh, and R. D. Inverson. A new approach to acoustic tomography using diffraction techniques. *Acoustical Imaging / Acoustic Holography*, 8:615–628, 1980.

[674] R. K. Mueller, M. Kaveh, and G. Wade. Reconstructive tomography and applications to ultrasonics. *Proceedings of the IEEE*, 67(4):567–587, April 1979.

[675] D. C. Munson, Jr. An introduction to strip-mapping synthetic aperture radar. In *Proceedings of the IEEE International Conference on Acoustics, Speech, and Signal Processing*, pages 2245–2248, Dallas, Texas, 6–9 April 1987.

[676] D. C. Munson, Jr and W. E. Higgins. Faster algorithms for computer-aided tomography and spotlight mode synthetic aperture radar. In *Proceedings of the 16st Asilomar Conference on Circuits, Systems and Computing*, pages 426–430, Pacific Grove, California, 1982.

[677] D. C. Munson, Jr and W. K. Jenkins. A common framework for spotlight mode synthetic aperture radar and computer-aided tomography. In *Proceedings of the 15th Asilomar Conference on Circuits, Systems and Computing*, page 217, Pacific Grove, California, 9–11 November 1981.

[678] D. C. Munson, Jr and J. D. O'Brien. An analysis of artifacts due to Fourier domain interpolation. In *Proceedings of the Twentieth Asilomar Conference on Signals, Systems and Computers*, pages 82–85, Washington, DC, 1987.

[679] D. C. Munson, Jr, J. D. O'Brien, and W. K. Jenkins. A tomographic formulation of spotlight-mode synthetic aperture radar. *Proceedings of the IEEE*, 17(8):917–925, August 1983.

[680] D. C. Munson, Jr and J. L. C. Sanz. Image reconstruction from frequency-offset Fourier data. *Proceedings of the IEEE*, 72(6):661–669, June 1984.

[681] D. C. Munson, Jr and J. L. C. Sanz. The importance of random phase for image reconstruction from frequency offset Fourier data. In *Proceedings of the IEEE International Conference on Acoustics, Speech and Signal Processing*, pages 158–161, March 1984.

[682] D. C. Munson, Jr, J. L. C. Sanz, W. K. Jenkins, G. Kakazu, and B. Mather. A comparison of algorithms for polar-to-Cartesian interpolation in spotlight mode SAR. In *Proceedings of the IEEE International Conference on Acoustics, Speech and Signal Processing*, Tampa, Florida, March 1985.

[683] D. C. Munson, Jr and R. L. Visentin. A signal processing view of strip-mapping synthetic aperture radar. *IEEE Transactions on Acoustics, Speech, and Signal Processing*, 37(12):2131–2147, December 1989.

[684] Y. Na, H. Sun, Y. H. Lee, L. C. Tai, and H. L. Chan. Performance evaluation of back-projection and range migration algorithms in foliage penetration radar imaging. In *Proceedings of the IEEE International Conference on Image Processing*, volume 1, pages 21–24, 2004.

[685] D. Nahamoo and A. C. Kak. Ultrasonic diffraction imaging. Technical report, School of Electrical Engineering, Purdue University, August 1982.

[686] D. Nahamoo, S. X. Pan, and A. C. Kak. Synthetic aperture diffraction tomography and its interpolation-free computer implementation. *IEEE Transactions on Sonics and Ultrasonics*, 31:218–229, 1984.

[687] K. Naidu and L. Lin. Data dome: Full k-space sampling data for high frequency radar research. In *Proceedings of SPIE, Algorithms for Synthetic Aperture Radar Imagery XI*, volume 5427, Orlando, Florida, 12–16 April 2004.

[688] M. Nannini and R. Scheiber. A time domain beamforming algorithm for SAR tomography. In *Proceedings of the European Conference on Synthetic Aperture Radar*, 2006.

[689] M. Nannini, R. Scheiber, and A. Moreira. Estimation of the minimum number of tracks for SAR tomography. *IEEE Transactions on Geoscience and Remote Sensing*, 47(2):531–543, February 2009.

[690] F. Natterer. *The Mathematics of Computerized Tomography.* Wiley, New York, 1986.

[691] F. Natterer, M. Cheney, and B. Borden. Resolution for radar and x-ray tomography. *Inverse Problems*, 19:55–63, 2003.

[692] F. Natterer and F. Wubbeling. A propagation-backpropagation method for ultrasound tomography. *Inverse Problems*, 11:1225–1232, 1995.

[693] Y. Nemoto, H. Nishino, M. Ono, H. Mizutamari, K. Nishikawa, and K. Tanaka. Japanese Earth resources satellite-1 synthetic aperture radar. *Proceedings of the IEEE*, 79(6):800–809, 1991.

[694] I. Newton. New theory about light and colours. *Philosophical Transactions*, 6(80):3075–3087, 19 February 1672.

[695] I. Newton. *Opticks: or a Treatise of the Reflections, Refractions, Infections and Colors of Light.* Oxford University, London, 1704. First published in 1704 (English), 1706 (Latin), 1717/18 (English). Now available under the same title, but based on the fourth posthumous edition of 1730, New York: Dover Publications, 1952.

[696] R. G. Newton. *Scattering Theory of Waves and Particles.* Dover Books on Physics, second edition, 1982.

[697] L. Nguyen, M. Ressler, and M. Soumekh. Enhancement of backprojection SAR imagery using digital spotlighting preprocessing. In *Proceedings of the IEEE International Radar Conference*, pages 53–58, 2004.

[698] L. Nguyen, M. Ressler, and M. Soumekh. Signal processing and image formation using low-frequency ultra-wideband radar data. In *Proceedings of SPIE, Detection and Remediation Technologies for Mines and Minelike Targets IX*, volume 5415, pages 1053–1064, 2004.

[699] L. H. Nguyen, T. Ton, D. Wong, and M. Soumekh. Adaptive coherent suppression of multiple wide-bandwidth RFI sources in SAR. In *Proceedings of SPIE, Algorithms for Synthetic Aperture Radar Imagery XI*, volume 5427, pages 1–16, Bellingham, Washington, 12–16 April 2004.

[700] N. Niki, T. Mizutani, Y. Takahashi, and T. Inouye. A high-speed computerized tomography image reconstruction using direct two-dimensional Fourier transform method. *Systems, Computers, Controls*, 14(3):56–65, 1983.

[701] S. Nilsson. Fast backprojection. Technical report, Linköping University, Sweden, July 4, 1996.

[702] S. Nilsson. *Application of Fast Backprojection Techniques for Some Inverse Problems of Integral Geometry.* PhD thesis, Department of Mathematics, Linköping University, S-58 183 Linköping, Sweden, September 1997.

[703] S. Nilsson and L.-E. Andersson. Application of fast backprojection techniques for some inverse problems of synthetic aperture radar. In *Proceedings of SPIE, Algorithms for Synthetic Aperture Radar Imagery V*, volume 3370, pages 62–72, Orlando, Florida, April 1998.

[704] C. J. Nolan and M. Cheney. Microlocal analysis of synthetic aperture radar imaging. *The Journal of Fourier Analysis and Applications*, 2002.

[705] C. J. Nolan and M. Cheney. Synthetic aperture inversion. *Inverse Problems*, 18:221–236, 2002.

[706] C. J. Nolan and M. Cheney. Synthetic aperture inversion for non-flat topography. In *Proceedings of the of EUSAR 2002 - 4th European Conference on Synthetic Aperture Radar*, volume 1, pages 105–108, 2002.

[707] C. J. Nolan and M. Cheney. Synthetic aperture inversion of arbitrary flight paths and nonflat topography. *IEEE Transactions on Image Processing*, 12(9):1035–1043, September 2003.

[708] C. J. Nolan and W. W. Symes. Global solution of a linearized inverse problem for acoustic wave equation. *Communications in Partial Differential Equations*, 22:919–952, 1997.

[709] D. O. North. An analysis of the factors which determine signal/noise discrimination in pulsed-carrier systems. *Proceedings of the IEEE*, 51:1016–1027 (reprint of RCA Technical Report PTR-6C, June 25, 1943), 1963.

[710] S. J. Norton. Reconstruction of a two-dimensional reflecting medium over a circular domain: Exact solution. *The Journal of the Acoustical Society of America*, 67:1266–1273, 1979.

[711] S. J. Norton and M. Linzer. Ultrasonic reflectivity in three dimensions: Reconstruction with spherical transducer arrays. *Ultrasonic Imaging*, 1:210–231, 1979.

[712] S. J. Norton and M. Linzer. Ultrasonic reflectivity tomography: Reconstruction with circular transducer arrays. *Ultrasonic Imaging*, 1:154–184, 1979.

[713] S. J. Norton and M. Linzer. Ultrasonic reflectivity imaging in three dimensions: Exact inverse scattering solutions for plane, cylindrical, and spherical apertures. *IEEE Transactions on Biomedical Engineering*, 28:202–220, February 1981.

[714] L. M. Novak, G. J. Owirka, and A. L. Weaver. Automatic target recognition using enhanced resolution SAR data. *IEEE Transactions on Aerospace and Electronic Systems*, 35:157–175, 1999.

[715] A. H. Nuttal. Some windows with very good sidelobe behavior. *IEEE Transactions on Acoustics, Speech and Signal Processing*, 29(1):84–91, 1981.

[716] S.-M. Oh, J. H. McClellan, and M. C. Cobb. Multi-resolution mixed-radix quadtree SAR image focusing algorithms. In *Proceedings of the Advanced Sensors Consortium Federated Laboratory Symposium*, pages 139–143, February 1999.

[717] G. P. Ohman. Getting high range resolution with pulse-compression radar. *Electronics*, October 1960.

[718] W. H. Oldendorf. Isolated flying spot detection of radiodensity discontinuities displaying the internal structural pattern of a complex object. *IRE Transactions on Bio-Medical Electronics*, 8:68–72, 1961.

[719] C. J. Oliver. Review article – synthetic aperture radar imaging. *Journal of Physics D: Applied Physics*, 25:871–890, 1989.

[720] C. J. Oliver, D. Belcher, A. Blake, and R. G. White. Algorithms for focused linear synthetic aperture radar imaging. In *Proceedings of SPIE, Algorithms for Synthetic Aperture Radar Imagery*, volume 2230, pages 60–71, Orlando, Florida, 4–8 April 1994.

[721] C. J. Oliver and S. Quegan. *Understanding Synthetic Aperture Radar Images*. Artech House, London, 1998.

[722] L. Onural. Impulse functions over curves and surfaces and their applications to diffraction. *Journal of Mathematical Analysis and Applications*, 322:18–27, 2006.

[723] L. Onural. Projection-slice theorem as a tool for mathematical representation of diffraction. *IEEE Signal Processing Letters*, 14(1):43–46, January 2007.

[724] A. V. Oppenheim, G. V. Frisk, and D. R. Martinez. An algorithm for the numerical evaluation of the Hankel transform. *Proceedings of the IEEE*, 66:264–265, 1978.

[725] A. V. Oppenheim and R. W. Schafer. *Digital Signal Processing*. Prentice–Hall, Englewood Cliffs, New Jersey, 1975.

[726] M. L. Oristaglio. Accuracy of the Born and Rytov approximations for reflection and refraction at a plane interface. *Journal of the Optical Society of America*, 2:1987–1993, 1985.

[727] J. D. O'Sullivan. A fast sinc function gridding algorithm for Fourier inversion in computer tomography. *IEEE Transactions on Medical Imaging*, 4(4):200–207, December 1985.

[728] R. Ottolini. Synthetic aperture radar data processing. *SEP Reports, Stanford University*, SEP-56:203–214, 1988.

[729] Y. Ozaki, H. Sumitani, T. Tomoda, and M. Tanaka. A new system for real-time synthetic aperture ultrasonic imaging. *IEEE Transactions on Ultrasonics, Ferroelectrics and Frequency Control*, 35(1):828, 1988.

[730] S. X. Pan and A. C. Kak. A computational study of reconstruction algorithms for diffraction tomography: Interpolation vs. filtered-backpropagation. *IEEE Transactions on Acoustics, Speech and Signal Processing*, 31(5):1262–1275, October 1983.

[731] X. Pan, E. Y. Sidky, and M. Vannier. Why do commercial CT scanners still employ traditional, filtered back-projection for image reconstruction? *Inverse Problems*, 25(123009):36pp, 2009.

[732] X. Pan, J. Siewerdsen, P. J. La Riviere, and W. A. Kalender. Anniversary paper: Development of x-ray computed tomography: The role of Medical physics and AAPM from the 1970s to present. *Medical Physics*, 35:3728–2739, 2008.

[733] A. Papoulis. *The Fourier Integral and its Applications.* McGraw-Hill, New York, 1962.

[734] A. Papoulis. *Systems and Transforms with Applications in Optics.* McGraw-Hill, New York, 1968.

[735] W. R. Paradella, R. D. Santos, P. Veneziani, and M. C. Morais. Synthetic aperture radar for geological applications in the moist tropics: Experiences from the Brazilian Amazon region. *Revista Brasileria de Geociencias*, 30(3):538–542, September 2000.

[736] D. Pastina, F. Colone, and P. Lombardo. Effects of apodization on SAR image understanding. *IEEE Transactions on Geoscience and Remote Sensing*, 45(11):3533–3551, November 2007.

[737] T. Payne. Phase analysis for the limitations of the tomographic paradigm on a 3D scene. In *IEEE Geoscience and Remote Sensing Symposium*, volume 7, pages 3030–3032, July 2001.

[738] H. Peng and H. Stark. Direct Fourier reconstruction in fan beam tomography. *IEEE Transactions on Medical Imaging*, 6(3):209–219, September 1987.

[739] R. P. Perry, R. C. DiPietro, A. Kozma, and J. J. Vaccaro. SAR image formation processing using planar subarrays. In *Proceedings of SPIE, Algorithms for Synthetic Aperture Radar Imagery*, volume 2230, pages 25–35, Orlando, Florida, 4–8 April 1994.

[740] R. P. Perry and H. W. Kaiser. Digital step transform approach to airborne radar processing. In *IEEE National Aerospace and Electronics Conference*, pages 280–287, May 1973.

[741] R. N. O. Petersson, A. G. Derneryd, and P. Ingvarson. Microstrip antennas for remote sensing satellites. In *Proceedings of the PIERS Conference*, Noordwikj, Netherlands, July 1994.

[742] J. D. Phillips. Potential-field continuation: Past practice vs. modern methods. In *66th meeting, SEG*, volume 2.5, pages 1411–1414, Denver, 1996.

[743] C. Pichot, G. Peronnet, and J. C. Bolomey. Active microwave imaging of inhomogeneous bodies. *IEEE Transactions on Antennas Propagation*, 33:416–425, 1985.

[744] J. E. B. Ponsonby, I. Morison, A. R. Birks, and J. K. Landon. Radar images of the Moon at 75 and 185 cm wavelengths. *The Moon*, 5:286–294, November/December 1972.

[745] R. P. Porter. Scattering wave inversion for arbitrary receiver geometry. *The Journal of the Acoustical Society of America*, 80:1220–1227, 1986.

[746] A. Potsis, A. Reigber, E. Alivizatos, A. Moreira, and N. K. Uzunoglou. Comparison of chirp scaling and wavenumber domain algorithms for airborne low-frequency SAR. In *Proceedings of SPIE, SAR Image Analysis, Modeling and Techniques V*, volume 4883, pages 25–36, March 2002.

[747] C. Prati. 3-D synthetic aperture radar surveys. In *Proceedings of the IEEE International Conference on Acoustics, Speech and Signal Processing*, pages 2376–2379, Glasgow, UK, 1989.

[748] C. Prati, A. M. Guarnieri, and F. Rocca. Spot mode SAR focusing with the ω-k technique. In *IEEE International Geoscience and Remote Sensing Symposium*, pages 631–634, Helsinki, Finland, June 3–6 1991.

[749] C. Prati and F. Rocca. 3-D synthetic aperture radar surveys. Technical report, Stanford Exportation Project, 1988.

[750] C. Prati and F. Rocca. Limits to the resolution of elevation maps from stereo SAR images. *International Journal of Remote Sensing*, 11(12):2215–2235, 1991.

[751] C. Prati and F. Rocca. Focusing SAR data with time-varying Doppler centroid. *IEEE Transactions on Geoscience and Remote Sensing*, 30(3):550–559, May 1992.

[752] C. Prati, F. Rocca, A. M. Guarnieri, and E. Damonti. Seismic migration for SAR focusing: Interferometrical applications. *IEEE Transactions on Geoscience and Remote Sensing*, 28(4):627–640, July 1990.

[753] M. Preiss, D. Gray, and N. Stacy. Space variant filtering of polar format spotlight SAR images for wavefront curvature correction and interferometric processing. In *Proceedings of the IEEE International Geoscience Remote Sensing Symposium*, Toronto, Ontario, 2002.

[754] M. Preiss and N. J. S. Stacy. Coherent change detection: Theoretical description and experimental results. Technical Report DSTO-TR-1851, Australian Government, Department of Defense, Defence Science and Technology Organisation, Edinburgh, South Australia, 2006.

[755] M. J. Prickett and C. C. Chen. Principles of inverse synthetic aperture radar (ISAR) imaging. In *IEEE EASCON Record*, page 340, 1980.

[756] J. Radon. Uber die bestimmung von funktionen durch ihre intergralwerte langs gewisser mannigfaltigheiten (On the determination of functions from their integrals along certain manifolds). *Berichte Saechsische Akademie der Wissenschaften (Leipzig), Math. Phys. Klass*, 29:262–277, 1917. See also F. John: *Plane Wave and Spherical Means Applied to Partial Differential Equations.*, New York: Wiley-Interscience, 1955.

[757] G. N. Ramachandran and A. V. Lakshiminarayanan. Three-dimensional reconstruction from radiographs and electron micrographs; I. Application of convolutions instead of Fourier transforms. *Indian Journal of Pure and Applied Physics*, (Raman Memorial Issue) 9:997–1003, 1982. Proceedings of the National Academy of Sciences, 68, 9, 2236–2240 (1971).

[758] G. N. Ramachandran and A. V. Lakshiminarayanan. Three-dimensional reconstruction from radiographs and electron micrographs: II. Application of convolutions instead of Fourier transforms. *Proceedings of the National Academy of Sciences*, 68(9):2236–2240, 1982.

[759] G. N. Ramachandran and A. V. Lakshiminarayanan. Three-dimensional reconstrution from radiographs and electron micrographs: III. Description and application of the convolution method. *Indian Journal of Pure and Applied Physics*, 9:997–1003, 1982.

[760] R. K. Raney. Synthetic aperture imaging radar and moving targets. *IEEE Transactions on Aerospace and Electronic Systems*, 7:499–505, May 1971.

[761] R. K. Raney. The Canadian RADARSAT program. In *Proceedings of the IEEE International Geoscience Remote Sensing Symposium*, 1982.

[762] R. K. Raney. Processing synthetic aperture radar data. *International Journal on Remote Sensing*, 3(2):243–257, 1982.

[763] R. K. Raney. Considerations for SAR image quantification unique to orbital systems. *IEEE Transactions on Geoscience and Remote Sensing*, 29(5):754–760, 1991.

[764] R. K. Raney. An exact wide field digital imaging algorithm. *International Journal of Remote Sensing*, 13(5):991–998, 1992.

[765] R. K. Raney. A new and fundamental Fourier transform pair. In *Proceedings of the IEEE International Geoscience Remote Sensing Symposium*, volume 1, pages 106–107, Clear Lake, Texas, May 1992.

[766] R. K. Raney and R. Bamler. A novel method for range migration correction for SAR. In *Proceedings of the IEEE Geoscience Remote Sensing Symposium*, page 1435, Espoo, Finland, July 1991.

[767] R. K. Raney and R. Bamler. A novel high precision SAR focusing algorithm based on chirp scaling. In *Proceedings of the IEEE Geoscience Remote Sensing Symposium*, volume 1, pages 372–375, Clear Lake, Texas, May 1992.

[768] R. K. Raney, I. G. Cumming, and F. H. Wong. *Synthetic Aperture Radar Processor to Handle Large Squint with High Phase and Geometric Accuracy.* United States Patent, Number 5,179,383, January, 1993.

[769] R. K. Raney, A. P. Luscombe, E. J. Langham, and S. Ahmed. RADARSAT. *Proceedings of the IEEE*, 79:839–849, June 1991.

[770] R. K. Raney, H. Runge, R. Bamler, I. G. Cumming, and F. W. Wong. Precision SAR processing using chirp scaling. *IEEE Transactions on Geoscience and Remote Sensing*, 32(4):786–799, July 1994.

[771] R. K. Raney and P. W. Vachon. A phase preserving SAR processor. In *Proceedings of the IEEE International Geoscience Remote Sensing Symposium*, volume 4, pages 2588–2591, Vancouver, July 1989.

[772] K. I. Ranney and M. Soumekh. Signal subspace change detection in averaged multilook SAR imagery. *IEEE Transactions on Geoscience and Remote Sensing*, 44(1):201–213, January 2006.

[773] T. Rastello, D. Vray, and J. Chatillon. Spatial undersampling for synthetic aperture ultrasonic imaging via digital spotlighting technique. *IEEE Transactions on Image Processing*, 8(9):1309–1313, September 1999.

[774] T. Rastello, D. Vray, P. Delachartre, and R. Prost. Fourier-based imaging applied to synthetic aperture sonar. In *Proceedings of the IEEE Oceans'06*, pages 1521–1526, 1996.

[775] P. A. Rattey and A. G. Lindgren. Sampling the 2-D Radon transform. *IEEE Transactions on Acoustics, Speech, and Signal Processing*, 29(5):994–1002, October 1981.

[776] C. Read, D. V. Arnold, D. M. Chabries, P. L. Jackson, and R. W. Christiansen. Synthetic aperture radar image formation from compressed data using a new computation technique. *IEEE Aerospace and Electronics Systems Magazine*, 3(1):3–10, October 1988.

[777] C. Read, D. M. Chabries, R. W. Christiansen, R. W. Jones, and V. Vaughn. A computation compression technique for SAR based on vector quantization. In *IEEE National Radar Conference*, pages 123–128, Ann Arbor, Michigan, 20–21 April 1988.

[778] A. Reigber. *Airborne Polarimetric SAR Tomography.* PhD thesis, University Stuttgart, Stuttgart, Germany, 2001.

[779] A. Reigber, E. Alivizatos, A. Potsis, and A. Moreira. Extended wavenumber-domain synthetic aperture radar focusing with integrated motion compensation. *IEE Proceedings on Radar, Sonar and Navigation*, 153(3):301–310, 2006.

[780] A. Reigber and A. Moreira. First demonstration of airborne SAR tomography using multibaseline L-band data. *IEEE Transactions on Geoscience and Remote Sensing*, 38(5):2142–2152, September 2000.

[781] A. Reigber, A. Potsis, E. Alivizatos, N. Uzunoglu, and A. Moreira. Wavenumber domain SAR focusing with integrated motion compensation. In *Proceedings of the Geoscience and Remote Sensing Symposium*, volume 3, pages 1465–1467, 2003.

[782] A. Reilly, G. Frazer, and B. Boashash. Analytic signal generation - tips and traps. *IEEE Transactions on Acoustics, Speech and Signal Processing*, 42(11):3241–3245, 1994.

[783] M. A. Richards. Iterative noncoherent angular superresolution. In *IEEE 1988 National Radar Conference, 3rd*, pages 100–105, Ann Arbor, Michigan, 20–21 April 1988.

[784] M. A. Richards. *Fundamentals of Radar Signal Processing.* McGraw-Hill, 2005.

[785] B. D. Rigling and R. L. Moses. Flight path strategies for 3-D scene reconstruction from bistatic SAR. *IEE Proceedings on Radar, Sonar and Navigation*, 151(3):149–157, 2004.

[786] B. D. Rigling and R. L. Moses. Taylor expansion of the differential range for monostatic SAR. *IEEE Transaction on Aerospace and Electronic Systems*, 41(1):60–64, 2005.

[787] E. Rignot, B. Hallet, and A. Fountain. Rock glacier surface motion in Beacon Valley, Antarctica, from synthetic-aperture radar interferometry. *Geophysical Research Letters*, 29(12):48–1–48–4, December 2002.

[788] A. W. Rihaczek. *Principles of High-Resolution Radar.* McGraw-Hill, 1969.

[789] E. Robinson. Image reconstruction in exploration geophysics. *IEEE Transactions on Sonics and Ultrasonics*, 31:259–270, July 1984.

[790] F. Rocca. Synthetic aperture radar: A new application for wave equation techniques. *Stanford Exploration Project SEP-56*, pages 167–189, 1987.

[791] F. Rocca, C. Cafforio, and C. Prati. Synthetic aperture radar: A new application for wave equation techniques. *Geophysical Prospecting*, 37:809–830, 1989.

[792] F. Rohrlich. *Classical Charged Particles.* Addison-Wesley Publishing, New York, 1965.

[793] K. D. Rolt and H. Schmidt. Azimuthal ambiguities in synthetic aperture sonar and synthetic aperture radar imagery. *IEEE Journal of Oceanic Engineering*, 17:73–79, 1992.

[794] V. G. Romanov. *Integral Geometry and Inverse Problems for Hyperbolic Equations.* Springer-Verlag, Berlin, 1974.

[795] O. Romer. Memories. *de l'Academie des Sciences*, 575:166–1699, 1676.

[796] J. H. Rose and M. Cheney. Self-consistent equations for variable velocity three-dimensional inverse scattering. *Physical Review Letters*, 59:954–957, 1987.

[797] J. H. Rose, M. Cheney, and B. DeFacio. Determination of the wavefield from scattering data. *Physical Review Letters*, 57:783–786, 1986.

[798] J. H. Rose, B. De Facio, and M. Cheney. Physical basis of three-dimensional inverse scattering for plasma wave equation. *Journal of the Optical Society of America*, 2A:1954–1957, 1996.

[799] S. Rosenbaum-Raz. On scatterer reconstruction from far-field data. *IEEE Transactions on Antennas Propagation*, 24:66–70, January 1976.

[800] W. L. V. Rossum, M. P. G. Otten, and R. J. P. Bree. Extended PGA for range migration algorithms. *IEEE Transactions on Aerospace and Electronic Systems*, 42(2):478–488, April 2006.

[801] M. S. Roulston and D. O. Muhleman. Synthesizing radar maps of polar regions with a Doppler-only method. *Applied Optics*, 36:3912–3919, 1997.

[802] P. D. Rowley. Quantitative interpretation of three-dimensional weakly refractive phase objects using holographic interferometry. *Journal on Optics Society of America*, 59:1496–1498, November 1969.

[803] W. Rudin. *Real and Complex Analysis.* McGraw-Hill, 1987.

[804] H. Rudolf, D. Tarchi, and A. J. Sieber. Combination of linear and circular SAR for 3-D features. In *Proceedings of the Geoscience and Remote Sensing Symposium*, volume 4, pages 1551–1553, 1997.

[805] H. Runge. Benefits of antenna yaw steering for SAR. In *Proceedings of the IEEE International Geoscience Remote Sensing Symposium*, pages 257–261, June 1991.

[806] H. Runge and R. Bamler. A novel high precision SAR focusing algorithm based on chirp scaling. In *Proceedings of the IEEE International Geoscience Remote Sensing Symposium,* volume 1, pages 372–375, Houston, Texas, May 1992.

[807] S. M. Rytov. Diffraction of light by ultrasonic waves. *Izvestiya Akademii Nauk SSR, Seriya Fizicheskaya (Proceedings of the USSR Academy of Sciences, Serie Physics)*, 2:223–259, 1937. In Russian.

[808] M. Sack and I. Cumming. Application of efficient linear FM matched filtering algorithms to synthetic aperture processing. *Proceedings of Institute of Electronic Engineering F*, 132(1):45–47, February 1985.

[809] R. W. Schafer and L. R. Rabiner. A digital signal processing approach to interpolation. *Proceedings of the IEEE*, 61:642, 1973.

[810] H. Scheffers. Vereinfachte ableitung der formeln fur die fraunhoferschen beugungserscheinungen. *Annalen der Physik*, 434(2–3):211–215, 1942.

[811] S. A. Schelkunoff. A mathematical theory of linear arrays. *The Bell System Technical Journal*, 22:80–107, 1943.

[812] T. E. Scheuer and F. H. Wong. Comparison of SAR processors based on a wave equation formulation. In *Proceedings of the IEEE International Geoscience Remote Sensing Symposium*, volume 2, pages 635–639, Helsinki, Finland, June 1991.

[813] W. A. Schneider. Integral formulation for migration in two and three dimensions. *Geophysics*, 43(1):49–76, February 1978.

[814] H. Schomberg and J. Timmer. The gridding method for image reconstruction by Fourier transformation. *IEEE Transactions on Medical Imaging*, 14(3):596–607, September 1985.

[815] J. Schulz-Stellenfleth and L. Lehner. Measurement of 2-D sea surface elevation fields using complex synthetic aperture radar data. *IEEE Transactions on Geoscience and Remote Sensing*, 42:1149–1160, June 2004.

[816] C. Scott. *The Spectral Domain Method in Electromagnetics*. Artech, Norwood, Massachusetts, 1989.

[817] H. J. Scudder. Introduction to computer-aided tomography. *Proceedings of the IEEE*, 66:628–637, June 1978.

[818] O. Seger, M. Herberthson, and H. Hellsten. Real-time SAR processing of low frequency ultra wide band radar data. In *Proceedings of EUSAR*, volume 2747, pages 489–492, Friedrichshafen, Germany, 25–27 May 1998.

[819] S. Y. Semenov. Three dimensional microwave tomography: Experimental prototype of the system and vector Born reconstruction method. *IEEE Transactions on Biomedical Engineering*, 46:937–946, 1999.

[820] M. Settle and J. Taranik. Use of the space shuttle for remote sensing research: Recent results and future prospects. *Science*, 218:993–995, 1982.

[821] C. E. Shannon. Communication in the presence of noise. *Proceedings of the IRE*, 49:10–21, January 1949.

[822] I. I. Shapiro, S. H. Zisk, A. E. E. Rogers, M. A. Slade, and T. W. Thompson. Lunar topography: Global determination by radar. *Science*, 178:939–948, 1972.

[823] Z. She, D. A. Gray, R. E. Bogner, and J. Homer. Image registration for multiple pass SAR. In *Proceedings of Information, Decision and Control*, pages 341–345, 1999.

[824] Z. She, D. A. Gray, R. E. Bogner, and J. Homer. Three-dimensional SAR imaging via multiple pass processing. In *Proceedings of the IEEE International Geoscience and Remote Sensing Symposium*, volume 5, pages 2389–2391, Hamburg, Germany, 1999.

[825] Z. She, D. A. Gray, R. E. Bogner, J. Homer, and I. D. Longstaff. Three-dimensional space-borne synthetic aperture SAR imaging with multiple pass processing. *International Journal Remote Sensing*, 23(20):4357–4382, 2002.

[826] D. M. Sheen. Foliage transmission measurements using a ground-based ultrawideband SAR system. *IEEE Transactions on Geoscience and Remote Sensing*, 32(1):118–130, January 1994.

[827] D. M. Sheen, D. L. Makin, and T. E. Hall. Three-dimensional millimeter-wave image imaging for concealed weapon detection. *IEEE Transactions on Microwave Theory and Techniques*, 49(9):1581–1592, September 2001.

[828] D. M. Sheen, D. L. Makin, and T. E. Hall. Near-field three-dimensional radar imaging techniques and applications. *Applied Optics*, 49(19):83–93, June 2010.

[829] D. M. Sheen, D. L. Makin, T. E. Hall, and R. H. Severtsen. *Real-Time Wideband Cylindrical Holographic Surveillance System*. United States Patent, Number 5,859,609, January 12, 1999.

[830] D. R. Sheen, C. M. Strawitch, and T. B. Lewis. UHF wideband SAR design and preliminary results. In *Proceedings of IEEE International Geoscience and Remote Sensing Symposium*, volume 1, pages 289–291, Pasadena, California, August 1994.

[831] L. A. Shepp and B. F. Logan. The Fourier reconstruction of a head section. *IEEE Transactions on Nuclear Science*, 21:21–43, 1974.

[832] C. W. Sherwin, J. P. Ruina, and R. D. Rawcliffe. Some early developments in synthetic aperture radar systems. *IRE Transactions on Military Electronics*, 6:111–115, April 1962.

[833] G. Shippey, S. Banks, and J. Pihl. SAS image reconstructions using fast polar backprojection: Comparisons with fast factorized backprojection and Fourier-domain imaging. In *Proceedings of Oceans 2005 – Europe*, volume 1, pages 96–101, Brest, France, June 2005.

[834] G. Shippey and T. Nordkvist. Phased array acoustic imaging in ground co-ordinates, with extension to synthetic aperture processing. *IEE Proceedings on Radar, Sonar, and Navigation*, 143(3):131–139, June 1996.

[835] S. Silver. On the integral equations of planar diffraction problems. *University of California, Antenna Laboratory*, Report No. 163, 1949.

[836] F. Simonetti and L. Huang. From beamforming to diffraction tomography. *Journal of Applied Physics*, 103:103–110, 2008.

[837] F. Simonetti and L. Huang. Synthetic aperture diffraction tomography for three-dimensional imaging. *Proceedings of the Royal Society A*, 465:2877–2895, July 2009.

[838] R. G. Simpson, H. H. Barrett, J. A Suback, and H. D. Fisher. Digital processing of annular coded-aperture imagery. *Optical Engineering*, 14(5):490–494, September-October 1975.

[839] T. K. Sjogren, V. T. Vu, and M. I. Pettersson. A comparative study of the polar version with the subimage version of fast factorized backprojection in UWB SAR. *International Radar Symposium*, pages 1–4, May 2008.

[840] M. I. Skolnik. *Radar Handbook*. McGraw–Hill, New York, New York, 1970.

[841] M. I. Skolnik. *Introduction to Radar Systems*. McGraw–Hill, New York, New York, second edition, 1980.

[842] M. I. Skolnik. Fifty years of radar. *Proceedings of the IEEE*, 73(2), February 1985.

[843] M. G. Slaney. *Imaging with Diffraction Tomography*. PhD thesis, Purdue University, May 1985.

[844] D. Slepian. On bandwidth. *Proceedings of the IEEE*, 64:292–300, 1976.

[845] D. Slepian. Some comments on Fourier analysis, uncertainty and modeling. *SIAM Review*, 25(3):379–393, July 1983.

[846] D. Slepian and H. O. Pollak. Prolate spheroidal wave functions, Fourier analysis and uncertainty–I. *The Bell System Technical Journal*, 40:43–63, 1961.

[847] D. Slepian and H. O. Pollak. Prolate spheroidal wave functions, Fourier analysis and uncertainty–II. *The Bell System Technical Journal*, 40:65–84, 1961.

[848] D. Slepian and H. O. Pollak. Prolate spheroidal wave functions, Fourier analysis and uncertainty–IV: Extensions to many dimensions; generalized prolate spheroidal functions. *The Bell System Technical Journal*, 43:3009–3057, 1964.

[849] A. M. Smith. A new approach to range-Doppler SAR processing. *International Journal on Remote Sensing*, 12:1702–1706, 1991.

[850] G. S. Smith. *An Introduction to Classical Electromagnetic Radiation*. Cambridge University Press, Cambridge, 1997.

[851] G. S. Smith. Comment on 'Rayleigh-Summerfeld diffraction and Poisson's spot'. *European Journal of Physics*, 27:21–23, 2006.

[852] H. M. Smith. *Principles of Holography*. Wiley and Sons, New York, 1969.

[853] P. R. Smith, T. M. Peters, and R. H. T. Bates. Image reconstruction from finite numbers of projections. *Journal of Physics A: Mathematical, Nuclear and General*, 6:361–382, 1973.

[854] J. Sok-Son, G. Thomas, and B. C. Flores. *Range-Doppler Radar Imaging and Motion Compensation*. Artech House, Norwood, Massachusetts, 2001.

[855] A. H. S. Solberg, C. Brekke, and P. Ove Husoy. Oil spill detection in Radarsat and Envisat SAR images. *IEEE Transactions on Geoscience and Remote Sensing*, 45(3):746–755, March 2007.

[856] D. C. Solomon. The x-ray transform. *Journal of Mathematical Analysis and Applications*, 56:61–83, 1976.

[857] M. Soumekh. Image reconstruction from limited projections using the angular periodicity of the Fourier transform of the Radon transform of an object. *Acoustical Imaging*, 13, 1983.

[858] M. Soumekh. Image reconstruction techniques in tomographic imaging systems. *IEEE Transactions on Acoustics, Speech, and Signal Processing*, 34(4):952–962, August 1986.

[859] M. Soumekh. Surface imaging via wave equation inversion. *Acoustical Imaging*, 16, 1987.

[860] M. Soumekh. Band-limited interpolation from unevenly spaced sampled data. *IEEE Transactions on Acoustics, Speech, and Signal Processing*, 36(1):110–122, January 1988.

[861] M. Soumekh. The effects of limited data in multifrequency reflection diffraction tomography. *NATO Advanced Science Institute Series: Signal Processing and Pattern Recognition in Nondestructive Evaluation of Materials*, 44:231–240, 1988.

[862] M. Soumekh. Phase reconstruction/unwrapping from amplitude for diffracted waves using a perturbation solution of the wave equation. *IEEE Transactions on Acoustics, Speech, and Signal Processing*, 36:1076, July 1988.

[863] M. Soumekh. Reconstruction from incomplete data in echo diffraction imaging. In *Proceedings of the IEEE International Conference on Acoustics, Speech, and Signal Processing*, pages 1304–1307, 1988.

[864] M. Soumekh. Echo imaging using physical and synthesized arrays. *Optical Engineering*, 29(5):545–554, May 1990.

[865] M. Soumekh. A system model and inversion for synthetic aperture radar imaging. In *Proceedings of the IEEE International Conference on Acoustics, Speech, and Signal Processing*, pages 1873–1876, 3–6 April 1990.

[866] M. Soumekh. Bistatic synthetic aperture radar inversion with application in dynamic object imaging. *IEEE Transactions on Signal Processing*, 39(9):2044–2055, September 1991.

[867] M. Soumekh. Array imaging with beam-steered data. *IEEE Transactions on Image Processing*, 1(3):379–390, July 1992.

[868] M. Soumekh. A system model and inversion for synthetic aperture radar. *IEEE Transactions on Image Processing*, 1(1):64–76, January 1992.

[869] M. Soumekh. Digital spotlighting and coherent subaperture formation for stripmap synthetic aperture radar. *IEEE Transactions on Image Processing*, 21(12):476–480, December 1994.

[870] M. Soumekh. Digital spotlighting and coherent subaperture image formation for stripmap SAR. In *Proceedings of the IEEE International Conference on Image Processing*, volume 1, pages 471–475, Austin, November 1994.

[871] M. Soumekh. *Fourier Array Imaging*. Prentice-Hall, Englewood Cliffs, New Jersey, 1994.

[872] M. Soumekh. Automatic aircraft landing using interferometric inverse synthetic aperture radar imaging. In *Proceedings of the IEEE International Conference on Image Processing*, pages 129–132, 1995.

[873] M. Soumekh. Reconnaissance with ultra wideband UHF synthetic aperture radar. *IEEE Signal Processing Magazine*, 12(4):21–40, July 1995.

[874] M. Soumekh. Automatic aircraft landing using interferometric inverse synthetic aperture radar imaging. *IEEE Transactions on Image Processing*, 5(9):1335–1345, September 1996.

[875] M. Soumekh. Reconnaissance with slant plane circular SAR imaging. *IEEE Transactions on Image Processing*, 5(8):1252–1265, August 1996.

[876] M. Soumekh. Super-resolution array processing in SAR. *IEEE Signal Processing Magazine*, 13:14–18, November 1996.

[877] M. Soumekh. Airborne synthetic aperture acoustic imaging. *IEEE Transactions on Image Processing*, 6(11):1545–1554, November 1997.

[878] M. Soumekh. Moving target detection in foliage using along track monopulse synthetic aperture radar imaging. *IEEE Transactions on Image Processing*, 6(8):1114–1118, August 1997.

[879] M. Soumekh. Phased array imaging of moving targets with randomized beam steering and area spotlighting. *IEEE Transactions on Image Processing*, 6(5):736–749, May 1997.

[880] M. Soumekh. Signal subspace fusion of uncalibrated sensors with application in SAR. In *Proceedings of IEEE International Conference on Image Processing*, pages 280–283, Santa Barbara, California, October 1997.

[881] M. Soumekh. Range stacking: An interpolation-free SAR reconstruction algorithm. In *Proceedings of SPIE, Algorithms for Synthetic Aperture Radar Imagery V*, volume 3370, pages 13–24, Orlando, Florida, April 1998.

[882] M. Soumekh. Wide-bandwidth continuous-wave monostatic/bistatic synthetic aperture radar imaging. In *Proceedings of the IEEE International Conference on Image Processing*, Chicago, October 1998.

[883] M. Soumekh. Depth-focused interior echo imaging. *IEEE Transactions on Image Processing*, 8(11):1608–1618, November 1999.

[884] M. Soumekh. Signal subspace fusion of uncalibrated sensors with application in SAR and diagnostic medicine. *IEEE Transactions on Image Processing*, 8(1):127–137, January 1999.

[885] M. Soumekh. *Synthetic Aperture Radar Signal Processing with MATLAB Algorithms.* John Wiley & Sons, 1999.

[886] M. Soumekh. UHF-band P-3 and X-band monopulse-GMTI SAR processing. In *Proceedings of European Conference on Synthetic Aperture Radar*, Munich, May 2000.

[887] M. Soumekh. Wavefront-based synthetic aperture radar signal processing. *EUSAR 2001 Special Issue on SAR, Frequenz Zeitschrift fur Telecommunication (Journal of Telecommunications)*, 55:99–113, March/April 2001.

[888] M. Soumekh. Moving target detection and imaging using an X band along-track monopulse SAR. *IEEE Transactions on Aerospace and Electronic Systems*, 38(1):315–333, January 2002.

[889] M. Soumekh. Time domain non-linear SAR processing. Final report, State University of New York, Buffalo, New York, January 2006.

[890] M. Soumekh. *Radiation Imaging Utilizing Data Reconstruction to Provide Transforms Which Accurately Reflect Wave Propagation Characteristics.* United States Patent, Number 5,170,170, December 8, 1992.

[891] M. Soumekh and J. Choi. Reconstruction in diffraction imaging. *IEEE Transactions on Ultrasonics, Ferroelectrics and Frequency Control*, 36(1):93–100, January 1989.

[892] M. Soumekh and J. Choi. Phase and amplitude-phase restoration in synthetic aperture radar imaging. *IEEE Transactions on Image Processing*, 1(2):229–242, April 1992.

[893] M. Soumekh, J. Choi, and W. Zwolinski. Target imaging using a synthesized bistatic radar. In *Proceedings of the IEEE International Conference on Systems Engineering*, pages 39–42, Dayton, Ohio, August 1994.

[894] M. Soumekh, G. Guenther, M. Linderman, and R. Kohler. Digitally-spotlighted subaperture SAR image formation using high performance computing. In *Proceedings of SPIE, Algorithms for Synthetic Aperture Radar Imagery VII*, volume 4053, pages 260–271, Orlando, Florida, April 2000.

[895] M. Soumekh and B. Himed. SAR-MTI processing of multi-channel airborne radar measurement (MCARM) data. In *Proceedings of IEEE International Conference on Radar Systems*, Long Beach, California, May 2002.

[896] M. Soumekh and M. Kaveh. Image reconstruction from frequency domain data on arbitrary contours. In *Proceedings of the IEEE International Conference on Acoustics, Speech, and Signal Processing*, pages 12A.2.1–12A.2.4, San Diego, California, 1984.

[897] M. Soumekh and M. Kaveh. A theoretic study of model approximation errors in diffraction tomography. *IEEE Transactions on Ultrasonics, Ferroelectrics and Frequency Control*, 33:10–20, 1986.

[898] M. Soumekh, M. Kaveh, and R. K. Mueller. Fourier reconstruction methods with applications to diffraction tomography. *Acoustical Imaging*, 13, 1984.

[899] M. Soumekh, D. Nobels, M. Wicks, and G. Genello. Signal processing of wide-bandwidth and wide-beamwidth P-3 SAR data. *IEEE Transactions on Aerospace and Electronic Systems*, 37(4):1122–1141, October 2001.

[900] M. Soumekh, S. Nugroho, S. Jones, R. Rysdyk, and S. Tran. ISAR imaging of an airborne DC-9. In *International Conference on Acoustics, Speech, and Signal Processing*, volume V, pages 465–469, Minneapolis, March 1993.

[901] M. Soumekh, S. Worrell, E. G. Zelnio, and B. Keaffaber. SAR wavefront reconstruction using motion compensated phase history (polar format) data and DPCA-based GMTI. In *Proceedings of SPIE, Algorithms for Synthetic Aperture Radar Imagery VII*, volume 4053, pages 64–75, Orlando, Florida, April 2000.

[902] I. Stakgold. *Green's Functions and Boundary Value Problems.* Wiley-Interscience, New York, 2nd edition, 1997.

[903] J. J. Stamnes. Waves, rays, and the method of stationary phase. *Optics Express*, 10(16):740–751, August 2002.

[904] H. C. Stankwitz, R. J. Dallaire, and J. R. Feinup. Nonlinear apodization for sidelobe control in SAR imagery. *IEEE Transactions on Aerospace and Electronic Systems*, 31(1):267–279, January 1995.

[905] H. Stark. Sampling theorems in polar coordinates. *Journal of the Optical Society of America*, 69:1519–1524, 1979.

[906] H. Stark, J. W. Woods, P. Indraneel, and R. Hingorani. An investigation of computerized tomography by direct Fourier inversion and optimum interpolation. *IEEE Transactions on Biomedical Engineering*, 28(7):496–505, July 1981.

[907] H. Stark, J. W. Woods, I. Paul, and R. Hingorani. Direct Fourier reconstruction in computer tomography. *IEEE Transactions on Acoustics, Speech and Signal Processing*, 29(2):237–245, April 1981.

[908] P. Stefanov and G. Uhlmann. Inverse backscattering for the acoustic equation. *SIAM Journal Mathematical Analysis*, 28:1191–1204, 1997.

[909] V. Stein. Physical optics method: Prediction of radar signatures. In L. B. Felsen, editor, *Theoretical Aspects of Target Classification*, volume 152 of *AGARD Lecture Series*, Rome, Munich, Norway, 1987. AGARD.

[910] B. D. Steinberg. *Principles of Aperture and Array System Design.* Wiley, New York, 1976.

[911] F. Stenger. Asymptomatic ultrasonics inversion based on using more than one frequency. *Acoustical Imaging*, pages 425–444, 1982.

[912] F. Stenger and S. A. Johnson. Ultrasonic transmission tomography based on the inversion of the Helmholtz equation for plane and spherical wave insonification. *Applied Mathematics Notes*, 4(3-4):102–127, December 1979.

[913] R. H. Stolt. Migration by Fourier transform. *Geophysics*, 43(1):23–48, February 1978.

[914] R. H. Stolt and A. B. Weglein. Migration and inversion of seismic data. *Geophysics*, 50(12):2548–2472, December 1985.

[915] J. A. Stratton. *Electromagnetic Theory.* McGraw-Hill, New York, 1941.

[916] J. A. Stratton and L. J. Chu. Diffraction theory of electromagnetic waves. *Physical Review*, 56:99–107, July 1939.

[917] N. Streibl. Fundamental restrictions for 3-D light distributions. *Optik*, 66:341, 1984.

[918] N. Streibl. Three-dimensional imaging by a microscope. *Journal of the Optical Society of America, A*, 2:121, 1985.

[919] J. W. Strohbehn. Line of sight wave propagation through turbulent atmosphere. *Proceedings of the IEEE*, 56:1301–1318, August 1968.

[920] J. W. Strohbehn. Optical propagation through the turbulent atmosphere. *Progress in Optics*, 9:75–122, 1971.

[921] G. W. Stroke. Image deblurring and aperture synthesis using 'a posteriori' processing by Fourier-transform holography. *Optica Acta*, 16:401–422, 1969.

[922] W. L. Stutzman and G. A. Thiele. *Antenna Theory and Design.* John Wiley & Sons, New York, 1997.

[923] B. Subiza, E. Gimeno-Nieves, J. M. Lopez-Sanchez, and J. Fortuny-Guasch. An approach to SAR imaging by means of non-uniform FFTs. In *Proceedings of the IEEE International Geoscience and Remote Sensing Symposium*, volume 6, pages 4089–4091, July 2003.

[924] N. S. Subotic and B. J. Thelen. Sequential processing of SAR phase history data for rapid detection. In *Proceedings of the IEEE International Conference on Image Processing*, pages 144–146, Washington, DC, October 1995.

[925] R. J. Sullivan. *Microwave Radar Imaging and Advanced Concepts.* Artech House, Boston, Massachusetts, 2000.

[926] X. B. Sun, T. S. Yeo, C. B. Zhang, Y. H. Lu, and P. S. Kooi. Time-varying step-transform algorithm for high squint SAR imaging. *IEEE Transactions on Geoscience and Remote Sensing*, 27:2668–2677, November 1999.

[927] D. W. Sweeney and C. M. Vest. Reconstruction of three-dimensional refractive index fields from multidirectional interferometric data. *Applied Optics*, 12(11):2649–2664, 1973.

[928] W. Tabbara, B. Duchene, C. Pichot, D. Lesselier, L. Chommeloux, and N. Joachimowicz. Diffraction tomography: Contribution to the analysis of some applications in microwaves and ultrasonics. *Inverse Problems*, 4:305–331, 1988.

[929] C.-T. Tai. A concise formulation of Huygens' principle for electromagnetic fields. *IEEE Transactions on Antennas and Propagation*, 8:634, November 1960.

[930] C.-T. Tai. Kirchhoff theory: Scalar, vector, or dyadic? *IEEE Transactions on Antennas and Propagation*, pages 114–115, January 1972.

[931] K. C. Tam and V. Perez-Mendez. Limited-angle three-dimensional reconstructions using Fourier transform iterations and Radon transform iterations. *Optical Engineering*, 20:586, 1981.

[932] K. C. Tam and V. Perez-Mendez. Tomographical imaging with limited-angle input. *Journal of the Optical Society of America*, 71:582, 1981.

[933] W. Tan, W. Hong, Y. Wang, Y. Lin, and Y. Wu. Synthetic aperture radar tomography sampling criteria and three-dimensional range migration algorithm with elevation digital spotlighting. *Science in China Series F: Information Sciences*, 52(1):100–114, January 2009.

[934] V. Tatarski. *Wave Propagation in a Turbulent Medium.* Dover, New York, 1967.

[935] O. Tedone. *Rendiconti*, volume 5, 1, pages 286–289. Accademia dei Lincei, 1917.

[936] A. R. Thompson and R. N. Bracewell. Interpolation and Fourier transformation of fringe visibilities. *Astronomical Journal*, 79(1):11–24, 1974.

[937] B. J. Thompson. Fundamentals of coherent optics. In *Proceedings of SPIE, Effective Utilization of Optics in Radar Systems*, volume 128, pages 30–45, Huntsville, Alabama, 27-29 September 1977.

[938] D. G. Thompson, , D. V. Arnold, D. G. Long, and G. F. Miner. YSAR: A compact, low-cost synthetic aperture radar. In *Proceedings of the International Geoscience and Remote Sensing Symposium*, pages 1892–1894, Lincoln, Nebraska, May 1996.

[939] D. G. Thompson, , D. V. Arnold, D. G. Long, G. F. Miner, and M. A. Jensen. YINSAR: A compact, low-cost interferometric synthetic aperture radar. In *Proceedings of the International Geoscience and Remote Sensing Symposium*, pages 1920–1922, 1998.

[940] J. H. Thomson and J. E. B. Ponsonby. Two-dimensional aperture synthesis in lunar radar astronomy. *Proceedings of the Royal Society A.*, 303:477–491, March 1968.

[941] K. Tomiyasu. Synthetic aperture radar in geosynchronous orbit. In *IEEE International Symposium (Digest) on Antennas and Propagation*, pages 42–45, College Park, Maryland, May 1978.

[942] K. Tomiyasu. Tutorial review of synthetic-aperture radar (SAR) with applications to imaging of the ocean surface. *Proceedings of the IEEE*, 66:563–583, May 1978.

[943] K. Tomiyasu. Conceptual performance of a satellite borne, wide swath synthetic aperture radar. *IEEE Transactions on Geoscience and Remote Sensing*, 19(2):108–116, April 1981.

[944] M. F. Toups, L. A. Bessett, and B. T. Binder. Foliage penetration data collections and investigations utilizing the P-3 UWB SAR. In *Proceedings of SPIE, Algorithms for Synthetic Aperture Radar Imagery III*, volume 2757, April 1996.

[945] R. Touzi. Extraction of point target response characteristics from complex SAR data. *IEEE Transactions on Geoscience and Remote Sensing*, 30(6):1158–1161, November 1992.

[946] O. Tretiak. Recovery of multi-dimensional signals from their projections. *Computer Graphics and Image Processing*, 1:179–195, October 1973.

[947] O. Tretiak, M. Eden, and M. Simon. Internal structures for three dimensional images. In *Proceedings of the 8th International Conference on Medical and Biological Engineering*, Chicago, Illinois, 1969.

[948] G. Tricoles and N. H. Farhat. Microwave holography: Applications and techniques. *Proceedings of the IEEE*, 65:108–121, January 1997.

[949] A. W. Trorey. A simple theory for seismic diffractions. *Geophysics*, 35:762–784, 1970.

[950] T. K. Truong, I. S. Reed, R. G. Lipes, A. L. Rubin, and S. A. Butman. Digital SAR processing using a fast polynomial transform. *IEEE Transactions on Acoustics, Speech, and Signal Processing*, 32(2):419–425, April 1984.

[951] S. V. Tsynkov. On SAR imaging through the Earth ionosphere. *SIAM Journal on Imaging Science*, 2:140–182, 2009.

[952] M. Tur, K. C. Chin, and J. W. Goodman. When is speckle noise multiplicative? *Applied Optics*, 21(7):1157–1159, April 1982.

[953] F. T. Ulaby and A. K. Fung. *Microwave Remote Sensing*. Artech House, 1986.

[954] L. M. H. Ulander. Approaching the wavelength resolution limit in ultra-wideband VHF-SAR. In *Proceedings of the PIERS Workshop on Advances in Radar Methods, Report EUR 18662 EN Space Applications Institute, Joint Research Centre of the European Commission*, pages 1540–1550, Beveno, Italy, July 1998.

[955] L. M. H. Ulander and P.-O. Frolund. Ultra-wideband SAR interferometry. *IEEE Transactions on Geoscience and Remote Sensing*, 36(5):1540–1550, September 1998.

[956] L. M. H. Ulander and H. Hellsten. A new formula for synthetic aperture radar resolution. *AEÜ, International Journal of Electronics and Communications*, 50(2):117–121, 1996.

[957] L. M. H. Ulander and H. Hellsten. Low-frequency ultra-wideband array-antenna SAR for stationary and moving target imaging. In *Proceedings of SPIE, Radar Sensor Technology IV*, volume 3704, pages 149–158, Orlando, Florida, 5 April 1999.

[958] L. M. H. Ulander, H. Hellsten, and G. Stenstrom. Synthetic-aperture radar processing using fast factorised backprojection. In *Proceedings of EUSAR*, pages 753–756, Munich, Germany, 23–25 May 2000.

[959] L. M. H. Ulander, H. Hellsten, and G. Stenstrom. Performance analysis of fast backprojection for synthetic-aperture radar processing. In *Proceedings of SPIE, Algorithms for Synthetic Aperture Radar Imagery VIII*, volume 4382, pages 13–21, 2001.

[960] L. M. H. Ulander, H. Hellsten, and G. Stenstrom. Synthetic-aperture radar processing using fast factorized back-projection. *IEEE Transactions on Aerospace and Electronic Systems*, 39(3):760–766, July 2003.

[961] P. W. Vachon and R. K. Raney. Estimation of the SAR system transfer function through process defocus. *IEEE Transactions on Geoscience and Remote Sensing*, 27(6):702–708, November 1989.

[962] B. K. Vainshtein and S. S. Orlov. Theory of the recovery of functions from their projections. *Soviet Physics-Crystallography*, 17(2):213–216, September-October 1972.

[963] M. Vandewal, R. Speck, and H. Suss. Efficient and precise processing for squinted spotlight SAR through a modified Stolt mapping. *EURASIP Journal on Advances in Signal Processing*, pages 1–7, 2007.

[964] M. R. Vant. A spatially variant autofocus synthetic aperture radar. In *Proceedings of the IEEE International Radar Conference*, pages 159–163, 1982.

[965] M. R. Vant, G. E. Haslam, and G. M. Royer. A digital signal processing approach for satellite-borne synthetic aperture radar. In *Proceedings of the IEEE International Radar Conference*, pages 251–256, Paris, France, December 1978.

[966] M. R. Vant, G. E. Haslam, and E. Shaw. Digital processing techniques for satellite-borne synthetic-aperture radar. *Canadian Journal of Remote Sensing*, 5:67–73, May 1979.

[967] M. R. Vant and K. H. Wu. A digital SAR processor based on the coherent subaperture addition technique. In *Proceedings of the IEEE International Radar Conference*, pages 425–429, 1984.

[968] T. Varslot, J. H. Morales, and M. Cheney. Synthetic-aperture imaging through dispersive media. *Inverse Problems*, 26(025008), 2010.

[969] S. M. Verbout and D. J. Blejer. Two- and three-dimensional radar imaging at close range to a synthetic aperture. In *Proceedings of the National Telesystems Conference*, Atlanta, Georgia, 1991.

[970] R. C. Waag, J. A. Campbell, J. Ridder, and P. R. Mesdag. Cross-sectional measurements and extrapolations of ultrasonic fields. *IEEE Transactions on Sonics and Ultrasonics*, 32(1):26–35, 1985.

[971] C. C. Wackerman and D. R. Lyzenga. Automatic focusing of synthetic aperture radar images of diffuse targets. In *IEEE International Geoscience and Remote Sensing Symposium*, pages 937–940, Ann Arbor, Michigan, 18–21 May 1987.

[972] D. E. Wahl, P. H. Eichel, D. C. Ghiglia, and C. V. Jakowatz. Phase gradient auto focus–A robust tool for high resolution SAR phase correction. *IEEE Transactions on Aerospace and Electronic Systems*, 30(3):827–835, July 1994.

[973] D. E. Wahl, P. H. Eichel, and C. V. Jakowatz. Implementation of the phase-gradient algorithm. In *Proceedings of SPIE, Advanced Signal Processing Algorithms, Architectures and Implementations*, volume 1348, pages 528–535, 1990.

[974] D. E. Wahl, C. V. Jakowatz, Jr, and P. A. Thompson. New approach to strip-map SAR auto focus. In *Sixth IEEE Digital Signal Processing Workshop*, pages 53–56, October 1994.

[975] D. E. Wahl, D. A. Yocky, and C. F. Jakowatz, Jr. An implementation of a fast backprojection image formation algorithm for spotlight-mode SAR. In *Proceedings of SPIE, Algorithms for Synthetic Aperture Radar Imagery XV*, volume 6970, Orlando, Florida, 16–20 March 2008.

[976] J. L. Walker. Range-Doppler imaging of rotating objects. *IEEE Transactions on Aerospace and Electronic Systems*, 16(1):23–52, January 1980.

[977] J. L. Walker, W. G. Carrara, and I. Cindrich. Optical processing of rotating-object radar data using a polar recording format. *Technical Report RADC-TR-73-136*, May 1973.

[978] I. Walterscheid, J. H. G. Ender, A. R. Brenner, and O. Loffeld. Bistatic SAR processing and experiments. *IEEE Transactions on Geoscience and Remote Sensing*, 44(10):2710–2717, October 2006.

[979] J.-N. Wang. Inverse backscattering problem for Maxwell's equations. *Mathematical Methods in the Applied Sciences*, 21:1441–1465, 1998.

[980] G. N. Watson. *The Theory of Bessel Functions.* Cambridge University Press, London, United Kingdom, 1952.

[981] J. Way and E. A. Smith. The evolution of synthetic aperture radar systems and their progression to the EOS SAR. *IEEE Transactions on Geoscience and Remote Sensing*, 29(6):962–985, November 1991.

[982] J. L. H. Webb and D. C. Munson, Jr. Radar imaging of three-dimensional surfaces using limited data. In *Proceedings of the IEEE International Conference on Image Processing*, pages 136–139, Washington, DC, October 1995.

[983] H. F. Weber. Ueber die integration der partiellen differentialgleichung. *Mathematische Annalen*, 1:1–36, 1869.

[984] D. R. Wehner. *High Resolution Radar.* Artech, New York, 1994.

[985] M. Weib, J. Ender, O. Peters, and T. Espeter. An airborne radar for three dimensional imaging and observation-technical realisation and status of ARTINO. In *EUSAR 2006, 6th European Conference on Synthetic Aperture Radar*, Dresden, Germany, May 2006.

[986] M. Weib, O. Peters, and J. Ender. A three dimensional SAR system on an UAV. In *Proceedings of the IEEE International Geoscience and Remote Sensing Symposium*, Barcelona, Spain, July 2007.

[987] W. G. Weis and W. K. Jenkins. Theory and simulation of digital FFT processors for spotlight mode synthetic aperture radar. Report lmsc/d566409, Lockheed Aircraft Corporation, December 1977.

[988] H. Weyl. Ausbreitung elektromagnetischer wellen Über einem ebenen leiter. *Annalen der Physik*, 60:481–500, 1919.

[989] R. White and M. L. Williams. Processing ISAR and spotlight SAR data to very high resolution. In *Proceedings of the IEEE Geoscience and Remote Sensing Symposium*, volume 1, pages 32–34, 1999.

[990] C. A. Wiley. Synthetic aperture radars: A paradigm for technology evolution. *IEEE Transactions on Aerospace and Electronic Systems*, 21(3):440–443, May 1985.

[991] P. R. Williamson and M. H. Worthington. Resolution limits in ray tomography due to wave behavior: Numerical experiments. *Geophysics*, 58:727–735, 1993.

[992] A. Witten, J. Tuggle, and R. Waag. A practical approach to ultrasonic imaging using diffraction tomography. *Journal of the Acoustical Society of America*, 83:1645, April 1988.

[993] E. Wolf. Three-dimensional structure determination of semi-transparent objects from holographic data. *Optics Communications*, 1(4):153–156, 1969.

[994] F. Wong and I. G. Cumming. Error sensitivities of a secondary range compression algorithm for processing squinted satellite SAR data. In *IEEE International Geoscience and Remote Sensing Symposium*, Vancouver, Canada, 1989.

[995] F. H. Wong and T. S. Yeo. New applications of nonlinear chirp scaling in SAR data processing. *IEEE Transactions on Geoscience and Remote Sensing*, 39(5):946–953, May 2001.

[996] J. W. Wood. The removal of azimuthal distortion in SAR images. *International Journal on Remote Sensing*, 9:1097–1107, 1988.

[997] P. M. Woodward. *Probability and Information Theory, with Applications to Radar.* McGraw-Hill, New York, 1953.

[998] S. Worrell and M. Soumekh. Automatic target recognition in SAR using digitally-spotlighted phase history data. In *Proceedings of SPIE, Algorithms for Synthetic Aperture Radar Imagery VII*, volume 4053, pages 344–353, Orlando, Florida, April 2000.

[999] C. Wu. A digital system to produce imagery from SAR data. In *AIAA Systems Design Driven by Sensors Conference*, volume Paper 76-968, Pasadena, California, 18–20 October 1976.

[1000] C. Wu. A digital fast correlation approach to produce SEASAT SAR imagery. In *Proceedings of the IEEE International Radar Conference*, Pasadena, California, June 1980.

[1001] C. Wu, K. Y. Liu, and M. Jin. Modeling and a correlation algorithm for space borne SAR signals. *IEEE Transactions on Aerospace and Electronic Systems*, 18(5):563–575, September 1982.

[1002] K. H. Wu and M. R. Vant. Extensions to the step transform SAR processing technique. *IEEE Transactions on Aerospace and Electronic Systems*, 21(3):338–344, May 1985.

[1003] S. Xiao and D. C. Munson, Jr. Spotlight-mode SAR imaging of a three-dimensional scene using spectral estimation techniques. In *Proceedings of International Geoscience Remote Sensing Symposium*, volume 2, pages 624–644, 1998.

[1004] S. Xiao, D. C. Munson, Jr, S. Basu, and Y. Bresler. An $n^2 \log n$ back-projection algorithm for SAR image formation. In *Proceedings of the 34th Asilomar Conference on Signals, Systems, and Computers*, pages 3–7, Pacific Grove, California, November 2000.

[1005] M. Xing, X. Jiang, R. Wu, F. Zhou, and Z. Bao. Motion compensation for UAV SAR based on raw radar data. *IEEE Transactions on Geoscience and Remote Sensing*, 47(8):2870–2883, August 2009.

[1006] X. Xu and R. M. Narayanan. Enhanced resolution in SAR/ISAR imaging using iterative sidelobe apodization. *IEEE Transactions on Image Processing*, 14(4):537–547, 2005.

[1007] H. Yang and M. Soumekh. Motion estimation and compensation in SAR/ISAR imaging. In *International Conference on Acoustics, Speech, and Signal Processing*, volume 3, pages 17–20, 23–26 March 1992.

[1008] H. Yang and M. Soumekh. Blind-velocity SAR/ISAR imaging of a moving target in a stationary background. *IEEE Transactions on Image Processing*, 2:80–95, January 1993.

[1009] B. Yazici, M. Cheney, and C. E. Yarman. Synthetic aperture inversion in the presence of noise and clutter. *Inverse Problems*, 22(5):1705–1729, 2006.

[1010] A. F. Yegulalp. Fast backprojection algorithm for synthetic aperture radar. In *Proceedings of the IEEE International Radar Conference*, pages 60–65, Waltham, Massachusetts, 20–22 April 1999.

[1011] J. L. Yen. On nonuniform sampling of bandlimited signals. *IRE Transactions on Circuit Theory*, 3:251–257, December 1956.

[1012] T. S. Yeo, N. L. Tan, C. B. Zhang, and Y. H. Lu. A new subaperture approach to high squint SAR processing. *IEEE Transactions on Geoscience and Remote Sensing*, 39(5):954–968, May 2001.

[1013] C. Yerkes and E. Webster. Implementation of ω-k synthetic aperture radar imaging algorithm on a massively parallel supercomputer. In *Proceedings of SPIE, Algorithms for Synthetic Aperture Radar Imagery*, volume 2230, pages 171–178, Orlando, Florida, 4–8 April 1994.

[1014] D. Yocky and C. V. Jakowatz, Jr. Automated wide-angle SAR stereo height extraction in rugged terrain using shift-scaling correlation. In *Proceedings of SPIE, Applications of Digital Image Processing XXVI*, volume 5203, pages 10–20, 2003.

[1015] D. Yocky and C. V. Jakowatz, Jr. Terrain elevation mapping results from airborne spotlight-mode coherent cross-track SAR stereo. *IEEE Transactions on Geoscience and Remote Sensing*, 42(2):301–308, February 2004.

[1016] D. Yocky, D. Wahl, and C. V. Jakowatz, Jr. Spotlight-mode SAR image formation utilizing the chirp Z-transform in two dimensions. In *Proceedings of the IEEE International Geoscience and Remote Sensing Symposium*, pages 4180–4182, July 2006.

[1017] H. Yonghong, M. Shiyi, and H. Wen. A fast approximate imaging algorithm for spaceborne synthetic aperture radar. *Journal of Electronics (China)*, 11(2):97–103, 1994.

[1018] D. C. Youla and H. Webb. Image restoration by the method of convex projections: Part 1. Theory. *IEEE Transactions on Medical Imaging*, 1:81–94, 1982.

[1019] S. Young, N. Nasrabadi, and M. Soumekh. SAR moving target detection and identification using stochastic gradient techniques. In *IEEE International Conference on Acoustics, Speech, and Signal Processing*, Detroit, Michigan, May 1995.

[1020] T. Young. Experiments demonstration of the general law of the interference of light. In *Philosophical Transactions of the Royal Society of London XCII*, volume 12, page 387, 1802. Miscellaneous works of the late Thomas Young, Vol I (London, J. Murray, 1885) pp. 140–170.

[1021] T. Young. Experiments demonstration of the general law of the interference of light. In *Philosophical Transactions of the Royal Society of London*, volume 94, 1804.

[1022] T. Young. *A Course of Lectures on Natural Philosophy and the Mechanical Arts*, volume 1. London, 1845.

[1023] R. Zahn and M. Schlott. Active antenna for X-band space SAR. In *Proceedings of the IEEE International Conference on Radar*, pages 36–41, Paris, France, 1994.

[1024] F. D. Zan and A. M. Guarnieri. TOPSAR: terrain observation by progressive scans. *IEEE Transactions on Geoscience and Remote Sensing*, 44(9):2352–2360, September 2006.

[1025] E. Z. Zaugg and D. G. Long. The BYU SAR: A small, student-built SAR for UAV operation. In *Proceedings of the IEEE Geoscience and Remote Sensing Symposium*, pages 411–414, 31 July – 4 August 2006.

[1026] E. Z. Zaugg and D. G. Long. Theory and application of motion compensation for LFM-CW SAR. *IEEE Transactions on Geoscience and Remote Sensing*, 46(10):2990–2998, October 2008.

[1027] E. Z. Zaugg and D. G. Long. Generalized frequency-domain SAR processing. *IEEE Transactions on Geoscience and Remote Sensing*, 47(11):3761–3773, November 2009.

[1028] H. A. Zebker and R. M. Goldstein. Topographic mapping from interferometric synthetic aperture radar observations. *Journal of Geophysical Research*, 91(B5):4993–4999, April 1986.

[1029] H. A. Zebker, S. N. Madsen, J. Martin, K. B. Wheeler, T. Miller Y. Lou, G. Alberti, S. Vetrella, and A. Cucci. The TOPSAR interferometric radar topographic mapping instrument. *IEEE Transactions on Geoscience and Remote Sensing*, 30(9):933–940, September 1992.

[1030] H. A. Zebker and J. Villasenor. Decorrelation in interferometric radar echoes. *IEEE Transactions on Geoscience and Remote Sensing*, 30(5):950–959, September 1992.

[1031] H. A. Zebker, C. L. Werner, P. A. Rosen, and S. Hensley. Accuracy of topographic maps derived from ERS–1 interferometric radar. *IEEE Transactions on Geoscience and Remote Sensing*, 32(4):823–836, July 1994.

[1032] X. X. Zhu and R. Bamler. Very high resolution spaceborne SAR tomography in urban environment. *IEEE Transactions on Geoscience and Remote Sensing*, 48(12):4296–4308, December 2010.

[1033] S. H. Zisk. Lunar topography: First radar-interferometer measurements of the Alphonsus-Ptolemaeus-Arzachel region. *Science*, 178:977–980, 1972.

[1034] S. H. Zisk. A new Earth-based radar technique for the measurement of lunar topography. *The Moon*, 4:296–306, 1972.

Index